Contemporary Mathematicians

Gian-Carlo Rota
Editor

R. P. Dilworth at retirement.

The Dilworth Theorems

Selected Papers of Robert P. Dilworth

Edited by
Kenneth P. Bogart, Ralph Freese,
and Joseph P. S. Kung

Birkhäuser
Boston • Basel • Berlin
1990

Kenneth P. Bogart
Department of Mathematics and Computer Science
Dartmouth College
Hanover, NH 03755
U.S.A.

Ralph Freese
Department of Mathematics
University of Hawaii
Honolulu, HI 96822
U.S.A.

Joseph P. S. Kung
Department of Mathematics
University of North Texas
Denton, TX 76203-5116
U.S.A.

Library of Congress Cataloging-in-Publication Data
The Dilworth theorems : selected papers of Robert P. Dilworth / edited
  by Kenneth P. Bogart, Ralph Freese, Joseph P. S. Kung.
      p.   cm. — (Contemporary mathematicians)
   Includes bibliographical references.
   ISBN 0-8176-3434-7 (alk. paper)
      1. Lattice theory.   2. Dilworth, Robert P. (Robert Palmer), 1914–
      I. Dilworth, Robert P. (Robert Palmer), 1914– . II. Bogart,
   Kenneth P. III. Freese, Ralph S., 1946– . IV. Kung, Joseph P. S.
   V. Series.
   QA171.5.D55   1990
   511.3'3—dc20   90-31372   90-31372

Printed on acid-free paper.

ISBN 0-8176-3434-7
ISBN 3-7643-3434-7

Camera-ready text provided by the editors.
Printed and bound by Edwards Brothers, Inc., Ann Arbor, Michigan.
Printed in the U.S.A.
9 8 7 6 5 4 3 2 1

Contents

List of Contributors — ix

Editors' Preface — xi

Biography — xv

Recollections of R. P. Dilworth   Peter Crawley — xix

Recollections of Professor Dilworth   Phillip Chase — xxi

Mathematical Publications of R. P. Dilworth — xxiii

Doctoral Students — xxvi

1   Chain Partitions in Ordered Sets

Background — 1

Reprinted Papers

   A Decomposition Theorem for Partially Ordered Sets — 7

   Some Combinatorial Problems on Partially Ordered Sets — 13

Articles

   K. Bogart, C. Greene, and J. Kung   *The Impact of the Chain Decomposition Theorem on Classical Combinatorics* — 19

   E. C. Milner   *Dilworth's Decomposition Theorem in the Infinite Case* — 30

   H. Kierstead   *Effective Versions of the Chain Decomposition Theorem* — 36

2 Complementation

Background 39
Reprinted Papers
   Lattices with Unique Complements 41
   On Complemented Lattices 73
  Articles
    M. Adams   *Uniquely Complemented Lattices* 79
    G. Kalmbach   *On Orthomodular Lattices* 85

3 Decomposition Theory

Background 89
Reprinted Papers
   Lattices with Unique Irreducible Decompositions 93
   The Arithmetical Theory of Birkhoff Lattices 101
   Ideals in Birkhoff Lattices 115
   Decomposition Theory for Lattices without Chain Conditions ... 145
    (with P. Crawley)
   Note on the Kurosch-Ore Theorem 167
   Structure and Decomposition Theory of Lattices 173
  Articles
    B. Jónsson   *Dilworth's Work on Decompositions in Semi-modular Lattices* ... 187
    B. Monjardet   *The Consequences of Dilworth's Work on Lattices with Unique Irreducible Decompositions* ... 192
    J. Kung   *Exchange Properties for Reduced Decompositions in Modular Lattices* ... 201
    M. Stern   *The Impact of Dilworth's Work on Semimodular Lattices on the Kurosch-Ore Theorem* ... 203

4 Modular and Distributive Lattices

Background 205
Reprinted Papers
   The Imbedding Problem for Modular Lattices (with M. Hall) ... 211
   Proof of a Conjecture on Finite Modular Lattices
   Distributivity in Lattices (with J. McLaughlin)
   Aspects of Distributivity
  Articles
    A. Day and R. Freese   *The Role of Gluing Constructions in Modular Lattice Theory* ... 251
    I. Rival   *Dilworth's Covering Theorem for Modular Lattices* ... 261

5 Geometric and Semimodular Lattices

Background 265
Reprinted Papers
   Dependence Relations in a Semi-modular Lattice 269
   A Counterexample to the Generalization of Sperner's Theorem 283
     (with C. Greene)
Articles
   U. Faigle *Dilworth's Completion, Submodular Functions, and* 287
     Combinatorial Optimization
   J. Kung *Dilworth Truncations of Geometric Lattices* 295
   J. Griggs *The Sperner Property in Geometric and Partition* 298
     Lattices

6 Multiplicative Lattices

Background 305
Reprinted Papers
   Abstract Residuation over Lattices 309
   Residuated Lattices (with M. Ward) 317
   Non-commutative Residuated Lattices 337
   Non-commutative Arithmetic 357
   Abstract Commutative Ideal Theory 369
Articles
   D. Anderson *Dilworth's Early Papers on Residuated and* 387
     Multiplicative Lattices
   E. Johnson *Abstract Ideal Theory: Principals and Particulars* 391
   D. Anderson *Representation and Embedding Theorems for* 397
     Noether Lattice and r-Lattices

7 Miscellaneous Papers

Background 403
Reprinted Papers
   The Structure of Relatively Complemented Lattices 407
   The Normal Completion of the Lattice of Continuous Functions 419
   A Generalized Cantor Theorem (with A. Gleason) 431
   Generators of Lattice Varieties (with R. Freese) 433

Contents

Articles

    G. McNulty   *Lattice Congruences and Dilworth's Decomposition*   439
      of Relatively Complemented Lattices

    G. Gierz   *The Normal Completion of the Lattice of Continuous*   445
      Functions

    J. Kung   *Cantor Theorems for Relations*   450

    J. B. Nation   *Ideal and Filter Constructions in Lattice Varieties*   451

8 Two Results from "Algebraic Theory of Lattices"

Background   455
Articles

    J. Kung   *Dilworth's Proof of the Embedding Theorem*   458

    G. Grätzer   *On the Congruence Lattice of a Lattice*   460

Permissions   465

List of Contributors

M. E. Adams, Department of Mathematics, State University of New York, New Platz, New York 12561, U. S. A.

Daniel D. Anderson, Department of Mathematics, University of Iowa, Iowa City, Iowa 52240, U. S. A.

Kenneth P. Bogart , Department of Mathematics, Dartmouth College, Hanover, New Hampshire 03755, U. S. A.

Phillip J. Chase , 8716 Oxwell Lane, Laurel, Maryland 20708, U. S. A.

Peter Crawley , Department of Mathematics, Brigham Young University, Provo, Utah 84602, U. S. A.

Alan Day, Department of Mathematics, Lakehead University, Thunder Bay, Ontario, Canada P7B 5E1.

Ulrich Faigle, Faculty of Applied Mathematics, University of Twente, P. O. Box 217, 7500 AE Enschede, the Netherlands.

Ralph Freese, Department of Mathematics, University of Hawaii, Honolulu, Hawaii 96822, U. S. A.

Gerhard Gierz, Department of Mathematics, University of California, Riverside, California 92521, U. S. A.

George A. Grätzer, Department of Mathematics, University of Manitoba, Winnipeg, Manitoba, Canada R3T 2N2.

Curtis Greene, Department of Mathematics, Haverford College, Haverford, Pennsylvania 19041, U. S. A.

Jerrold R. Griggs, Department of Mathematics, University of South Carolina, Columbia, South Carolina 29208, U. S. A.

Eugene W. Johnson, Department of Mathematics, University of Iowa, Iowa City, Iowa 52240, U. S. A.

Bjarni Jónsson, Department of Mathematics, Vanderbilt University, Nashville, Tennessee 37235, U. S. A.

Gudrun Kalmbach, Mathematische Institut, Universität Ulm, D-7900 Ulm, Federal Republic of Germany.

Henry A. Kierstead, Department of Mathematics, University of South Carolina, Columbia, South Carolina 29208, U. S. A.

Joseph P. S. Kung, Department of Mathematics, University of North Texas, Denton, Texas 76203, U. S. A.

George F. McNulty, Department of Mathematics, University of South Carolina, Columbia, South Carolina 29208, U. S. A.

E. C. Milner, Department of Mathematics, University of Calgary, Calgary, Alberta, Canada T2N 1N4.

Bernard Monjardet, Université Paris V and Centre d'Analyse et de Mathématique Sociale, 54 Boulevard Raspail, 75 270 Paris Cedex 06, France.

James B. Nation, Department of Mathematics, University of Hawaii, Honolulu, Hawaii 96822, U. S. A.

Ivan Rival, Department of Computer Science, University of Ottawa, Ottawa, Canada K1N 6N5.

Manfred Stern, Sektion Mathematik, Martin-Luther-Universität, DDR-4010 Halle, German Democratic Republic.

Editors' Preface

Lattice theory is one of those rare subjects to which it is possible to assign a precise birthday. This birthday occurred in 1897 with the publication of Dedekind's paper [2]. Not much was done before the subject was revived in the 1930's by Birkhoff, Mac Lane, Ore, Ward, and others. At that time, lattice theory was thought of as a subject whose interest depends on other areas of mathematics and the emergence of lattice theory as a subject with ideas, theorems, and problems of its own did not really occur till the fifties. Among the many many mathematicians whose work contributed to this evolution, Robert P. Dilworth is one of the most influential. This maturing of lattice theory is best described by quoting from Dilworth's preface to the proceedings [6] of the symposium on partially ordered sets and lattice theory held at Monterey in 1959:

... on April 15, 1938, the first general symposium on lattice theory was held in Charlottesville in conjunction with a regular meeting of the American Mathematical Society. The three principal addresses on that occasion were entitled: Lattices and their Applications, On the Application of Structure Theory to Groups, and The Representation of Boolean Algebras. It is interesting to observe that the first and last of these titles appear again as section titles for the present Symposium. ... Nevertheless there have been major changes in emphasis and interest during the intervening years and thus some general comments concerning the present state of the subject and its relationship to other areas of mathematics appear to be appropriate.

The theory of groups provided much of the motivation and many of the technical ideas in the early development of lattice theory. Indeed, it was the hope of many of the early researchers that lattice-theoretic methods would lead to the solution of some of the important problems in group theory. Two decades later, it seems to be a fair judgement that, while this hope has not been realized, lattice theory has provided a useful framework for the formulation of certain topics in the theory of groups ... and has produced some interesting and difficult group-theoretic problems ...

On the other hand, the fundamental problems of lattice theory have, for the most part, not come from this source but have arisen from attempts to answer intrinsically natural questions concerning lattices and partially ordered sets; namely, questions concerning the decompositions, representations, imbedding, and free structure, of such systems. ... As the study of these basic questions has progressed, there has come into being a sizable body of technical ideas and methods which are peculiarly lattice-theoretic in nature. These conceptual tools are intimately related to the underlying order relation and are particularly appropriate for the study of general lattice structure. At the 1938 Symposium, lattice theory was described as a "vigorous and promising younger brother of group theory". In the intervening years it has developed into a full-fledged member of the algebraic family with an extensive body of knowledge and a collection of exciting problems all of its own. Such outstanding problems as the construction of a set of structure invariants for certain classes of Boolean algebras, the characterization of the lattice of congruence relations of a lattice, the imbedding of finite lattices in finite partition lattices, the word problem for free modular lattices, and the construction of a dimension theory for continuous, non-complemented, modular lattices, have an intrinsic interest independent of the problems associated with other algebraic systems. Furthermore, these and other current problem are sufficiently difficult that imaginative and ingenious methods will be required in their solutions. ...

Rather than change his subject through building new foundations or developing theory as other important figures in contemporary mathematics have, Dilworth's approach to research in lattice theory was to attack the deepest unsolved problems. His work brought the subject to a level not reached before. He emphasized specific results, but basic archetypical results whose solutions require methods of general applicability. An example of this is the paper [4] in which he settled a problem posed by E. V. Huntington in 1904 by showing that a lattice in which every element has a unique complement is not necessarily a Boolean algebra. To do this, Dilworth developed several constructions which have turned out to be central in lattice theory and universal algebra. Similarly, the chain decomposition theorem [5], abstracted from a structural analysis of distributive lattices, has been the foundation on which entire theories in combinatorics and ordered sets rest. To use a fanciful image, Dilworth is a pioneer who discovers trails, mountain passes and river crossings, and points out the way to later, more systematic settlement. George McNulty, in his article on "The Structure of Relatively Complemented Lattices," expresses succinctly a view which we believe characterizes all of Dilworth's work.

Reading it some forty years after it was written, it strikes this reader as if it were a fresh result, conveyed with simplicity in the most current notation and speaking to issues now at the breaking edge of research in lattice theory. This is certainly a testament to the lasting impact that

Dilworth's contributions to lattice theory have had. Today we are speaking and writing and thinking about lattices in the manner of Robert Dilworth.

Dilworth started on lattice theory in the 1930's by reading Dedekind's papers [2,3] in Dedekind's collected works. He encouraged his students to read them as well, improving their German as they improved their understanding of the origins of lattice theory. He remarked that while Dedekind's papers were excellent introductions, the motivation behind them was unclear. For this reason, we have organized this volume differently. The book is organized into chapters covering different subject areas of Dilworth's work. Each chapter begins with a background, written by Dilworth, for his papers reprinted in the chapter. In these backgrounds, he discusses how and why he approached the problems he solved. Following the background in each chapter are the reprints of Dilworth's papers, followed by articles about these papers we have solicited from experts in the appropriate field. At Dilworth's suggestion, these articles are not just commentaries on his papers, but are general surveys on the subsequent work in the area, bringing the reader up to the current state of the subject.

Some of Dilworth papers have had such a broad influence on certain areas that we have more than one article dealing with the paper. For example, there are three articles dealing with various aspects of Dilworth's chain decomposition paper in Chapter 1. On the other hand, some of the articles deal with more than one of Dilworth's papers. Jónsson's article in Chapter 3, for example, deals with all of Dilworth's work on decompositions into irreducibles in semimodular lattices.

We are very pleased with the quality of the articles in this volume and wish to thank the authors for their contributions. We hope that, with these articles, Dilworth's backgrounds and his papers, this volume, like Dedekind's collected works before it, will serve as an inspiration to a new generation of lattice theorists.

This volume contains most of the papers Dilworth wrote in the theory of lattices and ordered sets. A complete chronological bibliography is given on pp. xxii to xxiv. Not included in this volume is the book [1] which Dilworth wrote with Peter Crawley. This book contains two influential results which Dilworth discovered in the 1940's but remained unpublished until they appeared in [1] in 1973. These two results are discussed in Chapter 8. Besides serving as an accessible and concise introductory text, [1] has also stimulated much research through the unsolved problems stated in it. A notable example is the problem of characterizing ordered sets with the fixed point property (see [7]). It would take us too far afield to adequately cover this aspect of Dilworth's ongoing influence.

We would like to thank Academic Press, the American Mathematical Society, Annals of Mathematics, Pacific Journal of Mathematics and Tôhoku Mathematical Journal for their generous response to our request for permission to reprint. We would like to thank Phillip Chase and Peter Crawley for their illuminating recollections of Dilworth, which give true insight to his character. In addition to those who have contributed articles to this volume, we would also like to thank F. Galvin, K. H. Hofmann, P. Johnstone, and R. S. Pierce. Finally, our special thanks go to

Garrett Birkhoff and J. B. Nation for their comments and suggestions throughout our editorial work and Gian-Carlo Rota for getting this project off to a running start.

REFERENCES

1. P. Crawley and R. P. Dilworth, "Algebraic Theory of Lattices," Prentice-Hall, Englewood Cliffs, New Jersey, 1973.
2. R. Dedekind, *Über Zerlegungen von Zahlen durch ihre grössten gemeinsamen Teiler*, Festschrift Technische Hochschule Braunschweig (1897). Reprinted in "Gesammelte Werke," Vol. 2, pp. 103–148.
3. R. Dedekind, *Über die von drei Moduln erzeugte Dualgruppe*, Math. Ann. **53** (1900), 371–403. Reprinted in "Gesammelte Werke," Vol. 2, pp. 236–271.
4. R. P. Dilworth, *Lattices with unique complements*, Trans. Amer. Math. Soc. **57** (1945), 123–154. Reprinted in Chapter 2 of this volume.
5. R. P. Dilworth, *A decomposition theorem for partially ordered sets*, Ann. of Math. (2) **51** (1950), 161–166. Reprinted in Chapter 1 of this volume.
6. R. P. Dilworth (ed.), "Lattice Theory," Proceedings of Symposia in Pure Mathematics, Vol. 2, Amer. Math. Soc., Providence, Rhode Island, 1961.
7. I. Rival, *The fixed point property*, Order **2** (1985), 219–221.

Biography

Robert Palmer Dilworth was born on December 2, 1914, in Hemet, California. He received his Bachelor of Science degree in 1936 at the California Institute of Technology. He stayed at Caltech for his graduate work under Morgan Ward and obtained the degree of doctor of philosophy in 1939. He was Sterling Research Fellow (1939-40) and Instructor in Mathematics (1940-43) at Yale. In 1943, he returned to join the mathematics faculty at Caltech. He was assistant professor (1943-45), associate professor (1945-50), professor (1950-82), and professor emeritus (1982-present).

Dilworth was one of the formative influences on the Caltech Mathematics Department. Through Morgan Ward, who was a student of Eric Temple Bell, he was linked directly to the beginnings of mathematics in America. (In fact, there was a period in the 50's when Bell, Ward, Dilworth, and Dilworth's student, Richard Pierce – four mathematical generations – were at Caltech at the same time.) Dilworth imparted through his teaching the precision and solidity in research as well as the concern for good teaching and exposition that characterized the best of early American mathematics. Dilworth had seventeen doctoral students at Caltech, but his impact extends beyond his "official" students. He served as the departmental graduate director during most of his tenure at Caltech. An idea of his teaching style can be gleaned from the following description by Ralph Freese and J. B. Nation: "When he lectured, he rarely used abbreviations and his handwriting was nearly perfect. Students had to write as fast as they could, using several abbreviations to keep up with him. When he got stuck he would step back from the blackboard, stare at the problem and whistle *Stars and Stripes Forever*."

In addition to his work at Caltech, Dilworth was active in the application of mathematics and mathematics education. One such activity is during his military service during the Second World War. In his own words,

In July 1944, I was appointed by John Harlan to be a member of a two-man operational analysis unit to be attached to the headquarters of the 1st Division of the 8th Air Force located in Brampton Park, England.

This unit was to serve as a liasion between the main operational analysis unit located at the headquarters of the 8th Air Force near London and the command of the 1st Air Division. This unit made regular reports to the 8th Air Force unit and forwarded the results of their studies to the appropriate personnel in the division headquarters. Eventually the unit carried out analyses desired by the commander of the 1st Air Division. After a few months, the second member of the unit returned to the States for health reasons and the unit became a one-man operation until the end of the war. In the spring of 1945, in collaboration with the division navigator, an elaborate experiment was carried out to evaluate the intrinsic accuracy of radar bombing. A special radar target placed in the Wash on east coast of Britain was used in this exercise.

Dilworth served as a consultant to the National Security Agency from 1955 to 1976. As he described it,

> I was appointed to the Mathematics Panel of the agency in August 1960. The mission of the panel was to review and suggest applications of mathematics to the problems of the agency. In February 1964, I was made chairman of the panel and was appointed to the Science Advisory Board of the agency. I continued to serve in this capacity until July 1970.

Dilworth also served as consultant to the Communications Research Division of the Institute for Defense Analysis through most of the 60's and 70's. He had a visiting appointment with the Communications Research Division for fifteen months in 1961-62.

Dilworth was also involved in mathematics education. He described his experience as follows:

> In June 1954, I was appointed to the College Board Advanced Mathematics Committee. This committee's mission was to set policy for the Advanced Mathematics Examination and to oversee its preparation for each academic year. I became chairman of the committee in 1957 and served in this capacity until June 1961.
>
> From 1962 through 1969, I was Director of Testing and Evaluation for the African Mathematics Program sponsored by Educational Services, Inc. and funded by the Agency for International Development. This program was established to train educational personnel from the newly independent countries of Africa in the techniques of mathematics curriculum development. The objective was to develop a core of mathematics educators in each of the participating countries who would be able to produce curriculum materials in mathematics which would be appropriate for the needs of each of the countries. During six summer sessions from 1962 to 1968,

the representatives of the African countries involved met with mathematics educators from the United States and Britain to develop specimen mathematics texts covering the primary and secondary years. It was the responsibility of the testing and evaluation group to see that there were African personnel in each of the countries trained in modern testing methods by developing tests and other evaluative materials to accompany the texts being written by the primary and secondary groups. The early summer sessions were held in Entebbe, Uganda, and the later sessions in Mombasa, Kenya.

In May 1962, I was appointed to the Graduate Record Examination Board of Examiners in Mathematics. This board supervised the construction and use of the Graduate Record Examination in Mathematics. I became chairman of the board in April 1964 and continued to serve in that capacity until June 1969.

I served on the School Mathematics Study Group Advisory Board from 1962 to 1965.

From 1968 to 1970, I was Director of Research for the Miller Mathematics Improvement Program. This program was designed to provide extensive in-service training for teachers of mathematics in the State of California. The research effort involved pre and post testing a very large sample of students of teachers participating in the program.

In the 70's, I served the National Science Foundation as Director of Evaluation for the Washington State Summer Workshop Program and the Teachers Center Project for San Diego.

At various times, I did consulting for the Radio Corporation of America, the Stanford Research Institute, and Automations Electronics.

Dilworth was actively involved in sports. In earlier years, he competed in the decathalon and was particularly good at pole-vaulting. Although bamboo poles had come into common use by then, several of Dilworth's poles broke while he was vaulting. He competed in several open track and field events in Los Angeles and did well. Later, he exercised by swimming at least half a mile at noon everyday.

Dilworth married Miriam White on December 23, 1940. They have two sons, Robert, Jr. and Gregory.

R. P. Dilworth at the start of his research career.

Recollections of R. P. Dilworth

PETER CRAWLEY

My memories and impressions of R. P. Dilworth focus on this central fact: he was an electrifying teacher and colleague. And apart from his intellectual power as a mathematician, I think this was primarily a product of two traits: Bob Dilworth loved a challenge, and he was tenacious in confronting one; and he had great mathematical taste.

Bob's love of the challenge was apparent when he interviewed me as a prospective Caltech freshman. He asked what interested me about mathematics, the order and beauty of it, or the fun of trying a problem. When I responded I liked to work problems, a twinkle came in his eyes, and I knew at that moment that attacking a mathematical problem was one of the great pleasures of the man interviewing me. Bob and Miriam would often invite the graduate students over to their house, and on such an occasion Bob would bring out his new puzzle. Each of us would have to take a turn at it; and each of us knew, though this was never explicitly announced, that the object of the game was to try to beat Dilworth's time. All of us who worked with him know that Dilworth never gave up on a problem. As an undergraduate I watched him labor over a certain extremely difficult problem, and then I watched him try alternative approaches ten or twelve years later. His catalogue of unsolved problems was always in his mind–Indeed I am certain that such a catalogue is in his mind at this moment.

Dilworth first taught me the idea of taste when I was an undergraduate in his lattice theory seminar. Like all of us in the seminar I tried my hand at a problem and one day I came to Dilworth with what I thought was a reasonable conjecture. His face darkened a bit, and I could see that my conjecture wasn't particularly satisfying to him. At that point he made a statement that profoundly affected me. Merely solving a problem doesn't necessarily make good mathematics, Dilworth observed. What counts, he went on, is the beauty, the depth, the surprising nature of the solution. Later I noticed how strictly he applied this in his professional life. All of us know of pieces of work which remained unpublished in his desk drawer, work

that we would have published without hesitation. But the theorem did not go far
enough, or the proof was not elegant enough, so the paper remained in Dilworth's
desk.

As his student, I don't ever recall Bob explicitly stating his expectations for
me. What I will never forget, however, is the implicit set of expectations that flowed
from his personality, his personal conduct, and his work as research mathematician.

Brigham Young University
Provo, UT 84602
U. S. A.

THE DILWORTH THEOREMS

Recollections of Professor Dilworth

PHILLIP J. CHASE

However sentimental or corny, I need to have the courage to say that my feelings of personal debt to Prof. Dilworth are of the same sort one ordinarily feels for a father. You will see why. During my graduate school days at Caltech I found myself following my nose, working on required courses, but otherwise not taking any particular initiative. But time was on the move. You can't really earn a Ph. D. in the mode I was in. At this point Prof. Dilworth walked into my office in Sloan and asked if I had considered who I wanted to work for. I said no. He asked "Well, how about me?" The real world thus suddenly having intruded, I quickly said yes. He asked what I wanted to work on. I said "I don't know." He said "Here are three ideas. Look them over and let me know if one of them appeals to you." Well, I did that, and picked out one of them. Not being particularly mature (but mature enough to know it), I explained I'd do better with fairly frequent consultations. "Fine. How about one o'clock on Fridays?", or some such time. The two following years were maybe that most wonderful of my life. Early on, maybe our first meeting, I showed up with an extended mathematical argument that I was proud of, one that didn't leave much white space on three sheets of narrow-lined paper, both sides. He listened a few minutes and said "Can you leave this with me?" About an hour later he came to my desk with just one sheet of paper, with only the top third written on, containing an elegant and beautiful equivalent of what I had given him. This made a powerful impact on me, and set a standard that remains with me.

Eventually there came a point where I had made it, and was to receive my degree in a few months. With my usual single-minded tunnel vision, I was transposing letters in my thesis, and scraping together dollars for more typed pages (one dollar per page). Prof. Dilworth walked in once again. This time "What are you going to do when you're finished?" As before "I don't know." "How about the National Security Agency?" I filled out the application, which asked a whole lot a questions, designed to find out to what degree I was a communist, and fired it off. A couple months later Prof. Dilworth asked what I had heard from NSA. "Nothing," I said.

Shocked, he had NSA on the phone immediately, with me still standing there. NSA had never heard of me! They'd apparently lost my application. Once again I waded through the application, got accepted, and am at NSA to this day. (Because of the lost application, NSA was unable to process me soon enough, so I wound up teaching two years at the College of Wooster.

Thus Prof. Dilworth in many respects dominated the course of my adult life (but I did pick out my own wife).

During my early years at NSA, I occasionally saw Prof. Dilworth there, where he served in an important scientific advisory capacity. I am told by some old-timers still remaining at NSA that before this he had been a star contributor to least one summer program, during which researchers from academia gathered to try to apply techniques from advanced mathematics to sensitive problems of importance to our community. Like myself, these old-timers remember Prof. Dilworth for his strength of character, powerful mathematical ability, and thorough professionalism.

I've said what I wanted to, but will close with a hodge-podge of miscellaneous recollections. Prof. Dilworth had an office next to Prof. Bohnenblust, whom he seemed very fond of, calling him "Boni." (He was boney, angular and tall.) Among the chatter that I occasionally heard between them, one particular session sticks in my mind. Prof. Dilworth had gotten some kind of a traffic ticket, apparently for failing to stop at a stop sign. He was incensed and refused to pay the fine, winding up in court. Prepared, as usual, he had photos showing that the stop sign was obscured by foliage. He won.

I also recall that Prof. Dilworth was a regular swimmer, commenting that running inflicted a beating on one's legs. He was always in trim physical condition, and this was well before the current fitness era. He never dawdled, but always walked with a spring in his step, and got wherever he was going very fast.

It was said, though I never knew first hand, that he had a study in his house that was sealed off using special soundproofing insulation.

He seemed particularly impressed by Marshall Hall, but who wasn't? Prof. Dilworth's own thesis advisor was Prof. Morgan Ward, who was closing out his own career, full of achievement during this time. My opinion is that Prof. Ward helped instill in Prof. Dilworth his profound respect for the teaching of mathematics, at all levels, even very elementary levels. Dilworth exhibited a sense of duty with respect to improving mathematics education. There were no bizarre "new math" distortions here, only sound stuff, always respect for problem solving and computational facility. He emphasized the value of quality exposition in mathematical writing. (Once I asked him which line the "=" should go on, if an equation wouldn't all fit on the current line. This gave him pause, he didn't really like either answer.)

8716 Oxwell Lane
Laurel, MD 20708
U. S. A.

                                             THE DILWORTH THEOREMS

Mathematical Publications of
Robert P. Dilworth

[1]   Abstract residuation over lattices, *Bulletin of the American Mathematical Society,* **44**(1938), 262–268.

[2]   (with Morgan Ward) Residuated lattices, *Proceedings of the National Academy of Sciences,* **24**(1938), 162–164.

[3]   (with Morgan Ward) Residuated lattices, *Transactions of the American Mathematical Society,* **45**(1939), 335–354.

[4]   Non-commutative residuated lattices, *Transactions of the American Mathematical Society,* **46**(1939), 426–444.

[5]   Non-commutative arithmetic, *Duke Mathematical Journal,* 5(1939), 270–280.

[6]   (with Morgan Ward) The lattice theory of ova, *Annals of Mathematics,* 40(1939), 600–608.

[7]   On complemented lattices, *Tôhoku Mathematical Journal,* **47**(1940) 18–23.

[8]   Lattices with unique irreducible decompositions, *Annals of Mathematics,* **41**(1940), 771–777.

[9]   Note on complemented modular lattices, *Bulletin of the American Mathematical Society,* **46**(1940), 74–76.

[10] The arithmetical theory of Birkhoff lattices, *Duke Mathematical Journal,* 8(1941), 286–299.

[11] Ideals in Birkhoff lattices, *Transactions of the American Mathematical Society,* **49** (1941), 325–353.

[12] Dependence relations in a semi-modular lattice, *Duke Mathematical Journal,* **11** (1944), 575–587.

[13] (with Marshall Hall, Jr.) The imbedding problem for modular lattices, *Annals of Mathematics,* **45**(1944), 450–456.

[14] Lattices with unique complements, *Transactions of the American Mathematical Society,* **57**(1945), 123–154.

[15] Note on the Kurosch-Ore theorem, *Bulletin of the American Mathematical Society,* **52**(1946), 659–663.

[16] (with Morgan Ward) Note on a paper by C. E. Rickart, *Bulletin of the American Mathematical Society,* **55**(1949), 1141.

[17] Note on the strong law of large numbers, *American Mathematical Monthly,* **54**(1949), 249–250.

[18] A decomposition theorem for partially ordered sets, *Annals of Mathematics,* **51**(1950), 161–166.

[19] The structure of relatively complemented lattices, *Annals of Mathematics,* **51**(1950), 348–359.

[20] The normal completion of the lattice of continuous functions, *Transactions of the American Mathematical Society,* **68**(1950), 427–438.

[21] (with Jack E. McLaughlin) Distributivity in lattices, *Duke Mathematical Journal,* **19**(1952), 683–694.

[22] Proof of a conjecture on finite modular lattices, *Annals of Mathematics,* **60**(1954), 359–364.

[23] (with Peter Crawley) Decomposition theory for lattices without chain conditions, *Transactions of the American Mathematical Society,* **60**(1960), 1–22.

[24] Some combinatorial problems on partially ordered sets, in R. Bellman and M. Hall, Jr., eds., *Combinatorial Analysis (Proceedings of the Tenth Symposium in Applied Mathematics, Columbia University, 1958),* American Mathematical Society, Providence, Rhode Island, 1960, pp. 85–90.

[25] Structure and decomposition theory of lattices, in R. P. Dilworth, ed., *Lattice Theory (Proceedings of Symposia in Pure Mathematics, Vol. 2),* American Mathematical Society, Providence, Rhode Island, 1961, pp. 3–16.

[26] Abstract commutative ideal theory, *Pacific Journal of Mathematics,* **12**(1962), 481–498.

[27] (with Andrew M. Gleason) A generalized Cantor theorem, *Proceedings of the American Mathematical Society,* **13**(1962), 704–705.

[28] (with Curtis Greene) A counterexample to the generalization of Sperner theorem, *Journal of Combinatorial Theory,* **10**(1971), 18–21.

[29] (with Peter Crawley) *Algebraic Theory of Lattices,* Prentice-Hall, Englewood Cliffs, New Jersey, 1973.

[30] (with Ralph Freese) Generators of lattice varieties, *Algebra Universalis,* **6**(1976), 263–267.

[31] The role of order in lattice theory, in I. Rival, ed., *Ordered Sets (Proceedings, NATO Advanced Study Institute, Banff, Alberta, 1981)*, Reidel, Dordrecht and Boston, 1982, pp. 333–353.

[32] Aspects of distributivity, *Algebra Universalis*, **18**(1984), 4–17.

Doctoral Students

Daniel T. Finkbeiner, 1949. *A General Dependence Relation and the Application to Lattice Imbeddings.*

Worthy L. Doyle, 1950. *An Arithmetical Theorem for Partially Ordered Sets.*

Jack E. McLaughlin, 1950. *Projectivities in Relatively Complemented Lattices.*

Richard B. Talmadge, 1951. *The Representation of Baire Functions.*

Richard S. Pierce, 1952. *Homomorphisms of Function Lattices.*

Don E. Edmondson, 1954. *Homomorphisms of a Modular Lattice.*

Juris Hartmanis, 1955. *Some Embedding Theorems of Lattices.*

John B. Johnston, 1955. *Universal Partial Orders.*

Peter L. Crawley, 1961. *A Decomposition Theory for Lattices without Chain Conditions.*

Alfred Hales, 1962. *On the Nonexistence of Free Complete Boolean Algebras.*

Phillip J. Chase, 1965. *Sublattices of Partition Lattices.*

Kenneth P. Bogart, 1968. *Structure Theorems for Local Noether Lattices.*

Curtis Greene, 1969. *Combinatorial Properties of Finite Geometric Lattices.*

Ralph S. Freese, 1972. *Varieties Generated by Modular Lattices of Width Four.*

James B. Nation, 1973. *Varieties of Algebras whose Congruence Lattices Satisfy Lattice Identities.*

John R. Stonesifer, 1973. *Combinatorial Inequalities for Geometric Lattices.*

Daniel Erickson, 1974. *Counting Zeroes of Polynomials over Finite Fields.*

Chain Partitions in Ordered Sets

Editors' note: The background for this chapter is based on a lecture intended for undergraduates.

Background

R. P. Dilworth

Let us consider the sets which can be made up from the numbers 1, 2, and 3. We have the singleton sets $\{1\}$, $\{2\}$, and $\{3\}$ made up of one number each. Then there are the two-element sets $\{1,2\}$, $\{1,3\}$, and $\{2,3\}$. Finally, there is the three-element set $\{1,2,3\}$. It is customary to include the empty set $\emptyset$ which has no numbers belonging to it. There is a natural relation between these sets, namely, $\{1\}$ is a subset of $\{1,2\}$ and $\{1,3\}$ is a subset of $\{1,2,3\}$. If A and B denote two of these sets, we write $A = B$ if A and B consist of the same numbers. Thus $\{1,2\} = \{2,1\}$. If every member of A is also a member of B we write $A \leq B$ and if, in addition, there are members of B which do not belong to A we write $A < B$. Thus $\{1,2\} < \{1,2,3\}$. It is convenient to diagram the containing relations existing between these sets as shown in Figure 1. Each set is represented by a small circle in the plane. Larger sets lie above smaller sets and the circles representing two sets are joined by a line if one contains the other and there is no set lying strictly between the two.

The subsets of a set also have an algebraic structure. If A and B are two sets, $A \vee B$ (A join B) will denote the set made up of the objects of A together with the objects of B. $A \wedge B$ (A meet B) will denote the set of objects belonging to both A and B. These operations are commutative, associative, distributive and idempotent ($A \vee A = A \wedge A = A$). There is also a unary operation of complementation. A' consists of the objects in the set which do not belong to A. In the example above,

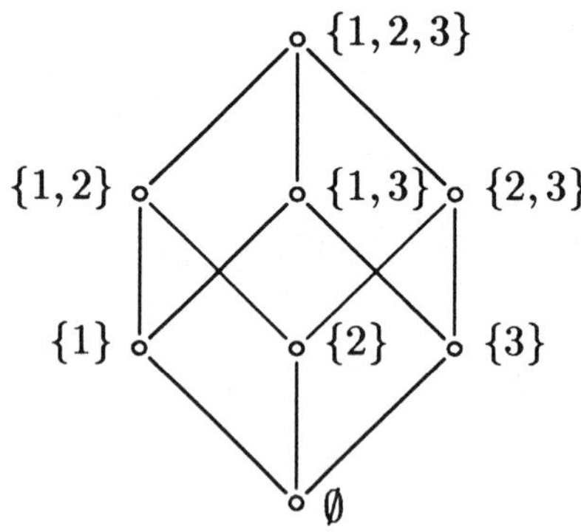

Figure 1

we have

$$\{1,2\} \vee \{3\} = \{1,2,3\} \qquad \{1,2\} \wedge \{2,3\} = \{2\} \qquad \{1,3\}' = \{2\}.$$

The subsets of the set $\{1,2,3\}$ under the operations of join, meet, and complementation provide a simple example of a Boolean algebra—named after George Boole—who studied such systems in the 1800's in connection with his work in symbolic logic.

The structure of a finite Boolean algebra is completely determined once the number of atoms (singleton sets) is known, namely each member of Boolean algebra is the unique join of the singleton sets contained in it. Now a singleton set together with the empty set $\emptyset$ make up a two-element Boolean algebra whose diagram is shown in Figure 2. Thus the basic structure result for finite Boolean algebras states that a finite Boolean algebra is a direct product of n two-element Boolean algebras where n is the number of atoms.

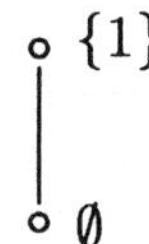

Figure 2

An important generalization of a Boolean algebra is a distributive lattice which has the join and meet operations with the same properties as in the case of Boolean algebras but without the complementation operation. For example, the sets $\emptyset$, $\{1\}$, $\{2\}$, $\{1,2\}$, $\{2,3\}$, $\{1,2,3\}$ form a distributive lattice having the diagram shown in Figure 3.

It is no longer true that every member of the lattice is a join of atoms, but it is true that every member is a join of join irreducibles. A join irreducible is an element of the lattice which cannot be represented as a join of elements distinct

                                   THE DILWORTH THEOREMS

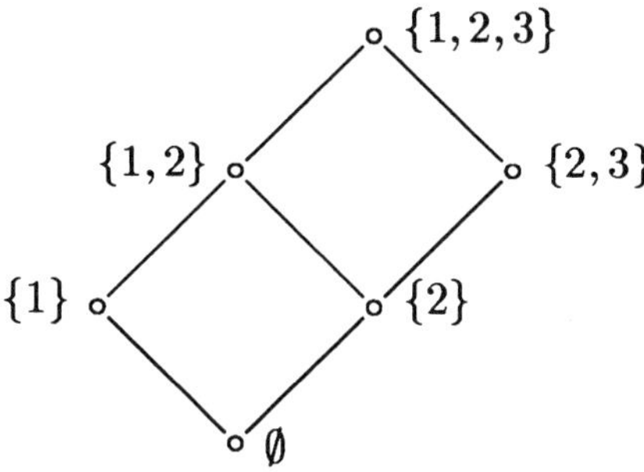

Figure 3

from itself. In the example, $\emptyset$, $\{1\}$, $\{2\}$, and $\{2,3\}$ are join irreducibles. Although $\{1,2,3\}$ can be represented as the join of $\{1\}$, $\{2\}$, and $\{2,3\}$ and also as the join of $\{1\}$ and $\{2,3\}$ the latter representation which is minimal is unique. Indeed, in an abstract finite distributive lattice, the minimal representations as joins of join irreducibles are unique and associating each element of the lattice with the set of join irreducibles less than or equal to it gives a representation of the lattice as a lattice of sets of join irreducibles. Conversely, if a finite ordered set is given, the subsets of the ordered set which are such that with each element in the subset also all elements less than or equal to it also belong to the subset form a distributive lattice. For example, consider the ordered set diagrammed in Figure 4.

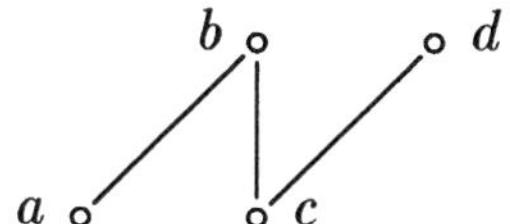

Figure 4

The sets $\emptyset$, $\{a\}$, $\{c\}$, $\{c,d\}$, $\{a,b,c\}$, $\{a,c,d\}$, $\{b,c,d\}$, $\{a,b,c,d\}$ form a distributive lattice diagrammed in Figure 5. The join irreducibles $\{a\}$, $\{a,b,c\}$, $\{c\}$, $\{c,d\}$ correspond to the elements a, b, c, d in the ordered set.

The correspondence between finite ordered sets and finite distributive lattices is one-to-one. Thus, as is frequently the case, questions concerning the structure of distributive lattices can be formulated in terms of questions concerning ordered sets. We also note from this correspondence that any representation of a finite distributive lattice as a lattice of subsets of some set requires that the set have as many members as there are join irreducibles in the distributive lattice. In other terms, the minimum number of two-element Boolean algebras whose direct product has a sublattice isomorphic to a given finite distributive lattice is the number of join irreducibles in the lattice.

When I completed my graduate work at Caltech and went to Yale University

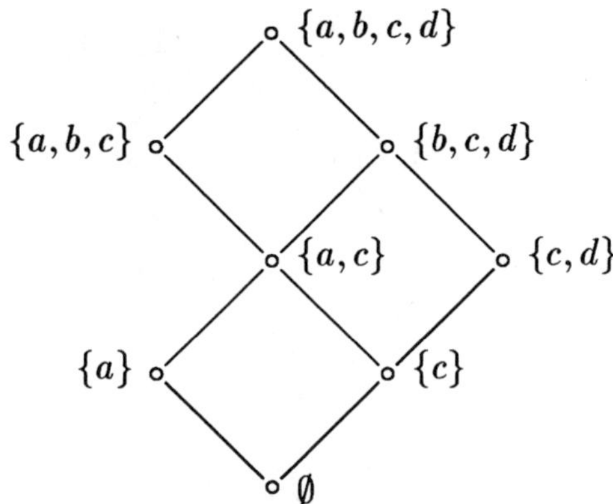

Figure 5

on a Sterling Research Fellowship, one of the most interesting problems in lattice theory at that time was the conjecture that a lattice in which every element had a unique complement was a Boolean algebra. The conjecture had been verified in several instances where one of a number of additional restrictions was imposed, namely modularity, orthocomplementation, and atomicity. The general conjecture was still open. Since the fellowship provided me with a full year in which to do research, I decided to concentrate on the conjecture. However, since it is difficult to work full time on one problem, I decided to fill the breaks by working on a problem concerning the representation of a distributive lattice as a sublattice of a direct product of totally ordered sets, *i.e.*, chains. Since it was easy to determine the least number of two-element chains whose direct product could contain the given distributive lattice as a sublattice, it seemed natural to ask for the least number of chains of arbitrary length whose direct product could contain the given distributive lattice as a sublattice.

Making use of the correspondence between finite distributive lattices and finite ordered sets it was quite straightforward to verify that imbedding a finite distributive lattice in a direct product of chains was equivalent to representing a finite ordered set as a set join of chains. One useful observation which turned up early in the investigation was the fact that the number of elements covering a given element in the distributive lattice was always a lower bound for the number of chains needed for a representation, since each covering element had to belong to a different chain. Again, making use of the correspondence between distributive lattices and ordered sets and examining many examples it became clear that the maximal number of elements covering an element in the distributive lattice was equal to the maximum number of non-comparable elements in the corresponding ordered set, *i.e.*, to the maximal number of elements in an antichain of the ordered set. It was also clear that the maximal number of elements in an antichain was a lower bound to the minimum number of chains needed to represent the ordered set as a set join of chains. Thus the problem would be solved if it could be shown that these two numbers were indeed equal.

If n is the maximal number of elements in an antichain of a finite ordered set, the most natural attack on the problem would be to take a maximal chain out of the ordered set and hope that the maximum size of an antichain in what is left is $n - 1$ or less. However, this doesn't work. In Figure 4, the maximum size of an antichain is 2. The elements b and c make up a maximal chain. The elements a and d which are left are an antichain of size 2.

Obviously, a good place to begin would be the case $n = 2$, since for $n = 1$, the ordered set is already a chain. By starting at the bottom of the ordered set and making an inductive argument it was easy to see that the ordered set is a set join of two chains when $n = 2$. This special technique fails for $n \geq 3$ so a new approach was required. Concentrating on $n = 3$, I tried removing maximal chains having various special properties with the hope that the remaining elements would form an ordered set with $n = 2$. Although this approach worked for ordered sets of relatively small orders, the required properties became more and more complicated as the size of the ordered set increased. After several months I abandoned this approach and turned to trying inductive arguments. Clearly, if the conjecture was true and a portion of the ordered set was split into n chains and x was an element not in any of the n chains, then it should be possible to include x and reorganize the chains so that the enlarged set was again a set join of n chains. Experimenting with a few examples made it clear that the reorganization should involve the portions of the chains lying above x and portions lying below x. For example, in Figure 4, the ordered set consisting of a, b, and c can be split into the two chains $\{a\}$ and $\{b, c\}$. The portion of $\{b, c\}$ lying below the element d is the element c. If c and d are put together as a chain, then the remaining elements a and b also form a chain, and the ordered set is represented as a join of two chains. If this method was to work in general, then it should be possible to select an upper portion of one of the chains and a lower portion of another chain so that when these elements are removed from the set join of the n chains, the remaining ordered set can be represented as set join of $n - 1$ chains. Assuming that this is not the case for each selection of an upper section and a lower section turns out to lead to a contradiction and thus the theorem follows by induction.

While I was working on the finite case it became clear that the result must hold for infinite ordered sets as long as the size of antichains remains bounded. However, by this time I had come back to Caltech as an assistant professor and was winding up my work on the unique complementation problem. Since it was now in the middle of World War II, I had been asked by John Harlan, who was then with the DOD, to be a member of one of the small operations analysis units which were being set up at the headquarters of each of the three divisions of the Eighth Air Force. I felt that I could not refuse, so until the end of the war I was in England doing an entirely different kind of research.

At the end of the war I returned to Caltech and eventually turned again to the decomposition problem for infinite ordered sets in which the size of antichains was bounded. It seemed me, at this time, that the problem now was more a set-

theoretic rather than a combinatorial problem. So again I went searching for chains
which when removed from the ordered set left an ordered set in which the maximum
size of an antichain was reduced. After several months of trying various restrictive
conditions, I decided to go to the strongest restriction possible, namely, to define a
chain to be 'strong' if its meet with any finite subset of the ordered set was a chain in
some decomposition of the finite subset into a minimal number of chains. Since the
theorem holds in the finite case, a chain consisting of a single element is 'strong' and
since 'strongness' is a finite property, there exists a maximal 'strong' chain. When
such a chain is removed from the ordered set, the size of a maximal antichain in the
remaining ordered set is reduced and theorem follows by induction.

The time had finally arrived when it was appropriate to write up the results
and send the manuscript off for publication.

Annals of Mathematics
Vol. 51, No. 1, January, 1950

A DECOMPOSITION THEOREM FOR PARTIALLY ORDERED SETS

By R. P. Dilworth

(Received August 23, 1948)

1. Introduction

Let P be a partially ordered set. Two elements a and b of P are *camparable* if either $a \geq b$ or $b \geq a$. Otherwise a and b are *non-comparable*. A subset S of P is *independent* if every two distinct elements of S are non-comparable. S is *dependent* if it contains two distinct elements which are comparable. A subset C of P is a *chain* if every two of its elements are comparable.

This paper will be devoted to the proof of the following theorem and some of its applications.

THEOREM 1.1. *Let every set of $k + 1$ elements of a partially ordered set P be dependent while at least one set of k elements is independent. Then P is a set sum of k disjoint chains.*[1]

It should be noted that the first part of the hypothesis of the theorem is also necessary. For if P is a set sum of k chains and S is any subset containing $k + 1$ elements, then at least one pair must belong to the same chain and hence be comparable.

Theorem 1.1 contains as a very special case the Radó-Hall theorem on representatives of sets (Hall [1]). Indeed, we shall derive from Theorem 1.1 a general theorem on representatives of subsets which contains the Kreweras (Kreweras [2]) generalization of the Radó-Hall theorem.

As a further application, Theorem 1.1 is used to prove the following imbedding theorem for distributive lattices.

THEOREM 1.2. *Let D be a finite distributive lattice. Let $k(a)$ be the number of distinct elements in D which cover a and let k be the largest of the numbers $k(a)$. Then D is a sublattice of a direct union of k chains and k is the smallest number for which such an imbedding holds.*

2. Proof of Theorem 1.1.

We shall prove the theorem first for the case where P is finite. The theorem in the general case will then follow by a transfinite argument. Hence let P be a finite partially ordered set and let k be the maximal number of independent elements. If $k = 1$, then every two elements of P are comparable and P is thus

---

[1] This theorem has a certain formal resemblance to a theorem of Menger on graphs (D. König, *Theorie der endlichen und unendlichen Graphen*, Leipzig, (1936)). Menger's theorem, however, is concerned with the characterization of the maximal number of disjoint, *complete* chains. Another type of representation of partially ordered sets in terms of chains has been considered by Dushnik and Miller [3] (see also Komm [4]). It can be shown that if n is the maximal number of non-comparable elements, then the dimension of P in the sense of Dushnik and Miller is at most n. Except for this fact, there seems to be little connection between the two representations.

a chain. Hence the theorem is trivial in this case and we may make an argument by induction. Let us assume, then, that the theorem holds for all finite partially ordered sets for which the maximal number of independent elements is less than k. Now it will be sufficient to show that if $C_1, \cdots, C_k$ are k disjoint chains of P and if a is an element belonging to none of the C_i, then $C_1 + \cdots + C_k + a$ is a set sum of k disjoint chains. For beginning with a set $a_1, \cdots, a_k$ of independent elements (which exist by hypothesis) we may add one new element at a time and be sure that at each stage we have a set sum of k disjoint chains. Since P is finite, we finally have P itself represented as a set sum of k chains.

Let, then, $C_1, \cdots, C_k$ be k disjoint chains and let a be an element not belonging to $C_1 + \cdots + C_k$. Let U_i be the set of all elements of C_i which contain a, let L_i be the set of all elements of C_i which are contained in a, and let N_i be the set of all elements of C_i which are non-comparable with a. Finally let

$$U = U_1 + \cdots + U_k$$
$$L = L_1 + \cdots + L_k$$
$$N = N_1 + \cdots + N_k$$
$$C = C_1 + \cdots + C_k.$$

Clearly $U_i + N_i + L_i = C_i$ and $U + N + L = C$.

We show now that for some m the maximal number of independent elements in $N + U - U_m$ is less than k. For suppose that for each j there exists a set S_j consisting of k independent elements of $N + U - U_j$. Since there are k elements in S_j and they belong to $C = C_1 + \cdots + C_k$, there is exactly one element of S_j in each of the chains C_i. Since S_j contains no elements of U_j it follows that S_j contains exactly one element of N_j. Thus $S = S_1 + \cdots + S_k$ contains at least one element of N_i for each i. Now let s_i be the minimal element of S which belongs to C_i. s_i exists since the intersection of S and C_i is a finite chain which we have proved to be non-empty. Furthermore, $s_i \, \epsilon \, N_i$ since there is at least one element of N_i which belongs to S and all of the elements of U_i properly contain all of the elements of N_i. Hence $s_1, \cdots, s_k \, \epsilon \, N$. Now if $s_i \geqq s_j$ for $i \neq j$, let $s_j \, \epsilon \, S_r$. Since S_r contains an element t_i belonging to C_i, we have from the definition of s_i that $t_i \geqq s_i \geqq s_j$ and $t_i \neq s_j$ since $t_i \, \epsilon \, C_i$ and $s_j \, \epsilon \, C_j$. But this contradicts our assumption that the elements of S_r are independent. Hence we must have $s_j \ngeqq s_j$ for $i \neq j$ and $s_1, \cdots, s_k$ form an independent set. But since s_i belongs to N, s_i is non-comparable with a and hence $a, s_1, \cdots, s_k$ is an independent set containing $k + 1$ elements. But this contradicts the hypothesis of the theorem and hence we conclude that for some m, the maximal number of independent elements in $N + U - U_m$ is less than k.

In an exactly dual manner it follows that for some l, the maximal number of independent elements in $N + L - L_l$ is less than k.

Now let T be an independent subset of $C - U_m - L_l$. If T contains an element x belonging to $U - U_m$ and an element y belonging to $L - L_l$, then $x \geqq a \geqq y$ contrary to the independence of T. Since

$$(N + U - U_m) + (N + L - L_l) = C - U_m - L_l$$

it follows that T is either a subset of $N + U - U_m$ or of $N + L - L_l$. Hence the number of elements in T is less than k and thus the maximal number of independent elements in $C - U_m - L_l$ is less than k. Since $U_m + L_l$ is a chain there is at least one independent set of $k - 1$ elements in $C - U_m - L_l$. Hence by the induction hypothesis $C - U_m - L_l = C_1' + \cdots + C_{k-1}'$ where $C_1', \cdots,$ C_{k-1}' are disjoint chains. Let C_k' be the chain $U_m + a + L_l$. Then

$$C + a = C_1' + \cdots + C_k'$$

and our assertion is proved.

We turn now to the proof of the general case. Again when $k = 1$ the theorem is trivial and we may proceed by induction. Hence let the theorem hold for all partially ordered sets having at most $k - 1$ independent elements and let P satisfy the hypotheses of the theorem. A subset C of P is said to be *strongly dependent* if for every finite subset S of P, there is a representation of S as a set sum of k disjoint chains such that all of the elements of C which belong to S are members of the same chain. Clearly any strongly dependent subset is a chain. Also from the theorem in the finite case it follows that a set consisting of a single element is always strongly dependent. Since strong dependence is a finiteness property it follows from the Maximal Principle that P contains a maximal strongly dependent subset C_1. Suppose that $P - C_1$ contains k independent elements $a_1, \cdots, a_k$. Then from the maximal property of C_1 we conclude that $C_1 + a_i$ is not strongly dependent for each i. Hence there exists a finite subset S_i such that in any representation as a set sum of k chains there are at least two chains which contain elements of $C_1 + a_i$. S_i must clearly contain a_i since C_1 is strongly dependent. Let $S = S_1 + \cdots + S_k$. By the strong dependence of C_1, $S = K_1 + \cdots + K_k$ where $K_1, \cdots, K_k$ are disjoint chains such that for some $n \leqq k$ we have $S \cdot C_1 \subseteqq K_n$. Since S contains $a_1, \cdots, a_k$ which are independent, for some $m \leqq k$ we have $a_m \, \epsilon \, K_n$. Let K_i' be the chain $S_m \cdot K_i$. Then $S_m = K_1' + \cdots + K_k'$ and $S_m \cdot C_1 \subseteqq S_m \cdot S \cdot C_1 \subseteqq S_m \cdot K_n = K_n'$. But by definition $a_m \, \epsilon \, S_m$ and $a_m \, \epsilon \, K_n$. Hence $S_m \cdot (C_1 + a_m) \subseteqq K_n'$ which contradicts the definition of S_m. We conclude that $P - C_1$ contains at most $k - 1$ independent elements. But since C_1 is a chain and P contains a set of k independent elements, it follows that $P - C_1$ contains a set of $k - 1$ independent elements. Thus by the induction hypothesis we have $P - C_1 = C_2 + \cdots + C_k$. Hence

$$P = C_1 + \cdots + C_k$$

and the proof of the theorem is complete.

3. Application to representatives of sets.

G. Kreweras has proved the following extension of the Radó-Hall theorem on representatives of sets:

Let $\mathfrak{A}$ and $\mathfrak{B}$ be two partitions of a set into n parts and let h be the smallest number such that for any r, r parts of $\mathfrak{A}$ contain at most $r + h$ parts of $\mathfrak{B}$. Let k be the smallest number such that $n + k$ elements serve to represent both partitions. Then $h = k$.

To show the power of Theorem 1.1 we shall prove an even more general theorem in which the partition requirement is dropped. Now if $\mathfrak{A}$ is any finite collection of subsets of a set S we shall say that a set of n elements (repetitions being counted) represents $\mathfrak{A}$ if there exists a one-to-one correspondence of the sets of $\mathfrak{A}$ onto a subset of the n elements such that each set contains its corresponding element. For example, the set $\{1, 1, 1\}$ represents the three sets $\{1, 2\}$, $\{1, 3\}$, and $\{1, 4\}$. The theorem can then be stated as follows:

THEOREM 3.1. *Let $\mathfrak{A}$ and $\mathfrak{B}$ be two finite collections of subsets of some set. Let $\mathfrak{A}$ and $\mathfrak{B}$ contain m and n sets respectively. Let h be the smallest number such that for every r, the union of any $r + h$ sets of $\mathfrak{A}$ intersects at least r sets of $\mathfrak{B}$. Let k be the smallest number such that $n + k$ elements serve to represent both collections $\mathfrak{A}$ and $\mathfrak{B}$. Then $h = k$.*

It can be easily verified that if $\mathfrak{A}$ and $\mathfrak{B}$ are partitions of a set, then h as defined in Theorem 2.1 is equivalent to the definition given in the theorem of Kreweras.

For the proof let $\mathfrak{A}$ consist of sets $A_1, \cdots, A_m$ and $\mathfrak{B}$ consist of sets $B_1, \cdots, B_n$. We make the sets $A_1, \cdots, A_m, B_1, \cdots, B_n$ into a partially ordered set P as follows:

$$A_i \geqq A_i \quad i = 1, \cdots, m$$

$$B_j \geqq B_j \quad j = 1, \cdots, n.$$

$$A_i \geqq B_j \text{ if and only if } A_i \text{ and } B_j \text{ intersect.}$$

It is obvious that P is a partially ordered set under this ordering. Now let w be the maximal number of independent elements of P. Since the union of any $r + h$ sets of $\mathfrak{A}$ intersects at least r sets of $\mathfrak{B}$, it follows that any independent subset of P can have at most $r + h + (n - r) = n + h$ elements. Hence $w \leqq n + h$. On the other hand for some r there are $r + h$ sets of $\mathfrak{A}$ whose union intersects precisely r sets of $\mathfrak{B}$. Hence these $r + h$ sets of $\mathfrak{A}$ and the remaining $n - r$ sets of $\mathfrak{B}$ form an independent subset of P containing $n + h$ elements. Thus $w = n + h$. By Theorem 1.1, P is the set sum of w chains $C_1, \cdots, C_w$. Now if a chain C_i contains two sets they have a non-null intersection by definition. Hence for each C_i there is an element a_i common to the sets of C_i. But since $A_1, \cdots, A_m$ are independent in P it follows that they belong to different chains and hence the w elements $a_1, \cdots, a_w$ represent $\mathfrak{A}$. Similarly, $a_1, \cdots, a_w$ represent $\mathfrak{B}$ and thus $n + k \leqq w$. But since P cannot be represented as a set sum of less than w chains, it follows that $n + k = w = n + h$. Hence $h = k$ and the theorem is proved.

4. Proof of Theorem 1.2.

Let us recall that an element q of a finite distributive lattice D is (*union*) *irreducible* if $q = x \cup y$ implies $q = x$ or $q = y$. It can be easily verified that if q is irreducible, then $q \leqq x \cup y$ implies $q \leqq x$ or $q \leqq y$. From the finiteness[2] of S it

---

[2] L is assumed to be finite for sake of simplicity. The theorem holds without this restriction. In the proof, "elements covered by a" must be replaced by "maximal ideals in a" and "irreducible elements" must be replaced by "prime ideals."

follows that every element of D can be expressed as a union of irreducible elements. From this fact we conclude that if $x > y$, there exists at least one irreducible q such that $x \geqq q$ and $y \not\geqq q$.

Now let P be the partially ordered set of union irreducible elements of D. Let a be such that $k = k(a)$. Then there are k elements $a_1, \cdots, a_k$ which cover a. Let q_i be an irreducible such that $a_i \geqq q_i$ and $a \not\geqq q_i$. Then if $q_i \geqq q_j$ where $i \neq j$ we have $a = a_i \cap a_j \geqq q_i \cap q_j \geqq q_j$ which contradicts $a \not\geqq q_j$. Hence $q_1, \cdots, q_k$ are an independent set of elements of P.

Next let $q_1', \cdots, q_l'$ be an arbitrary independent subset of P. Let $a' = q_1' \cup \cdots \cup q_l'$ and for each i let $p_i' = q_1' \cup \cdots \cup q_{i-1}' \cup q_{i+1}' \cup \cdots \cup q_l'$. Now if $p_i' = a'$ for some i, then

$$q_i' = q_i' \cap a' = q_i' \cap p_i'$$
$$= (q_i' \cap q_1') \cup \cdots \cup (q_i' \cap q_{i-1}') \cup (q_i' \cap q_{i+1}') \cup \cdots \cup M\,(q_i' \cap q_l')$$

and hence $q_i' = q_i' \cap q_j'$ for some $j \neq i$. But then $q_j' \geqq q_i'$ contrary to independence. Thus $a' > p_i'$ for each i and $p_i' \cup p_j' = a'$ for $i \neq j$. Let $a = p_1' \cap \cdots \cap p_l'$ and for each i let $p_i = p_1' \cap \cdots \cap p_{i-1}' \cap p_{i+1}' \cap \cdots \cap p_l'$. If $p_i = a$, then $p_i' = p_i' \cup a = p_i' \cup p_i = (p_i' \cup p_1') \cap \cdots \cap (p_i' \cup p_{i-1}') \cap (p_i' \cup p_{i+1}') \cap \cdots \cap (p_i' \cup p_l') = a'$ which contradicts $p_i' < a'$. Hence $p_i > a$ and $p_i \cap p_j = a$ for $i \neq j$. Let $p_i \geqq a_i$ where a_i covers a. Then $a \leqq a_i \cap a_j \leqq p_i \cap p_j = a$ for $i \neq j$ and hence $a_i \cap a_j = a$, $i \neq j$. Thus $a_1, \cdots, a_l$ are distinct elements of D covering a. It follows that $l \leqq k$ and hence k is the maximal number of independent elements of P.

Now by Theorem 1.1 P is the set sum of k disjoint chains $C_1, \cdots, C_k$. We adjoin the null element z of D to each of the chains C_i. Then for each $x \in D$, there is a unique maximal element x_i in C_i which is contained in x. Now suppose $x > x_1 \cup \cdots \cup x_k$ in D. Then there exists an irreducible q such that $x \geqq q$ and $x_1 \cup \cdots \cup x_k \not\geqq q$. But $q \in C_i$ for some i and hence $x_1 \cup \cdots \cup x_k \geqq x_i \geqq q$ contrary to the definition of q. Hence $x = x_1 \cup \cdots \cup x_k$. Consider the mapping of D into the direct union of $C_1, \cdots, C_k$ given by

$$x \to \{x_1, \cdots, x_k\}.$$

Now if $x_i = y_i$ for $i = 1, \cdots, k$, then $x = x_1 \cup \cdots \cup x_k = y_1 \cup \cdots \cup y_k = y$ and the mapping is thus one-to-one. Since $x \cup y \geqq x_i \cup y_i$ we have $(x \cup y)_i \geqq x_i \cup y_i$. But since $(x \cup y)_i$ is union irreducible we get $x \cup y \geqq (x \cup y)_i \to x \geqq (x \cup y)_i$ or $y \geqq (x \cup y)_i \to x_i \geqq (x \cup y)_i$ or $y_i \geqq (x \cup y)_i \to x_i \cup y_i \geqq (x \cup y)_i$. Thus $(x \cup y)_i = x_i \cup y_i$ and we have

$$x \cup y \to \{x_1 \cup y_1, \cdots, x_k \cup y_k\}.$$

Similarly $x \cap y \geqq x_i \cap y_i \to (x \cap y)_i \geqq x_i \cap y_i$. But $x \geqq x \cap y \to x_i \geqq (x \cap y)_i$ and $y \geqq x \cap y \to y_i \geqq (x \cap y)_i$. Hence $x_i \cap y_i \geqq (x \cap y)_i$. Thus $(x \cap y)_i = x_i \cap y_i$ and we have

$$x \cap y \to \{x_1 \cap y_1, \cdots, x_k \cap y_k\}.$$

This completes the proof that D is isomorphic to a sublattice of a direct union of k chains.

Now suppose that D is a sublattice of the direct union of l chains $C'_1, \cdots, C'_l$ where $l < k$. Again let a be such that $k(a) = k$ and let $a_1, \cdots, a_k$ be the k distinct elements covering a. Define $a' = a_1 \cup \cdots \cup a_k$ and let $a'_i = a_1 \cup \cdots \cup a_{i-1} \cup a_{i+1} \cup \cdots \cup a_k$ for each i. Now $a'_i = q'_1 \cup \cdots \cup q'_i$ where $q'_i \in C'_i$. And if $q'_i = x' \cup y'$, then $q'_i = x'_i \cup y'_i$ where $x'_i, y'_i \in C'_i$. But then either $q'_i = x'_i \cup y'_i = x'_i$ or $q'_i = x'_i \cup y'_i = y'_i$ and hence either $q'_i = x'$ or $q'_i = y'$. Thus each q'_i is union irreducible. But $a_1 \cup \cdots \cup a_k = a' \geq q'_i$ for $i = 1, \cdots, l$. Thus for each $i \leq l$ there is a j such that $a_j \geq q'_i$. Since $l < k$ there is some r such that $a'_r \geq q'_i \cup \cdots \cup q'_l = a' \geq a_r$. But then $a_r = a'_r \cap a_r = a$ which contradicts the fact that a_r covers a. Hence $l \geq k$ and we conclude that k is the least number of chains whose direct union contains D as a sublattice. This completes the proof of Theorem 1.2.

YALE UNIVERSITY
CALIFORNIA INSTITUTE OF TECHNOLOGY

REFERENCES

1. P. HALL. *On representatives of subsets.* J. London Math. Soc. 10 (1935), 26–30.
2. G. KREWERAS. *Extension d'un théorème sur les répartitions en classes.* C. R. Acad. Sci. Paris 222 (1946), 431–432.
3. B. DUSHNIK AND E. W. MILLER. *Partially ordered sets.* Amer. J. of Math. vol. 63 (1941), 600–610.
4. H. KOMM. *On the dimension of partially ordered sets.* Amer. J. of Math. vol. 20 (1948), 507–520.

SOME COMBINATORIAL PROBLEMS ON
PARTIALLY ORDERED SETS

BY

R. P. DILWORTH

1. Introduction. This paper is concerned with some combinatorial problems related to the following theorem on partially ordered sets:

The minimal number of chains in the representation of a finite partially ordered set P as a set union of chains is equal to the maximal number of mutually non-comparable elements of P.

The theorem was first formulated and proved in connection with a problem on subdirect union representations of distributive lattices (Dilworth [1]). It was well known that any distributive lattice could be represented as a subdirect union of chains and the problem concerned the minimal number of chains required for such a representation. Now it is easily shown that subdirect union representations of a finite distributive lattice in terms of chains correspond to the decomposition of the partially ordered set of join irreducibles into a set union of chains. On the other hand, a mutually non-comparable set of n join irreducibles leads directly to an element s of the distributive lattice which covers exactly n elements of the distributive lattice. These elements are the *cover set* of s. Conversely, every cover set of an element of the lattice produces a collection of non-comparable join irreducibles having the same number of elements. Thus the theorem on partially ordered sets gives the following theorem for distributive lattices.

The minimal number of chains in the representation of a finite distributive lattice as a subdirect union of chains is equal to the maximal number of elements in the cover sets of the lattice.

As a by-product of the investigation, it was observed that a wide variety of theorems on representatives of sets could be derived directly from the theorem by applying it to the partially ordered set obtained from a collection of subsets together with the elements themselves by ordering each subset to the elements which it contains. In particular, the Rado-Hall theorem on representatives of sets is an immediate consequence of the theorem.

The next development in connection with the theorem was the discovery by Dantzig and Hoffman [2] that the problem of finding the minimal decomposition of a finite partially ordered set into disjoint chains can be formulated as a transportation type linear programming problem and that the theorem follows from the duality theorem of linear inequality theory. If $P_1, \cdots, P_n$ are the elements of a finite partially ordered set Dantzig and Hoffman consider the array $\{x_{ij}\}$, $i, j = 0, 1, \cdots, n$ and require that x_{00} be a maximum subject to the restrictions

$$\sum_{i=0}^{n} x_{ij} = \begin{cases} n, & j = 0, \\ 1, & j \neq 0; \end{cases}$$

$$\sum_{j=0}^{n} x_{ij} = \begin{cases} n, & i = 0, \\ 1, & i \neq 0; \end{cases}$$

$$x_{ij} \geqq 0; \quad x_{ij} = 0 \quad \text{if} \quad P_i \nleq P_j \quad \text{or if} \quad i = j \neq 0.$$

There are always integers x_{ij} which give the required maximum. The maximum value for x_{00} is $n - m$ where m is the minimal number of chains in the representation of the partially ordered set as a set union of chains. A chain of the minimal representation may be obtained as follows: Select P_j so that $x_{0j} = 1$. Then there is exactly one element x_{jk} in the jth row which is equal to 1. If $k \neq 0$, $P_j \geqq P_k$. Similarly there is exactly one element x_{kl} in the kth row which is equal to 1. If $l \neq 0$, then $P_j \geqq P_k \geqq P_l$. This process continues until an $x_{r0} = 1$ is obtained in which case P_r terminates the chain. One of the important consequences of the work of Dantzig and Hoffman is the fact that the techniques of linear programming may be used to construct the chains of a minimal representation.

Now it had earlier been observed by Birkhoff that there is a certain formal resemblance between the partially ordered set theorem and a theorem of König on linear graphs. If V denotes the set of vertices of a linear graph, let $V = V_1 + V_2$ be a partition of V into two disjoint subsets. A *cut* of the graph is a collection of vertices such that every edge joining a vertex of V_1 to a vertex of V_2 has at least one of its end points in the collection. A *join* is a collection of edges joining vertices of V_1 to vertices V_2 such that no two edges of the collection have a vertex in common. König's theorem asserts that *the minimal number of vertices forming a cut of the graph is equal to the maximal number of edges making up a join of the graph.*

The König theorem can also be obtained as an application of the duality theorem of linear programming. Furthermore, the formulation of the problem in linear programming terms is so closely similar to the transportation problem described above that Fulkerson [3] succeeded in deducing the partially ordered set theorem from König's theorem and conversely.

The method employed by Fulkerson is as follows: If $P_1, \cdots, P_n$ are the elements of the partially ordered set, let V be the set of $2n$ vertices $a_1, \cdots, a_n$, $b_1, \cdots, b_n$ where a_i is joined to b_j by an edge if and only if $P_i > P_j$. Let $V_1 = \{a_1, \cdots, a_n\}$ and $V_2 = \{b_1, \cdots, b_n\}$. If D is a decomposition of the partially ordered set into chains, the collection of edges $a_i b_j$ where P_i covers P_j in one of the chains clearly forms a join J of the graph. Conversely every join leads to a decomposition of the partially ordered set. It is easily shown that $n(D) + n(J) = n$. Furthermore if U is a maximal non-comparable subset of the partially ordered set, then each $P_i \notin U$ is such that either $P_i > u$ or $u > P_i$ for some $u \in U$. Set $a_i \in C$ if $P_i > a$ and $b_i \in C$ if $u > P_i$. Then C is a cut of the graph and $n(U) + n(C) = n$. The equivalence of the two theorems can clearly be deduced from these formulas.

As noted above, the Rado-Hall theorem on representatives of sets is an immediate consequence of the decomposition theorem for partially ordered sets. On the other hand, it is well known that the König theorem on graphs can be easily derived from the Rado-Hall theorem. Thus, making use of Fulkerson's correspondence between partially ordered sets and graphs, the partially ordered set theorem can be proved from the Rado-Hall theorem. However such a derivation is clearly indirect and moreover requires the introduction of a somewhat artificial auxiliary partially ordered set. Actually, there is a much more intimate relation between the partially ordered set theorem and the Rado-Hall theorem. I will develop this relationship in the following section and will also show that there is still another connection between the theorem and properties of distributive lattices.

2. **The theory of maximal non-comparable sets.** Let us consider first a class of partially ordered sets for which the decomposition theorem is immediately equivalent to the Rado-Hall theorem. Let $A = \{a_1, \cdots, a_n\}$, $B = \{b_1, \cdots, b_n\}$ be two sets each of which contain n distinct elements. A class of pairs (a_i, b_j) such that $a_i, b_j \notin A \cap B$ is selected. We write $a_i > b_j$ if (a_i, b_j) belongs to the selected class. Denote this partially ordered set by $P(A, B)$. For each a_i let S_{a_i} denote the set of b_j such that $a_i \geq b_j$.

LEMMA 2.1. *The maximal number of mutually non-comparable elements in $P(A, B)$ is n if and only if the union of any k of the subsets S_{a_i} contains at least k elements.*

For if $S_{a_1} \vee \cdots \vee S_{a_k} = S$ contains less than k elements, $(B - S) \vee \{a_1, \cdots, a_k\}$ is a non-comparable subset of $P(A, B)$ with more than n elements. Conversely, if C is a non-comparable subset of $P(A, B)$ with more than n elements let $C \cap A = \{a_1, \cdots, a_k\}$. Then $C \cap B$ contains more than $n - k$ elements. But since $(S_{a_1} \vee \cdots \vee S_{a_k}) \wedge C = \emptyset$, it follows that $S_{a_1} \vee \cdots \vee S_{a_k}$ contains less than k elements.

LEMMA 2.2. *The minimal number of chains whose union contains $P(A, B)$ is n if and only if there exists a set of distinct representatives for the sets $S_{a_1}, \cdots, S_{a_n}$.*

For if $P(A, B)$ is the set union of n chains, then each chain contains exactly one element a_i of A and b_j of B. The b_j provide a set of distinct representatives for the sets S_{a_i}. Clearly a set of distinct representatives for the S_{a_i} gives a representation of $P(A, B)$ as a set union of n chains.

From Lemmas 2.1, 2.2, and the Rado-Hall theorem it follows that if the maximal number of mutually non-comparable elements in $P(A, B)$ is n, then $P(A, B)$ is the set union of n distinct chains. These chains, in turn, define a 1-1 mapping of A onto B, $a_i \to b_j$ such that $a_i \geq b_j$.

Now let P be a finite partially ordered set in which n is the maximal number of mutually non-comparable elements. The proof that P is the set union of n chains can be easily reduced to the case in which each element of P belongs to at least one non-comparable set having n elements. For let P' be the union of all such non-comparable sets in P. Then if P' is the set union of n chains, let C be one of these chains. Suppose $P - C$ contains a non-comparable subset with n elements. Then the elements of this set belong to P' and at least one of them belongs to C contrary to the assumption that the elements belong to $P - C$. Thus the maximal number of mutually non-comparable elements in $P - C$ is $n - 1$. Induction on n then completes the proof for the partially ordered set P.

We shall assume now that each element of P belongs to at least one non-comparable set of n elements.

Let L denote the collection of non-comparable subsets of P having n elements. We shall call these sets n-sets. If Q and Q' are n-sets, a partial ordering on L is defined by the relation $Q \leq Q'$ if and only if for each $q \in Q$, there exists $q' \in Q'$ such that $q \leq q'$. The following lemma shows that this relation is self dual.

LEMMA 2.3. $Q \leq Q'$ *if and only if* $q' \in Q'$ *implies that there exists* $q \in Q$ *such that* $q \leq q'$.

For let $Q \leq Q'$ and let q' be an arbitrary element of Q'. Then since $\{q'\} \vee Q$ is no longer non-comparable, there exists $q \in Q$ which is comparable with q'. If $q \leq q'$, the lemma follows. If $q' \leq q$, then it follows from $Q \leq Q'$ that $q'' \in Q'$ exists such that $q \leq q''$. But then $q' \leq q''$ and since Q' is a non-comparable set we must have $q' = q''$. Hence $q \leq q'$. Thus the necessity of the condition is proved and a dual argument gives the sufficiency.

If Q and Q' are arbitrary n-sets of P, let M denote the maximal elements of $Q \vee Q'$ and let N denote the minimal elements of $Q \vee Q'$. Clearly M and N are non-comparable subsets of P. We shall show that they are n-sets of P.

LEMMA 2.4. $M \vee N = Q \vee Q'$.

Clearly $M \vee N \leq Q \vee Q'$. Suppose that $s \in Q \vee Q'$ but $s \notin M \vee N$. If $s \in Q$, then since $s \notin N$ there exists $s_1 \in Q \vee Q'$ such that $s_1 < s$. Similarly since $s \notin M$ there exists $s_2 \in Q \vee Q'$ such that $s < s_2$. Since s_1 and s are comparable, it follows that $s_1 \in Q'$. Similarly we find that $s_2 \in Q'$. But $s_1 < s_2$ contrary to the non-comparability of Q'. If $s \in Q'$, a similar contradiction can be obtained for Q. Thus s must belong to $M \vee N$ and the lemma is proved.

LEMMA 2.5. $M \wedge N = Q \wedge Q'$.

For let $s \in M \wedge N$. Then if s is comparable with an element t of $Q \vee Q'$ we must have $s = t$ since s is both maximal and minimal. But s is comparable with at least one element of Q and hence $s \in Q$. Similarly $s \in Q'$

and thus $s \in Q \wedge Q'$. Conversely, if $s \in Q \wedge Q'$ then $s = q = q'$ where $q \in Q$ and $q' \in Q'$. Let $x \geq s$. Then if $x \in Q$ we have $x \geq q \Rightarrow x = q = s$. If $x \in Q'$, then $x \geq q' \Rightarrow x = q' = s$. Thus $x \geq s \Rightarrow x = s$ and hence $s \in M$. A dual argument shows that $s \in N$ and hence that $s \in M \wedge N$.

LEMMA 2.6. *M and N are n-sets of P.*

For let $n(A)$ denote the number of elements in A. Then $n(M) + n(N) = n(M \vee N) + n(M \wedge N) = n(Q \vee Q') + n(Q \wedge Q') = n(Q) + n(Q') = 2n$. But since M and N are sets consisting of mutually non-comparable elements, we have $n(M) \leq n$ and $n(N) \leq n$. But then $n(M) = n(N) = n$ and M and N are n-sets.

LEMMA 2.7. *M, N are respectively the l.u.b. and g.l.b. of Q and Q' under the partial ordering of L.*

For $q \in Q$ implies $q \leq s$ where s is a maximal element of $Q \vee Q'$. Hence $s \in M$ and thus $Q \leq M$. Similarly $Q' \leq M$. Now let $Q \leq R$, $Q' \leq R$ where R is an n-set. Let $s \in M$. Then $s \in Q$ or $s \in Q'$. If $s \in Q$ there exists $r \in R$ such that $s \leq r$. A similar argument holds if $s \in Q'$. Hence $M \leq R$ and M is the l.u.b. of Q and Q' in L. A dual argument shows that N is the g.l.b. and the proof of the lemma is thus complete.

It follows from Lemma 2.7, that L is a lattice under the partial ordering of n-sets. In fact, a much stronger result holds.

THEOREM 2.1. *L is a distributive lattice.*

Proof. Let Q_1, Q_2, and Q_3 be n-sets of P. Let $s \in Q_1 \cap (Q_2 \cup Q_3)$. Then s is a minimal element of $Q_1 \vee (Q_2 \cup Q_3)$. Let us suppose first that $s \in Q_1$. Then since $Q_1 \cap (Q_2 \cup Q_3) \leq Q_2 \cup Q_3$, it follows that $s \leq t$ where $t \in Q_2 \cup Q_3$. Then $t \in Q_2 \vee Q_3$ and, by symmetry it will suffice to consider the case $t \in Q_2$. Now suppose that s is not a minimal element of $Q_1 \vee Q_2$. Then $r < s$ where $r \in Q_1 \vee Q_2$. Since $s \in Q_1$ it follows that $r \notin Q_1$ and hence $r \in Q_2$. But then $r < t$ where $r, t \in Q_2$ contrary to the non-comparability of Q_2. Thus $s \in Q_1 \cap Q_2$. But $Q_1 \cap Q_2 \leq (Q_1 \cap Q_2) \cup (Q_2 \cap Q_3)$ implies that $s \leq s'$ where $s' \in (Q_1 \cap Q_2) \cup (Q_1 \cap Q_3)$. Next let us suppose that $s \notin Q_1$. Then $s \in Q_2 \cup Q_3$ and s is a maximal element of $Q_2 \vee Q_3$. Since $Q_1 \cap (Q_2 \cup Q_3) \leq Q_1$, it follows that there exists $t \in Q_1$ such that $s \leq t$. Again, by symmetry, we may suppose that $s \in Q_2$. If s is not a minimal element of $Q_1 \vee Q_2$ then $r < s$ where $r \in Q_1$. But then $r < t$ contrary to the non-comparability of Q_1. Thus $s \in Q_1 \cap Q_2$ and hence we again have $s \leq s'$ where $s' \in (Q_1 \cap Q_2) \cup (Q_1 \cap Q_3)$. Thus $Q_1 \cap (Q_2 \cup Q_3) \leq (Q_1 \cap Q_2) \cup (Q_1 \cap Q_3)$ and L is distributive.

For the proof of the next theorem the following lemma is required.

LEMMA 2.8. *If $q < q'$ where $q \in Q$ and $q' \in Q'$, then $q \in Q \cap Q'$ and $q' \in Q \cup Q'$.*

For if q is not a minimal element of $Q \vee Q'$, there exists $q'' \in Q \vee Q'$ such that $q'' < q$. But then $q'' \in Q'$ and $q'' < q'$ contrary to the non-comparability of Q'. Thus $q \in Q \cap Q'$ and a similar argument shows that $q' \in Q \cup Q'$.

THEOREM 2.2. *Let $Q_1 \leqq Q_2 \leqq \cdots \leqq Q_m$ be a maximal chain of L. Then $P = Q_1 \vee Q_2 \vee \cdots \vee Q_m$.*

Proof. Let $q \in P$. Then q is comparable with at least one element of Q_i for each i. Now by hypothesis $q \in Q$ for some $Q \in L$. Since $Q_1 \leqq \cdots \leqq Q_m$ is a maximal chain, Q_1 is the minimal n-set of L and hence $Q_1 \leqq Q$. Thus $q_1 \leqq q$ for some $q_1 \in Q_1$. Let k be maximal such that $q_k \leqq q$ for some $q_k \in Q_k$. If $k = m$, then since Q_m is the maximal n-set of L, we have $q \leqq q'_m$ for some $q'_m \in Q_m$. But then $q_m \leqq q \leqq q'_m$ and thus $q = q_m = q'_m \in Q_m \subseteq Q_1 \vee \cdots \vee Q_m$. If $k < m$, then q is comparable with an element of Q_{k+1} and hence by the maximal property of k we must have $q < q_{k+1}$ where $q_{k+1} \in Q_{k+1}$. Now suppose that $q_k < q$. Then by Lemma 2.8, $q \in Q_k \cup Q$. Again by Lemma 2.8 since $q < q_{k+1}$ we have $q \in (Q_k \cup Q) \cap Q_{k+1}$. But $Q_k \leqq (Q_k \cup Q) \cap Q_{k+1} \leqq Q_{k+1}$ and by the maximal property of the chain we must have $Q_k = (Q_k \cup Q) \cap Q_{k+1}$ or $(Q_k \cup Q) \cap Q_{k+1} = Q_{k+1}$. Hence either $q \in Q_k$ or $q \in Q_{k+1}$. But $q_k < q$ contradicts $q \in Q_k$ and $q < q_{k+1}$ contradicts $q \in Q_{k+1}$. Thus we must have $q = q_k$. Hence $q \in Q_k \subseteq Q_1 \vee \cdots \vee Q_m$.

Making use of this theorem we can now apply the Rado-Hall theorem to give a simple direct construction of a representation of P as a set union of n-chains. For the maximal number of non-comparable elements in $P(Q_2, Q_1)$ is n. Thus, by the Rado-Hall theorem there is a one-to-one mapping of Q_1 onto Q_2 such that if $q_1 \to q_2$ then $q_1 \leqq q_2$. Similarly there is a one-to-one mapping of Q_2 onto Q_3 such that if $q_2 \to q_3$ then $q_2 \leqq q_3$. Continuing in this manner we get a chain $q_1 \leqq q_2 \leqq \cdots \leqq q_m$. The n possible choices for q_1 give the n chains of the representation. Clearly the union of the chains is $Q_1 \vee \cdots \vee Q_m$ which by Theorem 2.2 is the partially ordered set P.

REFERENCES

1. R. P. Dilworth, *A decomposition theorem for partially ordered sets*, Ann. of Math. vol. 51 (1950) pp. 161–166.

2. G. B. Dantzig and A. Hoffman, *On a theorem of Dilworth*, Contributions to linear inequalities and related topics, Annals of Mathematics Studies, no. 38.

3. D. R. Fulkerson, *Note on Dilworth's decomposition for partially ordered sets*, Proc. Amer. Math. Soc. vol. 7 (1956) pp. 701–702.

4. P. Hall, *On representatives of subsets*, J. London Math. Soc. vol. 10 (1935) pp. 26–30.

5. D. König, *Theorie der Graphen*, New York, Chelsea, 1950.

CALIFORNIA INSTITUTE OF TECHNOLOGY,
PASADENA, CALIFORNIA

The Impact of the Chain Decomposition Theorem
on Classical Combinatorics

KENNETH P. BOGART, CURTIS GREENE
AND JOSEPH P. S. KUNG

1. Introduction. Dilworth's chain decomposition has become such a standard concept for specialists in ordered sets that at the NATO Advanced Study Institute on Ordered Sets held in Banff in 1981 participants would use the phrase "Dilworth-type theorem" when they meant "minimax theorem" or when they meant "partition theorem." Harper and Rota, in their foundational survey of matching theory [30], describe a number of equivalent matching theorems which can be stated as minimax theorems and are all derivable from each other. Among these, they choose Dilworth's theorem as "perhaps the most elegant." Mirsky and Perfect regard Dilworth's theorem as perhaps "the most fundamental among the finite results" of this type in matching theory [46].

The impact of Dilworth's theorem on classical finite combinatorics has had several directions. The directions we shall discuss are

- Alternate proofs of Dilworth's theorem
- Extensions and generalizations of Dilworth's theorem
- Relationships with operations research
- Applications to dimension theory.

Each of these areas is a major chapter of classical combinatorics which has sprung from Dilworth's work. Connections of Dilworth's theorem with recursive function theory and infinite set theory are taken up in other articles in this chapter.

There are many other applications of Dilworth's theorem, too many to enumerate in a short survey. We shall give one example here. An important special case of Dilworth's theorem is the case when the ordered set P is constructed from a finite sequence $(a_1, a_2, ..., a_n)$ of positive integers in the following way: P is the set of ordered pairs (a_i, i), ordered by the product order: $(a_i, i) < (a_j, j)$ if $a_i \leq a_j$ and $i < j$. In P, a chain is a nondecreasing subsequence and an antichain is a nonincreasing subsequence. From this, one can derive the following theorem of Erdős and Szekeres [16] by a simple counting argument: *Any finite sequence of length $n^2 + 1$ contains a monotone subsequence of length $n + 1$.*

Graham, Rothschild, and Spencer have observed in [26, p.17] that Dilworth's theorem can be interpreted as a "Ramsey theorem." To see this, observe that the Erdős-Szekeres theorem is in the spirit of Ramsey theory, in the sense that it asserts, in the words of Burkill and Mirsky [7], "that every system of a certain class possesses a large subsystem with a higher degree of organization than the original system." With the Erdős-Szekeres theorem as a model, Dilworth's theorem can be recast as follows:

> *Any ordered set P of size at least $ab + 1$ contains either a chain of length of $a + 1$ or an antichain of size $b + 1$.*

The lower bound $ab + 1$ is best possible and thus, Dilworth's theorem allows the calculation of an exact Ramsey function.

2. Alternative proofs. Since 1950, many alternative proofs of Dilworth's decomposition theorem have appeared. These proofs can be divided roughly into two groups: direct elementary arguments and indirect arguments deriving Dilworth's theorem from other theorems in combinatorics.

In addition to Dilworth's own alternative proof given in [15], direct arguments were given by Perles [49], Tverberg [57], and Pretzel [53]. We shall sketch these proofs. The *width* of a finite ordered set P is the maximum size of an antichain or independent subset in P.

Perles argues by induction on the number of elements. Let P be a finite ordered set of width m. First suppose that there exists a maximum-sized antichain $\{a_1, a_2, ..., a_m\}$ in P not equal to the set of maximal elements or the set of minimal elements. Define P^+ to be the subset $\{x : x \geq a_i \text{ for some } i\}$ and P^- to be the subset $\{x : x \leq a_i \text{ for some } i\}$. The subsets P^+ and P^- are proper subsets of P with union P and intersection $\{a_1, a_2, ..., a_m\}$. By induction, P^+ and P^- can be decomposed into m chains. Patching together these chains at the elements a_i yields a decomposition of P in m chains. One can now suppose that a maximum-sized antichain in P must the set of maximal elements or the set of minimal elements. Let a be a maximal element and b a minimal element such that $a \geq b$. (It is possible that $a = b$.) The width of $P - \{a, b\}$ is $m - 1$ and by induction $P - \{a, b\}$ can be decomposed into $m - 1$ chains. These chains, together with the chain $\{a, b\}$, yield a decomposition of P into m chains.

Tverberg's argument is also by induction. If there exists a maximal chain C whose removal decreases the width of P, then C together with $m - 1$ chains in a decomposition of $P - C$ yield a decomposition of P into m chains. Otherwise, proceed as in the first case of Perles' proof.

Both Perles and Tverberg made implicit use of a saturated chain, that is, a chain intersecting all maximum-sized antichains. Pretzel's proof starts with a method for

                                                         THE DILWORTH THEOREMS

constructing saturated chains. Dilworth [15] has proved that the maximum-sized antichains of P form a lattice with the join (respectively, meet) of two maximum-sized antichains A and B defined to be the set of maximal (respectively, minimal) elements of the subset $A \cup B$. Using this fact, we can construct a saturated chain $a_1 < a_2 < a_3 < \cdots$ recursively by choosing a_{i+1} to be an element in the meet of all the maximum-sized antichains not intersecting $\{a_1, a_2, ..., a_i\}$ such that $a_{i+1} > a_i$. Since the width of $P - C$, where C is a saturated chain, is smaller than the width of P, Dilworth's theorem follows by induction.

Freese [21] has given another proof of the theorem in Dilworth [15] that the maximum-sized antichains form a distributive lattice. He used this fact to prove a result of Kleitman, Edelberg, and Lubell [37] that there exists a maximum-sized antichain which is a union of orbits of the group of automorphisms of the ordered set.

Dilworth's theorem is one of the first examples of a minimax theorem in combinatorics and can be derived from several such theorems. In fact, Dilworth's theorem can also be derived from results in matching theory, linear programming, network flow theory, and graph theory. As described in more detail in the next section, Fulkerson [22] derived it from König's maximum-matching minimum-vertex-cover theorem. Accounts of how Dilworth's theorem fits into matching theory can be found in [30,43,45,50]. Ford and Fulkerson [19] and Gallai [23] derived it from the Ford-Fulkerson theorem relating a maximum flow and minimum cut in a network. Dantzig and Hoffman [12] obtained it as a consequence of the duality theorem of linear programming. Using the fact that an incidence matrix associated with the ordered set is totally unimodular (that is, has all its subdeterminants equal to 0, $+1$, or -1), Hoffman [32] derived Dilworth's theorem from the duality and integrality theorems for totally unimodular linear programs. [33] and [56, p. 275] contain excellent summaries of how linear programming can be used to prove Dilworth's theorem.

Now let G be a directed graph and $\beta_0(G)$ be the maximum size of a set of vertices, no two of which are connected by an edge. Gallai and Milgram [24] proved that G can be decomposed into $\beta_0(G)$ paths such that every vertex is contained in exactly one of the paths. From this, they derived Dilworth's theorem. An edge-covering analogue of the Gallai-Milgram theorem, and hence, of Dilworth's theorem, can be found in [39]. Deming [13] showed that for G an undirected graph, $\beta_0(G)$ equals the maximum over all acyclic orientations ω of the size of a minimum chain decomposition of the directed graph G oriented according to ω. Dilworth's theorem follows from this result. Finally, Dilworth's theorem is equivalent to the theorem that the incomparability graph of an ordered set (that is, the graph on the elements of the ordered set with two elements joined by an edge whenever they are incomparable) is perfect (that is, its chromatic number equals the size of a maximum-sized clique). Thus, it can be proved using Lovász' perfect graph theorem and the easy fact that the comparability graph (that is, the complement of the incomparability graph) is perfect [41,42].

3. Extensions. Since Dilworth's theorem deals with partitioning an ordered set into a union of chains, it is a natural misconception that it tells us the minimum number of chains needed to partition the ordering (rather than the underlying set) or at least the minimum number of chains needed to determine the ordering. However, even with three elements, it is possible to give examples of two non-isomorphic orderings of a set and a partition of the set into classes which are a minimum-sized chain decomposition for each ordering. Thus it is natural to ask for the minimum number of chains needed to write the ordering as the union of the orderings of the chains or at least as the transitive closure of this union. To answer this question, we define two covering pairs (x, y) and (z, w) (with x covered by y and z covered by w) to be incomparable if one of x or y is incomparable with one of z or w. Bogart [4] showed that the minimum number of chains such that the transitive closure of the union of their orderings is the ordering of an ordered set is the maximum size of a set of mutually incomparable covers. (If we work with reflexive rather than strict orderings, then the minimum number of chains needed is increased by the number of isolated elements.) Dilworth suggested the idea behind the following proof. Order the covering pairs by saying that $(x, y) < (z, w)$ if $y \leq z$. Each chain of this ordered set corresponds to a chain of the original ordered set in an obvious way. Further, each chain C of an ordered set corresponds to the chain in the covering ordering whose covers are the covers both of whose elements are in the chain C. However, since an ordering is the transitive closure of its covering relation, each chain decomposition of the covering ordered set gives a chain covering of the original ordered set such that the transitive closure of the union of the chains is the ordering. Conversely, each chain covering of the ordered set such that the transitive closure of the union of its orderings is the ordering must have each covering pair in its transitive closure. Since a covering pair cannot be the consequence of a nontrivial application of the transitive law, each covering pair must be a pair in one of the orderings of one of the chains and thus the associated chains of the covering ordering must cover the set of covers. Thus the minimum number of chains needed is the size of a maximum antichain in the covering ordering. This ordered set of covering pairs has been studied in more detail in Behrendt [3] and in unpublished work of Bonin [6]. Both Behrendt and Bonin have characterized the ordered sets that can arise as covering orderings and have shown that except for the ordering relations of maximal and minimal elements, orderings with isomorphic covering orderings are isomorphic. In particular, two orderings with universal bounds are isomorphic if and only if their covering orderings are.

Dilworth's theorem has been extended to more general subsets of ordered sets than chains and antichains in a variety of ways, giving other packing and covering theorems for chains and antichains in posets. One such theorem, which has itself led to much subsequent work, is due to Greene and Kleitman [29]. A *k-antichain* $A \subseteq P$ is a subset which meets every chain in at most k elements (equivalently, A is the union of k antichains). The Greene-Kleitman result is a minimax theorem which uses chain partitions to bound the maximum size of a k-antichain.

Clearly, if C is any chain, then $|A \cap C| \leq \min\{k, |C|\}$. Hence for any chain

THE DILWORTH THEOREMS

partition of P, a bound on $|A|$ is obtained by summing $\min\{k, |C|\}$ over each chain C in the partition. The theorem states that this bound is exact, that is, equality occurs for some chain partition. The case $k = 1$ is Dilworth's theorem.

Partitions which achieve the bound for a particular k are called *k-saturated*. For each fixed k, there always exist partitions which are simultaneously k- and $(k+1)$-saturated, but there need not exist partitions which are k-saturated for all k.

The original proof by Greene and Kleitman used lattice theoretic methods, generalizing Dilworth's observation [15] in the case $k = 1$ that the maximum-sized antichains form a distributive lattice. It can be shown that the k-antichains of maximum size also have the structure of a distributive lattice. It is perhaps interesting to note that the family of *all k*-antichains also forms a lattice, but it is not distributive for $k > 1$. The lattice structure is used to show that in the nontrivial cases there exists an element common to all maximum-sized k-antichains, and the proof continues by induction.

Other proofs have been given by Saks [54], Hoffman and Schwartz [36], Frank [20], and Fomin [18]. Saks gives a direct argument in which the result is obtained by applying Dilworth's theorem to $P \times \mathbf{k}$, the product of P with a k-element chain. Perfect [51] has used this argument to prove the existence of partitions which are simultaneously k- and $(k + 1)$-saturated. The Hoffman-Schwartz proof is based on linear programming duality and the ideas in [12] which were used to prove Dilworth's original theorem. The other proofs mentioned above are based on network flows. Cameron and Edmonds [10] explored other connections with linear programming duality, and gave an algorithm for finding minimum-weight k-antichains. Extensions of the theorem to directed graphs have been obtained by several authors, including Linial [40], Hoffman [34], and Cameron [9].

If the words "chain" and "antichain" are reversed in the statement of Dilworth's theorem, the result remains true, but the proof is trivial (just partition P with antichains P_k consisting of the elements of rank k). That this result is "equivalent" to the harder version of Dilworth's theorem is a consequence of Lovász' perfect graph theorem [41,42]. The Greene-Kleitman theorem also remains true if the notions of chain and antichain are interchanged, but the proof is nontrivial [27]. The argument rests on a "symmetrical" relation between the portion of P coverable by h chains and the portion coverable by k antichains, for all h and k. The result states that there exists a partition $\{\lambda_1 \geq \lambda_2 \geq \cdots \geq \lambda_Q\}$ of the integer $|P|$ such that for each k, the maximum size of a k-antichain is $\lambda_1 + \lambda_2 + \cdots + \lambda_k$, and the maximum size of an h-chain (*i.e.*, a union of h chains) is $\lambda_1^* + \lambda_2^* + \cdots + \lambda_h^*$, where λ^* denotes the partition conjugate to λ. In fact this result includes both versions of Dilworth's theorem as well as both versions of the Greene-Kleitman theorem as a special case. For more details as well as a discussion of the connection with perfect graphs and Young tableaux, see [28]. Other related results can be found in Saks [55] and Woodall [62].

For an account of other variations and extensions of the Greene-Kleitman theorem, see the excellent survey article by West [60]. The considerable body of work presented here can be viewed as the expansion and development of ideas suggested by Dilworth's theorem, and by some of its early proofs.

4. Decompositions and Operations Research. At about the time that Dilworth was led to his decomposition theorem for structural reasons, a group of people in operations research began analyzing a scheduling problem whose answer would eventually follow from the theorem. In a Rand Corporation Memorandum dated 1950, Robinson and Walsh discussed a problem involving the routing of empty oil tankers to locations where they would be needed in order to meet a fixed transportation schedule. In an Institute for Naval Analysis Memorandum dated 1952, Tomkins gave a combinatorial description of the problem of meeting a fixed transportation schedule with tankers, full or empty. In the first published reference to the problem, Dantzig and Fulkerson [11] described the problem of moving oil tankers between pickup points and discharge points so that a fixed delivery schedule is met with a minimum number of tankers. They described the problem as a linear programming problem and showed how to find the minimum number of tankers and the assignment of tankers to routes.

On seeing this work, Hoffman recognized that the problem of determining the minimum number of tankers that could be used was an instance of the problem Dilworth had solved [35]. The set being ordered is the set of delivery trips required—a trip is specified by the time and location where the cargo is to be loaded and the time and location where it is to be delivered. A trip from location i to location j is ordered before a trip from location i' to location j' if the time required to move a tanker from location j to location i' and load it is less than or equal to time available between the end of the trip from i to j and the beginning of the trip from i' to j'. Two trips are thus comparable if and only if they can be served by the same tanker, so that for any chain of the ordering, all its trips may be served by the same tanker. Thus the minimum number of tankers needed is the minimum number of chains in a chain decomposition. This application demonstrates the importance of actually finding a chain decomposition. Based on this work, Dantzig and Hoffman recognized that the problem of finding a chain decomposition of an arbitrary ordered set could be formulated as a linear programming problem and in [12] used this formulation to derive Dilworth's theorem as a consequence of the duality theorem of linear programming.

Dilworth's theorem is formally similar to König's theorem that the maximum size of a matching in a bipartite graph is the minimum size of a vertex cover of the edges, an observation Dilworth [15] credits to Birkhoff. It is straightforward to derive König's theorem from Dilworth's since the matching edges correspond to two-element chains in a bipartite ordered set. Fulkerson saw a clever way to derive Dilworth's theorem from the König theorem [22]. In more picturesque language than Fulkerson's, we construct a bipartite graph with two vertices, one in each part, for each member of our ordered set. We call one vertex associated with x the "bottom" of x and the other the "top" of x. We draw a line between the top of x and the bottom of y if x is less than or equal to y in the ordering. (The parts of the bipartite graph are, of course, the bottoms and the tops.) After finding a matching in this graph, we add edges from the bottom of x to the top of x and the connected components of the resulting "matching or equal to" graph correspond to the chains of a chain decomposition of the ordering. In 1958, Ford and Fulkerson showed how to derive Hall's theorem on systems of distinct representatives from the max-flow min-cut theorem for networks, giving an indirect proof of Dilworth's theorem from network flow techniques (relying on the equivalence of Hall's and König's theorems). In their 1962 book [19], Ford and Fulkerson showed how to apply network flow methods to a network derived from an ordering in essentially the same way that Fulkerson's bipartite graph is derived from an ordering, establishing a direct and elegant proof of Dilworth's theorem from network flow techniques. Their construction also allowed the use of network flow algorithms to find minimum-sized chain decompositions, just as the Fulkerson graph allows us to use bipartite matching algorithms to find minimum-sized chain decompositions. It is interesting that the flow problem that corresponds to Dilworth's theorem is finding a minimum flow subject to certain required flow amounts rather than finding a maximum flow subject to certain capacities.

Because techniques to (efficiently) find matchings in bipartite graphs and flows in networks were known and because, with little additional work, these techniques also locate antichains of maximum size, no direct work was done on algorithms for finding maximum-sized antichains or chain decompositions in ordered sets. Bogart and Magagnosc [5,44] showed that an order-theoretic approach to these problems leads to an algorithm similar to those used for matching or flow problems; in fact, when properly stated, the natural order-theoretic algorithm is an instance of the others. The central idea of the order-theoretic algorithm is to begin with the trivial decomposition of the ordered set into chains of size one and refine it by joining two chains to make one if all elements of one are over all elements of the other, joining three chains to make two if the elements of one chain may be divided into elements below the second chain and elements above the third and so on. In retrospect, one can see that their approach is implicit in Dilworth's original proof in [14] of the decomposition theorem.

In their 1962 book [19], Ford and Fulkerson also gave a simple algorithm [which is easily seen to be $0(n^2)$] for finding chain decompositions of the ordered sets studied extensively by Fishburn and now known as interval orders [17].

Typical timetables for railroads, airlines, bus companies, etc., repeat on a daily or weekly basis. This adds a bit of complexity to the problem of assigning vehicles since a vehicle which has completed servicing one route must be available either to repeat that route or begin another one. The version of the problem that does not allow deadheading (moving empty vehicles) was solved by Bartlett and Charnes [2] without reference to chain decompositions; Orlin [48] outlines work of Dantzig, Simpson, Wollmer, and Orlin which leads to a chain-decomposition style solution to the problem of minimizing the number of vehicles needed to service a repeating fixed schedule. In particular, Orlin introduces the idea of a periodic ordering and shows how its chain decompositions relate to the problem at hand. Although the ordered set is infinite, the solution takes a polynomial number of steps in the number of elements in the finite ordered set given by the timetable. Interestingly, if we require that each route be served by the same vehicle in each iteration of the timetable, then we convert the problem of deciding whether the timetable can be served by k vehicles into the NP-complete problem of determining whether a circular-arc graph has a coloring with k colors.

Minimizing the number of vehicles used to meet the timetable does not necessarily minimize the cost of meeting the schedule. Among the assignments of vehicles, some may involve moving empty vehicles over greater distances than others. Thus we wish to find a chain decomposition of minimum total cost or a chain decomposition of minimum total cost among those using a minimum number of vehicles. Algorithms for finding minimum cost flows or matchings of maximum value or size convert directly to algorithms for finding minimum-cost chain decompositions. Although these algorithms assume a finite network or graph, the relationship of Orlin's and Wollmer's solutions to the concept of a circulation in a network shows that algorithms for finding minimum-cost circulations may be applied to periodic orderings to minimize the cost of assigning vehicles to meet periodic timetables as well. For examples of such algorithms, see [25].

Of course it may also be of interest to find a minimum-cost antichain of maximum size; by using the Fulkerson graph we may translate this into the problem of finding a minimum-cost vertex-cover of the edges, a problem first solved by Norman and Rabin [47]. Cameron and Edmonds [10] used linear programming duality and Cameron [8] used a direct approach to convert the minimum-weight antichain (or more generally k-antichain) problem to a dual transportation problem, a problem which may be solved by standard algorithmic techniques.

5. Dimension Theory. The dimension of an ordered set is the smallest number of linear orderings of the set whose intersection is the given ordering or, alternatively, the smallest number of linear orderings in whose product the ordered set may be imbedded. In a footnote in [14], Dilworth points out that the dimension of an ordered set with finite width is no more than its width.

In fact, Dilworth's proof in [14] that a distributive lattice is a sublattice of a product of k chains if and only if k is the maximum size of a set of distinct covers of an element of the lattice contains the kernel of an elegant proof of this bound on the dimension. Namely, if the ordered set is a union of a set of chains then it may be imbedded in the product of these chains, each augmented by a bottom element. The embedding theorem also shows that *the dimension of a distributive lattice is equal to the width of the ordered set of its join-irreducible elements.*

Baker, Fishburn, and Roberts [1] and Wille [61] rekindled an interest in dimension theory by outlining its relation with certain problems of social science interest, and by characterizing modular lattices of dimension two, respectively, leading to an outpouring of papers in the 1970's. One theme of these papers was the relationship between dimension and width; another was the relationship between dimension and size. These themes intersect in an idea originally conjectured by Curtis Greene. Not only is the dimension bounded by the size of a largest antichain, it is also bounded by the size of the complement of a largest antichain as long as that size is two or more. This result was proved independently by Kimble [38] and Trotter [59]; except

THE DILWORTH THEOREMS

for the need for careful analysis of the situation in which the complement has size 2, the result follows by applying Hiraguchi's result [31] that the dimension of an ordered set decreases by at most one when we remove one element.

It is now natural to ask how the removal of an arbitrary antichain affects the dimension and how the dimension relates to the width of what remains after the removal of an antichain. Trotter showed that the dimension is at most one more than the width of the ordered set we get by removing all the maximal (or all the minimal) elements [59] and at most one more than twice the width of the ordered set that results from removing an arbitrary antichain [58]. Pretzel introduced the concept of a double antichain $\langle A, B \rangle$ consisting of antichains A and B with each a in A below each b in B. He related the double width (*i.e.*, size of a double antichain) to decompositions of ordered sets into two different families of chains which are, in a combinatorial sense, orthogonal and showed that the dimension of an ordered set is no more than 1 plus the double width of the ordered set that results from removing an antichain [52].

REFERENCES

1. K. A. Baker, P. C. Fishburn, and F. S. Roberts, *Partial orders of dimension 2*, Networks **2** (1971), 11–28.
2. T. E. Bartlett and A. Charnes, *Cyclic scheduling and combinatorial topology: Assignment and routing of motive power to meet scheduling and maintenance requirements; II. Generalization and analysis*, Naval Res. Logist. Quart. **4** (1957), 207–220.
3. G. Behrendt, *Covering posets*, Discrete Math. (1988), 189–195.
4. K. P. Bogart, *Decomposing partial orderings into chains*, J. Combin. Theory **9** (1976), 97–99.
5. —————, "Introductory Combinatorics," Harcourt Brace Jovanovich, San Diego, 1990.
6. J. Bonin, *Personal communication, 1988*.
7. H. Burkill and L. Mirsky, *Monotonicity*, J. Math. Anal. Appl. **41** (1973), 391–410.
8. K. Cameron, *Antichain sequences*, Order **2** (1985), 249–255.
9. —————, *On k-optimum dipath partitions and partial k-colourings of acyclic diagraphs*, European J. Combin. **7** (1986), 115–118.
10. K. Cameron and J. Edmonds, *Algorithms for optimum antichains*, in "Proceedings of the 10th Southeastern Conference on Combinatorics, Graph Theory and Computing," F. Hoffman, ed., Utilitas Mathematica, Winnipeg, Manitoba, pp. 229–240.
11. G. B. Dantzig and D. R. Fulkerson, *Minimizing the number of tankers to meet a fixed schedule*, Naval Res. Logist. Quart. **1** (1954), 217–222.
12. G. B. Dantzig and A. J. Hoffman, *Dilworth's theorem on partially ordered sets*, in "Linear Inequalities and Related Systems," Annals of Mathematics Studies No. 38, H. W. Kuhn and A. W. Tucker, eds., Princeton Univ. Press, Princeton, New Jersey, 1956, pp. 207–214.
13. R. W. Deming, *Acyclic orientations of a graph and chromatic and independence number*, J. Combin. Theory Ser. B **26** (1979), 101–110.
14. R. P. Dilworth, *A decomposition theorem for partially ordered sets*, Ann. of Math. (2) **51** (1950), 161–166. Reprinted in Chapter 1 of this volume.
15. —————, *Some combinatorial problems on partially ordered sets*, in "Combinatorial Analysis," Proceedings of the Tenth Symposium in Applied Mathematics, New York, 1958, R. Bellman and M. Hall, Jr., eds., Amer. Math. Soc., Providence, Rhode Island, 1960, pp. 85–90. Reprinted in Chapter 1 of this volume.
16. P. Erdős and G. Szekeres, *A combinatorial problem in geometry*, Compositio Math. **2** (35), 464–470.

17. P. C. Fishburn, "Interval Graphs and Interval Orders," Wiley, New York, 1985.

18. S. V. Fomin, *Finite partially ordered sets and Young diagrams*, Soviet Math. Dokl. **19** (1976), 1510–1514.

19. L. R. Ford and D. R. Fulkerson, "Flows in Networks," Princeton Univ. Press, Princeton, New Jersey, 1962.

20. A. Frank, *On chain and antichain families of a partially ordered set*, J. Combin. Theory Ser. B **29** (1980), 176–184.

21. R. Freese, *An application of Dilworth's lattice of maximal antichains*, Discrete Math. **7** (1974), 107–109.

22. D. R. Fulkerson, *Note on Dilworth's decomposition theorem for partially ordered sets*, Proc. Amer. Math. Soc. **7** (1956), 701–702.

23. T. Gallai, *Maximum-Minimum Satzes über Graphen*, Acta Math. Acad. Sci. Hung. **9** (1958), 395–434.

24. T. Gallai and A. N. Milgram, *Verallgemeinerung eines graphentheoretischen Satzes von Rédei*, Acta Sci. Math. (Szeged) **21** (1960), 181–186.

25. A. V. Goldberg, S. A. Plotkin, and E. Tardos, *Combinatorial algorithms for the generalized circulation problem*, in "Twenty-ninth Annual Symposium on Foundations of Computer Science," IEEE, Washington, D.C., 1988.

26. R. L. Graham, B. L. Rothschild, and J. H. Spencer, "Ramsey Theory," Wiley, New York, 1980.

27. C. Greene, *Sperner families and partitions of a partially ordered set*, in "Combinatorics, Part 2: Graph Theory, Foundations, Partitions, and Combinatorial Geometry," M. Hall, Jr. and J. H. van Lint, eds., Math. Centrum, Amsterdam, 1974, pp. 91–106.

28. __________, *Some partitions associated with a partially ordered set*, J. Combin. Theory Ser. A **20** (1976), 69-79.

29. C. Greene and D. Kleitman, *The structure of Sperner k-families*, J. Combin. Theory Ser. A **20** (1976), 41–68.

30. L. H. Harper and G.-C. Rota, *Matching theory, an introduction*, in "Advances in Probability, Vol. 1," P. Ney, ed., Marcel Dekker, New York, 1971, pp. 171–215.

31. T. Hiraguchi, *On the dimension of partially ordered sets*, Sci. Rep. Kanazawa Univ. **1** (1951), 77–94.

32. A. J. Hoffman, *Some recent applications of the theory of linear inequalities to extremal combinatorial analysis*, in "Combinatorial Analysis," Proceedings of Symposia on Applied Mathematics, Vol. 10, R. Bellman and M. Hall, Jr., eds., Amer. Math. Soc., Providence, Rhode Island, 1960, pp. 113–127.

33. __________, *Ordered sets and linear programming*, in "Ordered Sets," Proceedings of the Banff Conference, 1982, I. Rival, ed., Reidel, Dordrecht and Boston, 1982, pp. 619–654.

34. __________, *Extending Greene's theorem to directed graphs*, J. Combin. Theory Ser. A **34** (1983), 102–107.

35. __________, *Personal communication, 1989.*

36. A. J. Hoffman and D. E. Schwartz, *On partitions of a partially ordered set*, J. Combin. Theory Ser. B **23** (1977), 3–13.

37. D. J. Kleitman, M. Edelberg, and D. Lubell, *Maximal sized antichains in partial orders*, Discrete Math. **1** (1971), 47–54.

38. R. J. Kimble, "Extremal Problems in Dimension Theory for Partially Ordered Sets," Ph. D. thesis, M. I. T., 1973.

39. N. Linial, *Covering digraphs by paths*, Discrete Math. **23** (1978), 257–272.

40. __________, *Extending the Greene-Kleitman theorem to directed graphs*, J. Combin. Theory Ser. A **30** (1981), 331–334.

41. L. Lovász, *Normal hypergraphs and the perfect graph conjecture*, Discrete Math. **2** (1972), 253–267.

  THE DILWORTH THEOREMS

42. __________, *Perfect graphs*, in "Selected Topics in Graph Theory 2," L. W. Bieneke and R. J. Wilson, eds., Academic Press, New York, 1983, pp. 55–87.

43. L. Lovász and M. D. Plummer, "Matching Theory," Ann. Discrete Math., Vol. 29, North-Holland, Amsterdam and New York, 1986.

44. D. Magagnosc, "Cuts and Decompositions: Structure and Algorithms," Ph. D. thesis, Dartmouth College, 1987.

45. L. Mirsky, "Transversal Theory," Academic Press, New York, 1971.

46. L. Mirsky and H. Perfect, *Systems of representatives*, J. Math. Anal. Appl. **15** (1966), 520–568.

47. R. Z. Norman and M. Rabin, *An algorithm for a minimum cover of a graph.*, Proc. Amer. Math. Soc. **10** (1959), 315–319.

48. J. B. Orlin, *Minimizing the number of vehicles to meet a fixed periodic schedule: an application of periodic posets*, Operations Research **30** (1982), 760–776.

49. M. A. Perles, *A proof of Dilworth's decomposition theorem for partially ordered sets*, Israel J. Math. **1** (1963), 105–107.

50. H. Perfect, *Remarks on Dilworth's theorem in relation to transversal theory*, Glasgow Math. J. **21** (1980), 19–22.

51. __________, *Addendum to: "A short proof of the existence of k-saturated partitions of partially ordered sets" [Adv. in Math. 33(1979), no. 3, 207–211; MR82c:06008] by M. Saks*, Glasgow Math. J. **25** (1984), 31–33.

52. O. Pretzel, *On the dimension of partially ordered sets*, J. Combin. Theory Ser. A **22** (1977), 146–152.

53. __________, *Another proof of Dilworth's decomposition theorem*, Discrete Math. **25** (1979), 91–92.

54. M. E. Saks, *A short proof of the existence of k-saturated partitions of partially ordered sets*, Adv. in Math. **33** (1979), 207–211.

55. __________, *Some sequences associated with combinatorial structures*, Discrete Math. **59** (1986), 135–166.

56. A. Schrijver, "Theory of Linear and Integer Programming," Wiley, Chichester and New York, 1986.

57. H. A. Tverberg, *A proof of Dilworth's decomposition theorem*, J. Combin. Theory **3** (1967), 305–306.

58. W. T. Trotter, *Irreducible posets with large height exist*, J. Combin. Theory Ser. A **17** (1974), 337–344.

59. __________, *Inequalities in dimension theory for posets*, Proc. Amer. Math. Soc. **52** (1975), 33–39.

60. D. B. West, *Extremal problems in partially ordered sets*, in "Ordered Sets," Proceedings of the Banff Conference, 1982, I. Rival, ed., Reidel, Dordrecht and Boston, 1982, pp. 473–521.

61. R. Wille, *On modular lattices of order dimension two*, Proc. Amer. Math. Soc. **43**, 287–292.

62. D. R. Woodall, *Menger and König systems*, in "Theory and Applications of Graphs," Y. Alavi and D. R. Lick, eds., Lecture Notes in Mathematics, Vol. 642, Springer-Verlag, Berlin and New York, 1978, pp. 620–635.

Dartmouth College
Hanover, NH 03755
U. S. A.

Haverford College
Haverford, PA 19041
U. S. A.

University of North Texas
Denton, TX 76203
U. S. A.

Dilworth's Decomposition Theorem in the Infinite Case

E. C. Milner

Now-a-days we should probably say that the infinite case of Dilworth's decomposition theorem [6] follows from the finite case by "a standard compactness argument." Depending upon one's upbringing, what we would have in mind is an application of Gödel's compactness theorem in logic (*cf.* Church [5]), Tychonoff's theorem on compact topological spaces [25], or (if one is a combinatorialist) Rado's selection lemma [19]. For example, we might argue as follows. Let $\mathcal{P} = \langle P, \leq \rangle$ be a (partially) ordered set with no antichain of size $k + 1$. Then, by the finite case of Dilworth's theorem, any finite subcollection of the set of sentences

$$S = \{p_{x1} \vee p_{x2} \vee \cdots \vee p_{xk} : x \in P\} \cup \{\neg(p_{xi} \wedge p_{yi}) : 1 \leq i \leq k, x \perp y\}$$

has a model (where $x \perp y$ means x and y incomparable in P, and p_{xi} is a propositional variable with the intended interpretation "x belongs to the ith chain"). Hence S has a model and P is the union of the k chains $C_i = \{x \in P : p_{xi}$ is true$\}$ $(1 \leq i \leq k)$.

However, in the 1940's mathematicians were not so familiar with such arguments, and it is interesting to read Dilworth's own account (in the background for this chapter) of how he eventually found the clever *ad hoc* proof of the infinite case of his theorem using Zorn's lemma. As witness to the fact that such compactness arguments were not then part of a mathematicians armoury let us recall two other famous results published at about the same time as Dilworth's paper [6]. The first is Marshall Hall's [11] extension of the marriage theorem to the case of infinitely many finite sets and the second is the theorem of de Bruijn and Erdös [4] which states that, for finite k, a graph is k-colourable if and only if every finite subgraph is k-colourable. At that time these results were certainly not considered to be "standard." In fact, it was because of applications to results like these and Dilworth's theorem that mathematicians gradually became aware of the effectiveness of compactness to bridge the gap between the finite and the infinite.

The *width*, $\mu(\mathcal{P})$, of an ordered set $\mathcal{P}$ is the smallest cardinal μ such that $|A| < 1 + \mu$ holds for any antichain $A \subseteq P$. We will write $\chi(\mathcal{P})$ to denote the smallest cardinal number χ such that P is a union of χ chains. Dilworth's theorem says that, if the width $\mu(\mathcal{P})$ is finite, then $\chi(\mathcal{P}) = \mu(\mathcal{P})$. Several people noticed that this is false if the width is infinite, but the first was Perles [17] who gave the following simple example. The direct product $\kappa \otimes \kappa$, where κ is an infinite cardinal and the elements are ordered componentwise (*i.e.*, $(\alpha,\beta) \leq (\gamma,\delta) \Leftrightarrow \alpha \leq \gamma$ and $\beta \leq \delta$), contains no infinite antichain but it is not the union of fewer than κ chains; in other words, if $\mu(\mathcal{P}) = \omega$, then $\chi(\mathcal{P})$ may be arbitrarily large and so cannot be described as a function of $\mu(\mathcal{P})$ alone.

Although Dilworth's theorem cannot be extended to general ordered sets of infinite width, the result does extend to the special case of trees. The ordered set $\mathcal{P} = \langle P, \leq \rangle$ is a *tree* if $\{x \in P : x < p\}$, the set of predecessors of any element p, is well-ordered by the induced ordering. Although the result seems to have been overlooked, Kurepa [13] proved a long time ago that the inequality $\chi(\mathcal{P}) \leq \mu(\mathcal{P})$ holds in the case when $\mathcal{P}$ is a tree; in fact, if $\mu(\mathcal{P}) = \mu \geq \omega$, then μ is regular (by the theorem of [7] stated below) and if the Suslin hypothesis $\text{SH}(\mu)$ holds, then $\chi(\mathcal{P}) < \mu$ (see [15]). This result was independently rediscovered by F. Galvin (unpublished) and others (see [15]). A proof of this and a generalization to a wider, but still rather special, class of ordered sets is given in Milner, Li and Wang [14].

A stronger assertion than the cardinal inequality $\chi(\mathcal{P}) \leq \mu(\mathcal{P})$ is the statement that there is a decomposition of $\mathcal{P}$ into disjoint chains $P = \bigcup\{C_i : i \in I\}$ and there is an antichain A such that $A \cap C_i \neq \emptyset$ for each $i \in I$; we call such a decomposition a *Dilworth decomposition* of $\mathcal{P}$. Of course, Dilworth's theorem says that any ordered set of finite width has such a decomposition. Oellrich and Steffens [16] proved that

$$(*) \qquad \textit{If } \mathcal{P} \textit{ has no infinite chain, it has a Dilworth decomposition.}$$

Actually in [16] this theorem is stated only for denumerable ordered sets, but what is really proved is an interesting connection with the following generalization of König's theorem on bipartite graphs. A *bipartite graph* is a pair $\mathcal{G} = (A \cup B, E)$, where A, B are non-empty disjoint sets and $E \subseteq A \times B$; the elements of $A \cup B$ are the *vertices* of $\mathcal{G}$ and E is the set of *edges*. Two edges (a,b) and (a',b') are disjoint if $a \neq a'$ and $b \neq b'$. A *matching* of $\mathcal{G}$ is a set of pairwise disjoint edges, and a *covering set* is a set $S \subseteq A \cup B$ such that $\{a,b\} \cap S \neq \emptyset$ for every edge $(a,b) \in E$. Let us say that the bipartite graph $\mathcal{G} = (A \cup B, E)$ is *fully matchable* if there is a matching M and a covering set $S \subseteq A \cup B$ such that $|\{a,b\} \cap S| = 1$ for every edge $(a,b) \in M$ and for any $x \in S$ there is an edge $(a,b) \in M$ such that $x \in \{a,b\}$. What Oellrich and Steffens really prove in [16] is that $(*)$ is equivalent to the assertion

$$(**) \qquad \textit{Any bipartite graph is fully matchable.}$$

The statement $(**)$ for finite bipartite graphs is the well-known theorem of König [12], and Podewski and Steffens [18] proved it for countable graphs. This was all that was known at the time when [16] was published which is why $(*)$ is stated only for countable ordered sets in [16]. Subsequently, Aharoni [2] proved that $(**)$ is true in general and hence so also is $(*)$. The condition in $(*)$ that $\mathcal{P}$ should not contain any infinite chain is rather too restrictive, and it would be of considerable interest if this could be relaxed in some way. In this connection, let us mention another interesting result from [16] which gives a necessary and sufficient condition for a tree to have a Dilworth decomposition. Call the node x of the tree $\mathcal{T}$ a *splitting node* if there are $y, z \geq x$ in $\mathcal{T}$ such that $y \perp z$. It is shown that [16] the tree $\mathcal{T}$ has a Dilworth decomposition if and only if for every node x there is a non-splitting node $y \geq x$.

Abraham [1] has shown that, in a certain sense, the Perles counter-example mentioned above is the simplest one possible. If $\mu(\mathcal{P}) = \omega$, then $\mathcal{A}(\mathcal{P})$, the set of all antichains of $\mathcal{P}$ ordered by reverse inclusion, is well-founded. Consequently, this ordered set has a height $h(\mathcal{A}(\mathcal{P}))$. Now the height of $\mathcal{A}(\omega_1 \otimes \omega_1)$ is ω_1^2, and what Abraham shows (by induction on this height function) is that, if $\mu(\mathcal{P}) = \omega$ and $h(\mathcal{A}(\mathcal{P})) < \omega_1^2$, then $\mathcal{P}$ is a union of countably many chains, *i.e.*, $\chi(\mathcal{P}) = \mu(\mathcal{P})$.

This result of Abraham is interesting since we know very few results giving sufficient conditions for a ordered set to be a union of countably many chains. Todorćević [22] suggests that one reason for this, apart from the Perles counter-example, is "our very poor understanding to which posets we can add an uncountable chain in a reasonable forcing extension." He then goes on to prove another result of this kind for a class of ordered sets which has already been quite well studied in the literature. To describe this class we need some new definitions. Let $\mathcal{M}(\mathcal{P})$ denote the set of all maximal chains in $\mathcal{P}$. A *cutset* of the ordered set $\mathcal{P}$ is a subset $X \subseteq P$ such that $X \cap C \neq \emptyset$ for every chain $C \in \mathcal{M}(\mathcal{P})$; a *cutset for an element* x is a set $F_x \subseteq x^\perp = \{y \in P : x \perp y\}$ such that $\{x\} \cup F_x$ is a cutset of $\mathcal{P}$. The ordered set $\mathcal{P}$ is said to have the *finite cutset property* if every cutset of $\mathcal{P}$ contains a finite cutset. This notion was introduced by Bell and Ginsburg [3] who were interested in the (still unsettled) question: when is a compact topological space representable, for some suitable $\mathcal{P}$, as the space $\mathcal{M}(\mathcal{P})$ in the product topology induced by 2^P? They showed that $\mathcal{M}(\mathcal{P})$ is closed (*i.e.*, compact) in the product space $2^P \iff \mathcal{P}$ has the finite cutset property $\iff$ there is a finite cutset for every element $x \in P$. The ordered set $\mathcal{P}$ is said to be *conditionally σ-chain complete* if every bounded countable chain in $\mathcal{P}$ has an infimum and a supremum in $\mathcal{P}$. As we remarked, the class of ordered sets which are conditionally σ-chain complete and have the finite cut set property have been well considered in the literature (*e.g.* [3, 8, 9, 10, 20]). For example, Ginsburg, Rival and Sands [9] showed that every antichain in such a ordered set is countable (*i.e.*, $\mu(\mathcal{P}) \leq \omega_1$), and they conjectured that, under these conditions, every uncountable subset of P contains an uncountable chain. This conjecture was settled by Todorćević [21], and he later proved the much stronger result referred to above [22]: *if P is conditionally σ-chain complete and has the finite cutset property, then P is a union of countably many chains.* An interesting key idea used in his proof of this uses a lemma of Sauer and Woodrow [20] which depends upon the fact that, under the stated conditions, the order relation $\preceq$ on

THE DILWORTH THEOREMS

$\mathcal{P}^2 = \{(x, y) : x < y \text{ in } \mathcal{P}\}$ defined by setting

$$(x, y) \preceq (a, b) \iff a \leq x < y \leq b \text{ and } x^{\perp} \cap y^{\perp} \not\subseteq a^{\perp} \cap b^{\perp},$$

is well-founded. Todorčević proves his theorem by a clever induction on the height of $\mathcal{P}^2$ under this ordering.

The elements x, y of an ordered set $\mathcal{P} = \langle P, \leq \rangle$ are *compatible* if there is an element z such that $x \leq z$ and $y \leq z$, and we say that $\mathcal{P}$ is directed if every pair of elements is compatible; a directed subset is a set $D \subseteq P$ such that $\mathcal{P}|_D$ is directed. A subset $A \subseteq P$ is a strong antichain if no two elements of A are compatible. Of course, a strong antichain is an antichain and a chain is a directed subset. By analogy with our earlier notation, let us write $\mu^*(\mathcal{P})$ to denote the least cardinal μ^* such that $\mathcal{P}$ contains no strong antichain of size $1 + \mu^*$, and let $\chi^*(\mathcal{P})$ be the smallest cardinal χ^* such that $\mathcal{P}$ is a union of χ^* directed sets. An older decomposition result than Dilworth's theorem is the fact that, if $\mu^*(\mathcal{P})$ is finite, then $\chi^*(\mathcal{P}) = \mu^*(\mathcal{P})$. This is almost obvious, for if S is any strong antichain of maximal (finite) size, then P is the union of the directed sets $D_x = \{y \in P : y \text{ is compatible with } x\}$ ($x \in S$). This simple fact was observed by Erdös and Tarski [7], but the main result of their paper is the theorem that the number $\mu^*(\mathcal{P})$ cannot be $\aleph_0$ or a singular cardinal (but every other value is possible). Thus, in particular, if P has no infinite strong antichain, then $\mu^*(\mathcal{P}) = k$ for some finite k and so $\chi^*(\mathcal{P}) = k$ also. The first interesting question is, what happens if $\mu^*(\mathcal{P}) = \omega_1$? Just as Dilworth's theorem fails badly when $\mu(\mathcal{P}) = \omega$, here too, for any infinite cardinal κ there is an ordered set $\mathcal{P}$ such that $\mu^*(\mathcal{P}) = \omega_1$ and $\chi^*(\mathcal{P}) > \kappa$. For an example of this due to Baumgartner, see [15].

Despite these negative results [15,17] showing that Dilworth's theorem and the Erdös-Tarski result both fail for higher cardinals, there is a decomposition theorem which is something of a mixture of the two results. Answering a question of F. Galvin, Milner and Prikry [15] showed that if μ is an infinite cardinal and $\mu(\mathcal{P}) = \mu$, then $\chi^*(\mathcal{P}) \leq \mu^{<\mu}$. It is interesting to note that the transitivity of the order relation is not really needed for this since Todorčević [23] has proved the following more general result. If $\langle A, R \rangle$ is any structure where R is a reflexive binary relation on A, then an *antichain* of $\langle A, R \rangle$ is a subset of R-unrelated elements of A, and a λ-*directed subset* is a set $D \subseteq A$ such that for any subset $X \subseteq D$ of cardinality $|X| < \lambda$, there is an element $d \in D$ such that $x \, R \, d$ holds for all $x \in X$. For cardinals κ, λ we write $\kappa \ll \lambda$ if $\rho^{\sigma} < \lambda$ whenever $\rho < \lambda$ and $\sigma < \kappa$. We can now state Todorčević's generalization of the Milner-Prikry result [23]: *if λ is a regular cardinal and $\kappa \ll \lambda$, then any binary structure $\langle A, R \rangle$ with no antichain of size κ is a union of fewer than λ of its λ-directed sets.* This is a strong result. For example, one immediate corollary is the well-known partition relation $\lambda \to (\lambda, \kappa)^2$ for $\kappa \ll \lambda = \mathrm{cf}\,\lambda$ (for a partition $K \cup K'$ of $[\lambda]^2 = \{X \subseteq \lambda : |X| = 2\}$, consider the binary relation $R = \{(\alpha, \beta) : \{\alpha, \beta\} \in K \text{ and } \alpha < \beta\}$).

If $\mu(\mathcal{P}) = \omega$, then by the Erdös-Tarski theorem $\mu^*(\mathcal{P})$ is finite and so P is a finite union of directed sets. However it remains an open question whether $\mu^{<\mu}$

can be replaced by any smaller cardinal in the Milner-Prikry theorem when μ is an uncountable cardinal. In particular, denote by κ_0 the smallest cardinal such that every ordered set having no uncountable antichain is the union of $\leq \kappa_0$ directed sets, then it is not known if $\kappa_0 \leq \omega_1$ can be proved without any additional set-theoretical assumption. Todorčevič [24] has shown, using the proper forcing axiom, that it is consistent that κ_0 can have the least possible value, $\kappa_0 = \omega$. (Of course, in this model there can be no Suslin tree since, if $\mathcal{T}$ is a Suslin tree then, $\mu(\mathcal{T}) = \chi(\mathcal{T}) = \omega_1$.)

References

1. U. Abraham, *A note on Dilworth's theorem in the infinite case*, Order **4** (1987), 107–125.
2. R. Aharoni, *König's duality theorem for infinite bipartite graphs*, J. London Math. Soc. (2) **29** (1984), 1–12.
3. M. Bell and J. Ginsburg, *Compact spaces and spaces of maximal complete subgraphs*, Trans. Amer. Math. Soc. **283** (1984), 329–338.
4. N. G. de Bruijn and P. Erdös, *A colour problem for infinite graphs and a problem in the theory of relations*, Indag. Math. **13** (1951), 369–373.
5. A. Church, "Introduction to Mathematical Logic," Vol. 1, Princeton Univ. Press, Princeton, New Jersey, 1956.
6. R. P. Dilworth, *A decomposition theorem for partially ordered sets*, Ann. of Math. (2) **51** (1950), 161–166. Reprinted in Chapter 1 of this volume.
7. P. Erdös and A. Tarski, *On families of mutually exclusive sets*, Ann. of Math. (2) **44** (1943), 315–329.
8. J. Ginsburg, *Compactness and subsets of ordered sets that meet all maximal chains*, Order **1** (1986), 143–157.
9. J. Ginsburg, I. Rival, and B. Sands, *Antichains and finite sets that meet all maximal chains*, Canad. J. Math. **38** (1986), 619–632.
10. P. A. Grillet, *Maximal chains and antichains*, Fund. Math. **65** (1969), 157–167.
11. M. Hall, Jr., *Distinct representatives of subsets*, Bull. Amer. Math. Soc. **54** (1948), 922–926.
12. D. König, "Theorie der Endlichen und Unendlichen Graphen," Akademischen Verlagsgesellschaft, Leipzig, 1936. (Reprinted, Chelsea, New York, 1950.)
13. D. Kurepa, *Ensembles ordonnés et ramifiés*, Publ. Math. Univ. Belgrade **4** (1935), 1–138.
14. E. C. Milner, B. Y. Li, and S. Z. Wang, *Some inequalities for partial orders*, Order **3** (1987), 369–382.
15. E. C. Milner and K. Prikry, *The cofinality of a partially ordered set*, Proc. London Math. Soc. (3) **46** (1983), 454–470.
16. H. Oellrich and K. Steffens, *On Dilworth's decomposition theorem*, Discrete Math. **15** (1976), 301–304.
17. M. A. Perles, *On Dilworth's theorem in the infinite case*, Israel J. Math. **1** (1963), 108–109.
18. K. P. Podewski and K. Steffens, *Injective choice functions for countable families*, J. Combin. Theory Ser. A **21** (1976), 40–46.
19. R. Rado, *Axiomatic treatment of rank in infinite sets*, Canad. J. Math. **1** (1949), 337–343.
20. N. Sauer and R. Woodrow, *Finite cutsets and finite antichains*, Order **1** (1984), 35–46.
21. S. Todorčevič, *Posets with finite cutsets*, preprint, March 1984.
22. —————, *A chain decomposition theorem*, J. Combin. Theory Ser. A **48** (1988), 65–76.
23. —————, *Directed sets and cofinal types*, Trans. Amer. Math. Soc. **290** (1985), 711–723.

24. __________, *A note on the proper forcing axiom*, in "Axiomatic Set Theory," (Boulder, Colorado, 1983) Contemporary Math., Vol. 31, Amer. Math. Soc., Providence, R.I., 1984, pp. 209–218.

25. A. Tychonoff, *Über topologische Erweiterung von Räumen*, Math. Ann. **102** (1930), 544–561.

University of Calgary
Calgary, Alberta T2N 1N4
Canada

Effective Versions of the Chain Decomposition Theorem

H. Kierstead

Because of the central role Dilworth's decomposition theorem plays in the structure theory of (partially) ordered sets, it is important to consider it from a computational point of view. The theorem is proved in two steps. The first step is to prove the result for finite ordered sets. Dilworth's original proof in [1] of the finite result is constructive, but the notion of computational complexity is not addressed. In 1956, Fulkerson [2] gave a network flow proof of Dilworth's theorem. This proof allows the calculation of a Dilworth decomposition using any of the fast network flow algorithms. In particular, a decomposition can be calculated in $O(n^3)$ time using the network flow algorithms in [3] or [8].

The second step of Dilworth's proof is to use the Maximal Principle and the finite result to prove the infinite case. The use of the Maximal Principle results in a non-constructive proof. In fact, it turns out that the proof is necessarily non-constructive in the following technical sense. An ordered set is said to be recursive provided there is an algorithm which calculates its order relation. A partition is said to be recursive provided there is an algorithm that calculates the characteristic function of each part of the partition. A constructive proof of Dilworth's theorem should produce a recursive partition into k chains, at least in the case that the original ordered set is recursive. But, as Schmerl pointed out, there exists a recursive ordered set with a maximum antichain of size two, which cannot be recursively partitioned into two, or even three chains.

Schmerl asked whether every recursive ordered set with a finite maximum antichain could be recursively partitioned into finitely many antichains. Kierstead [4] gave a positive answer to Schmerl's question with the following constructive version of Dilworth's theorem.

THEOREM 1. *Every recursive ordered set with a maximum finite antichain of size k can be recursively partitioned into $(5^k - 1)/4$ chains. Moreover, for every finite $k > 1$, there exist recursive ordered sets with a maximum antichain of size k which cannot be recursively partitioned into $4(k - 1)$ chains.*

Let $D(k)$ be the least integer m such that any recursive ordered set with a maximum antichain of size k can be recursively partitioned into m chains. By Theorem 1,

$$4(k - 1) \le D(k) \le (5^k - 1)/4.$$

The problem of determining good bounds on $D(k)$ remains open; in particular it is not known whether $D(k)$ is bounded above by a polynomial in k. The only improvement has been made by Szemerédi and Trotter [9], who showed that the lower bound is at least quadratic. One special case of interest is interval orders. Kierstead and Trotter [7] showed that any recursive interval order with a finite maximum antichain of size k can be recursively partitioned into $3k - 2$ chains, and that this number is best possible. Kierstead, McNulty, and Trotter [6] used Theorem 1 to prove results about the recursive dimension of ordered sets in the same way that Dilworth's theorem is used to prove results about the ordinary dimension of ordered sets.

The proof of Theorem 1 has a finite interpretation. An on-line algorithm for partitioning an ordered set is an algorithm that considers the points of the set in some externally determined presentation order and assigns the nth point to a part of the partition using only information about the ordered set induced by the first n points. It follows from the proofs in [4] and [9] that there is an on-line algorithm that partitions every ordered set of width k into $(5^k - 1)/4$ chains, regardless of presentation order and that such an algorithm must use at least k^2 chains. Recently Kierstead [5] has used the on-line version of the result for interval orders mentioned above to give an improved polynomial time approximation algorithm for dynamic storage allocation.

REFERENCES

1. R. P. Dilworth, *A decomposition theorem for partially ordered sets*, Ann. of Math. (2) **51** (1950), 161–166. Reprinted in Chapter 1 of this volume.
2. D. R. Fulkerson, *Note on Dilworth's decomposition theorem for partially ordered sets*, Proc. Amer. Math. Soc. **7** (1956), 701–702.
3. A. V. Karazanov, *Determining the maximal flow in a network with the method of Preflows*, Soviet Math. Dokl. **15** (1974), 434–437.
4. H. A. Kierstead, *An effective version of Dilworth's Theorem*, Trans. Amer. Math. Soc. **268** (1981), 63–77.
5. H. A. Kierstead, *A polynomial time approximation algorithm for Dynamic Storage Allocation.*
6. H. A. Kierstead, G. F. McNulty, and W. T. Trotter, *A theory of recursive dimension for ordered sets*, Order 1 (1984), 67–82.

7. H. A. Kierstead and W. T. Trotter, *An extremal problem in recursive combinatorics*, Congressus Numerantium **33** (1981), 143–153.
8. V. M. Malhotra, M. P. Kumar, and S. N. Maheshwari, *An $O(|v|^3)$ algorithm for finding maximum flows in networks*, Information Proc. Letters **7** (1978), 277–278.
9. E. Szemerédi and W. T. Trotter, *private communication.*

University of South Carolina
Columbia, SC 29208
U. S. A.

Complementation

Background

R. P. Dilworth

Unique complementation. As was pointed out in Chapter 1, upon completing my graduate work at Caltech, I was granted a Sterling Research Fellowship at Yale University. An outstanding problem at that time was the conjecture that a lattice with unique complements was a Boolean algebra. The truth of the conjecture had been verified under any one of a number of additional restrictions, *i.e.,* modularity, orthocomplementation, and atomicity. I was determined to make a major effort to settle the conjecture.

The better part of the first year was spent in trying a variety of approaches to proving the conjecture. Clearly, the key to a proof would involve using the complements of some of the elements of the lattice to construct complements of other elements. Then the uniqueness of complements would imply some structural restrictions on the lattice. However, none of these approaches came even close to producing new complements. This being the case, I decided it would probably be more profitable to try to show that the conjecture was not true. The problem of constructing a counterexample would not be easy since any finiteness restriction would likely make the lattice a Boolean algebra. If it could be done, constructing the free lattice with unique complements would surely settle the conjecture since it would either be a Boolean algebra, in which case the conjecture would be proved, or it, itself, would be a counterexample.

Now it was clear that if a lattice had unique complements this complementation would be a unary operation on the system. Thus a reasonable first step would be to start with a given lattice and construct a free lattice with unary operator which had the given lattice as a sublattice. Hopefully, each application of the unary operation would produce an element which, from a lattice point of view, would be equivalent to adjoining a free element to the lattice. If this were the case, then the further reductions required to make the operation a complement would not disturb the lattice structure already present. Now, in a lattice with unique complements, each element must be the unique complement of its complement. Hence the next step was to construct a free lattice with reflexive unary operator having the given lattice as a sublattice. This was accomplished by selecting an appropriate sublattice of the free lattice with unary operator. The final step consisted of constructing a homomorphic image of the free lattice with reflexive unary operator in which the image of the operator became a complementation. A careful examination of the properties of the free lattice with unary operator was required in order to show that this complementation indeed produced a unique complement for each element.

Whitman had shown much earlier that the lattice generated by three unordered elements contains as a sublattice the free lattice generated by a denumerable set of unordered elements. It seemed natural to ask if there were similar results for the free systems constructed in this investigation. The free lattice with unary operator generated by one element turned out to contain as a sublattice the free lattice with unary operator generated by a denumerable set of elements. This result also held for lattices with reflexive unary operator. Furthermore, the free lattice with unique complements generated by two unordered elements again contains a sublattice generated by a denumerable set of unordered elements. Since the Boolean algebra generated by a finite set of elements is finite, this shows clearly how far lattices with unique complements may differ from Boolean algebras.

On complemented lattices. This investigation arose from reading a paper by Kodi Husimi entitled 'Studies on the foundations of quantum mechanics'. In this paper Husimi conjectured that a lattice with a negation is modular if all sublattices closed with respect to relative negation have the property that all maximal chains have the same length. It seemed to me that this conjecture could not be true so I set about trying to construct a counterexample. After several false starts it occurred to me that modifying the lattice of the seven-point Fano plane might be a good way to begin. By adding and subtracting incidences between points and lines, I was finally able to get a lattice, shown in Figure 1, with a negation which was neither upper nor lower semimodular but in which every sublattice closed with respect to relative negation had all maximal chains of equal length.

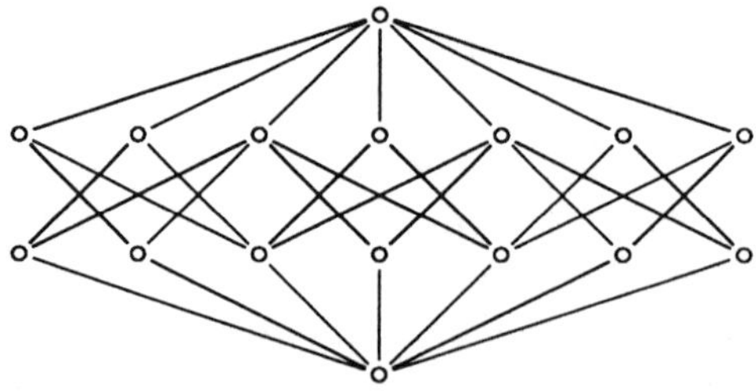

Figure 1

Along the lines of the Husimi conjecture, I was able to show that a lattice with negation is modular if every sublattice closed with respect to lower relative complementation has maximal chains all of the same length.

LATTICES WITH UNIQUE COMPLEMENTS

BY

R. P. DILWORTH

Introduction. For several years one of the outstanding problems of lattice theory has been the following: *Is every lattice with unique complements a Boolean algebra?* Any number of weak additional restrictions are sufficient for an affirmative answer. For example, if a lattice is modular (G. Bergman [1]([1])) or ortho-complemented (G. Birkhoff [1]) or atomic (G. Birkhoff and M. Ward [1]), then unique complementation implies distributivity and the lattice is a Boolean algebra.

In spite of these results, I shall show here that the theorem is *not* true in general. Indeed, the following counter theorem is proved:

Every lattice is a sublattice of a lattice with unique complements.

Thus any nondistributive lattice is a sublattice of a lattice with unique complements which a fortiori is *not* a Boolean algebra.

The actual construction gives a somewhat more general result; namely, that every partially ordered set P can be imbedded in a lattice with unique complements in such a way that least upper bounds and greatest lower bounds, whenever they exist, of pairs of elements are preserved. In particular, if P is unordered the construction yields the free lattice with unique complements generated by P.

The initial step consists in imbedding P in a lattice L so that bounds, whenever they exist, of pairs of elements of P are preserved. L is the free lattice generated by P in the sense that the only containing relations in L are those which follow from lattice postulates and preservation of bounds. Thus this imbedding represents the other extreme from the usual bound preserving imbedding by means of normal subsets. The methods employed are an extension of those used by Whitman [1, 2] in the study of free lattices. Indeed, if P is unordered, L is precisely the free lattice studied by Whitman.

Next, the lattice L is extended to a lattice O with *unary operator*, that is, a lattice over which an operation a^* is defined with the property

$$(\alpha) \qquad\qquad a = b \quad \text{implies} \quad a^* = b^*.$$

It is to be emphasized that the equality symbol denotes lattice equality which is not necessarily logical identity. Again the only containing relations in O are those which follow from lattice postulates, preservation of bounds, and from (α). Curiously, the main difficulties in obtaining an imbedding lattice with unique complements occur in connection with the structure of O.

Presented to the Society, April 24, 1943; received by the editors December 13, 1943.

([1]) Numbers in brackets refer to the references cited at the end of the paper.

123

In the third step, a sublattice N is selected from O over which a new operation a^* is defined for which (α) holds and also having the property

$$(\beta) \qquad\qquad (a^*)^* = a.$$

Thus N is a lattice with *reflexive, unary operator*. N is again free in the sense that the only containing relations in N are those which follow from lattice postulates, preservation of bounds, and the two properties (α) and (β).

Finally, a homomorphic image M of N is constructed in which the operation a^* becomes a complementation. It follows from the structure theorems of O that this complementation is unique. Furthermore, M contains P and is indeed the free lattice with unique complements generated by P.

At each stage, necessary and sufficient conditions are determined that a sublattice of the free lattice with operator be free. When the results are applied to the free lattice with unique complements having two generators, one gets the following theorem:

The free lattice with unique complements generated by two elements contains as a sublattice the free lattice with unique complements generated by a denumerable set of elements.

Since the free Boolean algebra generated by a finite number of elements is always finite, this theorem shows clearly how far lattices with unique complements may differ from Boolean algebras.

1. The free lattice generated by a partially ordered set. We begin with a fixed, but arbitrary, partially ordered set P of elements $a, b, c, \cdots$ and inclusion relation $\geq$. If $a \geq b$ and $b \geq a$, we write $a = b$ where the equality is in general not logical identity. $a > b$ denotes proper inclusion. If two elements a and b have a least upper bound or greatest lower bound in P, it will be denoted by l.u.b.(a, b) or g.l.b.(a, b) respectively.

In the construction which follows we shall use as building stones the three formal operation symbols $\cup$, $\cap$, and $*$.

DEFINITION 1.1. *Operator polynomials* over P are defined inductively as follows:

(1) The elements $a, b, c, \cdots$ of P are operator polynomials over P.

(2) If A and B are operator polynomials over P, then $A \cup B$, $A \cap B$, and A^* are operator polynomials over P.

In short, the operator polynomials over P are all finite, formal expressions which can be obtained from the symbols $a, b, c, \cdots$ by the operation symbols $\cup$, $\cap$, and $*$. The set of all operator polynomials will be denoted by O.

Those operator polynomials which are obtained from the symbols $a, b, c, \cdots$ by means of the two operations $\cup$ and $\cap$ are called *lattice polynomials*. Thus the symbol $*$ does not occur in a lattice polynomial. The set of all lattice polynomials will be denoted by L.

DEFINITION 1.2. The *rank* $r(A)$ of an operator polynomial is defined inductively as follows:

(1) $r(A) = 0$ if $A \in P$.

(2) $r(A \cup B) = r(A \cap B) = r(A) + r(B) + 1$ and $r(A^*) = r(A) + 1$.

In short, $r(A)$ is simply the number of times any of the symbols $\cup$, $\cap$, and $*$ occur in A. It is clear from definition 1.2 that $r(A) = 0$ if and only if A is an element of P.

DEFINITION 1.3. Two operator polynomials A and B are *identical* (in symbols $A \equiv B$) if, inductively,

(1) A and B have rank zero and represent the same element of P,

(2) A and B have rank $n > 0$ and either (i) A and B have the forms $A_1 \cup A_2$ and $B_1 \cup B_2$ respectively with $A_1 \equiv B_1$ and $A_2 \equiv B_2$ or (ii) A and B have the forms $A_1 \cap A_2$ and $B_1 \cap B_2$ respectively with $A_1 \equiv B_1$ and $A_2 \equiv B_2$ or (iii) A and B have the forms A_1^* and B_1^* respectively with $A_1 \equiv B_1$.

Stated less precisely, two operator polynomials are identical if and only if they look exactly alike.

The identity relation is clearly reflexive, symmetric, transitive, and preserves the operations $\cup$, $\cap$, and $*$.

Before going further, we must make precise what is meant by "the free lattice generated by a partially ordered set P." Now it is clear that if the lattice is to be of any use in imbedding problems it must be more restrictive than the free lattice generated by P as an unordered set. Indeed, it is desirable that the lattice properties of P be preserved[2]. Moreover, this can be done most simply by requiring that least upper bounds and greater lower bounds of pairs of elements of P shall be preserved whenever they exist[3]. Hence, by "the free lattice generated by P" we shall mean *the free lattice*[4] *generated by P and preserving bounds, whenever they exist, of pairs of elements of P.*

---

[2] It might be suggested that the free lattice generated by P should preserve only the order in P. However, this seems to be too general for most purposes since even if P were a lattice, union and crosscut in the free lattice would be distinct from the union and crosscut in P. This suggests that a better term for the free lattice generated by P and preserving order would be "the *completely* free lattice generated by P."

[3] Another possibility would be the requirement that finite bounds be preserved whenever they exist. This would, however, introduce a great many complications into the notation while all of the essential difficulties seem to occur in the case of bounds of pairs. Let us notice that if P is unordered, any of these requirements yield the free lattice generated by P in the usual sense.

[4] We shall frequently have to consider the most general lattice (sometimes with operator) generated by a set S and satisfying certain additional restrictions. The existence of such a lattice, which we shall call the *free* lattice (with operator) generated by S and satisfying the given restrictions, follows from general existence theorems on free algebras. Namely, let A be an algebra consisting of a set of operations 0, a set of relations R, and a set of postulates P. A *polynomial* over S is any formal expression in elements of S obtained by finite application of the operations 0. A *formula* consists of two polynomials connected by a relation R. We assume further that the postulates P are either formulas or implications between formulas. Then the set of all polynomials p over S can be made into an algebra A by defining $p_1 R p_2$ if and only if the formula $p_1 R p_2$ can be deduced as a formula from the postulates P. Furthermore A is the most general algebra generated by S in the sense that any other algebra generated by S and satisfying the postulates P is a homomorphic image of A.

Since this section treats only lattices generated by P, we may restrict our attention to the set L of lattice polynomials. Those lattice polynomials which are significant in P can be characterized as follows:

DEFINITION 1.4. The lattice polynomial A of L has a *value* $v(A)$ in P if and only if, inductively,

(1) $A \in P$, in which case $v(A) \equiv A$,

(2) $A \equiv A_1 \cup A_2$ where $v(A_1)$, $v(A_2)$, and l.u.b.$(v(A_1), v(A_2))$ exist, in which case[5] $v(A) = $l.u.b.$(v(A_1), v(A_2))$; or $A \equiv A_1 \cap A_2$ where $v(A_1)$, $v(A_2)$ and g.l.b.$(v(A_1), v(A_2))$ exist, in which case $v(A) = $g.l.b.$(v(A_1), v(A_2))$.

From Definition 1.4 follows immediately:

LEMMA 1.1. *If $v(A)$ exists, then $v(A) \in P$.*

The next definition introduces the basic containing relation in L.

DEFINITION 1.5. If $A, B \in L$ let us set

(i) $A \geq B$ (1) if $A \equiv B$ or if $v(A)$, $v(B)$ exist and $v(A) \geq v(B)$ in P.

(ii) $A \geq B(n)$ where $n > 1$ if and only if one of the following hold:

(1) $A \geq C(n-1)$ and $C \geq B(n-1)$ for some $C \in L$.

(2) $A \equiv A_1 \cup A_2$ where $A_1 \geq B(n-1)$ or $A_2 \geq B(n-1)$.

(3) $A \equiv A_1 \cap A_2$ where $A_1 \geq B(n-1)$ and $A_2 \geq B(n-1)$.

(4) $B \equiv B_1 \cup B_2$ where $A \geq B_1(n-1)$ and $A \geq B_2(n-1)$.

(5) $B \equiv B_1 \cap B_2$ where $A \geq B_1(n-1)$ or $A \geq B_2(n-1)$.

(iii) $A \geq B$ if and only if $A \geq B(n)$ for some n.

LEMMA 1.2. $A \geq B(n)$ *implies* $A \geq B(k)$ *for all* $k \geq n$.

It is clearly sufficient to show that $A \geq B(n)$ implies $A \geq B(n+1)$. If $A \geq B(1)$, then since $B \equiv B$ we have $B \geq B(1)$ and $A \geq B(2)$ by (1) of (ii). Let us suppose that it has been shown that $A \geq B(n)$ implies $A \geq B(n+1)$ for all $n < m$. Let $A \geq B(m)$. Then $B \geq B(m)$ by the induction assumption. Hence $A \geq B(m+1)$ by (1) of (ii). The lemma follows by induction.

THEOREM 1.1. L *is a lattice under the containing relation* $A \geq B$.

Proof. $A \geq A$ since $A \equiv A$ implies $A \geq A(1)$ by (i). Let $A \geq B$ and $B \geq C$. Then $A \geq B(m)$ and $B \geq C(n)$ for some m and n by (iii). But then $A \geq B(k)$ and $B \geq C(k)$ where $k = \max (m, n)$ by Lemma 1.2. Hence $A \geq C(k+1)$ by (1) of (ii) and $A \geq C$ by (iii). Thus L is partially ordered by the relation $A \geq B$. Now $A \cup B \geq A, B$ since $A \cup B \geq A, B(2)$ by (2) of (ii). Similarly $A, B \geq A \cap B$. If $X \geq A, B$ then $X \geq A(m)$ and $X \geq B(n)$ for some m and n by (iii). Hence by Lemma 1.2, $X \geq A(k)$ and $X \geq B(k)$ where $k = \max (m, n)$. But then $X \geq A \cup B(k+1)$ by (4) of (ii). Hence $X \geq A \cup B$ by (iii). In a similar manner $A, B \geq X$ implies $A \cap B \geq X$ and L is thus a lattice under $A \geq B$.

---

[5] Since equality need not be logical identity, $v(A)$ may be a multivalued function from L to P. However, the various values of $v(A)$ are equal in P.

THEOREM 1.2. *L is the free lattice generated by P. That is, L is the most general lattice generated by P and preserving bounds, if they exist, or pairs of elements of P.*

Proof. L is clearly generated by P. Now let l.u.b.(a, b) exist in P. Then $v(a \cup b) =$ l.u.b.(a, b) and $v($l.u.b.$(a, b)) =$ l.u.b.(a, b) by Definition 1.4. Hence $a \cup b \geq$ l.u.b.$(a, b)(1)$ and l.u.b.$(a, b) \geq a \cup b(1)$. By (iii) we get $a \cup b =$ l.u.b.(a, b) in L. Similarly if g.l.b.(a, b) exists in P, then $a \cap b =$ g.l.b.(a, b) in L. Hence l.u.b. and g.l.b., if they exist, of pairs of elements of P are preserved in L.

We show next that if $v(A)$ exists, then $A = v(A)$ in the free lattice generated by P. If A is of rank zero, then $v(A) \equiv A$ and hence $v(A) = A$ in the free lattice. We proceed by induction. If $A \equiv A_1 \cup A_2$ and $v(A)$ exists, then $v(A_1)$, $v(A_2)$, and l.u.b.$(v(A_1), v(A_2))$ exist and $v(A) =$ l.u.b.$(v(A_1), v(A_2))$. But then by the induction assumption $v(A_1) = A_1$ and $v(A_2) = A_2$ in the free lattice generated by P and since bounds of pairs of elements are preserved $A \equiv A_1 \cup A_2 =$ l.u.b.$(v(A_1), v(A_2)) = v(A)$. A similar argument holds if $A \equiv A_1 \cap A_2$. Thus the above statement follows by induction. Also since bounds are preserved, $a \geq b$ in P implies $a \geq b$ in the free lattice. Hence $v(A) \geq v(B)$ in P implies $A = v(A) \geq v(B) = B$ in the free lattice generated by P. Thus $A \geq B(1)$ implies $A \geq B$ in the free lattice. But since (1), (2), (3), (4), and (5) of (ii) follow from lattice properties, it is clear that $A \geq B(n)$ implies $A \geq B$ in the free lattice. Hence L is isomorphic to the free lattice generated by P.

It follows from Theorem 1.2 that $a \geq b$ in P implies $a \geq b$ in L. However, if L is to contain P as a sub-partially ordered set we must verify that $a \sim \geq b$ in P implies $a \sim \geq b$ in L ($\sim \geq$ means "does not contain"). Now it is well known that the normal subsets of P form a lattice which preserves the ordering of P and all bounds which exist. In particular, it preserves the bounds of pairs of elements of P. Also $a \sim \geq b$ in P implies $a \sim \geq b$ in the lattice of normal subsets. Hence if we take that sublattice of the lattice of normal subsets which is generated by P we have a lattice containing P, preserving bounds of pairs whenever they exist, and such that $a \sim \geq b$ whenever $a \sim \geq b$ in P. But since L is the free lattice generated by and preserving bounds of pairs, it follows that $a \sim \geq b$ in P implies $a \sim \geq b$ in L.

This proof, though short, is non-constructive and it seems worthwhile to give a constructive proof which at the same time exhibits something of the structure of the lattice L.

THEOREM 1.3. *L contains P as a sub-partially ordered set.*

Proof. From Theorem 1.2 it follows that $a \geq b$ in P implies $a \geq b$ in L. Before proving the converse we need a result on finite vectors with whole number components. Consider the vector $\mu = \{m_1, \cdots, m_k\}$ where m_i is a positive integer. μ is said to undergo a *reduction* if some m_i is omitted or is replaced by a vector $\{m_{i1}, \cdots, m_{ie}\}$ where $m_{ij} < m_i$. The set of finite vectors is partially ordered by defining a vector to be contained in μ if it is obtained from μ by a series of reductions. *The set of finite vectors so ordered satisfies the descending chain condition.* To prove this we must show that after a finite number of reductions we always reach a vector which can be reduced no farther; that is, a vector of the form $\{1\}$. Let us make an induction on $m = \max(m_1, \cdots, m_k)$. If $m = 1$, the result is trivial. Now suppose it holds for all vectors whose maximum is less than m. If $k = 1$, then any reduction gives a vector whose maximum is less than m and the result follows by the induction assumption. Now let us make a second induction upon k and assume that the statement holds for all vectors of maximum m and length less than k. If m_i is untouched, a series of reductions must end since it is a series of reductions on $\{m_1, \cdots, m_{i-1}, m_{i+1}, \cdots, m_k\}$ whose length is less than k. Hence after a finite number of reductions each m_i has either been omitted or replaced by smaller numbers so that the resulting vector has a maximum which is less than m. For this vector the above statement holds by the induction assumption. Hence the result holds for all vectors of maximum m by induction on k, and induction on m completes the proof.

Now consider a chain

$$(1) \qquad A_1 \geq A_2(m_1), A_2 \geq A_3(m_2), \cdots, A_k \geq A_{k+1}(m_k),$$

where $v(A_1)$ and $v(A_{k+1})$ exist. *We shall show that $v(A_1) \geq v(A_{k+1})$ in P.* We may clearly suppose that $A_i \neq A_{i+1}$ since otherwise A_i or A_{i+1} may be omitted from the chain. Suppose $m_1 = m_2 = \cdots = m_k = 1$. Then by (i), $v(A_1) \geq v(A_2) \geq \cdots \geq v(A_{k+1})$ in P and the result follows. Let us associate with the chain (1) the vector $\mu = \{m_1, \cdots, m_k\}$ and suppose that the conclusion holds for all chains whose vector is properly contained in μ. Now if $A_i \geq C(m_i - 1)$, $C \geq A_{i+1}(m_i - 1)$ for some i, by substituting C into the above chain we get a chain whose vector is $\{m_1, \cdots, m_{i-1}, m_i - 1, m_i - 1, m_{i+1}, \cdots, m_k\}$ and which is properly contained in μ. Hence $v(A_1) \geq v(A_{k+1})$ in P by assumption. Thus we may assume that (1) of (ii) holds for none of the containing relations of (1). Also if $v(A_i)$ exists where $1 < i < k+1$ then again by the induction assumption $v(A_1) \geq v(A_i) \geq v(A_{k+1})$ in P. Hence we can assume that $m_i \neq 1$, $i = 1, \cdots, k$. Suppose next that $A_{k+1} \equiv B_{k+1} \cup C_{k+1}$ where $A_k \geq B_{k+1}(m_k - 1)$ and $A_k \geq C_{k+1}(m_k - 1)$. Then since the chains from A_1 to B_{k+1} and C_{k+1} have vectors contained in μ, we have $v(A_1) \geq v(B_{k+1}), v(C_{k+1})$ in P. Hence by Definition 1.4, $v(A_1) \geq \mathrm{l.u.b.}(v(B_{k+1}), v(C_{k+1})) = v(A_{k+1})$. If $A_{k+1} \equiv B_{k+1} \cap C_{k+1}$ where either $A_{k+1} \geq B_{k+1}(m_k - 1)$ or $A_{k+1} \geq C_{k+1}(m_k - 1)$ then as before either $v(A_1) \geq v(B_{k+1})$ or $v(A_1) \geq v(C_{k+1})$ in P. Hence $v(A_1) \geq \mathrm{g.l.b.}(v(B_{k+1}), v(C_{k+1})) = v(A_{k+1})$ in P. Thus since $m_k > 1$ we can assume that either (2) or (3) of (ii) in Definition 1.5 holds for $A_k \geq A_{k+1}(m_k)$. Now let A_r be the first member of the chain such that (2) or (3) of (ii) holds for $A_r \geq A_{r+1}(m_r)$. Suppose $r = 1$. If (2) holds, then $A_1 \equiv B_1 \cup C_1$, where $B_1 \geq A_2(m_1 - 1)$ or $C_1 \geq A_2(m_1 - 1)$. Thus by assumption $v(B_1) \geq v(A_{k+1})$ or $v(C_1) \geq v(A_{k+1})$. Hence $v(A_1) = \mathrm{l.u.b.}(v(B_1), c(C_1)) \geq v(A_{k+1})$ in P. If (3) holds, a similar argument shows that $v(A_1)$

$\geq v(A_{k+1})$ in P. Thus we have only to consider the case $r > 1$. But then (4) or (5) of (ii) must hold for $A_{r-1} \geq A_r(m_{r-1})$. Thus either $A_r \equiv B_r \cup C_r$ or $A_r \equiv B_r \cap C_r$ and either $A_{r-1} \geq B_r(m_{r-1} - 1)$, $B_r \geq A_{r+1}(m_r - 1)$ or $A_{r-1} \geq C_r(m_{r-1} - 1)$, $C_r \geq A_{r+1}(m_r - 1)$. Hence if we replace A_r by B_r or C_r, as the case may be, we get a chain whose vector is properly contained in μ. Thus $v(A_1) \geq v(A_{k+1})$ in P by the induction assumption. But since the descending chain condition holds in the set of vectors it follows that $v(A_1) \geq v(A_{k+1})$ for every chain (1) for which $v(A_1)$ and $v(A_{k+1})$ exist.

Now let $a \geq b(m)$ where $a, b \in P$. Then $v(a) \equiv a$ and $v(b) \equiv b$ and by the result we have just proved we conclude that $a \geq b$ in P. Hence $a \sim \geq b$ in P implies $a \sim \geq b$ in L and the proof of the theorem is complete.

P will frequently be specialized to two extreme cases.

COROLLARY 1. *If P is a lattice under the partial ordering, then L is isomorphic to P.*

COROLLARY 2. *If P is an unordered([6]) set S, then L is the free lattice generated by S.*

In connection with sublattices of the free lattice generated by an unordered set, Whitman [2] has proved the following quite surprising theorem.

THEOREM 1.4. *The free lattice generated by three elements contains as a sublattice the free lattice generated by a countable set of elements.*

We shall give here a new proof of Whitman's result since similar methods will be used later in proving analogous theorems for lattices with operators.

LEMMA 1.3. *Let $\mathfrak{S}$ consisting of elements $A, B, C, \cdots$ be a subset of the free lattice generated by an unordered set S. Then the sublattice $L_\mathfrak{S}$ generated by $\mathfrak{S}$ is isomorphic to the free lattice generated by $\mathfrak{S}$ as an unordered set if and only if*
   (1) $A \geq B$ *implies* $A \equiv B$ *if* $A, B \in \mathfrak{S}$,
   (2) $\mathfrak{A} \cup \mathfrak{B} \geq A$ *implies* $\mathfrak{A} \geq A$ *or* $\mathfrak{B} \geq A$ *if* $\mathfrak{A}, \mathfrak{B} \in L_\mathfrak{S}$ *and* $A \in \mathfrak{S}$,
   (3) $A \geq \mathfrak{A} \cap \mathfrak{B}$ *implies* $A \geq \mathfrak{A}$ *or* $A \geq \mathfrak{B}$ *if* $\mathfrak{A}, \mathfrak{B} \in L_\mathfrak{S}$ *and* $A \in \mathfrak{S}$.

The necessity of (1) is obvious. In view of footnote 5, the necessity of (2) and (3) follows from the fact that (2) and (3) of (ii), Definition 1.5, are the only possibilities which can occur respectively in these two cases. On the other hand if (1), (2) and (3) are satisfied, let $L_\mathfrak{S}'$ be the free lattice generated by $\mathfrak{S}$ as an unordered set. Let $\mathfrak{A} \geq \mathfrak{B}$ in $L_\mathfrak{S}$. If $\mathfrak{A}, \mathfrak{B} \in \mathfrak{S}$, then $\mathfrak{A} \geq \mathfrak{B}$ in $L_\mathfrak{S}'$ by (1). Now make an induction on the sum of the ranks of $\mathfrak{A}$ and $\mathfrak{B}$. If $\mathfrak{A} \equiv \mathfrak{A}_1 \cap \mathfrak{A}_2$, then $\mathfrak{A}_1 \geq \mathfrak{B}$ and $\mathfrak{A}_2 \geq \mathfrak{B}$ imply $\mathfrak{A}_1 \geq \mathfrak{B}$ and $\mathfrak{A}_2 \geq \mathfrak{B}$ in $L_\mathfrak{S}'$ imply $\mathfrak{A} \geq \mathfrak{B}$ in $L_\mathfrak{S}'$. A similar argument holds if $\mathfrak{B} \equiv \mathfrak{B}_1 \cup \mathfrak{B}_2$. Hence we can suppose that $\mathfrak{A} \in \mathfrak{S}$ or

---

[6] If P is unordered, then $A \geq B$ (1) if and only if $A \equiv B$. In this case, as Whitman [1] has shown, (1) of (ii) may be replaced by (1)' $A \geq B(n-1)$. For a general partially ordered set, however, the transitivity of the new partial ordering cannot be proved on the basis of (1)'. On the other hand, it is (1) of (ii) which makes the proof of Theorem 1.3 difficult.

$\mathfrak{A} = \mathfrak{A}_1 \cup \mathfrak{A}_2$ and $\mathfrak{B} \in \mathfrak{S}$ or $\mathfrak{B} = \mathfrak{B}_1 \cap \mathfrak{B}_2$. If $\mathfrak{A} = \mathfrak{A}_1 \cup \mathfrak{A}_2$ and $\mathfrak{B} \in \mathfrak{S}$, then $\mathfrak{A} \geq \mathfrak{B}$ in $L_\mathfrak{S}'$ by (2). If $\mathfrak{A} \in \mathfrak{S}$ and $\mathfrak{B} = \mathfrak{B}_1 \cap \mathfrak{B}_2$ then $\mathfrak{A} \geq \mathfrak{B}$ in $L_\mathfrak{S}'$ by (3). Finally if $\mathfrak{A} = \mathfrak{A}_1 \cup \mathfrak{A}_2$ and $\mathfrak{B} = \mathfrak{B}_1 \cap \mathfrak{B}_2$ then one of $\mathfrak{A} \geq \mathfrak{B}_1$, $\mathfrak{A} \geq \mathfrak{B}_2$, $\mathfrak{A}_1 \geq \mathfrak{B}$, $\mathfrak{A}_2 \geq \mathfrak{B}$ holds by (ii), Definition 1.5. Hence by the induction assumption $\mathfrak{A} \geq \mathfrak{B}$ in $L_\mathfrak{S}'$. Thus $\mathfrak{A} \geq \mathfrak{B}$ in $L_\mathfrak{S}$ implies $\mathfrak{A} \geq \mathfrak{B}$ in $L_\mathfrak{S}'$ and since $L_\mathfrak{S}'$ is the free lattice generated by $\mathfrak{S}$ as an unordered set it follows that $L_\mathfrak{S}$ and $L_\mathfrak{S}'$ are isomorphic.

Now let L be the free lattice generated by the three unordered elements a, b, c. Let us set $x_0 \equiv a$ and define inductively

$$x_n \equiv a \cup (b \cap (c \cup (a \cap (b \cup (c \cap x_{n-1}))))), \qquad n = 1, 2, \cdots,$$

$$x_{-n} \equiv a \cap (b \cup (c \cap (a \cup (b \cap (c \cup x_{-n+1}))))), \qquad n = 1, 2, \cdots.$$

LEMMA 1.4. *The following relations hold in L:*
(1) $x_n \cup b \sim \geq c$; $c \sim \geq x_{-n} \cap b$.
(2) $a, b \sim \geq x_n \cap c$; $x_{-n} \cup c \sim \geq a, b$.

For $x_n \cup b \geq c \rightarrow x_n \geq c \rightarrow a \cup b \geq c$ and $a \geq x_n \cap c \rightarrow a \geq b \cap c$. But both of these conclusions are impossible. Dual proofs give the other relations.

LEMMA 1.5. *If $a \cup c \geq X$, $Y \geq c$, then $a \cup (b \cap X) \geq a \cup (b \cap Y) \rightarrow b \cap X \geq b \cap Y$. If $c \geq X$, $Y \geq a \cap c$, then $a \cap (b \cup X) \geq a \cap (b \cup Y) \rightarrow b \cup X \geq b \cup Y$.*

For $a \cup (b \cap X) \geq a \cup (b \cap Y) \rightarrow a \cup (b \cap X) \geq b \cap Y$. Now $a \cup (b \cap X) \geq b \rightarrow a \cup X \geq b \rightarrow a \cup c \geq b$ which is impossible. $a \cup (b \cap X) \geq Y \rightarrow a \cup b \geq c$ which is impossible. $a \geq b \cap Y \rightarrow a \geq b \cap c$ which again is impossible. Hence the only possibility according to (ii), Definition 1.5, is $b \cap X \geq b \cap Y$.

It is clear that Lemma 1.5 holds under cyclic permutations of a, b, c.

LEMMA 1.6. $\cdots < x_{-n-1} < x_{-n} < \cdots < x_{-1} < x_0 < x_1 < \cdots < x_n < x_{n+1} < \cdots$.

Clearly $x_1 \geq x_0$. Suppose we have shown that $x_n \geq x_{n-1}$, then since the operations $\cup$ and $\cap$ preserve order we have $x_{n+1} \geq x_n$. Similarly $x_{-n-1} \leq x_{-n}$. Now suppose $x_n \geq x_{n+1}$. Then by successive application of Lemma 1.5 we get $x_{n-1} \geq x_n$. Hence eventually we have $a \geq x_1 \geq b \cap c$ which is impossible. Similarly $x_{-n-1} \sim \geq x_{-n}$. Thus the containing relations in the chain are all proper.

The elements of the countable generating set are defined explicitly as follows:

$$a_n \equiv b \cup (x_n \cap (x_{-n} \cup c)), \qquad n = 1, 2, \cdots.$$

Clearly $a_n \sim \geq c$ since $a \cup b \geq x_n \cup b \geq a_n$.

LEMMA 1.7. $a_n \geq a_m$ *implies* $m = n$.

For $a_n \geq a_m \rightarrow b \cup (x_n \cap (x_{-n} \cup c)) \geq x_m \cap (x_{-m} \cup c)$. Now $a_n \geq x_{-m} \cup c \rightarrow a_n \geq c$ which is impossible. Also $a_n \geq x_m \rightarrow b \cup x_{-n} \cup c \geq x_m \geq a \rightarrow x_{-n} \cup c \geq a$ and $b \geq x_m \cap (x_{-m} \cup c) \rightarrow b \geq x_m \cap c$, both conclusions contradicting Lemma 1.4. Hence

$x_n \cap (x_{-n} \cup c) \geq x_m \cap (x_{-m} \cup c)$. But then $x_n \geq x_m \cap (x_{-m} \cup c)$ and $x_{-n} \cup c \geq x_m \cap (x_{-m} \cup c)$. Now $x_n \geq x_{-m} \cup c \rightarrow x_n \geq c$ which contradicts Lemma 1.4 and since $a, b \sim \geq x_m \cap c$ we must have $x_n \geq x_m$. Whence $n \geq m$. Also $x_{-n} \geq x_m \cap (x_{-m} \cup c)$ $\rightarrow a \geq x_m \cap c$ which contradicts Lemma 1.4 and $c \geq x_m \cap (x_{-m} \cup c) \rightarrow c \geq x_{-m}$ which is also impossible. Since $x_{-n} \cup c \sim \geq x_m$ we have $x_{-n} \cup c \geq x_{-m} \cup c \geq x_{-m}$. But since $x_{-n} \cup c \sim \geq a, b$ and $c \sim \geq x_{-m}$ we have $x_{-n} \geq x_{-m}$. Whence $n \leq m$ and thus $m = n$.

Let $\mathfrak{S}$ denote the set of elements $a_1, a_2, \cdots$. Then it follows from Lemma 1.7 that (1) of Lemma 1.3 holds for the set $\mathfrak{S}$.

Let $\mathfrak{A}, \mathfrak{B}, \mathfrak{C}, \cdots$ denote lattice polynomials over $\mathfrak{S}$.

LEMMA 1.8. $\mathfrak{A} \sim \geq c$; $\mathfrak{A} \geq b$.

For $a_n \sim \geq c$ and $\mathfrak{A} \cup \mathfrak{B} \geq c \rightarrow \mathfrak{A} \geq c$ or $\mathfrak{B} \geq c$ while $\mathfrak{A} \cap \mathfrak{B} \geq c \rightarrow \mathfrak{A} \geq c$ and $\mathfrak{B} \geq c$. Hence the first relation follows by induction on the rank of $\mathfrak{A}$. Now $a_n \geq b$ for all n. Hence if $\mathfrak{A}, \mathfrak{B} \geq b$, then $\mathfrak{A} \cup \mathfrak{B} \geq b$ and $\mathfrak{A} \cap \mathfrak{B} \geq b$ and the result follows by induction.

LEMMA 1.9. $\mathfrak{A} \cup \mathfrak{B} \geq x_n$ *implies* $\mathfrak{A} \geq x_n$ *or* $\mathfrak{B} \geq x_n$.

For $\mathfrak{A} \cup \mathfrak{B} \geq x_n \rightarrow \mathfrak{A} \cup \mathfrak{B} \geq a \rightarrow \mathfrak{A} \geq a$ or $\mathfrak{B} \geq a \rightarrow \mathfrak{A} \geq a \cup b$ or $\mathfrak{B} \geq a \cup b \rightarrow \mathfrak{A} \geq x_n$ or $\mathfrak{B} \geq x_n$.

LEMMA 1.10. $\mathfrak{A} \cup \mathfrak{B} \geq a_n$ *implies* $\mathfrak{A} \geq a_n$ *or* $\mathfrak{B} \geq a_n$.

If $\mathfrak{A} \cup \mathfrak{B} \geq a_n$, then $\mathfrak{A} \cup \mathfrak{B} \geq x_n \cap (x_{-n} \cup c)$. If $\mathfrak{A} \geq x_n \cap (x_{-n} \cup c)$ then $\mathfrak{A} \geq b \cup (x_n \cap (x_{-n} \cup c))$ and $\mathfrak{A} \geq a_n$. If $\mathfrak{B} \geq x_n \cap (x_{-n} \cup c)$ then $\mathfrak{B} \geq a_n$. Now $\mathfrak{A} \cup \mathfrak{B} \sim \geq x_{-n} \cup c$. Hence the only remaining possibility is $\mathfrak{A} \cup \mathfrak{B} \geq x_n$ and hence $\mathfrak{A} \geq x_n$ or $\mathfrak{B} \geq x_n$ by Lemma 1.9. Thus either $\mathfrak{A} \geq x_n \cap (x_{-n} \cup c)$ or $\mathfrak{B} \geq x_n \cap (x_{-n} \cup c)$ and hence either $\mathfrak{A} \geq a_n$ or $\mathfrak{B} \geq a_n$.

LEMMA 1.11. $b \sim \geq \mathfrak{A}$.

For $b \geq a_n \rightarrow b \geq b \cup (x_n \cap (x_{-n} \cup c)) \rightarrow b \geq x_n \cap (x_{-n} \cup c) \rightarrow b \geq x_n \cap c$ which contradicts Lemma 1.4. Induction on the rank of $\mathfrak{A}$ completes the proof.

LEMMA 1.12. $a_n \geq \mathfrak{A} \cap \mathfrak{B}$ *implies* $a_n \geq \mathfrak{A}$ *or* $a_n \geq \mathfrak{B}$.

For suppose that $a_n \geq \mathfrak{A} \cap \mathfrak{B}$ but $a_n \sim \geq \mathfrak{A}$ and $a_n \sim \geq \mathfrak{B}$. But then $b \cup (x_n \cap (x_{-n} \cup c)) \geq \mathfrak{A} \cap \mathfrak{B}$ and $b \sim \geq \mathfrak{A} \cap \mathfrak{B} \rightarrow x_n \cap (x_{-n} \cup c) \geq \mathfrak{A} \cap \mathfrak{B} \rightarrow x_{-n} \cup c \geq \mathfrak{A} \cap \mathfrak{B} \geq b$ which contradicts Lemma 1.4.

Lemmas 1.10 and 1.12 give conditions (2) and (3) of Lemma 1.3 and Theorem 1.4 follows.

2. **Lattices with unary operator.** In the previous section, the containing relation in P has been extended to the set L of lattice polynomials. We now extend the containing relation in L to the set O of all operator polynomials. In order to connect the operator polynomials with the set of lattice polynomials another definition is required.

DEFINITION 2.1. An operator polynomial A has an *upper cover* $\overline{A}$ if and only if, inductively,

(1) $A \in P$ in which case $\overline{A} \equiv A$,

(2) $A \equiv A_1 \cup A_2$ where A_1 and A_2 have upper covers $\overline{A}_1$ and $\overline{A}_2$ respectively in which case $\overline{A} \equiv \overline{A}_1 \cup \overline{A}_2$,

(3) $A \equiv A_1 \cap A_2$ where either A_1 or A_2 has an upper cover. If $\overline{A}_1$ exists and $\overline{A}_2$ does not exist, then $\overline{A} \equiv \overline{A}_1$. If $\overline{A}_2$ exists and $\overline{A}_1$ does not exist, then $\overline{A} \equiv \overline{A}_2$. If both $\overline{A}_1$ and $\overline{A}_2$ exist, then $\overline{A} \equiv \overline{A}_1 \cap \overline{A}_2$.

Although $\overline{A}$ is defined for all lattice polynomials, it should be noted that it is also defined for some operator polynomials which are not lattice polynomials. For example, $\overline{a \cap b^*}$ is defined and indeed $\overline{a \cap b^*} \equiv a$.

In an exactly dual manner, the *lower cover* $*A$ is defined. If $A \equiv A_1 \cup A_2$, then $*A$ exists if and only if $*A_1$ or $*A_2$ exists while if $A \equiv A_1 \cap A_2$, then $*A$ exists if and only if both $*A_1$ and $*A_2$ exist.

LEMMA 2.1. $\overline{A}(*A)$, *if it exists, is a lattice polynomial.*

For if A belongs to P, then $\overline{A} \equiv A$ and $\overline{A}$ is a lattice polynomial trivially. If $A \equiv A_1 \cup A_2$, then $\overline{A} \equiv \overline{A}_1 \cup \overline{A}_2$ and if $\overline{A}_1$ and $\overline{A}_2$ are lattice polynomials so also is $\overline{A}$. If $A \equiv A_1 \cap A_2$ then $\overline{A} \equiv \overline{A}_1$, $\overline{A}_2$, or $\overline{A}_1 \cap \overline{A}_2$ and again $\overline{A}$ is a lattice polynomial if $\overline{A}_1$ or $\overline{A}_2$ or both are lattice polynomials. If $A \equiv A_1^*$ then, by Definition 2.1, $\overline{A}$ does not exist. A dual argument holds for $*A$.

LEMMA 2.2. *If A is a lattice polynomial, then $\overline{A}$ and $*A$ exist and $\overline{A} \equiv *A \equiv A$.*

For if $A \in P$ the lemma is trivial. By induction, if $A \equiv A_1 \cup A_2$; then A_1 and A_2 are lattice polynomials and $\overline{A}_1 \equiv A_1$ and $\overline{A}_2 \equiv A_2$. By Definition 2.1, $\overline{A} \equiv \overline{A}_1 \cup \overline{A}_2 \equiv A_1 \cup A_2 \equiv A$. If $A \equiv A_1 \cap A_2$, then A_1 and A_2 are lattice polynomials, whence both $\overline{A}_1$ and $\overline{A}_2$ exist by the induction assumption and $\overline{A}_1 \equiv A_1$, $\overline{A}_2 \equiv A_2$. But then $\overline{A} \equiv \overline{A}_1 \cap \overline{A}_2 \equiv A_1 \cap A_2 \equiv A$. A dual argument holds for $*A$.

LEMMA 2.3. *If $\overline{A}$ and $*A$ exist, then $\overline{A} \geq *A$ in L.*

For if $A \in P$, then $\overline{A} \equiv *A \equiv A$. Whence $\overline{A} \geq *A$ by Definition 1.5. Now by induction, if $A \equiv A_1 \cup A_2$ we have $\overline{A} \equiv \overline{A}_1 \cup \overline{A}_2$ and $*A \equiv *A_1$ or $*A_2$ or $*A_1 \cup *A_2$. But by the induction assumption $\overline{A}_1 \geq *A_1$ or $\overline{A}_2 \geq *A_2$ according as $*A_1$ or $*A_2$ exists. Thus $\overline{A} \equiv \overline{A}_1 \cup \overline{A}_2 \geq *A$. If $A \equiv A_1 \cap A_2$ the dual of the above argument shows that $\overline{A} \geq *A$. If $A \equiv A_1^*$ then $\overline{A}$ and $*A$ do not exist and the lemma holds vacuously.

The containing relation in O is defined in a manner analogous to that in L.

DEFINITION 2.2. If $A, B \in O$ let us set

(i) $A \supseteq B(1)$ if $A \equiv B$ or if $*A$ and $\overline{B}$ exist with $*A \geq \overline{B}$ in L,

(ii) $A \supseteq B(n)$ where $n > 1$ if and only if one of the following hold:

(1) $A \supseteq B(n-1)$,

(2) $A \equiv A_1 \cup A_2$ where $A_1 \supseteq B(n-1)$ or $A_2 \supseteq B(n-1)$,

(3) $A \equiv A_1 \cap A_2$ where $A_1 \supseteq B(n-1)$ and $A_2 \supseteq B(n-1)$,
(4) $B \equiv B_1 \cup B_2$ where $A \supseteq B_1(n-1)$ and $A \supseteq B_2(n-1)$,
(5) $B \equiv B_1 \cap B_2$ where $A \supseteq B_1(n-1)$ or $A \supseteq B_2(n-1)$,
(6) $A \equiv A_1^*$ and $B \equiv B_1^*$ where $A_1 \supseteq B_1(n-1)$ and $B_1 \supseteq A_1(n-1)$,
(iii) $A \supseteq B$ if and only if $A \supseteq B(n)$ for some n.

LEMMA 2.4. $\overline{A} \supseteq A$.

For $A \equiv A$ implies $A \supseteq A(1)$.

LEMMA 2.5. *Let* $A \supseteq B$. *Then if* $\overline{A}$ *exists,* $\overline{B}$ *also exists and* $\overline{A} \geq \overline{B}$ *in* L.

First, let $A \supseteq B(1)$. If $A \equiv B$, the lemma is trivial. If $*A$ and $\overline{B}$ exist with $*A \geq \overline{B}$ in L, then the first part of the lemma is immediate. But by Lemma 2.3, $\overline{A} \geq *A \geq \overline{B}$ in L. This gives the second part of the lemma. Now proceed by induction and suppose the lemma holds if $A \supseteq B(n-1)$. Let $A \supseteq B(n)$. We have six possibilities: If $A \supseteq B(n-1)$ the lemma holds by assumption. If $A \equiv A_1 \cup A_2$ where $A_1 \supseteq B(n-1)$ or $A_2 \supseteq B(n-1)$ then if $\overline{A}$ exists, $\overline{A}_1$ and $\overline{A}_2$ also exist and hence $\overline{B}$ exists by the induction assumption. Furthermore, either $\overline{A}_1 \geq \overline{B}$ or $\overline{A}_2 \geq \overline{B}$ in L. Hence $\overline{A} \equiv \overline{A}_1 \cup \overline{A}_2 \geq \overline{B}$ in L. If $A \equiv A_1 \cap A_2$ where $A_1 \supseteq B(n-1)$ and $A_2 \supseteq B(n-1)$, then if $\overline{A}$ exists, either $\overline{A}_1$ or $\overline{A}_2$ exists and $\overline{A} \equiv \overline{A}_1$ or $\overline{A}_2$ or $\overline{A}_1 \cap \overline{A}_2$. But again by the induction assumption $\overline{B}$ exists and $\overline{A} \equiv \overline{A}_1$ or $\overline{A}_2$ or $\overline{A}_1 \cap \overline{A}_2 \geq \overline{B}$ in L. If $B \equiv B_1 \cup B_2$ where $A \supseteq B_1(n-1)$ and $A \supseteq B_2(n-1)$, then if $\overline{A}$ exists both $\overline{B}_1$ and $\overline{B}_2$ exist by the induction assumption and $\overline{A} \geq \overline{B}_1$, $\overline{B}_2$ in L. But then $\overline{B} \equiv \overline{B}_1 \cup \overline{B}_2$ exists and $\overline{A} \geq \overline{B}$ in L. If $B \equiv B_1 \cap B_2$ where $A \supseteq B_1(n-1)$ or $A \supseteq B_2(n-1)$ then if $\overline{A}$ exists, either $\overline{B}_1$ or $\overline{B}_2$ exists by induction and $\overline{A} \geq \overline{B}_1$ or $\overline{A} \geq \overline{B}_2$. But then $\overline{B} \equiv \overline{B}_1$ or $\overline{B}_2$ or $\overline{B}_1 \cap \overline{B}_2$ and hence $\overline{A} \geq \overline{B}$ in L. Finally if $A \equiv A_1^*$ and $B \equiv B_1^*$, then $\overline{A}$ does not exist and the lemma holds vacuously. Induction on n completes the proof.

LEMMA 2.6. *Let* $A \supseteq B$. *Then if* $*B$ *exists,* $*A$ *also exists and* $*A \geq *B$ *in* L.

The proof is the exact dual of that of Lemma 2.5.

LEMMA 2.7. $A \supseteq B$ *and* $B \supseteq C$ *imply* $A \supseteq C$.

We shall show first that $A \supseteq B(m)$ and $B \supseteq C(n)$ imply $A \supseteq C$ by making an induction on $l = m + n$. Suppose $m = 1$ so that $A \supseteq B(1)$. If $A \equiv B$, then $B \supseteq C(n)$ implies $A \supseteq C(n)$ implies $A \supseteq C$. If $*A$ and $\overline{B}$ exist with $*A \geq \overline{B}$ in L, then since $B \supseteq C$ it follows from Lemma 2.5 that $\overline{C}$ exists and $\overline{B} \geq \overline{C}$ in L. But then $*A \geq \overline{C}$ in L and $A \supseteq C(1)$. Whence $A \supseteq C$ by Definition 2.2. Thus the result holds if $m = 1$ and a dual argument gives the case where $n = 1$. We may thus suppose that m, $n > 1$. Let us assume that the lemma holds if $m + n < l$ and let $m + n = l$. Since $A \supseteq B(m)$ and $m > 1$ we have six possibilities:
(1) $A \supseteq B(m-1)$. But then $A \supseteq C$ by induction.
(2) $A \equiv A_1 \cup A_2$ where $A_1 \supseteq B(m-1)$ or $A_2 \supseteq B(m-1)$. But then $A_1 \supseteq C$ or $A_2 \supseteq C$ by the induction assumption. That is $A_1 \supseteq C(k)$ or $A_2 \supseteq C(k)$ for some k.

But then $A \equiv A_1 \cup A_2 \supseteq C(k+1)$ by (2) of (ii). Hence $A \supseteq C$ by Definition 2.2.

(3) $A \equiv A_1 \cap A_2$ where $A_1 \supseteq B(m-1)$ and $A_2 \supseteq B(m-1)$. But then $A_1 \supseteq C$ and $A_2 \supseteq C$ by induction. That is $A_1 \supseteq C(i)$ and $A_2 \supseteq C(j)$ for some i and j. But then $A_1 \supseteq C(k)$ and $A_2 \supseteq C(k)$ where $k = \max(i, j)$. Hence $A \equiv A_1 \cap A_2 \supseteq C(k+1)$ by (3) of (ii). Thus $A \supseteq C$ by (iii).

We leave possibilities (4), (5), and (6) for the moment and consider the six possibilities on $B \supseteq C(n)$. If $B \supseteq C(n-1)$, then $A \supseteq C$ by induction. If $C \equiv C_1 \cup C_2$ with $B \supseteq C_1(n-1)$ and $B \supseteq C_2(n-1)$ then $A \supseteq C$ by the exact dual of the argument in (3) above. If $C \equiv C_1 \cap C_2$ with either $B \supseteq C_1(n-1)$ or $B \supseteq C_2(n-1)$ then $A \supseteq C$ by the exact dual of (2) above.

Now consider possibility (6). $A \equiv A_1^*$ and $B \equiv B_1^*$ where $A_1 \supseteq B_1(m-1)$ and $B_1 \supseteq A_1(m-1)$. Since $B \supseteq C(n)$ and the possibilities $B \supseteq C(n-1)$, $C \equiv C_1 \cup C_2$, and $C \equiv C_1 \cap C_2$ have been treated, we must have $C \equiv C_1^*$ where $B_1 \supseteq C_1(n-1)$ and $C_1 \supseteq B_1(n-1)$. But then by induction $A_1 \supseteq C_1$ and $C_1 \supseteq A_1$. That is $A_1 \supseteq C_1(i)$ and $C_1 \supseteq A_1(j)$ for some i and j. But then $A_1 \supseteq C_1(k)$ and $C_1 \supseteq A_1(k)$ where $k = \max(i, j)$. Thus $A_1^* \supseteq C_1^*(k+1)$ by (6) of (ii). Hence $A \supseteq C$ by (iii) of Definition 2.2.

The only remaining possibilities are

(4) $B \equiv B_1 \cup B_2$ with $A \supseteq B_1(m-1)$ and $A \supseteq B_2(m-1)$ and either $B_1 \supseteq C(n-1)$ or $B_2 \supseteq C(n-1)$.

(5) $B \equiv B_1 \cap B_2$ with either $A \supseteq B_1(m-1)$ or $A \supseteq B_2(m-1)$ and $B_1 \supseteq C(n-1)$, $B_2 \supseteq C(n-1)$.

But in both (4) and (5), $A \supseteq C$ by induction. It follows that $A \supseteq B(m)$ and $B \supseteq C(n)$ imply $A \supseteq C$. But if $A \supseteq B$ and $B \supseteq C$ then $A \supseteq B(m)$ for some m and $B \supseteq C(n)$ for some n. Hence $A \supseteq C$ and the proof of the lemma is complete.

THEOREM 2.1. *The set O of operator polynomials forms a lattice under the relation $A \supseteq B$.*

Proof. Lemmas 2.4 and 2.7 show that O is partially ordered by the relation $A \supseteq B$. Now $A \cup B \supseteq A, B$ since $A \cup B \supseteq A, B(2)$ by (2) of (ii). Similarly $A, B \supseteq A \cap B$. Now let $X \supseteq A$ and $X \supseteq B$. Then $X \supseteq A(n)$ and $X \supseteq B(n)$ for some n. But then $X \supseteq A \cup B(n+1)$ by (4). Hence $X \supseteq A \cup B$ by (iii) of Definition 2.2. Similarly if $A, B \supseteq X$, then $A \cap B \supseteq X$. Hence $A \cup B$ and $A \cap B$ are least upper bound and greatest lower bound respectively of A and B. This completes the proof.

DEFINITION 2.3. $A \simeq B$ if and only if $A \supseteq B$ and $B \supseteq A$.

The relation $A \simeq B$ is reflexive, symmetric, transitive, and preserves the operations of union and crosscut.

THEOREM 2.2. *$A \supseteq B$ in O if and only if one of the following holds:*
(1) $A \equiv B$ *or* $^*A$ *and* $\overline{B}$ *exist with* $^*A \geq \overline{B}$ *in* L.
(2) $A \equiv A_1 \cup A_2$ *with* $A_1 \supseteq B$ *or* $A_2 \supseteq B$.
(3) $A \equiv A_1 \cap A_2$ *with* $A_1 \supseteq B$ *and* $A_2 \supseteq B$.
(4) $B \equiv B_1 \cup B_2$ *with* $A \supseteq B_1$ *and* $A \supseteq B_2$.

(5) $B \equiv B_1 \cap B_2$ *with* $A \supseteq B_1$ *or* $A \supseteq B_2$.

(6) $A \equiv A_1^*$ *and* $B \equiv B_1^*$ *with* $A_1 \backsim B_1$.

Proof. The proof is clear from Definitions 2.2 and 2.3.

THEOREM 2.3. L *is a sublattice of* O.

Proof. L is clearly a subset of O and furthermore the operations of union and crosscut in L are the same in O. Hence we have only to show that $A \supseteq B$ where $A, B \in L$ implies $A \geq B$ and conversely. First, let $A \supseteq B(1)$. If $A \equiv B$, then $A \geq B$ trivially. If $*A$ and $\overline{B}$ exist with $*A \geq \overline{B}$, then since A and B are lattice polynomials, $*A \equiv A$ and $\overline{B} \equiv B$, whence $A \geq B$. We make an induction and let $A \supseteq B(n)$. If $A \supseteq B(n-1)$, there is nothing to prove. If $A \equiv A_1 \cup A_2$ where $A_1 \supseteq B(n-1)$ or $A_2 \supseteq B(n-1)$, then A_1 and A_2 are lattice polynomials and by the induction assumption $A_1 \geq B$ or $A_2 \geq B$. Hence $A \equiv A_1 \cup A_2 \geq B$. If $A \equiv A_1 \cap A_2$ where $A_1 \supseteq B(n-1)$ and $A_2 \supseteq B(n-1)$, then A_1 and A_2 are lattice polynomials and by the induction assumption $A_1 \geq B$ and $A_2 \geq B$. Hence $A \equiv A_1 \cap A_2 \geq B$. Cases $B \equiv B_1 \cup B_2$ and $B \equiv B_1 \cap B_2$ are treated similarly. Since A is a lattice polynomial, $A \equiv A_1^*$ cannot occur. Induction on n shows that $A \supseteq B(n)$ implies $A \geq B$. But $A \supseteq B$ where $A, B \in L$ implies $A \supseteq B(n)$ for some n implies $A \geq B$. Conversely let $A \geq B$. Since A and B are lattice polynomials we have $*A \equiv A$ and $\overline{B} \equiv B$ and thus $*A \geq \overline{B}$. Hence $A \supseteq B(1)$ and $A \supseteq B$ by (iii) of Definition 2.2. This completes the proof.

COROLLARY. O *contains* P *as a sub-partially ordered set and preserves* l.u.b. *and* g.l.b. *of pairs whenever they exist.*

The corollary follows from Theorems 2.3 and 1.2.

The remaining theorems of this section will develop the general structure of the lattice O.

THEOREM 2.4. $\overline{A} \supseteq A \supseteq *A$ *whenever the covers exist.*

Proof. We make an induction on the rank of A. If $r(A) = 0$, then $A \in P$ and $\overline{A} \equiv A \equiv *A$ and the theorem holds. Now let $r(A) = n$ where $n > 0$. If $A \equiv A_1 \cup A_2$ and $\overline{A}$ exists, then $\overline{A_1}$ and $\overline{A_2}$ exist and $\overline{A} \equiv \overline{A_1} \cup \overline{A_2}$. But since $r(A_1) < n$ and $r(A_2) < n$ we have $\overline{A_1} \supseteq A_1$ and $\overline{A_2} \supseteq A_2$. Hence $\overline{A} \equiv \overline{A_1} \cup \overline{A_2} \supseteq A_1 \cup A_2 \equiv A$. If $A \equiv A_1 \cap A_2$ and $\overline{A}$ exists, then $\overline{A_1}$ or $\overline{A_2}$ exists and $\overline{A} \equiv \overline{A_1}$ or $\overline{A_2}$ or $\overline{A_1} \cap \overline{A_2}$. But by induction $\overline{A_1} \supseteq A$ or $\overline{A_2} \supseteq A$. Hence $\overline{A} \supseteq A_1 \cap A_2 \equiv A$. If $A \equiv A_1^*$, then $\overline{A}$ does not exist and the theorem holds vacuously. Thus by induction $\overline{A} \supseteq A$ whenever $\overline{A}$ exists. A dual proof gives the second inclusion.

THEOREM 2.5. $A^* \supseteq B^*$ *if and only if* $A \backsim B$.

Proof. The only possibilities of Theorem 2.2 which can occur are $A \equiv B$ and $A \backsim B$. In either case $A \backsim B$.

COROLLARY. $A \backsim B$ *implies* $A^* \backsim B^*$.

It should be noted that as a consequence of Theorem 2.5, the polynomials of the form A^* form an unordered set. This is in marked contrast to ortho-complementation where $A \supseteq B$ implies $B' \supseteq A'$.

Theorem 2.6. *$A^* \supseteq B \cap C$ if and only if $A^* \supseteq B$ or $A^* \supseteq C$.*

Proof. Since $^*A$ does not exist, the only possibility of Theorem 2.2 is (5). That is, $A^* \supseteq B$ or $A^* \supseteq C$.

Theorem 2.7. *$B \cup C \supseteq A^*$ if and only if $B \supseteq A^*$ or $C \supseteq A^*$.*

Proof. The proof is the dual of that of Theorem 2.6.

Theorem 2.8. *If $A \cap B \simeq C^*$, then either $A \simeq C^*$ or $B \simeq C^*$.*

Proof. If $A \cap B \simeq C^*$, then $C^* \supseteq A \cap B$ and $C^* \supseteq A$ or $C^* \supseteq B$ by Theorem 2.6. But $A \supseteq C^*$ and $B \supseteq C^*$. Hence either $A \simeq C^*$ or $B \simeq C^*$.

Theorem 2.9. *If $A \cup B \simeq C^*$, then either $A \simeq C^*$ or $B \simeq C^*$.*

Proof. The proof is the dual of that of Theorem 2.8.

Theorem 2.10. *$A^* \sim\!\supseteq a$ and $a \sim\!\supseteq A^*$ if $a \in P$ ($\sim\!\supseteq$ means "does not contain").*

Proof. None of the possibilities of Theorem 2.2 can occur.

Theorems 2.4–2.10 are quite elementary and follow immediately from the definition of the containing relation in O. To get at the deeper theorems, however, we shall need the more detailed structure of operator polynomials.

Definition 2.4. A is a *component* of B if one of the following holds: $B \equiv A \cup X$, $B \equiv X \cup A$, $B \equiv A \cap X$, $B \equiv X \cap A$, $B \equiv A^*$.

Lemma 2.8. *If A is a component of B, then $r(A) < r(B)$.*

Definition 2.5. A is a *sub-polynomial* of B if there is a chain of operator polynomials $A \equiv A_1, A_2, \cdots, A_n \equiv B$ where A_i is a component of A_{i+1}. If $A \not\equiv B$, then A is a *proper* sub-polynomial of B.

Lemma 2.9. *If A is a sub-polynomial of B, then $r(A) \leq r(B)$. If A is a proper sub-polynomial of B, then $r(A) < r(B)$.*

Lemma 2.10. *If A is a sub-polynomial of B and B is a sub-polynomial of C, then A is a sub-polynomial of C.*

Lemma 2.11. *If A is a proper sub-polynomial of B, then A is a sub-polynomial of a component of B.*

For A is a sub-polynomial of A_{n-1} which is a component of B.

Theorem 2.11. *If $A \supseteq B^*$, there is a sub-polynomial B_1^* of A such that $B_1 \simeq B$.*

Proof. If $A \supseteq B^*(1)$, then since $\overline{B}^*$ does not exist we must have $A \equiv B^*$. But then $B_1^* \equiv B^*$ is a sub-polynomial of A such that $B_1 \simeq B$. Now suppose that it has been shown that $A \supseteq B^*(k)$ implies that a sub-polynomial B_1^* of A exists such that $B_1 \simeq B$ for all $k < n$. Let $A \supseteq B^*(n)$. If $A \supseteq B^*(n-1)$, the existence of B_1^* follows by induction. If $A \equiv A_1 \cup A_2$ where either $A_1 \supseteq B^*(n-1)$ or $A_2 \supseteq B^*(n-1)$, then by induction either A_1 or A_2 contain a sub-polynomial B_1^* such that $B_1 \simeq B$. But since A_1 and A_2 are components of A, B_1^* is also a sub-polynomial of A and $B_1 \simeq B$. If $A \equiv A_1 \cap A_2$ where $A_1 \supseteq B^*(n-1)$ and $A_2 \supseteq B^*(n-1)$, then by induction A_1 contains a sub-polynomial B_1^* such that $B_1 \simeq B$. But B_1^* is then a sub-polynomial of A and $B_1 \simeq B$. Since possibilities (4) and (5) cannot occur, the only other possibility is $A \equiv A_1^*$ where $A_1 \supseteq B(n-1)$ and $B \supseteq A_1(n-1)$. But then $A_1 \simeq B$ and $B_1^* \equiv A_1^*$ is a sub-polynomial of A such that $B_1 \simeq B$. Thus the conclusion of the theorem holds for $k = n$ and by induction for all k. But if $A \supseteq B^*$, then $A \supseteq B^*(k)$ for some k and hence a sub-polynomial B_1^* of A exists such that $B_1 \simeq B$.

THEOREM 2.12. *If $B^* \supseteq A$, there is a sub-polynomial B_1^* of A such that $B_1 \simeq B$.*

Proof. The proof is the dual of that of Theorem 2.11.

THEOREM 2.13. *If $A^* \cup B \supseteq C$ and $B \sim \supseteq C$, then a sub-polynomial A_1^* of C exists such that $A_1 \simeq A$.*

Proof. Let $A^* \cup B \supseteq C(1)$. If $A^* \cup B \equiv C$, then $A_1^* \equiv A^*$ is a sub-polynomial of C such that $A_1 \simeq A$. On the other hand if $*A^* \cup *B$ and $\overline{C}$ exist such that $*A^* \cup *B \geq \overline{C}$ in L, then since $*A^*$ does not exist, $*B$ must exist and $*A^* \cup *B \equiv *B$. But then $*B \geq \overline{C}$ in L and $B \supseteq C(1)$. Thus $B \supseteq C$ contrary to hypothesis. Hence the theorem holds in this case.

Now suppose we have shown that $A^* \cup B \supseteq C(k)$ and $B \sim \supseteq C$ imply that a sub-polynomial A_1^* of C exists such that $A_1 \simeq A$ for all $k < n$. Let $A^* \cup B \geq C(n)$ and $B \sim \supseteq C$. If $A^* \cup B \supseteq C(n-1)$ the existence of A_1^* follows from the induction assumption. If $A^* \supseteq C(n-1)$, then $A^* \supseteq C$ and by Theorem 2.12 there is a sub-polynomial A_1^* of C such that $A_1 \simeq A$. The possibility $B \supseteq C(n-1)$ cannot occur since otherwise $B \supseteq C$. If $C \equiv C_1 \cup C_2$ where $A^* \cup B \supseteq C_1(n-1)$ and $A^* \cup B \supseteq C_2(n-1)$, then since $B \sim \supseteq C$ we must have $B \sim \supseteq C_1$ or $B \sim \supseteq C_2$. Hence by the induction assumption either C_1 or C_2 must contain a sub-polynomial A_1^* such that $A_1 \simeq A$. But then A_1^* is a sub-polynomial of C such that $A_1 \simeq A$. If $C \equiv C_1 \cap C_2$ where either $A^* \cup B \supseteq C_1(n-1)$ or $A^* \cup B \supseteq C_2(n-1)$, then since $B \sim \supseteq C$ we have $B \sim \supseteq C_1$ and $B \sim \supseteq C_2$. Hence by assumption a sub-polynomial A_1^* of either C_1 or C_2 exists such that $A_1 \simeq A$. A_1^* is clearly a sub-polynomial of C. Possibility (6) cannot occur since $A^* \cup B \not\equiv X^*$ for every X. Thus by induction $A^* \cup B \supseteq C(n)$ and $B \sim \supseteq C$ implies that a sub-polynomial A_1^* of C exists such that $A_1 \simeq A$. If $A^* \cup B \supseteq C$ and

$B \sim \supseteq C$, then $A^* \cup B \supseteq C(n)$ for some n and the conclusion of the theorem follows.

Theorem 2.14. *If $C \supseteq A^* \cap B$ and $C \sim \supseteq B$, then a sub-polynomial A_1^* of C exists such that $A_1 \simeq A$.*

Proof. The proof is the dual of that of Theorem 2.13.

Definition 2.6. The *length $l(A)$* of an operator polynomial A is the least value of $r(X)$ for $X \simeq A$.

Lemma 2.12. $r(A) \geq l(A)$.

Theorem 2.15. *If $A \supseteq B^*$, then $l(A) > l(B)$.*

Proof. By Theorem 2.11, if $X \supseteq B^*$ then X contains a sub-polynomial B_1^* such that $B_1 \simeq B$. Now let $X \simeq A$. Then $X \supseteq B^*$ and $r(X) \geq r(B_1^*) = r(B_1) + 1 > r(B_1) \geq l(B)$. Thus $r(X) > l(B)$ for all $X \simeq A$. Hence $l(A) > l(B)$.

Theorem 2.16. *If $B^* \supseteq A$, then $l(A) > l(B)$.*

Proof. The proof is the dual of that of Theorem 2.15.

Theorem 2.17. $l(A^*) = l(A) + 1$.

Proof. Now by Definition 2.6, X exists such that $X \simeq A$ and $r(X) = l(A)$. But then by Theorem 2.5, corollary, $X^* \simeq A^*$ and $r(X^*) = r(X) + 1 = l(A) + 1$. Hence $l(A^*) \leq r(X^*) = l(A) + 1$. But since $A^* \supseteq A^*$ by Theorem 2.15 we have $l(A^*) > l(A)$. Now $l(A) < l(A^*) \leq l(A) + 1$ implies $l(A^*) = l(A) + 1$.

Theorem 2.18. *If $A^* \cup B \supseteq C$ and $B \sim \supseteq C$, then $l(C) > l(A)$.*

Proof. Let $X \simeq C$. Then if $A^* \cup B \supseteq C$ and $B \sim \supseteq C$ we also have $A^* \cup B \supseteq X$ and $B \sim \supseteq X$. But then by Theorem 2.13 a sub-polynomial A_1^* of X exists such that $A_1 \simeq A$. Thus $r(X) \geq r(A_1^*) = r(A_1) + 1 > r(A_1) \geq l(A)$. Thus $r(X) > l(A)$ for all $X \simeq C$ and hence $l(C) > l(A)$.

Theorem 2.19. *If $C \supseteq A^* \cap B$ and $C \sim \supseteq B$, then $l(C) > l(A)$.*

Proof. The proof is the dual of that of Theorem 2.18.

Theorem 2.20. *If $A^* \cup B \supseteq A$, then $B \supseteq A$.*

Proof. Let $A^* \cup B \supseteq A$. If $B \sim \supseteq A$, then by Theorem 2.18, $l(A) > l(A)$ which is impossible. Hence $B \supseteq A$.

Theorem 2.21. *If $A \supseteq A^* \cap B$, then $A \supseteq B$.*

Proof. The proof is the dual of Theorem 2.20.

Theorems 2.20 and 2.21 are particularly important in the construction of lattices with unique complementation.

In order to characterize the lattice O another definition is needed.

DEFINITION 2.7. A lattice has a *unary operator* if to each x is ordered an element x^* such that

(α)
$$x = y \quad \text{implies} \quad x^* = y^*.$$

In view of Theorem 2.5, corollary, we have the following theorem.

THEOREM 2.22. *O is a lattice with unary operator.*

As in the previous section, by the "free lattice with unary operator generated by P" we shall mean the free lattice with unary operator generated by P and preserving bounds, if they exist, of pairs of elements of P.

THEOREM 2.23. *O is the free lattice with unary operator generated by P.*

Proof. Clearly the free lattice with unary operator generated by P consists of all operator polynomials over P. Furthermore since O is a lattice with unary operator, $A \supseteq B$ in the free lattice implies $A \supseteq B$ in O. Now if $A \supseteq B(1)$, then either $A \equiv B$, in which case $A \supseteq B$ in the free lattice with unary operator, or $*A$ and $\overline{B}$ exist and $*A \geq \overline{B}$ in L. But since L is the free lattice generated by P we have $*A \supseteq \overline{B}$ in the free lattice with unary operator. From Definition 2.1 it follows that $\overline{A} \supseteq A \supseteq *A$ in the free lattice with unary operator whenever the covers exist. But then $A \supseteq *A \supseteq \overline{B} \supseteq B$. Now suppose we have shown that $A \supseteq B(n-1)$ implies $A \supseteq B$ in the free lattice. Let $A \supseteq B(n)$. If any of the possibilities (1), $\cdots$, (5) occur, then $A \supseteq B$ in the free lattice follows from lattice properties. On the other hand, if $A \equiv A_1^*$, $B \equiv B_1^*$ where $A_1 \supseteq B_1(n-1)$ and $B_1 \supseteq A_1(n-1)$ then $A_1 \supseteq B_1$, $B_1 \supseteq A_1$ in the free lattice and hence $A_1 \simeq B_1$. But then $A \equiv A_1^* \simeq B_1^* \equiv B$ by (α). Hence $A \supseteq B$ in the free lattice with unary operator. Now if $A \supseteq B$ in O, then $A \supseteq B(n)$ for some n and thus $A \supseteq B$ in the free lattice with unary operator generated by P.

In concluding this section, we give two theorems on the free lattice with unary operator generated by an unordered set S. The first theorem answers the question: When is a sublattice of the free lattice again a free lattice?

THEOREM 2.24. *Let O be the free lattice with unary operator generated by the unordered set S. Let A, B, C, $\cdots$ be elements of a subset $\mathfrak{S}$ of O and let $\mathfrak{A}$, $\mathfrak{B}$, $\mathfrak{C}$, $\cdots$ be the elements of $O_{\mathfrak{S}}$, the operator sublattice of O generated by the elements of $\mathfrak{S}$. Then $O_{\mathfrak{S}}$ is isomorphic to the free lattice with unary operator generated by $\mathfrak{S}$ as an unordered set if and only if*
  (1) *$A \supseteq B$ implies $A \equiv B$ if A, $B \in \mathfrak{S}$,*
  (2) *$\mathfrak{A} \cup \mathfrak{B} \supseteq A$ implies $\mathfrak{A} \supseteq A$ or $\mathfrak{B} \supseteq A$ if $A \in \mathfrak{S}$,*
  (3) *$A \supseteq \mathfrak{A} \cap \mathfrak{B}$ implies $A \supseteq \mathfrak{A}$ or $A \supseteq \mathfrak{B}$ if $A \in \mathfrak{S}$,*
  (4) *$\mathfrak{A}^* \sim \supseteq A$ if $A \in \mathfrak{S}$,*
  (5) *$A \sim \supseteq \mathfrak{A}^*$ if $A \in \mathfrak{S}$.*

Proof. Since S is unordered, $v(A)$ exists only if $A \in S$ and hence the second part of (i), Definition 1.5, can be omitted. But then the second part of (i),

Definition 2.2, can be omitted and hence also the second part of (1), Theorem 2.2. Now conditions (1), (2), (3), (4), (5) of the theorem are clearly necessary in view of Definition 2.2. We shall show the sufficiency by proving that $\mathfrak{A} \supseteq \mathfrak{B}$ if and only if one of the conditions of Theorem 2.2 holds.

First of all, let $\mathfrak{A}$ be of rank zero; that is, $\mathfrak{A} \equiv A$ where $A \in \mathfrak{S}$. Now if $\mathfrak{B} \equiv B$ is in $\mathfrak{S}$, then $\mathfrak{A} \supseteq \mathfrak{B}$ implies $\mathfrak{A} \equiv \mathfrak{B}$ by (1) and hence the first case of Theorem 2.2 occurs. If $\mathfrak{B} \equiv \mathfrak{B}_1 \cup \mathfrak{B}_2$ then by lattice properties $A \supseteq \mathfrak{B}_1$, and $A \supseteq \mathfrak{B}_2$ and hence case (4) of Theorem 2.2 occurs. If $\mathfrak{B} \equiv \mathfrak{B}_1 \cap \mathfrak{B}_2$, then either $A \supseteq \mathfrak{B}_1$ or $A \supseteq \mathfrak{B}_2$ by (3) and hence case (2) of Theorem 2.2 occurs. Now $\mathfrak{B} \equiv \mathfrak{B}_1{}^*$ cannot occur by (5). Thus the theorem holds if $\mathfrak{A}$ is of rank zero and clearly $\mathfrak{B}$ of rank zero is treated similarly. Hence we may suppose that both $\mathfrak{A}$ and $\mathfrak{B}$ are of positive rank. But then Theorem 2.2 itself applies and $\mathfrak{A} \supseteq \mathfrak{B}$ if and only if one of the conditions of Theorem 2.2 holds. This completes the proof.

Application of Theorem 2.24 to the free operator lattice generated by a single element gives a particularly interesting conclusion.

THEOREM 2.25. *The free lattice with unary operator generated by a single element contains as a sublattice the free lattice with unary operator generated by a countable set of elements.*

Proof. Let O be generated by the single element a. A set of operator polynomials is constructed inductively as follows: $A_1 \equiv a \cup a^*, A_{n+1} \equiv a \cup (a^* \cup A_n^*)^*$. Now suppose that $A_i \supseteq A_1$ where $i > 1$. Then $a \cup (a^* \cup A_{i-1}^*)^* \supseteq a \cup a^* \supseteq a^*$. Since $a \sim \supseteq a^*$ we have $a^* \cup A_{i-1}^* \supseteq a^*$ and hence $a^* \cup A_{i-1}^* \simeq a$. But then $a \supseteq a^*$, which contradicts Theorem 2.10. Hence $A_i \sim \supseteq A_1$, $i > 1$, and similarly $A_1 \sim \supseteq A_i$, $i > 1$. Suppose $A_i \supseteq A_j$ where $i, j > 1$ and $i \neq j$. Then $a \cup (a^* \cup A_{i-1}^*)^* \supseteq a \cup (a^* \cup A_{j-1}^*)^* \supseteq (a^* \cup A_{j-1}^*)^*$ and $(a^* \cup A_{i-1}^*)^* \supseteq (a^* \cup A_{j-1}^*)^*$ by Theorem 2.7. But then $a^* \cup A_{i-1}^* \simeq a^* \cup A_{j-1}^* \supseteq A_{j-1}^*$. If $a^* \supseteq A_{j-1}^*$, then $a = A_{j-1}$ and $a \supseteq a^*$ or $a \supseteq (a^* \cup A_{j-2}^*)^*$, both of which are impossible. Hence $a^* \sim \supseteq A_{j-1}^*$ and thus $A_{i-1}^* \supseteq A_{j-1}^*$ by Theorem 2.7. But then $A_{i-1} \supseteq A_{j-1}$ and successive applications lead to one of the previous cases which we have shown to be impossible. Thus $A_i \supseteq A_j$ implies $i = j$ and condition (1) of Theorem 2.24 holds.

Let us consider next the operator polynomials generated by the A_i. If $\mathfrak{A}$ is any such polynomial, then $a \sim \supseteq \mathfrak{A}$. For if $\mathfrak{A} \equiv A_i$ for some i, then $a \supseteq \mathfrak{A}$ implies $a \supseteq a^*$ or $a \supseteq (a^* \cup A_{i-1}^*)^*$ both of which are impossible by Theorem 2.10. Now, using induction, if $\mathfrak{A} \equiv \mathfrak{A}_1 \cup \mathfrak{A}_2$ or $\mathfrak{A} \equiv \mathfrak{A}_1 \cap \mathfrak{A}_2$ then $a \supseteq \mathfrak{A}$ implies $a \supseteq \mathfrak{A}_1$ or $a \supseteq \mathfrak{A}_2$, which are impossible by assumption. If $\mathfrak{A} \equiv \mathfrak{A}_1{}^*$ then $a \supseteq \mathfrak{A}$ cannot occur by Theorem 2.10. Hence $a \sim \supseteq \mathfrak{A}$ follows by induction.

We show next that $\mathfrak{A} \supseteq a^*$ implies $\mathfrak{A} \supseteq a$. Since $A_i \supseteq a$, all i, this is trivial if $\mathfrak{A} \equiv A_i$. Using induction, if $\mathfrak{A} \equiv \mathfrak{A}_1 \cup \mathfrak{A}_2$ or $\mathfrak{A} \equiv \mathfrak{A}_1 \cap \mathfrak{A}_2$, then $\mathfrak{A} \supseteq a^*$ implies $\mathfrak{A}_1 \supseteq a^*$ or $\mathfrak{A}_2 \supseteq a^*$ or both. But then $\mathfrak{A}_1 \supseteq a$ or $\mathfrak{A}_2 \supseteq a$ or both and hence $\mathfrak{A} \supseteq a$. If $\mathfrak{A} \equiv \mathfrak{A}_1{}^*$ then $\mathfrak{A} \supseteq a^*$ implies $\mathfrak{A}_1 \simeq a$ contrary to $a \sim \supseteq \mathfrak{A}_1$. Hence the conclusion follows by induction.

In a similar manner one shows that $\mathfrak{A} \supseteq (a^* \cup A_i^*)^*$ implies $\mathfrak{A} \supseteq a$. For if $\mathfrak{A} \equiv A_j$ for some j, the result is trivial. Furthermore the cases $\mathfrak{A} \equiv \mathfrak{A}_1 \cup \mathfrak{A}_2$ and $\mathfrak{A} \equiv \mathfrak{A}_1 \cap \mathfrak{A}_2$ are treated as before. If $\mathfrak{A} \equiv \mathfrak{A}_1^*$, then $\mathfrak{A} \supseteq (a^* \cup A_i^*)^*$ implies $\mathfrak{A}_1 \simeq a^* \cup A_i^*$. But then $\mathfrak{A}_1 \supseteq a^*$ implies $\mathfrak{A}_1 \supseteq a$ by what we have just shown. Thus $a^* \cup A_i^* \supseteq a$ which is impossible since $a^* \sim \supseteq a$ and $A_i^* \sim \supseteq a$. Hence $\mathfrak{A} \supseteq (a^* \cup A_i^*)^*$ implies $\mathfrak{A} \supseteq a$ vacuously in this case. Induction gives the result.

Now let $\mathfrak{A} \cup \mathfrak{B} \supseteq A_i$. If $i = 1$, then $\mathfrak{A} \cup \mathfrak{B} \supseteq a \cup a^* \supseteq a^*$ and $\mathfrak{A} \supseteq a^*$ or $\mathfrak{B} \supseteq a^*$ by Theorem 2.7. Hence $\mathfrak{A} \supseteq a \cup a^*$ or $\mathfrak{B} \supseteq a \cup a^*$ by the above result. Thus $\mathfrak{A} \supseteq A_1$ or $\mathfrak{B} \supseteq A_1$. If $i > 1$, then $\mathfrak{A} \cup \mathfrak{B} \supseteq a \cup (a^* \cup A_{i-1}^*)^* \supseteq (a^* \cup A_{i-1}^*)^*$ implies $\mathfrak{A} \supseteq (a^* \cup A_{i-1}^*)^*$ or $\mathfrak{B} \supseteq (a^* \cup A_{i-1}^*)^*$. But then $\mathfrak{A} \supseteq a \cup (a^* \cup A_{i-1}^*)^*$ or $\mathfrak{B} \supseteq a \cup (a^* \cup A_{i-1}^*)^*$ and $\mathfrak{A} \supseteq A_i$ or $\mathfrak{B} \supseteq A_i$. Hence condition (2) of Theorem 2.24 holds.

To prove condition (3), let $A_i \supseteq \mathfrak{A} \cap \mathfrak{B}$. Then $a \cup (a^* \cup A_{i-1}^*)^* \supseteq \mathfrak{A} \cap \mathfrak{B}$. If $A_i \supseteq \mathfrak{A}$ or $A_i \supseteq \mathfrak{B}$ we have nothing to prove. Otherwise we must have $a \supseteq \mathfrak{A} \cap \mathfrak{B}$ or $(a^* \cup A_{i-1}^*)^* \supseteq \mathfrak{A} \cap \mathfrak{B}$. But then by Theorem 2.2 one of the following must occur: $a \supseteq \mathfrak{A}$, $a \supseteq \mathfrak{B}$, $(a^* \cup A_{i-1}^*)^* \supseteq \mathfrak{A}$, $(a^* \cup A_{i-1}^*)^* \supseteq \mathfrak{B}$. However any one of these possibilities implies $A_i \supseteq \mathfrak{A}$ or $A_i \supseteq \mathfrak{B}$ and (3) follows.

$\mathfrak{A}^* \sim \supseteq A_i$ since $\mathfrak{A}^* \supseteq a$ by Theorem 2.10. Hence condition (4) holds.

Finally if $A_i \supseteq \mathfrak{A}^*$ where $i > 1$, then $a \cup (a^* \cup A_{i-1}^*)^* \supseteq \mathfrak{A}^*$ and since $a \sim \supseteq \mathfrak{A}^*$ we have $(a^* \cup A_{i-1}^*)^* \supseteq \mathfrak{A}^*$ But then $\mathfrak{A} \simeq a^* \cup A_{i-1}^* \supseteq a^*$. Hence $\mathfrak{A} \supseteq a$ and thus $a^* \cup A_{i-1}^* \supseteq a$ which is impossible. A similar proof holds for $i = 1$. Hence $A_i \sim \supseteq \mathfrak{A}^*$ and condition (5) hold.

Since (1), $\cdots$, (5) of Theorem 2.24 have been verified it follows that the operator sublattice generated by A_1, A_2, $\cdots$ is the free lattice with unary operator generated by a countable set of symbols.

It may be noted that a^*, $(a \cup a^*)^*$, $(a \cup (a \cup a^*)^*)^*$, $\cdots$ is a sequence of operator polynomials which also generates an operator sublattice which is a free lattice with a countable set of generators. However, this sequence is too special for use in later work.

3. The free lattice with reflexive unary operator. We begin defining a reflexive operator.

DEFINITION 3.1. A lattice with unary operator is *reflexive* if

(β) $(A^*)^* \simeq A.$

It will also be convenient to speak of reflexive elements of O.

DEFINITION 3.2. An element A of O is *reflexive* if $A \simeq (X^*)^*$ for some $X \in O$.

Now let us denote by N the set of all operator polynomials of O which contain *no* reflexive sub-polynomials.

LEMMA 3.1. *N contains P.*

For $a \simeq (X^*)^*$ implies $a \supseteq (X^*)^*$ which is impossible by Theorem 2.2.

LEMMA 3.2. *If $A \in N$, then every sub-polynomial of A is in N.*

Theorem 3.1. *N is a sublattice of O.*

Proof. Let A and B belong to N. Now a proper sub-polynomial of $A \cup B$ is either a sub-polynomial of A or of B and hence is not reflexive by the definition of N. If $A \cup B$ is reflexive we have $A \cup B \simeq (X^*)^*$ for some X, which implies $A \simeq (X^*)^*$ or $B \simeq (X^*)^*$ by Theorem 2.9 and hence either A or B is reflexive contrary to assumption. Thus $A \cup B$ is not reflexive and hence $A \cup B \in N$. A similar proof gives $A \cap B \in N$.

It is clear that although N is a sublattice of O, it is not closed under the operation $*$. However it is possible to define an operation A' over N such that A' agrees with A^* if $A^* \in N$ and also has the property α.

Definition 3.3. Let A be an operator polynomial of N.

(i) If $A^* \in N$, let $A' \equiv A^*$.

(ii) If $A^* \notin N$, then A' is defined inductively as follows:

(1) $A \equiv A_1 \cup A_2$. Since $A \in N$ and $A^* \notin N$, A^* is reflexive and $A^* \simeq (X^*)^*$ for some X. Hence $A \equiv A_1 \cup A_2 \simeq X^*$. By Theorem 2.9 we have the three possibilities $A_1 \simeq X^* \simeq A$, $A_2 \sim\simeq X^*$; or $A_2 \simeq X^* \simeq A$, $A_1 \sim\simeq X^*$; or $A_1 \simeq A_2 \simeq X^* \simeq A$ ($\sim\simeq$ means "not equivalent to"). Let us set, respectively, $A' \equiv A_1'$, or $A' \equiv A_2'$, or $A' \equiv A_1' \cup A_2'$ for the three possibilities.

(2) $A \equiv A_1 \cap A_2$. As in (1), $A \simeq X^*$ for some X and hence we set $A' \equiv A_1'$, or $A' \equiv A_2'$, or $A' \equiv A_1' \cap A_2'$ according as $A_1 \simeq X^* \simeq A$, $A_2 \sim\simeq X^*$; or $A_2 \simeq X^* \simeq A$, $A_1 \sim\simeq X^*$; or $A_1 \simeq A_2 \simeq X^* \simeq A$.

(3) $A \equiv A_1^*$. In this case we set $A' \equiv A_1$.

Let us note that (ii) of Definition 3.3 is independent of the choice of the operator polynomial X. Since if $A^* \simeq (Y^*)^*$ then $Y^* \simeq X^*$ and A_1 or $A_2 \simeq Y^*$ if and only if A_1 or $A_2 \simeq X^*$ respectively.

Lemma 3.3. *If $A \in N$, then $A' \in N$.*

For if $A^* \in N$, then $A' \equiv A^*$ is in N. If $A^* \notin N$, then cases (1) and (2) give A' in N by induction. In case (3), A' is a sub-polynomial of A and hence is in N.

Lemma 3.4. *If $A \in N$ and $A^* \in N$, then $A' \equiv A^*$.*

Lemma 3.5. *If $A, B \in N$ and $A \simeq B$, then $A' \simeq B'$.*

Let us note first that $A^* \in N$ implies $B^* \in N$. For if $B^* \notin N$, then B^* is reflexive, that is, $B^* \simeq (X^*)^*$ and $A^* \simeq B^* \simeq (X^*)^*$, whence A^* is reflexive contrary to $A^* \in N$. Hence if either $A^* \in N$ or $B^* \in N$, then $A' \equiv A^* \simeq B^* \equiv B'$. We proceed by induction. If $r(A) = r(B) = 0$, then A^* and B^* belong to N and the lemma holds by the remark above.

Now suppose the lemma has been proved for all A and B such that $r(A) < n$ and $r(B) < n$. Let $r(A) = n$. If $r(B) = 0$, then $B^* \in N$ and $A' \simeq B'$ as before. Let us suppose we have shown that $A' \simeq B'$ if $r(B) < k$ where $k < n$. We shall show that the result also holds if $r(B) = k$. Consider first $A \equiv A_1 \cup A_2$. If $A^* \in N$ the lemma has been proved. If $A^* \notin N$, we have three possibilities:

(1) $A' \equiv A_1'$ where $A \simeq A_1$, $A \sim \simeq A_2$. But then $A_1 \simeq B$ and $r(A_1) < r(A) = n$, $r(B) = k < n$. Hence by the induction assumption $A_1' \simeq B'$. But then $A' \equiv A_1' \simeq B'$. (2) $A' \equiv A_2'$ where $A \simeq A_2$, $A \sim \simeq A_1$. But then $A_2 \simeq B$ and $r(A_2) < r(A) = n$, $r(B) = k < n$. Whence $A_2' \simeq B'$ and $A' \equiv A_2' \simeq B'$ by Definition 3.3. (3) $A' \equiv A_1' \cup A_2'$ where $A \simeq A_1 \simeq A_2$. But then $r(A_1) < r(A) = n$, $r(A_2) < r(A) = n$ and $r(B) = k < n$. Hence since $A_1 \simeq B$ we have $A_1' \simeq A_2'$ and $A_1' \simeq B'$. Thus $A' \equiv A_1' \cup A_2' \simeq A_1' \simeq B'$. Hence $A' \simeq B'$ follows if $A \equiv A_1 \cup A_2$. An exactly dual proof handles the case $A \equiv A_1 \cap A_2$. Next let $B \equiv B_1 \cup B_2$. Again we have three possibilities since the case $B^* \in N$ has been treated above: (1) $B' \equiv B_1'$ where $B \simeq B_1$, $B \sim \simeq B_2$. But then $A \simeq B_1$ and $r(B_1) < r(B) = k$. Hence $A' \simeq B_1' \equiv B'$ by the second induction assumption. (2) $B \equiv B_2'$ where $B \simeq B_2$, $B \sim \simeq B_1$ is treated similarly. (3) $B \equiv B_1' \cup B_2'$ where $B \simeq B_1 \simeq B_2$. But then $r(B_1) < r(B) = k < n$ and $r(B_2) < r(B) = k < n$. Since $A \simeq B_1$ we have by induction $A' \simeq B_1'$ and $B_1' \simeq B_2'$. Thus $A' \simeq B_1' \simeq B_1' \cup B_2' \equiv B'$. Hence the lemma follows if $B \equiv B_1 \cup B_2$. Again an exactly dual proof handles the case $B \equiv B_1 \cap B_2$. Now we are left with only the possibilities $A \equiv A_1^*$ and $B \equiv B_1^*$ where $A' \equiv A_1$ and $B' \equiv B_1$. But $A \simeq B$ implies $A_1 \simeq B_1$ implies $A' \simeq B'$. Thus the lemma holds for $r(B) = k$ and by induction it follows for $r(A) \leq n$ and $r(B) < n$. By symmetry the lemma holds if $r(A) < n$ and $r(B) \leq n$.

Now let $r(A) = n$ and $r(B) = n$. We may assume that $A^* \notin N$ and $B^* \notin N$. Let $A \equiv A_1 \cup A_2$. We have three possibilities: (1) $A' \equiv A_1'$ where $A \simeq A_1$, $A \sim \simeq A_2$. But then $A_1 \simeq B$ and $r(A_1) < r(A) = n$ while $r(B) = n$. Hence $A_1' \simeq B'$ and $A' \equiv A_1' \simeq B'$. (2) $A' \equiv A_2'$ where $A \simeq A_2$, $A \sim \simeq A_1$ is treated similarly. (3) $A' \equiv A_1' \cup A_2'$ where $A \simeq A_1 \simeq A_2$. But then $A_1 \simeq B$ and $r(A_1) < r(A) = n$, $r(A_2) < r(A) = n$, $r(B) = n$ and by assumption $A_1' \simeq B'$, $A_1' \simeq A_2'$. Hence $A' \equiv A_1' \cup A_2' \simeq A_1' \simeq B'$. A dual argument holds for $A \equiv A_1 \cap A_2$. Now by symmetry it follows that the lemma holds if $B \equiv B_1 \cup B_2$ or $B \equiv B_1 \cap B_2$. Hence we have only the possibility $A \equiv A_1^*$, $B \equiv B_1^*$. But then $A' \equiv A_1 \simeq B_1 \equiv B'$. Hence the lemma has been shown to hold if $r(A) \leq n$ and $r(B) \leq n$. A final induction on n gives the lemma for all A and B. The proof is then complete.

LEMMA 3.6. *If $A \in N$ and $A^* \notin N$, then $A \simeq (A')^*$ and $(A')^* \in N$.*

Let us make an induction on $r(A)$. If $r(A) = 0$, then $A^* \in N$ and the lemma holds vacuously. Now suppose the lemma is true for all A with $r(A) < n$ and let $r(A) = n$. If $A \equiv A_1 \cup A_2$ according to Definition 3.3 we have three possibilities: (1) $A \equiv A_1'$ where $A \simeq A_1$ and $A \sim \simeq A_2$. Since A_1 is a subpolynomial of A, $A_1 \in N$. If $A_1^* \in N$, then A_1^* is not reflexive and hence $A^* \simeq A_1^*$ is not reflexive contrary to $A^* \notin N$. Thus $A_1^* \notin N$ and since $r(A_1) < r(A) = n$ we have by assumption $A_1 \simeq (A_1')^*$ and $(A_1')^* \in N$. But since $A' \equiv A_1'$ we have $A \simeq A_1 \simeq (A')^*$ and $(A')^* \in N$. (2) $A' \equiv A_2'$ where $A \simeq A_2$, $A \sim \simeq A_1$ is treated similarly. (3) $A' \equiv A_1' \cup A_2'$ where $A \simeq A_1 \simeq A_2$. But then $A_1, A_2 \in N$ while $A_1^*, A_2^* \notin N$. Furthermore $A_1' \simeq A_2'$ by Lemma 3.5. Since $r(A_1) < r(A) = n$ and $r(A_2) < r(A) = n$ we have by assumption $A_1 \simeq (A_1')^*$,

$A_2 \simeq (A_2')^*$ where $(A_1')^* \in N$ and $(A_2')^* \in N$. But then $(A')^* \equiv (A_1' \cup A_2')^*$ $\simeq (A_1')^* \simeq (A_1')^* \cup (A_2')^* \simeq A_1 \cup A_2 \equiv A$. Also $(A')^*$ belongs to N since otherwise $(A_1' \cup A_2')^*$ is reflexive and thus $(A_1')^* \simeq (A_1' \cup A_2')^*$ is reflexive contrary to $(A_1')^* \in N$. Thus the lemma holds if $A \equiv A_1 \cup A_2$ and an exactly dual proof gives the case $A \equiv A_1 \cap A_2$. Now let $A \equiv A_1^*$. Then $A' \equiv A_1$ and $(A')^* \equiv A_1^* \equiv A$. Hence $A \simeq (A')^*$ and $(A')^* \in N$. Thus the lemma holds if $r(A) = n$ and induction upon n completes the proof.

LEMMA 3.7. *If $A \in N$, then $(A')' \simeq A$.*

First let $A^* \in N$. Then $A' \equiv A^*$. But then $(A')^*$ is not in N and since $A' \equiv A^*$ we have $(A')' \equiv A$. Hence $(A')' \simeq A$ in this case. If $A^* \notin N$, then by Lemma 3.6, $A \simeq (A')^*$ where $(A')^* \in N$. But then $(A')' \equiv (A')^*$ by Definition 3.3. Hence $(A')' \simeq A$.

THEOREM 3.2. *N is a lattice with reflexive unary operator.*

Proof. Lemmas 3.5 and 3.7 show that (α) and (β) hold for the operation A'.

In agreement with our previous usage of the word "free," by the *free lattice with reflexive unary operator generated by the partially ordered set P* we shall mean the free lattice with reflexive unary operator generated by P and preserving bounds, whenever they exist, of pairs of elements of P.

THEOREM 3.3. *N is the free lattice with reflexive unary operator generated by P.*

Proof. Let N' denote the free lattice with reflexive unary operator generated by P. Let us note that N' consists of the set of operator polynomials over P. Furthermore the relations between these polynomials are determined by the lattice postulates and (α), (β). Also, if any relation holds among the polynomials as elements of O, it must also hold as elements of N' since O is the free lattice under lattice postulates and (α) by Theorem 2.23. Now since N is a sublattice of O, if $A \supseteq B$ in N we have $A \supseteq B$ in O and hence $A \supseteq B$ in N'. Thus to complete the proof we have only to show that every operator polynomial is equivalent in N' to an operator polynomial in N. Let us note first that this is trivially true for polynomials of rank zero. Suppose it has been shown for all operator polynomials of rank less than n. Let $r(A) = n$. If $A \equiv A_1 \cup A_2$, then by the induction assumption $A_1 \simeq B_1$ and $A_2 \simeq B_2$ where B_1 and B_2 are in N and the equivalence is in N'. But then by lattice postulates $A \equiv A_1 \cup A_2 \simeq B_1 \cup B_2$ in N' and $B_1 \cup B_2$ is in N by Theorem 3.1. A similar argument holds if $A \equiv A_1 \cap A_2$. Now let $A \equiv A_1^*$ and $A_1 \simeq B_1$ where $B_1 \in N$. If $B_1^* \in N$, then by (α), $A \equiv A_1^* \simeq B_1^*$ where $B_1^* \in N$. If $B_1^* \notin N$, then by Lemma 3.6, $B_1 \simeq (B_1')^*$ and $(B_1')^* \in N$ where the equivalence is in O and hence holds also in N'. By (α) and (β) we have $A \simeq B_1^* \simeq ((B_1')^*)^* \simeq B_1'$ and B_1' is in N. Thus by induction every operator polynomial is equivalent in N' to an operator polynomial in N. It is also clear from the above argument that $A \simeq B$

and $B \in N$ implies $A^* \simeq B'$ in N'. Hence it follows that N is isomorphic to N'.

In the previous section we answered the question: When is a sublattice of the free lattice with unary operator generated by an unordered set again free? We turn now to the similar problem for lattices with reflexive unary operator. We shall need a new tool for the investigation.

DEFINITION 3.4. If $A \in O$, the operation $f(A)$ is defined inductively as follows:

(i) If $r(A) = 0$, then $f(A) \equiv A$.

(ii) If $r(A) = n > 0$ we have three cases. (1) $A \equiv A_1 \cup A_2$. $f(A)$ exists if and only if $f(A_1), f(A_2)$, or both $f(A_1)$ and $f(A_2)$ exist in which case $f(A) \equiv f(A_1)$, $f(A) \equiv f(A_2)$, or $f(A) \equiv f(A_1) \cup f(A_2)$ respectively. (2) $A \equiv A_1 \cap A_2$. $f(A)$ exists if and only if $f(A_1)$ and $f(A_2)$ exist in which case $f(A) \equiv f(A_1) \cap f(A_2)$. (3) $A \equiv A_1^*$. $f(A)$ exists if and only if $f(A_1)$ exists, $f(A_1) \simeq A_1$ and $[f(A_1)]^* \in N$ in which case $f(A) \equiv [f(A_1)]^*$.

There is clearly a dual operation $g(A)$.

Since N is a sublattice of O, Definition 3.4 gives the following lemma.

LEMMA 3.8. *If $f(A)$ exists, then $f(A) \in N$.*

LEMMA 3.9. *If $f(A)$ exists, then $A \supseteq f(A)$.*

LEMMA 3.10. *If $A \in N$, then $f(A)$ exists and $f(A) \equiv A$.*

If $A \in P$ the lemma follows from Definition 3.4. Making an induction upon $r(A)$, if $A \equiv A_1 \cup A_2$ and $A \in N$, then $A_1, A_2 \in N$ and hence $f(A_1) \equiv A_1$, $f(A_2) \equiv A_2$ by assumption. But then $f(A) \equiv f(A_1) \cup f(A_2) \equiv A_1 \cup A_2 \equiv A$. Similarly if $A \equiv A_1 \cap A_2$, then $f(A) \equiv f(A_1) \cap f(A_2) \equiv A_1 \cap A_2 \equiv A$. Finally if $A \equiv A_1^*$, then $f(A_1) \equiv A_1$ by the induction assumption. But then $f(A_1)$ exists, $f(A_1) \simeq A_1$ and $[f(A_1)]^* \equiv A_1^* \in N$. Hence $f(A)$ exists and $f(A) \equiv [f(A_1)]^* \equiv A_1^* \equiv A$.

LEMMA 3.11. *If $*A$ exists, then $f(A)$ exists and $f(A) \supseteq *A$.*

The lemma is trivial if $A \in P$ since $f(A) \equiv A \equiv *A$. By induction, if $A \equiv A_1 \cup A_2$ and $*A$ exists, then $*A_1$ or $*A_2$ exist. If $*A_1$ exists and $*A_2$ does not exist, then $f(A_1)$ exists by assumption and $f(A) \equiv f(A_1)$ or $f(A_1) \cup f(A_2)$. Hence $f(A) \supseteq f(A_1) \supseteq *A_1 \equiv *A$. If $*A_2$ exists and $*A_1$ does not exist a similar argument holds. If $*A_1$ and $*A_2$ both exist, then $f(A_1)$ and $f(A_2)$ both exist and by the induction assumption $f(A) \equiv f(A_1) \cup f(A_2) \supseteq *A_1 \cup *A_2 \equiv *A$. Next if $A \equiv A_1 \cap A_2$, then $*A_1$ and $*A_2$ exist and hence $f(A_1)$ and $f(A_2)$ exist. But then $f(A) \equiv f(A_1) \cap f(A_2) \supseteq *A_1 \cap *A_2 \equiv *A$. Finally if $A \equiv A_1^*$, then $*A$ does not exist and the lemma holds vacuously.

LEMMA 3.12. *If $A \supseteq B$ and $f(B)$ exists, then $f(A)$ exists and $f(A) \supseteq f(B)$.*

Let us suppose first that $A \supseteq B(1)$. If $A \equiv B$, then $f(B)$ exists if and only if $f(A)$ exists and $f(A) \equiv f(B)$. If $*A$ and $\overline{B}$ exist with $*A \supseteq \overline{B}$, then $B \supseteq f(B)$ by Lemma 3.9 and $\overline{f(B)}$ exists with $\overline{B} \supseteq f(B)$ by Lemma 2.5. Since $*A$ exists,

$f(A)$ exists and $f(A) \supseteq *A$ by Lemma 3.11. But then $*f(A)$ exists and $*f(A) \supseteq *A$ by Lemmas 2.2 and 2.6. Hence $*f(A) \supseteq *A \supseteq \overline{B} \supseteq f(B)$ and $f(A) \supseteq f(B)$ by Definition 2.2.

Now let us suppose we have shown that the lemma holds if $A \supseteq B(n-1)$ and let $A \supseteq B(n)$. If $A \equiv A_1 \cup A_2$ where $A_1 \supseteq B(n-1)$ or $A_2 \supseteq B(n-1)$ then by induction $f(A_1)$ or $f(A_2)$ exists and $f(A_1) \supseteq f(B)$ or $f(A_2) \supseteq f(B)$. Hence $f(A) \equiv f(A_1)$, $f(A_2)$, or $f(A_1) \cup f(A_2) \supseteq f(B)$. A similar argument holds if $A \equiv A_1 \cap A_2$. If $B \equiv B_1 \cup B_2$ with $A \supseteq B_1(n-1)$ and $A \supseteq B_2(n-1)$, then either $f(B_1)$ or $f(B_2)$ exists according to Definition 3.4 and hence $f(A)$ exists by the induction assumption. But also $f(A) \supseteq f(B_1)$ or $f(A) \supseteq f(B_2)$. Hence $f(A) \supseteq f(B)$. If $B \equiv B_1 \cap B_2$ with $A \supseteq B_1(n-1)$ or $A \supseteq B_2(n-1)$, then $f(B_1)$ and $f(B_2)$ exist and hence $f(A)$ exists and $f(A) \supseteq f(B_1)$ or $f(A) \supseteq f(B_2)$. Thus $f(A) \supseteq f(B_1) \cap f(B_2) \equiv f(B)$. Finally if $A \equiv A_1^*$ and $B \equiv B_1^*$ with $A_1 \supseteq B_1(n-1)$ and $B_1 \supseteq A_1(n-1)$ then since $f(B)$ exists, $f(B_1)$ also exists and $f(B_1) \simeq B$. By induction $f(A_1)$ exists and $f(A_1) \supseteq f(B_1)$, $f(B_1) \supseteq f(A_1)$. Hence $f(A_1) \simeq f(B_1) \simeq B_1 \simeq A_1$. If $[f(A_1)]^*$ is not in N, since $f(A_1) \in N$, $[f(A_1)]^*$ is reflexive. But then $f(B) \equiv [f(B_1)]^* \simeq [f(A_1)]^*$ is reflexive contrary to $f(B) \in N$. Hence $[f(A_1)]^* \in N$ and $f(A) \equiv [f(A_1)]^* \supseteq f(B)$. Induction upon n completes the proof.

Let us restrict P to be an unordered set S. Then N is the free lattice with reflexive unary operator generated by S. Let $\mathfrak{S}$ consisting of operator polynomials A, B, $\cdots$ be a subset of N. We desire necessary and sufficient conditions that the operator sublattice generated by $\mathfrak{S}$ be isomorphic to the free lattice with reflexive unary operator generated by $\mathfrak{S}$ as an unordered set. Now we may clearly assume that $A^* \in N$ for each $A \in \mathfrak{S}$, since otherwise we replace A by A' and the resulting set generates the same operator sublattice while $(A')^* \in N$ by Lemma 3.6. $\mathfrak{S}$ is said to be *regular* if it has this property. We have then the following theorem.

THEOREM 3.4. *The operator sublattice $N_{\mathfrak{S}}$ of N generated by a regular subset $\mathfrak{S}$ is isomorphic to the free lattice with reflexive unary operator generated by $\mathfrak{S}$ as an unordered set if and only if the operator sublattice of O generated by $\mathfrak{S}$ is isormorphic to the free lattice with unary operator generated by $\mathfrak{S}$ as an unordered set.*

Proof. Let A, B, C, $\cdots$ denote the operator polynomials of $\mathfrak{S}$. We shall show first that if $A, B, C, \cdots$ generate a free lattice with reflexive unary operator in N, then in O they generate a free lattice with unary operator. It is sufficient to show that properties (1)–(5) of Theorem 2.24 hold. (1) is trivial since $\mathfrak{S}$ is unordered in N and hence in O. Now if $\mathfrak{A}$ is any operator polynomial over $\mathfrak{S}$, then $f(\mathfrak{A})$ is an operator polynomial over $\mathfrak{S}$. For if $\mathfrak{A} \in \mathfrak{S}$, then $\mathfrak{A} \in N$ and $f(\mathfrak{A}) \equiv \mathfrak{A}$ and $f(\mathfrak{A})$ is a polynomial over $\mathfrak{S}$. By induction, if $\mathfrak{A} \equiv \mathfrak{A}_1 \cup \mathfrak{A}_2$ and $f(\mathfrak{A})$ exists, then $f(\mathfrak{A}) \equiv f(\mathfrak{A}_1)$, $f(\mathfrak{A}_2)$, or $f(\mathfrak{A}_1) \cup f(\mathfrak{A}_2)$. Hence if $f(\mathfrak{A}_1)$ or $f(\mathfrak{A}_2)$ is a polynomial over $\mathfrak{S}$, then $f(\mathfrak{A})$ is also. If $\mathfrak{A} \equiv \mathfrak{A}_1 \cap \mathfrak{A}_2$, then $f(\mathfrak{A}) \equiv f(\mathfrak{A}_1) \cap f(\mathfrak{A}_2)$

and again $f(\mathfrak{A})$ is a polynomial generated by $\mathfrak{S}$ if the same holds for $f(\mathfrak{A}_1)$ and $f(\mathfrak{A}_2)$. If $\mathfrak{A} \equiv \mathfrak{A}_1^*$ and $f(\mathfrak{A})$ exists, then $f(\mathfrak{A}) \equiv [f(\mathfrak{A}_1)]^*$ and if $f(\mathfrak{A}_1)$ is a polynomial over $\mathfrak{S}$, then $f(\mathfrak{A})$ is also.

Now let $\mathfrak{A} \cup \mathfrak{B} \supseteq A$ where $A \in \mathfrak{S}$ and $\mathfrak{A}$ and $\mathfrak{B}$ are operator polynomials over $\mathfrak{S}$. Since $A \in N$, $f(A)$ exists by Lemma 3.10 and hence $f(\mathfrak{A} \cup \mathfrak{B})$ exists by Lemma 3.12. But then $f(\mathfrak{A} \cup \mathfrak{B}) \supseteq f(A) \equiv A$. If $f(\mathfrak{A} \cup \mathfrak{B}) \equiv f(\mathfrak{A})$, then $\mathfrak{A} \supseteq f(\mathfrak{A}) \supseteq A$. If $f(\mathfrak{A} \cup \mathfrak{B}) \equiv f(\mathfrak{B})$, then $\mathfrak{B} \supseteq f(\mathfrak{B}) \supseteq A$. If $f(\mathfrak{A} \cup \mathfrak{B}) \equiv f(\mathfrak{A}) \cup f(\mathfrak{B})$, then since $f(\mathfrak{A}), f(\mathfrak{B}) \in N$ and are operator polynomials over $\mathfrak{S}$, we have either $f(\mathfrak{A}) \supseteq A$ or $f(\mathfrak{B}) \supseteq A$ since by hypothesis $A, B, C, \cdots$ generate a free lattice with reflexive unary operator in N. But then either $\mathfrak{A} \supseteq f(\mathfrak{A}) \supseteq A$ or $\mathfrak{B} \supseteq f(\mathfrak{B}) \supseteq A$. Hence $\mathfrak{A} \cup \mathfrak{B} \supseteq A$ implies $\mathfrak{A} \supseteq A$ or $\mathfrak{B} \supseteq A$ and condition (2) is satisfied.

A similar proof using the dual operation $g(A)$ gives (3).

Let us suppose that $\mathfrak{A}^* \supseteq A$ where $A \in \mathfrak{S}$ and $\mathfrak{A}$ is an operator polynomial over $\mathfrak{S}$. Then since $f(A)$ exists, $f(\mathfrak{A}^*)$ exists and $f(\mathfrak{A}^*) \equiv [f(\mathfrak{A})]^* \supseteq f(A) \equiv A$. Since $f(\mathfrak{A})$ is a polynomial over $\mathfrak{S}$ and $f(\mathfrak{A})$, $[f(\mathfrak{A})]^*$ belong to N, this contradicts the fact that $\mathfrak{S}$ generates a free lattice with reflexive unitary operator in N. Thus $\mathfrak{A}^* \sim \supseteq A$ and similarly $A \sim \supseteq \mathfrak{A}^*$. Hence (4) and (5) of Theorem 2.24 hold and the proof of the necessity is complete.

To prove the sufficiency let us suppose that the polynomials $A, B, C, \cdots$ of $\mathfrak{S}$ generate in O an operator sublattice isomorphic to the free lattice with unary operator generated by $\mathfrak{S}$ as an unordered set. Then by Theorem 3.3, this lattice contains a sublattice $N_{\mathfrak{S}}'$ isomorphic to the free lattice with reflexive unary operator generated by $\mathfrak{S}$ as an unordered set. Hence we have only to show that $N_{\mathfrak{S}}$ is isomorphic to $N_{\mathfrak{S}}'$. Now each element of $N_{\mathfrak{S}}$ is an operator polynomial over $\mathfrak{S}$ whose sub-polynomials are non-reflexive in O. Hence the sub-polynomials which are polynomials over $\mathfrak{S}$ are certainly non-reflexive in a sublattice of O and thus the elements of $N_{\mathfrak{S}}$ belong to $N_{\mathfrak{S}}'$. Now let $\mathfrak{A}$ be a polynomial over $\mathfrak{S}$ and let $\mathfrak{A} \simeq X^*$ where $X \in O$. We shall show that $\mathfrak{A} \simeq \mathfrak{X}^*$ where $\mathfrak{X}$ is a polynomial over $\mathfrak{S}$. For if $\mathfrak{A} \in \mathfrak{S}$, then $\mathfrak{A} \simeq X^*$ implies $\mathfrak{A}^*$ is reflexive, contrary to the regularity of $\mathfrak{S}$. Hence the statement holds vacuously in this case. By induction, if $\mathfrak{A} \equiv \mathfrak{A}_1 \cup \mathfrak{A}_2$, then $\mathfrak{A} \simeq X^*$ implies $\mathfrak{A}_1 \simeq X^*$ or $\mathfrak{A}_2 \simeq X^*$. Hence $\mathfrak{A}_1 \simeq \mathfrak{X}^*$ or $\mathfrak{A}_2 \simeq \mathfrak{X}^*$. But then $\mathfrak{A} \simeq X^* \simeq \mathfrak{A}_1$ or $\mathfrak{A}_2 \simeq \mathfrak{X}^*$. A similar argument holds if $\mathfrak{A} \equiv \mathfrak{A}_1 \cap \mathfrak{A}_2$. If $\mathfrak{A} \equiv \mathfrak{A}_1^*$ we need only pick $\mathfrak{X} \equiv \mathfrak{A}_1$. The statement above follows by induction. From this result follows an even sharper result, namely, $\mathfrak{A} \simeq (X^*)^*$ in O implies $\mathfrak{A} \simeq (\mathfrak{X}^*)^*$ where $\mathfrak{X}$ is an operator polynomial over $\mathfrak{S}$. If $\mathfrak{A} \in \mathfrak{S}$ the statement holds vacuously. By induction, if $\mathfrak{A} \equiv \mathfrak{A}_1 \cup \mathfrak{A}_2$ or $\mathfrak{A} \equiv \mathfrak{A}_1 \cap \mathfrak{A}_2$ we get $\mathfrak{A} \simeq (\mathfrak{X}^*)^*$ as before. Finally if $\mathfrak{A} \equiv \mathfrak{A}_1^*$, then $\mathfrak{A}_1^* \simeq (X^*)^*$ and $\mathfrak{A}_1 \simeq X^*$. By the previous result $\mathfrak{A}_1 \simeq \mathfrak{X}^*$ and $\mathfrak{A} \equiv \mathfrak{A}_1^* \simeq (\mathfrak{X}^*)^*$ which completes the proof of the statement. Now if $\mathfrak{A}$ is an operator polynomial of $N_{\mathfrak{S}}'$, then every sub-polynomial $\mathfrak{A}_1$ of $\mathfrak{A}$ considered as a polynomial over $\mathfrak{S}$ is non-reflexive and hence by the result just proved is non-reflexive over O. But now any sub-polynomial of $\mathfrak{A}$ is either a sub-polynomial of some $A \in \mathfrak{S}$ and hence is non-reflexive in O or is a polynomial over $\mathfrak{S}$ in which case it is again non-reflexive in O. Thus $\mathfrak{A}$ belongs to N and is a polynomial over $\mathfrak{S}$. Hence $\mathfrak{A} \in N_{\mathfrak{S}}$ and $N_{\mathfrak{S}}$ and $N'_{\mathfrak{S}}$ consist of the same operator polynomials of O. Since both are sublattices of O, they are clearly lattice isomorphic.

We have still to show that the unary operation is preserved. But if $\mathfrak{A} \in N_{\mathfrak{S}}$, then $\mathfrak{A}^* \in N_{\mathfrak{S}}$ if and only if $\mathfrak{A}^* \in N'_{\mathfrak{S}}$ since $N_{\mathfrak{S}}$ and $N'_{\mathfrak{S}}$ are identical. But in both $N_{\mathfrak{S}}$ and $N'_{\mathfrak{S}}$, $\mathfrak{A}' \equiv \mathfrak{A}^*$ if $\mathfrak{A}^*$ belongs to the set. Hence the unary operation is preserved in this case. Now by induction, if $\mathfrak{A} \equiv \mathfrak{A}_1 \cup \mathfrak{A}_2$ then in either $N_{\mathfrak{S}}$ or $N'_{\mathfrak{S}}$, $\mathfrak{A}' \equiv \mathfrak{A}'_1$, or $\mathfrak{A}'_2$, or $\mathfrak{A}'_1 \cup \mathfrak{A}'_2$ according as $\mathfrak{A} \simeq \mathfrak{A}_1$, or $\mathfrak{A} \simeq \mathfrak{A}_2$, or both. Hence again the operation is preserved. $\mathfrak{A} \equiv \mathfrak{A}_1 \cap \mathfrak{A}_2$ is treated similarly. If $\mathfrak{A} \equiv \mathfrak{A}_1^*$, then in both cases $\mathfrak{A}' \equiv \mathfrak{A}_1$ and hence the unary operation is the same in both $N_{\mathfrak{S}}$ and $N'_{\mathfrak{S}}$. Thus $N_{\mathfrak{S}}$ and $N'_{\mathfrak{S}}$ are isomorphic and the proof of the theorem is complete.

As a consequence of Theorem 3.4 one proves the following theorem.

THEOREM 3.5. *The free lattice with reflexive unary operator generated by a single element contains as a sublattice the free lattice with reflexive unary operator generated by a denumerable set of elements.*

Proof. If a is the single generator let us define $A_1 \equiv a \cup a^*$, $A_{n+1} \equiv a \cup (a^* \cup A_n^*)^*$ as in the proof of Theorem 2.25. Since $A_1, A_2, \cdots$ generate a free lattice with unary operator in O according to Theorem 3.4 it is only necessary to prove that $A_1, A_2, \cdots$ is a regular set. But $A_i \simeq X^*$ implies $X^* \supseteq a$ which is impossible by Theorem 2.10. Hence A_i is regular for each i and $A_1, A_2, \cdots$ generate a free lattice with reflexive unary operator on a denumerable set of elements.

4. The free lattice with unique complements. In order to construct the free lattice with unique complements generated by P, the lattice N must be still further restricted.

DEFINITION 4.1. An operator polynomial $A \in N$ is *union singular* if $A \supseteq X, X'$ where $X \in N$. A is *crosscut singular* if $X, X' \supseteq A$ where $X \in N$. A is *singular* if it is either union or crosscut singular.

LEMMA 4.1. $A \in N$ *is union singular if and only if* $A \supseteq X, X^*$ *where* $X, X^* \in N$.

For if A is union singular, then $A \supseteq X, X'$ where $X \in N$. If $X^* \in N$, then $X' \equiv X^*$ and $A \supseteq X, X^*$ with $X, X^* \in N$. If $X^* \notin N$, then $X \simeq (X')^*$ and $(X')^* \in N$. But then $A \supseteq X'$ and $A \supseteq X \supseteq (X')^*$ where X' and $(X')^*$ are in N. The sufficiency is obvious. Dualizing, one gets the following lemma.

LEMMA 4.2. $A \in N$ *is crosscut singular if and only if* $X, X^* \supseteq A$ *where* $X, X^* \in N$.

Now let us denote by M the set of all operator polynomials of N containing no singular sub-polynomials together with the two symbols u and z. The operator polynomials of M are clearly partially ordered by the relation $A \supseteq B$.

We further define $u \supseteq A \supseteq z$ for all polynomials $A \in M$. M is thus a partially ordered set with unit element u and null element z. It is also convenient to set $u' \equiv z$ and $z' \equiv u$.

THEOREM 4.1. *M is a lattice. Furthermore if $A \vee B$ and $A \wedge B$ denote union and crosscut in M, then $A \vee B \equiv A \cup B$ if $A \cup B$ is nonsingular and $A \wedge B \equiv A \cap B$ if $A \cap B$ is nonsingular.*

Proof. Let $X \supseteq A$ and $X \supseteq B$ where A, $B \in M$. Then $X \supseteq A \cup B$. If $A \cup B \in M$, then $A \cup B$ is a l.u.b. of A and B in M and we may take $A \vee B \equiv A \cup B$. If $A \cup B \notin M$, then $A \cup B$ must contain a singular sub-polynomial. But since a proper sub-polynomial of $A \cup B$ is a sub-polynomial of either A or B, $A \cup B$ itself must be singular. Now $A \cup B$ cannot be crosscut singular since Y, $Y' \supseteq A \cup B$ implies Y, $Y' \supseteq A$ contrary to $A \in M$. Hence $A \cup B$ is union singular and $A \cup B \supseteq Y$, Y' where $Y \in N$. But then $X \supseteq Y$, Y' and if $X \in M$ we must have $X \equiv u$. Thus $A \vee B \equiv u$ in this case. A dual argument shows that $A \wedge B \equiv A \cap B$ or z according as $A \cap B \in M$ or not.

LEMMA 4.3. *If $A \in M$, then $A' \in M$.*

The lemma is trivial if $A \equiv u$ or z so we may suppose that A is an operator polynomial. Let us treat first the case where $A^* \in N$. If $A^* \notin M$, then A^* must contain a singular sub-polynomial. But since every proper sub-polynomial of A^* is a sub-polynomial of A and $A \in M$ it follows that A^* itself is singular. If A^* is union singular, then by Lemma 4.1, $A^* \supseteq X$, X^* where X, $X^* \in N$. But then by Theorem 2.5, $A^* \simeq X^*$ and $X^* \supseteq X$. Hence by Theorem 2.16, $l(X) > l(X)$ which is impossible. Similarly A^* is not crosscut singular. Thus A^* is not singular and hence $A^* \in M$. But then $A' \equiv A^* \in M$ and the lemma holds in this case.

We proceed with an induction on $r(A)$. If $r(A) = 0$, then $A^* \in N$ and the lemma holds as above. Suppose that the lemma holds for all A such that $r(A) < n$. Let $r(A) = n$. Now we may suppose that $A^* \notin N$ since the case $A^* \in N$ has been treated above. If $A \equiv A_1 \cup A_2$ we have three possibilities: (1) $A \equiv A_1'$ where $A \simeq A_1$ and $A \frown \simeq A_2$. But then $r(A_1) < r(A) = n$ and since A_1 is a sub-polynomial of A, $A_1 \in M$. Hence by the induction assumption $A' \equiv A_1'$ belongs to M. (2) $A \equiv A_2'$ where $A \simeq A_2$ and $A \frown \simeq A_1$. As before $A' \equiv A_2' \in M$. (3) $A' \equiv A_1' \cup A_2'$ where $A \simeq A_1 \simeq A_2$. Now $A_1, A_2 \in M$ since they are sub-polynomials of A, hence by the induction assumption A_1' and A_2' belong to M. Suppose that $A_1' \cup A_2' \notin M$. Then $A_1' \cup A_2'$ is union singular and $A_1' \cup A_2' \supseteq X$, X' where $X \in N$. But by Lemma 3.5, $A_1' \simeq A_2'$ and hence $A_1' \simeq A_1' \cup A_2' \supseteq X$, X' and A_1' is singular contrary to $A_1' \in M$. Thus $A_1' \cup A_2' \in M$ and hence $A' \in M$. If $A \equiv A_1 \cap A_2$ an exactly dual proof gives $A' \in M$. Finally let $A \equiv A_1^*$. Then since A_1 is a sub-polynomial of A, $A_1 \in M$ and hence $A' \equiv A_1$ belongs to M. Thus if $r(A) = n$ we have $A' \in M$ and the lemma follows by induction.

C>COROLLARY. *M is a lattice with reflexive unary operator.*

For $(u')' \simeq u$ and $(z')' \simeq z$ and for operator polynomials the property follows from Theorem 3.2.

THEOREM 4.2. *Each element A of M has the unique complement A'.*

Proof. Since u and z are the unit and null elements respectively of M, it follows that $u' \equiv z$ is the unique complement of u and $z' \equiv u$ is the unique complement of z. Thus we may devote our attention to the operator-polynomials of M. Clearly A' is a complement of A since $A \cup A' \supseteq A$, A' implies $A \vee A' \equiv u$ and $A, A' \supseteq A \cap A'$ implies $A \wedge A' \equiv z$.

Now let $A \vee B \simeq u$ and $A \wedge B \simeq z$ where $B \in M$. But then $B \sim \simeq u, z$ and hence is an operator polynomial of M. Since $A \vee B \simeq u$, by Theorem 4.1, $A \cup B$ is union singular and hence by Lemma 4.1, $A \cup B \supseteq X$, X^* where X and X^* are in N. Similarly $A \cap B$ is crosscut singular and hence by Lemma 4.2, $Y, Y^* \supseteq A \cap B$ where Y and Y^* are in N. Since $A \cup B \supseteq X^*$ by Theorem 2.7, $A \supseteq X^*$ or $B \supseteq X^*$. Also since $Y^* \supseteq A \cap B$ by Theorem 2.6 either $Y^* \supseteq A$ or $Y^* \supseteq B$. Hence we have four possibilities.

(1) $A \supseteq X^*$ and $Y^* \supseteq A$. But in this case $Y^* \supseteq X^*$ and $X \simeq Y$ by Theorem 2.5. Hence $X^* \simeq Y^*$ and $Y^* \supseteq A \supseteq X^*$ implies $X^* \simeq A$. Thus $X^* \cup B \simeq A \cup B \supseteq X$. But then $B \supseteq X$ by Theorem 2.20. Also $Y \supseteq A \cap B \simeq Y^* \cap B$ implies $Y \supseteq B$ by Theorem 2.21. Hence $Y \supseteq B \supseteq X$ and $X \simeq Y \simeq B$. Since $X^* \in N$ we have $X' \equiv X^* \simeq A$. But then $B \simeq X \simeq (X')' \simeq A'$ by Theorem 3.2 and Lemma 3.5. Hence $B \simeq A'$ in this case.

(2) $A \supseteq X^*$, $Y^* \supseteq B$. Now $Y \supseteq A \cap B \supseteq X^* \cap B$ and $Y^* \cup A \supseteq B \cup A \supseteq X$. Clearly $Y \sim \supseteq B$. Since if $Y \supseteq B$, then $Y, Y^* \supseteq B$ and B is singular contrary to $B \in M$. Similarly $A \sim \supseteq X$. But then by Theorem 2.19, $l(Y) > l(X)$ and from Theorem 2.18 we get $l(X) > l(Y)$. This is impossible and hence this case cannot occur.

(3) $B \supseteq X^*$, $Y^* \supseteq A$. As in case (2), this leads to a contradiction by an exactly similar argument.

(4) $B \supseteq X^*$, $Y^* \supseteq B$. In this case as in (1) we get $X^* \simeq Y^* \simeq B$ and $X \simeq Y \simeq A$. Since $X^* \in N$ we have $X' \equiv X^*$ and thus $B \simeq X^* \simeq X' \simeq A'$.

Hence in every case $B \simeq A'$ and the proof is complete.

THEOREM 4.3. *M contains P as a sub-partially ordered set and preserves bounds of pairs of elements of P whenever the bounds exist.*

Proof. If $a \in P$, then $a \sim \supseteq A^*$ and $A^* \sim \supseteq a$ for every operator polynomial A. Hence a is nonsingular and belongs to M. Also $a \geq b$ if and only if $a \supseteq b$ in O and hence $a \geq b$ if and only if $a \supseteq b$ in M. Thus P is a sub-partially ordered set of M. Now let $c = $ l.u.b.(a, b) exist in P. Then $c \simeq a \cup b$ in O by the corollary to Theorem 2.3. But since a and b are nonsingular, $a \cup b$ is nonsingular and hence $a \cup b \in M$. Thus $c \simeq a \vee b$ in M and least upper bound is preserved in M if the bound exists. A similar argument holds for the greatest lower bound.

If P is a lattice, Theorem 4.3 gives as a corollary the theorem mentioned in the introduction.

THEOREM 4.4. *Every lattice is a sublattice of a lattice with unique complements.*

It is clear that the unit and null element of the imbedding lattice will be different from the unit and null element respectively of the original lattice. The lattice M may be further characterized as follows:

THEOREM 4.5. *M is the free lattice with unique complements generated by P and preserving bounds, whenever they exist, of pairs of elements of P.*

Proof. Let M' denote the free lattice[7] with unique complements generated by P and preserving bounds, whenever they exist, of pairs of elements of P. M' clearly consists of all operator polynomials over P. Furthermore since complements are unique we have (α) $A \simeq B$ in M' implies $A' \simeq B'$ and (β) $(A')' \simeq A$ in M'. Hence M' is a lattice with reflexive unitary operator. But then $A \supseteq B$ in M implies $A \supseteq B$ in N implies $A \supseteq B$ in M' since N is the free lattice with reflexive unitary operator generated by P. Now clearly $A \supseteq B$ in M' with $A, B \in M$ implies $A \supseteq B$ in M since M' is the free lattice with unique complements generated by P. Hence we have only to show that each operator polynomial $A \in O$ is equivalent in M' to an operator polynomial of M or to u or z. We make an induction on the rank of A. If $r(A) = 0$, then $A \in M$ and there is nothing to be proved. Let $r(A) = n$. If $A \notin N$, then A contains a sub-polynomial B which is reflexive, that is, $B \simeq (X^*)^*$. But since $B \equiv B_1 \cup B_2$ implies $B_1 \simeq (X^*)^*$ or $B_2 \simeq (X^*)^*$ and similarly for $B \equiv B_1 \cap B_2$, B contains a sub-polynomial B_1^* such that $B_1^* \simeq (X^*)^* \simeq B$. Hence $B_1 \simeq X^*$. But then B_1 contains a sub-polynomial B_2^* such that $B_2^* \simeq X^*$. Hence $B \simeq (X^*)^* \simeq (B_2^*)^*$. But then $B \simeq (B_2^*)^* \simeq B_2$ in M'. Hence replacing B by B_2 in A we obtain an operator polynomial A_1 of smaller rank such that $A \simeq A_1$ in M'. But by the induction assumption A_1 is equivalent in M' to $A_2 \in M$. Hence $A \simeq A_2$ in M'. If $A \in N$ but $A \notin M$, then A contains a sub-polynomial C which is singular. Hence $C \simeq u$ or z in M'. Hence replacing C by u or z respectively and using the relations $u \cup X \simeq u$, $u \cap X \simeq X$, $z \cup X \simeq X$, $z \cap X \simeq z$, $u' \simeq z$, $z' \simeq u$ we obtain A_1 which is either u, z or an operator polynomial of O of smaller rank. But $A \simeq A_1$ in M' and by induction $A_1 \simeq A_2$ in M' where $A_2 \in M$. Hence $A \simeq A_2$ in M'. Finally if $A \in M$, then $A \simeq A$ where $A \in M$. Thus every $A \in O$ is equivalent in M' to an operator polynomial of M or to u or z and hence the proof is complete.

COROLLARY. *If P is a lattice L, then M is the free lattice with unique complements generated by L.*

---

[7] The existence of M' follows from general existence theorems on free algebras (cf. footnote 4).

As in the previous sections, we shall determine conditions under which an operator sublattice of the free lattice with unique complements generated by an unordered set S is again free.

THEOREM 4.6. *Let O be the set of operator polynomials over the unordered set S. Let $\mathfrak{S}$ be a regular subset of M which generates in O a free lattice with unary operator. Then the operator sublattice $M_\mathfrak{S}$ of M generated by $\mathfrak{S}$ is isomorphic to the free lattice with unique complements generated by $\mathfrak{S}$ as an unordered set if and only if the following two conditions hold:*
(1) $\mathfrak{A}\cup\mathfrak{B}\supseteq X, X^*$ *where* $\mathfrak{A}, \mathfrak{B}\in M_\mathfrak{S}, X\in O\rightarrow\mathfrak{A}\cup\mathfrak{B}\supseteq\mathfrak{X}, \mathfrak{X}^*$ *where* $\mathfrak{X}^*\in M_\mathfrak{S}$.
(2) $X, X^*\supseteq\mathfrak{A}\cap\mathfrak{B}$ *where* $\mathfrak{A}, \mathfrak{B}\in M_\mathfrak{S}, X\in O\rightarrow\mathfrak{X}, \mathfrak{X}^*\supseteq\mathfrak{A}\cap\mathfrak{B}$ *where* $\mathfrak{X}^*\in M_\mathfrak{S}$.

Proof. Let us suppose first that $M_\mathfrak{S}$ is isomorphic to the free lattice with unique complements generated by $\mathfrak{S}$ on an unordered set. Then if $\mathfrak{A}\cup\mathfrak{B}\supseteq X$, X^* where $\mathfrak{A}, \mathfrak{B}\in M_\mathfrak{S}, X\in O$, we have $\mathfrak{A}\vee\mathfrak{B}\simeq u$ in $M_\mathfrak{S}$ and hence $\mathfrak{A}\vee\mathfrak{B}\simeq u$ in the free lattice with unique complements generated by $\mathfrak{S}$. Since $\mathfrak{A}, \mathfrak{B}\in M_\mathfrak{S}$, it follows that $\mathfrak{A}\cup\mathfrak{B}\supseteq\mathfrak{X}, \mathfrak{X}^*$ where $\mathfrak{X}$ is a polynomial over $\mathfrak{S}$. But by Theorem 2.11 we can take $\mathfrak{X}^*$ to be a sub-polynomial of $\mathfrak{A}\cup\mathfrak{B}$. Now $\mathfrak{A}\cup\mathfrak{B}\neq\mathfrak{X}^*$. Hence $\mathfrak{X}^*$ is a sub-polynomial of either $\mathfrak{A}$ or $\mathfrak{B}$ and hence belongs to $M_\mathfrak{S}$. Thus (1) holds and a dual proof gives (2).

To prove the sufficiency, let O' be the sublattice of O generated by $\mathfrak{S}$ and let $M'_\mathfrak{S}$ be the subset of O' containing no polynomials having reflexive or singular sub-polynomials. Then $M'_\mathfrak{S}$ is the free lattice with unique complements generated by $\mathfrak{S}$ as an unordered set. Under the assumption of (1) and (2) we have to show that $M_\mathfrak{S}$ is isomorphic to $M'_\mathfrak{S}$. Now since the containing relations in O and O' are the same, the operator polynomials in $M_\mathfrak{S}$ clearly belong to $M'_\mathfrak{S}$. Hence we have only to show that the elements of $M'_\mathfrak{S}$ belong to $M_\mathfrak{S}$ and that the unary operations correspond. But since the unary operation is unique complementation this follows from the lattice isomorphism. Thus we have only to show that the elements of $M'_\mathfrak{S}$ belong to $M_\mathfrak{S}$. Now if $\mathfrak{A}\in\mathfrak{S}$, then trivially $\mathfrak{A}\in M_\mathfrak{S}$ and we may use induction upon the rank of $\mathfrak{A}$ over $\mathfrak{S}$. Since $\mathfrak{A}\in M'_\mathfrak{S}$ we have $\mathfrak{A}\in N'_\mathfrak{S}$ and hence $\mathfrak{A}\in N_\mathfrak{S}$ by Theorem 3.4. Let $\mathfrak{A}\equiv\mathfrak{A}_1\cup\mathfrak{A}_2$. By the induction assumption $\mathfrak{A}_1$ and $\mathfrak{A}_2$ belong to $M_\mathfrak{S}$. Hence if $\mathfrak{A}\notin M_\mathfrak{S}$, $\mathfrak{A}$ is union singular; that is, $\mathfrak{A}\equiv\mathfrak{A}_1\cup\mathfrak{A}_2\supseteq X, X^*$ where $X\in O$. But then $\mathfrak{A}_1\cup\mathfrak{A}_2\supseteq\mathfrak{X}, \mathfrak{X}^*$ where $\mathfrak{X}^*\in M_\mathfrak{S}$ by (1). Thus $\mathfrak{A}\equiv\mathfrak{A}_1\cup\mathfrak{A}_2$ is singular over O' contrary to $\mathfrak{A}\in M'_\mathfrak{S}$. Hence $\mathfrak{A}\in M_\mathfrak{S}$ in this case. $\mathfrak{A}\equiv\mathfrak{A}_1\cup\mathfrak{A}_2$ is treated similarly. If $\mathfrak{A}\equiv\mathfrak{A}_1^*$, then $\mathfrak{A}_1^*\supseteq X^*\rightarrow\mathfrak{A}_1^*\simeq X^*\rightarrow X^*\supseteq X$ which is impossible. The proof is thus complete.

It is an interesting fact that, contrary to the case of lattices with reflexive unary operator, a regular set $A, B, C, \cdots$ of M may generate a free lattice with unique complements as a sublattice of M and yet *not* generate a free lattice with unary operator as a sublattice of O. Indeed, consider the operator polynomials $A\equiv a\cup(a\cup b^*)^*, B\equiv a\cup b^*$. It can be verified that A and B generate a free lattice with unique complements in M. However, since $B\cup B^*\supseteq A$,

A and B do not generate a free lattice with unary operator in O. The statement of both necessary and sufficient conditions (in terms of the containing relation in O) that a regular subset of M generate a free lattice with unique complements seems to be quite difficult.

Now it is clear that the free lattice with unique complements generated by a single element a consists of the four elements a, a', u, and z. Hence there is no theorem for lattices with unique complements analogous to Theorems 2.25 and 3.5. However, there is a similar theorem for lattices with two generators. We shall need the following lemma.

LEMMA 4.4. *Let X be an operator polynomial generated by the polynomials $A_1, \cdots, A_n$. Then if Y is a sub-polynomial of X, either Y is a sub-polynomial of A_i for some i or Y is a polynomial over $A_1, \cdots, A_n$.*

If $X \equiv A_i$ for some i, the lemma is trivial. Now use induction on the rank of X over $A_1, \cdots, A_n$. If $X \equiv X_1 \cup X_2$, then either Y is a sub-polynomial of X_1 or X_2 in which case the lemma holds by hypothesis or $Y \equiv X$ in which case the lemma is trivially true. If $X \equiv X_1 \cap X_2$ a similar argument holds. If $X \equiv X_1^*$, then either $Y \equiv X$ or Y is a sub-polynomial of X_1 and the lemma holds by hypothesis. Induction on the rank of X over $A_1, \cdots, A_n$ completes the proof.

THEOREM 4.7. *The free lattice with unique complements generated by two elements contains as a sublattice the free lattice with unique complements generated by a countable set of elements.*

Proof. Let M be the free lattice with unique complements generated by the two elements a and b. Let $A_1 \equiv a \cup b^*$ and define inductively $A_{n+1} \equiv a \cup (a^* \cup A_n^*)^*$. It follows from the proof of Theorem 2.25 that $A_1, A_2, \cdots$ generate in O a free lattice with unitary operator. By Theorem 4.6 we have only to show that conditions (1) and (2) hold.

Let us note first that $\mathfrak{A} \sim \supseteq a^*$ and $\mathfrak{A} \sim \supseteq b$ for every operator polynomial over $A_1, A_2, \cdots$. For $A_n \supseteq a^* \to a \cup (a^* \cup A_{n-1}^*)^* \supseteq a^* \to a^* \cup A_{n-1}^* \simeq a \to a \supseteq a^*$ which is impossible and $A_n \supseteq b \to (a^* \cup A_{n-1}^*)^* \supseteq b$ which contradicts Theorem 2.10. An easy induction gives the result.

Now let $\mathfrak{A} \cup \mathfrak{B} \supseteq X$, X^* where $\mathfrak{A}$ and $\mathfrak{B}$ are operator polynomials over $A_1, A_2, \cdots$ and belong to M. Then $\mathfrak{A} \cup \mathfrak{B}$ contains a sub-polynomial $\mathfrak{X}^*$ such that $\mathfrak{X} \sim X$ by Theorem 2.11. But then $\mathfrak{X}^*$ is a sub-polynomial of either $\mathfrak{A}$ or $\mathfrak{B}$ and hence $\mathfrak{X}^* \in M$. Finally, if $\mathfrak{X}^*$ is not a polynomial over $A_1, A_2, \cdots$, then by Lemma 4.4, $\mathfrak{X}^*$ is a sub-polynomial of some A_i. But then $\mathfrak{X}^* \equiv a^*$, $(a^* \cup A_n^*)^*$, or b^*. But since $\mathfrak{A} \cup \mathfrak{B} \sim \supseteq a^*$, b none of these possibilities can occur. Hence $\mathfrak{X}^*$ is an operator polynomial over $A_1, A_2, \cdots$ belonging to M and thus (1) holds.

Since $a^* \supseteq A_n \to a^* \supseteq a$ and $b \supseteq A_n \to b \supseteq a$ it follows that $a^* \sim \supseteq A_n$ and $b \sim \supseteq A_n$. But then the dual of the argument of the previous paragraph gives condition (2). Hence by Theorem 4.6 the operator sublattice of M generated by A_1, A_2, $\cdots$ is isomorphic to the free lattice with unique complements generated by A_1, A_2, $\cdots$ as an unordered set. The proof is thus complete.

Theorem 4.7 shows with particular clarity how far lattices with unique complements differ from Boolean algebras. For the free Boolean algebra generated by n symbols contains 2^{2^n} elements and hence does not contain as a sublattice the Boolean algebra generated by k symbols for $k > n$. On the other hand, the free lattice with unique complements generated by just *two* symbols contains as a sublattice the free lattice with unique complements generated by n symbols for any positive integer n.

REFERENCES

G. BERGMAN
1. *Zur Axiomatic der Elementargeometrie*, Monatschrift für Mathematik und Physik vol. 36 (1929) pp. 269–284.

G. BIRKHOFF
1. *Lattice theory*, Amer. Math. Soc. Colloquium Publication, vol. 25, 1940.

G. BIRKHOFF and M. WARD
1. *A characterization of Boolean algebras*, Ann. of Math. vol. 40 (1939) pp. 609–610.

P. M. WHITMAN
1. *Free lattices*, Ann. of Math. vol. 42 (1941) pp. 325–330.
2. *Free lattices*. II, Ann. of Math. vol. 43 (1942) pp. 104–115.

YALE UNIVERSITY,
    NEW HAVEN, CONN.
CALIFORNIA INSTITUTE OF TECHNOLOGY,
    PASADENA, CALIF.

On Complemented Lattices,

by

R. P. Dilworth, Pasadena, Calif. U.S.A..

1. *Introduction.* In a paper on the foundation of quantum mechanics, Kôdi Husimi[1] conjectured that a lattice with a negation is modular if the chain law holds for every sublattice closed with respect to relative negation. Although the theorem in this form does not hold, as we show by an example, we prove a theorem of a similar nature for relatively complemented lattices.

We also show that any complemented, non-modular lattice of finite dimensions has a complemented non-modular sublattice of order five. This theorem is the analogue for complemented lattices of the theorem of Dedekind that any non-modular lattice contains a non-modular sublattice of order five. As an application, we give a new proof of the theorem due to G. Birkhoff and M. Ward[2] that a lattice of finite dimensions is a Boolean algebra if and only if every element has a unique complement.

2. *Notation and terminology.* We denote the fixed lattice of elements $a, b, c, \ldots$ by $\mathfrak{S}$. $(\,,\,), [\,,\,], \supset$ denote union, cross-cut, and lattice division respectively. German capitals will denote sublattices of $\mathfrak{S}$ and subsets of $\mathfrak{S}$ which are not necessarily sublattices will be denoted by latin capitals. If $a \supset x \supset b$, $a \neq b$ implies $x = a$ or $x = b$ we say that a "covers" b and write $a > b$. Elements which cover the null element z of a lattice are called *points* and elements which are covered by the unit element i are said to be *simple*.

A lattice $\mathfrak{S}$ is said to satisfy the *ascending chain condition* if every chain $a_1 \subset a_2 \subset a_3 \subset \ldots$ has only a finite number of distinct members. Similarly $\mathfrak{S}$ is said to satisfy the descending chain condition if every chain $a_1 \supset a_2 \supset a_3 \supset \ldots$ has only a finite number of distinct members. If both the ascending and descending chain conditions hold, $\mathfrak{S}$ is said to have finite dimensions. A chain $a = a_0 \supset a_1 \supset a_2 \supset \ldots \supset a_n = b$ joining two elements a and b is said to be *complete*

---

[1] K. Husimi, *Studies on the foundations of quantum mechanics.* Proc. of the Physico-Math. Soc. of Japan, **19** (1937), pp. 766–789.

[2] G. Birkhoff and M. Ward, Bull. of the Amer. Math. Soc., abstract (45-1-78).

if $a_i > a_{i+1}$. A lattice of finite dimensions is said to satisfy the chain law if every complete chain joining any two elements a and b with $a \supset b$ has the same length.

An element a' is said to be a complement of a if $(a, a') = i$ and $[a, a'] = z$. If every element of $\mathfrak{S}$ has a complement, then $\mathfrak{S}$ is said to be complemented. $\mathfrak{S}$ is said to be relatively complemented if $a \supset b$ implies there exists an element b_1 such that $(b, b_1) = a$ and $[b, b_1] = z$.

An involutory automorphism $a \longleftrightarrow a'$ of $\mathfrak{S}$ is called a *negation* if $(a, a') = i$ and $[a, a'] = z$. A lattice with a negation is clearly complemented. If $\mathfrak{S}$ has a negation, then a sublattice $\mathfrak{A}$ of $\mathfrak{S}$ is said to be closed with respect to relative negation if with a and b, $a \supset b$ it contains $[a, b']$ and $a = (b, [a, b'])$. A lattice closed with respect to relative negation is clearly relatively complemented.

Definition 2. 1. A lattice $\mathfrak{S}$ is said to be a *Birkhoff* lattice if :

$$(1) \qquad\qquad a > [a, b] \text{ implies } (a, b) > b.$$

$\mathfrak{S}$ is said to be a dual Birkhoff lattice if

$$(2) \qquad\qquad (a, b) > a \text{ implies } b > [a, b].$$

Condition (1) and (2) are closely connected with modularity[1] as is shown by the following lemma proved by Garrett Birkhoff[2]:

Lemma 2.1. *A finite dimensional lattice $\mathfrak{S}$ is modular if and only if it is both a Birkhoff and dual Birkhoff lattice.*

3. *Relatively complemented lattices.* We are now ready to prove the first theorem mentioned in the introduction.

Theorem 3.1. *Let $\mathfrak{S}$ be a relatively complemented lattice of finite dimensions. Then if every relatively complemented sublattice satisfies the chain law, $\mathfrak{S}$ is a dual Birkhoff lattice.*

Proof. If $\mathfrak{S}$ is not a dual Birkhoff lattice there is an element x such that there exist two elements x_1 and x_2 for which $x > x_1$, $x \supset x_2$, $x_1 \not\supset x_2$, $x_2 \not> [x_1, x_2]$. For clearly $(x_1, x_2) = x > x_1$. Now let S be the set of all such elements x. Then since the descending chain condition holds in $\mathfrak{S}$, S must have at least one minimal element a. Since $a \in S$ there exist elements a_1 and y such that $a > a_1$, $a \supset y$, $a_1 \not\supset y$, $y \not> [a_1, y]$. For a and a_1 fixed let T be the set of all such elements y. Then T must have at least one minimal element b.

---

[1] A lattice $\mathfrak{S}$ is modular if $a \supset b$ implies $[a, (b, c)] = (b, [a, c])$.

[2] Garrett Birkhoff. *On combination of subalgebra*, Proc. of the Cambridge Phil. Soc., **29** (1933), pp. 441-464.

Hence $a > a_1$, $a \supset b$, $a_1 \not\supset b$, $b \not\gg [a_1, b]$. Since $b \not\gg [a_1, b]$ there exists an x such that $b > x \supset [a_1, b]$, $x \neq [a_1, b]$. But if $x \not\gg [a_1, b]$, then x belongs to T which contradicts the minimal property of b. Thus $b > x > [a_1, b]$. We will now show that $[a_1, b] = z$. Suppose that $[a_1, b] \neq z$. Since $\mathfrak{S}$ is relatively complemented, there exists an element y such that $(y, [a_1, b]) = b$, $[y, [a_1, b]] = z$. Since $[a_1, b] \neq z$, we have $b \neq y$ and there exists an element m such that $b > m \supset y$. Then $m \neq x$ since otherwise $x \supset y$ and $x \supset [a_1, b]$. Whence $x \supset (y, [a_1, b]) = b$ contradicting $b > x$. Also $[m, x] \neq [a_1, b]$ since otherwise $m \supset (y, [a_1, b]) = b$ which contradicts $b > m$. Now $b = (m, x) > x$ and hence $m > [m, x]$ by the minimal property of a. Similarly $x > [m, x]$. Now $x = ([m, x], [a_1, b]) > [a_1, b]$. Hence $[m, x] > [[m, x], [a_1, b]] = [m, a_1, b] = [m, a_1]$ by the minimal property of a. But then $m > [m, x] > [m, a_1]$. Hence $a \supset m$, $a_1 \not\supset m$, $m \not\gg [m, a_1]$ and m then belongs to T. This however contradicts the minimal property of b. Hence we have $[a_1, b] = z$.

Since $b \supset x$ there exists an element x_1 such that $(x, x_1) = b$, $[x, x_1] = z$. Now $a \supset (a_1, x) \supset a_1$ and since $a > a_1$, either $a = (a_1, x)$ or $a_1 \supset x$. But if $a_1 \supset x$, then since $b \supset x$ we have $z = [a_1, b] \supset x$ and hence $x = z$ which contradicts $x > [a_1, b]$. Hence $(a_1, x) = a$ and similarly $(a_1, x_1) = a$. But $[a_1, (x, x_1)] = [a_1, b] = z$. Hence $\{a, a_1, b, x, x_1, z\}$ is a sublattice closed with respect to relative complement in which the chain law does *not hold*. This contradicts the hypothesis of the theorem and $\mathfrak{S}$ is thus a dual Birkhoff lattice.

Corollary: *Let $\mathfrak{S}$ satisfy the hypotheses of theorem 3.1. Then if $\mathfrak{S}$ has a negation, $\mathfrak{S}$ is modular.*

For $\mathfrak{S}$ is a dual Birkhoff lattice by theorem 3.1 and since $\mathfrak{S}$ has a dual automorphism $\mathfrak{S}$ is also a Birkhoff lattice. Hence by lemma 2.1 $\mathfrak{S}$ is modular.

It will be noted that the converse of theorem 3.1 does not hold in general; that is, in a relatively complemented dual Birkhoff lattice every relatively complemented sublattice need *not* satisfy the chain law. Consider for example the lattice diagramed in Fig. 1.

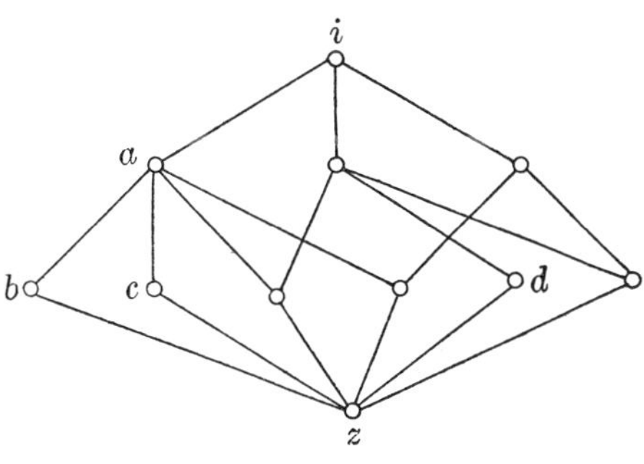

Fig. 1

The sublattice $\{i, a, b, c, d, z\}$ does not satisfy the chain law.

We conclude this section with an example of a lattice in which the theorem conjectured by H u s i m i does not hold.

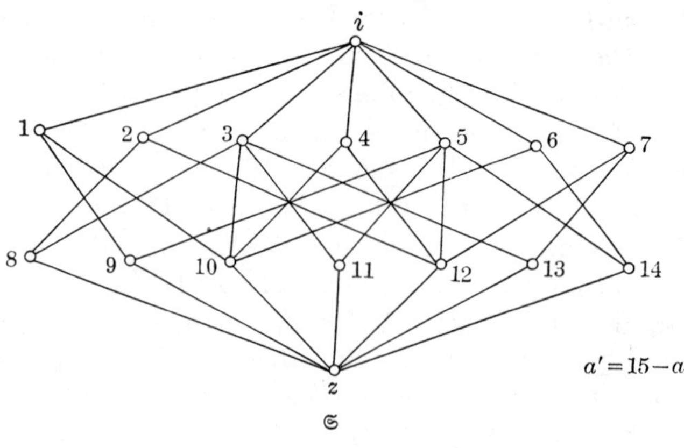

Fig. 2

℈ has a negation and is closed with respect to relative negation. Furthermore every sublattice closed with respect to relative negation satisfies the chain law. However ℈ is neither a B i r k h o f f lattice nor a dual B i r k h o f f lattice and hence is non-modular by lemma 2.1.

4. *Complemented non-modular lattices.* We prove now the second theorem mentioned in the introduction.

Theorem 4.1. *Every complemented, non-modular lattice of finite dimensions contains a complemented non-modular sublattice of order five.*

Proof. Let q be a cross-cut irreducible([1]) of the lattice ℈. Then if q is not a simple element, there exists an element $q_1 \neq i$ such that $q_1 > q$. Let q_1' be the complement of q_1. Then $(q_1, q_1') = (q, q_1') = i$ and $[q_1, q_1'] = [q, q_1'] = z$. Hence $\{i, q_1, q_1', q, z\}$ is a sublattice of the desired type. We may thus assume that the only cross-cut irreducibles are simple elements and similarly that the only union irreducibles are points.

We show now that if ℈ contains no complemented, non-modular

([1]) An element q is said to be cross-cut irreducible if $q = [a, b]$ implies either $q = a$ or $q = b$. If the lattice satisfies the descending chain condition q is cross-cut irreducible if and only if there is only one element covering q. Similarly p is union irreducible if $p = (a, b)$ implies either $p = a$ or $p = b$. p then covers only one element of ℈ if ℈ satisfies the ascending chain condition.

sublattice of order five, then $\mathfrak{S}$ is a Birknoff lattice. Dualizing the proof then shows that $\mathfrak{S}$ is a dual Birkhoff lattice and hence is modular by lemma 2.1 thus contradicting the hypothesis of the theorem.

If $\mathfrak{S}$ is not a Birkhoff lattice there is an element x with the property that $(x, p) \gtrdot x$, $(x, p) \neq x$ for some point p. For by definition 2.1 there exists an element x_1 such that $x_1 > [x, x_1]$ but $(x, x_1) \gtrdot x$. Now by the first paragraph we may assume that each element is a union of points. Hence there exists a point p such that $x_1 = ([x, x_1], p)$. But then $(x, p) = (x, [x, x_1], p) = (x, x_1) \gtrdot x$. Let S be the set of all such elements x and let a be a maximal element of S. Since a is in S there is an element a_1 such that $(a, p) \supset a_1 > a$. If $(a, p) \gtrdot a_1$, then a_1 is in S contradicting the maximal property of a. Hence $(a, p) > a_1 > a$. We show now that a_1 is simple and hence $(a, p) = i$. If a_1 is not simple, there exists an element y such that $y > a_1$, $y \neq (a, p)$. Now let $a_2 = [y, (a, a_1')]$. Then $y \supset a_2 \supset a$. Suppose that $y \supset a_2 \supset a_1$. Then either $y = a_2$ or $a_2 = a_1$ since $y > a_1$. If $y = a_2$, then $(a, a_1') \supset y$ and hence $(a, a_1') = (a, a_1', a_1') \supset (y, a_1') \supset (a_1, a_1') = i$. Thus $(a, a_1') = (a_1, a_1') = i$ and $[a, a_1'] = [a_1, a_1'] = z$. But then $\{i, a, a_1, a_1', z\}$ is a complemented non-modular sublattice of order five which contradicts our assumption. If $a_2 = a_1$, then $(a, a_1') \supset a_1$ and hence $(a, a_1') = (a, a_1', a_1') \supset (a_1, a_1') = i$. Thus $\{i, a, a_1, a_1', z\}$ is again a complemented non-modular sublattice of order five which contradicts our assumption. Hence $y \supset a_2 \supset a_1$ does not hold. Suppose now that $a_1 \supset a_2 \supset a$. Since $a_1 > a$ and $a_1 \neq a_2$ we must have $a_2 = a$ and $a \supset [y, (a, a_1')] \supset (a, [y, a_1']) \supset a$. Hence $(a, [y, a_1']) = a$ and $a \supset [a_1', y]$. But then $z = [a_1, a_1'] \supset [a, a_1'] \supset [y, a_1', a_1'] = [y, a_1']$. Thus $[y, a_1'] = z = [a_2, a_1']$. Also $(y, a_1') = (a_1, a_1') = i$. Thus $\{i, y, a_1, a_1', z\}$ is a complemented non-modular sublattices of order five which contradicts our assumption. Hence $a_1 \supset a_2 \supset a$ does not hold. Since $y > a_1 > a$ we have $(a_1, a_2) = y$, $[a_1, a_2] = a$. Now $(p, a_2) \neq (a, p)$. For if $(p, a_2) = (a, p)$, thene $a_1 = [y, (a, p)] = [y, (p, a_2)] = [y, ((a, p), a_2)] \supset (a_2, [y, (a, p)]) = (a_2, a) = y$ which contradicts $y > a_1$. Also $(y, p) > (a, p)$. For if $(y, p) \gtrdot (a, p)$, let $y = (a_1, p_1)$. Then $(p_1, (a, p)) = (p_1, (a_1, p)) = ((p_1, a), p) = (y, p)$. Also $(a, p) \gtrdot p_1$ since otherwise $(a, p) \supset (a_1, p_1, p) = (y, p) \supset y$ which is impossible. But then $(p_1, (a, p)) \gtrdot (a, p)$, $(p_1, (a, p)) \neq (a, p)$ and (a, p) is a proper divisor of a. But then (a, p) is in S which contradicts the maximal property of a.

We have $(y, p) \supset (a_2, p) \supset (a, p)$. Hence by the result we have just obtained $(y, p) = (a_2, p)$.   Thus $(a, p) \supset y \supset a_2$, $(a_2, p) \neq y$ since otherwise $y \supset p$ implies $y \supset (a, p) \supset a_1$ implies $y = (a, p)$ which contradicts the definition of y.   Also $y \neq a_2$ as has already been shown. Hence $(a_2, p) \not> a_2$ and $(a_2, p) \neq a_2$.   But $a_2 \supset a$ and $a_2 \neq a$.   This contradicts the maximal property of a.   Thus a_1 is simple and $(a, p) = i$. But then $\{i, a_1, a, p, z\}$ is a complemented non-modular sublattice of order five which contradicts our assumptions.   Hence $\mathfrak{S}$ is a Birkhoff lattice and the theorem is proved.

Theorem 4.1 may be used to give a new proof of the following theorem due to G. Birkhoff and M. Ward.

Theorem 4.2. *A lattice of finite dimensions is a Boolean algebra if and only if every element has a unique complement.*

For if every element of a lattice $\mathfrak{S}$ has a unique complement, then $\mathfrak{S}$ must be modular by theorem 4.1.   But it is well known[1] that a modular lattice with unique complement is a Boolean algebra. This completes the proof.

California Institute of Technology,
Pasadena, Calif..

(Received September 12, 1939).

---

[1] See for example, Huntington, Trans. Amer. Math. Soc., **5** (1904), p. 288 ; Skolem, Videnskapsselskepets Skrifter (1919); Bergman, Monatshefte f. Math. u. Phys., **36** (1929).

Uniquely Complemented Lattices

M. E. Adams

At the end of the nineteenth century, a number of mathematicians were concerned with the axiomatization of Boolean algebras. These considerations led Huntington [34] to what became known as Huntington's problem: is every uniquely complemented lattice distributive? Since a lattice is Boolean if and only if it is distributive and every element has a complement, an affirmative answer to Huntington's problem would have shown a lattice to be Boolean if and only if it was uniquely complemented.

Evidence mounted over time in support of a positive solution. For example, Bergmann [10] (later superseded by Birkhoff's characterization of distributivity) showed that, for any relatively complemented lattice, if relative complements are unique, then the lattice is distributive. Other known constraints on a uniquely complemented lattice sufficient to force distributivity included that the lattice be finite dimensional (Birkhoff and Ward; see Dilworth [21]), complete, atomic, and dually atomic (Birkhoff and Ward [13]), orthocomplemented (Birkhoff [11]), or modular (Birkhoff and von Neumann; see Birkhoff [12]).

It was a surprise when, in Dilworth [22], not only was a negative solution provided to Huntington's problem, but it was achieved by embedding an arbitrary lattice into one with unique complements. The proof is a four step *tour-de-force* all the more amazing as the conventional wisdom of the time dictated the opposite outcome.

Dilworth's proof is in the spirit of Whitman's fundamental paper on free lattices [49]. First of all he extended Whitman's solution of the word problem for free lattices to the lattice $F(P)$ freely generated by an ordered set P under the requirement (applied also in the subsequent steps) that least upper bounds and greatest lower bounds of pairs of elements of P be preserved whenever they exist. He showed that the order in P is preserved in $F(P)$. Using the results of the first step, Dilworth proceeded to give a solution to the word problem for a lattice $F^*(P)$ with a unary operation $*$ freely generated by an ordered set P. It was shown that the identity on P extends to an isomorphism of F(P) onto a sublattice of F*(P), so that the order in P is preserved in F*(P). In the third step, Dilworth selected a suitable sublattice of F*(P) containing P over which he defined a unary operation $'$ such that, for any element a of the sublattice, a' is a^* whenever a^* is in the sublattice. He demonstrated that F$'$(P), the sublattice together with the unary operation $'$, is a lattice with a reflexive unary operation freely generated by the ordered set P. In the final step Dilworth chose a suitable quotient F^c(P) of F$'$(P) shown to be a uniquely complemented lattice freely generated by P in which complements are prescribed by the unary operation $'$. Having established that the order in P is preserved by the quotient map, Dilworth concluded that any lattice L can be embedded in one with unique complements, for example, the lattice F^c(L).

In [50], Whitman had given necessary and sufficient conditions that a subset of a free lattice generates a free lattice over the subset. This had enabled him to show that F($\aleph_0$) may be embedded in F(3). (Note that, when P is a cardinal, the free object generated by P is simply the free object generated by an unordered set of the appropriate size.) Dilworth obtained analogous results for free lattices with a unary operation, a reflexive unary operation, and unique complementation. Moreover, he showed that lattices F*(1), F$'$(1), and F^c(2) contain copies of F*($\aleph_0$), F$'$($\aleph_0$), and F^c($\aleph_0$), respectively. The latter results are sharp since F^c(1) is clearly a 4-element lattice.

With this we conclude the description of Dilworth's landmark paper, the influence of which is discussed below.

With [22], Dilworth became the first person to consider amalgamation for lattices. He also explored the notion of a completely free lattice CF(P) generated by an ordered set P where only the order in P is required to be preserved. This notion was not developed in the paper as it was not suitable for the work at hand. Ten years later, Dean [18] proceeded to investigate the structure of completely free lattices generated by ordered sets and gave a solution to the word problem; see also Crawley and Dean [16] and Dean [19]. In a related vein, those ordered sets P for which CF(P) is finite were characterized first, in the special case that P is a disjoint union of chains, by Sorkin [46] and then, in general, by Wille [51]. In subsequent work, those ordered sets P for which CF(P) does not contain a sublattice isomorphic to F(3) were characterized, initially when P is a disjoint union of chains, by Rolf [40] and then, generally, by Rival and Wille [39]. Other authors have considered questions of a different ilk. For example, generalizing a result of Galvin and Jónsson [23] for free lattices, Adams and Kelly [1] have shown that, for a (necessarily) uncountable regular cardinal κ, if P has no chains of cardinality κ, then neither does CF(P).

Following a more general approach, Dean [20] has given a construction for a lattice FL(P) freely generated by an ordered set P where it is required that least upper bounds and greatest lower bounds of prescribed finite sets be preserved whenever they exist in P. Unlike F(P) and CF(P), the algorithm for deciding whether two polynomials are comparable in FL(P) is not necessarily finitistic.

In the second step of his proof, Dilworth had inductively defined a lower cover

        THE DILWORTH THEOREMS

and an upper cover (whenever they exist) for each polynomial. The notions of lower
and upper covers have had far reaching consequences. In particular, they facilitate
an elegant solution to the word problem for free products of lattices due to Grätzer,
Lakser, and Platt [29]. In turn, this has yielded a large number of interesting results
for free products; for further information, see Grätzer [28]. The first indication that
lower and upper covers had a role to play in varieties of lattices other than the
variety of all lattices was given in Jónsson [35]. This role is illustrated by the proof,
due to Grätzer and Sichler [32], that any two representations of a lattice in any
non-trivial variety V as a V-free product have a common refinement.

In [15], Chen and Grätzer gave a new proof of Dilworth's theorem. Although
their proof uses many of the ideas contained in [22], it also avoids some of the difficulties. Further refinement in Grätzer [26] and [27], led to the notion of C-reduced
free products: a free product of bounded lattices subject to the requirement that
complements in component lattices be preserved and that each member of a prescribed family of 2-element subsets, whose elements belong to distinct components,
be a pair of complements. In [27], Grätzer gave a solution to the word problem
for C-reduced free products which was then used to locate all their complements.
It was then a straightforward matter to show that every bounded lattice in which
any element has at most one complement is isomorphic to a $(0,1)$-sublattice of a
uniquely complemented lattice.

Since this work, other applications for C-reduced free products have been found.
For example, in [30], Grätzer and Sichler showed that every monoid is isomorphic to
the $(0,1)$-endomorphism monoid of a bounded lattice. This result was later strengthened, using similar techniques, in Adams and Sichler [3]. Another application, also
due to Grätzer and Sichler [31], was motivated by a result of H. Neumann: they
showed that a free product of hopfian lattices need not be hopfian (whether or not
the free product be bounded). Later, in showing that every finite monoid is isomorphic to the $(0,1)$-endomorphism monoid of a finite lattice, Adams and Sichler [2],
[4], (cf. [6]) observed that, for a variety V of lattices, the presence of certain testing
lattices, dubbed cover set lattices, was sufficient to keep track of complements in
C-reduced V-free products. Simultaneously, they introduced the more general notion of an R-reduced V-free product, which made it possible to strengthen some of
the earlier mentioned results. This approach was developed still further by Koubek
in [36].

In addition to the directions indicated above, uniquely complemented lattices
continued to draw the interest of mathematicians. In [38], Ogasawara and Sasaki
strengthened the result of Birkhoff and Ward by showing that any uniquely complemented atomic lattice is distributive. In conjunction with the result of Birkhoff and
von Neumann, this was then derived as a corollary to McLaughlin's proof [37] that
any complemented atomic lattice with unique comparable complements is modular.
Other restrictions on uniquely complemented lattices which force distributivity were
also discovered. For example, it is sufficient that the mapping which sends each element to its complement be order-inverting (Birkhoff [12]; cf. Szász [48]), or that,
for every element, there is a prime ideal that does not contain it (Chen [14]), or that

the lattice be relatively complemented (Szász [47]), initially complemented (Beran [9]), 0-modular (Grillet and Varlet [33]), 0-semi-modular (Saliĭ [43]), algebraic (Saliĭ [41]), weakly atomic (Bandelt and Padmanabhan [8]), or continuous (Bandelt [7] and, independently, Saliĭ [42]). This list is not exhaustive and, for a fuller account, the reader is urged to consult Saliĭ's monograph [45].

In [17], Crawley and Dilworth gave another proof of Dilworth's theorem. Nevertheless, the feeling that non-distributive uniquely complemented lattices are somewhat pathological in nature persisted, and still does. How can this feeling be made more concrete? In this context, Adams and Sichler [5] showed that there are continuum many varieties of lattices for which Dilworth's theorem holds: for each such variety $\mathcal{V}$, every lattice in $\mathcal{V}$ may be embedded in a uniquely complemented lattice which also belongs to $\mathcal{V}$. While answering a question of Grätzer, this also demonstrated non-distributive uniquely complemented lattices to be more common than had been expected.

At the time of this writing, only the tip of the uniquely complemented iceberg has been sighted. Many questions remain unanswered, some old, some new (see Grätzer [28] and Saliĭ [45]). One can still ask whether there exists a natural example of a non-distributive uniquely complemented lattice; this is clearly one of those questions that will be better formulated once it has been answered. The variety of p-modular lattices was first studied by Gedeonová in [24]. Presently, this is the smallest known non-distributive variety of lattices to satisfy Dilworth's theorem. Is it in fact the smallest such variety? Is there a variety that contains a non-distributive uniquely complemented lattice but fails to satisfy Dilworth's theorem? It is known that there exist non-distributive uniquely complemented lattices that are locally finite, but does there exist a finitely generated one, or, related to the above, does there exist a non-distributive locally finite variety of lattices that satisfies Dilworth's theorem? While Saliĭ [44] has shown that every complete uniquely complemented lattice is isomorphic to a direct product of a complete atomic Boolean lattice and a complete atomless uniquely complemented lattice, it is not yet known whether there exists a complete non-distributive uniquely complemented lattice. Nor is it known whether the MacNeille completion of a uniquely complemented lattice is necessarily uniquely complemented, apart from Glivenko's result [25] that the MacNeille completion of a Boolean lattice is Boolean.

These questions attest to the longstanding and vibrant contribution made by Dilworth's work on lattices with unique complements. With this one paper written in 1945, Dilworth made a rich and multifaceted contribution to lattice theory. In addition to solving Huntington's problem, Dilworth's insight has inspired investigation in diverse areas of lattice theory and has substantially influenced both the field and an ever widening circle of mathematicians.

Acknowledgement: It is a pleasure to acknowledge comments and suggestions made by R. Freese, M. Gould, G. Grätzer, and J. Sichler.

REFERENCES

1. M. E. Adams and D. Kelly, *Chain conditions in free products of lattices*, Algebra Universalis **7** (1977), 235–244.
2. M. E. Adams and J. Sichler, *Bounded endomorphisms of lattices of finite height*, Canad. J. Math. **29** (1977), 1254–1263.
3. M. E. Adams and J. Sichler, *Homomorphisms of bounded lattices with a given sublattice*, Arch. Math. (Basel) **30** (1978), 122–128.
4. M. E. Adams and J. Sichler, *Cover set lattices*, Canad. J. Math. **32** (1980), 1177–1205.
5. M. E. Adams and J. Sichler, *Lattices with unique complementation*, Pacific J. Math. **92** (1981), 1–13.
6. M. E. Adams and J. Sichler, *Refinement property of reduced free products in varieties of lattices*, in "Contributions to Lattice Theory," Colloq. Math. Soc. János Bolyai (Szeged 1980), vol. 33, 1983, pp. 19–53.
7. H. J. Bandelt, *Complemented continuous lattices*, Arch. Math. (Basel) **36** (1981), 474–475.
8. H. J. Bandelt and R. Padmanabhan, *A note on lattices with unique comparable complements*, Abh. Math. Sem. Univ. Hamburg **48** (1979), 112–113.
9. L. Beran, *Über die Charakterisierung von Boole-Verbänden*, Praxis Math. **17** (1975), 98–103.
10. G. Bergmann, *Zur Axiomatic der Elementargeometrie*, Monatsh. Math. Phys. **36** (1929), 269–284.
11. G. Birkhoff, "Lattice Theory," first edition, Colloquium Publications, Amer. Math. Soc., Providence, R.I., 1940.
12. G. Birkhoff, "Lattice Theory," rev. ed., Colloquium Publications, Amer. Math. Soc., Providence, R.I., 1948.
13. G. Birkhoff and M. Ward, *A characterization of Boolean algebras*, Ann. of Math. **40** (1939), 609–610.
14. C. C. Chen, *On uniquely complemented lattices*, J. Nanyang Univ. **3** (1969), 380–384.
15. C. C. Chen and G. Grätzer, *On the construction of complemented lattices*, J. Algebra **11** (1969), 56–63.
16. P. Crawley and R. A. Dean, *Free lattices with infinite operations*, Trans. Amer. Math. Soc. **92** (1959), 35–47.
17. P. Crawley and R. P. Dilworth, "Algebraic Theory of Lattices," Prentice-Hall, Englewood Cliffs, NJ, 1973.
18. R. A. Dean, *Completely free lattices generated by partially ordered sets*, Trans. Amer. Math. Soc. **83** (1956), 238–249.
19. R. A. Dean, *Sublattices of free lattices*, in "Lattice Theory," Proc. Symp. Pure Math. II, R. P. Dilworth, editor, Amer. Math. Soc., Providence, Rhode Island, 1961, pp. 31–42.
20. R. A. Dean, *Free lattices generated by partially ordered sets and preserving bounds*, Canad. J. Math. **16** (1964), 136–148.
21. R. P. Dilworth, *On complemented lattices*, Tôhoku Math. J. **47** (1940), 18–23. Reprinted in Chapter 2 of this volume.
22. R. P. Dilworth, *Lattices with unique complements*, Trans. Amer. Math. Soc. **57** (1945), 123–154. Reprinted in Chapter 2 of this volume.
23. F. Galvin and B. Jónsson, *Distributive sublattices of a free lattice*, Canad. J. Math. **13** (1961), 265–272.
24. E. Gedeonová, *Jordan-Hölder theorem for lines*, Mat. Časopis Sloven. Akad. Vied. **22** (1972), 177–198.
25. V. I. Glivenko, "Théorie Générale des Structures," Actualités Sci. Indust., Hermann, Paris, 1938.
26. G. Grätzer, *A reduced free product of lattices*, Fund. Math. **73** (1971), 21–27.
27. G. Grätzer, *Free products and reduced free products of lattices*, Proc. Univ. Houston Lattice Theory Conf. (1973), 539–563.

28. G. Grätzer, "General Lattice Theory," Series on Pure and Applied Mathematics, Academic Press, New York, N.Y.; Mathematische Reihe, Band 52, Birkhauser Verlag, Basel; Akademie Verlag, Berlin., 1978.

29. G. Grätzer, H. Lakser, and C. R. Platt, *Free products of lattices*, Fund. Math. **69** (1970), 233–240.

30. G. Grätzer and J. Sichler, *On the endomorphism semigroup (and category) of bounded lattices*, Pacific J. Math. **35** (1970), 639–647.

31. G. Grätzer and J. Sichler, *Free products of hopfian lattices*, J. Austral. Math. Soc. **27** (1974), 234–245.

32. G. Grätzer and J. Sichler, *Free decompositions of a lattice*, Canad. J. Math. **27** (1975), 276–285.

33. P. A. Grillet and J. C. Varlet, *Complementedness conditions in lattices*, Bull. Soc. Roy. Sci. Liège **36** (1967), 628–642.

34. E. V. Huntington, *Sets of independent postulates for the algebra of logic*, Trans. Amer. Math. Soc. **5** (1904), 288–309.

35. B. Jónsson, *Relatively free products of lattices*, Algebra Universalis 1 (1971), 362–373.

36. V. Koubek, *Towards minimal binding varieties of lattices*, Canad. J. Math. **36** (1984), 263–285.

37. J. E. McLaughlin, *Atomic lattices with unique comparable complements*, Proc. Amer. Math. Soc. **7** (1956), 864–866.

38. T. Ogasawara and U. Sasaki, *On a theorem in lattice theory*, J. Sci. Hiroshima Univ. Ser. A **14** (1949), p. 13.

39. I. Rival and R. Wille, *Lattices freely generated by partially ordered sets: which can be "drawn"?*, J. Reine Angew. Math. **310** (1979), 56–80.

40. H. L. Rolf, *The free lattice generated by a set of chains*, Pacific J. Math. **8** (1958), 585–595.

41. V. N. Saliĭ, *A compactly generated lattice with unique complements is distributive*, (Russian), Mat. Zametki **12** (1972), 617–620.

42. V. N. Saliĭ, *A continuous uniquely complemented lattice is distributive*, (Russian), in "Fifth All-Union Conf. Math. Logic, Abstracts of Reports," Inst. Mat. Sibirsk. Otdel. Akad. Nauk SSSR, Novosibirsk, 1979, p. 134.

43. V. N. Saliĭ, *Some conditions for distributivity of a lattice with unique complements*, (Russian), Izv. Vyss. Ucebn. Zaved. Matematika **5** (1980), 47–49.

44. V. N. Saliĭ, *Regular elements in complete uniquely complemented lattices*, (Russian), in "Universal Algebra and Applications (Warsaw 1978)," Banach Center Publications, 1982, pp. 15–19.

45. V. N. Saliĭ, "Lattices with Unique Complements," Translations of the Amer. Math. Soc., Amer. Math. Soc., Providence, R. I., 1988.

46. J. I. Sorkin, *Free unions of lattices*, (Russian) Mat. Sbornik **30** (1952), 677–694.

47. G. Szász, *On complemented lattices*, Acta Sci. Math. (Szeged) **19** (1958), 77–81.

48. G. Szász, *On the de Morgan formulae and the antitony of complements in lattices*, Czechoslovak Math. J. **28 (103)** (1978), 406–440.

49. Ph. M. Whitman, *Free lattices*, Ann. of Math. (2) **42** (1941), 325–330.

50. Ph. M. Whitman, *Free lattices II*, Ann. of Math. (2) **43** (1942), 104–115.

51. R. Wille, *On lattices freely generated by finite partially ordered sets*, in "Contributions to Universal Algebra," Colloq. Math. Soc. János Bolyai (Szeged 1975), vol. 17, 1977, pp. 581–593.

State University of New York
New Paltz, New York 12561
U. S. A.

On Orthomodular Lattices

Gudrun Kalmbach

The paper *On complemented lattices* was the third paper in the new theory of orthomodular lattices which started in 1936 with Birkhoff and von Neumann's idea of developing a new many-valued logic for quantum mechanics by using the lattice of closed subspaces $C(\mathcal{H})$ of a Hilbert space $\mathcal{H}$ as the valuation lattice.

In this article we concern ourselves only with those aspects of Dilworth's paper which are related to orthomodular lattices. The list of 260 authors in the field of orthomodular lattices and their papers is too long to be quoted here. The interested reader can consult [5, 6].

The second paper on orthomodular lattices was written by Husimi and contains the orthomodular law which holds in all $C(\mathcal{H})$. We write $U \vee V$ for the smallest closed subspace of $\mathcal{H}$ containing U and V in $C(\mathcal{H})$, $U \wedge V$ for their intersection and U' for the orthocomplement of U. The deep and broad investigation of the orthomodular law: for $V, W \in C(\mathcal{H})$,

$$\text{(1)} \qquad W \subseteq V \quad \text{implies} \quad V = W \vee (W' \wedge V),$$

was initiated by Dilworth's discovery of the first finite (nonmodular) orthomodular lattice D_{16} which is still one of the first examples one checks in looking for counterexamples to a conjecture in orthomodular lattice theory. This lattice is constructed as follows: Take three copies B_1, B_2, and B_3, $i = 1, 2, 3$ of the Boolean algebra 2^3 with the zero element 0_i, the three atoms x_i, x in $\{a, b, c\}$, the three coatoms x_i' and the unit element 1_i. Paste all zero elements together to form a new zero $0 = \{0_1, 0_2, 0_3\}$, paste all unit elements together to form a new unit $1 = \{1_1, 1_2, 1_3\}$ and paste atoms or coatoms together as follows: $d_1 = \{a_2, c_1\}$, $d_1' = \{a_2', c_1'\}$, $d_2 = \{a_3, c_2\}$, $d_2' = \{a_3', c_2'\}$. Figure 1 gives what is known as the Greechie diagram, see [5]. The actual lattice is given in Figure 1 of the Background.

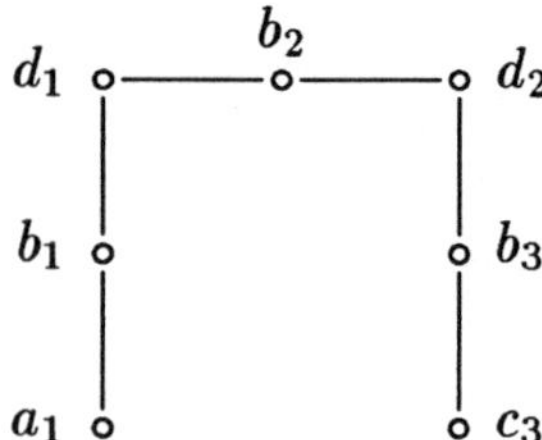

Figure 1

The resulting nonmodular, orthomodular lattice has a negation $'$, is finite-dimensional and satisfies the chain law. D_{16} was not expected by the authors of the first and second paper of the orthomodular theory.

Birkhoff and von Neumann had suggested that in quantum mechanics one needs a new kind of $C(\mathcal{H})$-valued logic which does not obey the distributive law for "AND" and "OR" of classical logic, but keeps a classical negation $'$. They had suggested the weaker modular law

$$(2) \qquad x < y \quad \text{implies} \quad x \vee (z \wedge y) = (x \vee z) \wedge y$$

as a substitute. It does not hold in an infinite-dimensional $C(\mathcal{H})$.

The lattice interpretation of "AND" by $\wedge$, of "OR" by $\vee$ and of "NOT" by $'$ is natural in logic. It is in general not clear what "IMPLIES", "$\rightarrow$", means. If the the members of the equational class of all orthomodular lattices is used as models of a many-valued orthomodular logic then again D_{16} makes it clear that only one meaningful choice for "$\rightarrow$" exists. $\rightarrow$ can be expressed as a polynomial-implication which reduces, if twice applied, to the classical negation:

$$x \rightarrow (x \rightarrow y) = x' \vee y.$$

The discovery of laws which hold in $C(\mathcal{H})$, but not in all orthomodular lattices and the axiomatic approach to "quantum logic" produced several deep and unexpected results. Keller, Künzi and Gross [4,7] found infinite-dimensional Hilbert spaces S whose orthocomplemented lattices $C(S)$ of $'$-closed subspaces obey a set of axioms (to be atomic, infinite-dimensional, irreducible, complete, orthomodular with the exchange axiom) which until the 1980's was conjectured to be a set of axioms characterizing classical Hilbert spaces. The new examples have a countable dimension and do not have an orthonormal basis of equal-length vectors. We are now back to asking experimental physicists: Is there a need in quantum mechanics for nonarchimedean Hilbert spaces where no two vectors of an orthonormal basis have equal length? The mathematical models are present in the form of Keller orthomodular spaces.

REFERENCES

1. G. Birkhoff and J. von Neumann, *The logic of quantum mechanics*, Ann. of Math. **37** (1936), 823–843.

2. R. P. Dilworth, *On complemented lattices*, Tôhoku Math. J. **47** (1940), 18–23. Reprinted in Chapter 2 of this volume.
3. K. Husimi, *Studies on the foundations of quantum mechanics, I*, Proc. Physics-Math. Soc. Japan **19** (1937), 766–789.
4. H. Gross and U.-M. Künzi, *On the class of orthomodular spaces*, L'Enseign. Math. **31** (1985), 187–212.
5. G. Kalmbach, "Orthomodular Lattices," Academic Press, London, 1983.
6. G. Kalmbach, "Measures and Hilbert Lattices," World Scientific, Singapore, 1986.
7. H. A. Keller, *Ein nicht-klassischer Hilbertscher Raum*, Math. Z. **172** (1980), 41–49.

Universität Ulm
D-7900 Ulm
West Germany

Decomposition Theory

Background

R. P. Dilworth

My interest in semimodular lattices began with a remark by Morgan Ward that he had observed that a lattice each of whose elements had a unique representation as a reduced meet of meet irreducibles was necessarily semimodular. Since Birkhoff had already shown that a modular lattice with unique decomposition into irreducibles was necessarily distributive, it suggested to me that lattices having unique meet decompositions could be characterized in terms of semimodularity and some weak form of distributivity. Furthermore, semimodular lattices seemed to be the appropriate domain in which to study decomposition questions of this type. I quickly came upon the lattice diagrammed in Figure 1 as the simplest example of a non-distributive lattice having unique reduced meet decompositions.

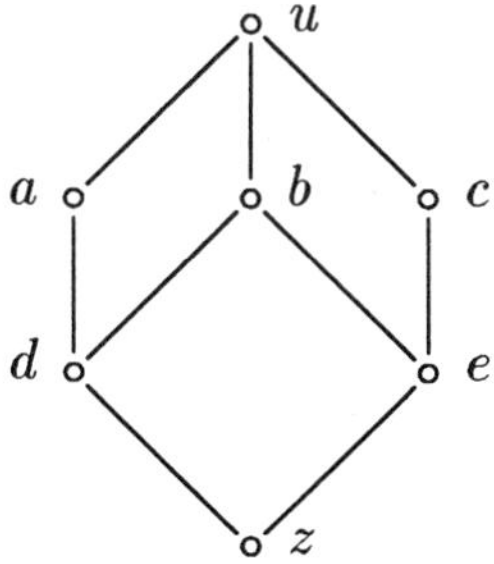

Figure 1

In this lattice d, e, and z have the unique representations

$$d = a \wedge b, \quad e = b \wedge c, \quad z = a \wedge c.$$

The lattice is not distributive since it contains the non-distributive sublattice $\{u, a, c, e, z\}$ of order five. It should be noted, however, that the quotient lattices generated by the elements covering an element of the lattice is in each case distributive. This example immediately suggests the conjecture that this form of local distributivity together with semimodularity is necessary and sufficient for unique irreducible decompositions in a lattice satisfying a suitable finiteness restriction. In order to insure the existence of decompositions and the existence of covering elements, it is sufficient that the lattice have a unit and every interval (quotient sublattice) is finite dimensional. It turned out that under these conditions the conjecture could be proved. This result also led to a second characterization suggested by the Birkhoff theorem, namely, that the local distributivity condition could be replaced by the condition that every modular sublattice is distributive.

Now Kurosch and Ore had already proved a decomposition theorem for modular lattices satisfying the ascending chain condition. For such a lattice the theorem asserts that if an element has two reduced decompositions into meet irreducibles, then the number of components are equal and any irreducible in one decomposition can be replaced by a suitably chosen irreducible in the other decomposition. For example, in the lattice diagrammed in Figure 2 the reduced decompositions of z are $z = a \wedge b = a \wedge c = b \wedge c.$

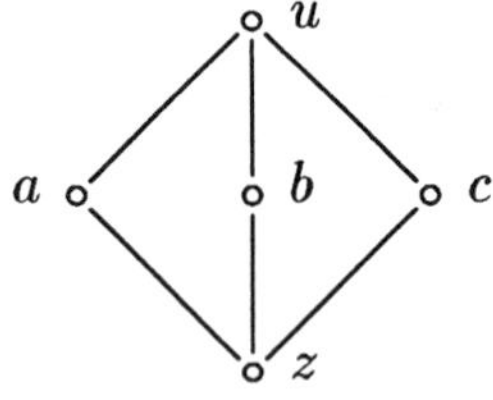

Figure 2

Since the uniqueness of decompositions can be characterized in terms of local conditions, it is reasonable to suppose that in a semimodular lattice uniqueness of the number of components in the decompositions of an element be expressed in terms of a local condition. In view of the Kurosch-Ore theorem the natural candidate for this local condition is modularity. Consider the following lattice diagrammed in Figure 3.

The lattice is semimodular but not modular. The reduced decompositions of the element z are given by

$$z = a \wedge c = a \wedge d = a \wedge e = b \wedge c = b \wedge d = b \wedge e = c \wedge e = d \wedge e.$$

Note that the quotient lattices generated by the elements covering a given element are all modular.

                    THE DILWORTH THEOREMS

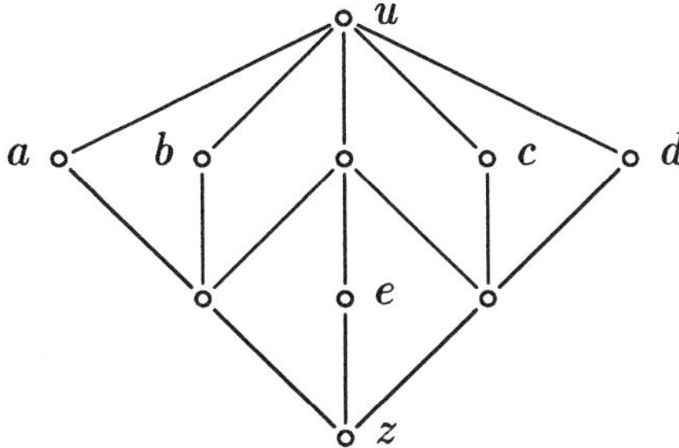

Figure 3

Let us then assume that the lattice is semimodular, has a unit element, and every interval is finite dimensional. Then decompositions into meet irreducibles exist and the existence of covering elements is assured. It can then be proved that the number of components is unique for each element of the lattice if and only if the intervals generated by the elements covering an element of the lattice are modular, *i.e.*, the lattice is locally modular. Furthermore, out of the uniqueness of the number follows the replacement property of Kurosch-Ore.

Now in order to insure that each element of a lattice has a reduced representation as a meet of a finite number of meet irreducibles, it suffices to have the ascending chain condition holding in the lattice. Thus it would be desirable to characterize lattices with unique decompositions and lattices with unicity of the number of components under just the ascending chain condition. In order to get an analogue of the finite dimensional case, it is necessary to go to the lattice of dual ideals (called 'ideals' then). Given an element a in the lattice, there always exists a dual ideal which covers the principal dual ideal (a). The interval in the lattice of dual ideals generated by the dual ideals covering (a) plays the role previously played by the interval in the lattice itself generated by the elements covering a. Consider the lattice in Figure 4.

The dual ideal consisting of the a's and c's covers the principal ideal (z) as does the dual ideal consisting of the b's and c's. The join of these two dual ideals is the dual ideal of the c's. This lattice is semimodular in a strong sense. Corresponding to the two minimal dual ideals, z has the unique representation $z = a_1 \wedge b_1$.

In general, a lattice is semimodular in the strong sense if P is a dual ideal covering (a) and Q is a dual ideal above (a) but not above P, then $P \vee Q$ covers Q. With this strong definition of semimodularity, a lattice satisfying the ascending chain condition has unique irreducible decompositions if and only if it is semimodular and is locally distributive. Furthermore, in this case the interval generated by the dual ideals covering (a) is a finite Boolean algebra. Likewise, a semimodular lattice satisfying the ascending chain condition has unicity of the number of components if and only if it is locally modular. Again, the interval generated by the dual ideals covering (a) is a finite dimensional modular lattice and the decompositions of a have

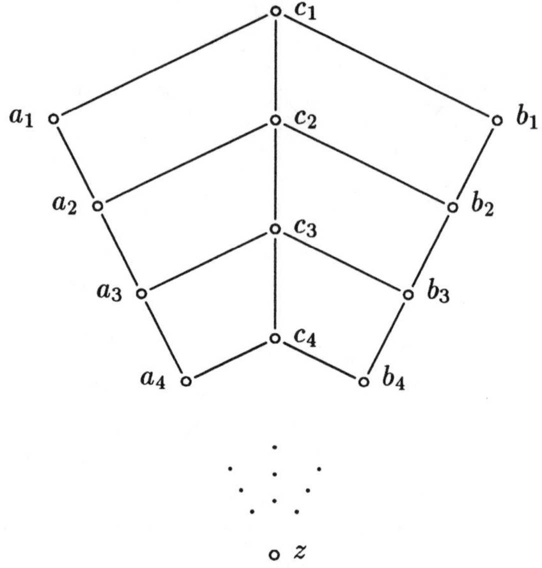

Figure 4

the replaceability property. The paper, "Ideals in Birkhoff Lattices," completed the work on the arithmetical properties of lattices satisfying the ascending chain condition.

Some twenty years later, Peter Crawley and I returned to these questions for lattices without a chain condition. We were motivated by examples from the theory of abelian groups, vector space lattices, and lattices of congruence relations. For this investigation the ascending chain condition was replaced by compact generation and the descending chain condition by atomicity. The decompositions were no longer finite but these assumptions together with semimodularity were sufficient to insure the existence of an irredundant representation of an element as a meet of completely meet irreducible elements. Analogues of the arithmetical theorems in the finite dimensional case could then be proved for these infinite decompositions.

ANNALS OF MATHEMATICS
Vol. 41, No. 4, October, 1940

LATTICES WITH UNIQUE IRREDUCIBLE DECOMPOSITIONS

By R. P. Dilworth

(Received June 6, 1939)

Consider a lattice $\mathfrak{S}$ in which the ascending chain condition holds. Then each element of $\mathfrak{S}$ has at least one reduced[1] representation as a cross-cut of irreducibles. Now it is well known that the requirement that this representation be unique considerably restricts the structure of the lattice. For example, Garrett Birkhoff [1] has proved that a modular lattice in which every element is uniquely expressible as a reduced cross-cut of irreducibles is distributive. Furthermore, Morgan Ward has shown that unicity of the irreducible decompositions implies that the lattice is a Birkhoff lattice.[2] These results suggest the interesting problem of characterizing a lattice in which every element has a unique reduced representation as a cross-cut of irreducibles in terms of the structure of the lattice. We give here a complete solution of this problem. We show, namely, that such lattices are simply those Birkhoff lattices in which every modular sublattice is distributive. The detailed statement of our theorem is as follows:

THEOREM 1.1. *Let $\mathfrak{S}$ be a lattice with unit element in which every quotient lattice is of finite dimensions. Then each element a of $\mathfrak{S}$ is uniquely expressible as a cross-cut of irreducibles if and only if $\mathfrak{S}$ is a Birkhoff lattice in which every modular sublattice is distributive. In this case, for each element covering a there is exactly one irreducible not dividing it but dividing the remaining elements covering a. These irreducibles are the components of a.*

1. Throughout the paper unless otherwise stated $\mathfrak{S}$ will denote a lattice of elements a, b, c, $\cdots$ and unit element i in which every quotient lattice[3] is of finite dimensions. a covers b will be written $a > b$. An element a is said to be *irreducible* if $a = [b, c]$ implies $a = b$ or $a = c$.

DEFINITION 1.1. $\mathfrak{S}$ is said to have *unique irreducible decompositions* if every element of $\mathfrak{S}$ has a unique reduced representation as a cross-cut of irreducibles.

DEFINITION 1.2. $\mathfrak{S}$ is said to be a Birkhoff lattice if $a > [a, b]$ implies that $(a, b) > b$.[4]

---

[1] A cross-cut representation is said to be "reduced" if no member of the representation is superfluous.

[2] This result was mentioned to the author in conversation. Birkhoff lattices are defined in definition 1.2.

[3] If $a \supset b$, the quotient lattice associated with a and b is the sublattice of elements x such that $a \supset x \supset b$ (Ore [1]).

[4] Cf. Klein (1).

771

If $\mathfrak{S}$ is a Birkhoff lattice, then there exists a rank function $\rho(a)$ (Birkhoff [2]) with the properties

(i) $\rho(i) = 0$.

(ii) $a > b$ implies $\rho b = \rho a + 1$.

(iii) $\rho[a, b] + \rho(a, b) \geqq \rho a + \rho b$.

DEFINITION 1.3. A set S of elements of $\mathfrak{S}$ is said to be *cross-cut independent* if the cross-cut of any finite number of elements of S is not divisible by any of the remaining elements of S. Similarly S is said to be union independent if the union of any finite set of elements of S divides none of the remaining elements of S.[5]

Our proof of theorem 1.1 rests on a series of lemmas. The first lemma proves the necessity while the remaining lemmas are devoted to proving the sufficiency.

LEMMA 1.1. *If $\mathfrak{S}$ has unique irreducible decompositions, then $\mathfrak{S}$ is a Birkhoff lattice in which every modular sublattice is distributive.*

PROOF. Let $a > [a, b]$ and suppose that $(a, b) \supset b_1 \supset b$ where $(a, b) \neq b_1$, and $b_1 \neq b$. Furthermore let $a = [q_1, \cdots, q_k]$ be the irreducible decomposition of a. Then since $b \not\supset b_1$ there is an irreducible component q of b such that $q \not\supset b_1$ and $q \not\supset a$. Hence $[a, b] = [q_{i_1}, \cdots, q_{i_l}, q]$ $l \leq k$ is an irreducible decomposition of $[a, b]$. Similarly let q' be an irreducible component of b_1 such that $q' \not\supset a$. Then $[a, b] = [q_{i_1}, \cdots, q_{i_m}, q']$ $m \leq k$ is an irreducible decomposition of $[a, b]$. But $q \neq q'$ and $q \neq q_i$ $(i = 1, \cdots, k)$. Thus $[a, b]$ has two irreducible decompositions which contradicts our assumption.

Now if there exists a modular sublattice which is not distributive, then $\mathfrak{S}$ contains a sublattice $\{u, a, b, c, d\}$ where $(a, b) = (b, c) = (a, c) = u$ and $[a, b] = [b, c] = [a, c] = d$ (Birkhoff [2]). Let $b = [q_1, \cdots, q_k]$ be the irreducible decomposition of b. Furthermore let $p_1, \cdots, p_n$ be the irreducible components of a which do not divide b and let $p'_1, \cdots, p'_m$ be the irreducible components of c which do not divide b. Then $d = [a, b] = [q_1, \cdots, q_k, p_1, \cdots, p_n]$ and dropping superfluous elements we get an irreducible decomposition $d = [q_{i_1}, \cdots, q_{i_s}, p_{i_1}, \cdots, p_{i_t}]$. In a similar manner we get an irreducible decomposition $d = [q_{j_1}, \cdots, q_{j_u}, p'_{j_1}, \cdots, p'_{j_v}]$. But $p_{i_1} \neq p'_{j_r} (r = 1, \cdots, v)$ since otherwise we would have $p_{i_1} \supset (a, c) \supset b$. Similarly $p_{i_1} \neq q_{j_r} (r = 1, \cdots, u)$. Hence d has two irreducible decompositions contrary to the hypothesis of the lemma.

Now let $\mathfrak{S}$ be a Birkhoff lattice in which every modular sublattice is distributive.

LEMMA 1.2. $a_1, a_2, \cdots, a_k > a$ *and* $(a_1, \cdots, a_k) \supset x \supset a, x \neq a$ *imply* $x \supset a_i$ *for some* i.

PROOF. We prove the lemma by induction and assume that the lemma is true for any a and $k - 1$ covering elements. Now $(a_1, \cdots, a_k) \supset (x, a_1) \supset a_1$. If $(x, a_1) = a_1$, then $a_1 \supset x \supset a$ and $x = a_1$ since $a_1 > a$. Hence $x \supset a_1$ and the lemma holds. We may thus assume that $(x, a_1) \neq a_1$. But then since $[a_1, a_2] = \cdots = [a_1, a_k] = a$ we have $(a_1, a_2), \cdots, (a_1, a_k) > a$, by definition

---

[5] Cf. MacLane (1), Whitney (1).

1.2. Hence by the induction assumption $(x, a_1) \supset (a_1, a_2)$ say. But then $(a_1, \cdots, a_k) \supset (x, a_1) \supset (a_1, a_2)$. If $(x, a_1) = (a_1, a_2)$ then $(a_1, a_2) \supset x \supset a$ and $x \supset a_1$ or $x \supset a_2$ by our induction assumption. We may thus assume that $(x, a_1) \neq (a_1, a_2)$. Now $(a_1, a_2, a_3), \cdots, (a_1, a_2, a_k) > (a_1, a_2)$. For $(a_1, a_2, a_i) > (a_1, a_2)$ or $(a_1, a_2, a_i) = (a_1, a_2)$ by the Birkhoff condition. But if $(a_1, a_2, a_i) = (a_1, a_2)$, then $(a_1, a_2) \supset a_i \supset a$, $a_i \neq a$ and $a_i \supset a_1$ or $a_i \supset a_2$ by the induction assumption. But this is impossible since $a_1, \cdots, a_k$ are distinct elements covering a. Now since $(a_1, \cdots, a_k) = ((a_1, a_2, a_3), \cdots, (a_1, a_2, a_k))$ we have $(x, a_1) \supset (a_1, a_2, a_3)$ say by assumption. Continuing in this manner we find that either the lemma is true or $(x, a_1) \supset (a_1, a_2, \cdots, a_{k-1})$. Hence $(a_1, \cdots, a_k) \supset (x, a_1) \supset (a_1, a_2, \cdots, a_{k-1})$ and $(a_1, \cdots, a_k) > (a_1, a_2, \cdots, a_{k-1})$. If $(x, a_1) = (a_1, \cdots, a_{k-1})$, then $(a_1, \cdots, a_{k-1}) \supset x \supset a$ and $x \supset a_j$ for some j by the induction assumption. Hence we may assume that $(x, a_1) = (a_1, \cdots, a_k) = u$. Let $a_1' = (a_2, \cdots, a_k)$. Clearly $(x, a_1') = u$ since otherwise $a_1' \supset x$ and $x \supset a_j$ by the induction assumption. Now $a_1' \supset [a_1', x] \supset a$. Hence if $[a_1', x] \neq a$, then $x \supset a_j$ by assumption. We may thus assume that $[a_1', x] = a$. Similarly we find $[a_1, x] = a$ and $[a_1, a_1'] = a$. But then $\{u, a_1, a_1', x, a\}$ is a modular, non-distributive sublattice which contradicts our hypothesis on $\mathfrak{S}$. Since the lemma is obviously true for $k = 1$, the proof is complete.

COROLLARY 1.1. *The elements covering any given element of $\mathfrak{S}$ are union independent.*

COROLLARY 1.2. *There are only a finite number of elements covering any given element of $\mathfrak{S}$.*

LEMMA 1.3. *The elements covering any element a of $\mathfrak{S}$ generate a finite Boolean algebra which is dense in $\mathfrak{S}$.*

PROOF. Let $a_1, \cdots, a_n$ denote the elements of $\mathfrak{S}$ covering a. Let $\mathfrak{B}$ be the sublattice generated by $a_1, \cdots, a_n$. Let $(a_1, \cdots, a_n) \supset x \supset a$, $x \supset a_1$, $\cdots, a_k$, and $x \not\supset a_{k+1}, \cdots, a_n$. Then $(a_1, \cdots, a_n) \supset x \supset (a_1, \cdots, a_k)$. Now $(a_1, \cdots, a_k, a_{k+1}), \cdots, (a_1, \cdots, a_k, a_n) > (a_1, \cdots, a_k)$ by definition 1.2 and corollary 1.1. Then if $x \neq (a_1, \cdots, a_k)$ we have $x \supset (a_1, \cdots, a_k, a_j) \supset a_j \ k + 1 \leq j \leq n$ by lemma 1.2. This contradicts $x \supset a_j \ k + 1 \leq j \leq n$. Hence $x = (a_1, \cdots, a_k)$. In particular, each element of $\mathfrak{B}$ is one union of the a_i which it divides. Set up the correspondence $x \leftrightarrow S_x$ where S_x is the set of elements a_i divisible by x. This correspondence is clearly $1 - 1$ by corollary 1.1. Furthermore $S_{(x,y)} = S_x + S_y$ and $S_{[x,y]} = S_x \wedge S_y$. Hence $\mathfrak{B}$ is a Boolean algebra.

LEMMA 1.4. *Let a Birkhoff lattice $\mathfrak{S}$ have the property that any three elements covering any given element of $\mathfrak{S}$ generate a Boolean algebra of order eight. Then $a, b > [a, b]; q_1, q_2 \supset a; q_1, q_2 \not\supset b$ where q_1 and q_2 are irreducibles, imply $q_1 = q_2$.*

PROOF. Suppose that $q_1 \neq q_2$. If $[q_1, q_2] \neq a$, then $[q_1, q_2] \supset a_1 > a$. But then $b_1 = (a, b) > a$ by definition 2.1 and $q_1, q_2 \supset a_1$, $q_1, q_2 \not\supset b_1$. If $[q_1, q_2] \neq a_1$, then $[q_1, q_2] \supset a_2 > a_1$ and $b_2 = (b_1, a_1) > b_1$; $q_1, q_2 \supset a_2$; $q_1, q_2 \not\supset b_2$. Hence by the ascending chain condition we may assume that $q_1, q_2 \supset a; q_1, q_2 \not\supset b; [q_1, q_2] = a$. Let $q_1 \supset a_1 > a$, $q_2 \supset a_2 > a$. This is always possible since if

$q_1 = a$ then $q_2 \neq a$ and $q_2 \supset (a, b) \supset b$ which contradicts $q_2 \not\supset b$. Now $a_1 \neq (a, b)$ since otherwise $q_1 \supset b$. Similarly $a_2 \neq (a, b)$. Hence $(a, b), a_1, a_2$ are three distinct elements covering a and thus generate a Boolean algebra by hypothesis. Let $a_2' = (a_1, a_2)$, $a_3' = (a_1, (a, b))$. Then $a_2' > a_1$, $a_3' > a_1$ and a_2' and a_3' are distinct. Now $q_1 \not\supset a_3'$ since otherwise $q_1 \supset b$, and $q_1 \not\supset a_2'$ since otherwise $a = [q_1, q_2] \supset a_2$. Hence there exists an element a_1' distinct from a_2' and a_3' such that $q_1 \supset a_1' > a_1$. But then a_1, a_2, a_3 generate a Boolean algebra. Hence a_1' is reducible and as before there exist elements a_1'', a_2'', a_3'' such that $a_2'' > a_1'$, a_2', $a_3'' > a_1'$, a_3' and $q_1 \supset a_1'' > a_1'$. But then $a \subset a_1 \subset a_1' \subset a_1'' \subset \cdots$ is an infinite ascending chain which contradicts the ascending chain condition

LEMMA 1.5. *Let $\mathfrak{S}$ satisfy the hypothesis of lemma 1.4. Then any set of elements covering a given element a of $\mathfrak{S}$ is union independent.*

PROOF. Let us suppose that the lemma is true for any a and any $n - 1$ elements covering a. Let $a_1, \cdots, a_n > a$ and suppose that $(a_1, \cdots, a_{n-1}) \supset a_n$. Then $(a_1, a_2), \cdots, (a_1, a_n) > a_1$ and $(a_1, a_2), \cdots, (a_1, a_n)$ are distinct since any three elements covering a generate a Boolean algebra by assumption. But then $((a_1, a_2), \cdots, (a_1, a_{n-1})) \supset (a_1, a_n)$ which contradicts the induction assumption. Hence $a_1, \cdots, a_n$ are union independent.

LEMMA 1.6. *Let $\mathfrak{S}$ satisfy the conditions of lemma 1.4 and let $a_1, \cdots, a_n$ be the elements covering a. Then for each a_i there is one and only one irreducible q_i such that $q_i \not\supset a_i$, $q_i \supset (a_1, \cdots, a_{i-1}, a_{i+1}, \cdots, a_n)$.*

PROOF. Let $a_i' = (a_1, \cdots, a_{i-1}, a_{i+1}, \cdots, a_n)$. Then $a_i' \not\supset a_i$ by lemma 1.5. Hence there exists an irreducible q_i such that $q_i \not\supset a_i$, $q_i \supset a_i'$. Suppose there were a second irreducible q_i' such that $q_i' \not\supset a_i$, $q_i' \supset a_i'$. Then $a_j, a_i > [a_j, a_i]$ $j \neq i$, $q_i, q_i' \supset a_j$ and $q_i, q_i' \not\supset a_i$. Hence $q_i = q_i'$ by lemma 1.4.

We will designate the irreducibles of lemma 1.6 as the irreducibles *belonging* to a. We have then

LEMMA 1.7. *Let $\mathfrak{S}$ satisfy the conditions of lemma 1.4. Then each element of $\mathfrak{S}$ is uniquely expressible as a reduced cross-cut of irreducibles. These irreducibles are simply the irreducibles belonging to the element.*

PROOF. If a is irreducible the lemma is trivial. Let a be reducible and let $a = [p_1, \cdots, p_k]$ be a reduced decomposition of a into irreducible components. Now if $p_i \not\supset (a_1, \cdots, a_n)$, then $p_i \not\supset a_j$ for some j. Also $p_i \supset a_k$ for some k since $p_i \supset a$. Now let q_j be the irreducible belonging to a_j according to lemma 1.6. Then $q_j \not\supset a_j$ and $q_j \supset a_k$. Hence $p_i = q_j$ by lemma 1.4. Thus either $p_i \supset (a_1, \cdots, a_n)$ or $p_i = q_j$ for some j. Furthermore for each q_l there is a p_i such that $q_l = p_i$ since otherwise $a = [p_1, \cdots, p_k] \supset a_l$ which is impossible. Hence if $p_i \supset (a_1, \cdots, a_n)$, p_i is superfluous in the decomposition of a which is contrary to assumption. Thus with suitable numbering $p_i = q_i$.

Lemmas 1.2, 1.3, and 1.7 together give theorem 1.1.

The following corollaries are a consequence of theorem 1.1.:

COROLLARY 1.3. *The number of irreducible components of an element a of a lattice having unique irreducible decompositions is equal to the number of distinct elements covering a.*

COROLLARY 1.4. *Let both the union and cross-cut decomposition of a lattice $\mathfrak{S}$ be unique. Then $\mathfrak{S}$ is distributive.*

For both $\mathfrak{S}$ and its dual satisfy the Birkhoff condition. Hence $\mathfrak{S}$ is modular and thus distributive by theorem 1.1.

2. This section will be devoted to some applications of theorem 1.1. We first give a new characterization of finite Boolean algebras.

THEOREM 2.1. *A complemented lattice of finite dimensions is a Boolean algebra if and only if every element is uniquely expressible as a reduced cross-cut of irreducibles.*

PROOF. Let u be the union of the points of $\mathfrak{S}$. If $u' \neq z$, then $u' \supset p$ and $[u, u'] \supset p$ which is impossible. Hence $u' = z$ and $u = i$. Thus the Boolean algebra generated by the points of $\mathfrak{S}$ according to lemma 1.3 is identical with $\mathfrak{S}$.

In a paper on the algebra of lattice functions, Morgan Ward has proved that a modular, non-distributive lattice satisfying the ascending chain condition always contains a complete,[6] modular, non-distributive sublattice of order five. Theorem 1.1 gives a generalization of this result.

THEOREM 2.2. *Let $\mathfrak{S}$ be a Birkhoff lattice. Then if $\mathfrak{S}$ contains a modular non-distributive sublattice it also contains a complete, modular non-distributive sublattice of order five.*

PROOF. Since $\mathfrak{S}$ contains a modular, non-distributive sublattice, not every element of $\mathfrak{S}$ is uniquely expressible as a reduced cross-cut of irreducibles by theorem 1.1. Hence by lemma 1.7 there exist three elements covering an element a of $\mathfrak{S}$ which do not generate a Boolean algebra. These three elements must then generate a complete, modular, non-distributive sublattice of order five.

In lemma 1.3 we found that any set of elements covering a given element of $\mathfrak{S}$ generate a Boolean algebra if $\mathfrak{S}$ has unique irreducible decompositions. This result may be generalized as follows:

THEOREM 2.3. *Let $\mathfrak{S}$ be a Birkhoff lattice. Then the sublattice generated by any union independent set of elements covering $a \ \epsilon \ \mathfrak{S}$ is a Boolean algebra.*

PROOF. Let $a_1, \cdots, a_n$ be a union independent set of elements covering a. Let U be the set of elements of $\mathfrak{S}$ which can be expressed as a union of the a_i. U is obviously closed with respect to union. Let now $a = (a_1, \cdots, a_k, a_{k+1},$ $\cdots, a_l)$, $b = (a_1, \cdots, a_k, a'_{k+1}, \cdots, a'_m)$ where $a_1, \cdots, a_l, a'_{k+1}, \cdots, a'_m$ are distinct. Now since $a_1, \cdots, a_n$ are independent, $r(a_1, \cdots, a_k) = k$ where $r(x) = \rho(a) - \rho(x)$. Hence $r[a, b] \leq ra + rb - r(a, b) = k + m - (k + m - l)$ $= l$. But $[a, b] \supset (a_1, \cdots, a_l)$. Hence $r[a, b] \geq l$. Thus $r[a, b] = l$ and $[a, b] = (a_1, \cdots, a_l)$. Hence U is also closed with respect to cross-cut and is the sublattice generated by $a_1, \cdots, a_n$. If we set up the correspondence $x \leftrightarrow S_x$ where $x \ \epsilon \ U$ and S_x is the set of elements a_i divisible by x, then the correspondence is $1 - 1$ and $S_{(x,y)} = S_x + S_y$, $S_{[x,y]} = S_x \wedge S_y$.

If $\mathfrak{S}$ is modular it can be shown that the Boolean algebra is dense in $\mathfrak{S}$. How-

---

[6] A sublattice $\mathfrak{S}'$ of $\mathfrak{S}$ is said to be *complete* if $a > b$ in $\mathfrak{S}'$ implies $a > b$ in $\mathfrak{S}$.

ever for Birkhoff lattices the Boolean algebra is in general not dense in $\mathfrak{S}$ as is easily shown by examples.

For lattices having unique irreducible decompositions the direct product decomposition may be characterized as follows:

THEOREM 2.4. *Let $\mathfrak{S}$ be a lattice satisfying the ascending chain condition in which every element is uniquely expressible as a reduced cross-cut of irreducibles. Then $\mathfrak{S}$ is the direct product of sublattices $\mathfrak{S}_1$ and $\mathfrak{S}_2$ if and only if the irreducibles of $\mathfrak{S}$ can be separated into two disjoint subsets A and B such that the set sum of any cross-cut independent set of A and any cross-cut independent set of B is again cross-cut independent.*

PROOF. Let A and B be sets of irreducibles having the property of the theorem and let $\mathfrak{S}_1$ and $\mathfrak{S}_2$ be the sublattices generated by A and B respectively. Then $x \in \mathfrak{S}$ has the representation $x = [p_1, \cdots, p_r, q_1, \cdots, q_s]$ where $p_1, \cdots, p_r \in A$ and $q_1, \cdots, q_s \in B$ and $p_1, \cdots, q_s$ are cross-cut independent. Hence $x = [a, b]$ where $a \in \mathfrak{S}_1$, $b \in \mathfrak{S}_2$. Let $x = [a', b']$ where $a' \in \mathfrak{S}_1$ and $b' \in \mathfrak{S}_2$. Let $a' = [p_1', \cdots, p_t']$ $b' = [q_1', \cdots, q_u']$ be the reduced representations of a' and b'. Then $p_i' \in A$ and $q_i' \in B$ and hence $p_1', \cdots, p_t', q_1', \cdots, q_u'$ are independent by assumption. Hence $[p_1', \cdots, q_u']$ is a reduced decomposition of $[a, b]$. Thus with suitable numbering $p_i = p_i'$, $q_j = q_j'$ and $a = a'$, $b = b'$. If $x = [a, b]$, $Y = [a_1, b_1]$, then clearly $[X, Y] = [[a, a_1], [b, b_1]]$. Also $(X, Y) = [(a, a_1), (b, b_1)]$. Hence $\mathfrak{S}$ is the direct product of $\mathfrak{S}_1$ and $\mathfrak{S}_2$.

On the other hand let $\mathfrak{S} = \mathfrak{S}_1 \times \mathfrak{S}_2$. Then each irreducible belongs either to $\mathfrak{S}_1$ or $\mathfrak{S}_2$. Let $p_1, \cdots, p_r$ be a cross-cut independent set of irreducibles of $\mathfrak{S}_1$ and $q_1, \cdots, q_s$ be a cross-cut independent set of irreducibles of $\mathfrak{S}_2$. Suppose that $p_1 \supset [p_2, \cdots, p_r, q_1, \cdots, q_s]$. Then $p_1 \supset [p_2, \cdots, p_r]$ contrary to the assumption that $p_1, \cdots, p_r$ are cross-cut independent. Hence $p_1, \cdots, p_r$, $q_1, \cdots, q_s$ are independent.

If $\mathfrak{S}$ has unique irreducible decompositions then $a \supset b$ does not necessarily imply that each component of a divides some component of b. We show that this holds if and only if $\mathfrak{S}$ is distributive.

THEOREM 2.5. *Let $\mathfrak{S}$ have unique irreducible decompositions. Then $\mathfrak{S}$ is distributive if and only if $a \supset b$ implies that each component of a divides some component of b.*

PROOF. Let $a \supset b$ imply that each component of a divides some component of b. Let $q \supset [a, b]$ where q is irreducible. Then $q \supset [p_1, \cdots, p_r, q_1, \cdots, q_s]$ where $p_1, \cdots, p_r \supset a$; $q_1, \cdots, q_s \supset b$ and $[p_1, \cdots, q_s]$ is the irreducible decomposition of $[a, b]$. Then by hypothesis $q \supset p_i$ or $q \supset q_j$ and hence $q \supset a$ or $q \supset b$. Thus every irreducible is a prime. Let $([a, b], [a, c]) = [p_1, \cdots, p_r]$. Then $p_i \supset [a, b]$, $p_i \supset [a, c]$. If $p_i \not\supset a$, then $p_i \supset b$, $p_i \supset c$. Hence either $p_i \supset a$ or $p_i \supset (b, c)$. Thus in either case $p_i \supset [a, (b, c)]$. Therefore $([a, b], [a, c]) \supset (a, [b, c])$. Hence $([a, b] [a, c]) = (a, [b, c])$ and $\mathfrak{S}$ is distributive.

Conversely if $\mathfrak{S}$ is distributive, then every irreducible is a prime (Ward [1]) and $a \supset b$ implies $p_i \supset [q_1, \cdots, q_s]$ implies $p_i \supset q_j$ for some j.

Theorem 1.1 gives finally the following result on the sublattice of a lattice having unique irreducible decompositions.

THEOREM 2.6. *Let $\mathfrak{S}$ have unique irreducible decompositions. Then a sublattice $\mathfrak{S}'$ of $\mathfrak{S}$ has unique irreducible decompositions if and only if it is a Birkhoff lattice.*

CALIFORNIA INSTITUTE OF TECHNOLOGY.

REFERENCES

G. Birkhoff   1. Duke Math. Journal, Vol. 3 (1933), pp. 443–454.
              2. Camb. Phil. Proc., Vol. 29 (1933), pp. 441–464.
F. Klein   1. Math. Zeit., Vol. 42 (1936), pp. 58–81.
S. MacLane   1. Amer. J. of Math., Vol. 58 (1936), pp. 236–240.
O. Ore   1. These Annals., Vol. 36 (1935), pp. 406–437.
M. Ward   1. These Annals., Vol. 39 (1938), pp. 558–568.
H. Whitney   1. Amer. J. of Math., Vol. 57 (1935), pp. 509–533.

Reprinted from Duke Mathematical Journal
Vol. 8, No. 2, June, 1941

THE ARITHMETICAL THEORY OF BIRKHOFF LATTICES

By R. P. Dilworth

1. **Introduction and summary.** In the development of lattice theory considerable work has been devoted to the study of the arithmetical properties of modular and distributive lattices. Indeed most of the decomposition theorems of abstract algebra have been extended to these more general domains. Nevertheless, there are lattices with very simple arithmetical properties which come under neither of these classifications. For example, the lattices with unique irreducible decompositions, which were studied by the author in a previous paper [3][1] satisfy the Birkhoff condition[2] which is even less restrictive than the modular axiom. Furthermore, there are important algebraic systems which give rise to non-modular, Birkhoff lattices. Thus, since every exchange lattice (Mac Lane [4]) is a Birkhoff lattice, the systems which satisfy Mac Lane's exchange axiom form lattices of the type in question. In this paper we shall study the arithmetical structure of general Birkhoff lattices and in particular determine necessary and sufficient conditions that certain important arithmetical properties hold.

In §§2–4 we characterize the irreducible decompositions in terms of the structure of the lattice and apply the results to determine necessary and sufficient conditions that the number of irreducible components be unique for each element of the lattice. The main result is the following:

Let $\mathfrak{S}$ be a Birkhoff lattice in which every quotient lattice is Archimedean. Then the number of irreducible components is unique for each element a of $\mathfrak{S}$ if and only if the sublattice generated by the elements covering a is a dense, modular sublattice of $\mathfrak{S}$.

§5 contains some methods for constructing Birkhoff lattices in which given arithmetical conditions hold. In §6 we treat the problem of determining the conditions that a set of irreducible components of an element must satisfy in order that it may be extended into a reduced representation. This problem is given a particularly simple solution in the case of a Birkhoff lattice in which the number of components is unique.

2. **Notation and definitions.** Throughout the paper $\mathfrak{S}$ will denote a lattice of elements a, b, c, $\cdots$ and unit element u in which every quotient lattice is Archimedean (Ore [5]). Union, cross-cut, and lattice division will be denoted by (,), [,], and $\supset$ respectively. a is said to *cover* b (in symbols $a > b$) if $a \supset b$, $a \neq b$ and $a \supset x \supset b$ implies $a = x$ or $x = b$. Elements covered by

Received October 9, 1940. The author is a Sterling Research Fellow at Yale University.

[1] Numbers in brackets refer to the references at the end of the paper.

[2] See Definition 2.1.

286

the unit element u are said to be *simple*. If $\mathfrak{S}$ has a null element z, the elements which cover z will be called *points*. $\mathfrak{S}$ is said to be *atomic* if every element is a cross-cut of simple elements. Similarly, if every element is a union of points, $\mathfrak{S}$ is said to be a *point* lattice.

An element x is said to be *cross-cut irreducible* (or simply *irreducible*) if $x = [x_1 , x_2]$ implies $x = x_1$ or $x = x_2$. x is said to be *union irreducible* if $x = (x_1 , x_2)$ implies $x = x_1$ or $x = x_2$. A set of elements $x_1 , \cdots , x_n$ is said to be *cross-cut independent* if $x_i \not\supset [x_1 , \cdots , x_{i-1} , x_{i+1} , \cdots , x_n]$ $(i = 1, \cdots , n)$. Similarly, $x_1 , \cdots , x_n$ are *union independent* if $(x_1 , \cdots , x_{i-1} , x_{i+1} , \cdots , x_n) \not\supset x_i$ $(i = 1, \cdots , n)$.

Since the ascending chain condition holds in $\mathfrak{S}$, each element a of $\mathfrak{S}$ may be expressed as a cross-cut of irreducibles $a = [q_1 , \cdots , q_k]$. If $q_1 , \cdots , q_k$ are cross-cut independent, the representation is said to be *reduced*. Clearly any representation can be reduced by dropping out suitable members. If an irreducible q occurs in a reduced decomposition of a, q is called a *component* of a.

Let $\mathfrak{A}_1 , \cdots , \mathfrak{A}_n$ be sublattices of $\mathfrak{S}$ with common null element a. The sublattice generated by $\mathfrak{A}_1 , \cdots , \mathfrak{A}_n$ is the *direct sum* of $\mathfrak{A}_1 , \cdots , \mathfrak{A}_n$ $(\mathfrak{A} = \mathfrak{A}_1 + \mathfrak{A}_2 + \cdots + \mathfrak{A}_n)$ if $x \,\epsilon\, \mathfrak{A}$ implies $x = (x_1 , \cdots , x_n)$, where $x_i \,\epsilon\, \mathfrak{A}_i$, $(x, y) = ((x_1 , y_1) , \cdots , (x_n , y_n))$, $[x, y] = ([x_1 , y_1], \cdots , [x_n , y_n])$ and $x \supset y$ if and only if $x_i \supset y_i$ $(i = 1, \cdots , n)$. If $\mathfrak{A}_2$ is the quotient lattice p/a where $p > a$, we write $\mathfrak{A} = \mathfrak{A}_1 + \mathfrak{A}_2 = \mathfrak{A}_1 + p$ and say that $\mathfrak{A}_1 + p$ is the direct sum of $\mathfrak{A}_1$ and the point p.

DEFINITION 2.1. A lattice $\mathfrak{S}$ is said to be a *Birkhoff*[3] lattice of type I (or simply *Birkhoff* lattice) if $a > [a, b]$ implies $(a, b) > b$. $\mathfrak{S}$ is said to be a *Birkhoff* lattice of type II if $(a, b) > b$ implies $a > [a, b]$.

If $\mathfrak{S}$ is a Birkhoff lattice of type I, a *rank* function ρa may be defined over $\mathfrak{S}$ with the properties (i) $\rho u = 0$, (ii) $\rho a = \rho b + 1$ if $b > a$, (iii) $\rho(a, b) + \rho[a, b] \geq \rho a + \rho b$. If $\mathfrak{S}$ is a Birkhoff lattice of type II, then in place of (iii) ρ satisfies (iii)$'$ $\rho(a, b) + \rho[a, b] \leq \rho a + \rho b$.

The following lemma proved by G. Birkhoff in [1] relates the Birkhoff conditions to modularity.

LEMMA 2.1. $\mathfrak{S}$ *is modular if and only if it is a Birkhoff lattice of type* I *and type* II.

3. Structure characterization of reduced decompositions.

We begin by proving a basic lemma from which most of the properties of Birkhoff lattices may be derived.

FUNDAMENTAL LEMMA. *Let $\mathfrak{S}$ be a Birkhoff lattice. Let $\mathfrak{A}$ be a complete*[4] *modular sublattice of $\mathfrak{S}$ with unit element v and null element a. Then if $p > a$ and $v \not\supset p$, the sublattice generated by $\mathfrak{A}$ and p is the direct sum $\mathfrak{A} + p$. $\mathfrak{A} + p$ is also complete in $\mathfrak{S}$.*

---

[3] G. Birkhoff (*Lattice Theory*, Amer. Math. Soc. Colloquium Publications, vol. 25, New York, 1940) uses the terms *upper* and *lower semi-modular* lattices for Birkhoff lattices of types I and II. The original terminology of Klein has been used here.

[4] A sublattice $\mathfrak{A}$ of $\mathfrak{S}$ is said to be complete in $\mathfrak{S}$ if $a > b$ in $\mathfrak{A}$ implies $a > b$ in $\mathfrak{S}$.

Proof. Let a_1 and a_2 be elements of $\mathfrak{A}$. If $(a_1, p) = (a_2, p)$, then $(a_1, p) \supset (a_1, a_2) \supset a_1$. Since $(a_1, p) > a_1$, either $(a_1, p) = (a_1, a_2)$ or $a_1 \supset a_2$. But if $(a_1, p) = (a_1, a_2)$, then $v \supset p$, and this is contrary to assumption. Hence $a_1 \supset a_2$ and similarly $a_2 \supset a_1$, $a_1 = a_2$. Thus $(a_1, x_1) = (a_2, x_2)$, $a_1, a_2 \,\epsilon\, \mathfrak{A}$, $x_1, x_2 \,\epsilon\, p/a$ implies $a_1 = a_2$, $x_1 = x_2$.

Clearly $((a_1, x_1), (a_2, x_2)) = ((a_1, a_2), (x_1, x_2))$.

Now $\rho[(a_1, p), (a_2, p)] \geqq \rho(a_1, p) + \rho(a_2, p) - \rho(a_1, a_2, p) = \rho a_1 + \rho a_2 - 2 - \rho(a_1, a_2) + 1 = \rho[a_1, a_2] - 1 = \rho([a_1, a_2], p)$ since $\mathfrak{A}$ is modular. Hence $\rho[(a_1, p), (a_2, p)] \geqq \rho([a_1, a_2], p)$. Since $[(a_1, p), (a_2, p)] \supset ([a_1, a_2], p)$, it follows that $[(a_1, p), (a_2, p)] = ([a_1, a_2], p)$. Also $([a_1, a_2], p) \supset [(a_1, p), a_2] \supset [a_1, a_2]$ and $a_2 \not\supset p$ imply $[(a_1, p), a_2] = [a_1, a_2]$. Hence $[(a_1, x_1), (a_2, x_2)] = ([a_1, a_2], [x_1, x_2])$, and the lemma is proved.

By repeated application of the Fundamental Lemma we have the following lemmas:

LEMMA 3.1. *Let $\mathfrak{S}$ be a Birkhoff lattice and let $a_1, \cdots, a_n > a$. Then each union independent set of the a_i is contained in a maximal union independent set.*

LEMMA 3.2. *Let $\mathfrak{S}$ be a Birkhoff lattice and let $a_1, \cdots, a_n > a$. Then every union independent set of the a_i generates a Boolean algebra.*

LEMMA 3.3. *Let $\mathfrak{S}$ be a Birkhoff lattice and let $a_1, \cdots, a_n > a$. Then any two maximal independent sets of the a_i have the same number of elements and any element of one set may be replaced by a suitably chosen element of the other.*

For if $a_1, \cdots, a_k$ and $a_1', \cdots, a_l'$ are the two maximal independent sets of the a_i, we may replace a_i by an a_j' such that $(a_1, \cdots, a_{i-1}, a_{i+1}, \cdots, a_n) \not\supset a_j'$.

LEMMA 3.4. *Let $\mathfrak{S}$ be a Birkhoff lattice and let $a > s_1, \cdots, s_k$ where $s_1, \cdots, s_k$ are cross-cut independent and $\rho[s_1, \cdots, s_k] - \rho(a) = k$. Then $s_1, \cdots, s_k$ generate a Boolean algebra of length k.*

For if $p_i = [s_1, \cdots, s_{i-1}, s_{i+1}, \cdots, s_k]$, then $p_i > [s_1, \cdots, s_k]$ and $p_1, \cdots, p_k$ are independent. Thus $s_i = (p_1, \cdots, p_{i-1}, p_{i+1}, \cdots, p_k)$ and hence the sublattice generated by $s_1, \cdots, s_k$ is simply the Boolean algebra generated by $p_1, \cdots, p_k$.

If a is an element of $\mathfrak{S}$, let u_a denote the union of the elements which cover a. We associate with each element a the quotient lattice $\mathfrak{S}_a$ of all elements x such that $u_a \supset x \supset a$.

DEFINITION 3.1. An element $c \neq u_a$ of $\mathfrak{S}_a$ is said to be *characteristic* if there exists an irreducible q such that $q \supset c$ and q divides exactly the same points of $\mathfrak{S}_a$ as c.

Each irreducible component of a divides at least one characteristic element c. For the union of the points of $\mathfrak{S}_a$ divisible by q clearly has the properties of Definition 3.1.

We show now that each reduced decomposition of a into irreducibles gives at least one representation of a as a reduced cross-cut of characteristic elements of $\mathfrak{S}_a$ and conversely.

THEOREM 3.1. *An element a of $\mathfrak{S}_a$ has a reduced decomposition $a = [q_1, \cdots, q_k]$ where $q_1, \cdots, q_k$ are irreducible, if and only if a has a reduced representation $a = [c_1, \cdots, c_k]$, where $c_1, \cdots, c_k$ are characteristic elements of $\mathfrak{S}_a$ such that $q_i \supset c_i$.*

Proof. Let $a = [q_1, \cdots, q_k]$ be an irreducible decomposition of a and let $c_1, \cdots, c_k$ be a set of associated characteristic elements. Then $a = [q_1, \cdots, q_k] \supset [c_1, \cdots, c_k] \supset a$. Hence $a = [c_1, \cdots, c_k]$. Now since $q_1, \cdots, q_k$ are cross-cut independent, $[q_1, \cdots, q_{i-1}, q_{i+1}, \cdots, q_k] \supset p_i$, where $p_i > a$. But then $[c_1, \cdots, c_{i-1}, c_{i+1}, \cdots, c_k] \supset p_i$ by Definition 3.1. Hence $c_i \not\supset [c_1, \cdots, c_{i-1}, c_{i+1}, \cdots, c_k]$ and the representation $a = [c_1, \cdots, c_k]$ is reduced.

Conversely, let $a = [c_1, \cdots, c_k]$ be a reduced representation of a in terms of characteristic elements and $q_1, \cdots, q_k$ be a set of associated irreducibles. If $[q_1, \cdots, q_k] \neq a$, then $[q_1, \cdots, q_k] \supset p > a$. Since $q_i \supset p$ implies $c_i \supset p$, we have $a = [c_1, \cdots, c_k] \supset p > a$, and this is impossible. Hence $a = [q_1, \cdots, q_k]$. If $q_i \supset [q_1, \cdots, q_{i-1}, q_{i+1}, \cdots, q_k]$, then $a = [q_1, \cdots, q_{i-1}, q_{i+1}, \cdots, q_k] \supset [c_1, \cdots, c_{i-1}, c_{i+1}, \cdots, c_k] \supset a$. This contradicts the assumption that $c_1, \cdots, c_k$ are cross-cut independent.

It will be noted that Theorem 3.1 is independent of the Birkhoff condition.

In view of Theorem 3.1 the structure characterization of the irreducible decompositions will be accomplished if we determine the characteristic elements of $\mathfrak{S}_a$ in terms of the lattice structure. We need the following lemma:

LEMMA 3.5. *$\mathfrak{S}_a$ is a complemented, atomic lattice.*

Proof. Let $b \in \mathfrak{S}_a$ and let $p_1, \cdots, p_k$ be a maximal union independent set of elements covering a and divisible by b. Imbed $p_1, \cdots, p_k$ in a maximal union independent set of points $p_1, \cdots, p_n$ (Lemma 3.1). Let $b' = (p_{k+1}, \cdots, p_n)$. Then clearly $(b, b') = u_a$. Suppose that $[b, b'] \neq a$. Then $[b, b'] \supset p > a$. But since $b \supset p$, we have $(p_1, \cdots, p_k) \supset p$ and $a = [(p_1, \cdots, p_k), (p_{k+1}, \cdots, p_n)] \supset p$ (Lemma 3.2), and this is impossible. Hence $[b, b'] = a$ and thus $\mathfrak{S}_a$ is complemented. Let now b be an irreducible element of $\mathfrak{S}_a$. Then $(b, p_{k+1}), \cdots, (b, p_n) > b$ and hence $(b, p_{k+1}) = \cdots = (b, p_n)$. But then $u_a = (b', b) > b$, and thus the only irreducible elements of $\mathfrak{S}_a$ are the simple elements.

If $b \in \mathfrak{S}_a$, an arbitrary complement of b in $\mathfrak{S}_a$ will be denoted by b'.

THEOREM 3.2. *Let $\mathfrak{S}$ be a Birkhoff lattice. Then an element $c \in \mathfrak{S}_a$ is characteristic if and only if there exists a divisor x of c such that $(c', x) > x$ and $[c', x] = a$ for every c'.*

Proof. Let us first assume that such an element x exists. Then $(u_a, x) = ((c, c'), x) = (c', x)$. Let q be an irreducible such that $q \supset x$, $q \not\supset (u_a, x)$. Since $q \supset x \supset c$, q divides every point of $\mathfrak{S}_a$ which c divides. Now let $q \supset p$. Then $x \supset p$ since if $x \not\supset p$ we would have $(c', x) \supset (x, p) \supset x$ and $(x, p) \neq x$. Hence $(c', x) = (x, p)$ and $q \supset (x, p) = (c', x)$, and this contradicts the definition of q. Now if $c \not\supset p$, then $c' \supset p$ for some c'. But then $a = [c', x] \supset p$, and this is impossible. Hence $q \supset p$ implies $c \supset p$ and c is thus characteristic.

On the other hand, let c be characteristic and let q be an irreducible associated with c. Then $(q, c') > q$ for every c'. For there is a point p such that $c' \supset p$, $c \not\supset p$ since otherwise we would have $c' = a$ and $c = u_a$. But then $(q, c') = (q, c, c') = (q, u_a) = (q, p) > q$. Now if $[c', q] \neq a$, then $[c', q] \supset p > a$, and hence $c' \supset p$, $q \supset p$. But then $c \supset p$ and hence $a = [c, c'] \supset p$; this is impossible. Thus $[c', q] = a$ for every c'.

COROLLARY 3.2. *Each simple element of $\mathfrak{S}_a$ is characteristic.*

For we may take x to be the element c itself.

COROLLARY 3.3. *If k is the maximal number of union independent elements covering a, then a has a reduced decomposition into irreducibles with k components.*

For by Theorem 3.1 and Corollary 3.2 a has a reduced representation as a cross-cut of k characteristic elements of $\mathfrak{S}_a$.

If $\mathfrak{S}$ is modular, then $\mathfrak{S}$ is a Birkhoff lattice of type II by Lemma 2.1 and hence by Theorem 3.2 we must have $c' > a$. But then $u_a > c$, and every characteristic element of $\mathfrak{S}_a$ is simple in $\mathfrak{S}_a$. Hence by the dual of Lemma 3.3 the number of components in any two irreducible decompositions is the same and each component of one decomposition may be replaced by a suitably chosen component of the other. Thus for modular lattices in which every quotient lattice is Archimedean we see that the Kurosch-Ore decomposition theorem rests on a familiar exchange property of independent bases.

4. Unicity of the number of components. In order to investigate the structure of Birkhoff lattices in which the number of components is unique we prove first a lemma which sharpens Corollary 3.3.

LEMMA 4.1. *Let $\mathfrak{S}$ be a Birkhoff lattice. Then a reduced representation $a = [q_1, \cdots, q_k]$ has the maximal number of components if and only if the characteristic elements belonging to the q_i are the simple elements of the Boolean algebra generated by a maximal union independent set of points of $\mathfrak{S}_a$.*

Proof. Clearly each such set of characteristic elements gives a reduced representation of a by Lemma 3.2 and Theorem 3.1. Now if $a = [q_1, \cdots, q_k]$ is a reduced representation of a, let $q_i' = [q_1, \cdots, q_{i-1}, q_{i+1}, \cdots, q_k]$. Then since $q_i' \neq a$, we must have $q_i' \supset p_i$ where $p_i > a$. $p_1, \cdots, p_k$ are union independent since if $(p_1, \cdots, p_{i-1}, p_{i+1}, \cdots, p_k) \supset p_i$, then $q_i \supset (q_1', \cdots, q_{i-1}', q_{i+1}', \cdots, q_k') \supset p_i$. Hence $a = [q_i, q_i'] \supset p_i$, and this contradicts $p_i > a$. Let $b_i = (p_1, \cdots, p_{i-1}, p_{i+1}, \cdots, p_k)$. Then $q_i \supset b_i$. Since $p_1, \cdots, p_k$ are union independent, $k \leq n$ where n is the maximal number of union independent points of $\mathfrak{S}_a$. But if the representation $a = [q_1, \cdots, q_k]$ has the maximal number of components, $k = n$ and $p_1, \cdots, p_k$ are a maximal union independent set of points of $\mathfrak{S}_a$ and hence generate a Boolean algebra with simple elements $b_1, \cdots, b_k$. $b_1, \cdots, b_k$ are clearly simple in $\mathfrak{S}_a$. Now let $c_1, \cdots, c_k$ be characteristic elements associated with $q_1, \cdots, q_k$. Then since $q_i \supset b_i$ we have

$u_a \supset c_i \supset b_i$ by Definition 3.1. But $u_a > b_i$ and $u_a \neq c_i$. Hence $c_i = b_i$ and the lemma is proved.

COROLLARY 4.1. *The maximal number of components in the reduced representations of a as a cross-cut of irreducibles is equal to the length of $\mathfrak{S}_a$.*

THEOREM 4.1. *Let $\mathfrak{S}$ be a Birkhoff lattice. Then the number of components in the reduced representations of a as a cross-cut of irreducibles is unique if and only if $\mathfrak{S}_a$ is modular and the characteristic elements are simple.*

Proof. If $\mathfrak{S}_a$ is modular and the only characteristic elements of $\mathfrak{S}_a$ are the simple elements, then since $\mathfrak{S}_a$ is a Birkhoff lattice of type II the number of elements in the representations of a as a cross-cut of characteristic elements is unique by the dual of Lemma 3.3. Hence the number of components in the irreducible decomposition of a is unique by Theorem 3.1.

On the other hand, if the number of components in the irreducible decompositions of a is unique, then each cross-cut independent set of simple elements of $\mathfrak{S}_a$ generates a Boolean algebra whose length is equal to the number of elements in the set. For if $u_a > s_1, \cdots, s_k$ and $s_1, \cdots, s_k$ do not generate a Boolean algebra, then $\rho[s_1, \cdots, s_k] - \rho u_a = l > k$ by Lemma 3.4. But then by adding $n - l$ simple elements $s_{l+1}, \cdots, s_n$ we have $a = [s_1, \cdots, s_k, s_{l+1}, \cdots, s_n]$, where n is the length of $\mathfrak{S}_a$. This, however, contradicts Lemma 4.1. Now let $b \in \mathfrak{S}_a$. Then $b = [s_1, \cdots, s_k]$ (Lemma 3.5), where $s_1, \cdots, s_k$ generate a Boolean algebra. If $s \not\supset b$, then $s, s_1, \cdots, s_k$ must generate a Boolean algebra of length $k + 1$ and hence $b > [s, b]$. Let $(x, y) > y$ where $x, y \in \mathfrak{S}_a$. Then by Lemma 3.5 there is a simple element s of $\mathfrak{S}_a$ such that $s \not\supset (x, y)$, $s \supset y$. Hence $y = [(x, y), s]$. Since $s \not\supset x$ we have $x > [x, s] = [x, (x, y), s] = [x, y]$. Thus $\mathfrak{S}_a$ is a Birkhoff lattice of type II and hence is modular by Lemma 2.1. The characteristic elements of $\mathfrak{S}_a$ are simple by Lemma 4.1.

To complete the proof of the theorem mentioned in the introduction we must prove first a lemma on modular sublattices of a lattice. Let $\mathfrak{A}$ and $\mathfrak{B}$ be quotient lattices of $\mathfrak{S}$ with unit elements a_1, b_1 and null elements a_2, b_2 respectively. Then if $a_1 \in \mathfrak{B}$ and $b_2 \in \mathfrak{A}$ we say that $\mathfrak{B}$ is an *extension* of $\mathfrak{A}$.

LEMMA 4.2. *Let $\mathfrak{A}_1, \cdots, \mathfrak{A}_k$ be quotient lattices of $\mathfrak{S}$ such that $\mathfrak{A}_{i+1}$ is an extension of $\mathfrak{A}_i$. Then the set sum $\mathfrak{S}_k$ of the lattice $\mathfrak{A}_1, \cdots, \mathfrak{A}_k$ is a sublattice of $\mathfrak{S}$ and is modular if and only if $\mathfrak{A}_1, \cdots, \mathfrak{A}_k$ are modular.*

Proof. The unit and null elements of $\mathfrak{A}_i$ will be denoted by a_i and b_i respectively. Let a and b be elements in the set sum $\mathfrak{S}_k$. Then $a \in \mathfrak{A}_i$ and $b \in \mathfrak{A}_j$ where we may assume $i \leq j$. Then $(a, b) = (a, b, b_j)$ since $b \supset b_j$. But then $a_j \supset (a, b, b_j) \supset b_j$ and hence $(a, b) \in \mathfrak{A}_j$. Similarly $[a, b] \in \mathfrak{A}_i$. Hence $\mathfrak{S}_k$ is a sublattice of $\mathfrak{S}$.

Now let $\mathfrak{A}_1, \cdots, \mathfrak{A}_k$ be modular and suppose that it has been shown that $\mathfrak{S}_{k-1}$ is modular. We prove then that $\mathfrak{S}_k$ is modular and the lemma follows by induction. Let a, b, c be any three elements of $\mathfrak{S}_k$ such that $a \supset b$. We have four non-trivial cases to consider.

(1) $a, c \in \mathfrak{A}_k$, $b \in \mathfrak{S}_{k-1}$. Then $(b, [a, c]) = (b, b_k, [a, c]) = ((b, b_k), [a, c]) = [a, (b, b_k, c)] = [a, (b, c)]$ since $\mathfrak{A}_k$ is modular.

(2) $a, b \in \mathfrak{A}_k$, $c \in \mathfrak{S}_{k-1}$. Then $[a, (b, c)] = [a, (b, b_k, c)] = [a, (b, (b_k, c))] = (b, [a, (b_k, c)]) = (b, [[a, a_{k-1}], (b_k, c)]) = (b, b_k, [a, a_{k-1}, c]) = (b, [a, c])$ since $\mathfrak{A}_k$ and $\mathfrak{S}_{k-1}$ are modular.

(3) $a \in \mathfrak{A}_k$, $b, c \in \mathfrak{S}_{k-1}$. Then $(b, [a, c]) = (b, [a, [a_{k-1}, c]]) = [a, a_{k-1}, (b, c)] = [a, (b, c)]$ since $\mathfrak{S}_{k-1}$ is modular.

(4) $c \in \mathfrak{A}_k$, $a, b \in \mathfrak{A}_{k-1}$. Then $(b, [a, c]) = (b, [a, [a_{k-1}, c]]) = [a, (b, [a_{k-1}, c])] = [a, (b, b_k, [a_{k-1}, c])] = [a, a_{k-1}, (b, b_{k-1}, c)] = [a, (b, c)]$ since $\mathfrak{A}_k$ and $\mathfrak{S}_{k-1}$ are modular.

Hence in any case $[a, (b, c)] = (b, [a, c])$ and $\mathfrak{S}_k$ is thus modular.

Theorem 4.2. *Let $\mathfrak{S}$ be a Birkhoff lattice. Then if $\mathfrak{S}_a$ contains a characteristic element which is not simple, there is an element $b \supset a$ such that $\mathfrak{S}_b$ is non-modular.*

Proof. Let c be a characteristic element of $\mathfrak{S}_a$ which is not simple. Then by Theorem 3.2 there is an element x and a complement c' such that $(c', x) > x$, $c' \not> a = [c', x]$. Let a_1 be the union of the points of $\mathfrak{S}_a$ which are divisible by $[u_a, x]$. Now if a_i has been defined, let a_{i+1} be the union of the points of $\mathfrak{S}_{a_i}$ which are divisible by $[u_{a_i}, x]$. Clearly $a_{i+1} \supset a_i$. Also a_{i+1} is distinct from a_i if $a_i \neq x$. Hence by the ascending chain condition $x = a_k$ for some k. Furthermore $a_{i+1} \in \mathfrak{S}_{a_i}$ since a_{i+1} is a union of points of $\mathfrak{S}_{a_i}$ and $u_{a_i} \in \mathfrak{S}_{a_{i+1}}$ since u_{a_i} is a union of points of $\mathfrak{S}_{a_{i+1}}$. Hence $\mathfrak{S}_{a_{i+1}}$ is an extension of $\mathfrak{S}_{a_i}$. Let $\mathfrak{S}_k$ be the set sum of $\mathfrak{S}_{a_1}, \cdots, \mathfrak{S}_{a_k}$. Then $\mathfrak{S}_k$ is a sublattice of $\mathfrak{S}$ and is non-modular since $(c', x) > x$ but $c' \not> [c', x]$ and $c', x \in \mathfrak{S}_k$. Hence by Lemma 4.2 $\mathfrak{S}_{a_j}$ is non-modular for some j. If we set $a_j = b$, the theorem follows.

Combining Theorems 4.1 and 4.2 we have

Theorem 4.3. *Let $\mathfrak{S}$ be a Birkhoff lattice. Then the number of irreducible components is unique in the reduced decompositions of each element of $\mathfrak{S}$ if and only if $\mathfrak{S}_a$ is modular for each element a.*

Proof. The necessity follows immediately from Theorem 4.1. If each $\mathfrak{S}_a$ is modular, then by Theorem 4.2 each characteristic element is simple and hence the number of components is unique by Theorem 4.1.

If $\mathfrak{S}_a$ is modular, let x be any element of $\mathfrak{S}_a$. Furthermore, let $p_1, \cdots, p_k$ be a maximal union independent set of points of $\mathfrak{S}_a$ divisible by x. Imbed $p_1, \cdots, p_k$ in a maximal union independent set of points $p_1, \cdots, p_n$. Then $x = [x, (p_1, \cdots, p_n)] = (p_1, \cdots, p_k, [x, (p_{k+1}, \cdots, p_n)])$. If $[x, (p_{k+1}, \cdots, p_n)] \neq a$, then $[x, (p_{k+1}, \cdots, p_n)] \supset p > a$ and hence $(p_1, \cdots, p_k) \supset p$. But then $a = [(p_1, \cdots, p_k), (p_{k+1}, \cdots, p_n)] \supset p$, and this is impossible. Thus $x = (p_1, \cdots, p_k)$ and hence $\mathfrak{S}_a$ is a point lattice. The statement of Theorem 4.3 is thus equivalent to that given in the introduction.

Kurosch has shown that in a modular lattice if $a = [q_1, \cdots, q_k] = [q_1', \cdots, q_l']$ are two irreducible decompositions of a, then each q_i may be replaced by a q_i', and conversely, without changing the representation. The unicity of the number of components, of course, follows from this replacement property. Now

Theorem 4.1 shows that the converse is true for Birkhoff lattices, namely, that if the number of components in each reduced representation of a is unique, then the Kurosch replacement property holds for the reduced representations of a. For $\mathfrak{S}_a$ must be modular and have only simple characteristic elements by Theorem 4.1. But then the replacement property follows from the dual of Lemma 3.2 and Theorem 3.1.

COROLLARY 4.1. *Let $\mathfrak{S}$ be an exchange lattice (Birkhoff point lattice). Then the number of components in the reduced decompositions of the elements of $\mathfrak{S}$ is unique if and only if $\mathfrak{S}$ is modular.*

5. Construction of Birkhoff lattices.

By definition u_a is never a characteristic element of $\mathfrak{S}_a$. Furthermore a is characteristic if and only if it is irreducible. The question then naturally arises whether there are any other elements of $\mathfrak{S}_a$ which can never be characteristic. In answer to this question we show that any $\mathfrak{S}_a$ may be imbedded in a Birkhoff lattice in which any given element of $\mathfrak{S}_a$ not equal to u_a or a is characteristic. The methods of construction are of considerable interest in themselves as they afford a means of obtaining Birkhoff lattices with various arithmetical properties.

Let us call two lattices $\mathfrak{S}_1$ and $\mathfrak{S}_2$ *compatible* if the set of common elements is a sublattice both of $\mathfrak{S}_1$ and $\mathfrak{S}_2$ and the operations of union and cross-cut in $\mathfrak{S}_1$ and $\mathfrak{S}_2$ agree in $\mathfrak{S}$.

LEMMA 5.1. *Let $\mathfrak{S}_1$ and $\mathfrak{S}_2$ be two compatible Birkhoff lattices satisfying the ascending chain condition. Moreover, let the common sublattice $\mathfrak{A}$ be complete in $\mathfrak{S}_1$. If a denotes the null element of $\mathfrak{S}_2$, let $a \in \mathfrak{S}_1$ and let $x \supset a$, $x \in \mathfrak{S}_1$ imply $x \in \mathfrak{S}_2$. Then the set sum of $\mathfrak{S}_1$ and $\mathfrak{S}_2$ can be made into a Birkhoff lattice $\mathfrak{S}$ with $\mathfrak{S}_1$ and $\mathfrak{S}_2$ as sublattices.*

Proof. Let $\mathfrak{S}$ denote the set sum of $\mathfrak{S}_1$ and $\mathfrak{S}_2$. In $\mathfrak{S}$ we define division as follows: If $x \in \mathfrak{S}_2$, then x divides y if and only if $x \supset (a, y)$, where the union is taken in $\mathfrak{S}_1$ or $\mathfrak{S}_2$ according as y is in $\mathfrak{S}_1$ or $\mathfrak{S}_2$. Since (a, y) is always in $\mathfrak{S}_2$, the division of the definition is in $\mathfrak{S}_2$. If $x \,\epsilon'\, \mathfrak{S}_2$, then x divides y if and only if $x \supset y$ in $\mathfrak{S}_1$. With respect to this division relation the union of two elements x and y is given by $((x, a), (y, a))$ if either x or y is in $\mathfrak{S}_2$ and is given by (x, y) in $\mathfrak{S}_1$ if $x, y \,\epsilon'\, \mathfrak{S}_2$. If $x \in \mathfrak{S}_2$, let u_x be the union of the elements of $\mathfrak{A}$ which are divisible by x. The union always exists by the ascending chain condition. Then the cross-cut relation in $\mathfrak{S}$ is given by $[u_x, y]$ if $x \,\epsilon'\, \mathfrak{S}_1$, $y \,\epsilon'\, \mathfrak{S}_2$ and otherwise by $[x, y]$ where the cross-cut relation is in $\mathfrak{S}_1$ or $\mathfrak{S}_2$ according as both x and y are in $\mathfrak{S}_1$ or $\mathfrak{S}_2$. It follows immediately that the union and cross-cut relations in $\mathfrak{S}$ reduce to those in $\mathfrak{S}_1$ and $\mathfrak{S}_2$ for elements in $\mathfrak{S}_1$ and $\mathfrak{S}_2$ respectively.

Now let $x > y$ in $\mathfrak{S}$. If $y \in \mathfrak{S}_2$, then $x \in \mathfrak{S}_2$ and hence $x > y$ in $\mathfrak{S}_2$. If $y \,\epsilon'\, \mathfrak{S}_2$, then $x \in \mathfrak{S}_1$. For if $x \,\epsilon'\, \mathfrak{S}_1$, then $x \supset a$ and $x \supset (y, a) \supset y$ in $\mathfrak{S}$. But then $x = (y, a)$ and $x \in \mathfrak{S}_1$, and this is contrary to assumption. Thus if $y \in \mathfrak{S}_1$, $x > y$ in $\mathfrak{S}_1$. Conversely, let $x > y$ in $\mathfrak{S}_2$. Then if $x \supset z \supset y$ in $\mathfrak{S}$, we have $z \in \mathfrak{S}_2$ and $x = z$ or $z = y$. Hence $x > y$ in $\mathfrak{S}$. Let $x > y$ in $\mathfrak{S}_1$.

Suppose $x \supset z \supset y$ in $\mathfrak{S}$. If $z \epsilon \mathfrak{S}_1$, then $x = z$ or $z = y$ since $x > y$ in $\mathfrak{S}_1$. If $z \epsilon' \mathfrak{S}_1$, then $z \supset a$ and hence $x \supset a$. But then $x \supset (a, y) \supset y$ in $\mathfrak{S}_1$. If $x = (a, y)$, then $x = z$. If $(a, y) = y$, then $y \supset a$ and $y \epsilon \mathfrak{S}_2$. But then $x \epsilon \mathfrak{S}_2$ and $x, y \epsilon \mathfrak{A}$. Since $x > y$ in $\mathfrak{S}_1$ we have $x > y$ in $\mathfrak{A}$. Since $\mathfrak{A}$ is complete in $\mathfrak{S}_2$, we have $x > y$ in $\mathfrak{S}_2$. Thus $x = z$ or $z = y$. We conclude then that $x > y$ in $\mathfrak{S}$ if and only if $x > y$ in $\mathfrak{S}_1$ or $x > y$ in $\mathfrak{S}_2$.

Now let $x, y > [x, y]$ in $\mathfrak{S}$. If $[x, y] \epsilon \mathfrak{S}_2$, then $x, y \epsilon \mathfrak{S}_2$ and hence $x, y > [x, y]$ in $\mathfrak{S}_2$. But then $(x, y) > x, y$ in $\mathfrak{S}_2$ and hence in $\mathfrak{S}$ since $\mathfrak{S}_2$ is a Birkhoff lattice. If $[x, y] \epsilon' \mathfrak{S}_2$, then $x, y \epsilon \mathfrak{S}_1$. Hence $x, y > [x, y]$ in $\mathfrak{S}_1$. But then since $\mathfrak{S}_1$ is a Birkhoff lattice, we have $(x, y) > x, y$ in $\mathfrak{S}_1$ and thus in $\mathfrak{S}$. Hence $\mathfrak{S}$ is a Birkhoff lattice and the lemma is proved.

LEMMA 5.2. *Let $\mathfrak{S}$ be a Birkhoff lattice satisfying the ascending chain condition and having a null element z. Let $\mathfrak{S}_p$ be the lattice obtained from $\mathfrak{S}$ by taking the direct sum of $\mathfrak{S}$ with a point p and deleting the simple elements. Then $\mathfrak{S}_p$ is a Birkhoff lattice containing $\mathfrak{S}$ as a complete sublattice.*

Proof. $\mathfrak{S}_p$ consists of two types of couples:
(1) $\{s, z\}$, $s \epsilon \mathfrak{S}$ and $s \neq u$,
(2) $\{s, p\}$, $s \epsilon \mathfrak{S}$, $u \not> s$.

If $\{x, y\}$ and $\{x_1, y_1\}$ are couples in $\mathfrak{S}_p$, we define $[\{x, y\}, \{x_1, y_1\}] = \{[x, x_1], [y, y_1]\}$. If $\{x, y\}$ and $\{x_1, y_1\}$ are $\mathfrak{S}_p$, then $\{[x, x_1], [y, y_1]\}$ is clearly in $\mathfrak{S}_p$. $(\{x, y\}, \{x_1, y_1\}) = \{(x, x_1), (y, y_1)\}$ if $\{(x, x_1), (y, y_1)\}$ is in $\mathfrak{S}_p$. Otherwise $(\{x, y\}, \{x_1, y_1\}) = \{u, p\}$. Then if $\{x, y\} \neq \{u, p\}$, $\{x, y\} > \{x_1, y_1\}$ if and only if $x > x_1, y = y_1$ or $x = x_1, y > y_1$. Hence the Birkhoff condition in $\mathfrak{S}_p$ follows from the Birkhoff condition in $\mathfrak{S}$. The correspondence $\{s, z\} \leftrightarrow s$, $s \neq u$ and $\{u, p\} \leftrightarrow u$ preserves union, cross-cut, and covering relations. Hence $\mathfrak{S}_p$ contains $\mathfrak{S}$ as a complete sublattice.

We shall refer to the process of Lemma 5.1 as the replacement of $\mathfrak{A}$ in $\mathfrak{S}_1$ by $\mathfrak{S}_2$. The process of Lemma 5.2 by which $\mathfrak{S}$ is imbedded in a lattice having the same unit and null elements will be called the imbedding of $\mathfrak{S}$ and $\mathfrak{S}_p$.

LEMMA 5.3. *Let $\mathfrak{S}$ be an Archimedean Birkhoff lattice in which the unit element is the union of the points of $\mathfrak{S}$. Then $\mathfrak{S}$ may be imbedded in a Birkhoff lattice $\mathfrak{S}'$ having the same unit and null elements but having a simple element s and a chain of elements $s > s_1 > s_2 > \cdots > z$ all of which are union irreducible.*

Proof. Imbed $\mathfrak{S}$ in $\mathfrak{S}_{p_1}$. Then p_1 is union irreducible and $p_1 \epsilon' \mathfrak{S}$. Let $\mathfrak{S}_1$ be the quotient lattice of elements of $\mathfrak{S}_{p_1}$ which divide p_1. Replace $\mathfrak{S}_1$ by $\mathfrak{S}_{1p_2}$ in $\mathfrak{S}$ to give a Birkhoff lattice $\mathfrak{S}_2$. Now p_2 is union irreducible in $\mathfrak{S}_2$. For if $p_2 > x$, then $x \epsilon \mathfrak{S}_{1p_2}$ since $p_2 \epsilon' \mathfrak{S}_{p_1}$. But then $x = p_1$ since p_2 is a point of $\mathfrak{S}_{1p_2}$. Continuing in this manner, we get a chain of union irreducibles $z < p_1 < p_2 < \cdots < p_k$ which by the ascending chain condition must lead to a simple element $s = p_k$. $\mathfrak{S}' = \mathfrak{S}_k$ is the desired lattice.

We apply Lemma 5.3 to an arbitrary quotient lattice.

THEOREM 5.1. *Let $\mathfrak{S}_a$ be the sublattice of a Birkhoff lattice $\mathfrak{S}$ belonging to the element a. Then if b is an arbitrary element of $\mathfrak{S}_a$ not the unit or null element,*

$\mathfrak{S}_a$ *may be imbedded in a lattice* $\mathfrak{S}'$ *in which* $\mathfrak{S}_a$ *is the quotient lattice belonging to* a *and* b *is characteristic.*

Proof. Let $\mathfrak{S}_b$ be the quotient lattice of elements of $\mathfrak{S}_a$ which divide b. Replace $\mathfrak{S}_b$ in $\mathfrak{S}_a$ by $\mathfrak{S}_b + b_1$, where b_1 is a point of $\mathfrak{S}_b + b_1$. Denote the lattice thus obtained by $\mathfrak{S}_{ab}$. Let $\mathfrak{S}_{b_1}$ be the quotient lattice of elements of $\mathfrak{S}_{ab}$ which divide b_1. Replace $\mathfrak{S}_{b_1}$ in $\mathfrak{S}_{ab}$ by $\mathfrak{S}'_{b_1}$ according to Lemma 5.3 to obtain a lattice $\mathfrak{S}'$. Then there is a chain of elements $b < b_1 < s_k < \cdots < s_1 < s$, where s is simple and $b_1, s_k, \cdots, s$ are union irreducible. Let b' be an arbitrary complement of b in $\mathfrak{S}_a$. Then $[b', s] = [b', s_1] = \cdots = [b', s_k] = [b', b_1] = [b, b'] = z$ and $(b', s) > s$. Hence b is characteristic by Theorem 3.1. Now let $\mathfrak{S}'_a$ be the quotient lattice associated with a in $\mathfrak{S}'$. Clearly $\mathfrak{S}_a \,\epsilon\, \mathfrak{S}'_a$. If $x \,\epsilon'\, \mathfrak{S}'_a$, then $x \supset b_1$ and $u_a \not\supset x$. Hence $x \,\epsilon'\, \mathfrak{S}'_a$. Thus $\mathfrak{S}_a = \mathfrak{S}'_a$.

6. Extension of reduced representations.

We turn now to the following problem: What are necessary and sufficient conditions that a given set of components of a may be extended to give a reduced decomposition of a? It is clearly necessary that the set of components be cross-cut independent. However, this is generally not sufficient if the elements of the lattice do not have unique irreducible decompositions. We obtain sufficient conditions by generalizing the notion of independence (Theorem 6.1). The above remark suggests the further problem of determining the conditions that a lattice must satisfy in order that every cross-cut independent set of components may be extended into a reduced decomposition. Theorem 6.2 gives a particularly simple answer, in case the number of components in the irreducible decompositions of the elements of the lattice is unique.

If $\mathfrak{S}_a$ is the quotient lattice associated with the element $a \,\epsilon\, \mathfrak{S}$, let $\mathfrak{P}_a$ denote the elements of $\mathfrak{S}_a$ which can be expressed as a union of points of $\mathfrak{S}_a$. Now $\mathfrak{P}_a$ is clearly closed with respect to union and hence may be made into a lattice by defining cross-cut in terms of union. If c is a characteristic element of $\mathfrak{S}_a$, then there is exactly one characteristic element of $\mathfrak{S}_a$ which is associated with the same irreducibles as c and which belongs to $\mathfrak{P}_a$. Inasmuch as this characteristic element is the cross-cut of all characteristic elements associated with the same irreducibles as c, we shall call it the minimal characteristic element associated with c.

DEFINITION 6.1. A set of characteristic elements of $\mathfrak{S}_a$ is said to be sub-independent if the associated minimal characteristic elements are cross-cut independent in $\mathfrak{P}_a$.

We note that if the number of components is unique in $\mathfrak{S}$, then $\mathfrak{S}_a = \mathfrak{P}_a$ (Theorem 4.3) and sub-independence becomes ordinary cross-cut independence.

THEOREM 6.1. *Let* $\mathfrak{S}$ *be a Birkhoff lattice. Then a set* $q_1, \cdots, q_k$ *of components of* a *can be extended into a reduced decomposition of* a *if and only if the associated characteristic elements* $c_1, \cdots, c_k$ *of* $\mathfrak{S}_a$ *are sub-independent.*

Proof. Let $a = [q_1, \cdots, q_k, q_{k+1}, \cdots, q_n]$ be a reduced representation of a as a cross-cut of irreducibles. If $c_1, \cdots, c_k$ are not sub-independent, then with suitable numbering $c_1 \supset \{c'_2, \cdots, c'_k\}$, where c'_i is the minimal characteristic element associated with c_i and $\{\ ,\ \}$ denotes cross-cut in $\mathfrak{P}_a$. Let $[q_2, \cdots, q_n] \supset p > a$. Then $[c'_2, \cdots, c'_n] \supset p$ and hence $\{c'_2, \cdots, c'_n\} \supset p$. Then $c_1 \supset p$ and hence $a = [q_1, \cdots, q_n] \supset [c_1, \cdots, c_n] \supset \{c'_1, \cdots, c'_n\} \supset p$. This contradicts $p > a$. Hence $c_1, \cdots, c_k$ are sub-independent.

Now let $c_1, \cdots, c_k$ be sub-independent. Let $p_1, \cdots, p_l$ be a maximal union independent set of elements covering a and divisible by $[q_1, \cdots, q_k]$ so that $\{c'_1, \cdots, c'_k\} = (p_1, \cdots, p_l)$. Since $c_1, \cdots, c_k$ are sub-independent, there exist points $p'_1, \cdots, p'_k$ such that $[c_1, \cdots, c_{i-1}, c_{i+1}, \cdots, c_k] \supset p'_i$, $c_i \not\supset p'_i$. Suppose $(p_2, \cdots, p_l, p'_1, \cdots, p'_k) \supset p_1$ say. Then since $(p_2, \cdots, p_l) \not\supset p_1$, there is a first i such that $(p_2, \cdots, p_l, p'_1, \cdots, p'_i) \supset p_1$. But then $(p_1, p_2, \cdots, p_l, p'_1, \cdots, p'_{i-1}) \supset p'_i$ by the Birkhoff condition. Hence with suitable numbering this case reduces to the case $(p_1, \cdots, p_l, p'_2, \cdots, p'_k) \supset p'_1$. But then $c_1 \supset (p_1, \cdots, p_l, p'_2, \cdots, p'_k) \supset p'_1$, and this contradicts $c_1 \not\supset p'_1$. Hence $p_1, \cdots, p_l, p'_1, \cdots, p'_k$ are union independent and may be imbedded in a maximal union independent set $p_1, \cdots, p_l, p'_1, \cdots, p'_k, \cdots, p'_n$. Let $s_1 = (p_2, \cdots, p'_n), \cdots, s_l = (p_1, \cdots, p_{l-1}, p'_1, \cdots, p'_n)$. Now if $[c_1, \cdots, c_k, s_1, \cdots, s_l] \neq a$, we have $[c_1, \cdots, c_k, s_1, \cdots, s_l] \supset p > a$. Then since $[c_1, \cdots, c_k] \supset p$, it follows that $(p_1, \cdots, p_l) \supset p$. Hence $a = [(p_1, \cdots, p_l), s_1, \cdots, s_l] \supset p$. This is impossible since $p_1, \cdots, p_l, p'_1, \cdots, p'_n$ generate a Boolean algebra. Thus $a = [c_1, \cdots, c_k, s_1, \cdots, s_l]$. We clearly cannot drop an s_i out of this representation since otherwise $a \supset p_i$ and this is impossible. Also we cannot drop out a c_i since then $c_i \supset p'_i$ and the definition of p'_i is contradicted. Hence $c_1, \cdots, c_k, s_1, \cdots, s_l$ is a cross-cut independent set of characteristic elements whose cross-cut is a. Thus by Theorem 3.1 there is a reduced representation of a containing $q_1, \cdots, q_k$.

COROLLARY 6.1. *Let $\mathfrak{S}$ be a Birkhoff lattice. Then an irreducible q is a component of a if and only if $q \supset a$ and $q \not\supset u_a$.*

COROLLARY 6.2. *Let $\mathfrak{S}$ be a Birkhoff lattice. Then if $q \supset b \supset a$ and q is a component of a, q is a component of b.*

For $u_b \supset u_a$. Hence if $q \supset u_b$, then $q \supset u_a$. This is impossible by Corollary 6.1.

COROLLARY 6.3. *I;'$_y$ and q' are components of a, then $q \supset q'$ implies $q = q'$.*

We are now ready to give a proof of Theorem 6.2 mentioned above using Theorem 6.1 and the Fundamental Lemma.

THEOREM 6.2. *Let $\mathfrak{S}$ be a Birkhoff lattice in which the number of components in the irreducible decompositions is unique for each element of $\mathfrak{S}$. Then each independent set of components of a can be extended into a reduced representation if and only if each characteristic element of $\mathfrak{S}_a$ belongs to exactly one irreducible component of a.*

Proof. Let $q_1, \cdots, q_k$ be an independent set of components of a. Let $c_1, \cdots, c_k$ be a set of characteristic elements associated with $q_1, \cdots, q_k$. We shall show that $c_1, \cdots, c_k$ are independent. If $c_1, \cdots, c_k$ are not independent, then $c_1 \supset [c_2, \cdots, c_k]$, say. Now $[q_2, \cdots, q_k] \neq [c_2, \cdots, c_k]$, since otherwise $q_1 \supset c_1 \supset [q_2, \cdots, q_k]$ and the independence of $q_1, \cdots, q_k$ is contradicted. Let $[q_2, \cdots, q_k] \supset x_1 > [c_2, \cdots, c_k]$. Let $\mathfrak{A}_1$ be the sublattice of elements of $\mathfrak{S}_a$ contained between u_a and $[c_2, \cdots, c_k]$. Now $u_a \not\supset x_1$, since otherwise $c_i = [u_a, q_i] \supset [q_i, x_1]$ (Theorem 4.1) and hence $[c_2, \cdots, c_k] \supset [q_2, \cdots, q_k, x_1] = [q_2, \cdots, q_k]$. This is impossible. Hence by the Fundamental Lemma, $\mathfrak{A}_1$ and x_1 generate a sublattice which is the direct sum of $\mathfrak{A}_1$ and x_1. Let $\mathfrak{A}_2$ be the sublattice of elements of the form (y, x_1), where $y \in \mathfrak{A}$. Then $\mathfrak{A}_2$ is isomorphic to $\mathfrak{A}_1$ and is thus modular. The unit element of $\mathfrak{A}_2$ is (u_a, x_1) and the null element is x_1. Furthermore $c_1' = (c_1, x_1), \cdots, c_k' = (c_k, x_1)$ belong to $\mathfrak{A}_2$. Clearly $c_1' \supset (x_1, [c_2, \cdots, c_k]) = [(x_1, c_2), \cdots, (x_1, c_k)] = [c_2', \cdots, c_k']$. Now if $q_1 \not\supset (c_1, x_1)$, there is a second component q_1' of a such that $q_1' \supset (c_1, x_1)$, $q_1' \not\supset u_a$. But this contradicts the hypothesis of the theorem. Hence $q_1 \supset c_1'$. Now $[q_2, \cdots, q_k] \neq [c_2', \cdots, c_k']$ since otherwise $q_1 \supset c_1' \supset [q_2, \cdots, q_k]$. This is contrary to the independence of $q_1, \cdots, q_k$. Thus $[q_2, \cdots, q_k] \supset x_2 > [c_2', \cdots, c_k']$ for some x_2. We note that $c_i' = [q_i, (u_a, x_1)]$. For $(u_a, x_1) \supset [q_i, (u_a, x_1)] \supset c_i'$, and since $u_a > c_i$ we have $(u_a, x_1) > (c_i, x_1) = c_i'$. Hence either $c_i' = [q_i, (u_a, x_1)]$ or $q_i \supset (u_a, x_1) \supset u_a$. But $q_i \not\supset u_a$ and thus $c_i' = [q_i, (u_a, x_1)]$. Now $(u_a, x_1) \not\supset x_2$ since otherwise $c_i' = [q_i, (u_a, x_1)] \supset x_2$ and $[c_2', \cdots, c_k'] \supset x_2$, and this contradicts $x_2 > [c_2', \cdots, c_k']$. Hence by the Fundamental Lemma, the sublattice generated by $\mathfrak{A}_2$ and x_2 is the direct sum of $\mathfrak{A}_2$ and x_2. Continuing this process we get an infinite ascending chain $x_1 < x_2 < x_3 < \cdots$ and the ascending chain condition is contradicted. Hence if $q_1, \cdots, q_k$ are cross-cut independent, then $c_1, \cdots, c_k$ are also cross-cut independent. But now since $\mathfrak{P}_a = \mathfrak{S}_a$, $c_1, \cdots, c_k$ are also sub-independent. Thus by Theorem 6.1 $q_1, \cdots, q_k$ may be extended into a reduced representation.

If there is a characteristic element of $\mathfrak{S}_a$ which has two irreducibles to which it belongs, then the two irreducibles are cross-cut independent by Corollary 6.3. However, they cannot be extended into a reduced representation by Theorem 3.1. Hence the theorem follows.

Let $\mathfrak{S}$ be a Birkhoff lattice. Let c be a characteristic element of $\mathfrak{S}_a$. Then there is at least one characteristic element c_1 of $\mathfrak{S}_c$ which does not divide u_a. Similarly, there is at least one characteristic element c_2 of $\mathfrak{S}_{c_1}$ which does not divide u_a. Continuing in this manner we eventually get a chain $c \subset c_1 \subset c_2 \subset \cdots \subset c_k$, where c_k is an irreducible component of a and c_i is a characteristic element of $\mathfrak{S}_{c_{i-1}}$ such that $c_i \not\supset u_a$.

Theorem 6.2 may then be stated as follows:

THEOREM 6.3. *Let $\mathfrak{S}$ be a Birkhoff lattice in which the number of components is unique. Then each cross-cut independent set of components of a can be extended into an irreducible decomposition if and only if for each characteristic element c of $\mathfrak{S}_a$ there is exactly one chain $c \subset c_1 \subset c_2 \subset \cdots \subset c_k$, where c_k is an irreducible such that $c_k \not\supset u_a$ and c_i is a characteristic element of $\mathfrak{S}_{c_{i-1}}$. Moreover, in this case $\mathfrak{S}_{c_i} = \mathfrak{S}_{c_i}' + (c_i, u_a)$, where $\mathfrak{S}_{c_i}'$ consists of those elements of $\mathfrak{S}_{c_i}$ which do not divide u_a.*

Proof. If in any $\mathfrak{S}_{c_i}$ there are two characteristic elements c_{i+1} and c'_{i+1} which do not divide u_a, let $q_{i+1} \supset c_{i+1}$, $q_{i+1} \not\supset u_a$, $q'_{i+1} \supset c'_{i+1}$, $q'_{i+1} \not\supset u_a$. Then q_{i+1} and q'_{i+1} are distinct components of a which divide c and hence cannot be extended into an irreducible representation. The condition is sufficient since c_k is the single irreducible component of a which divides c.

Since $\mathfrak{S}_{c_i}$ is modular, to show that $\mathfrak{S}_{c_i} = \mathfrak{S}_{c_i} + (c_i, u_a)$ we have only to show that $x, y \not\supset u_a$, $x, y \in \mathfrak{S}_{c_i}$ implies $(x, y) \not\supset u_a$. But now c_{i+1} is the single simple element of $\mathfrak{S}_{c_i}$ which does not divide u_a. Let $\mathfrak{A}_i$ be the set of elements of $\mathfrak{S}_{c_i}$ which divide u_a. Then since $\mathfrak{S}_{c_i}$ is an atomic lattice $\mathfrak{S}_{c_i} = \mathfrak{A}_i \times c_{i+1}$ by the dual of the Fundamental Lemma. Then since $x, y \in \mathfrak{A}_i$, $c_{i+1} \supset x, y$ and hence $c_{i+1} \supset (x, y)$. Thus $(x, y) \not\supset u_a$. Hence $\mathfrak{S}_{c_i}$ is a sublattice of $\mathfrak{S}$ and the direct sum decomposition follows from the Fundamental Lemma.

CoROLLARY 6.4. *Let $\mathfrak{S}$ be a modular lattice in which every complemented quotient lattice is irreducible. Then each cross-cut independent set of components of a can be extended into a reduced representation of a for all a if and only if $\mathfrak{S}$ is a chain of projective geometries connected by simple chains.*

The statement of Theorem 6.2 suggests the possibility of weakening the hypothesis that the number of components be unique. However, such a weakening will probably be very artificial since by using the methods of §5 the writer has constructed an example of a Birkhoff lattice $\mathfrak{S}$ in which each $\mathfrak{S}_a$ is a point lattice and each characteristic element is simple and has but one irreducible component as divisor. Furthermore, $\mathfrak{S}$ has only one element for which the number of components is not unique. Nevertheless not every cross-cut independent set of components can be extended into a reduced decomposition.

In §4 we pointed out that unicity of the number of components in a Birkhoff lattice implies the Kurosch replacement property. We conclude with a theorem which is a sort of converse result.

THEOREM 6.4. *Let $\mathfrak{S}$ be a Birkhoff lattice. Then if the Kurosch replacement property holds for the maximal representations of each element a, each representation of a is maximal, i.e., the number of components in the representations of a is unique.*

Proof. Let a be a simple element of $\mathfrak{S}_a$ which can be expressed as a union of points. Let b be any other element of $\mathfrak{S}_a$ which can be expressed as a union of points and such that $a \not\supset b$. Let $[a, b] \supset (p_1, \cdots, p_l)$, where $p_1, \cdots, p_l$ are a maximal union independent set of points of $\mathfrak{S}_a$ divisible by $[a, b]$. Let $a = (p_1, \cdots, p_l, p_{l+1}, \cdots, p_m)$ and $b = (p_1, \cdots, p_l, p'_{l+1}, \cdots, p'_k)$. Then there is a point of b, say p'_{l+1}, such that $p_1, \cdots, p_l, p_{l+1}, \cdots, p_m, p'_{l+1}$ form a maximal union independent set of points of $\mathfrak{S}_a$. Also $p_1, \cdots, p_l, p'_{l+1}, \cdots, p'_k$ can be imbedded in a maximal independent set $p_1, \cdots, p_l, p'_{l+1}, \cdots, p'_k, p'_{k+1}, \cdots, p'_r$. Let $a_1 = (p_2, \cdots, p_l, p_{l+1}, \cdots, p_m, p'_{l+1}), \cdots,$ $a_{m+1} = (p_1, \cdots, p_l, p_{l+1}, \cdots, p_m)$. Similarly, let $a'_1 = (p_2, \cdots, p_l, p'_{l+1}, \cdots, p'_r), \cdots, a'_{m+1} = (p_1, \cdots, p_l, p'_{l+1}, \cdots, p'_{r-1})$. Now $[a'_1, \cdots, a'_l, a'_{l+2}, \cdots, a'_{m+1}] = p'_{l+1}$ by Lemma 3.2. Hence the only element of $a_1, \cdots, a_{m+1}$ which can replace a'_{l+1} is a_{m+1}. But then $a'_1, \cdots, a'_l, a_{m+1}, a'_{l+2}, \cdots, a'_{m+1}$ must be simple elements of a Boolean algebra by Lemma 4.1. But $a = a_{m+1}$ and $b = [a'_{k+1}, \cdots, a'_{m+1}]$. Hence since a and b are elements of a Boolean algebra, we have $[a, b] = (p_1, \cdots, p_l)$ and $b > [a, b]$. Now let x and y be any two elements of $\mathfrak{S}_a$ which can be represented as a union of points of $\mathfrak{S}_a$. Then $x = [a_1, \cdots, a_k]$, where $a_1, \cdots, a_k$ are simple elements of $\mathfrak{S}_a$ which are unions of points. Hence $[x, y] = [a_1, \cdots, a_k, y]$ and hence $[x, y]$ is a union of points of $\mathfrak{S}_a$ by successive application of the result we have just obtained. But then the lattice generated by the points of $\mathfrak{S}_a$ is a modular sublattice of $\mathfrak{S}_a$ and hence is equal to $\mathfrak{S}_a$. $\mathfrak{S}_a$ is thus modular for each element a and the theorem follows from Theorem 4.3.

References

1. GARRETT BIRKHOFF, *On the combination of subalgebras*, Proceedings of the Cambridge Philosophical Society, vol. 29(1933), pp. 441–464.
2. GARRETT BIRKHOFF, *Abstract linear dependence and lattices*, American Journal of Mathematics, vol. 57(1935), pp. 800–804.
3. R. P. DILWORTH, *Lattices with unique irreducible decompositions*, Annals of Mathematics, vol. 41(1940), pp. 771–777.
4. S. MAC LANE, *A lattice formulation for transcendence degrees and p-bases*, this Journal, vol. 4(1938), pp. 435–468.
5. O. ORE, *On the foundation of abstract algebra*, I, Annals of Mathematics, vol. 36(1935), pp. 406–437.
6. H. WHITNEY, *On the abstract properties of linear dependence*, American Journal of Mathematics, vol. 57(1935), pp. 509–533.

YALE UNIVERSITY.

IDEALS IN BIRKHOFF LATTICES

BY

R. P. DILWORTH[1]

Introduction. In previous papers by the author (Dilworth [1, 2])[2] methods were developed for studying the arithmetical properties of Birkhoff lattices, that is, the properties of irreducibles and decompositions into irreducibles. These methods, however, required the assumption of both the ascending and descending chain conditions. In this paper we give a new technique which is applicable in general and which under the assumption of merely the ascending chain condition gives results quite as good as those of the previous work. Now the descending chain condition is equivalent to the requirement that every ideal[3] be principal. Hence if the descending chain condition does not hold we find it convenient to relate the arithmetical properties of the lattice to the structure of its lattice of ideals. Furthermore since the Birkhoff condition itself may lose much of its force if the descending chain condition does not hold, a lattice is defined to be a Birkhoff lattice if every element satisfies the Birkhoff condition[4] in the lattice of ideals. Hence if the descending chain condition holds, this definition reduces to that used in the previous papers. In the lattice of ideals, the existence of sufficient covering ideals to make the Birkhoff conditions effective can be proved.

In D1 and D2 it was shown that the arithmetical behavior of an element a was closely related to the structure of the quotient lattice $\mathfrak{S}_a$ generated by the elements covering a. Here we make a similar correlation with the structure of the quotient lattice of ideals $\mathfrak{L}_a$ generated by the ideals covering a. The important properties of $\mathfrak{S}_a$ follow from its finite dimensionality. $\mathfrak{L}_a$ on the other hand is in general *not* finite dimensional and thus one of the essential problems of the present treatment is the proof of the archimedean character of $\mathfrak{L}_a$ in the cases of interest.

If the descending chain condition holds, the Birkhoff condition is equivalent to Mac Lane's point-free exchange axiom E_5 (Mac Lane [1]). Now E_5 is independent of covering conditions, which suggests that it should be closely related to the Birkhoff condition in the lattice of ideals. We show that the Birkhoff condition in the lattice of ideals always implies E_5 and, if each principal ideal is covered by only a finite number of ideals, the two conditions are equivalent.

Presented to the Society, December 27, 1939; received by the editors May 11, 1940.

[1] Sterling Research Fellow, Yale University.

[2] These papers will be referred to as D1 and D2.

[3] An ideal is a sublattice which contains with each element all of its divisors. G. Birkhoff (Birkhoff [1]) uses the term *dual ideal* for such a sublattice.

[4] See §1, Conditions B1 and B1'.

325

In D1 it was shown that a lattice of finite dimensions has unique irreducible decompositions if and only if it is a Birkhoff lattice in which every modular sublattice is distributive. This result no longer holds if we drop the descending chain condition as we show by an example. However, by strengthening slightly the condition that every modular sublattice be distributive, we have the following theorem:

THEOREM 6.6. *Let $\mathfrak{S}$ satisfy the ascending chain condition. Then every element of $\mathfrak{S}$ is uniquely expressible as a reduced crosscut of irreducibles if and only if the following conditions hold.*

E_5. (Mac Lane's point-free exchange axiom.) $a \supset b \supset a \cap c$, $c \neq a \cap c$ *implies that* $c_1 \neq a \cap c$ *exists such that* $c \supset c_1 \supset a \cap c$ *and* $b = a \cap (b \cup c_1)$.

A. $a \cup b \supset x \supset a \cap b$, $a \cap x = b \cap x = a \cap b$ *implies* $x = a \cap b$.

If we go over to the lattice of ideals, E_5 may be replaced by the condition that $\mathfrak{S}$ be a Birkhoff lattice, and A, by the requirement that the ideals covering a principal ideal generate a Boolean algebra.

In D2, Birkhoff lattices in which the number of components in the irreducible decompositions of each element is unique were characterized in terms of the structure of the quotient lattices $\mathfrak{S}_a$. We prove here:

THEOREM 5.1. *Let $\mathfrak{S}$ be a Birkhoff lattice satisfying the ascending chain condition and let $\mathfrak{L}$ denote its lattice of ideals. Then the number of components in the irreducible decompositions of each element of the lattice $\mathfrak{S}$ is unique if and only if the ideals covering any principal ideal of the lattice $\mathfrak{L}$ generate a dense, modular sublattice of $\mathfrak{L}$.*

By means of ideal methods we give a new proof of the Kurosch-Ore decomposition theorem for modular lattices in its most general form. The proof rests on the fact that if an element of a modular lattice has a decomposition into irreducibles then the sublattice generated by the ideals covering the element is of finite dimensions.

Finally §§7 and 8 contain examples which show the complications which may arise when the descending chain condition does not hold.

1. **Notation and definitions.** The fixed lattice of elements a, b, c, $\cdots$ will be denoted by $\mathfrak{S}$. $\cup$ and $\cap$ will denote union and cross-cut in place of the symbols $(,)$ and $[,]$ used in D1 and D2. $\supset$ denotes lattice division. $a = b$ is defined by the two formulas $a \supset b$, $b \supset a$. If $a \supset b$, $a \neq b$ and $a \supset x \supset b$ implies $a = x$ or $x = b$, we say that a *covers* b and write $a > b$. Elements which cover the null element z of a lattice are called *points* and elements covered by the unit element u are said to be *simple*.

A lattice $\mathfrak{S}$ satisfies the ascending (descending) chain condition if every chain $a_1 \subset a_2 \subset a_3 \subset \cdots$ $(a_1 \supset a_2 \supset a_3 \supset \cdots)$ has only a finite number of distinct elements. If both the ascending and descending chain conditions hold, $\mathfrak{S}$ is said to be *archimedean* or of *finite dimensions*.

Throughout the paper we shall be particularly interested in lattices which satisfy the following weak form of the modular axiom.

B1. $a > a \cap b \to a \cup b > b$ [5].

Another form of B1 is the following:

B1'. $b > a, c \supset a, c \not\supset b \to b \cup c > c$.

If B1' is satisfied for a given a and any b and c we say that a satisfies the *Birkhoff condition* in $\mathfrak{S}$. Hence B1 holds in $\mathfrak{S}$ if and only if each element of $\mathfrak{S}$ satisfies the Birkhoff condition.

We state now some lemmas on elements satisfying the Birkhoff condition which are refinements of Lemmas 3.1–3.3 of D2.

LEMMA 1.1. *Let a satisfy the Birhoff condition in $\mathfrak{S}$ and let $a_1, \cdots, a_k > a$. Then each union independent*[6] *set of the a_i is contained in a maximal independent set.*

The usual proof is valid under the weaker hypotheses of the lemma.

LEMMA 1.2. *Let a satisfy the Birkhoff condition and let $a_1, \cdots, a_k > a$. Then each union independent set of the a_i generates a Boolean algebra.*

We note that the usual proof (for example Theorem 2.3 of D1) is not valid in this case since it depends upon the existence of a rank function. Under the hypotheses of the lemma, complete chains need not have the same length and hence a rank function will in general not exist.

Now let A and B be two arbitrary subsets of the set $\{a_1, \cdots, a_k\}$. Let $\Sigma(A)$ denote the union of the elements of A and denote the set-theoretic union and cross-cut of A and B by $A \cup B$ and $A \cap B$ respectively. We shall show that

$$(1) \qquad \Sigma(A) \cap \Sigma(B) = \Sigma(A \cap B).$$

Let $\mu(A)$ denote the number of elements in A and set $\nu(A) = k - \mu(A)$. If $\nu(A \cap B) = 0$, then $\mu(A \cap B) = k$ and $A = B$. Hence (1) holds. If $\nu(A \cap B) = 1$, then either $A \supset B$ or $B \supset A$ and again (1) holds. Now let (1) hold for all A and B such that $\nu(A \cap B) < l$. Let $\nu(A \cap B) = l$ for some A and B. Then $\mu(A \cap B) = k - l = r$. Hence $A = \{a_1, \cdots, a_r, a_{r+1}, \cdots, a_s\}$ and $B = \{a_1, \cdots, a_r, a'_{r+1}, \cdots, a'_t\}$. Since (1) is trivial if $B \supset A$, we may assume that $s > r$. Let $B' = \{a_1, \cdots, a_r, a_{r+1}, a'_{r+1}, \cdots, a'_t\}$. Now $\mu(A \cap B') = r + 1$ and hence $\nu(A \cap B') = k - (r+1) = l - 1 < l$. By the induction assumption $\Sigma(A \cap B') = \Sigma(A) \cap \Sigma(B')$. Thus $\Sigma(A \cap B') = \Sigma(A) \cap \Sigma(B') \supset \Sigma(A) \cap \Sigma(B) \supset \Sigma(A \cap B)$.

---

[5] $\to$ denotes formal implication.

[6] A set of elements $x_1, \cdots, x_n$ is said to be *union independent* or simply *independent* if $x_1 \cup \cdots \cup x_{i-1} \cup x_{i+1} \cup \cdots \cup x_n \not\supset x_i, i = 1, \cdots, n$. Similarly the set is said to be *cross-cut independent* if $x_i \not\supset x_1 \cap \cdots \cap x_{i-1} \cap x_{i+1} \cap \cdots \cap x_n, i = 1, \cdots, n$.

Since $a_1, \cdots, a_k$ are independent we have $\Sigma(A \cap B) \not\supset a_{r+1}$ and hence $\Sigma(A \cap B') = a_{r+1} \cup \Sigma(A \cap B) > \Sigma(A \cap B)$. If $\Sigma(A \cap B') = \Sigma(A) \cap \Sigma(B)$, then $\Sigma(B) \supset a_{r+1}$ contrary to the independence of $a_1, \cdots, a_k$. Hence $\Sigma(A) \cap \Sigma(B) = \Sigma(A \cap B)$. Thus (1) holds for $\nu(A \cap B) = l$ and by induction (1) holds for all A and B. Clearly $\Sigma(A) \cup \Sigma(B) = \Sigma(A \cup B)$. If $\Sigma(A) = \Sigma(B)$, then $A = B$ by the independence of $a_1, \cdots, a_k$. Hence the elements which can be expressed as a union of the a_i are isomorphic to the subsets of $a_1, \cdots, a_k$ under union and cross-cut and thus $a_1, \cdots, a_k$ generate a Boolean algebra. This completes the proof of the lemma.

LEMMA 1.3. *Let a satisfy the Birkhoff condition and let $a_1, \cdots, a_k > a$. Then any two maximal union independent sets of the a_i have the same number of elements and any element of one set may be replaced by a suitably chosen element of the other without altering the maximal property.*

The usual proof is valid in this case.

LEMMA 1.4. *Let a satisfy the Birkhoff condition and let $a_1, \cdots, a_k > a$. Then any chain joining $a_1 \cup \cdots \cup a_k$ to a has not more than $k+1$ distinct members.*

We may clearly suppose that $a_1, \cdots, a_k$ are independent. Let $a = b_0 \subset b_1 \subset b_2 \subset \cdots \subset b_{l-1} \subset b_l = a_1 \cup \cdots \cup a_k$ be a chain joining $a_1 \cup \cdots \cup a_k$ to a having $l+1$ distinct members and let us assume that $l > k$. Clearly $b_0 < b_0 \cup a_1 < \cdots < b_0 \cup a_1 \cup \cdots \cup a_{k-1} < a_1 \cup \cdots \cup a_k$ by the Birkhoff condition. Now suppose that it has been shown that $a \subset b_1 \subset \cdots \subset b_i < b_i \cup a_1 < \cdots < b_i \cup a_1 \cup \cdots \cup a_{k_i-1} < a_1 \cup \cdots \cup a_k$ where $k_i \leq k - i$ and $i < k$. Consider the chain $a \subset b_1 \subset \cdots \subset b_i \subset b_{i+1} \subset b_{i+1} \cup a_1 \subset \cdots \subset b_{i+1} \cup a_1 \cup \cdots \cup a_{k_i-1} \subset a_1 \cup \cdots \cup a_k$. Let us assume that all of the members of this chain are distinct. If $b_i \cup a_1 \cup \cdots \cup a_{k_i-1} \not\supset b_{i+1}$, then $a_1 \cup \cdots \cup a_k \supset e_{i+1} \cup a_1 \cup \cdots \cup a_{k_i-1} \supset b_i \cup a_1 \cup \cdots \cup a_{k_i-1}$ and $b_{i+1} \cup a_1 \cup \cdots \cup a_{k_i-1} \neq b_i \cup a_i \cup \cdots \cup a_{k_i-1}$. But $a_1 \cup \cdots \cup a_k > b_i \cup a_1 \cup \cdots \cup a_{k_i-1}$ and hence $a_1 \cup \cdots \cup a_k = b_{i+1} \cup a_1 \cup \cdots \cup a_{k_i-1}$ contrary to our assumption. Thus $b_i \cup a_1 \cup \cdots \cup a_{k_i-1} \supset b_{i+1}$. If $b_i \cup a_1 \cup \cdots \cup a_{k_i-2} \not\supset b_{i+1}$, we have $b_i \cup a_1 \cup \cdots \cup a_{k_i-1} \supset b_{i+1} \cup a_1 \cup \cdots \cup a_{k_i-2} \supset b_i \cup a_1 \cup \cdots \cup a_{k_i-2}$. But $b_i \cup a_1 \cup \cdots \cup a_{k_i-1} > b_i \cup a_1 \cup \cdots \cup a_{k_i-2}$ and hence $b_i \cup a_1 \cup \cdots \cup a_{k_i-1} = b_{i+1} \cup a_1 \cup \cdots \cup a_{k_i-2}$ contrary to our assumption. Thus $b_i \cup a_1 \cup \cdots \cup a_{k_i-2} \supset b_{i+1}$. Continuing in this manner we eventually have $b_i \cup a_1 \supset b_{i+1}$. But then $b_i \cup a_1 \supset b_{i+1} \supset b_i$ and $b_{i+1} \neq b_i$. Hence $b_{i+1} = b_i \cup a_1 = b_{i+1} \cup a_1$ which contradicts our assumption. We conclude, then, that at least two members of the above chain are equal. Thus (renumbering the a's if necessary) using the Birkhoff condition we have $a \subset b_1 \subset \cdots \subset b_i \subset b_{i+1} < b_{i+1} \cup a_1 < \cdots < b_{i+1} \cup a_1 \cup \cdots \cup a_{k_{i+1}-1} < a_1 \cup \cdots \cup a_k$ where $k_{i+1} \leq k_i - 1 \leq k - (i+1)$. By induction, we get $a = b_0 \subset b_1 \subset \cdots \subset b_{r-1} < a_1 \cup \cdots \cup a_k$ where $r \leq k$. But then $b_r = a_1 \cup \cdots \cup a_k$ and hence $r = l$ which contradicts $l > k$. Thus $l \leq k$ and the lemma follows.

The dual of condition B1 is the condition

B2. $a \cup b > b \rightarrow a > a \cap b$.

G. Birkhoff (Birkhoff [2]) has proved the following lemma which relates B1 and B2 to modularity.

LEMMA 1.5. *An archimedean lattice $\mathfrak{S}$ is modular if and only if* B1 *and* B2 *are satisfied.*

2. **Lattice ideals.** A sublattice $\mathfrak{a}$ of $\mathfrak{S}$ is said to be an *ideal* if $x \supset a$, $a \in \mathfrak{a}$ implies $x \in \mathfrak{a}$. If $\mathfrak{a}$ consists of all elements x such that $x \supset a$ for a fixed a, then $\mathfrak{a}$ is said to be a *principal* ideal and we write $\mathfrak{a} = (a)$. Now suppose that $\mathfrak{S}$ satisfies the descending chain condition. Then the set of elements in $\mathfrak{a}$ has a cross-cut which can be expressed as a cross-cut of a finite number of them and hence belongs to $\mathfrak{a}$. Thus $\mathfrak{a}$ consists of all divisors of a fixed element of $\mathfrak{S}$ and hence is principal. Conversely, if every ideal of $\mathfrak{S}$ is principal, then a descending chain $a_1 \supset a_2 \supset \cdots$ generates an ideal $\mathfrak{a}$ which consists of all x such that $x \supset a_k$ for some k. But then $\mathfrak{a} = (a)$ and $a \supset a_k$ for some k. Hence $a = a_k = a_{k+1} = \cdots$ and every descending chain has only a finite number of distinct elements. We thus have

LEMMA 2.1. $\mathfrak{S}$ *satisfies the descending chain condition if and only if every ideal is principal.*

The set of ideals of $\mathfrak{S}$ will be denoted by $\mathfrak{L}$.

DEFINITION 2.1. *The union $\mathfrak{a} \cup \mathfrak{b}$ of two ideals $\mathfrak{a}$ and $\mathfrak{b}$ is the set of all elements x such that $x \supset a \cup b$ for some $a \in \mathfrak{a}$ and $b \in \mathfrak{b}$. Similarly the cross-cut $\mathfrak{a} \cap \mathfrak{b}$ is the set of all elements y such that $y \supset a \cap b$ for some $a \in \mathfrak{a}$ and $b \in \mathfrak{b}$.*

It is readily verified that the union and cross-cut so defined are ideals and that $\mathfrak{L}$ is a lattice under these operations. The union $\mathfrak{a} \cup \mathfrak{b}$ is simply the set-theoretic cross-cut of $\mathfrak{a}$ and $\mathfrak{b}$.

The definition of cross-cut may be readily extended to any subset S of $\mathfrak{L}$. $\Pi(S)$ consists of all elements of $\mathfrak{S}$ which belong to the cross-cut of a finite number of ideals of S. If $\mathfrak{S}$ has a unit element u, the union $\Sigma(S)$ is also defined and is simply the set-theoretic cross-cut of the ideals of S.

If $\mathfrak{a}$ and $\mathfrak{b}$ are principal ideals $\mathfrak{a} = (a)$ and $\mathfrak{b} = (b)$, then by Definition 2.1 $\mathfrak{a} \cup \mathfrak{b} = (a \cup b)$ and $\mathfrak{a} \cap \mathfrak{b} = (a \cap b)$. Hence the set of principal ideals forms a sublattice of $\mathfrak{L}$ which is isomorphic to $\mathfrak{S}$ and we may thus consider $\mathfrak{S}$ as a sublattice of $\mathfrak{L}$.

LEMMA 2.2. $\mathfrak{L}$ *is a modular (distributive) if and only if $\mathfrak{S}$ is modular (distributive).*

Since $\mathfrak{S}$ is a sublattice of $\mathfrak{L}$, the modularity (distributivity) of $\mathfrak{L}$ implies the modularity (distributivity) of $\mathfrak{S}$.

Now let $\mathfrak{S}$ be distributive and let $x \in \mathfrak{a} \cup (\mathfrak{b} \cap \mathfrak{c})$. Then $x \supset a \cup (b \cap c)$ where $a \in \mathfrak{a}$, $b \in \mathfrak{b}$ and $c \in \mathfrak{c}$ by Definition 2.1. But then $x \supset a \cup (b \cap c) \supset (a \cup b) \cap (a \cup c)$ since $\mathfrak{S}$ is distributive and hence $x \in (\mathfrak{a} \cup \mathfrak{b}) \cap (\mathfrak{a} \cup \mathfrak{c})$. Thus $\mathfrak{a} \cup (\mathfrak{b} \cap \mathfrak{c}) \supset (\mathfrak{a} \cup \mathfrak{b}) \cap (\mathfrak{a} \cup \mathfrak{c})$. But $(\mathfrak{a} \cup \mathfrak{b}) \cap (\mathfrak{a} \cup \mathfrak{c}) \supset \mathfrak{a} \cup (\mathfrak{b} \cap \mathfrak{c})$ trivially. Hence $\mathfrak{L}$ is distributive. Now let $\mathfrak{S}$ be modular. Suppose $\mathfrak{a} \supset \mathfrak{b}$ and $x \in \mathfrak{b} \cup (\mathfrak{a} \cap \mathfrak{c})$. Then $x \supset b \cup (a \cap c)$ where $a \in \mathfrak{a}$, $b \in \mathfrak{b}$ and $c \in \mathfrak{c}$ by Definition 2.1. Now since $\mathfrak{a} \supset \mathfrak{b}$ we have $a \supset b_1$ where $b_1 \in \mathfrak{b}$. But then $x \supset (b \cap b_1) \cup (a \cap c)$ and $a \supset b \cap b_1$ where $b_1 \cap b \in \mathfrak{b}$. Hence $x \supset a \cap ((b \cap b_1) \cup c)$ since $\mathfrak{S}$ is modular and $x \in \mathfrak{a} \cap (\mathfrak{b} \cup \mathfrak{c})$. Thus $\mathfrak{b} \cup (\mathfrak{a} \cap \mathfrak{c}) \supset \mathfrak{a} \cap (\mathfrak{b} \cup \mathfrak{c})$ and since $\mathfrak{a} \cap (\mathfrak{b} \cup \mathfrak{c}) \supset \mathfrak{b} \cup (\mathfrak{a} \cap \mathfrak{c})$ trivially, $\mathfrak{L}$ is modular. This completes the proof.

LEMMA 2.3. *Let* $\mathfrak{a} \supset \mathfrak{b} \supset \cdots \supset \mathfrak{u} \supset \cdots$ *be a chain of ideals such that* $\mathfrak{u} \supset (a)$ *and* $\mathfrak{u} \neq (a)$ *for all ideals of the chain. Then if* $\mathfrak{p}$ *is the cross-cut of the ideals of the chain,* $\mathfrak{p} \supset (a)$ *and* $\mathfrak{p} \neq (a)$.

We note that $\mathfrak{p}$ is the set-theoretic union of the elements of the ideals $\mathfrak{a}, \mathfrak{b}, \cdots, \mathfrak{u}, \cdots$. For if $x \in \mathfrak{p}$, then x divides a finite cross-cut of the ideals of the chain and hence divides some ideal of the chain. Now suppose $\mathfrak{p} = a$. Then $a \in \mathfrak{p}$ and $a \in \mathfrak{u}$ for some $\mathfrak{u}$. But then $\mathfrak{u} = (a)$ contrary to assumption. Hence $\mathfrak{p} \neq a$.

The results so far have been independent of the well ordering hypothesis. However, to prove the fundamental property of the ideals we must assume that the elements of $\mathfrak{S}$ can be well ordered. This will be assumed through the remainder of the paper.

THEOREM 2.1. *Let* $\mathfrak{b} \supset (a)$ *and* $\mathfrak{b} \neq (a)$. *Then there exists an ideal* $\mathfrak{p}$ *such that* $\mathfrak{b} \supset \mathfrak{p} > (a)$.

Proof. Let U be the set of all elements x such that $x \supset a$. Let U be well ordered, $U = \{x_\nu\}$, $\nu < \sigma$. Define $\mathfrak{a}_0 = \mathfrak{b}$. Now suppose that $\mathfrak{a}_\mu$ has been defined for all $\mu < \nu$ in such a way that $\mathfrak{a}_\mu \neq (a)$, $\mathfrak{a}_\mu \supset \mathfrak{a}_{\mu'}$ if $\mu \leq \mu'$, and $\mathfrak{a}_\mu \cap x_\mu = \mathfrak{a}_\mu$ or $\mathfrak{a}_\mu \cap x_\mu = a$. Let $\mathfrak{c}_\nu$ be the cross-cut of all $\mathfrak{a}_\mu$ with $\mu < \nu$. Then $\mathfrak{c}_\nu \neq (a)$ by Lemma 2.3. If $\mathfrak{c}_\nu \cap x_\nu \neq (a)$, let $\mathfrak{a}_\nu = \mathfrak{c}_\nu \cap x_\nu$; otherwise let $\mathfrak{a}_\nu = \mathfrak{c}_\nu$. Then $\mathfrak{a}_\nu \neq (a)$ and $\mathfrak{a}_\mu \supset \mathfrak{a}_\nu$, all $\mu < \nu$. Clearly $\mathfrak{a}_\nu \cap x_\nu = a$ or $\mathfrak{a}_\nu$. Now let $\mathfrak{p} = \Pi_{\nu < \sigma} \mathfrak{a}_\nu$. Then $\mathfrak{p} \neq a$ by Lemma 2.3 and $\mathfrak{b} \supset \mathfrak{p}$. If $\mathfrak{p} \supset \mathfrak{a} \supset (a)$ and $\mathfrak{p} \neq \mathfrak{a}$, there exists an element $x \in \mathfrak{a}$ such that $x \notin \mathfrak{p}$. Since $x \supset a$ we have $x = x_\nu$ for some ν. But then $\mathfrak{a}_\nu \cap x = a$ since otherwise $x_\nu \supset \mathfrak{a}_\nu \supset \mathfrak{p}$ which contradicts $x \notin \mathfrak{p}$. Thus $a = \mathfrak{a}_\nu \cap x \supset \mathfrak{p} \cap \mathfrak{a} = \mathfrak{a} \supset a$ and $\mathfrak{a} = (a)$. Hence $\mathfrak{p} > a$.

In the special instances of Boolean algebras and distributive lattices, Theorem 2.1 gives respectively the existence of the prime ideals of Stone (Stone [1]) and the maximal collections of Wallman (Wallman [1]).

We next prove a theorem which enables us to pass from ideal relations to the corresponding element relations. The following lemma is required.

LEMMA 2.4. *Let* $\mathfrak{a} = \mathfrak{a}(\mathfrak{a}_1, \cdots, \mathfrak{a}_n)$ *be an ideal obtained from the ideals*

$\mathfrak{a}_1, \cdots, \mathfrak{a}_n$ *by forming a finite number of unions and cross-cuts. Then if $x \in \mathfrak{a}$, there exist elements $a_1, \cdots, a_n, a_i \in \mathfrak{a}_i$, such that $x \supset \mathfrak{a}(a_1, \cdots, a_n)$.*

For let $n(\mathfrak{a})$ denote the number of union and cross-cut symbols in the expression $\mathfrak{a}(\mathfrak{a}_1, \cdots, \mathfrak{a}_n)$. Suppose that the lemma is true for all expressions $\mathfrak{a}$ for which $n(\mathfrak{a}) < k$. Let $n(\mathfrak{a}) = k$. Then $\mathfrak{a} = \mathfrak{a}_1 \circ \mathfrak{a}_2$ where $\circ$ is either $\cap$ or $\cup$ and $n(\mathfrak{a}_1) < k$, $n(\mathfrak{a}_2) < k$. Now if $x \in \mathfrak{a}$ we have $x \supset x_1 \circ x_2$ where $x_1 \in \mathfrak{a}_1$ and $x_2 \in \mathfrak{a}_2$ by the definition of union and cross-cut. But then by the induction assumption elements $a_1', \cdots, a_n'$ and $a_1'', \cdots, a_n''$ exist such that $x_1 \supset \mathfrak{a}_1(a_1', \cdots, a_n')$, $x_2 \supset \mathfrak{a}_2(a_1', \cdots, a_n'')$. Let $a_i = a_i' \cap a_i''$. Then $x \supset x_1 \circ x_2 \supset \mathfrak{a}_1(a_1', \cdots, a_n')$ $\circ \mathfrak{a}_2(a_1', \cdots, a_n'') \supset \mathfrak{a}_1(a_1, \cdots, a_n) \circ \mathfrak{a}_2(a_1, \cdots, a_n) = \mathfrak{a}(a_1, \cdots, a_n)$ and a_i is clearly in $\mathfrak{a}_i$. Since the lemma is trivially true when $n(\mathfrak{a}) = 1$ by Definition 2.1, the proof is complete.

THEOREM 2.2. *Let $(a) = \mathfrak{a}(\mathfrak{a}_1, \cdots, \mathfrak{a}_n)$ where $\mathfrak{a}$ is obtained from $\mathfrak{a}_1, \cdots, \mathfrak{a}_n$ by forming a finite number of union and cross-cuts. Then $(a) = \mathfrak{a}(a_1, \cdots, a_n)$ where $a_i \in \mathfrak{a}_i$.*

Proof. By Lemma 2.4 $a \supset \mathfrak{a}(a_1, \cdots, a_n)$ where $a_i \in \mathfrak{a}_i$. But then $\mathfrak{a}(a_1, \cdots, a_n) \supset \mathfrak{a}(\mathfrak{a}_1, \cdots, \mathfrak{a}_n) = (a)$. Hence $(a) = \mathfrak{a}(a_1, \cdots, a_n)$.

As an example, if $a = \mathfrak{a}_1 \cap \cdots \cap \mathfrak{a}_n$ then elements $a_i \in \mathfrak{a}_i$ exist such that $a = a_1 \cap \cdots \cap a_n$.

We conclude this section with two useful lemmas on irreducibles[7].

LEMMA 2.5. *If q is irreducible in $\mathfrak{S}$, then q is irreducible in $\mathfrak{L}$.*

For if q is reducible in $\mathfrak{L}$, then $q = \mathfrak{a} \cap \mathfrak{b}$, $\mathfrak{a}$, $\mathfrak{b} \neq q$. But then $q = a \cap b$, $a \in \mathfrak{a}$, $b \in \mathfrak{b}$ by Theorem 2.2. Clearly $a \neq q$ and $b \neq q$. Hence q is reducible in $\mathfrak{S}$. Inverting the logic gives the lemma.

LEMMA 2.6. *Let every element of $\mathfrak{S}$ be expressible as a cross-cut of irreducibles. Then if $\mathfrak{a} \supset \mathfrak{b}$, $\mathfrak{a} \neq \mathfrak{b}$, there exists an irreducible q of $\mathfrak{S}$ such that $q \supset \mathfrak{b}$, $q \not\supset \mathfrak{a}$.*

For since $\mathfrak{a} \neq \mathfrak{b}$, b exists such that $b \in \mathfrak{b}$, $b \notin \mathfrak{a}$. Let $b = q_1 \cap \cdots \cap q_k$. If $q_i \in \mathfrak{a}$ for every i then $b \in \mathfrak{a}$ contrary to assumption. Hence $q_i \notin \mathfrak{a}$ for some i. But then $q_i \supset b \supset \mathfrak{b}$.

3. **Birkhoff lattices.** In D1 and D2 a lattice satisfying B1 was defined to be a Birkhoff lattice. Since both the ascending and descending chain conditions were assumed to hold, B1 was never satisfied trivially. Now in a sufficiently general lattice no covering relations may exist and B1 will hold vacuously. Hence we formulate a more general definition which reduces to that used in D1 and D2 if the descending chain condition holds.

DEFINITION 3.1. *A lattice $\mathfrak{S}$ is said to be a Birkhoff lattice if each element of $\mathfrak{S}$ satisfies the Birkhoff condition in the lattice of ideals.*

---

[7] An element q is said to be *cross-cut irreducible* or simply *irreducible* if $q = a \cap b \rightarrow q = a$ or $q = b$. q is said to be *union irreducible* if $q = a \cup b \rightarrow q = a$ or $q = b$.

A lattice $\mathfrak{S}$ is never vacuously a Birkhoff lattice since by Theorem 2.1 covering ideals always exist. Furthermore if the descending chain condition holds, then every ideal is principal and $\mathfrak{S}$ is a Birkhoff lattice if and only if B1 holds in $\mathfrak{S}$.

Now if $\mathfrak{S}$ has a unit element u and a is any element of $\mathfrak{S}$, then the union of the ideals covering a exists and will be denoted by $\mathfrak{u}_a$. Let $\mathfrak{L}_a$ denote the quotient lattice of all ideals of $\mathfrak{L}$ which are divisible by $\mathfrak{u}_a$ and which divide a. Then $\mathfrak{L}_a$ is a dense sublattice of $\mathfrak{L}$ and every proper divisor of a in $\mathfrak{L}$ divides some point ideal of $\mathfrak{L}_a$ by Theorem 2.1. Clearly $\mathfrak{L}_a$ reduces to the sublattice $\mathfrak{S}_a$ of the previous papers if the descending chain condition holds. The essential properties of $\mathfrak{S}_a$ followed from its finite dimensionality. But $\mathfrak{L}_a$ is in general *not* finite dimensional. However we now prove a theorem which insures the archimedean character of $\mathfrak{L}_a$ in most cases of arithmetical interest. We need the following lemma:

LEMMA 3.1. *Let $\mathfrak{S}$ be a Birkhoff lattice. Then if $\mathfrak{p}_1, \cdots, \mathfrak{p}_k$ is a maximal independent set of point ideals of $\mathfrak{L}_a$, the length of any chain of $\mathfrak{L}_a$ is not greater than k.*

Since the length of any chain is one less than the number of distinct members of the chain, the lemma follows immediately from Lemma 1.4 and Definition 3.1.

According to Lemma 3.1, $\mathfrak{L}_a$ is archimedean if and only if $\mathfrak{u}_a$ can be expressed as a union of a finite number of point ideals of $\mathfrak{L}_a$.

THEOREM 3.1. *Let $\mathfrak{S}$ be a Birkhoff lattice in which every element may be represented as a cross-cut of irreducibles. Then $\mathfrak{L}_a$ is archimedean if and only if the number of components in the irreducible decompositions of a is bounded.*

Proof. Let the number of components in the irreducible decompositions of a be bounded, say less than n. Then if $\mathfrak{L}_a$ is not archimedean, by Lemmas 1.2 and 3.1 there are n union independent point ideals $\mathfrak{p}_1, \cdots, \mathfrak{p}_n$ of $\mathfrak{L}_a$ which generate a Boolean algebra. Let $\mathfrak{a}_i = \mathfrak{p}_1 \cup \cdots \cup \mathfrak{p}_{i-1} \cup \mathfrak{p}_{i+1} \cup \cdots \cup \mathfrak{p}_n$. Then $a = \mathfrak{a}_1 \cap \cdots \cap \mathfrak{a}_n$. Hence by Theorem 2.2 $a = a_1 \cap \cdots \cap a_n$ where $a_i \in \mathfrak{a}_i$. Now let $a_i = q_{i1} \cap \cdots \cap q_{ik_i}$ where $q_{i1}, \cdots, q_{ik_i}$ are irreducibles of $\mathfrak{S}$. Then $a = q_{11} \cap q_{12} \cap \cdots \cap q_{nk_n}$ and this representation may be reduced([8]) by dropping our superfluous irreducibles. However not all of the irreducibles belonging to any one a_i may be dropped out since otherwise $a = q_{11} \cap \cdots \cap q_{nk_n} \supset a_1 \cap \cdots \cap a_{i-1} \cap a_{i+1} \cap \cdots \cap a_n \supset \mathfrak{a}_1 \cap \cdots \cap \mathfrak{a}_{i-1} \cap \mathfrak{a}_{i+1} \cap \cdots \cap \mathfrak{a}_n \supset \mathfrak{p}_i$ contrary to $\mathfrak{p}_i > a$. Hence a has a decomposition having at least n components. But this contradicts our assumption that the number of components is less than n. Hence $\mathfrak{L}_a$ is archimedean and of length less than n.

On the other hand let the number of components be unbounded. Then for

---

([8]) A representation $a = a_1 \cap a_2 \cap \cdots \cap a_n$ is said to be reduced if $a_1, \cdots, a_n$ are cross-cut independent.

every k there is an irreducible decomposition $a = q_1 \cap \cdots \cap q_n$ with $n \geq k$. Let $q_i' = q_1 \cap \cdots \cap q_{i-1} \cap q_{i+1} \cap \cdots \cap q_n$. Then $q_i' \supset a$ and $q_i' \neq a$ since the representation is reduced. Hence $q_i' \supset \mathfrak{p}_i > a$ by Theorem 2.1. Suppose $\mathfrak{p}_1 \cup \cdots \cup \mathfrak{p}_{i-1} \cup \mathfrak{p}_{i+1} \cup \cdots \cup \mathfrak{p}_n \supset \mathfrak{p}_i$. Then $q_i \supset q_1' \cup \cdots \cup q_{i-1}' \cup q_{i+1}' \cup \cdots \cup q_n' \supset \mathfrak{p}_1 \cup \cdots \cup \mathfrak{p}_{i-1} \cup \mathfrak{p}_{i+1} \cup \cdots \cup \mathfrak{p}_n \supset \mathfrak{p}_i$ and $a = q_i \cap q_i' \supset \mathfrak{p}_i$ which contradicts $\mathfrak{p}_i > a$. Thus $\mathfrak{p}_1, \cdots, \mathfrak{p}_n$ are union independent. Hence for every k there are more than k union independent point ideals of $\mathfrak{L}_a$ and $\mathfrak{L}_a$ is *not* archimedean.

If $\mathfrak{L}_a$ is archimedean it has some simple structure properties which follow from the Birkhoff condition.

Theorem 3.2. *Let $\mathfrak{S}$ be a Birkhoff lattice. Then if $\mathfrak{L}_a$ is archimedean, it is complemented and every ideal can be expressed as a cross-cut of simple ideals.*

Proof. Let $a \in \mathfrak{L}_a$ and let $\mathfrak{p}_1, \cdots, \mathfrak{p}_k$ be a maximal independent set of point ideals of $\mathfrak{L}_a$ divisible by a. Imbed $\mathfrak{p}_1, \cdots, \mathfrak{p}_k$ in a maximal independent set $\mathfrak{p}_1, \cdots, \mathfrak{p}_n$. Let $a' = \mathfrak{p}_{k+1} \cup \cdots \cup \mathfrak{p}_n$. Then $a \cup a' \supset \mathfrak{p}_1 \cup \cdots \cup \mathfrak{p}_n \supset \mathfrak{u}_a$. Hence $a \cup a' = \mathfrak{u}_a$. Now suppose that $a \cap a' \neq a$. Then $a \cap a' \supset \mathfrak{p} > a$ by Theorem 2.1. Since $a \supset \mathfrak{p}$ we have $\mathfrak{p}_1 \cup \cdots \cup \mathfrak{p}_k \supset \mathfrak{p}$ by the maximal property of $\mathfrak{p}_1, \cdots, \mathfrak{p}_k$ and $a = (\mathfrak{p}_1 \cup \cdots \cup \mathfrak{p}_k) \cap a' \supset \mathfrak{p}$, which contradicts $\mathfrak{p} > a$. Hence $a \cap a' = a$ and $\mathfrak{L}_a$ is complemented.

Now let q be irreducible in $\mathfrak{L}_a$. Let $\mathfrak{p}_1, \cdots, \mathfrak{p}_k$ be a maximal independent set of point ideals of $\mathfrak{L}_a$ divisible by q and let this set be imbedded in a maximal independent set $\mathfrak{p}_1, \cdots, \mathfrak{p}_k, \cdots, \mathfrak{p}_n$. Then $q \not\supset \mathfrak{p}_{k+1}, \cdots, \mathfrak{p}_n$ and hence $q \cup \mathfrak{p}_i > q$, $i = k+1, \cdots, n$, by B1$'$. But since q is irreducible in $\mathfrak{L}_a$ we have $q \cup \mathfrak{p}_{k+1} = \cdots = q \cup \mathfrak{p}_n$. Hence $\mathfrak{u}_a = q \cup \mathfrak{u}_a = q \cup \mathfrak{p}_{k+1} \cup \cdots \cup q \cup \mathfrak{p}_n = q \cup \mathfrak{p}_{k+1} > q$. Thus each ideal which is irreducible in $\mathfrak{L}_a$ is a simple ideal of $\mathfrak{L}_a$ and since $\mathfrak{L}_a$ is archimedean each ideal of $\mathfrak{L}_a$ can be represented as a cross-cut of simple ideals.

If $\mathfrak{L}_a$ is not archimedean it will in general neither be complemented nor will every ideal be expressible as a cross-cut of simple ideals[9]. In the archimedean case an arbitrary complement of a in $\mathfrak{L}_a$ will be denoted by a'.

Definition 3.2. *An ideal $c \neq \mathfrak{u}_a$ of $\mathfrak{L}_a$ is said to be characteristic if there exists an irreducible q of $\mathfrak{S}$ which divides exactly the same point ideals of $\mathfrak{L}_a$ as c.*

Theorem 3.3. *An element $a \in \mathfrak{S}$ has a reduced representation $a = q_1 \cap \cdots \cap q_n$ where $q_1, \cdots, q_n$ are irreducibles if and only if a has a reduced representation $a = c_1 \cap \cdots \cap c_n$ where $c_1, \cdots, c_n$ are characteristic ideals of $\mathfrak{L}_a$ such that $q_i \supset c_i$.*

Proof. Let $a = q_1 \cap \cdots \cap q_n$ be a reduced representation of a as a cross-cut of irreducibles. If $q_i \supset \mathfrak{u}_a$ for some i, then $q_1 \cap \cdots \cap q_{i-1} \cap q_{i+1} \cap \cdots \cap q_n \supset \mathfrak{p}_i > a$ and hence $a = q_1 \cap \cdots \cap q_n \supset \mathfrak{p}_i > a$, which is impossible. Thus $q_i \not\supset \mathfrak{u}_a$. Let c_i be a characteristic ideal associated with q_i. There is always at least one

---

[9] See §7 for an example.

such ideal, namely, the union of the point ideals of $\mathfrak{L}_a$ divisible by q_i. Now $a = q_1 \cap \cdots \cap q_n \supset c_1 \cap \cdots \cap c_n \supset a$ implies $a = c_1 \cap \cdots \cap c_n$. Suppose $c_i \supset c_1 \cap \cdots \cap c_{i-1} \cap c_{i+1} \cap \cdots \cap c_n$. Then $q_i \cap \cdots \cap q_{i-1} \cap q_{i+1} \cap \cdots \cap q_n \supset \mathfrak{p}_i > a$ implies $c_1 \cap \cdots \cap c_{i-1} \cap c_{i+1} \cap \cdots \cap c_n \supset \mathfrak{p}_i$. But then $a = c_i \cap c_1 \cap \cdots \cap c_{i-1} \cap c_{i+1} \cap \cdots \cap c_n \supset \mathfrak{p}_i$ which is impossible. Hence the representation $a = c_1 \cap \cdots \cap c_n$ is reduced.

Now let $a = c_1 \cap \cdots \cap c_n$ where $c_1, \cdots, c_n$ are characteristic ideals and the representation is reduced. Let $q_1, \cdots, q_n$ be associated irreducibles. Suppose $q_1 \cap \cdots \cap q_n \supset \mathfrak{p} > a$. Then $a = c_1 \cap \cdots \cap c_n \supset \mathfrak{p} > a$ which is impossible. Hence $a = q_1 \cap \cdots \cap q_n$. It follows easily that this representation is reduced.

The characteristic ideals of $\mathfrak{L}_a$ can be characterized in terms of the structure of $\mathfrak{L}$ as follows:

THEOREM 3.4. *Let $\mathfrak{S}$ be a Birkhoff lattice in which each element can be expressed as a cross-cut of irreducibles. Then if $\mathfrak{L}_a$ is archimedean, c is characteristic if and only if there exists an ideal $\mathfrak{x} \in \mathfrak{L}$ such that $\mathfrak{x} \supset c$, $c' \cup \mathfrak{x} > \mathfrak{x}$ and $c' \cap \mathfrak{x} = a$ for every c'.*

Proof. Let us first assume that such an ideal $\mathfrak{x}$ exists. Then $\mathfrak{u}_a \cup \mathfrak{x} = c \cup c' \cup \mathfrak{x} = c' \cup \mathfrak{x}$. Let q be an irreducible such that $q \supset \mathfrak{x}$, $q \not\supset \mathfrak{u}_a \cup \mathfrak{x}$ (Lemma 2.6). Since $q \supset \mathfrak{x} \supset c$, q divides every point ideal of $\mathfrak{L}_a$ which c divides. Now let $q \supset \mathfrak{p}$. Then if $\mathfrak{x} \not\supset \mathfrak{p}$ we have $c' \cup \mathfrak{x} = \mathfrak{u}_a \cup \mathfrak{x} \supset \mathfrak{p} \cup \mathfrak{x} \supset \mathfrak{x}$ and $\mathfrak{p} \cup \mathfrak{x} \neq \mathfrak{x}$. Hence $c' \cup \mathfrak{x} = \mathfrak{p} \cup \mathfrak{x}$ and $q \supset \mathfrak{p} \cup \mathfrak{x} \supset c' \cup \mathfrak{x}$ which contradicts the definition of q. Hence $\mathfrak{x} \supset \mathfrak{p}$. Now if $c \not\supset \mathfrak{p}$, then $c' \supset \mathfrak{p}$ for some c'. But then $a = c' \cap \mathfrak{x} \supset \mathfrak{p}$ which is impossible. Hence $q \supset \mathfrak{p}$ implies $c \supset \mathfrak{p}$ and c is thus characteristic.

On the other hand let c be characteristic and let q be an irreducible associated with c. Then $q \cup c' > q$ for every c'. For there is a point ideal $\mathfrak{p}$ such that $c' \supset \mathfrak{p}$, $c \not\supset \mathfrak{p}$ since otherwise we would have $c' = a$ and $c = \mathfrak{u}_a$ contrary to the definition of a characteristic ideal. Now $q \cup \mathfrak{p} = q \cup \mathfrak{u}_a > q$ since q is irreducible in $\mathfrak{L}$ by Lemma 2.5. Hence $q \cup \mathfrak{u}_a = q \cup c' = q \cup \mathfrak{p} > q$. Now if $c' \cap q \neq a$, then $c' \cap q \supset \mathfrak{p} > a$ and hence $c' \supset \mathfrak{p}$, $q \supset \mathfrak{p}$ by Theorem 2.2. But then $c \supset \mathfrak{p}$ and hence $a = c \cap c' \supset \mathfrak{p}$ which is impossible. Thus $c' \cap q = a$ for every c'.

COROLLARY 3.1. *Each simple ideal of $\mathfrak{L}_a$ is characteristic.*

We may take $\mathfrak{x}$ to be the simple ideal itself.

THEOREM 3.5. *Let $\mathfrak{S}$ be a Birkhoff lattice in which each element can be expressed as a cross-cut of irreducibles. Then if $\mathfrak{L}_a$ is archimedean, each characteristic ideal c of $\mathfrak{L}_a$ occurs in a reduced representation $a = c \cap c_1 \cap \cdots \cap c_k$ where k is the number of maximal independent point ideals divisible by c and $c_1, \cdots, c_k$ are characteristic ideals of $\mathfrak{L}_a$.*

Proof. Let $\mathfrak{p}_1, \cdots, \mathfrak{p}_k$ be a maximal independent set of point ideals of $\mathfrak{L}_a$ divisible by c. Imbed $\mathfrak{p}_1, \cdots, \mathfrak{p}_k$ in a maximal independent set $\mathfrak{p}_1, \cdots, \mathfrak{p}_k, \cdots, \mathfrak{p}_n$. Let $c_i = \mathfrak{p}_1 \cup \cdots \cup \mathfrak{p}_{i-1} \cup \mathfrak{p}_{i+1} \cup \cdots \cup \mathfrak{p}_k \cup \cdots \cup \mathfrak{p}_n$, $i = 1, \cdots, k$. If $c \cap c_1$

$\cap \cdots \cap c_k \neq a$ we have $c \cap c_1 \cap \cdots \cap c_k \supset \mathfrak{p} > a$ and $c \supset \mathfrak{p}$ implies $\mathfrak{p}_1 \cup \cdots \cup \mathfrak{p}_k \supset \mathfrak{p}$. But then $a = (\mathfrak{p}_1 \cup \cdots \cup \mathfrak{p}_k) \cap c_1 \cap \cdots \cap c_k \supset \mathfrak{p}$ which is impossible. Hence $a = c \cap c_1 \cap \cdots \cap c_k$. Also since $c \cap c_1 \cap \cdots \cap c_{i-1} \cap c_{i+1} \cap \cdots \cap c_k \supset \mathfrak{p}_i$ the representation is reduced. Since $c_1, \cdots, c_k$ are simple ideals of $\mathfrak{L}_a$, they are characteristic by Corollary 3.1.

COROLLARY 3.2. *Let $\mathfrak{S}$ be a Birkhoff lattice in which every element can be expressed as a cross-cut of irreducibles. Then if $\mathfrak{L}_a$ is archimedean of length k, a has a reduced decomposition into irreducibles with k components.*

For by Lemma 1.2 and Theorem 3.4, a has a reduced representation as a cross-cut of k characteristic ideals of $\mathfrak{L}_a$.

LEMMA 3.2. *Let $\mathfrak{S}$ be a Birkhoff lattice and let $\mathfrak{L}_a$ be archimedean for some a. Then $\mathfrak{L}_a$ is modular if and only if it satisfies B2.*

For let $\mathfrak{L}_a$ satisfy B2 and let $\mathfrak{q}$ be a union irreducible ideal of $\mathfrak{L}_a$. If $\mathfrak{s} \not\supset \mathfrak{q}$ and $\mathfrak{s}$ is a simple ideal of $\mathfrak{L}_a$ we have $\mathfrak{q} > \mathfrak{q} \cap \mathfrak{s}$ by B2. Hence since $\mathfrak{q}$ is union irreducible we have $\mathfrak{q} \cap \mathfrak{s} = \mathfrak{q} \cap \mathfrak{s}'$ for any two simple ideals $\mathfrak{s}$ and $\mathfrak{s}'$ which do not divide $\mathfrak{q}$. Let $a = \mathfrak{s}_1 \cap \cdots \cap \mathfrak{s}_n$ where $\mathfrak{s}_1, \cdots, \mathfrak{s}_l \supset \mathfrak{q}$; $\mathfrak{s}_{l+1}, \cdots, \mathfrak{s}_n \not\supset \mathfrak{q}$. Then $a = \mathfrak{q} \cap a = \mathfrak{q} \cap \mathfrak{s}_1 \cap \cdots \cap \mathfrak{s}_n = (\mathfrak{q} \cap \mathfrak{s}_{l+1}) \cap \cdots \cap (\mathfrak{q} \cap \mathfrak{s}_n) = \mathfrak{q} \cap \mathfrak{s}_{l+1} < \mathfrak{q}$. Hence $\mathfrak{q}$ is a point of $\mathfrak{L}_a$ and every ideal of $\mathfrak{L}_a$ is a union of point ideals. Now let $\mathfrak{a} > \mathfrak{a} \cap \mathfrak{b}$ in $\mathfrak{L}_a$. Then since every ideal is a union of point ideals, there exists a point ideal $\mathfrak{p}$ such that $\mathfrak{a} \supset \mathfrak{p}$, $\mathfrak{a} \cap \mathfrak{b} \not\supset \mathfrak{p}$. But then $\mathfrak{a} = (\mathfrak{a} \cap \mathfrak{b}) \cup \mathfrak{p}$. Hence $\mathfrak{a} \cup \mathfrak{b} = (\mathfrak{a} \cap \mathfrak{b}) \cup \mathfrak{p} \cup \mathfrak{b} = \mathfrak{p} \cup \mathfrak{b} > \mathfrak{b}$ since $\mathfrak{S}$ is a Birkhoff lattice. Thus B1 and B2 hold in $\mathfrak{L}_a$ and $\mathfrak{L}_a$ is modular by Lemma 1.5. Conversely, if $\mathfrak{L}_a$ is modular, then B2 is satisfied by Lemma 1.5. This completes the proof.

According to Theorem 3.1, if every element of a lattice $\mathfrak{S}$ has a decomposition into irreducibles and the number of components in the decompositions of a is bounded, then $\mathfrak{L}_a$ is archimedean. This result can be sharpened considerably if $\mathfrak{S}$ is modular.

LEMMA 3.3. *Let $\mathfrak{S}$ be a modular lattice. Then if an element a has a decomposition into irreducibles, $\mathfrak{L}_a$ is archimedean.*

For let $a = q_1 \cap \cdots \cap q_k$ where $q_1, \cdots, q_k$ are irreducible. Since $\mathfrak{S}$ is modular, $\mathfrak{L}$ is modular by Lemma 2.2. Now if $q_i \not\supset \mathfrak{p}$ where $\mathfrak{p} > a$, we have $q_i \cup \mathfrak{p} > q_i$ and hence $q_i \cup \mathfrak{u}_a > q_i$ since q_i is irreducible. But then $\mathfrak{u}_a > \mathfrak{u}_a \cap q_i$ since $\mathfrak{L}$ is modular. Thus each irreducible q_i divides a simple characteristic ideal $c_i = q_i \cap \mathfrak{u}_a$. Since $\mathfrak{L}$ is modular, we have $\mathfrak{u}_a > c_1 > c_1 \cap c_2 > \cdots > c_1 \cap \cdots \cap c_k = a$. Hence $\mathfrak{L}_a$ is archimedean and the lemma is proved.

If $\mathfrak{S}$ is modular and a has two reduced decompositions into irreducibles, then by Lemma 3.3, $\mathfrak{L}_a$ is archimedean and a has two reduced representations as a cross-cut of simple ideals. Now by Lemma 3.2, B2 holds in $\mathfrak{L}_a$ and hence by the dual of Lemma 1.3 any two reduced representations of a as a cross-cut of simple ideals have the same number of components and any simple ideal of one decomposition may be replaced by a suitably chosen simple ideal of the other. Thus by Theorem 3.3 and Corollary 3.1 we have the

KUROSCH-ORE DECOMPOSITION THEOREM. *Let an element of a modular lattice have two reduced decompositions into irreducibles. Then the number of components in the two decompositions is the same and any component in one decomposition may be replaced by a suitably chosen component of the other.*

4. Lattices with unique decompositions. This section will be devoted to the proof of the following theorem:

THEOREM 4.1. *Let $\mathfrak{S}$ satisfy the ascending chain condition. Then each element of $\mathfrak{S}$ has a unique representation as a reduced cross-cut of irreducibles if and only if $\mathfrak{S}$ is a Birkhoff lattice and $\mathfrak{L}_a$ is a Boolean algebra for each a.*

We begin with a series of lemmas, the first of which proves the necessity of the conditions of the theorem.

LEMMA 4.1. *Let $\mathfrak{S}$ satisfy the ascending chain condition and let each element have a unique representation as a reduced cross-cut of irreducibles. Then $\mathfrak{S}$ is a Birkhoff lattice and $\mathfrak{L}_a$ is a Boolean algebra for each a.*

For let $\mathfrak{b} > a$, $\mathfrak{c} \supset a$ and $\mathfrak{c} \not\supset \mathfrak{b}$. If $\mathfrak{b} \cup \mathfrak{c} \not> \mathfrak{c}$ we have $\mathfrak{b} \cup \mathfrak{c} \supset \mathfrak{b} \supset \mathfrak{c}$ where $\mathfrak{b} \cup \mathfrak{c} \neq \mathfrak{b} \neq \mathfrak{c}$. Since $\mathfrak{b} \not\supset \mathfrak{b}$, there exists a $d \in \mathfrak{b}$ such that $d \not\supset \mathfrak{b}$. Since $\mathfrak{b} \neq \mathfrak{c}$, there exists a c such that $c \in \mathfrak{c}$, $d \supset c$, and $c \not\supset \mathfrak{b}$. Furthermore since $c \not\supset \mathfrak{b}$ there exists an irreducible q_c such that $q_c \supset c$, $q_c \not\supset \mathfrak{b}$ (Lemma 2.6). But then $\mathfrak{b} \supset \mathfrak{b} \cap q_c \supset a$ and if $\mathfrak{b} = \mathfrak{b} \cap q_c$ we have $q_c \supset \mathfrak{b} \cup \mathfrak{c} \supset \mathfrak{b}$ which contradicts $q_c \not\supset \mathfrak{b}$. Hence $a = \mathfrak{b} \cap q_c$. Similarly there exists an irreducible q_d such that $q_d \supset d$ and $a = \mathfrak{b} \cap q_d$. By Theorem 2.2 we have $a = b_c \cap q_c$ and $a = b_d \cap q_d$ where b_c, $b_d \in \mathfrak{b}$. Let $b = b_c \cap b_d$. Then $b \in \mathfrak{b}$ and $a = b \cap q_c = b \cap q_d$. Let $b = q_1 \cap \cdots \cap q_k$. Then a has two reduced representations $a = q_{i_1} \cap \cdots \cap q_{i_l} \cap q_c = q_{j_1} \cap \cdots \cap q_{j_m} \cap q_d$. Now $q_c \neq q_d$ since otherwise $q_c \supset \mathfrak{b}$ and $q_c \neq q_{j_r}$ since otherwise $q_c \supset \mathfrak{b} \cup \mathfrak{c} \supset \mathfrak{b}$ contrary to $q_c \not\supset \mathfrak{b}$. Hence a has two distinct reduced representations as a cross-cut of irreducibles which contradicts our hypothesis. Thus $\mathfrak{b} \cup \mathfrak{c} > \mathfrak{c}$ and hence each element of $\mathfrak{S}$ satisfies the Birkhoff condition in the lattice of ideals.

Now since each element has a unique decomposition into irreducibles, the number of components is obviously bounded and hence $\mathfrak{L}_a$ is archimedean by Theorem 3.1. Let $\mathfrak{p}_1, \cdots, \mathfrak{p}_k$ be a maximal independent set of point ideals of $\mathfrak{L}_a$. Then $\mathfrak{p}_1, \cdots, \mathfrak{p}_k$ generate a Boolean algebra with simple ideals $\mathfrak{s}_1, \cdots, \mathfrak{s}_k$. $\mathfrak{s}_1, \cdots, \mathfrak{s}_k$ are clearly simple ideals of $\mathfrak{L}_a$ and hence are characteristic ideals by Corollary 3.2. Thus a has a decomposition $a = q_1 \cap \cdots \cap q_k$ where $q_i \supset \mathfrak{s}_i$ (Theorem 3.3). Now suppose there is a simple ideal $\mathfrak{s}$ distinct from $\mathfrak{s}_1, \cdots, \mathfrak{s}_k$. Let $q \supset \mathfrak{s}$, $q \not\supset \mathfrak{u}_a$. Then q is a component of a by Theorem 3.5 and hence $q = q_i$ for some i since a has but one reduced decomposition into irreducibles. But then $q \supset \mathfrak{s} \cup \mathfrak{s}_i = \mathfrak{u}_a$ which is impossible. Hence $\mathfrak{s}_1, \cdots, \mathfrak{s}_k$ are all of the simple ideals of $\mathfrak{L}_a$ and since each ideal of $\mathfrak{L}_a$ can be expressed as a union cross-cut of simple ideals, $\mathfrak{L}_a$ is simply the Boolean algebra generated by $\mathfrak{p}_1, \cdots, \mathfrak{p}_k$.

LEMMA 4.2. *If $\mathfrak{L}_a$ is a Boolean algebra, then it is archimedean.*

For if $\mathfrak{L}_a$ has an infinite number of point ideals, let $\mathfrak{p}_1, \mathfrak{p}_2, \mathfrak{p}_3, \cdots$ be a denumerable sequence of point ideals. Let $\mathfrak{p}_i' = \mathfrak{p}_1 \cup \mathfrak{p}_2 \cup \cdots \cup \mathfrak{p}_{i-1} \cup \mathfrak{p}_{i+1} \cup \cdots$. Then since $\mathfrak{L}_a$ is a Boolean algebra we have $a = \mathfrak{p}_1' \cap \mathfrak{p}_2' \cap \cdots$. But since the cross-cut of an infinite number of ideals consists of all elements contained in finite cross-cuts $a = \mathfrak{p}_1' \cap \mathfrak{p}_2' \cap \cdots \cap \mathfrak{p}_k'$ for some k. Then $a \supset \mathfrak{p}_{k+1}$ which contradicts $\mathfrak{p}_{k+1} > a$. Hence $\mathfrak{L}_a$ has only a finite number of point ideals and thus is archimedean.

LEMMA 4.3. *Let $\mathfrak{S}$ be a Birkhoff lattice in which each $\mathfrak{L}_a$ is archimedean. Then if every three ideals covering a principal ideal generate a Boolean algebra of order eight, $\mathfrak{L}_a$ is a Boolean algebra for each a.*

For let the hypotheses of the lemma be satisfied and let every three ideals covering a principal ideal generate a Boolean algebra. We show first that the ideals of any finite set of ideals covering a principal ideal are independent. Suppose that for any a every $k-1$ ideals covering a are independent. Let $\mathfrak{p}_1, \cdots, \mathfrak{p}_k$ be k distinct ideals covering a. If $\mathfrak{p}_1, \cdots, \mathfrak{p}_k$ are not independent let $\mathfrak{p}_1 \cup \mathfrak{p}_2 \cup \cdots \cup \mathfrak{p}_{k-1} \supset \mathfrak{p}_k$ say. Now $\mathfrak{p}_1 \cup \mathfrak{p}_i \not\supset \mathfrak{p}_2, \cdots, \mathfrak{p}_{i-1}, \mathfrak{p}_{i+1}, \cdots, \mathfrak{p}_k$ $(i=2, \cdots, k)$ since every three ideals covering a generate a Boolean algebra. Hence elements $x_{ij} \in \mathfrak{p}_1$ exist such that $x_{ij} \cup \mathfrak{p}_i \not\supset \mathfrak{p}_j$ $(j=2, \cdots, i-1, i+1, \cdots, k;$ $i=2, \cdots, k)$. Let $x = x_{23} \cap x_{24} \cap \cdots \cap x_{k\ k-1}$. Then $x \in \mathfrak{p}_1$ and $x \cup \mathfrak{p}_i \not\supset \mathfrak{p}_2, \cdots, \mathfrak{p}_{i-1}, \mathfrak{p}_{i+1}, \cdots, \mathfrak{p}_k$ $(i=2, \cdots, k)$. Clearly $x \not\supset \mathfrak{p}_2, \cdots, \mathfrak{p}_k$. Hence $\mathfrak{p}_2' = x \cup \mathfrak{p}_2 > x, \cdots, \mathfrak{p}_k' = x \cup \mathfrak{p}_k > x$ and $\mathfrak{p}_2', \cdots, \mathfrak{p}_k'$ are distinct. Thus by the induction assumption $\mathfrak{p}_2', \cdots, \mathfrak{p}_k'$ are independent. But $\mathfrak{p}_2' \cup \cdots \cup \mathfrak{p}_{k-1}' \supset x \cup \mathfrak{p}_2 \cup \cdots \cup \mathfrak{p}_{k-1} \supset x \cup \mathfrak{p}_1 \cup \cdots \cup \mathfrak{p}_{k-1} \supset x \cup \mathfrak{p}_k = \mathfrak{p}_k'$ which is contrary to the independence. Hence the independence of any finite set of covering elements follows by induction.

Now let $a \in \mathfrak{L}_a$ and let $\mathfrak{p}_1, \cdots, \mathfrak{p}_k$ be a maximal independent set of point ideals of $\mathfrak{L}_a$ divisible by a. Imbed $\mathfrak{p}_1, \cdots, \mathfrak{p}_k$ in a maximal independent set $\mathfrak{p}_1, \cdots, \mathfrak{p}_k, \cdots, \mathfrak{p}_n$. Set $b = \mathfrak{p}_1 \cup \cdots \cup \mathfrak{p}_k$. Then $a \supset b$. If $b \not\supset a$, there exists an element $b_1 \in b$ such that $b_1 \not\supset a$. Now $b \cup \mathfrak{p}_{k+i} \not\supset \mathfrak{p}_{k+j}$ $(j=k+1, \cdots, k+i-1, k+i+1, \cdots, n; i=1, \cdots, n-k)$. Hence as above there exists an element $b_2 \in b$ such that $b_2 \cup \mathfrak{p}_{k+i} \not\supset \mathfrak{p}_{k+j}, i \neq j$. Also $a = a \cup b \not\supset \mathfrak{p}_{k+1}, \cdots, \mathfrak{p}_n$. Hence an element $b_3 \in b$ exists such that $a \cup b_3 \not\supset \mathfrak{p}_{k+1}, \cdots, \mathfrak{p}_n$. Set $b = b_1 \cap b_2 \cap b_3$. Then $b \in b, b \not\supset a, b \cup \mathfrak{p}_{k+i} \not\supset \mathfrak{p}_{k+j}, i \neq j$, and $a \cup b \not\supset \mathfrak{p}_{k+1}, \cdots, \mathfrak{p}_n$. Clearly $b \not\supset \mathfrak{p}_{k+1}, \cdots, \mathfrak{p}_n$. Hence $\mathfrak{p}_{k+1}' = b \cup \mathfrak{p}_{k+1} > b, \cdots, \mathfrak{p}_n' = b \cup \mathfrak{p}_n > b$ and $\mathfrak{p}_{k+1}', \cdots, \mathfrak{p}_n'$ are distinct. Let $b \cup a \supset \mathfrak{p} > b$. Then $\mathfrak{p}$ is distinct from $\mathfrak{p}_{k+1}', \cdots, \mathfrak{p}_n'$. For if $\mathfrak{p} = \mathfrak{p}_{k+i}'$, then $b \cup a \supset \mathfrak{p}_{k+i}$ contrary to the definition of b. Thus by the result of the above paragraph $\mathfrak{p}, \mathfrak{p}_{k+1}', \cdots, \mathfrak{p}_n'$ are independent. But $\mathfrak{p}_{k+1}' \cup \cdots \cup \mathfrak{p}_n' = b \cup \mathfrak{p}_{k+1} \cup \cdots \cup \mathfrak{p}_n = b \cup \mathfrak{p}_1 \cup \cdots \cup \mathfrak{p}_n \supset b \cup a \supset \mathfrak{p}$ which is impossible. Hence $a = b$ and

$\mathcal{L}_a$ is a point lattice. But then the point ideals of $\mathcal{L}_a$ are independent and generate $\mathcal{L}_a$. Thus $\mathcal{L}_a$ is a Boolean algebra by Lemma 1.2.

Lemma 4.4. *Let $\mathfrak{S}$ be a Birkhoff lattice satisfying the ascending chain condition in which every three ideals covering a principal ideal generate a Boolean algebra. Let q be an irreducible of $\mathfrak{S}$ such that $q \supset a$; $\mathfrak{b}, \mathfrak{c} > a$ and $\mathfrak{b} \neq \mathfrak{c}$. Then either $q \supset \mathfrak{b}$ or $q \supset \mathfrak{c}$.*

Let us suppose that for some a we have $q \supset a$; $\mathfrak{b}, \mathfrak{c} > a$, $\mathfrak{b} \neq \mathfrak{c}$, $q \not\supset \mathfrak{b}$ and $q \not\supset \mathfrak{c}$. We shall show that a proper divisor a' of a exists with the same properties and hence the lemma follows from the ascending chain condition. Now $q \neq a$ since otherwise $q = \mathfrak{b} \cap \mathfrak{c}$ contrary to the irreducibility of q. Hence $q \supset \mathfrak{p} > a$ by Theorem 2.1. Clearly $\mathfrak{p} \neq \mathfrak{b}, \mathfrak{c}$ since otherwise $q \supset \mathfrak{b}$ or $q \supset \mathfrak{c}$. Hence $\mathfrak{p}, \mathfrak{b}$ and $\mathfrak{c}$ generate a Boolean algebra. Since $\mathfrak{p} \cup \mathfrak{b} \not\supset \mathfrak{c}$ there exists an element $p \in \mathfrak{p}$ such that $p \cup \mathfrak{b} \not\supset \mathfrak{c}$. Since $q \supset \mathfrak{p}$, there exists an element $p' \in \mathfrak{p}$ such that $q \supset p'$. Let $a' = p \cap p'$. Then $a' \in \mathfrak{p}$ and hence $a' \neq a$. Clearly $q \supset a'$. Let $\mathfrak{b}' = a' \cup \mathfrak{b}$, $\mathfrak{c}' = a' \cup \mathfrak{c}$. Then $\mathfrak{b}' > a'$ and $\mathfrak{c}' > a'$ by the Birkhoff condition. If $\mathfrak{b}' = \mathfrak{c}'$, then $p \cup \mathfrak{b} \supset a' \cup \mathfrak{b} \supset \mathfrak{c}$, which contradicts $p \cup \mathfrak{b} \not\supset \mathfrak{c}$. Hence $\mathfrak{b}' \neq \mathfrak{c}'$. Since $q \not\supset \mathfrak{b}$, $q \not\supset \mathfrak{c}$ we have $q \not\supset \mathfrak{b}'$, $q \not\supset \mathfrak{c}'$. Thus a' is a proper divisor of a with the desired properties.

Lemma 4.5. *Let $\mathfrak{S}$ be a Birkhoff lattice satisfying the ascending chain condition in which every three ideals covering a principal ideal generate a Boolean algebra. Then if a has a reduced representation $a = q_1 \cap \cdots \cap q_k$, $\mathcal{L}_a$ is archimedean of length k and each q_i divides a simple ideal of $\mathcal{L}_a$.*

For let $\mathfrak{a}_i$ be the union of the point ideals of $\mathcal{L}_a$ which are divisible by q_i. Then $\mathfrak{a}_i \neq \mathfrak{u}_a$ since $\mathfrak{a}_i$ is a characteristic ideal of $\mathcal{L}_a$. Now let $\mathfrak{p}, \mathfrak{p}'$ be any two point ideals of $\mathcal{L}_a$ which are not divisible by $\mathfrak{a}_i$. Then $\mathfrak{a}_i \cup \mathfrak{p} > \mathfrak{a}_i$ and $\mathfrak{a}_i \cup \mathfrak{p}' > \mathfrak{a}_i$ by the Birkhoff condition. Now suppose that $\mathfrak{a}_i \cup \mathfrak{p} \not\supset \mathfrak{p}'$. Then there exists an element $a_1 \in \mathfrak{a}_i$ such that $a_1 \cup \mathfrak{p} \not\supset \mathfrak{p}'$. Since $q_i \supset \mathfrak{a}_i$, there exists an element $a_2 \in \mathfrak{a}_i$ such that $q_i \supset a_2$. Let $a_i = a_1 \cap a_2$. Then $q_i \supset a_i$ and $a_i \cup \mathfrak{p} \not\supset \mathfrak{p}'$. Clearly $a_i \not\supset \mathfrak{p}'$. If $a_i \supset \mathfrak{p}$, then $q_i \supset \mathfrak{p}$ and $\mathfrak{a}_i \supset \mathfrak{p}$ contrary to assumption. Hence $a_i \cup \mathfrak{p} > a_i$, $a_i \cup \mathfrak{p}' > a_i$ and $a_i \cup \mathfrak{p} \neq a_i \cup \mathfrak{p}'$. Since $q_i \supset a_i$ by Lemma 4.4 we have either $q_i \supset a_i \cup \mathfrak{p}$ or $q_i \supset a_i \cup \mathfrak{p}'$. Hence $q_i \supset \mathfrak{p}$ or $q_i \supset \mathfrak{p}'$. But then $\mathfrak{a}_i \supset \mathfrak{p}$ or $\mathfrak{a}_i \supset \mathfrak{p}'$ contrary to assumption. Thus $\mathfrak{a}_i \cup \mathfrak{p} \supset \mathfrak{p}'$ and $\mathfrak{a}_i \cup \mathfrak{p} = \mathfrak{a}_i \cup \mathfrak{p}'$ for every pair of point ideals of $\mathcal{L}_a$ not divisible by $\mathfrak{a}_i$. But then $\mathfrak{a}_i \cup \mathfrak{p} = \mathfrak{a}_i \cup \mathfrak{u}_a = \mathfrak{u}_a$ and $\mathfrak{u}_a > \mathfrak{a}_i$. Hence $\mathfrak{a}_i$ is simple and each q_i divides a simple ideal of $\mathcal{L}_a$.

Now let $\mathfrak{b}_0 = \mathfrak{u}_a$ and let $\mathfrak{b}_i$ denote the union of the point ideals of $\mathcal{L}_a$ which are divisible by $q_1, \cdots, q_i$. Then $\mathfrak{b}_1 = \mathfrak{a}_1$ and $\mathfrak{b}_0 > \mathfrak{a}_1$ by the result we have just obtained. Clearly $\mathfrak{b}_{l-1} \supset \mathfrak{b}_l$. If $\mathfrak{b}_{l-1} = \mathfrak{b}_l$, let $q_1 \cap \cdots \cap q_{l-1} \cap q_{l+1} \cap \cdots \cap q_k \supset \mathfrak{p}_l > a$. $\mathfrak{p}_l$ exists since the representation is reduced. Now $q_1 \cap \cdots \cap q_{l-1} \supset \mathfrak{p}_l$ and hence $\mathfrak{b}_{l-1} \supset \mathfrak{p}_l$. But then $\mathfrak{b}_l \supset \mathfrak{p}_l$ and hence $q_l \supset \mathfrak{p}_l$. Thus $a = q_1 \cap \cdots \cap q_k \supset \mathfrak{p}_l$ which is impossible. Hence $\mathfrak{b}_{l-1} \neq \mathfrak{b}_l$. Now let $\mathfrak{p}$ and $\mathfrak{p}'$ be two point ideals divisible by $\mathfrak{b}_{l-1}$ but not by $\mathfrak{b}_l$. If $\mathfrak{b}_l \cup \mathfrak{p} \neq \mathfrak{b}_l \cup \mathfrak{p}'$ there exists an element $b_l \in \mathfrak{b}_l$ such that $q_l \supset b_l$, $b_l \cup \mathfrak{p} > b_l$, $b_l \cup \mathfrak{p}' > b_l$ and $b_l \cup \mathfrak{p} \neq b_l \cup \mathfrak{p}'$. But then $q_l \supset b_l \cup \mathfrak{p}$

or $q_l \supset \mathfrak{b}_l \cup \mathfrak{p}'$ by Lemma 4.4. Hence either $\mathfrak{b}_l \supset \mathfrak{p}$ or $\mathfrak{b}_l \supset \mathfrak{p}'$ which is contrary to assumption. Hence $\mathfrak{b}_l \cup \mathfrak{p} = \mathfrak{b}_l \cup \mathfrak{p}'$ for every two point ideals of $\mathfrak{b}_{l-1}$ which are not divisible by $\mathfrak{b}_l$. Thus $\mathfrak{b}_l \cup \mathfrak{p} = \mathfrak{b}_l \cup \mathfrak{b}_{l-1} = \mathfrak{b}_{l-1}$ and $\mathfrak{b}_{l-1} > \mathfrak{b}_l$ by the Birkhoff condition. Hence we have the chain $\mathfrak{u}_a > \mathfrak{b}_1 > \mathfrak{b}_2 > \cdots > \mathfrak{b}_k$. But $\mathfrak{b}_k = a$ and the lemma follows from Lemma 3.1.

LEMMA 4.6. *Let $\mathfrak{S}$ be a Birkhoff lattice satisfying the ascending chain condition in which every three ideals covering a principal ideal generate a Boolean algebra. Then each $a \in \mathfrak{S}$ has a unique reduced representation $a = q_1 \cap \cdots \cap q_k$ where $q_1, \cdots, q_k$ are irreducibles. $\mathfrak{L}_a$ is a Boolean algebra of order 2^k and each q_i divides a simple ideal of $\mathfrak{L}_a$.*

It follows from Lemmas 4.3 and 4.5 that $\mathfrak{L}_a$ is a Boolean algebra of order 2^k. q_i divides a simple ideal $\mathfrak{s}_i$ of $\mathfrak{L}_a$ by Lemma 4.5. Now let $a = q_1' \cap \cdots \cap q_l'$ be a reduced decomposition of a. By Lemma 4.5, $l = k$ and q_i' divides a simple ideal $\mathfrak{s}_j$. Let $b = q_i' \cap q_j$. Then $b \supset \mathfrak{s}_j$, and $b \not\supset \mathfrak{u}_a$. Let $\mathfrak{s}_j \not\supset \mathfrak{p} > a$. Then $\mathfrak{u}_a = \mathfrak{s}_j \cup \mathfrak{p}$ and $b \cup \mathfrak{p} = b \cup \mathfrak{s}_j \cup \mathfrak{p} = b \cup \mathfrak{u}_a > b$ by the Birkhoff condition. If $q_j \neq b$, we have $q_j \supset \mathfrak{p}_j > b$ and $\mathfrak{p}_j \neq \mathfrak{p} \cup b$ since otherwise $q_j \supset \mathfrak{p} \cup \mathfrak{s}_j = \mathfrak{u}_a$. Hence by Lemma 4.4, either $q_i' \supset \mathfrak{p}_j$ or $q_i' \supset \mathfrak{p}$. But if $q_i' \supset \mathfrak{p}_j$, then $b = q_i' \cap q_j \supset \mathfrak{p}_j > b$ which is impossible. Hence $q_i' \supset \mathfrak{p}$ and $q_i' \supset \mathfrak{u}_a$ which is impossible. Thus $q_j = b$ and similarly $q_i' = b$. Hence q_i' is equal to q_j and the two representations are identical. This completes the proof of the lemma.

Lemma 4.1 and Lemma 4.6 together give Theorem 4.1.

In view of Lemma 4.6, lattices with unique irreducible decompositions may be characterized in terms of the local properties of the lattice of ideals as follows:

THEOREM 4.2. *Let $\mathfrak{S}$ satisfy the ascending chain condition. Then each element of $\mathfrak{S}$ has a unique reduced decomposition into irreducibles if and only if $\mathfrak{S}$ is a Birkhoff lattice in which every three ideals covering an element of $\mathfrak{S}$ are independent.*

As a corollary to Lemma 4.6 we have

COROLLARY 4.1. *Let $\mathfrak{S}$ satisfy the ascending chain condition and let every element of $\mathfrak{S}$ have a unique reduced decomposition into irreducibles. Then the number of irreducible components of a is equal to the number of ideals covering a.*

COROLLARY 4.2. *Let $\mathfrak{S}$ be a Birkhoff lattice satisfying the ascending chain condition. Then if $\mathfrak{S}$ contains a modular, non-distributive sublattice, the lattice of ideals of $\mathfrak{S}$ contains a complete*([10]) *modular, non-distributive sublattice of order five.*

For if $\mathfrak{S}$ contains a modular, non-distributive sublattice of order five, at

---

([10]) A sublattice $\mathfrak{L}'$ of $\mathfrak{L}$ is said to be *complete* if $\mathfrak{a} > \mathfrak{b}$ in $\mathfrak{L}'$ implies $\mathfrak{a} > \mathfrak{b}$ in $\mathfrak{L}$.

least one element of $\mathfrak{S}$ does *not* have a unique decomposition into irreducibles. But then there are three ideals covering a principal ideal which are dependent. These three ideals generate a complete, modular, non-distributive sublattice of $\mathfrak{L}$ of order five.

5. **Unicity of the number of components.** In the previous section lattices with unique irreducible decompositions were completely characterized as Birkhoff lattices with certain special properties. Simple examples show that a similar characterization of lattices in which the *number* of components is unique will require lattices that are considerably more general than Birkhoff lattices. Hence we shall restrict ourselves to the characterization of Birkhoff lattices having the number of components unique. We prove the following theorem:

THEOREM 5.1. *Let $\mathfrak{S}$ be a Birkhoff lattice satisfying the ascending chain condition. Then the number of components in the reduced decompositions of each element into irreducibles is unique if and only if $\mathfrak{L}_a$ is modular for each a.*

As in §4, the proof rests on a series of lemmas.

LEMMA 5.1. *Let $\mathfrak{S}$ be a Birkhoff lattice satisfying the ascending condition. Then the number of components in the irreducible decompositions of a is unique if and only if $\mathfrak{L}_a$ is archimedean, modular, and every characteristic ideal of $\mathfrak{L}_a$ is simple.*

Since the ascending chain condition holds each element of $\mathfrak{S}$ has a decomposition into irreducibles. Now if the number of components in the irreducible decompositions of a is unique it is certainly bounded and hence $\mathfrak{L}_a$ is archimedean by Theorem 3.1. Now let $\mathfrak{c}$ be a characteristic ideal of $\mathfrak{L}_a$ and let $\mathfrak{p}_1, \cdots, \mathfrak{p}_k$ be a maximal independent set of point ideals divisible by $\mathfrak{c}$. Imbed $\mathfrak{p}_1, \cdots, \mathfrak{p}_k$ in a maximal independent set $\mathfrak{p}_1, \cdots, \mathfrak{p}_k, \cdots, \mathfrak{p}_n$. By Theorem 3.5 and Theorem 3.3, a has an irreducible decomposition having $k+1$ components. But by Corollary 3.2 a has a decomposition having n components. Hence if the number of components is unique we have $n=k+1$. But then $\mathfrak{p}_1\cup \cdots \cup\mathfrak{p}_k$ is a simple ideal of $\mathfrak{L}_a$ and $\mathfrak{u}_a\supset\mathfrak{c}\supset\mathfrak{p}_1\cup \cdots \cup\mathfrak{p}_k$, $\mathfrak{u}_a\neq\mathfrak{c}$. Hence $\mathfrak{c}=\mathfrak{p}_1\cup \cdots \cup\mathfrak{p}_k$ and $\mathfrak{c}$ is a simple ideal of $\mathfrak{L}_a$.

Now let $\mathfrak{s}$ be an arbitrary simple ideal of $\mathfrak{L}_a$ and let $\mathfrak{a}$ be any ideal of $\mathfrak{L}_a$ such that $\mathfrak{s}\supset\mathfrak{a}$. By Theorem 3.2, $\mathfrak{a}$ has a reduced representation $\mathfrak{a}=\mathfrak{s}_1\cap \cdots \cap\mathfrak{s}_l$ where $\mathfrak{s}_1, \cdots, \mathfrak{s}_l$ are simple ideals of $\mathfrak{L}_a$. If $\mathfrak{a}\cap\mathfrak{s}\neq\mathfrak{a}$, by Theorem 3.2 there exists a simple ideal $\mathfrak{s}_{l+2}$ such that $\mathfrak{s}_{l+2}\supset\mathfrak{a}\cap\mathfrak{s}$. Similarly if $\mathfrak{a}\cap\mathfrak{s}\cap\mathfrak{s}_{l+2}\neq\mathfrak{a}$, there exists a simple ideal $\mathfrak{s}_{l+3}$ such that $\mathfrak{s}_{l+3}\supset\mathfrak{a}\cap\mathfrak{s}\cap\mathfrak{s}_{l+2}$. Thus we eventually have $\mathfrak{a}\cap\mathfrak{s}\cap\mathfrak{s}_{l+2}\cap \cdots \cap\mathfrak{s}_m=\mathfrak{a}$. Then $\mathfrak{a}=\mathfrak{s}_1\cap \cdots \cap\mathfrak{s}_l\cap\mathfrak{s}\cap\mathfrak{s}_{l+2}\cap \cdots \cap\mathfrak{s}_m$ and since each simple ideal is characteristic this decomposition gives a decomposition into irreducibles with the same number of terms. Hence if the number of components in the irreducible decompositions of a is unique we have $m\geq n$ where n is the length of $\mathfrak{L}_a$. But $\mathfrak{u}_a\supset\mathfrak{s}_1\supset\mathfrak{s}_1\cap\mathfrak{s}_2\supset \cdots \supset\mathfrak{s}_1\cap \cdots \cap\mathfrak{s}_l\supset\mathfrak{s}_1$

$\cap \cdots \cap \mathfrak{s}_l \cap \mathfrak{s} \supset \mathfrak{s}_1 \cap \cdots \cap \mathfrak{s}_l \cap \mathfrak{s} \cap \mathfrak{s}_{l+2} \supset \cdots \supset \mathfrak{s}_1 \cap \cdots \cap \mathfrak{s}_m = a$ and the ideals of this chain are distinct. Hence $m \leq n$ by Lemma 3.1. Thus $m = n$ and each ideal of the chain covers the ideal which immediately follows. Hence $\mathfrak{a} > \mathfrak{a} \cap \mathfrak{s}$. Now let $\mathfrak{a}$ and $\mathfrak{b}$ be any two ideals of $\mathfrak{L}_a$ such that $\mathfrak{a} \cup \mathfrak{b} > \mathfrak{b}$. By Theorem 3.2 and ideal $\mathfrak{s}$ exists such that $\mathfrak{s} \supset \mathfrak{b}$, $\mathfrak{s} \not\supset \mathfrak{a} \cup \mathfrak{b}$. But then $\mathfrak{b} = (\mathfrak{a} \cup \mathfrak{b}) \cap \mathfrak{s}$. Hence $\mathfrak{a} > \mathfrak{a} \cap \mathfrak{s} = \mathfrak{a} \cap (\mathfrak{a} \cup \mathfrak{b}) \cap \mathfrak{s} = \mathfrak{a} \cap \mathfrak{b}$. Thus $\mathfrak{a} \cup \mathfrak{b} > \mathfrak{b}$ implies $\mathfrak{a} > \mathfrak{a} \cap \mathfrak{b}$ and B2 holds in $\mathfrak{L}_a$. But then $\mathfrak{L}_a$ is modular by Lemma 3.2.

On the other hand let $\mathfrak{L}_a$ be archimedean, modular, and every characteristic ideal be simple. Let $a = q_1 \cap \cdots \cap q_k$ be a reduced decomposition into irreducibles. By Theorem 3.3, a has a reduced representation $a = c_1 \cap \cdots \cap c_k$ where c_i is a characteristic ideal of $\mathfrak{L}_a$. But then c_i is a simple ideal of $\mathfrak{L}_a$ by assumption. Thus $\mathfrak{u}_a > c_1 > c_1 \cap c_2 > \cdots > c_1 \cap \cdots \cap c_k = a$ since B2 holds in $\mathfrak{L}_a$ by Lemma 1.5. Hence k is simply the length of $\mathfrak{L}_a$ and every reduced decomposition of a into irreducibles has the same number of components. This completes the proof of the lemma.

LEMMA 5.2. *Let $\mathfrak{S}$ be a Birkhoff lattice satisfying the ascending chain condition. Then if $\mathfrak{L}_a$ is modular for each a, every characteristic ideal of $\mathfrak{L}_a$ is simple.*

Let every characteristic ideal of $\mathfrak{L}_b$ be simple for every proper divisor b of a. We shall show that every characteristic ideal of $\mathfrak{L}_a$ is simple and the lemma follows by the ascending chain condition.

If c is a characteristic ideal of $\mathfrak{L}_a$ which is not simple, let q be an associated irreducible. If x is any element of $\mathfrak{S}$ divisible by q, let q_x denote the union of the point ideals of $\mathfrak{L}_x$ divisible by q. Then since c is a characteristic ideal associated with q we have $c \supset q_a$ and hence q_a is not a simple ideal of $\mathfrak{L}_a$. Now suppose that for every two point ideals $\mathfrak{p}$ and $\mathfrak{p}'$ such that $q_a \not\supset \mathfrak{p}, \mathfrak{p}'$ we have $q_a \cup \mathfrak{p} = q_a \cup \mathfrak{p}'$. Then $\mathfrak{u}_a = q_a \cup \mathfrak{u}_a = q_a \cup \mathfrak{p} > q_a$ and q_a is simple contrary to assumption. Hence there are two point ideals $\mathfrak{p}$ and $\mathfrak{p}'$ such that $q_a \not\supset \mathfrak{p}, q_a \not\supset \mathfrak{p}'$, and $q_a \cup \mathfrak{p} \neq q_a \cup \mathfrak{p}'$. Now $q \cap (q_a \cup \mathfrak{p} \cup \mathfrak{p}') = (q \cap \mathfrak{u}_a) \cap (q_a \cup \mathfrak{p} \cup \mathfrak{p}') = q_a \cup (q \cap \mathfrak{u}_a \cap (\mathfrak{p} \cup \mathfrak{p}')) = q_a \cup (q \cap (\mathfrak{p} \cup \mathfrak{p}'))$ since $\mathfrak{L}_a$ is modular. If $q \cap (\mathfrak{p} \cup \mathfrak{p}') \neq a$, we have $q \cap (\mathfrak{p} \cup \mathfrak{p}') \supset \mathfrak{p}_1 > a$. If $\mathfrak{p}' = \mathfrak{p}_1$, we have $q \supset \mathfrak{p}'$ and hence $q_a \supset \mathfrak{p}'$ contrary to hypothesis. Thus $\mathfrak{p}_1 \neq \mathfrak{p}$ and $\mathfrak{p}_1 \neq \mathfrak{p}'$. Now $\mathfrak{p} \cup \mathfrak{p}' \supset \mathfrak{p} \cup \mathfrak{p}_1 \supset \mathfrak{p}$ and $\mathfrak{p} \cup \mathfrak{p}_1 \neq \mathfrak{p}$. Hence $\mathfrak{p} \cup \mathfrak{p}' = \mathfrak{p} \cup \mathfrak{p}_1$ by the Birkhoff condition. Since $q \supset \mathfrak{p}_1$ we have $q_a \supset \mathfrak{p}_1$ and hence $q_a \cup \mathfrak{p} \supset \mathfrak{p}_1 \cup \mathfrak{p} \supset \mathfrak{p}'$. But then $q_a \cup \mathfrak{p} = q_a \cup \mathfrak{p}'$ which contradicts the definition of $\mathfrak{p}$ and $\mathfrak{p}'$. Thus $q \cap (\mathfrak{p} \cup \mathfrak{p}') = a$ and $q \cap (q_a \cup \mathfrak{p} \cup \mathfrak{p}') = q_a$.

Now suppose that q_a is *not* principal. Let X be the set of all elements x such that $q \supset x \supset q_a$, $q \neq x$. If $x \in X$, let $\mathfrak{p}_x = q \cap (x \cup \mathfrak{p} \cup \mathfrak{p}')$. Clearly X generates q_a. We shall show

    (1) There exists an $x_0 \in X$ such that $x \cup \mathfrak{p} \cup \mathfrak{p}' > \mathfrak{p}_x > x$ for all $x \in X$, $x_0 \supset x$.

    (2) The set of ideals $\mathfrak{p}_x$, $x \in X$, $x_0 \supset x$, generates q_a.

    (1) Since $q_a \cup \mathfrak{p} \not\supset \mathfrak{p}'$ and X generates q_a, there exists an element $x_0 \in X$ such that $x_0 \cup \mathfrak{p} \not\supset \mathfrak{p}'$. Let $x_0 \supset x$, $x \in X$ and suppose that $x = \mathfrak{p}_x$. Since x is a proper divisor of a we have $\mathfrak{u}_x > q_x$. By the Birkhoff condition $x \cup \mathfrak{p} > x$,

$x \cup \mathfrak{p}' > x$ and hence $x \cup \mathfrak{p}$, $x \cup \mathfrak{p}'$ belong to $\mathfrak{L}_x$. Now $\mathfrak{q}_x \not\supset x \cup \mathfrak{p} \cup \mathfrak{p}'$ since otherwise $q \supset \mathfrak{p}$. Hence by the modularity of $\mathfrak{L}_x$ we have $x \cup \mathfrak{p} \cup \mathfrak{p}' > \mathfrak{q}_x \cap (x \cup \mathfrak{p} \cup \mathfrak{p}')$. Then $x \cup \mathfrak{p} \cup \mathfrak{p}' \supset q \cap (x \cup \mathfrak{p} \cup \mathfrak{p}') \supset \mathfrak{q}_x \cap (x \cup \mathfrak{p} \cup \mathfrak{p}')$ and $q \cap (x \cup \mathfrak{p} \cup \mathfrak{p}') \neq x \cup \mathfrak{p} \cup \mathfrak{p}'$. Thus $x = \mathfrak{p}_x = q \cap (x \cup \mathfrak{p} \cup \mathfrak{p}') = \mathfrak{q}_x \cap (x \cup \mathfrak{p} \cup \mathfrak{p}')$ and hence $x \cup \mathfrak{p} \cup \mathfrak{p}' > x$. But then $x \cup \mathfrak{p} \cup \mathfrak{p}' \supset x \cup \mathfrak{p} \supset x$ and if $x = x \cup \mathfrak{p}$ we have $q \supset \mathfrak{p}$ which is impossible. Thus $x \cup \mathfrak{p} = x \cup \mathfrak{p} \cup \mathfrak{p}' \supset \mathfrak{p}'$ and $x_0 \cup \mathfrak{p} \supset x \cup \mathfrak{p} \supset \mathfrak{p}'$ contrary to the definition of x_0. Hence $x \neq \mathfrak{p}_x$. Let $\mathfrak{p}_x \supset \mathfrak{p}_x' > x$. Clearly $x \cup \mathfrak{p} \cup \mathfrak{p}' \supset x \cup \mathfrak{p} \cup \mathfrak{p}_x' \supset x \cup \mathfrak{p}$ and $x \cup \mathfrak{p} \cup \mathfrak{p}' > x \cup \mathfrak{p}$ by the Birkhoff condition. If $x \cup \mathfrak{p} \cup \mathfrak{p}_x' = x \cup \mathfrak{p}$ we have $x \cup \mathfrak{p} \supset \mathfrak{p}_x'$ and $x = q \cap (x \cup \mathfrak{p}) \supset \mathfrak{p}_x'$ which contradicts $\mathfrak{p}_x' > x$. Hence $x \cup \mathfrak{p} \cup \mathfrak{p}' = x \cup \mathfrak{p} \cup \mathfrak{p}_x'$. But then $x \cup \mathfrak{p} \cup \mathfrak{p}_x' \supset \mathfrak{p}_x \supset \mathfrak{p}_x'$ and $x \cup \mathfrak{p} \cup \mathfrak{p}_x' > \mathfrak{p}_x'$. If $x \cup \mathfrak{p} \cup \mathfrak{p}_x' = \mathfrak{p}_x$, then $q \supset x \cup \mathfrak{p} \cup \mathfrak{p}'$ which is impossible. Hence $\mathfrak{p}_x = \mathfrak{p}_x'$ and $x \cup \mathfrak{p} \cup \mathfrak{p}' > \mathfrak{p}_x > x$.

(2) Clearly $\mathfrak{p}_x \supset \mathfrak{q}_a$ for every x since $\mathfrak{p}_x \supset x \supset \mathfrak{q}_a$. Now let $a_1 \in \mathfrak{q}_a$. Then $a_1 \supset \mathfrak{q}_a = q \cap (\mathfrak{q}_a \cup \mathfrak{p} \cup \mathfrak{p}')$ and hence $a_1 \supset q \cap (a_2 \cup \mathfrak{p} \cup \mathfrak{p}')$ where $a_2 \in \mathfrak{q}_a$ by Theorem 2.2. Let $x = x_0 \cap a_2$. Then $x \in X$ and $a_1 \supset q \cap (x \cup \mathfrak{p} \cup \mathfrak{p}') = \mathfrak{p}_x$. Hence each element of $\mathfrak{q}_a$ divides some $\mathfrak{p}_x$ and thus the ideals $\mathfrak{p}_x$ generate $\mathfrak{q}_a$.

Now let y be an arbitrary element of $\mathfrak{q}_a \cup \mathfrak{p}$. Then $y \supset \mathfrak{q}_a$ and hence $y \supset \mathfrak{p}_x$ where $x_0 \supset x$ by (2). But then by (1) $x \cup \mathfrak{p} \cup \mathfrak{p}' > \mathfrak{p}_x > x$ and $\mathfrak{p}_x \not\supset \mathfrak{p}$ since otherwise $q \supset \mathfrak{p}$. Now $x \cup \mathfrak{p} \cup \mathfrak{p}' \supset \mathfrak{p}_x \cup \mathfrak{p} \supset \mathfrak{p}_x$ and $\mathfrak{p}_x \cup \mathfrak{p} \neq \mathfrak{p}_x$. Hence $x \cup \mathfrak{p} \cup \mathfrak{p}' = \mathfrak{p}_x \cup \mathfrak{p}$ which gives $\mathfrak{p}_x \cup \mathfrak{p} \supset \mathfrak{p}'$. Thus $y \cup \mathfrak{p} \supset \mathfrak{p}'$ for every y and hence $\mathfrak{q}_a \cup \mathfrak{p} \supset \mathfrak{p}'$. The assumption that $\mathfrak{q}_a$ is *not* principal has thus led to a contradiction and we conclude that $\mathfrak{q}_a$ is principal, say $\mathfrak{q}_a = (a_1)$. Since a_1 is a proper divisor of a, by hypothesis we have $\mathfrak{u}_{a_1} > \mathfrak{q}_{a_1}$. Hence $\mathfrak{q}_{a_1} \cup (a_1 \cup \mathfrak{p} \cup \mathfrak{p}') > \mathfrak{q}_{a_1}$. But $\mathfrak{q}_{a_1} \cap (a_1 \cup \mathfrak{p} \cup \mathfrak{p}') = q \cap (a_1 \cup \mathfrak{p} \cup \mathfrak{p}') = a_1$ and $a_1 \cup \mathfrak{p} \cup \mathfrak{p}' \not> a_1$. Hence $\mathfrak{L}_{a_1}$ is non-modular contrary to assumption. Thus $\mathfrak{q}_a$ is simple and hence $\mathfrak{c}$ is a simple ideal of $\mathfrak{L}_a$.

LEMMA 5.3. *Let $\mathfrak{S}$ be a Birkhoff lattice satisfying the ascending chain condition. Then if $\mathfrak{L}_a$ is modular for every a, $\mathfrak{L}_a$ is archimedean.*

For let $a = q_1 \cap \cdots \cap q_n$ be a reduced decomposition of a into irreducibles. Then a has the reduced representation $a = \mathfrak{c}_1 \cap \cdots \cap \mathfrak{c}_k$ where $\mathfrak{c}_i$ is a characteristic ideal associated with q_i. By Lemma 5.2, $\mathfrak{c}_i$ is a simple ideal of $\mathfrak{L}_a$. Hence since $\mathfrak{L}_a$ is modular we have $\mathfrak{u}_a > \mathfrak{c}_1 > \mathfrak{c}_1 \cap \mathfrak{c}_2 > \cdots > \mathfrak{c}_1 \cap \cdots \cap \mathfrak{c}_k = a$. Thus $\mathfrak{L}_a$ is archimedean of length k.

Lemmas 5.1–5.3 together give Theorem 5.1.

COROLLARY 5.1. *Let $\mathfrak{S}$ be a Birkhoff lattice satisfying the ascending chain condition. Let the number of components in the reduced decompositions of an element a be unique. Then in any two reduced decompositions of a, each component of one decomposition may be replaced by a suitably chosen component of the other.*

For by Lemma 5.1, the two decompositions give two reduced representations of a as a cross-cut of simple ideals of $\mathfrak{L}_a$. However, since $\mathfrak{L}_a$ is modular, B2 is satisfied and the replacement property follows from the dual of Lemma 1.3.

Corollary 5.2. *Let $\mathfrak{S}$ be a Birkhoff lattice satisfying the ascending chain condition. Then if the number of components in the decompositions of an element a is unique, that number is simply the length of $\mathfrak{L}_a$.*

Corollary 5.3. *Let $\mathfrak{S}$ be a complemented Birkhoff lattice in which every element can be expressed as a cross-cut of a finite number of irreducibles. Then the number of components in the reduced decompositions of the null element z is unique if and only if $\mathfrak{S}$ is a complemented modular lattice of finite dimensions.*

For since $\mathfrak{S}$ is complemented, $\mathfrak{L}_z$ is simply $\mathfrak{L}$, the lattice of ideals.

6. **The Mac Lane exchange axiom.** In order to free condition B1 of the covering properties, Mac Lane (Mac Lane [1]) formulated the following axiom.

E_5. If $a \supset b \supset a \cap c$ and $c \neq a \cap c$, then there exists an element $c_1 \neq a \cap c$ such that $c \supset c_1 \supset a \cap c$ and $b = a \cap (b \cup c_1)$.

Mac Lane showed that E_5 is equivalent to a transposition property of chains and in case covering elements exist, that is, if $b \supset a$, $b \neq a$, implies b' exists such that $b \supset b' > a$, it reduces to B1. Thus both E_5 and the requirement that each element satisfy the Birkhoff condition in the lattice of ideals are generalizations of B1. We shall be particularly interested in the conditions under which they are equivalent.

Theorem 6.1. *Every Birkhoff lattice satisfies E_5.*

Proof. Let $a \supset b \supset a \cap c$ and $c \neq a \cap c$. Then by Theorem 2.1 and ideal $\mathfrak{p}$ exists such that $c \supset \mathfrak{p} > a \cap b$. Now $b \not\supset \mathfrak{p}$ since otherwise $a \cap c \supset \mathfrak{p}$ which is impossible. Hence $b \cup \mathfrak{p} > b$ by the Birkhoff condition. But then $b \cup \mathfrak{p} \supset a \cap (b \cup \mathfrak{p}) \supset b$ and if $b \cup \mathfrak{p} = a \cap (b \cup \mathfrak{p})$ we have $a \supset \mathfrak{p}$ which is impossible. Hence $b = a \cap (b \cup \mathfrak{p})$. Thus by Theorem 2.2 an element $p \in \mathfrak{p}$ exists such that $b = a \cap (b \cup p)$. Let $c_1 = c \cap p$. Then $c \supset c_1 \supset \mathfrak{p} > a \cap c$ and hence $c \supset c_1 \supset a \cap c$, $c_1 \neq a \cap c$. Also $b = a \cap (b \cup p) \supset a \cap (b \cup c_1) \supset b$. Thus $b = a \cap (b \cup c_1)$ and c_1 satisfies the requirements of E_5.

Theorem 6.2. *Let $\mathfrak{S}$ satisfy E_5 and have the property that each element is covered by only a finite number of covering ideals. Then $\mathfrak{S}$ is a Birkhoff lattice.*

Proof. Let $\mathfrak{a} \supset a$, $\mathfrak{p} > a$ and $\mathfrak{a} \not\supset \mathfrak{p}$. Let $\mathfrak{p}, \mathfrak{p}_1, \cdots, \mathfrak{p}_n$ be the finite number of ideals covering a. Now if $\mathfrak{p} \cup \mathfrak{a} \not> \mathfrak{a}$, we have $\mathfrak{p} \cup \mathfrak{a} \supset c \supset \mathfrak{a}$, $\mathfrak{p} \cup \mathfrak{a} \neq c \neq \mathfrak{a}$. Since $\mathfrak{p} \supset \mathfrak{p} \cap c \supset a$ and $c \not\supset \mathfrak{p}$ we have $\mathfrak{p} \cap c = a$ and hence by Theorem 2.2 elements $p' \in \mathfrak{p}$ and $c \in c$ exist such that $p' \cap c = a$. Since $c \neq \mathfrak{a}$, there exists an element $b' \in \mathfrak{a}$ such that $b' \not\supset c$. Let $b = c \cap b'$. Then $b \not\supset c$ and $c \supset b \supset c \cap p'$. Now since $\mathfrak{p} \not\supset \mathfrak{p}_i$ $(i = 1, \cdots, n)$ elements p_i' exist such that $p_i' \in \mathfrak{p}$ and $p_i' \not\supset \mathfrak{p}_i$ $(i = 1, \cdots, n)$. Set $p = p' \cap p_1' \cap \cdots \cap p_n'$. Then $p \in \mathfrak{p}$, $p' \supset p$ and $p \not\supset \mathfrak{p}_i$ $(i = 1, \cdots, n)$. Clearly $a = c \cap p' \supset c \cap p \supset a$ implies $c \cap p = a$. Hence $c \supset b \supset c \cap p$ and $p \neq c \cap p$. Thus by E_5 an element p_1 exists such that $p_1 \neq a$, $p \supset p_1 \supset a$ and

$b = c \cap (b \cup p_1)$. Now $p_1 \supset \mathfrak{p}' > a$ and $\mathfrak{p}' \neq \mathfrak{p}_i$ $(i = 1, \cdots, n)$ since otherwise $p \supset p_1 \supset \mathfrak{p}_i$ contrary to the definition of p. Hence $\mathfrak{p}' = \mathfrak{p}$. But then $b = c \cap (b \cup p_1) \supset c \cap (\mathfrak{a} \cup \mathfrak{p}) \supset \mathfrak{c}$ which contradicts $b \not\supset c$. Hence $\mathfrak{p} \cup \mathfrak{a} > \mathfrak{a}$ and $\mathfrak{S}$ is a Birkhoff lattice.

Now by Lemma 4.6, if $\mathfrak{S}$ is a Birkhoff lattice in which the ascending chain condition holds and every three ideals covering a principal ideal generate a Boolean algebra, then each a is covered by only a finite number of ideals. However, Theorem 6.2 does not enable us to replace the Birkhoff condition in the lemma by E_5 since the proof of the finiteness required the Birkhoff condition. To carry out this replacement we first replace the condition that every three ideals covering a principal ideal generate a Boolean algebra by an equivalent condition.

THEOREM 6.3. *Let $\mathfrak{S}$ be a Birkhoff lattice satisfying the ascending chain condition. Then every three ideals covering a principal ideal generate a Boolean algebra if and only if $a \cup b \supset q > a \cap b$ implies $a \supset q$ or $b \supset q$.*

Proof. Let every three ideals covering a principal ideal generate a Boolean algebra and suppose that $a \cup b \supset q > a \cap b$ but $a \not\supset q$, $b \not\supset q$. Then $a, b \neq a \cap b$. For if $a = a \cap b$, then $b = a \cup b \supset q$ contrary to assumption. Now with b fixed let a be maximal such that $a \cup b \supset q > a \cap b$ and $a \not\supset q$, $b \not\supset q$ for some q. Let $b \supset \mathfrak{p} > a \cap b$. Suppose $\mathfrak{a} = \mathfrak{p} \cup a \not\supset q$. Then $\mathfrak{a} \cup b = a \cup \mathfrak{p} \cup b = a \cup b \supset q$ and $\mathfrak{a} \cap b \supset a \cap b$. Now $\mathfrak{a} \neq a$ since otherwise $a \supset \mathfrak{p}$ and $a \cap b \supset \mathfrak{p} > a \cap b$ which is impossible. Let a_1' be an element of $\mathfrak{a}$ such that $a_1' \not\supset q$. Now $a \cup b \supset b \supset \mathfrak{p}$ and hence $a \cup b \supset \mathfrak{p} \cup a = \mathfrak{a}$. Thus $a_1 = (a \cup b) \cap a_1' \supset \mathfrak{a}$ and $a \cup b \supset a_1$. But $a \cup b \supset a_1 \cup b \supset a \cup b$. Hence $a_1 \cup b = a \cup b \supset q$. Also $a_1 \not\supset q$ since otherwise $a_1' \supset q$. Now $q \supset (a_1 \cap b) \cap q \supset a \cap b$ and $q \neq (a_1 \cap b) \cap q$ since otherwise $b \supset q$. Hence $q > a \cap b = (a_1 \cap b) \cap q$. By the Birkhoff condition we have $q_1 = q \cup (a_1 \cap b) > a_1 \cap b$. Since $a_1 \cup b \supset q$ we have $a_1 \cup b \supset q_1 > a_1 \cap b$. Clearly $a_1 \not\supset q_1$, $b \not\supset q_1$, and $a_1 \neq a$. This contradicts the maximal property of a. Hence we have $\mathfrak{p} \cup a \supset q$.

Now let $a \supset \mathfrak{p}_1 > a \cap b$. $\mathfrak{p}_1 \neq q$ since otherwise $a \supset q$ and $\mathfrak{p}_1 \neq \mathfrak{p}$ since otherwise $a \cap b \supset \mathfrak{p} > a \cap b$. Hence $\mathfrak{p}_1 \cup \mathfrak{p} \not\supset q$ since every three ideals covering $a \cap b$ generate a Boolean algebra. Thus $p_1 \cup \mathfrak{p} \not\supset q$ for some $p_1 \in \mathfrak{p}_1$. Set $x = p_1 \cap a$. Then $a \supset x \supset a \cap b$ and $x \neq a \cap b$, $x \cup \mathfrak{p} \not\supset q$. Let a_2 be a maximal such x. Then $a_2 \neq a$ since $a \cup \mathfrak{p} \supset q$. Hence $a \supset \mathfrak{p}_2 > a_2$ for some ideal $\mathfrak{p}_2$. Then if $\mathfrak{p}_2 \cup \mathfrak{p} \not\supset q$ we have $p_2 \cup \mathfrak{p} \not\supset q$ for some $p_2 \in \mathfrak{p}_2$. Let $a_2' = a \cap p_2$. Then $a \supset a_2' \supset a \cap b$, $a_2' \neq a \cap b$ and $a_2' \cup \mathfrak{p} \not\supset q$ contrary to the maximal property of a_2. Hence $\mathfrak{p}_2 \cup \mathfrak{p} \supset q$. Let $q_2 = a_2 \cup q$ and $\mathfrak{p}_3 = a_2 \cup \mathfrak{p}$. We have $a_2 \not\supset \mathfrak{p}$ since $a \not\supset \mathfrak{p}$ and hence $\mathfrak{p}_2, q_2, \mathfrak{p}_3 > a_2$ by the Birkhoff condition. Clearly $\mathfrak{p}_2 \cup \mathfrak{p}_3 = \mathfrak{p}_2 \cup a_2 \cup \mathfrak{p} \supset a_2 \cup q = q_2$. Now $\mathfrak{p}_2 \neq q_2$ since otherwise $a \supset q$. Also $q_2 \neq \mathfrak{p}_3$ since otherwise $a_2 \cup \mathfrak{p} \supset q$ contrary to the definition of a_2. Finally $\mathfrak{p}_2 \neq \mathfrak{p}_3$ since $a \not\supset \mathfrak{p}$. Thus $\mathfrak{p}_2, q_2, \mathfrak{p}_3$ do not generate a Boolean algebra, which contradicts our hypothesis. Hence either $a \supset q$ or $b \supset q$.

On the other hand if three ideals $\mathfrak{a}$, $\mathfrak{b}$, $\mathfrak{c}$ covering d do not generate a Boolean algebra, then $\mathfrak{a} \cup \mathfrak{b} \supset \mathfrak{c}$ say. Since $d = \mathfrak{a} \cap \mathfrak{b}$, elements $a \in \mathfrak{a}$ and $b \in \mathfrak{b}$

exist such that $d = a \cap b$. Hence $a \cup b \supset a \cup \mathfrak{b} \supset \mathfrak{c} > a \cap b = d$. If $a \supset \mathfrak{c}$, then $a \supset a \cup \mathfrak{c} \supset \mathfrak{b}$ and $d = a \cap b \supset \mathfrak{b}$ which is impossible. Hence $a \not\supset \mathfrak{c}$ and $b \not\supset \mathfrak{c}$. Thus $a \cup b \supset \mathfrak{c} > a \cap b$ but $a \not\supset \mathfrak{c}$ and $b \not\supset \mathfrak{c}$. This completes the proof of the theorem.

We show now that if the ascending chain condition holds and the condition of Theorem 6.3 is satisfied, then E_5 is equivalent to the Birkhoff condition. A preliminary lemma is required.

LEMMA 6.1. $\mathfrak{S}$ is a Birkhoff lattice if and only if $b \supset a$, $b \not\supset \mathfrak{p} > a$ implies $\mathfrak{p} \cup b > b$.

The necessity of the condition is obvious. To prove the sufficiency let $b > a$, $\mathfrak{c} \supset a$ and $\mathfrak{c} \not\supset b$. If $b \cup \mathfrak{c} \not> \mathfrak{c}$ we have $b \cup \mathfrak{c} \supset \mathfrak{b} \supset \mathfrak{c}$ where $b \cup \mathfrak{c} \neq \mathfrak{b} \neq \mathfrak{c}$. Since $\mathfrak{b} \not\supset b$ an element $d \in \mathfrak{b}$ exists such that $d \not\supset b$. Since $d \supset \mathfrak{c}$ and $\mathfrak{c} \not\supset b$ there exists an element $c \in \mathfrak{c}$ such that $d \supset c$ and $c \not\supset b$. Now $c \supset a$, $c \not\supset b > a$. Hence $c \cup b > c$. But $c \cup b \supset d \cap (c \cup b) \supset c$ and $c \cup b \neq d \cap (c \cup b)$ since $d \not\supset b$. Thus $c = d \cap (c \cup b) \supset \mathfrak{b} \cap (c \cup b) = \mathfrak{b}$ which contradicts $c \not\supset b$. Hence $b \cup \mathfrak{c} > \mathfrak{c}$ and $\mathfrak{S}$ is a Birkhoff lattice.

THEOREM 6.4. Let $\mathfrak{S}$ satisfy E_5, the ascending chain condition, and let $a \cup b \supset \mathfrak{q} > a \cap b$ imply $a \supset \mathfrak{q}$ or $b \supset \mathfrak{q}$. Then $\mathfrak{S}$ is a Birkhoff lattice.

Proof. Let X be the set of all elements x such that $y \supset x$, $\mathfrak{p} > x$, $y \not\supset \mathfrak{p}$, and $\mathfrak{p} \cup y \not> y$ for some y and $\mathfrak{p}$. If $\mathfrak{S}$ is not a Birkhoff lattice, then X is non-empty by Lemma 6.1. Let a be a maximal element of X. Then b and $\mathfrak{p}$ exist such that $b \supset a$, $b \not\supset \mathfrak{p} > a$ and $b \cup \mathfrak{p} \supset \mathfrak{c} > b$, $b \cup \mathfrak{p} \neq \mathfrak{c}$. Now $\mathfrak{p} \supset \mathfrak{p} \cap \mathfrak{c} \supset a$ and $\mathfrak{p} \neq \mathfrak{p} \cap \mathfrak{c}$ since otherwise $b \cup \mathfrak{p} = \mathfrak{c}$. Hence $\mathfrak{p} \cap \mathfrak{c} = a$. Let $p \in \mathfrak{p}$, $c \in \mathfrak{c}$ such that $p \cap c = a$. Then $p \cap c = p \cap b = a$ and $c \neq b$. Hence by E_5 an element p_1 exists such that $p \supset p_1 \supset a$, $p_1 \neq a$ and $b = c \cap (b \cup p_1)$. Now $b \cup p_1 \not\supset \mathfrak{c}$ since otherwise $b = c \cap (b \cup p_1) \supset \mathfrak{c}$ which contradicts $\mathfrak{c} > b$. Now suppose that we have found $p_1, \cdots, p_k$ such that $b \cup p_1 \cup p_2 \cup \cdots \cup p_k \not\supset \mathfrak{c}$ and $(b \cup p_1 \cup \cdots \cup p_i) \cap p_{i+1} = a$ $(i = 1, \cdots, k-1)$. Since $b \cup p_1 \cup p_2 \cup \cdots \cup p_k \not\supset \mathfrak{c}$ we have $b \cup p_1 \cup \cdots \cup p_k \not\supset \mathfrak{p}$ and hence $(b \cup p_1 \cup \cdots \cup p_k) \cap \mathfrak{p} = a$. Thus $p'_{k+1} \in \mathfrak{p}$ exists such that $(b \cup p_1 \cup \cdots \cup p_k) \cap p'_{k+1} = a$. Let $p''_{k+1} = p'_{k+1} \cap p$. Then $(b \cup p_1 \cup \cdots \cup p_k) \cap p''_{k+1} = a$ and $c \cap p''_{k+1} = b \cap p''_{k+1} = a$, $p''_{k+1} \neq a$. Hence by E_5 an element p_{k+1} exists such that $p''_{k+1} \supset p_{k+1} \supset a$, $p_{k+1} \neq a$ and $b = c \cap (b \cup p_{k+1})$. Then $b \cup p_{k+1} \not\supset \mathfrak{c}$ since otherwise $b = c \cap (b \cup p_{k+1}) \supset \mathfrak{c}$. Now $\mathfrak{c} \supset c \cap ((b \cup p_1 \cup \cdots \cup p_k) \cap (b \cup p_{k+1})) \supset b$ and $\mathfrak{c} \neq c \cap ((b \cup p_1 \cup \cdots \cup p_k) \cap (b \cup p_{k+1}))$ since otherwise $b \cup p_{k+1} \supset (b \cup p_1 \cup \cdots \cup p_k) \cap (b \cup p_{k+1}) \supset \mathfrak{c}$. Hence since $\mathfrak{c} > b$ we have $\mathfrak{c} > c \cap ((b \cup p_1 \cup \cdots \cup p_k) \cap (b \cup p_{k+1})) = b$. But now since b is a proper divisor of a, by the maximal property of a we must have $c \cup ((b \cup p_1 \cup \cdots \cup p_k) \cap (b \cup p_{k+1})) > (b \cup p_1 \cup \cdots \cup p_k) \cap (b \cup p_{k+1})$. Now suppose that $b \cup p_1 \cup \cdots \cup p_{k+1} \supset \mathfrak{c}$. Then $(b \cup p_1 \cup \cdots \cup p_k) \cup (b \cup p_{k+1}) \supset c \cup ((b \cup p_1 \cup \cdots \cup p_k) \cap (b \cup p_{k+1})) > (b \cup p_1 \cup \cdots \cup p_k) \cap (b \cup p_{k+1})$. Hence by hypothesis either $b \cup p_1 \cup \cdots \cup p_k \supset \mathfrak{c}$ or $b \cup p_{k+1} \supset \mathfrak{c}$ both of which are impossible. Thus $b \cup p_1 \cup \cdots \cup p_{k+1} \not\supset \mathfrak{c}$. By induction we get an infinite chain $b \subset b \cup p_1 \subset b \cup p_1 \cup p_2 \subset \cdots \subset b \cup p_1 \cup \cdots$

$\cup p_i \subset \cdots$ and $b \cup p_1 \cup \cdots \cup p_i \neq b \cup p_1 \cup \cdots \cup p_{i+1}$ since $(b \cup p_1 \cup \cdots \cup p_i) \cap p_{i+1} = a$. This chain contradicts the ascending chain condition and hence $\mathfrak{S}$ is a Birkhoff lattice.

The condition $a \cup b \supset q > a \cap b$ implies $a \supset q$ or $b \supset q$, may be given a purely combinatorial statement as follows:

THEOREM 6.5. *$a \cup b \supset q > a \cap b$ implies $a \supset q$ or $b \supset q$ if and only if*
(A) *$a \cup b \supset x \supset a \cap b$, $a \cap x = b \cap x = a \cap b$ implies $x = a \cap b$.*

Proof. Let $a \cup b \supset x \supset a \cap b$, $a \cap x = b \cap x = a \cap b$. If $x \neq a \cap b$, let $x \supset q > a \cap b$. Then $a \cup b \supset q > a \cap b$ and hence $a \supset q$ say. But then $a \cap b = a \cap x \supset q > a \cap b$ which is impossible. Hence $x = a \cap b$.

On the other hand, let $a \cup b \supset q > a \cap b$. If $a \not\supset q$ and $b \not\supset q$ we have $a \cap q = b \cap q = a \cap b$. Hence for some $x \in q$ we have $a \cap x = b \cap x = a \cap b$ and $a \cup b \supset x$ by Theorem 2.2. But then $a \cup b \supset x \supset a \cap b$, $a \cap x = b \cap x = a \cap b$ and $x \neq a \cap b$ which contradicts condition A.

Lemma 4.6 with Theorems 6.1–6.5 give

THEOREM 6.6. *Let $\mathfrak{S}$ satisfy the ascending chain condition. Then every element of $\mathfrak{S}$ is uniquely expressible as a reduced cross-cut of irreducibles if and only if conditions E_5 and A are satisfied.*

Theorem 6.6 has the following interesting corollary:

COROLLARY 6.1. *Let $\mathfrak{S}$ satisfy the ascending chain condition and let each element of $\mathfrak{S}$ have a unique reduced decomposition into irreducibles. Then a sublattice $\mathfrak{S}'$ of $\mathfrak{S}$ has unique irreducible decompositions if and only if E_5 holds in $\mathfrak{S}'$.*

Axiom A is clearly a slightly stronger form of the requirement that every modular sublattice be distributive. In D1 it was shown that under the assumption of both the ascending and descending chain conditions, this weaker condition and B1 were necessary and sufficient for unique decomposition into irreducibles. But A *cannot* be replaced by the requirement that every modular sublattice be distributive in Theorem 6.6 as the example of Figure 1 shows.

The non-principal ideals of $\mathfrak{S}$ are the ideals $\mathfrak{a}$, generated by $a_1, a_2, a_3, \cdots$; $\mathfrak{b}$, generated by $b_1, b_2, b_3, \cdots$; and $\mathfrak{c}$, generated by $c_1, c_2, c_3, \cdots$. Clearly $\mathfrak{a} > z$, $\mathfrak{b} > b$, and $\mathfrak{c} > z$. Now $b \cup a_{2n} = d_n > a_{2n}$ and $b \cup a_{2n+1} = b_{2n+1} > a_{2n+1}$. Hence $b \cup a_i > a_i$. Similarly $b \cup c_i > c_i$. Now let x be any element of $\mathfrak{S}$ not equal to b or z. Then b_1/x is an archimedean lattice and B1 is readily verified in b_1/x since each element has at most two covering elements. Thus we have only to verify the Birkhoff condition for non-principal ideals. Clearly $\mathfrak{b} > \mathfrak{a}, b, \mathfrak{c} > z$. Hence $\mathfrak{a} \cup \mathfrak{b} > \mathfrak{b}, \mathfrak{a}; \mathfrak{a} \cup \mathfrak{c} > \mathfrak{a}, \mathfrak{c}; \mathfrak{b} \cup \mathfrak{c} > \mathfrak{b}, \mathfrak{c}$. $\mathfrak{a} \cup c_{2n} = e_n > c_{2n}$ and $\mathfrak{a} \cup c_{2n+1} = b_{2n+2} > c_{2n+1}$. Hence $\mathfrak{a} \cup c_i > c_i$. Similarly $\mathfrak{c} \cup a_i > a_i$. Thus every element of $\mathfrak{S}$ satisfies the Birkhoff condition in the lattice of ideals and hence $\mathfrak{S}$ is a Birkhoff lattice. By Theorem 6.1, E_5 holds in $\mathfrak{S}$. Now if $\mathfrak{S}$ contains a modular, non-distributive sublattice it also contains one of the form $\{u, v, w, x, y\}$ where $v\cup w = w\cup x = v\cup x = u$ and $v\cap w = w\cap x = v\cap x = y$. Since every element not equal to b or z is covered by at most two elements we must have $y = z$. But then $v = a_i$, $w = b$, $x = c_j$ and $v\cup w = a_i\cup b \neq b\cup c_j = w\cup x$ which contradicts $v\cup w$

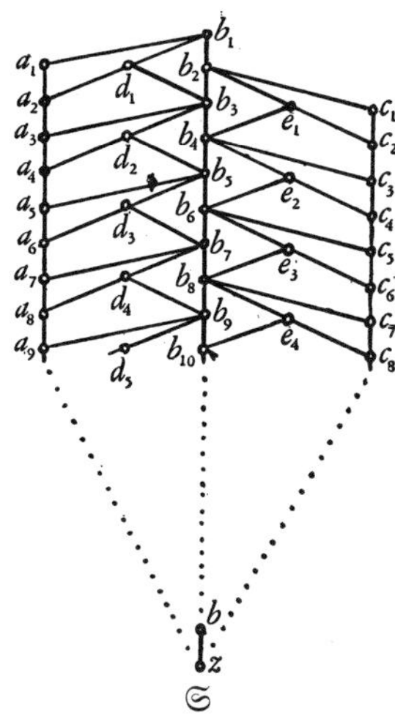

FIG. 1

$= x\cup w$. Hence every modular sublattice of $\mathfrak{S}$ is distributive. However $\mathfrak{S}$ does *not* have unique irreducible decompositions since $z = a_1\cap b = b\cap c_1 = a_1\cap c_1$ and a_1, b, c are irreducibles of $\mathfrak{S}$. Axiom A does not hold since $a_i\cup c_j \supset b \supset a_i\cap c_j$ and $a_i\cap b = c_j\cap b = a_i\cap c_j$ but $b \neq a_i\cap c_j = z$.

Since $\mathfrak{a}$, b, $\mathfrak{c}$ generate a modular lattice, $\mathfrak{L}_x$ is modular for every x. Hence by Theorem 5.1, the number of components in the reduced decompositions of each element must be unique. This can be readily verified.

According to Theorem 6.2, if every element of a lattice $\mathfrak{S}$ is covered by only a finite number of ideals, then E_5 implies that $\mathfrak{S}$ is a Birkhoff lattice. We prove now an even stronger theorem, namely, under this restriction E_5 implies that B1 holds in the lattice of ideals. We begin with necessary lemmas.

LEMMA 6.2. B1 *holds in the lattice of ideals of* $\mathfrak{S}$ *if and only if* $\mathfrak{a} > x\cap\mathfrak{a}$ *implies* $x\cup\mathfrak{a} > x$ *for every* $\mathfrak{a}\in\mathfrak{L}$ *and* $x\in\mathfrak{S}$.

For if B1 holds in $\mathfrak{L}$, then clearly $\mathfrak{a} > x\cap\mathfrak{a}$ implies $x\cup\mathfrak{a} > x$. Now let $\mathfrak{a} > x\cap\mathfrak{a}$ imply $\mathfrak{a}\cup x > x$ for each $\mathfrak{a}$ and x. Suppose that B1 does not hold in $\mathfrak{L}$. Then ideals $\mathfrak{a}$ and $\mathfrak{b}$ exist such that $\mathfrak{a} > \mathfrak{a}\cap\mathfrak{b}$ but $\mathfrak{a}\cup\mathfrak{b} \supset \mathfrak{c} \supset \mathfrak{b}$, $\mathfrak{a}\cup\mathfrak{b} \neq \mathfrak{c} \neq \mathfrak{b}$. Let $x_1\supset\mathfrak{b}$, $x_1\not\supset\mathfrak{a}$. Such an x_1 always exists since $\mathfrak{b}\not\supset\mathfrak{a}$. Also since $\mathfrak{b}\not\supset\mathfrak{c}$, an element x_2 exists such that $x_2\supset\mathfrak{b}$, $x_2\not\supset\mathfrak{c}$. Finally since $\mathfrak{b}\cup\mathfrak{c}\not\supset\mathfrak{a}$ there is an element x_3 such that $x_3\supset\mathfrak{b}$, $x_3\cup\mathfrak{c}\not\supset\mathfrak{a}$. Let $x = x_1\cap x_2\cap x_3$. Then $x\supset\mathfrak{b}$, $x\not\supset\mathfrak{a}$, and $x\cup\mathfrak{c}\not\supset\mathfrak{a}$. Now $\mathfrak{a}\supset\mathfrak{a}\cap x\supset\mathfrak{a}\cap\mathfrak{b}$ and $\mathfrak{a}\neq\mathfrak{a}\cap x$. Hence $\mathfrak{a}\cap x = \mathfrak{a}\cap\mathfrak{b}$ and thus $\mathfrak{a} > \mathfrak{a}\cap x$. By hypothesis then $x \cup \mathfrak{a} > x$. Now $x \cup \mathfrak{a} \supset c \cup x \supset x$ and $c \cup x \neq x$. Hence $x \cup \mathfrak{a} = c \cup x$ which implies $x \cup c \supset \mathfrak{a}$ contrary to the definition of x. Hence B1 holds in $\mathfrak{L}$.

LEMMA 6.3. *Let $\mathfrak{S}$ be a Birkhoff lattice. Then if $\mathfrak{a} > x \cap \mathfrak{a}$ and $x \cup \mathfrak{a} \not> x$, each $x \cap a$, $a \in \mathfrak{a}$ is covered by an infinite number of ideals.*

For let $\mathfrak{a} > x \cap \mathfrak{a}$ and $x \cup \mathfrak{a} \not> x$. Then clearly $a \cap x \neq a$ for every $a \in \mathfrak{a}$ since otherwise $x \supset a \supset \mathfrak{a}$ and $\mathfrak{a} \not> x \cap \mathfrak{a}$. Hence $a \supset \mathfrak{p}_a > a \cap x$ for some ideal $\mathfrak{p}_a$ by Theorem 2.1. Let S_a denote the set of all ideals $\mathfrak{p}_a$. Now $x \not\supset \mathfrak{p}_a$, since otherwise $a \cap x \supset \mathfrak{p}_a > a \cap x$ which is impossible. Thus $\mathfrak{p}_a \supset x \cap \mathfrak{p}_a \supset a \cap x$ and $\mathfrak{p}_a \neq x \cap \mathfrak{p}_a$. Hence $x \cap \mathfrak{p}_a = a \cap x$ and $\mathfrak{p}_a > x \cap \mathfrak{p}_a$ where $x \cap \mathfrak{p}_a$ is a principal ideal of $\mathfrak{L}$. Since $\mathfrak{S}$ is a Birkhoff lattice we have $x \cup \mathfrak{p}_a > x$ for every $\mathfrak{p}_a$.

Now in S_a we set $\mathfrak{p}_a \sim \mathfrak{p}_a'$ if and only if $x \cup \mathfrak{p}_a = x \cup \mathfrak{p}_a'$. Then $\sim$ is an equivalence relation which separates S_a into mutually exclusive sets of ideals. Let B_a denote an arbitrary equivalence class and let $\mathfrak{b}_a = \Sigma(B_a)$. If $\mathfrak{p}_a \in B_a$, then $x \cup \mathfrak{p}_a = x \cup \mathfrak{p}_a'$ for every other ideal $\mathfrak{p}_a'$ of B_a and hence $x \cup \mathfrak{p}_a = x \cup \mathfrak{b}_a$. Thus $x \cup \mathfrak{b}_a > x$. Let T_a denote the set of ideals $\mathfrak{b}_a$. Now if $a \supset a_1 \supset \mathfrak{a}$ and $\mathfrak{b}_{a_1} \in T_{a_1}$, let $\mathfrak{b}_{a_1} \supset \mathfrak{p}_{a_1} > x \cap a_1$. Then $\mathfrak{p}_{a_1} \supset (x \cap a) \cap \mathfrak{p}_{a_1} \supset x \cap a_1$ and $\mathfrak{p}_{a_1} \neq (x \cap a) \cap \mathfrak{p}_{a_1}$ since otherwise $x \supset x \cap a \supset \mathfrak{p}_{a_1}$ which contradicts $x \not\supset \mathfrak{p}_{a_1}$. Hence $\mathfrak{p}_{a_1} > (x \cap a) \cap \mathfrak{p}_{a_1} = x \cap a_1$ and $(x \cap a) \cup \mathfrak{p}_{a_1} > x \cap a$ by the Birkhoff condition. Also $a \supset a \cup a_1 \supset (x \cap a) \cup \mathfrak{p}_{a_1} > x \cap a$ and hence $\mathfrak{p}_a = (x \cap a) \cup \mathfrak{p}_{a_1}$ belongs to S_a. Now $x \cup \mathfrak{p}_a = x \cup (x \cap a) \cup \mathfrak{p}_{a_1} = x \cup \mathfrak{p}_{a_1}$. Let $\mathfrak{b}_{a_1} \supset \mathfrak{p}_{a_1}' > x \cap a_1$. Then $x \cup \mathfrak{p}_a' = x \cup \mathfrak{p}_{a_1}'$ where $\mathfrak{p}_a' = (x \cap a) \cup \mathfrak{p}_{a_1}'$. Thus $x \cup \mathfrak{p}_a = x \cup \mathfrak{p}_{a_1} = x \cup \mathfrak{p}_{a_1}' = x \cup \mathfrak{p}_a'$ and $\mathfrak{p}_a \sim \mathfrak{p}_a'$ in S_a. Hence $\mathfrak{b}_{a_1} \subset \mathfrak{b}_a$ where $\mathfrak{b}_a$ is an ideal of T_a. Now suppose that T_{a_1} contains a second ideal $\mathfrak{b}_{a_1}'$, which is divisible by $\mathfrak{b}_a$. Let $\mathfrak{b}_{a_1}' \supset \mathfrak{p}_{a_1}'' > x \cap a_1$. Then $x \cup \mathfrak{p}_{a_1} = x \cup \mathfrak{p}_a = x \cup \mathfrak{b}_a \supset x \cup \mathfrak{b}_{a_1}' \supset x \cup \mathfrak{p}_{a_1}'' \supset x$. Since $x \cup \mathfrak{p}_{a_1}'' \neq x$, we have $x \cup \mathfrak{p}_{a_1} = x \cup \mathfrak{p}_{a_1}''$ and $\mathfrak{p}_{a_1} \sim \mathfrak{p}_{a_1}''$ in S_{a_1} contrary to assumption. Hence $\mathfrak{b}_{a_1}$ is the only ideal of T_{a_1} divisible by $\mathfrak{b}_a$. Next suppose that T_a contains another ideal $\mathfrak{b}_a'$ such that $\mathfrak{b}_a' \supset \mathfrak{b}_{a_1}$. Then $x \cup \mathfrak{b}_a' \supset x \cup \mathfrak{b}_{a_1} = x \cup \mathfrak{b}_a$ and hence $\mathfrak{b}_a = \mathfrak{b}_a'$. We thus conclude that *each ideal $\mathfrak{b}_{a_1}$ of T_{a_1} is divisible by exactly one ideal $\mathfrak{b}_a$ of T_a and $\mathfrak{b}_{a_1}$ is the only ideal of T_{a_1} which is divisible by $\mathfrak{b}_a$.*

Let n_a denote the cardinal number of the set T_a. If $x \cap \mathfrak{a}$ is covered by only a finite number of ideals for some a, then n_a is finite for some a and hence has a minimal value for some a_0. If $a_0 \supset a \supset \mathfrak{a}$, then $n_a \leq n_{a_0}$ since distinct ideals of T_a are divisible by distinct ideals of T_{a_0}. But since n_{a_0} is minimal we have $n_a = n_{a_0}$. Hence each ideal of T_{a_0} divides exactly one ideal of T_a. Let $\mathfrak{b}_0$ be an ideal of T_{a_0} and let $\mathfrak{b}_a$ denote the ideal of T_a divisible by $\mathfrak{b}_0$. In general, for any $a \in \mathfrak{a}$, let $\mathfrak{b}_a$ be the ideal of T_a divisible by $x \cup \mathfrak{b}_0$. Such an ideal $\mathfrak{b}_a$ always exists since $\mathfrak{b}_0 \supset \mathfrak{b}_a'$ where $\mathfrak{b}_a' \in T_{a \cap a_0}$ and hence $\mathfrak{b}_a \in T_a$ exists such that $\mathfrak{b}_a \supset \mathfrak{b}_a'$. Clearly $x \cup \mathfrak{b}_0 = x \cup \mathfrak{b}_a' = x \cup \mathfrak{b}_a \supset \mathfrak{b}_a$ and $\mathfrak{b}_a$ is unique as shown above. If $a \supset a'$, we have $\mathfrak{b}_a \supset \mathfrak{b}_{a'}$ since $x \cup \mathfrak{b}_a = x \cup \mathfrak{b}_{a'}$. Hence $\mathfrak{b}_{a_1} \cap \mathfrak{b}_{a_2} \cap \cdots \cap \mathfrak{b}_{a_n} \supset \mathfrak{b}_{a_1 \cap a_2 \cap \cdots \cap a_n}$.

Now let $\mathfrak{a}_1 = \prod_{a \in \mathfrak{a}} (\mathfrak{b}_a)$. Since $a \supset \mathfrak{b}_a \supset \mathfrak{a}_1$, we have $\mathfrak{a} \supset \mathfrak{a}_1 \supset x \cap \mathfrak{a}$. If $\mathfrak{a}_1 = x \cap \mathfrak{a}_1$ then $x \cap \mathfrak{a}$ divides the cross-cut of a finite number of the ideals $\mathfrak{b}_a$ and hence

$x \cap a \supset \mathfrak{b}_{a'}$ for some a'. But then $(x \cap a) \cap a' \supset \mathfrak{b}_{a'} \cap \mathfrak{b}_{a'} \supset \mathfrak{b}_{a'} \supset \mathfrak{b}_{a \cap a'} \supset x \cap (a \cap a')$ and $\mathfrak{b}_{a \cap a'} = x \cap (a \cap a')$ contrary to the definition of $\mathfrak{b}_{a \cap a'}$. Thus $a_1 \neq x \cap a$ and hence $a = a_1$ since $a > x \cap a$. But then $x \cup \mathfrak{b}_0 \supset x \cup a_1 \supset x \cup a \supset x$. Since $x \cup \mathfrak{b}_0 > x$ and $x \not\supset a$ we have $x \cup a = x \cup \mathfrak{b}_0 > x$ which contradicts $x \cup a \not> x$. Hence each

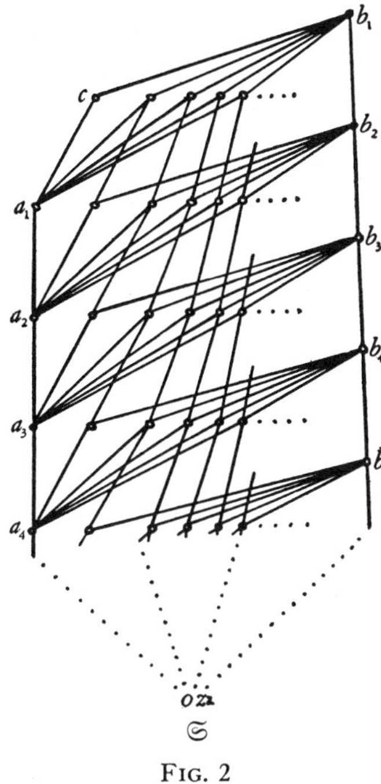

FIG. 2

$x \cap a$ is covered by an infinite number of ideals. The proof is thus complete.
Lemmas 6.2 and 6.3 and Theorem 6.2 give immediately

THEOREM 6.7. *Let each element of $\mathfrak{S}$ be covered by at most a finite number of ideals. Then the following conditions are equivalent:*
　(1) E_5 *holds in $\mathfrak{S}$.*
　(2) $\mathfrak{S}$ *is a Birkhoff lattice.*
　(3) B1 *holds in the lattice of ideals.*

If $\mathfrak{S}$ is a Birkhoff lattice in which each element is *not* covered by at most a finite number of ideals, then even though the ascending chain condition holds in $\mathfrak{S}$ B1 need not be satisfied in the lattice of ideals. For example, consider the lattice diagramed in Figure 2.

All of the elements distinct from z form an ideal a which is generated by $a_1, a_2, a_3, \cdots . b_1, b_2, b_3, \cdots$ clearly form an ideal $\mathfrak{b}$ which divides a. Now let $\mathfrak{b} \supset c \supset a$, $\mathfrak{b} \neq c$. Let $y \in c$, $y \notin \mathfrak{b}$. Then there exists a b_i such that $b_i > x \supset y$, $x \notin \mathfrak{b}$. But by the method of construction there exists an integer j such that

$x \cap b_k = a_k$ all $k \geq j$. Hence $\mathfrak{a} \supset x \cap \mathfrak{b} \supset y \cap \mathfrak{b} \supset \mathfrak{c} \cap \mathfrak{b} = \mathfrak{c}$. Thus $\mathfrak{c} = \mathfrak{a}$ and $\mathfrak{b} > \mathfrak{a}$. Clearly $\mathfrak{b} \cap a_1 = \mathfrak{a}$ and $\mathfrak{b} \cup a_1 = b_1$. But then $\mathfrak{b} > a_1 \cap \mathfrak{b}$ and $a_1 \cup \mathfrak{b} \not\succ a_1$. Hence B1 does *not* hold in $\mathfrak{L}$. On the other hand it is readily verified that $\mathfrak{S}$ is a Birkhoff lattice since $\mathfrak{a}$ is the only non-principal ideal which covers a principal ideal and every element distinct from z divides $\mathfrak{a}$.

The number of ideals covering an element of a lattice is closely related to the number of decompositions of the element into irreducibles. We prove

THEOREM 6.8. *Let $\mathfrak{S}$ be a Birkhoff lattice in which each element can be represented as a cross-cut of irreducibles. Then if an element a has a finite number of decompositions into irreducibles, $\mathfrak{L}_a$ is finite.*

Proof. Since a has only a finite number of decompositions into irreducibles, the number of components in the irreducible decompositions of a is bounded. Hence $\mathfrak{L}_a$ is archimedean by Theorem 3.1. Let $\mathfrak{p}_1, \cdots, \mathfrak{p}_k$ be a maximal independent set of point ideals of $\mathfrak{L}_a$. Let $\mathfrak{s}_i = \mathfrak{p}_1 \cup \cdots \cup \mathfrak{p}_{i-1} \cup \mathfrak{p}_{i+1} \cup \cdots \cup \mathfrak{p}_k$ and suppose that $\mathfrak{L}_a$ has an infinite sequence $\mathfrak{s}_1, \cdots, \mathfrak{s}_k, \mathfrak{s}_{k+1}, \cdots$ of simple ideals. Now if for each i there are only a finite number of simple ideals of the sequence which do not divide $\mathfrak{p}_i$, we have $\mathfrak{s}_n \supset \mathfrak{p}_i$ for all $n \geq l_i$ for some l_i. Let $n \geq \max (l_1, \cdots, l_k)$. Then $\mathfrak{s}_n \supset \mathfrak{p}_i$, $i = 1, \cdots, k$, and $\mathfrak{s}_n \supset \mathfrak{u}_a$, which is impossible. Hence for some $\mathfrak{p}_i$, say $\mathfrak{p}_k$, there are an infinity of ideals in the sequence $\mathfrak{s}_1, \mathfrak{s}_2, \cdots$ which do not divide $\mathfrak{p}_k$. We may assume that $\mathfrak{s}_k, \mathfrak{s}_{k+1}, \cdots$ do not divide $\mathfrak{p}_k$. Let $q_i \supset \mathfrak{s}_i$, $q_i \not\supset \mathfrak{u}_a$. Then $q_i \neq q_j$, $i \neq j$, since otherwise $q_i = q_i \cup q_j \supset \mathfrak{s}_i \cup \mathfrak{s}_j \supset \mathfrak{u}_a$. Now $a = \mathfrak{p}_k \cap \mathfrak{s}_{k+l} = \mathfrak{s}_1 \cap \cdots \cap \mathfrak{s}_{k-1} \cap \mathfrak{s}_{k+l}$ $(l = 0, 1, 2, \cdots)$ implies $a = q_1 \cap q_2 \cap \cdots \cap q_{k-1} \cap q_{k+l}$ $(l = 0, 1, 2, \cdots)$ by Theorem 3.3. If this representation is reduced for each l, q_{k+l} always remains since otherwise $a = q_1 \cap \cdots \cap q_{k-1} \supset \mathfrak{p}_k$. Hence a has an infinite number of irreducible decompositions, which contradicts our hypothesis. Thus $\mathfrak{L}_a$ has only a finite number of simple ideals. But by Theorem 3.2 every ideal of $\mathfrak{L}_a$ can be expressed as a cross-cut of simple ideals. Hence $\mathfrak{L}_a$ is finite.

Theorems 6.7 and 6.8 give

THEOREM 6.9. *Let $\mathfrak{S}$ be a lattice in which every element has at least one and at most a finite number of decompositions into irreducibles. Then $\mathfrak{S}$ is a Birkhoff lattice if and only if B1 is satisfied in the lattice of ideals*[11].

7. **Example of a Birkhoff lattice.** In §3 we have shown that the existence of a decomposition into irreducibles for an element a of a modular lattice implies that $\mathfrak{L}_a$ is archimedean. Hence if the ascending chain condition holds,

---

[11] Various considerations suggest that the finiteness of the number of irreducible decompositions of an element always implies the finiteness of the number of ideals covering the element, in which case E_5 and the finiteness of the number of decompositions would imply that $\mathfrak{S}$ is a Birkhoff lattice. However, I have been unable to prove this. I have also been unable to prove that E_5 is equivalent to the Birkhoff condition under the assumption of the ascending chain condition although this seems quite likely.

$\mathfrak{L}_a$ is archimedean for each a. Also if the ascending chain condition holds in a Birkhoff lattice and the number of components is bounded for an element a, then $\mathfrak{L}_a$ is archimedean. We shall construct in this section a Birkhoff lattice satisfying the ascending chain condition but containing an element a such that $\mathfrak{L}_a$ is not archimedean. By the above remark the number of components in the irreducible decompositions of a must be unbounded.

Latin capitals A, B, C, $\cdots$ will denote finite subsets of the set of positive integers $1, 2, 3, \cdots$. If A is such a set, let $n(A)$ denote the number of elements in A. Small Latin letters a, b, c, $\cdots$ will denote positive integers. $A \cup B$ and $A \cap B$ will denote set-theoretic union and cross-cut respectively and $a \cup b$, $a \cap b$ are respectively the maximum and minimum of a and b. Let $\mathfrak{S}$ be the set of all ordered couples $\alpha = \{A, a\}$ where $n(A) < a$ together with the elements u and z. In $\mathfrak{S}$ we define

$$\alpha \cup \beta = \{A \cup B, a \cup b\} \quad \text{if} \quad n(A \cup B) < a \cup b,$$
$$= u \quad \text{if} \quad n(A \cup B) \geq a \cup b,$$
$$\alpha \cap \beta = \{A \cap B, a \cap b\} \quad \text{if} \quad A \cap B \text{ is not null},$$
$$= z \quad \text{if} \quad A \cap B \text{ is null},$$
$$\alpha \cup z = \alpha \cap u = \alpha, \quad \alpha \cup u = u, \quad \alpha \cap z = z.$$

If $A \cap B$ exists, then $n(A \cap B) \leq n(A) < a \leq a \cap b$. Hence $\alpha \cap \beta$ is in $\mathfrak{S}$ if α and β are in $\mathfrak{S}$. Now it can be readily verified that the union and cross-cut so defined in $\mathfrak{S}$ are idempotent, commutative and associative. Consider $\alpha \cap (\alpha \cup \beta)$. If $\alpha \cup \beta = u$, then $\alpha \cap (\alpha \cup \beta) = \alpha$. If $\alpha \cup \beta \neq u$, then $\alpha \cap (\alpha \cup \beta) = \{A \cap (A \cup B), a \cap (a \cup b)\} = \{A, a\} = \alpha$. Hence $\alpha \cap (\alpha \cup \beta) = \alpha$ in all cases. Similarly $\alpha \cup (\alpha \cap \beta) = \alpha$. Hence $\mathfrak{S}$ is a lattice under the union and cross-cut operations defined above. Clearly $\alpha \supset \beta$ if and only if $A \supset B$ and $a \leq b$.

$\mathfrak{S}$ *satisfies the ascending chain condition.* For let $\alpha_1 \subset \alpha_2 \subset \alpha_3 \subset \cdots$ be an infinite ascending chain. We may assume that $\alpha_1 \neq z$ so that $\alpha_1 = \{A_1, a_1\}$. But then $A_1 \subset A_2 \subset A_3 \subset \cdots$ and $a_1 \geq a_2 \geq a_3 \geq \cdots$. Now $n(A_i) < a_i \leq a_1$. Hence the chain $A_1 \subset A_2 \subset \cdots$ has only a finite number of distinct sets. Clearly the chain $a_1 \geq a_2 \geq \cdots$ has only a finite number of distinct members. Thus the chain $\alpha_1 \subset \alpha_2 \subset \cdots$ has only a finite number of distinct members.

Now let $\mathfrak{a}$ be an ideal of $\mathfrak{S}$ with elements β, γ, δ, $\cdots$. Then the set of integers is either bounded or unbounded. If bounded, let a be the largest of them. Now suppose that $B \cap C \cap D \cap \cdots$ is null. Then there exist a finite number of them B, C, $\cdots$, L whose cross-cut is null. But then $\mathfrak{a}$ contains z since $\mathfrak{a}$ is closed with respect to finite cross-cut. Hence $\mathfrak{a} = \mathfrak{S}$ in this case. If $B \cap C \cap D \cap \cdots$ is not null, let $A = B \cap C \cap D \cap \cdots$. Then $A = B \cap C \cap \cdots \cap L$ for a finite number of the sets and hence $\alpha' = \{A, a'\}$ and $\alpha'' = \{A', a\}$ are in $\mathfrak{a}$. But then $\mathfrak{a}$ contains $\alpha = \alpha' \cap \alpha'' = \{A, a\}$ and $\beta \supset \alpha$ for every $\beta \in \mathfrak{a}$. Hence if the integers of the ideal are bounded, $\mathfrak{a}$ is a principal ideal. If the integers of the ideal are unbounded, then as before either $\mathfrak{a} = \mathfrak{S}$

or there exists a set A such that $\{A, a\}$ is in $\mathfrak{a}$ for each positive integer a and $\beta \supset \{A, a\}$ for some a if $\beta \in \mathfrak{a}$. Hence the ideals of $\mathfrak{S}$ have the form $\mathfrak{a} = \{A, \infty\}$ if $\mathfrak{a}$ is not principal. ∞ denotes the ideal of all positive integers. Clearly if a and b are positive integers or ∞

$$
\begin{aligned}
\mathfrak{a} \cap \mathfrak{b} &= \{A \cap B, a \cap b\} && \text{if } A \cap B \text{ is not null,} \\
&= z && \text{if } A \cap B \text{ is null,} \\
\mathfrak{a} \cup \mathfrak{b} &= \{A \cup B, a \cup b\} && \text{if } n(A \cup B) < a \cup b, \\
&= u && \text{if } n(A \cup B) \geq a \cup b.
\end{aligned}
$$

THEOREM 7.1. *$\mathfrak{S}$ is a Birkhoff lattice.*

Proof. We shall show that B1 holds in the lattice of ideals. Let $\mathfrak{a} > \mathfrak{a} \cap \mathfrak{b}$. If $\mathfrak{a} \cap \mathfrak{b} = z$, then $A \cap B$ is null. Since $\mathfrak{a} > z$ we have $\mathfrak{a} = \{(a), \infty\}$. But then $\mathfrak{a} \cup \mathfrak{b} = \{(a) \cup B, \infty \cup b\} = \{(a) \cup B, b\}$ and $(a) \cup B > B$. Thus $\mathfrak{a} \cup \mathfrak{b} > \mathfrak{b}$. If $\mathfrak{a} \cap \mathfrak{b} \neq z$, then $\{A, a\} > \{A \cap B, a \cap b\}$ and hence either $A > A \cap B, a = a \cap b$ or $A = A \cap B, a = (a \cap b) - 1$. In the first case $A > A \cap B \rightarrow A \cup B > B$ and $a = a \cap b$ implies $b = a \cup b$. Hence $\mathfrak{a} \cup \mathfrak{b} = \{A \cup B, a \cup b\} = \{A \cup B, b\} > \{B, b\} = \mathfrak{b}$. If $n(A \cup B) \geq b$, then $n(B) = b - 1$ and hence $u > B$. In the second case $A = A \cap B \rightarrow A \cup B = B$ and $a > a \cap b \rightarrow a = b - 1$. Hence $\mathfrak{a} \cup \mathfrak{b} = \{B, a\} > \mathfrak{b}$. Thus $\mathfrak{a} > \mathfrak{a} \cap \mathfrak{b}$ implies $\mathfrak{a} \cup \mathfrak{b} > \mathfrak{b}$ and B1 holds in the lattice of ideals of $\mathfrak{S}$.

The point ideals of $\mathfrak{L}_z$ are clearly the ideals $\mathfrak{p}_i = \{(i), \infty\}$. Now $\mathfrak{p}_1 \cup \mathfrak{p}_2 \cup \cdots = u$ and hence $\mathfrak{L}_z = \mathfrak{L}$. The ascending chain $\mathfrak{p}_1 \subset \mathfrak{p}_1 \cup \mathfrak{p}_2 \subset \mathfrak{p}_1 \cup \mathfrak{p}_2 \cup \mathfrak{p}_3 \subset \cdots$ has distinct members and $\mathfrak{L}_z$ is thus *not* archimedean. The point ideals $\mathfrak{p}_1, \mathfrak{p}_2, \cdots$ are *not* independent since the union of any infinite set is u. However, every finite set of the $\mathfrak{p}_i$ is independent and thus generates a Boolean algebra. $\mathfrak{L}_z$ is *not* complemented. For if $\mathfrak{p}_i \cup \mathfrak{a} = u$, then $\mathfrak{a} = u$ and $\mathfrak{a} \supset \mathfrak{p}_i$. The simple ideals of $\mathfrak{L}_z$ are the elements of the form $\{A, a\}$ where $n(A) = a - 1$. Clearly $\mathfrak{p}_i$ cannot be represented as a cross-cut of simple ideals. Hence it is *not* true that every ideal of $\mathfrak{L}_z$ may be represented as a cross-cut of simple ideals. The irreducibles of $\mathfrak{S}$ are the simple elements of $\mathfrak{L}_z$, namely, those elements $\{A, a\}$ with $n(A) = a - 1$. Now let A_i be the set $\{1, 2, \cdots, i-1, i+1, \cdots, k\}$ $(i = 1, \cdots, k)$. Then $\alpha_i = \{A_i, k\}$ is simple for each i and $z = \alpha_1 \cap \alpha_2 \cap \cdots \cap \alpha_k$ since $A_1 \cap A_2 \cap \cdots \cap A_k$ is null. This representation of z is clearly reduced. Hence for any positive integer $k > 1$, z has a reduced decomposition with k components.

This example clearly indicates the complications that may arise if $\mathfrak{L}_a$ is *not* archimedean even though the ascending chain condition holds in $\mathfrak{S}$.

8. **Example of a lattice satisfying E_5 which is not a Birkhoff lattice.** Let S be the set of elements $p_1, p_2, p_3, \cdots$. From the set of all subsets of S omit those infinite sets which contain either p_1 or p_2 but not both. Denote this set of subsets by $\mathfrak{S}$. $\mathfrak{S}$ is clearly closed under infinite cross-cut. Since $\mathfrak{S}$ contains a unit element, the union of any set of sets of $\mathfrak{S}$ may be defined in terms

of the cross-cut operation. $\mathfrak{S}$ is thus a continuous lattice in which every element is a union of points. Now consider $A \cup p$ where $A \in \mathfrak{S}$ and p is any element of S. If A is finite, then clearly $A \cup p = A + p$ where $+$ indicates set-theoretic union. Also if A is infinite and contains both p_1 and p_2, $A \cup p = A + p$. Now if A is infinite and does not contain p_1 or p_2, then $A \cup p = A + p$ if $p \neq p_1, p_2$ and $A \cup p = A + p_1 + p_2$ if $p = p_1$ or $p = p_2$. Now let $A + p_1 + p_2 \supset B \supset A$. If $B \neq A$, then B contains p_1 or p_2 and hence contains both p_1 and p_2 by the definition of $\mathfrak{S}$. Thus $B = A + p_1 + p_2$. Hence in every case $A \cup p > A$ if $p \notin A$.

Now let $A \supset B \supset A \cap C$ and $C \neq A \cap C$ where A, B, C are in $\mathfrak{S}$. Since every element of $\mathfrak{S}$ is a union of points and $C \neq A \cap C$, there exists a point p such that $C \supset p$, $A \cap C \not\supset p$. Set $C_1 = (A \cap C) \cup p$. Then $C \supset C_1 \supset A \cap C$ and $C_1 \neq A \cap C$. Now $B \cup p \supset A \cap (B \cup C_1) \supset B$ and $B \cup p \neq A \cap (B \cup C_1)$ since otherwise $A \supset B \cup p \supset p$ and $A \cap C \supset p$ which contradicts $A \cap C \not\supset p$. Since $B \cup p > B$ we thus have $B = A \cap (B \cup C_1)$ and hence E_5 holds in $\mathfrak{S}$.

In $\mathfrak{S}$ let $\mathfrak{a}$ be the ideal generated by the sets $A_k = \{p_k, p_{k+1}, \cdots\}$ $(k = 3, 4, 5, \cdots)$. Then by Theorem 2.1, there exists a point ideal $\mathfrak{p}$ such that $\mathfrak{a} \supset \mathfrak{p} > z$. Every set of $\mathfrak{S}$ occurring in $\mathfrak{p}$ contains an infinite number of elements. For suppose that $Q \in \mathfrak{p}$ and Q contains only a finite number of elements. Let k be the largest subscript occurring among the elements of Q. Then $z = Q \cap A_{k+1} \supset \mathfrak{p} \cap \mathfrak{a} \supset \mathfrak{p}$ which contradicts $\mathfrak{p} > z$. If $Q \in \mathfrak{p}$, then $Q \cup p_1 \supset p_2$ by the definition of $\mathfrak{S}$. Hence $\mathfrak{p} \cup p_1 \supset p_1 \cup p_2 \supset p_1$ where $\mathfrak{p}'$ $'p_1 \neq p_1 \cup p_2$ and $p_1 \cup p_2 \neq p_1$. Thus $\mathfrak{p} \cup p_1 \not> p_1$ and hence $\mathfrak{S}$ is not a Birkhoff lattice. If $A \supset B$ and $A \neq B$, let $A \supset p$, $B \not\supset p$. Then $A \supset B_1 > B$ where $B_1 = B \cup p$. Thus $\mathfrak{S}$ is an example of a continuous point lattice in which covering elements exist and E_5 holds, but which is *not* a Birkhoff lattice. Whether or not an exchange lattice, i.e., a continuous point lattice satisfying E_5 and a finite dependence axiom (Mac Lane [1]) is a Birkhoff lattice is an open question.

REFERENCES

G. BIRKHOFF
1. *Rings of sets*, Duke Mathematical Journal, vol. 3 (1937), pp. 443-454.
2. *On combination of subalgebras*, Proceedings of the Cambridge Philosophical Society, vol. 29 (1933), pp. 441–464.

R. P. DILWORTH
1. *Lattices with unique irreducible decompositions*, Annals of Mathematics.
2. *The arithmetical theory of Birkhoff lattices*, submitted to Duke Mathematical Journal.

S. MAC LANE
1. *A lattice formulation for transcendence degrees and p-bases*, Duke Mathematical Journal, vol. 24 (1938), pp. 455-468.

M. H. STONE
1. *The theory of representations of Boolean algebra*, these Transactions, vol. 40 (1936), pp. 37–111.

H. WALLMAN
1. *Lattices and topological spaces*, Annals of Mathematics, (2), vol. 39 (1938), pp. 112–126.

YALE UNIVERSITY,
NEW HAVEN, CONN.

DECOMPOSITION THEORY FOR LATTICES WITHOUT CHAIN CONDITIONS

BY

R. P. DILWORTH AND PETER CRAWLEY

1. Introduction. The classical structure theorems of algebraic systems usually assume some type of finiteness condition. The most common finiteness restriction is a chain condition. Thus the proofs of the fundamental structure and decomposition theorems for lattices have customarily required the ascending chain condition. Moreover, these theorems generally fail to hold in arbitrary lattices. Nevertheless, there are important examples of decomposition theorems for lattices associated with abelian groups and rings in which the ascending chain condition does not hold. These lattices have in common another distinctive property—they are compactly generated. Namely, the lattice is generated by a collection of elements which are finitely dependent in the sense that any such element is contained in the union of a set of lattice elements if and only if it is contained in the union of a finite subset. The compact elements of the lattice of ideals of a ring are the finitely generated ideals. Likewise the compact elements of a lattice of congruence relations are the congruence relations generated by collapsing a finite collection of element pairs. More generally, the lattice of congruence relations and the lattice of subsystems of a universal algebra are compactly generated. Since structure theorems for an algebraic system correspond to decomposition theorems in the lattice of congruence relations, this strongly suggests that compactly generated lattices are the appropriate domain in which to study decomposition theory. Furthermore, since every lattice satisfying the ascending chain condition is trivially compactly generated, it follows that the classical case will be subsumed in the more general theory.

In the present paper we shall make the further restriction that the lattice is atomic, that is, that every quotient contains minimal elements. Thus this case generalizes the finite dimensional theory (Dilworth [2; 3]) in that the ascending chain condition is replaced by compact generation and the descending chain condition, by atomicity. Now the basic technique in the classical case consisted in establishing a relationship between the properties of the decompositions of an element a of the lattice and the structure of the finite dimensional quotient lattice generated by the elements covering a. In the present case, the quotient lattice is no longer finite dimensional and the decompositions are no longer finite. Nevertheless, compact generation and atomicity imply sufficient regularity in the structure of the quotient lattices

---

Received by the editors September 1, 1959.

and the decompositions that a relationship can be established which preserves many of the properties of the finite dimensional case.

It should be noted that there are many important examples of compactly generated atomic lattices which do not satisfy the ascending chain condition. Some of these examples are the lattice of subgroups of an infinite torsion abelian group, the lattice of congruences of a weakly atomic modular lattice, the lattices of subspaces of an infinite dimensional vector space, and any infinite dimensional exchange lattice.

The decomposition theory for nonatomic compactly generated lattices requires a quite different approach and will be treated by one of us elsewhere.

2. **Preliminaries.** Lattice elements will be denoted by lower case latin letters while sets of lattice elements will be denoted by latin capitals. If a is an element of the lattice L and S is a subset of L, $a \cap S$ will denote the set of elements $a \cap s$ where $s \in S$. $a \cup S$ is similarly defined. The covering relation in L will be denoted by $a \succ b$. $S \wedge T$ and $S \vee T$ will denote set-intersection and union respectively.

DEFINITION 2.1. An element c of a lattice L is said to be *compact* if $c \leq \cup S$ implies $c \leq \cup S'$ for a finite subset S' of S.

If c_1 and c_2 are compact, then $c_1 \cup c_2 \leq \cup S$ implies $c_1 \leq \cup S$ and $c_2 \leq \cup S$. Hence there exist finite subsets S_1 and S_2 of S such that $c_1 \leq \cup S_1$ and $c_2 \leq \cup S_2$. But then $c_1 \cup c_2 \leq \cup (S_1 \vee S_2)$. Thus we have the following lemma.

LEMMA 2.1. *The compact elements of a lattice are closed under finite union.*

The compact elements of L will be denoted by $C(L)$.

DEFINITION 2.2. A lattice L is *compactly generated* if L is complete and $a = \cup \{ c \in C(L) \mid c \leq a \}$ for all $a \in L$.

A complete lattice is thus compactly generated if the compact elements generate the lattice under unrestricted joins. If every element is compact, the lattice is trivially compactly generated.

LEMMA 2.2. *Let A be an ideal of a complete lattice L. Then if $\cup A$ is compact, A is principal.*

For if $c = \cup A$ is compact, then $c \leq \cup S'$ where S' is a finite subset of A. But then $c \in A$ and hence $A = (c)$.

COROLLARY. *Every element of a complete lattice L is compact if and only if L satisfies the ascending chain condition.*

The following lemmas develop some of the properties of lattices satisfying the ascending chain condition which also hold in compactly generated lattices.

LEMMA 2.3. *Every compactly generated lattice is join continuous; that is, $a \cap \cup B = \cup (a \cap B)$ for every ideal B of L.*

For let $c \leq a \cap \cup B$ where $c \in C(L)$. Then $c \leq \cup B$ and hence $c \leq \cup B'$ where

B' is a finite subset of B. But then $c \in B$ and hence $c = a \cap c \leq \cup(a \cap B)$. Thus $a \cap \cup B = \cup\{c \in C(L) \mid c \leq a \cap \cup B\} \leq \cup(a \cap B)$. Since $a \cap \cup B \geq \cup(a \cap B)$ holds trivially, it follows that $a \cap \cup B = \cup(a \cap B)$.

DEFINITION 2.3. An element q of a lattice L is *completely meet irreducible* if for all $S \subseteq L$, $q = \cap S$ implies $q \in S$.

Clearly every completely meet irreducible element is meet irreducible. For atomic([1]) lattices the two concepts coincide.

LEMMA 2.4. *Every meet irreducible element of a complete atomic lattice is completely meet irreducible.*

For if q is a meet irreducible element, let $q = \cap S$ and suppose that $s > q$ all $s \in S$. Then by atomicity there exist p such that $p > q$. If $s \not\geq p$ for some $s \in S$, then $s \cap p = q$ contrary to the meet irreducibility of q. Thus $s \geq p$ all $s \in S$ and hence $q = \cap S \geq p$ contrary to $p > q$. It follows that $q = s \in S$ for some s and hence q is completely meet irreducible.

LEMMA 2.5. *If L is a compactly generated lattice and a, b are elements of L such that $a \not\geq b$, then there exists a completely meet irreducible element q such that $a \leq q$ and $b \not\leq q$.*

For since $a \not\geq b$ and $b = \cup\{c \in C(L) \mid c \leq b\}$, there exists a compact element $c \leq b$ such that $c \not\leq a$. But then $c \notin (a)$ where (a) is the principal ideal generated by a. By the Maximal Principle there exists a maximal ideal A such that $a \in A$ and $c \notin A$. Let $q = \cup A$. If $c \leq q$, then $c \leq \cup A$ and hence $c \leq \cup A'$ where A' is a finite subset of A, and thus $c \in A$ contrary to assumption. It follows that $c \not\leq q$ and hence $b \not\leq q$. If $q = \cap S$ and $s > q$ all $s \in S$, then since $A \subset (s)$ we have $c \leq s$ all s by the maximal property of A. Hence $c \leq \cap S = q$ contrary to $c \not\leq q$. Thus $q = s \in S$ for some s and q is completely meet irreducible.

COROLLARY. *Each element of a compactly generated lattice can be represented as a meet of completely meet irreducible elements.*

For if b is the meet of all completely meet irreducibles containing a and $b \not\leq a$, then according to Lemma 2.5 there would exist a completely meet irreducible q such that $a \leq q$ and $b \not\leq q$ contrary to the definition of b.

It is evident from the corollary that a compactly generated lattice contains sufficiently many completely meet irreducibles to give substance to a decomposition theory. Indeed it is likely that the additional assumption of atomicity is sufficient to insure the existence of irredundant decompositions. However, we shall begin our study with the consideration of semimodular lattices([2]).

([1]) A lattice is atomic if $a > b$ implies $a \geq c > b$ for some $c \in L$.

([2]) At the present time, an adequate decomposition theory does not exist for lattices more general than semimodular lattices even in the finite dimensional case.

3. Semimodular lattices. Throughout this section L will denote a compactly generated atomic lattice. Completely meet irreducible elements of L will simply be called "irreducibles." For each $a \in L$ let P_a denote the set of elements covering a. We set $u_a = \cup P_a$, and define $u_a/a = \{x \mid a \leq x \leq u_a\}$.

LEMMA 3.1. u_a/a *is a compactly generated atomic lattice. Furthermore the elements of* P_a *are compact elements of* u_a/a.

For if $x \in u_a/a$, then $x = \cup\{c \in C(L) \mid c \leq x\}$. But then
$$x = \cup \{a \cup c \mid c \in C(L), c \leq x\}.$$
$a \cup c$ is compact in u_a/a since if $a \cup c \leq \cup S$ where $S \subseteq u_a/a$, then $c \leq \cup S$ and hence $c \leq \cup S'$ where S' is a finite subset of S. But then $a \cup c \leq \cup S'$. Since u_a/a is a quotient lattice of an atomic lattice, it is atomic. Finally if $p > a$ and $p \geq c$, where $a \not\geq c$ and $c \in C(L)$, then $p = a \cup c$ and hence p is compact in u_a/a.

DEFINITION 3.1. An element $s \in u_a/a$ is a *relative (meet) irreducible* of u_a/a if there exists a completely meet irreducible q of L such that $q \geq a$ and $q \cap u_a = s$.

COROLLARY. *Each irreducible of u_a/a is a relative irreducible.*

For if s is an irreducible of u_a/a, by the corollary to Lemma 2.5, $s = \cap Q$ where Q is a set of irreducible of L. But then $s = \cap(Q \cap u_a)$ and $Q_a \cap u_a \subseteq u_a/a$. Hence $s = q \cap u_a$ for some q and thus s is a relative irreducible.

The decompositions of an element a into irreducibles in L are closely related to the decompositions of a into relative irreducibles in u_a/a, as indicated in the following lemma. The proof is left to the reader.

LEMMA 3.2. *To each decomposition $a = \cap Q$ of the element a into irreducibles in L corresponds the decomposition $a = \cap(Q \cap u_a)$ into relative irreducibles in u_a/a. The decomposition $a = \cap Q$ is irredundant if and only if the decomposition $a = \cap(Q \cap u_a)$ is irredundant. Furthermore each decomposition of a into relative irreducibles in u_a/a can be obtained in this manner from a decomposition into irreducibles in L.*

By means of the correspondence of Lemma 3.2 it is frequently possible to reduce the study of decompositions in L to the study of decompositions in u_a/a. This approach is particularly appropriate when the relative irreducibles of u_a/a coincide with the irreducibles of u_a/a.

Semimodularity is defined in the usual way.

DEFINITION 3.2. A lattice is *(upper) semimodular* if $a > a \cap b$ implies $a \cup b > b$.

A lattice is lower semimodular if its dual is upper semimodular. For finite dimensional lattices, upper semimodularity is equivalent to the following weaker condition.

DEFINITION 3.3. A lattice is *weakly (upper) semimodular* if $a, b > a \cap b$ imply $a \cup b > a, b$.

THE DILWORTH THEOREMS

We will now show that these two conditions are equivalent for compactly generated atomic lattices.

LEMMA 3.3. *A weakly semimodular, compactly generated, atomic lattice is semimodular.*

For, let $a > a \cap b$ and define a subset W as follows:

$$W = \{x \mid b \geq x \geq a \cap b,\ a \cup x > x\}.$$

The set W is nonempty since it contains $a \cap b$. Now let X be a chain of W and let $y = \cup X$. Then $b \geq y \geq a \cap b$. Also $a \cup y > y$, since otherwise $b \geq y = a \cup y \geq a$ contrary to $a > a \cap b$. Let $a \cup y \geq z > y$. Since the lattice is compactly generated there exists a compact element c such that $z \geq c$, $y \not\geq c$. Then $c \leq a \cup y = a \cup (\cup X)$. Since c is compact there exist $x_1, \cdots, x_n$ in X such that $c \leq a \cup x_1 \cup \cdots \cup x_n$. Since X is a chain, $x_1 \cup \cdots \cup x_n = x \in X$. Hence $c \leq a \cup x$ and thus $x \leq c \cup x \leq a \cup x$. If $x = c \cup x_1$ then $y \geq x = c \cup x \geq c$ contrary to $y \not\geq c$. Thus $x < c \cup x \leq a \cup x$. Since $a \cup x > x$ we have $c \cup x = a \cup x \geq a$. But then $z \geq c \cup y \geq c \cup x \geq a$ and hence $z \geq a \cup y$. Thus $z = a \cup y$ and hence $a \cup y > y$. It follows that $\cup X = y \in W$. By the Maximal Principle W contains a maximal element x_0. Suppose $b > x_0$. Then by atomicity, $b \geq p > x_0$ for some p. Since $b \not\geq a$ it follows that $p \neq a \cup x_0$. Also $a \cup x_0 > x_0$ since $x_0 \in W$. Hence $p \cap (a \cup x_0) = x_0$ and by weak semimodularity we have $a \cup p = (a \cup x_0) \cup p > p$. But $b \geq p \geq a \cap b$ and hence $p \in W$ contrary to the maximal property of x_0. Thus $b = x_0$ and hence $a \cup b = a \cup x_0 > x_0 = b$.

The following generalization of a well known characterization of finite dimensional modular lattices will be needed.

LEMMA 3.4. *A compactly generated atomic lattice which is both upper and lower semimodular is modular.*

For let $a \geq b$ and let V be defined as follows:

$$V = \{x \mid b \cup (a \cap c) \geq x \geq a \cap c,\ x = a \cap (x \cup c)\}.$$

V is nonempty since it contains $a \cap c$. Now let X be a chain of V and let $y = \cup X$. Clearly $b \cup (a \cap c) \geq y \geq a \cap c$. Also, $a \cap (y \cup c) = a \cap \cup(X \cup c) = \cup(a \cap (X \cup c)) = \cup X = y$ since the lattice is continuous. Hence $y \in V$ and thus every chain in V has a bound in V. By the maximal principle V contains a maximal element x_0. If $b \cup (a \cap c) > x_0$, there exists p such that $b \cup (a \cap c) \geq p > x_0$. Now $x_0 \cup c \not\geq p$ since otherwise $x_0 = a \cap (x_0 \cup c) \geq p$. Thus $p \cap (x_0 \cup c) = x_0 < p$. By upper semimodularity we have $p \cup c = p \cup (x_0 \cup c) > x_0 \cup c$. Now $x_0 \cup c \not\geq a \cap (p \cup c)$ since otherwise $x_0 = a \cap (x_0 \cup c) \geq a \cap (p \cup c) \geq p$. Hence $[a \cap (p \cup c)] \cup (x_0 \cup c) = p \cup c > x_0 \cup c$. By lower semimodularity we have

$$a \cap (p \cup c) > a \cap (p \cup c) \cap (x_0 \cup c) = a \cap (x_0 \cup c) = x_0.$$

But $a \cap (p \cup c) \geq p > x_0$ and hence $a \cap (p \cup c) = p$. Thus $p \in V$ contrary to the maximal property of x_0. It follows that $x_0 = b \cup (a \cap c)$ and hence

$$b \cup (a \cap c) = x_0 = a \cap (x_0 \cup c) = a \cap (b \cup (a \cap c) \cup c) = a \cap (b \cup c).$$

We thus conclude that the lattice is modular.

DEFINITION 3.4. *A subset N of P_a is independent if* $p \cap \cup (N - p) = a$ *for all* $p \in N$.

LEMMA 3.5. *A subset N of P_a is independent if and only if every finite subset of N is independent.*

If N is not independent, then for some $p \in N$ we have $p \geq p \cap \cup (N - p) > a$ and hence $p = p \cap \cup (N - p) \leq \cup (N - p)$. Since p is compact in u_a/a we have $p \leq \cup (N' - p)$ where N' is a finite subset of N containing p. Thus N' is not independent. Conversely, it is clear that any finite subset of an independent set is independent.

LEMMA 3.6. *Let a be an element of a semimodular, compactly generated, atomic lattice and let N be an independent subset of P_a. Then N is a maximal independent subset of P_a if and only if $\cup N = u_a$.*

If $\cup N = u_a$, then N is clearly a maximal independent subset of P_a. Now let N be maximal and let $p \in P_a$. Then either $\cup N \geq p$ or there exists $p' \in N$ such that $\cup (N - p') \cup p \geq p'$. In the second case $\cup (N - p') \not\geq p$ since otherwise $\cup (N - p') \geq p'$ contrary to the independence of N. Thus we have $\cup (N - p') \cup p > \cup (N - p')$ by semimodularity. But $\cup (N - p') \cup p \geq \cup (N - p') \cup p' = \cup (N) > \cup (N - p')$. Thus $\cup (N) = \cup (N - p') \cup p \geq p$. Hence $\cup N \geq p$ in either case. Since p was an arbitrary element of P_a we have $\cup N \geq \cup P_a = u_a$.

LEMMA 3.7. *Let a be an element of a semimodular, compactly generated, atomic lattice L and let N be an independent subset of P_a. Then the elements of L which are joins of subsets of N form a complete sublattice of L which is isomorphic with the lattice of all subsets of S.*

Consider the mapping $S \rightarrow \cup S$. If $\cup S_1 = \cup S_2$, then $p_1 \leq \cup S_1 \leq \cup S_2$ all $p_1 \in S_1$. Since N is independent it follows that $p_1 \in S_2$ all $p_1 \in S_1$. Thus $S_1 \subseteq S_2$ and similarly $S_2 \subseteq S_1$. The mapping is thus one-to-one. Now let S_α be a collection of subsets of N. Clearly $\cup_\alpha (\cup S_\alpha) = \cup (\vee_\alpha S_\alpha)$ and hence the mapping preserves joins. Let $b = \cup (\wedge_\alpha S_\alpha)$ and let $T_\alpha = \{ b \cup p \mid p \in S_\alpha - \wedge_\alpha S_\alpha \}$. Then $p' > b$ for each $p' \in T_\alpha$ since N is independent and L is semimodular. Also $\cup T_\alpha = b \cup \cup (S_\alpha - \wedge_\alpha S_\alpha) = \cup (\wedge_\alpha S_\alpha) \cup \cup (S_\alpha - \wedge_\alpha S_\alpha) = \cup (S_\alpha)$. Now $b \cup p_1 = b \cup p_2$ where $p_1, p_2 \in S_\alpha - \wedge_\alpha S_\alpha$ implies that $p_1 = p_2$ since N is independent. Thus $\wedge_\alpha T_\alpha = \varnothing$. Let us suppose that $\cap_\alpha (\cup T_\alpha) > b$. Then $\cap_\alpha (\cup T_\alpha) \geq r > b$ by atomicity and hence $r \leq \cup T_\alpha$ all α. But r is compact in u_b/b and hence $r \leq \cup T_\alpha'$ for some finite subset T_α' of T_α for each α. Now pick a fixed α, then by semimodularity there exists $p' \in T_\alpha'$ such that $p' \leq \cup (T_\alpha' - p') \cup r$. Since $\wedge_\alpha T_\alpha = \varnothing$ there exists β such that $p' \notin T_\beta$. But then $r \leq \cup T_\beta$ implies $p' \leq \cup (T_\alpha' - p') \cup \cup T_\beta \leq \cup (T_\alpha \vee T_\beta - p')$. If $p' = b \cup p$, then we have $p \leq \cup (S_\alpha \vee S_\beta - p)$ contrary to

                                                    THE DILWORTH THEOREMS

the independence of N. Hence we conclude that $\bigcap_\alpha (\bigcup S_\alpha) = \bigcap_\alpha (\bigcup T_\alpha) = b = \bigcup(\bigwedge_\alpha S_\alpha)$. Thus the mapping preserves meets and the proof of the lemma is complete.

COROLLARY. *If a is an element of a semimodular, compactly generated, atomic lattice, then u_a/a is complemented and each element of u_a/a is a meet of elements covered by u_a.*

For if $b \in u_a/a$, let S be a maximal independent set of points of u_a/a contained in b. Extend S to a maximal independent set of points P. Then $b \cup \bigcup(P-S) = \bigcup P = u_a$, and clearly $b \cap \bigcup(P-S) = a$ since $\bigcap S \cap \bigcap (P-S) = a$. Now $p \cup b \succ b$ for each $p \in P - S$. Let $P_1 \subseteq P - S$ be such that the set $\{p \cup b \,|\, p \in P_1\}$ is maximal independent. Then if $p_1 \in P_1$ we have $u_a \succ \bigcup\{p\cup b \,|\, p\in P_1 - p_1\}$ and $\bigcap_p \bigcup\{p\cup b \,|\, p\in P_1 - p\} = b$ by Lemma 3.7.

In order to prove a sufficiently general existence theorem on irredundant decompositions it is necessary to have a criterion for subsets of such decompositions. Let then $a = \bigcap Q$ be an irredundant decomposition of a into irreducibles and let $R \subseteq Q$. If $q \in R$, we have $\bigcap(R-q) \geq \bigcap(Q-q) > a$. Hence there exists $p \in P_a$ such that $\bigcap(R-q) \geq \bigcap(Q-q) \geq p$. Clearly $q \nleq p$ since $q \cap \bigcap(Q-q) = \bigcap Q = a$. We thus make the following definition:

DEFINITION 3.5. *A subset R of L is irredundant over a if for each $r \in R$, there exists $p \succ a$ such that $\bigcap(R-r) \geq p$ and $r \nleq p$.*

We can now state and prove the fundamental existence theorem on irredundant decompositions.

THEOREM 3.1. *Let a be an element of a semimodular, compactly generated, atomic lattice and let R be a set of completely meet irreducible elements containing a. Then R can be extended to an irredundant decomposition of a if and only if R is irredundant over a.*

Proof. We have proved the necessity in the paragraph preceding Definition 3.5. To prove the sufficiency let R be irredundant over a. Then for each $q \in R$, there exists an element $p_q \in P_a$ such that $q \nleq p_q$ and $\bigcap(R-q) \geq p_q$. Let $K_1 = \{p_q \,|\, q \in R\}$ and let K_2 be a maximal independent subset of $\{p \in P_a \,|\, p \leq \bigcap R\}$. Let $K = K_1 \vee K_2$. Now suppose that $\bigcup(K-p_q) \geq p_q$ for some $p_q \in K_1$. Then since $q \geq \bigcap R \geq \bigcup K_2$ and $q \geq \bigcap(R-q') \geq p_{q'}$ all $q' \in R-q$ we have $q \geq \bigcup(K_1-p_q) \cup \bigcup K_2 \geq \bigcup(K-p_q) \geq p_q$ contrary to $q \nleq p_q$. Thus $\bigcup(K-p_q) \nleq p_q$ all $p_q \in K_1$. Next suppose that $\bigcup(K-p) \geq p$ for some $p \in K_2$. Since p is compact in u_a/a there exist finite sets $K_1' \subseteq K_1$ and $K_2' \subseteq K_2 - p$ such that $\bigcup(K_1' \vee K_2') \geq p$. Since K_2 is independent, K_2' is nonempty. Replacing K_2' by a smaller set if necessary we may assume that for some $p_q \in K_1'$, $\bigcup(K_2' \vee K_1' - p_q) \nleq p$. By semimodularity we have $p_q \leq \bigcup(K_2' \vee K_1' - p_q) \cup p \leq \bigcup(K-p_q)$ contrary to $p_q \nleq \bigcup(K-p_q)$. Thus $\bigcup(K-p) \nleq p$ all $p \in K_2$ and K is thus an independent subset of P_a.

Now for each $p \in K_2$, let q_p be an irreducible such that $q_p \geq \bigcup(K-p)$,

$q_p \not\geq p$. Such irreducibles exist by Lemma 2.5. Let $Q' = \{q_p \mid p \in K_2\}$. Since $q_{p'} \geq \cup\{K - p'\} \geq p$ all $p' \neq p$ in K_2 we have $\cap(Q' - q_p) \geq p$. Now let $Q = Q' \vee R$. If $q \in R$, then $\cap(Q - q) = \cap(R - q) \cap \cap Q' \geq \cap(R - q) \cap \cup K_1 \geq p_q$. If $p \in K_2$, then $\cap(Q - q_p) = \cap R \cap \cap(Q' - q_p) \geq \cup K_2 \cap \cap(Q' - q_p) \geq p$. Thus if we show that $\cap Q = a$ then this decomposition is a fortiori irredundant. Suppose that $\cap Q > a$. Then $\cap Q \geq p'$ for some $p' \in P_a$. Since $\cap R \geq \cap Q$ we have $p' \leq \cap R$ and hence $p' \leq \cap K_2$ by Lemma 3.6. Furthermore $p' \leq \cap Q' \leq q_p$ for all $p \in K_2$. If $p' \not\leq \cup(K_2 - p)$ for some $p \in K_2$, then $q_p \geq \cup(K_2 - p) \cup p' = \cup K_2 \geq p$ contrary to $q_p \not\geq p$. Thus $p' \leq \cup(K_2 - p)$ all $p \in K_2$. But then $p' \leq \cap_p \cup(K_2 - p) = \cup(\wedge_p (K_2 - p)) = \cup(\varnothing) = a$ by Lemma 3.7 contrary to $p' > a$. Thus $\cap Q = a$ and the proof of the theorem is complete.

Since the null set is trivially irredundant we have

CoROLLARY 1. *Every element of a semimodular, compactly generated, atomic lattice has an irredundant decomposition into completely meet irreducibles.*

Let q be an irreducible containing a such that $q \not\leq u_a$. If R consists of the single element q, then R is irredundant since $\cap(R - q) = u \geq q$. Hence we have

CoROLLARY 2. *If a is an element of a semimodular, compactly generated, atomic lattice and q is a completely meet irreducible element such that $q \not\leq u_a$, then there exists an irredundant decomposition $a = \cap Q$ such that $q \in Q$.*

We conclude this section with the following theorem on the cardinality of maximal independent sets of P_a.

THEOREM 3.2. *Let L be a semimodular, compactly generated, atomic lattice. Then any two maximal independent subsets of P_a have the same cardinality.*

Proof. Let M and N be two maximal independent subsets of P_a. We may suppose that $|M| \leq |N|$, where $|X|$ denotes the cardinality of a set X. If N is finite, then u_a/a is finite dimensional and the number of elements in any maximal independent set is simply the dimension of u_a/a. Hence $|M| = |N|$. If N is infinite, we have $\cup M = \cup N = u_a$ and hence for each $p \in M$, $p \leq \cup N$. But then there exists a finite subset $N_p \subseteq N$ such that $p \leq \cup N_p$. Let $S = \{N_p \mid p \in M\}$ and let $N' = \vee_{p \in M} N_p$. Then N' as a subset of N is independent. Moreover, $p \leq \cup N'$ for each $p \in M$ implies that $\cup N' = \cup P_a = u_a$. Hence N' is a maximal independent subset of P_a and thus $N' = N$. Now for each p, N_p is a finite subset of N, and since S is infinite, it follows that $|S| = |N'| = |N|$. But the mapping $p \to N_p$ is single valued and hence $|M| \geq |S| = |N|$. Thus $|M| = |N|$.

4. Modular lattices. Throughout this section L will denote a modular, compactly generated, atomic lattice.

For modular lattices satisfying the ascending chain condition, the principal results on irreducible decompositions concern the replacement of elements in two irredundant decompositions (Dilworth [5]). Since the decompositions are finite it follows from these replacement theorems, that the number of elements in an irredundant decomposition is unique. In this section we shall extend some of these replacement theorems to compactly generated modular lattices. We shall see that there are many different classes of irredundant decompositions and that replacement properties are dependent upon the classes of decompositions involved. Moreover, the unicity of the number of elements in irredundant decompositions does not generalize completely. Thus, an example of a compactly generated modular lattice is constructed in which an element has two irredundant decompositions with different cardinalities.

We begin by showing that in modular compactly generated lattices the existence of irredundant decompositions is equivalent to atomicity.

THEOREM 4.1. *Every element of a modular compactly generated L lattice has an irredundant decomposition into completely irreducible elements if and only if L is atomic.*

Proof. We need only show that if every element of L has an irredundant decomposition, then L is atomic.

Suppose $a = \cap Q$ is an irredundant decomposition of a into completely irreducible elements. Let $q \in Q$ and $q^* = \cap \{ x \in L \mid x > q \} > q$. Then $q \cup \cap (Q - q) > q$ since $\cap (Q - q) > a$, and hence $q \cup \cap (Q - q) \geqq q^*$. By modularity we have $q \cup [q^* \cap \cap (Q - q)] = q^* \cap [q \cup \cap (Q - q)] = q^* > q$, and hence $q^* \cap \cap (Q - q) > q \cap q^* \cap \cap (Q - q) = a$. Hence if every element of L has an irredundant decomposition, then every element of L is covered by some element.

Now suppose $b > a$. From the above, $p \in L$ exists such that $p > a$. Suppose $b \not\geqq p$. Then $b \cap p = a$. Let $\{ x_\alpha \}$ be a chain of elements of L such that $b \cap x_\alpha = a$ for every index α. Then by Lemma 2.4, $b \cap \cup_\alpha x_\alpha = \cup_\alpha b \cap x_\alpha = a$. Thus by the Maximal Principle it follows that a maximal element m exists such that $m \cap b = a$. Let $s \in L$ be such that $s > m$. Then $b \cap s > a$. Hence $(b \cap s) \cup m = s > m$, whence $b \cap s > b \cap s \cap m = a$. Thus under any circumstances $r \in L$ exists such that $b \geqq r > a$, and hence L is atomic.

Our next theorem is a direct generalization of the classical replacement theorem.

THEOREM 4.2. *If a is an element of a compactly generated, atomic, modular lattice and $a = \cap Q = \cap Q'$ are two decompositions of a, then for each $q \in Q$ there exists $q' \in Q'$ such that $a = \cap (Q - q) \cap q'$. If the decomposition $a = \cap Q$ is irredundant, then the decomposition $a = \cap (Q - q) \cap q'$ is also irredundant.*

Proof. Let $q \in Q$. For each $q' \in Q'$, define $r_{q'}$ by

$$r_{q'} = \cap (Q - q) \cap q'.$$

Then $a = \cap_{q' \in Q'} r_{q'}$, and $a \leqq r_{q'} \leqq \cap (Q - q)$ for each $q' \in Q'$. Now since L is modular, the quotient sublattices $q \cup (\cap (Q - q))/q$ and $\cap (Q - q)/\cap (Q - q) \cap q$

$=\bigcap(Q-q)/a$ are isomorphic. q is completely irreducible in L and hence q is completely irreducible in the quotient $q\cup(\bigcap(Q-q))/q$. Thus a is completely irreducible in the quotient $\bigcap(Q-q)/a$. But $a=\bigcap_{q'\in Q'} r_{q'}$ is a representation of a as a meet of elements of $\bigcap(Q-q)/a$, and hence for some $q'\in Q'$, $a=r_{q'}=\bigcap(Q-q)\cap q'$.

Suppose the decomposition $a=\bigcap Q$ is irredundant. Then $\bigcap(Q-q)>a$, so that if $a=\bigcap(Q-q)\cap q'$ is a redundant decomposition, there exists an element $q_1\in Q-q$ such that $a=\bigcap(Q-\{q, q_1\})\cap q'$. Now in this decomposition, q' can be replaced by some $q_2\in Q$ giving a decomposition of a. But then either $a=\bigcap(Q-q)$ or $a=\bigcap(Q-q_1)$, contrary to the irredundance of the decomposition $a=\bigcap Q$. Hence the decomposition $a=\bigcap(Q-q)\cap q'$ is also irredundant.

We note that the theorem holds in any complete modular lattice.

The corollary to Lemma 3.7 can be sharpened in the case of modular lattices to give

LEMMA 4.1. *The quotient u_a/a is complemented point lattice.*

For since u_a/a is complemented and modular, it is also relatively complemented. Hence if $x\in u_a/a$ and y is the union of the points contained in x, then the relative complement of y in x/a must be a and thus $x=y$.

The following lemma relating irreducibles containing a to the elements covering a will be needed.

LEMMA 4.2. *Let $a\in L$ and let p, p' be two distinct elements covering a. If q is an irreducible such that $q\geq a$ and $q\not\geq p$, p', then $q\cap(p\cup p')>a$.*

For since $q\not\geq p$, p' we have $p\cap q=p'\cap q=a$ and hence $p\cup q>q$ and $p'\cup q>q$. Since q is irreducible it follows that $p\cup q=p'\cup q=p\cup p'\cup q$. Thus $(p\cup p')\cup q>q$ and hence $p\cup p'>q\cap(p\cup p')$. Since $p\cup p'>p>a$ it follows that $q\cap(p\cup p')>a$.

DEFINITION 4.1. If $a=\bigcap Q$ is a decomposition of a, let

$$H_Q = \{p\in P_a \mid \bigcap(Q-q)\geq p \text{ some } q\in Q\} \quad \text{and} \quad h_Q = \bigcup H_Q.$$

LEMMA 4.3. *Let $a=\bigcap Q$ be a decomposition of a into irreducibles. Then*
(1) H_Q is an independent subset of P_a.
(2) H_Q contains at most one element p such that $\bigcap(Q-q)\geq p$.
(3) If $a=\bigcap Q$ is irredundant, then H_Q contains exactly one element p such that $\bigcap(Q-q)\geq p$.

In order to see that (2) holds, let $\bigcap(Q-q)\geq p,p'$ where p and p' are distinct. Since $a=q\cap\bigcap(Q-q)$ it follows that $q\not\geq p,p'$ and hence $q\cap(p\cup p')>a$ by Lemma 4.2 which contradicts $a=q\cap\bigcap(Q-q)\geq q\cap(p\cup p')$. Now if $\bigcap(Q-q)\geq p$ and $p'\in H_Q-p$, then $\bigcap(Q-q)\not\geq p'$ and hence $\bigcap(Q-q')\geq p'$ where $q'\neq q$. But then $q\geq\bigcap(Q-q')\geq p'$. Thus $q\geq\bigcup(H_Q-p)$. Since $q\not\geq p$ it follows that $\bigcup(H_Q-p)\not\geq p$. Thus H_Q is independent and (1) holds. (3) follows immediately from (2).

According to Theorem 3.1 each element a has at least one irredundant decomposition into irreducibles. For modular lattices a much stronger existence theorem holds.

THEOREM 4.3. *Let a be an element of a compactly generated, atomic, modular lattice and let $a = \bigcap Q$ be an irredundant decomposition of a into irreducibles. Then if J is an independent subset of P_a such that $J \supseteq H_Q$, there exists an irredundant decomposition $a = \bigcap Q'$ such that $H_{Q'} = J$.*

Proof. By a trivial application of the Maximal Principle J can be extended to a maximal independent set $M \subseteq P_a$. Let $J_1 = J - H_Q$ and $M_1 = M - J_1$. If we set $b = \bigcup J_1$ and $c = \bigcup M_1$, then by Lemma 3.6, $b \cup c = \bigcup M = u_a$ and by Lemma 3.7, $b \cap c = a$. For each $p \in H_Q$, let q_p be the unique element of Q such that $\bigcap (Q - q_p) \geq p$. Then $q_p \not\geq p$ and hence $q_p \cup p \succ q_p$. Since q_p is irreducible, $q_p \cup p_1 = q_p \cup p$ for any other $p_1 \in M_1$ such that $q_p \not\geq p_1$. Thus $q_p \cup c = q_p \cup \bigcup M_1 = q_p \cup p \succ q_p$. By modularity we have $c \succ c \cap q_p$. Now $b \cup (c \cap q_p) \neq u_a$, since otherwise $c = c \cap u_a = c \cap (b \cup (c \cap q_p)) = (c \cap b) \cup (c \cap q_p) = c \cap q_p$ contrary to $c \succ c \cap q_p$. Thus $u_a = b \cup c \succ b \cup (c \cap q_p)$. Let us set $s_p = b \cup (c \cap q_p)$ so that s_p is a maximal element of u_a / a for each $p \in H_Q$.

Now for each $p \in J_1$ let us set $s_p = \bigcup (J_1 - p) \cup c$. Clearly $u_a \succ s_p$ for each $p \in J_1$. Furthermore, since J_1 is an independent subset of P_a it follows from Lemma 3.7 that $\bigcap_p \bigcup (J_1 - p) = a$. Since L is modular, the mapping $x \to x \cup c$ maps the quotient $b/a = b/b \cap c$ isomorphically onto $b \cup c / c = u_a / c$. Hence $\bigcap_{p \in J_1} s_p = c$. Thus

$$\bigcap_J s_p = \bigcap_{J_1} s_p \cap \bigcap_{H_Q} S_p = c \cap \bigcap_{H_Q} (b \cup (c \cap q_p)) = \bigcap_{H_Q} (c \cap (b \cup (c \cap q_p)))$$

$$= \bigcap_{H_Q} ((c \cap b) \cup (c \cap q_p)) = \bigcap_{H_Q} (c \cap q_p) = c \cap \bigcap Q = a.$$

It follows immediately from the definition that if $p \in J_1$, then $s_p \geq c \geq \bigcup H_Q$ and if $p \in H_Q$, then $s_p \geq b \geq \bigcup J_1$. On the other hand, if $p \in J_1$, then $s_p \geq \bigcup (J_1 - p)$ and if $p \in H_Q$, then $s_p \geq c \cap q_p \geq \bigcup (H_Q - p)$. Thus for each $p \in J$ we have $s_{p'} \geq p$ all $p' \neq p$ and hence $\bigcap \{ s_{p'} \mid p' \neq p, \ p \in J \} \geq p$. By Lemma 2.5 for each $p \in J$ there exists a completely meet irreducible element q_p' such that $q_p' \geq s_p$ and $q_p' \not\geq u_a$. Let $Q' = \{ q_p' \mid p \in J \}$. Then $u_a \cap q_p' = s_p$ and hence by Lemma 3.2, $\bigcap Q' = a$ is an irredundant decomposition of a. Clearly $H_{Q'} = J$.

For finite irreducible decompositions it is easily verified that $\bigcup_q \bigcap (Q - q) \geq u_a$. This property no longer holds for general decompositions. In fact, we shall show that $\bigcup_q \bigcap (Q - q) \cap u_a = h_Q$. A preliminary lemma is needed.

LEMMA 4.4. *Let $a \in L$ and let $a = \bigcap Q$ be an irredundant decomposition of a into irreducibles. Furthermore let $Q' = \{ q_1, \cdots, q_n \}$ be a finite subset of Q such that $\bigcap (Q - q_i) \geq p_i \succ a$, for each i. Then $\bigcap (Q - Q') \geq p$ implies $p_1 \cup \cdots \cup p_n \geq p$ for each $p \in P_a$.*

For $n = 1$, the lemma follows immediately from Lemma 4.3. Now suppose that the lemma holds for $n=k-1$ and let $\cap(Q-\{q_1, \cdots, q_k\}) \geq p > a$. If $q_k \geq p$, then $\cap(Q-\{q_1, \cdots, q_{k-1}\}) \geq p$ and hence by the induction hypothesis $p_1 \cup \cdots \cup p_k \geq p_1 \cup \cdots \cup p_{k-1} \geq p$. If $q_k \not\geq p$, then since $q_k \not\geq p_k$ we have $q_k \cap (p \cup p_k) > a$ by Lemma 4.2. Now $\cap(Q-\{q_1, \cdots, q_{k-1}\}) \geq q_k \geq q_k \cap (p \cup p_k)$ and by the induction hypothesis we have $p_1 \cup \cdots \cup p_{k-1} \geq q_k \cap (p \cup p_k)$. Then

$$p_1 \cup \cdots \cup p_k \geq p_k \cup (q_k \cap (p \cup p_k)) = (p_k \cup q_k) \cap (p \cup p_k)$$

$$= (p \cup q_k) \cap (p \cup p_k) \geq p.$$

The lemma follows by induction.

Theorem 4.4. *If $a \in L$ and $a = \cap Q$ is an irredundant decomposition of a, then $\cup_a \cap (Q-q) \cap u_a = h_Q$.*

Proof. Let $\cup_q \cap (Q-q) \geq p > a$. Then since p is compact in u_a/a, there exists a finite subset $\{q_1, \cdots, q_n\}$ of Q such that $\cup_i \cap (Q - q_i) \geq p$. Thus $\cap(Q-\{q_1, \cdots, q_n\}) \geq \cup_i \cap (Q-q_i) \geq p$. Let $\cap(Q-q_i) \geq p_i > a$. Then $p_i \in H_Q$ and by Lemma 4.4, $p_1 \cup \cdots \cup p_n \geq p$. Hence $h_Q = \cup H_Q \geq p$ all $p \leq \cup_q \cap (Q-q)$. Since u_a/a is a point lattice we have $h_Q \geq \cup_q \cap (Q-q) \cap u_a$. But $\cup_q \cap (Q-q) \cap u_a \geq h_Q$ trivially.

In the theorems which follow it will be shown that the replacement properties of irredundant decompositions are determined by the order properties of the elements h_Q.

Theorem 4.5. *Let $a = \cap Q$ be an irredundant decomposition of $a \in L$ and let $a = \cap Q'$ where Q' is obtained from Q by replacing $q \in Q$ by an irreducible q'. Then $h_{Q'} = h_Q$.*

Proof. According to Theorem 4.2 the decomposition $a = \cap Q'$ is irredundant. Let p be the unique element of H_Q such that $\cap(Q-q) \geq p$. Now for each $p^* \in H_Q - p$ let q^* be the unique element of Q such that $\cap(Q-q^*) \geq p^*$ and let $\bar{p}^*$ be the unique element of $H_{Q'}$ such that $\cap(Q'-q^*) \geq \bar{p}^*$. Clearly the correspondence $p^* \to \bar{p}^*$ is a one-to-one mapping of $H_Q - p$ onto $H_{Q'} - p$.

Now let $p^* \in H_Q - p$ be such that $p^* \neq \bar{p}^*$ and let q^* be the unique element of Q for which $\cap(Q-q^*) \geq p^*$. Then $q' \not\geq p^*$ since otherwise $\cap(Q'-q^*) = \cap(Q-\{q, q^*\}) \cap q' \geq p^*$ and hence $p^* = \bar{p}^*$ contrary to assumption. Also $q' \not\geq p$ and hence by Lemma 4.2, $q' \cap (p \cup p^*) > a$. But then $\cap(Q'-q^*) = \cap(Q-\{q, q^*\}) \cap q' \geq q' \cap (p \cup p^*)$ and thus $\bar{p}^* = q' \cap (p \cup p^*)$. It follows that $p \cup \bar{p}^* = p \cup (q' \cap (p \cup p^*)) = (p \cup q') \cap (p \cup p^*) = p \cup p^*$. But $p \cup \bar{p}^* = p \cup p^*$ holds trivially if $p^* = \bar{p}^*$. Thus for all $p^* \in H_Q - p$ we have $p \cup \bar{p}^* = p \cup p^*$. Thus

$$h_{Q'} = \cup H_{Q'} = \bigcup_{p^*} (p \cup \bar{p}^*) = \bigcup_{p^*} (p \cup p^*) = \cup H_Q = h_Q.$$

This completes the proof of the theorem.

From Lemma 3.6 we get the following corollary to Theorem 4.5.

Corollary. *Under the hypotheses of Theorem 4.5, if H_Q is a maximal independent set of P_a, then $H_{Q'}$ is also a maximal independent set of P_a.*

If the ascending chain condition is satisfied, then $h_Q = u_a$ for every irredundant decomposition $a = \cap Q$. Thus it is not surprising that an additional hypothesis is required for a simultaneous replacement theorem in the more general case.

Theorem 4.6. *If a is an element of a compactly generated, atomic, modular, lattice and $a = \cap Q = \cap Q'$ are two irredundant decompositions of a with $h_Q \geq h_{Q'}$, then for each $q' \in Q'$ there exists $q \in Q$ such that $a = \cap(Q-q) \cap q' = \cap(Q'-q') \cap q$.*

Proof. Suppose that the irredundant decompositions $a = \cap Q = \cap Q'$ are such that $h_Q \geq h_{Q'}$, and that $q' \in Q'$. Let $p' \in H_{Q'}$ be such that $\cap(Q'-q') \geq p'$. Since $\cup H_Q \geq \cup H_{Q'} \geq p'$, there exists a finite subset $\{p_1, \cdots, p_n\} \subseteq K_Q$ such that $p_1 \cup \cdots \cup p_n \geq p'$. We may assume that $\{p_1, \cdots, p_n\}$ is a minimal such subset. Then if $q_i \in Q$ is such that $\cap(Q - q_i) \geq p_i$ $(i = 1, \cdots, n)$, $\cap(Q - \{q_1, \cdots, q_n\}) \geq p_1 \cup \cdots \cup p_n \geq p'$. Moreover, if for some i, $q_i \geq p'$, then $\cap(Q - \{q_1, \cdots, q_{i-1}, q_{i+1}, \cdots, q_n\}) \geq p'$, and hence it follows from Lemma 4.4 that $p_1 \cup \cdots \cup p_{i-1} \cup p_{i+1} \cup \cdots \cup p_n \geq p'$, contrary to the minimality of $\{p_1, \cdots, p_n\}$. Thus for each $i = 1, \cdots, n$, $q_i \not\geq p'$. Now $q' \not\geq p'$, and hence there must exist some $p_j \in \{p_1, \cdots, p_n\}$ such that $q' \not\geq p_j$. Suppose $\cap(Q - q_j) \cap q' > a$. Then since L is atomic, $p \in P_a$ exists such that $\cap(Q - q_j) \cap q' \geq p$. But then $\cap(Q - q_j) \geq p$, p_j, contradicting Lemma 4.3. Hence $\cap(Q - q_j) \cap q' = a$. It is also true that $q_j \not\geq p'$, and hence by a similar argument, $\cap(Q' - q') \cap q_j = a$. The irredundancy of the decompositions $a = \cap(Q - q_j) \cap q' = \cap(Q' - q') \cap q_j$ follows from Theorem 4.1.

Corollary. *Let $a \in L$ and $a = \cap Q$ be an irredundant decomposition of a. Then for each irredundant decomposition $a = \cap Q'$ and each $q' \in Q'$, there exists $q \in Q$ such that $a = \cap(Q-q) \cap q'$, if and only if $h_Q = u_a$.*

Proof. If $h_Q = u_a$, then for any irredundant decomposition $a = \cap Q'$, $h_Q = u_a \geq h_{Q'}$, and hence Theorem 4.6 holds for the decompositions $a = \cap Q = \cap Q'$. Suppose $h_Q < u_a$. Then H_Q is not a maximal independent subset of P_a, and hence there exists an independent subset $J \subseteq P_a$ such that $J \supset H_Q$. By Theorem 4.2, there exists an irredundant decomposition $a = \cap Q'$ such that $H_{Q'} = J$. Let $p \in H_{Q'} - H_Q$, and let $q' \in Q'$ be such that $\cap(Q'-q') \geq p$. Then $q' \geq \cup H_Q$, and hence $\cap(Q-q) \cap q' > a$ for all $q \in Q$. Thus q' can replace none of the irreducibles in the decomposition $a = \cap Q$.

The replacement property of Theorem 4.2 can be considerably sharpened. In order to simplify the statement of the theorem we shall introduce the notion of Q-equivalence.

Let $a = \cap Q$ be a decomposition of a, and let $S \subseteq Q$. A set T of completely irreducible elements of L is said to be *Q-equivalent to S* if there is a one-one mapping ϕ of S onto T such that for each $q \in Q$, $a = \cap(Q-q) \cap \phi(q)$.

THEOREM 4.7. *If a is an element of a compactly generated, atomic, modular lattice and $a = \cap Q = \cap Q'$ are two irredundant decompositions of a, then for each finite subset $S \subseteq Q$ there exists a subset $S' \subseteq Q'$ such that S' is Q-equivalent to S. If $h_Q \geq h_{Q'}$, then for each finite subset $S' \subseteq Q'$ there is a subset $S \subseteq Q$ such that S' is Q-equivalent to S.*

Proof. For $n = 1$, the first statement of this theorem is just that of Theorem 4.1. Let $\{q_1, \cdots, q_n\}$ be a finite subset of Q. For each $i = 1, \cdots, n$, let $S_i = \{q' \in Q \mid a = \cap(Q - q_i) \cap q'\}$. Suppose for some k-element subset $\{q_{i_1}, \cdots, q_{i_k}\} \subseteq \{q_1, \cdots, q_n\}$, $k \leq n$, $S_{i_1} \vee \cdots \vee S_{i_k}$ contains $m < k$ elements, say $q_1', \cdots, q_m'$. Let $p_{i_j} \in H_Q$ be such that $\cap(Q - q_{i_j}) \geq p_{i_j}$ $(j = 1, \cdots, k)$, and let $p_j' \in H_{Q'}$ be such that $\cap(Q' - q_j') \geq p_j'$ $(j = 1, \cdots, m)$. Then for each $q' \in Q' - \{q_1', \cdots, q_m'\}$, $\cap(Q - q_i) \cap q' > a$, and hence, since $\cap(Q - q_i)$ contains only one element of P_a, $q' \geq p_i$ for each $i = 1, \cdots, k$. Thus

$$\cap(Q' - \{q_1', \cdots, q_m'\}) \geq p_1 \cup \cdots \cup p_k,$$

and it follows from Lemma 4.4 that $p_1' \cup \cdots \cup p_m' \geq p_1 \cup \cdots \cup p_k$. But this is impossible, since the dimension m of the quotient $p_1' \cup \cdots \cup p_m'/a$ is less than the dimension k of the quotient $p_1 \cup \cdots \cup p_k/a$. Hence for each k-element subset $\{q_{i_1}, \cdots, q_{i_k}\} \subseteq \{q_1, \cdots, q_n\}$, $S_{i_1} \vee \cdots \vee S_{i_k}$ contains at least k distinct elements. It now follows from P. Hall's theorem on representatives of subsets that there are n distinct elements $q_1', \cdots, q_n' \in Q'$ such that $q_i' \in S_i$ $(i = 1, \cdots, n)$, completing the proof of the first part.

Now let $h_Q \geq h_{Q'}$. Let $\{q_1', \cdots, q_n'\}$ be a finite n-element subset of Q', and let $T_i = \{q \in Q \mid a = \cap(Q - q) \cap q_i'\}$. In view of the preceding paragraph, to prove the second part of the theorem it suffices to show that $T_1 \vee \cdots \vee T_n$ contains at least n distinct elements. Suppose $T_1 \vee \cdots \vee T_n$ contains $m < n$ elements, say $q_1, \cdots, q_m$. For each $i = 1, \cdots, m$, let $p_i \in H_Q$ be such that $\cap(Q - q_i) \geq p_i$. Then if $p \in H_Q - \{p_1, \cdots, p_m\}$, $q_i' \geq p$ for each $i = 1, \cdots, n$, and hence $q_1' \cap \cdots \cap q_n' \geq \cup(H_Q - \{p_1, \cdots, p_m\})$. Since $h_Q \geq \cup H_{Q'}$, and since for each i, q_i' is covered by only one element, $h_Q \cup q_i' > q_i'$ and hence $h_Q > h_Q \cap q_i'$. Furthermore if $k < n$, then $(h_Q \cap q_1' \cap \cdots \cap q_k') \cup q_{k+1}' > q_{k+1}'$, and hence $h_Q \cap q_1' \cap \cdots \cap q_k' > h_Q \cap q_1' \cap \cdots \cap q_{k+1}'$. Thus the quotient sublattice $h_Q/h_Q \cap q_1' \cap \cdots \cap q_n'$ is of dimension n. But now we have a contradiction, since the quotient $h_Q/\cup(H_Q - \{p_1, \cdots, p_m\})$ is of dimension $m < n$. Hence $T_1 \vee \cdots \vee T_n$ contains at least n distinct elements, and the proof is complete.

COROLLARY. *If a is an element of a compactly generated, atomic, modular lattice and $a = \cap Q = \cap Q'$ are two irredundant decompositions of a with $h_Q \leq h_{Q'}$, then there exists a subset $S' \subseteq Q'$ such that S' is Q-equivalent to Q.*

Proof. For each $q \in Q$, let $S_q = \{q' \in Q' \mid a = \cap(Q - q) \cap q'\}$. Then from Theorem 4.7 it follows that for any finite subset $\{q_1, \cdots, q_n\} \subseteq Q$, $S_{q_1} \vee \cdots \vee S_{q_n}$ contains at least n distinct elements. Now suppose $q \in Q$ and $p \in H_Q$

THE DILWORTH THEOREMS

is such that $\cap(Q-q)\geqq q$. Since $h_Q\leqq h_{Q'}$, $p\leqq\cup H_Q$, and hence $p\leqq p_1'\cup\cdots\cup p_k'$ for some finite subset $\{p_1',\cdots,p_k'\}\subseteq Q'$. If q_i' is such that $\cap(Q'-q_i')\geqq p_i'$ $(i=1,\cdots,k)$, then $\cap(Q'-\{q_1',\cdots,q_k'\})\geqq p$, and thus $\{q_1',\cdots,q_k'\}\supseteq S_q$. Hence for each $q\in Q$, S_q is finite. The corollary now follows from the Marshall Hall theorem (Hall [6]) on representatives of subsets.

COROLLARY. *If $a=\cap Q=\cap Q'$ are two irredundant decompositions of $a\in L$ such that $h_Q=h_{Q'}$, then Q and Q' have the same cardinality.*

We conclude this section with an example of a compactly generated, atomic, modular lattice in which the null element has two irredundant decompositions into irreducibles of different cardinalities.

For each integer i, let A_i be a group isomorphic with the additive group of integers modulo a fixed prime p, and let G be the *complete direct sum* of the groups A_i, that is, the set of all functions f on the integers such that $f(i)\in A_i$, with addition defined componentwise. Then G is an (additive) abelian group every element of which has order p. Let L be the lattice of subgroups of G. L is then compactly generated and modular, and since every element of G has finite order, L is also atomic. For each i, let Q_i be that subgroup of G consisting of all functions $f\in G$ for which $f(i)$ is the zero element of A_i. Then G/Q_i is isomorphic with A_i and hence Q_i is a maximal subgroup of G. Thus Q_i is a completely irreducible element of L for each i. Since G is the complete direct sum of the A_i, it follows that $0=\cap_i Q_i$ (0 denoting the zero subgroup of G). Moreover, for each i, $Q_i\cup(\cap_{j\neq i} Q_j)=G$, and hence the decomposition $0=\cap_i Q_i$ is irredundant. Now G can be considered as a vector space over the field of integers modulo p, and accordingly, G has a basis $\{f_\alpha\}$. Since G has cardinality $2^{\aleph_0}$, the number of f_α must also be $2^{\aleph_0}$. For each index α, let Q_α' be that subgroup of G generated by the set $\{f_\beta|\beta\neq\alpha\}$. Then each Q_α' is a maximal subgroup of G, and just as above, it follows that $0=\cap_\alpha Q_\alpha'$ is an irredundant decomposition of 0. Thus $0\in L$ has two irredundant decompositions with different cardinalities.

5. **Distributive lattices.** Under the hypothesis of the ascending chain condition, each element of a distributive lattice has a unique irredundant decomposition into irreducibles (Birkhoff [1, p. 142]). This decomposition is necessarily finite. In this section, we shall show that there exists a unique (though possibly infinite) irredundant decomposition for each element of a compactly generated atomic distributive lattice.

We begin with a theorem concerning irreducibles in semimodular lattices.

THEOREM 5.1. *Let a be an element of a semimodular, compactly generated, atomic lattice. Then an irreducible q appears in every irredundant decomposition of a if and only if there exists $p\succ a$ with $p\nleqq q$ such that $p\cap(x\cup y)=(p\cap x)\cup(p\cap y)$ for all $x,y\geqq a$.*

Proof. Suppose such a p exists. Let q' be an irreducible with $q'\nleqq p$. Then $p\cap q=p\cap q'=a$, and hence $a=(p\cap q)\cup(p\cap q')=p\cap(q\cup q')$. Now if $q\neq q'$, then $q\cup q'\geq u_a\geq p$ contrary to $p\nleq q\cup q'$. Thus $q=q'$, and q is the only irreducible not containing p. Hence q appears in every decomposition of a.

Suppose q appears in every irredundant decomposition of a. Let p be an element covering a such that $q\nleq p$. Let x, $y\geq a$, and suppose $x\cap p=y\cap p=a$. By Lemma 2.5 an irreducible element q_x exists such that $q_x\nleq p$ and $q_x\geq x$. Similarly an irreducible element q_y exists. Thus by Theorem 3.1 there are irredundant decompositions of a which contain q_x and q_y, respectively. But q appears in every irredundant decomposition of a and $q\nleq p$, whence it follows that $q=q_x=q_y$. Hence $q\geq x\cup y$, so that $x\cup y\nleq p$. Thus $p\cap(x\cup y)=(p\cap x)\cup(p\cap y)$, and the theorem follows.

Combining Theorems 3.1 and 5.1 we have the following theorem.

THEOREM 5.2. *Every element of a compactly generated, atomic, distributive lattice has a unique irredundant decomposition into irreducibles.*

For distributive lattices satisfying the ascending chain condition it is easy to show that $q'\geq a$ implies $q'\geq q$ where q belongs to the unique irredundant decomposition of a. This property does not generalize to compactly generated, atomic, distributive lattices. Consider, for example, the collection of all subsets S of the set I of positive integers such that either $S\subseteq I-\{1\}$ or $I-S$ is finite. It can be easily verified that this collection is closed under arbitrary union and finite intersection and thus is a complete distributive lattice. The compact elements of this lattice are the finite subsets of $I-\{1\}$ and the sets S such that $1\in S$. These sets clearly generate the lattice under arbitrary union and hence the lattice is compactly generated. The lattice is obviously atomic. Now the unique irredundant decomposition of the null set is $\varnothing=\cap_{n=2}^{\infty}Q_n$ where $Q_n=I-\{n\}$. Note that this lattice meet is *not* set intersection since $\wedge_{n=2}^{\infty}Q_n=\{1\}$. On the other hand, $Q_1=I-\{1\}$ is an irreducible containing $\varnothing$ such that $Q_1\not\supseteq Q_n$ for every $n\neq 1$.

6. **Unique decompositions.** In the previous section we have shown that each element of a distributive, compactly generated, atomic lattice has a unique irredundant decomposition into irreducibles. This section will be devoted to a characterization of lattices having unique irredundant decompositions into irreducibles. The characterization will be analogous to the finite dimensional case (Dilworth [2]), though quite different techniques of proof are required. We begin with the definition of the relevant concepts.

DEFINITION 6.1. An atomic lattice L is *locally distributive* (*locally modular*) if u_a/a is distributive (modular) for each $a\in L$.

Clearly a locally distributive or locally modular lattice is weakly semimodular. Hence from Lemma 3.3 we have

LEMMA 6.1. *A locally distributive or locally modular, compactly generated, atomic lattice is semimodular.*

We first show that for compactly generated atomic lattices, unique irredundant decompositions imply that the lattice is semimodular.

THE DILWORTH THEOREMS

LEMMA 6.2. *Let L be a compactly generated atomic lattice. Then if every element of L has a unique irredundant decomposition into irreducibles, L is semimodular.*

For let $p > a$, $b \geq a$, and $b \nleq p$. Suppose $p \cup b$ does not cover b. Then x exists such that $p \cup b > x > b$. By Lemma 2.5 there exists an irreducible q such that $q \geq b$ and $q \ngeq x$. Similarly there exists an irreducible q' such that $q' \geq x$ and $q' \ngeq p \cup b$. Now $a = p \cap q = p \cap q'$. By the Maximal Principal elements m and m' exist which are maximal such that $m \cap q = a$, $m \geq p$, and $m' \cap q' = a$, $m' \geq p$. By assumption m and m' have irredundant decompositions $m = \cap M$ and $m' = \cap M'$. Clearly $q \notin M'$ and $q' \notin M$. However, $a = q \cap \cap M = q' \cap \cap M'$ are two different irredundant decompositions of a, contrary to uniqueness. Thus $p \cup b > b$, and L is semimodular.

The next lemma is a direct consequence of Theorem 5.1 and Lemma 3.7 (and its corollary).

LEMMA 6.3. *Let L be a semimodular, compactly generated, atomic lattice. If $a \in L$ has a unique irredundant decomposition into irreducibles, then u_a/a is distributive.*

It should be noted that in any compactly generated atomic lattice, u_a/a will be a complete atomic, Boolean algebra if it is distributive. We now turn to the converse of the above lemma.

LEMMA 6.4. *Let L be a locally distributive, compactly generated, atomic lattice. Then each element of L has a unique irredundant decomposition into irreducibles.*

From Lemma 6.1 it follows that L is semimodular. We will show first that if $p > a$ and if q_1 and q_2 are irreducibles such that $q_1, q_2 \geq a$ and $q_1, q_2 \ngeq p$, then $q_1 = q_2$. If $q_1 \neq q_2$, then $q_1 \ngeq q_2$ since otherwise $q_1 \geq q_2 \cup p \geq p$ contrary to $q_1 \ngeq p$. Thus $q_2 > q_1 \cap q_2$ and hence there exists b_2 such that $q_2 \geq b_2 > q_1 \cap q_2$. Since $q_2 \ngeq p$ we have $b_2 \ngeq p$. Now let $S = \{s \mid q_1 \geq s \geq q_1 \cap q_2, s \cup b_2 \ngeq p\}$. S is nonempty since $q_1 \cap q_2 \in S$. Furthermore, since p is compact in q_1/a, S is inductive. Let s^* be a maximal element of S. If $q_1 > s^*$, then there exists w_1 such that $q_1 \geq w_1 > s^*$. Since $s^* \ngeq p$ and $s^* \geq a$ we have $w_2 = s^* \cup p > s^*$ by semimodularity. Also $s^* \ngeq b_2$, since otherwise $q_1 \geq s^* \geq b_2$ and hence $q_1 \cap q_2 \geq b_2$ contrary to $b_2 > q_1 \cap q_2$. Since $s^* \geq q_1 \cap q_2$ we have $w_3 = s^* \cup b_2 > s^*$ by semimodularity. Now $w_1 \neq w_2$ since $q_1 \ngeq p$; $w_1 \neq w_3$ since $q_1 \ngeq b_2$; and $w_2 \neq w_3$ since $s^* \cup b_2 \ngeq p$. By local distributivity $w_1 \cup w_3 \ngeq w_2$. But then $w_1 \cup b_2 = w_1 \cup s^* \cup b_2 = w_1 \cup w_3 \ngeq p$ and hence $w_1 \in S$ contrary to the maximal property of s^*. Thus we must have $q_1 = s^*$ and hence $q_1 \cup b_2 \ngeq p$. But then $q_1 \cup b_2 = q_1$, since otherwise $q_1 \cup b_2 > q_1$, $q_1 \cup p > q_1$ and $q_1 \cup b_2 \neq q_1 \cup p$ contrary to the irreducibility of q_1. Thus $q_1 \geq b_2$, and hence $q_1 \cap q_2 \geq b_2$ contrary to $b_2 > q_1 \cap q_2$. It follows that our original assumption is untenable and hence $q_1 = q_2$.

Now let q be irredundant in the decomposition $a = \cap Q$ and let $q \nleq \cap(Q' - q')$

in the decomposition $a = \cap Q'$. Then $q \not\geq p$ some $p \in P_a$. Now if $q'' \neq q$ all $q'' \in Q'$ we have $q'' \geq p$ all $q'' \in Q'$ by the result of the preceding paragraph. But then $a = \cap Q' \geq p$ contrary to $p > a$. Hence $q = q''$ some $q'' \in Q'$. But if $q'' \neq q'$ we have $q \geq \cap (Q' - q')$. Thus $q = q'$ and a has a unique decomposition into irreducibles. This completes the proof of the lemma.

Combining Lemmas 6.1-6.4 we have

THEOREM 6.1. *A compactly generated, atomic lattice has unique irreducible decompositions if and only if it is locally distributive.*

An examination of the proof of Lemma 6.4 shows that the only conditions required for the sufficiency argument are weak semimodularity and the fact that every three distinct elements covering an element of the lattice are independent. Thus we have the following corollary.

COROLLARY. *A compactly generated atomic lattice is locally distributive if and only if every three distinct covering elements generate a dense Boolean algebra of order eight.*

A further characterization analogous to that of Theorem 1.1 of [2] is the following:

COROLLARY. *A semimodular, compactly generated, atomic lattice is locally distributive if and only if every modular sublattice is distributive.*

If every modular sublattice is distributive then every three distinct covering elements must be independent and hence by the first corollary to Theorem 6.1, the lattice is locally distributive.

Now let L be locally distributive and let M be a modular sublattice which is not distributive. We may clearly suppose that M is the modular, nondistributive lattice of order five. Hence there exist elements a, b, c, d, e such that $b \cup c = b \cup d = c \cup d = e$ and $b \cap c = b \cap d = c \cap d = a$. It now follows (see Lemma 7.2 below) that there exist two irredundant decompositions $a = \cap Q = \cap Q'$ such that for each $q \in Q$ either $q \geq b$ or $q \geq c$, and for each $q' \in Q'$ either $q' \geq b$ or $q' \geq d$. Let $q \in Q$ be such that $q \not\geq b$. By Theorem 6.1, $q \in Q'$ and hence $q \geq d$. But then $q \geq c \cup d \geq b$, a contradiction. Thus every modular sublattice of L is distributive.

7. **Locally modular lattices.** The problem of the characterization of lattices in which the number of components in the irredundant decompositions of elements is unique is much more complex than the uniqueness problem treated in §6. In fact it is easy to give examples of nonsemimodular lattices in which the number of components is unique. Since a decomposition theory for lattices which are not semimodular has not yet been developed we will restrict our discussion to the semimodular case.

When a semimodular lattice satisfies the ascending chain condition the decompositions are finite and the basic result states that the number of components in the irredundant representations is unique if and only if L is locally modular (Dilworth [4]). For infinite decompositions, the example given at the end of §4 seems to indicate that questions concerning the invariance of the cardinality of decompositions are primarily set theoretic in nature. On the other hand, in the finite case, the invariance of the number of components is always obtained in terms of replacement properties and these properties are of a lattice theoretic character. Accordingly in this section we shall investigate the structure of semimodular, compactly generated, atomic lattices in which the fundamental replacement property of Theorem 4.2 holds. It will turn out that the lattices satisfying this replacement property will be precisely the locally modular lattices.

DEFINITION 7.1. A lattice is said to have *replaceable decompositions* if for every element a of the lattice, each irreducible in one irredundant decomposition of a can be replaced by a suitable irreducible in any other irredundant decomposition of a.

The following lemma and its corollary show the sufficiency of local modularity for replaceability.

LEMMA 7.1. *Let L be a locally modular, compactly generated, atomic lattice. Then $q \geqq a$ and $q \not\leqq u_a$ imply $u_a \succ q \cap u_a$ for each irreducible q.*

For suppose that $u_a \succ q \cap u_a$ does not hold. Then there exist two distinct elements p_1 and p_2 in P_a such that $(q \cap u_a) \cap (p_1 \cup p_2) = a$ and hence $q \cap (p_1 \cup p_2) = a$. Now let X be a chain of elements x such that $q \geqq x \geqq a$, $q \cap (x \cup p_1 \cup p_2) = x$, $x \cup p_1 \not\geqq p_2$. Then $q \geqq \bigcup X \geqq a$ and by continuity $\bigcup X = q \cap \bigcup_x (x \cup p_1 \cup p_2) = q \cap (\bigcup X \cup p_1 \cup p_2)$. Also $\bigcup X \cup p_1 \not\geqq p_2$ since p_2 is compact in u_a/a. Thus $\bigcup X$ satisfies the condition on x and since a trivially satisfies the conditions, it follows from the Maximal Principle that there is a maximal element m for which $q \geqq m \geqq a$. $q \cap (m \cup p_1 \cup p_2) = m$, $m \cup p_1 \not\geqq p_2$. Now clearly $m \cup p_1 \neq m \cup p_2$ and $m \cup p_1 \succ m$, $m \cup p_2 \succ m$. Thus $m = (m \cup p_1) \cap (m \cup p_2)$ and hence is reducible. It follows that $q > m$ and by atomicity we have $q \geqq m_1 \succ m$. If $m_1 \cup p_1 \geqq p_2$, then $m_1 \cup p_1 = m \cup p_1 \cup p_2$ and $m = q \cap (m \cup p_1 \cup p_2) = q \cap (m_1 \cup p_1) \geqq m_1$ contrary to $m_1 \succ m$. Thus $m_1 \cup p_1 \not\geqq p_2$. Since $q \geqq m_1 \geqq a$, by the maximal property of m we must have $q \cap (m_1 \cup p_1 \cup p_2) > m$. From $m_1 \cup p_1 \cup p_2 \succ m_1 \cup p_1 \succ m_1$ we conclude that $m_1 \cup p_1 \cup p_2 \succ q \cap (m_1 \cup p_1 \cup p_2)$. But $m_1 \cup p_1 \cup p_2 \in u_m/m$ and since u_m/m is modular by hypothesis, we have $m \cup p_1 \cup p_2 \succ q \cap (m \cup p_1 \cup p_2) = m$. Hence $m \cup p_1 = m \cup p_1 \cup p_2 \geqq p_2$ contrary to $m \cup p_1 \not\geqq p_2$. It follows from this final contradiction that $u_a \succ q \cap u_a$.

COROLLARY. *A locally modular, compactly generated, atomic lattice has replaceable decompositions.*

For, by Lemma 7.1, each irreducible containing a is such that $q \cap u_a$ is an irreducible element of u_a/a. But by Theorem 4.2 decompositions in u_a/a are replaceable. Hence by Lemma 3.2 decompositions in L are replaceable.

The necessity of local modularity is a considerably deeper result. We begin with two preliminary lemmas which will be needed later in the proof. The second lemma concerning finite dimensional semimodular lattices is of interest as a generalization of a well known classical theorem.

LEMMA 7.2. *Let L be a compactly generated, atomic lattice in which each element has an irredundant decomposition into irreducibles. Then if $a = a_1 \cap \cdots \cap a_n$ is an irredundant finite representation of a, there exists an irredundant decomposition $a = \cap Q$ such that for each $q \in Q$, $q \geq a_i$ for some i.*

For since L is continuous there exists a maximal element $m_1 \geq a_1$ such that $a = m_1 \cap a_2 \cap \cdots \cap a_n$. Similarly there exists a maximal element $m_2 \geq a_2$ such that $a = m_1 \cap m_2 \cap a_3 \cap \cdots \cap a_n$. By induction we get $a = m_1 \cap m_2 \cap \cdots \cap m_n$ where m_i is maximal such that $a = m_1 \cap \cdots \cap m_i \cap a_{i+1} \cap \cdots \cap a_n$. Thus $m_i' > m_i$ implies $m_1 \cap \cdots \cap m_{i-1} \cap m_i' \cap m_{i+1} \cap \cdots \cap m_n > a$. Now let $m_i = \cap Q_i$ be an irredundant decomposition of m_i into irreducibles and let $Q = \mathsf{V}_i \, Q_i$. Then $\cap Q = \cap_i \cap Q_i = \cap_i m_i = a$. Let $q_i \in Q_i$ then $\cap (Q_i - q_i) > m_i$. Hence $\cap (Q - q_i) \geq m_1 \cap \cdots \cap m_{i-1} \cap (Q_i - q_i) \cap m_{i+1} \cap \cdots \cap m_n > a$. Thus the decomposition $a = \cap Q$ is irredundant.

LEMMA 7.3. *Let L be a finite dimensional semimodular lattice in which the unit element u is a join of points. Let B denote the set of elements of L which are joins of points. Then if $u > s$, $s \not\geq b$ imply $b > s \cap b$ for all $b \in B$, L is modular.*

Proof. Let us suppose that L contains an element which is not a join of points. Then by finite dimensionality there exists a minimal element $b \in B$ such that b/z contains an element x which is not a join of points. Let $b > y \geq x$. By the minimal property of b, y is not a join of points. Let b_1 be the union of points in y. Then $b > y > b_1$. Since every element of L distinct from u is a meet of maximal elements, it follows that there exists s such that $u > s$, $s \not\geq b$ and $s \geq y$. Thus $s \cap b = y$. Now let S be a maximal independent set of points in b_1. Then $b_1 = \cup S$ by Lemma 3.6. Let S be extended to a maximal independent set T of points in b. Then $b = \cup T$. $S - T$ clearly contains at least two distinct points since otherwise $b = \cup T > \cup S = b_1$ contrary to $b > y > b_1$. Let p, q be distinct points of $S - T$. Then $s \not\geq p, q$ since otherwise $y = b \cap s \geq p, q$ and $b_1 = \cup S \geq p, q$, contrary to the independence of T. By the hypothesis of the lemma $p \cup q > s \cap (p \cup q) = s \cap b \cap (p \cup q) = y \cap (p \cup q)$. Thus $y \cap (p \cup q)$ is a point contained in y. Hence $b_1 \geq y \cap (b \cup q)$. But then $y \cap (p \cup q) \leq y \cap b_1 \cap (p \cup q) = \cup(S) \cap (p \cup q) = z$ since T is independent contrary to $y \cap (p \cup q) > z$. In view of this contradiction we conclude that every element of L is a join of points. Now let $a \cup b > b$ for $a, b \in L$. Let $u > s$, $s \geq b$, $s \not\geq a \cup b$. Then $s \cap (a \cup b) = b$. Clearly $s \not\geq a$ and since a is a join of points we have $a > s \cap a = s \cap (a \cup b) \cap a = a \cap b$. Then L is lower semimodular and since it is upper semimodular by hypothesis, it is modular. This completes the proof of the lemma.

It will now be shown that the modularity of u_a/a follows from the replacement property for the irredundant decompositions of a.

LEMMA 7.4. *Let L be a semimodular, compactly generated, atomic lattice such that the replacement property holds for the irredundant decompositions of a. Then u_a/a is modular.*

It will first be shown that $u_a > s$, $s \not\geq b$, *where b is a join of elements of P_a imply that $b > s \cap b$.*

Let us suppose that this is not the case and hence that $u_a > s$, $s \not\geq b$ while $b > s \cap b$ does not hold for some b which is a join of elements of P_a. Now let T be a maximal independent set of elements of P_a contained in $s \cap b$. Extend T to a maximal independent set S of elements of P_a contained in b. Then by Lemma 3.6, $\cup S = b$. Finally extend S to a maximal independent set P of points in P_a. For each $p \in P$ let q_p be an irreducible such that $q_p \geq \cup (P - p)$, $q_p \not\geq u_a$. Set $Q = \{q_p \mid p \in P\}$. Then by Lemma 3.7, $a = \cap Q$ is an irredundant decomposition of a into irreducibles. Now consider the case when $s \cap b = a$. Then by Lemma 7.2 there exists an irredundant decomposition $a = \cap Q'$ such that $q' \geq s$ or $q' \geq b$ for each $q' \in Q'$. Since $b \neq a$, there exists at least one $q' \in Q'$ such that $q' \geq s$. Suppose that there is another $q'' \in Q'$ such that $q'' \geq s$. Then $u_a \cap q' = s = u_a \cap q''$ and hence $u_a \cap \cap (Q' - q') = u_a \cap q'' \cap \cap (Q' - q') = u_a \cap q' \cap \cap (Q' - q') = u_a \cap \cap Q' = a$. Whence $\cap (Q' - q') = a$ contrary to the irredundancy of Q'. Thus there is exactly one $q' \in Q'$ such that $q' \geq s$ and hence $\cap (Q' - q') \geq b$. According to the hypothesis of the lemma q' can be replaced by q_p for some $p \in P$. Thus $a = q_p \cap \cap (Q' - q')$. Hence $a \geq q_p \cap b = \cup (P - p) \cap \cup (S) = \cup (S \wedge (P - p))$ and thus $S \wedge (P - p) = \varnothing$. Since $S \subseteq P$ it follows that $S = \{p\}$ and thus $b = \cup (S) = p > a = s \cap b$ contrary to hypothesis. Hence we may suppose that $s \cap b \neq a$. Next let $t^* = \cup (P - T)$ and suppose that $s \cap t^* = a$. Again there exists an irredundant decomposition $a = \cap Q'$ such that $q' \geq s$ or $q' \geq t^*$ for each $q' \in Q'$. Also there is a unique q' such that $q' \geq s$. For this q' we have $\cap (Q' - q') \geq t^*$. Since by hypothesis q' can be replaced by q_p for some $p \in P$ we get $a = \cup (P - p) \cap \cup (P - T) = \cup ((P - p) \wedge (P - T))$. Thus $(P - p) \wedge (P - T) = \varnothing$ and hence $T = P - p$. But then $u_a > \cup (P - p) = \cup T$ and hence $s = s \cap b$ contrary to $s \not\geq b$. We may thus assume that $s \cap b \neq a$ and $s \cap t^* \neq a$. Furthermore $b \cap t^* \neq a$. For since $\cup S = b > s \cap b \geq \cup (T)$ it follows that $\cup (S - T) > a$ and hence $b \cap t^* \geq \cup (S - T) > a$. On the other hand $a = s \cap b \cap t^*$. Since if $s \cap b \cap t^* \geq p > a$, we have $s \cap b \geq p$ and hence $\cup T \geq p$ which implies $a = \cup T \cap \cup (P - T) \geq p$ contrary to $p > a$. Thus the representation $a = s \cap b \cap t^*$ is irredundant and by Lemma 7.2, there exists an irredundant decomposition into irreducibles $a = \cap Q'$ such that either $q' \geq s$, $q' \geq b$, or $q' \geq t^*$ for each $q' \in Q'$. By the argument given above there is a unique q' in Q' such that $q' \geq s$. By the replacement property we have $a = q_p \cap \cap (Q' - q)$ for some $p \in P$. But then

$$a \geq \cup (P - p) \cap (b \cap t^*) = \cup (P - p) \cap \cup (S \wedge (P - T))$$

$$= \cup ((P - p) \wedge [S \wedge (P - T)]) = \cup (S - (T \vee p)).$$

Thus $S - T \vee p = \varnothing$ and thus $S = T \vee p$. Hence $b = \cup (S) = \cup T \cup p > \cup T$. But

$b > s \cap b \geq \cup T$. Thus $b > s \cap b$ contrary to hypothesis. The proof of the above statement is thus complete.

Now let x be an arbitrary element of u_a/a, and let c be a compact element such that $x \geq c$. Then $\cup P_a \geq x \geq c$ and hence there exists a finite set $\{p_1, \cdots, p_n\}$ of elements of P_a such that $p_1 \cup \cdots \cup p_n \geq c$. We show next that $p_1 \cup \cdots \cup p_n/a$ satisfies the conditions of Lemma 7.3. It is clearly a finite dimensional semimodular lattice in which the unit element is a join of points. Now let $p_1 \cup \cdots \cup p_n > t$, $p_1 \cup \cdots \cup p_n \geq w$ and $t \not\geq w$ where w is a join of elements of P_a. By Lemma 3.7, there exists v such that $u_a > v$, $v \geq t$ and $v \not\geq w$. By the first part of the proof we have $w > v \cap w = v \cap (p_1 \cup \cdots \cup p_n) \cap w = t \cap w$. Thus all of the conditions of Lemma 3.7 are satisfied and hence $p_1 \cup \cdots \cup p_n/a$ is modular. Since $a \leq a \cup c \leq p_1 \cup \cdots \cup p_n$ we conclude that $a \cup c$ is a join of points of $p_1 \cup \cdots \cup p_n/a$ and hence is a join of the elements of P_a. But since L is compactly generated, $x = \cup \{c \mid c \leq x\} = \cup \{a \cup c \mid c \leq x\}$ and thus x is a join of elements of P_a. It follows that u_a/a is a point lattice. Now if $x \cup y > y$ in u_a/a, by Lemma 3.7 there exists s such that $u_a > s$, $s \geq y$, and $s \not\geq x \cup y$. Then $s \not\geq x$ and since x is a join of points, the first part of the proof implies that $x > x \cap s$. But then $x > x \cap s = x \cap (x \cup y) \cap s = x \cap y$. Thus u_a/a is lower semimodular and we conclude from Lemma 3.4 that u_a/a is modular. Hence the proof of the lemma is complete.

Lemmas 7.1-7.4 give the following theorem.

THEOREM 7.1. *A semimodular, compactly generated, atomic lattice has replaceable decompositions if and only if it is locally modular.*

REFERENCES

1. G. Birkhoff, *Lattice theory*, rev. ed., Amer. Math. Soc. Colloquium Publications, vol. 25, New York, 1948.

2. R. P. Dilworth, *Lattices with unique irreducible decompositions*, Ann. of Math. (2) vol. 41 (1940) pp. 771–777.

3. ———, *The arithmetical theory of Birkhoff lattices*, Duke Math. J. vol. 8 (1941) pp. 286–299.

4. ———, *Ideals in Birkhoff lattices*, Trans. Amer. Math. Soc. vol. 49 (1941) pp. 325–353.

5. ———, *Note on the Kurosh-Ore theorem*, Bull. Amer. Math. Soc. vol. 52 (1946) pp. 659–663.

6. M. Hall, *Distinct representatives of subsets*, Bull. Amer. Math. Soc. vol. 54 (1948) pp. 922–926.

7. J. Hashimoto, *Direct, subdirect decompositions and congruence relations*, Osaka Math. J. vol. 9 (1957) pp. 87–112.

CALIFORNIA INSTITUTE OF TECHNOLOGY,
PASADENA, CALIFORNIA

NOTE ON THE KUROSCH-ORE THEOREM

R. P. DILWORTH

1. **Introduction.** The Kurosch-Ore theorem[1] asserts that if an element of a modular lattice has two decompositions into irreducibles, then each irreducible of one decomposition may be replaced by a suitably chosen irreducible from the other decomposition. It follows that the number of irreducibles in the two decompositions is the same.

The purpose of the present note is to study the manner in which the irreducibles of two decompositions can replace one another. Now from the Kurosch-Ore theorem it is not even clear that each irreducible of one decomposition is suitable for replacing some irreducible of the other decomposition. However, this follows from the following precise theorem:

THEOREM 1. *Let a be an element of a modular lattice and let $a = q_1 \cap \cdots \cap q_n = q_1' \cap \cdots \cap q_n'$ be two reduced decompositions into irreducibles. Then the q's may be renumbered in such a way that*

$$a = q_1 \cap \cdots \cap q_{i-1} \cap q_i' \cap q_{i+1} \cap \cdots \cap q_n, \qquad i = 1, \cdots, n.$$

Along the same line of ideas, the following theorem on simultaneous replacement is also proved.

THEOREM 2. *Let a be an element of a modular lattice and let $a = q_1 \cap \cdots \cap q_n = q_1' \cap \cdots \cap q_n'$ be two reduced decompositions into irreducibles. Then for each q_i, there exists q_j' such that q_j' can replace q_i in the first decomposition and q_i can replace q_j' in the second decomposition.*

On the other hand, an example is given which shows that, in general, it is impossible to renumber the q's in such a way that simultaneously q_i may replace q_i' and q_i' replace q_i.

As the principal tool in the investigation we introduce the concept of a *superdivisor* r of an element a. r has the fundamental property that its crosscut with any proper divisor of a is never equal to a. The superdivisors of a are closed under crosscut and indeed form a dual-ideal r_a which properly divides a.

A surprising by-product of the investigation is the fact that in a

---

Presented to the Society, September 8, 1942, under the title *On the decomposition theory of modular lattices*; received by the editors April 9, 1946.

[1] A simple proof is given in Birkhoff [1, p. 54]. Numbers in brackets refer to the references cited at the end of the paper.

659

modular lattice satisfying the ascending chain condition, $\mathfrak{r}_a$ can be used to prove the existence of covering ideals. Thus, in this case, the customary use of transfinite induction can be avoided.

2. Properties of superdivisors. Let M denote a modular lattice of elements a, b, c, $\cdots$. $a \supseteq b$ will denote ordinary lattice inclusion while $a \supset b$ will denote proper inclusion. We recall that an element q of M is (crosscut) irreducible if $q = x \cap y$ implies either $q = x$ or $q = y$.

DEFINITION 1. *A divisor r of a is a* superdivisor *of a if $r \cap x = a$ implies $x = a$ for all x in M.*

The following lemmas give the basic properties of superdivisors.[2]

LEMMA 1. *If r is a superdivisor of a and $s \supseteq r$, then s is a superdivisor of a.*

For $s \cap x = a$ implies $r \cap x = a$ implies $x = a$.

LEMMA 2. *If r and s are superdivisors of a, then $r \cap s$ is a superdivisor of a.*

For $(r \cap s) \cap x = a$ implies $r \cap (s \cap x) = a$ implies $s \cap x = a$ implies $x = a$.

COROLLARY. *The superdivisors of a form a dual-ideal $\mathfrak{r}_a$ of M.*

LEMMA 3. *If q is an irreducible divisor of a and $x \supset q$, then x is a superdivisor of a.*

For if $x \cap y = a$, then $q = q \cup a = q \cup (x \cap y) = x \cap (q \cup y)$ by the modular law. Since q is irreducible and $q \neq x$, it follows that $q = q \cup y$. Hence $y = q \cap y = q \cap x \cap y = q \cap a = a$. Thus x is a superdivisor of a.

Now if $a = q_1 \cap \cdots \cap q_n$ is a reduced decomposition of a into irreducibles, we shall set $Q_i = q_1 \cap \cdots \cap q_{i-1} \cap q_{i+1} \cap \cdots \cap q_n$. Clearly $a = q_i \cap Q_i$ and $Q_i \neq a$.

LEMMA 4. *Let $a = q_1 \cap \cdots \cap q_n$ be a reduced decomposition into irreducibles. Then if r is a superdivisor of a, $q_i \cup (r \cap Q_i)$ is also a superdivisor of a.*

By Lemma 3 if $q_i \cup (r \cap Q_i)$ is not a superdivisor of a, then $q_i \supseteq r \cap Q_i$. But then $r \cap Q_i = r \cap q_i \cap Q_i = a$. Hence $Q_i = a$ which contradicts $Q_i \neq a$.

LEMMA 5. *Let $a = q_1 \cap \cdots \cap q_n$ be a reduced decomposition into ir-*

---

[2] If the descending chain condition holds, it is easy to show that r is a superdivisor if and only if $r \supseteq u_a$ where u_a is the union of the elements covering a. Cf. Dilworth [2, p. 288].

reducibles and let $x \supseteq r_1 \cap Q_1, r_2 \cap Q_2, \cdots, r_i \cap Q_i$ where $r_1, \cdots, r_i$ are superdivisors of a. Then $x \supseteq r \cap q_{i+1} \cap \cdots \cap q_n$ where r is a superdivisor of a.

Now clearly $x \supseteq r \cap q_1 \cap \cdots \cap q_n$ for any superdivisor r of a. Let k be maximal such that $x \supseteq r \cap q_k \cap \cdots \cap q_n$ for some superdivisor r. Suppose $k \leq i$. Then by the hypothesis of the lemma $x \supseteq r_k \cap Q_k$. Let $r' = r \cap r_k$. Then $x \supseteq (r' \cap q_k \cap \cdots \cap q_n) \cup (r' \cap Q_k) = r' \cap q_{k+1} \cap \cdots \cap q_n \cap (q_k \cup (r' \cap Q_k))$. But $q_k \cup (r' \cap Q_k)$ is a superdivisor of a by Lemma 4. Hence $r' \cap (q_k \cup (r' \cap Q_k)) = r''$ is a superdivisor of a by Lemma 2. But then $x \supseteq r'' \cap q_{k+1} \cap \cdots \cap q_n$ contrary to the maximal property of k. Thus $k > i$ and the conclusion of the lemma follows.

3. Decomposition theory. The application of superdivisors to decomposition problems rests on the following lemma:

LEMMA 6. *Let $a = q_1 \cap \cdots \cap q_n$ be a reduced decomposition of a. Then q_i may be replaced by an irreducible divisor q of a if and only if $q \supseteq r \cap Q_i$ is false for every superdivisor r of a.*

Let us suppose that q can replace q_i. Then $a = q \cap Q_i$. Hence if $q \supseteq r \cap Q_i$ for some superdivisor r, then $r \cap Q_i = r \cap q \cap Q_i = r \cap a = a$ and $Q_i = a$ which is impossible. Thus $q \supseteq r \cap Q_i$ fails for every superdivisor r. Conversely suppose $q \supseteq r \cap Q_i$ holds for no superdivisors r. Then

$$q \supseteq q \cap Q_i = (q \cap Q_i) \cup (q_i \cap Q_i) = [q_i \cup (q \cap Q_i)] \cap Q_i.$$

Hence $q_i \cup (q \cap Q_i)$ is not a superdivisor of a and by Lemma 3 we have $q_i \supseteq q \cap Q_i$. Thus $q \cap Q_i = q \cap q_i \cap Q_i = a$ and q can replace q_i in the decomposition.

The theorems stated in the introduction can now be proved.

PROOF OF THEOREM 1. Let S_i' denote the set of irreducibles of the second decomposition which can replace q_i in the first decomposition. Now suppose that there are k of the sets S_i' which together contain less than k irreducibles. Renumbering if necessary, we can suppose that $S_1', \cdots, S_k'$ are composed of the irreducibles $q_1', \cdots, q_l'$ where $l < k$. It follows that q_j' cannot replace q_i if $j > l$ and $i \leq k$. Hence by Lemma 6, $q_j' \supseteq r_{ji} \cap Q_i$ for some superdivisor r_{ji} of a if $j > l$ and $i \leq k$. From Lemma 5 we conclude that $q_j' \supseteq r_j \cap q_{k+1} \cap \cdots \cap q_n$ for some superdivisor r_j of a if $j > l$. Thus $q_{l+1}' \cap \cdots \cap q_n' \supseteq r \cap q_{k+1} \cap \cdots \cap q_n$ where $r = r_{l+1} \cap \cdots \cap r_n$ is a superdivisor of a. But then

$$a = q_1' \cap \cdots \cap q_l' \cap q_{l+1}' \cap \cdots \cap q_n' \supseteq r \cap q_1' \cap \cdots \cap q_l' \cap q_{k+1} \cap \cdots \cap q_n \supseteq a.$$

Hence $a = r \cap q_1' \cap \cdots \cap q_l' \cap q_{k+1} \cap \cdots \cap q_n$. Since r is a superdivisor of a, we have

$$a = q_1' \cap \cdots \cap q_l' \cap q_{k+1} \cap \cdots \cap q_n.$$

Since $l < k$, the number of components in this decomposition is less than n, contrary to the Kurosch-Ore theorem. Thus every k of the sets S_i' contain at least k irreducibles. It follows from the Radó-Hall theorem on representatives of sets that there exists a distinct set of representatives for the sets $S_1', \cdots, S_n'$. Renumbering if necessary, we may suppose that these representatives are $q_1', \cdots, q_n'$. But then q_i' can replace q_i and the theorem is proved.

PROOF OF THEOREM 2. Renumbering if necessary, we may suppose that $q_1', \cdots, q_l'$ can replace q_i while the others cannot. According to Lemma 6, $q_j' \supseteq r_j \cap Q_i$, $j = l+1, \cdots, n$, where r_j is s superdivisor of a. Now suppose that q_i can replace none of the irreducibles $q_1', \cdots, q_l'$ in the second decomposition. Again by Lemma 6 we have $q_i \supseteq r_j' \cap Q_j'$, $j = 1, \cdots, l$. From Lemma 5 we conclude that $q_i \supseteq r' \cap q_{l+1}' \cap \cdots \cap q_n'$ for some superdivisor r'. Now $q_{l+1}' \cap \cdots \cap q_n' \supseteq r_{l+1} \cap \cdots \cap r_n \cap Q_i = r \cap Q_i$ where r is a superdivisor of a. Hence $q_i \supseteq r' \cap r \cap Q_i$ where $r' \cap r$ is a superdivisor of a. But then $r' \cap r \cap Q_i = r' \cap r \cap Q_i \cap q_i = a$ and $Q_i = a$ contrary to hypothesis. Hence q_j' can be replaced by q_i for some $j \leq l$. Thus q_i and q_j' can replace one another.

In order to see that a sharper theorem on simultaneous replacement cannot be proved in general, consider the lattice of subspaces of the seven-point projective plane. If $1, \cdots, 7$ denote the points, let the lines (and the points they contain) be denoted by $l_1(124)$, $l_2(235)$, $l_3(346)$, $l_4(457)$, $l_5(156)$, $l_6(267)$, $l_7(137)$. $l_1, \cdots, l_7$ are the irreducibles of the lattice. Let us consider the decompositions of the null space z. We have

$$z = l_1 \cap l_2 \cap l_3 = l_5 \cap l_6 \cap l_7.$$

Now the possible sets of replacements of l_1, l_2, l_3 respectively are (l_5, l_6, l_7), (l_6, l_5, l_7), and (l_6, l_7, l_5). But l_1, l_2, l_3 is a possible set of replacements only for (l_5, l_7, l_6), (l_7, l_5, l_6), and (l_7, l_6, l_5). Hence it is not possible in this case to renumber the irreducibles in such a way that corresponding irreducibles can replace one another.

4. **Existence of covering ideals.** It is well known that the lattice of dual-ideals of a lattice M contains M as the sublattice of principal ideals.[3] Hence if $\mathfrak{a}$ is a dual-ideal, by $\mathfrak{a} \supseteq a$ we shall mean $\mathfrak{a} \supseteq (a)$ where

---

[3] For the general properties of dual-ideals used in this paper see Dilworth [3, pp. 329–331].

(a) is the principal ideal generated by a. Also $\mathfrak{a} > a$ ($\mathfrak{a}$ "covers" a) means $\mathfrak{a} \supset a$ and no ideal exists which properly contains $\mathfrak{a}$ and is properly contained in (a). We give a proof of the existence of covering ideals which does not require transfinite induction.

THEOREM 3. *Let an element a of a modular lattice M have a decomposition into irreducibles. Then if $\mathfrak{a} \supseteq a$, there exists a dual-ideal $\mathfrak{p}$ such that $\mathfrak{a} \supseteq \mathfrak{p} > a$.*

PROOF. Let $\mathfrak{a}' = \mathfrak{a} \cap \mathfrak{r}_a$ where $\mathfrak{r}_a$ is the dual-ideal of all superdivisors of a. Then $\mathfrak{a}' \neq (a)$. For if $\mathfrak{a}' = (a)$, then $x \cap r = a$ where $x \in \mathfrak{a}$ and r is a superdivisor of a. But then $x = a$ and $\mathfrak{a} = (a)$ contrary to $\mathfrak{a} \supset a$. Now let $a = q_1 \cap \cdots \cap q_n$ be a reduced decomposition of a into irreducibles. Clearly $\mathfrak{r}_a \cap q_1 \cap \cdots \cap q_{i-1} \supseteq \mathfrak{r}_a \cap q_1 \cap \cdots \cap q_i$. Suppose $\mathfrak{r}_a \cap q_1 \cap \cdots \cap q_{i-1} = \mathfrak{r}_a \cap q_1 \cap \cdots \cap q_i$. Then $\mathfrak{r}_a \cap q_1 \cap \cdots \cap q_{i-1} \cap q_{i+1} \cap \cdots \cap q_n = a$ and hence $r \cap Q_i = a$ where r is a superdivisor of a. Thus $Q_i = a$ contrary to assumption. Next suppose that $\mathfrak{r}_a \cap q_1 \cap \cdots \cap q_{i-1} \supset \mathfrak{b} \supset \mathfrak{r}_a \cap q_1 \cap \cdots \cap q_i$. Then since the lattice of dual-ideals is modular we have $\mathfrak{b} = \mathfrak{r}_a \cap q_1 \cap \cdots \cap q_{i-1} \cap (\mathfrak{b} \cup q_i)$. Let $b \in \mathfrak{b}$. If $q_i \supseteq b$, then $q_i \supseteq \mathfrak{b}$ and $\mathfrak{b} = \mathfrak{r}_a \cap q_1 \cap \cdots \cap q_{i-1} \cap q_i$ contrary to hypothesis. Hence $q_i \cup b$ is a proper divisor of q_i for every $b \in \mathfrak{b}$. By Lemma 3, $q_i \cup b$ is a superdivisor of a for every $b \in \mathfrak{b}$. Hence $q_i \cup b \in \mathfrak{r}_a$ for every $b \in \mathfrak{b}$. Thus $q_i \cup \mathfrak{b} \supseteq \mathfrak{r}_a$. But then $\mathfrak{b} = \mathfrak{r}_a \cap q_1 \cap \cdots \cap q_{i-1}$ contrary to assumption. Thus $\mathfrak{r}_a \cap q_1 \cap \cdots \cap q_{i-1} > \mathfrak{r}_a \cap q_1 \cap \cdots \cap q_i$. But then $\mathfrak{r}_a > \mathfrak{r}_a \cap q_1 > \cdots > \mathfrak{r}_a \cap q_1 \cap \cdots \cap q_{n-1} > a$ is a finite complete chain joining $\mathfrak{r}_a$ to a. By the general theory of modular lattices (Birkhoff [1]) it follows that the quotient lattice $\mathfrak{r}_a/a$ is of finite dimension. Since $\mathfrak{a}' \in \mathfrak{r}_a/a$ we have $\mathfrak{a}' \supseteq \mathfrak{p} > a$ for some dual-ideal $\mathfrak{p}$ and the theorem is proved.

Now if the ascending chain condition holds in a modular lattice M, then every element has a decomposition into irreducibles and hence, by Theorem 3, there exist dual-ideals covering a for every a not the unit of M.

REFERENCES

1. G. Birkhoff, *Lattice theory*, Amer. Math. Soc. Colloquium Publications, vol. 25, 1940.

2. R. P. Dilworth, *Arithmetical theory of Birkhoff lattices*, Duke Math. J. vol. 8 (1941) pp. 286–299.

3. ———, *Ideals in Birkhoff lattices*, Trans. Amer. Math. Soc. vol. 49 (1941) pp. 325–353.

CALIFORNIA INSTITUTE OF TECHNOLOGY

STRUCTURE AND DECOMPOSITION THEORY OF LATTICES

BY

R. P. DILWORTH

1. **Introduction.** One of the most natural problems which arise in the investigation of an abstract algebraic system is that of representing the elements of the system in terms of a canonical subset by means of the operations of the system. Thus for a polynomial domain over a field with the operation that of ordinary polynomial multiplication it is the problem of representing polynomials as products of irreducible polynomials. For lattices there are two operations with respect to which we may consider the representation of the elements of the lattice. Since the operations are dual it suffices to consider representations with respect to one of the operations. Thus we shall treat only meet representations. Now an element which cannot be expressed as the meet of elements distinct from itself clearly has only trivial representations. Furthermore these elements must surely be included in any reasonable canonical set. Thus we shall be particularly concerned with meet representations in terms of meet irreducible elements.

A second natural problem which arises in any algebraic investigation is that of representing the system as a whole in terms of certain distinguished subsystems by means of canonical constructions. The most familiar of these constructions is the direct union (direct product) representation. Again, if possible, it is desirable to represent the system as a direct or subdirect union of systems which cannot be further decomposed, i.e., indecomposable systems.

For most algebraic systems, these two problems are quite different in character and the results in one case may have very little connection with the results in the other. For lattices, on the other hand, these two problems are intimately related. For direct and subdirect union representations of lattices, or indeed of any algebraic system, can be described in terms of the structure of the congruence relations on the algebraic system. But the congruence relations in a very natural way form a lattice. Furthermore, the meet representations of the null congruence relation correspond to the subdirect representations of the algebraic system. Hence decomposition theorems for the elements of a class of lattices will immediately give representation theorems for those algebraic systems having congruence lattices belonging to the given class of lattices. In particular, decomposition theorems which hold for the null element of the lattice of congruence relations of a lattice lead to subdirect (or direct) union representation theorems for the lattice itself. Structure theorems for lattices are greatly simplified by the fact that the lattice of congruence relations on a lattice is always distributive. Thus it suffices to study the decomposition theory of distributive

3

lattices in developing the structure theory of arbitrary lattices. However, applications to the structure of other algebraic systems require a decomposition theory for lattices of a much more general type. This paper will be devoted to a description of the relationship between structure and decomposition theorems followed by an account of the development of decomposition theory up to present. The final section contains a discussion of some of the outstanding problems of current interest in this area of lattice theory.

2. **Direct and subdirect unions.** Consider a collection of lattices L_α where α belongs to an index set A. $\prod_\alpha L_\alpha$ will denote the "Cartesian product" of the lattices L_α; that is the set of functions f on A such that $f(\alpha) \in L_\alpha$ for each $\alpha \in A$. $\prod_\alpha L_\alpha$ is a lattice if we define

$$(f \cup g)(\alpha) = f(\alpha) \cup g(\alpha),$$
$$(f \cap g)(\alpha) = f(\alpha) \cap g(\alpha).$$

A lattice L is "imbedded" in $\prod_\alpha L_\alpha$ if L is isomorphic to a sublattice of $\prod_\alpha L_\alpha$. L is a *subdirect union* of the lattices L_α if L is imbedded in $\prod_\alpha L_\alpha$ and for each $a_\alpha \in L_\alpha$ there exists a function $f \in \prod_\alpha L_\alpha$ corresponding to an element of L such that $f(\alpha) = a_\alpha$. Finally L is a *direct union* of the lattices L_α if L is imbedded in $\prod_\alpha L_\alpha$ and for every finite set of indices $\alpha_1, \cdots, \alpha_n$ and elements $a_{\alpha_1}, \cdots, a_{\alpha_n}$ in $L_{\alpha_1}, \cdots, L_{\alpha_n}$ respectively, there exists f corresponding to an element of L such that $f(\alpha_i) = a_{\alpha_i}, i = 1, \cdots, n$.

If L is imbedded in $\prod_\alpha L_\alpha$, this imbedding induces a natural set of congruence relations on L; namely

$$a\ \theta_\alpha\ b \text{ if and only if } a_\alpha = b_\alpha.$$

For the study of lattice structure it is useful to have a characterization of imbedding, subdirect unions, and direct unions in terms of properties of this set of congruence relations. We begin by pointing out that the set $\Theta(L)$ of congruence relations on L can be partially ordered by the relation

$$\theta \leq \phi \text{ if and only if } a\ \theta\ b \Rightarrow a\ \phi\ b.$$

$\Theta(L)$ has a unique maximal element ι, namely the congruence relation which identifies all elements of L. $\Theta(L)$ likewise has a unique minimal element ω, namely the equality relation on L. Under the above partial ordering $\Theta(L)$ is a complete lattice. If $\Phi \subseteq \Theta(L)$ then the meet and join of the subset Φ may be characterized as follows:

$$a \bigcap \Phi\ b \text{ if and only if } a\ \phi\ b \text{ for all } \phi \in \Phi;$$

$$a \bigcup \Phi\ b \text{ if and only if } a = a_0, a_1, \cdots, a_m = b \text{ exist such that } a_{i-1}\ \phi_i\ a_i \text{ for }$$
$\phi_i \in \Phi.$

In addition to the lattice operations, a permutability relation plays an important role in structure theorems. Congruence relations θ and ϕ are said to *permute* if $a\ \theta\ b$ and $b\ \phi\ c$ imply the existence of d such that $a\ \phi\ d$ and $d\ \theta\ c$.

Permutability is preserved by the lattice operations. If θ permutes with both ϕ and ψ, then θ permutes with $\phi \cup \psi$ and $\phi \cap \psi$. Furthermore if θ permutes with all of the congruence relations belonging to Φ, then θ permutes with $\bigcup \Phi$. It is not difficult to show that any two congruence relations on a relatively complemented lattice permute. Another useful result asserts that any lattice of permuting congruence relations on an arbitrary algebraic system is modular. For lattices there is the stronger theorem of Funayama [**9**] that $\Theta(L)$ is always distributive.

Now if θ is a congruence relation on L, the congruence classes form a lattice which is a homomorphic image of L. The lattice of congruence classes will be denoted by $\theta(L)$. If L is imbedded in $\prod_\alpha L_\alpha$ and θ_α is the congruence relation determined by the component L_α, then each congruence class of θ_α determines a unique $a_\alpha \in L_\alpha$ and $\theta_\alpha(L)$ is thus a sublattice of L_α. Furthermore if $a \, \theta_\alpha \, b$ for all $\alpha \in A$, then $a_\alpha = b_\alpha$ all $\alpha \in A$ and hence $a = b$. Thus $\bigcap_\alpha \theta_\alpha = \omega$. If L is a subdirect union of the lattices L_α, then $\theta_\alpha(L) = L_\alpha$. Thus subdirect union representations are characterized by meet decompositions of the minimal congruence relation. Clearly subdirect union representations in terms of subdirectly irreducible lattices correspond to decompositions of ω into indecomposable congruence relations.

We next observe that if L is a direct union of the lattices L_α and α, β are two distinct elements of A, then for any pair of elements $a, b \in L$, there exists $c \in L$ such that $c_\alpha = a_\alpha$ and $c_\beta = b_\beta$. Thus $a \, \theta_\alpha \, c$ and $c \, \theta_\beta \, b$. It follows that θ_α and θ_β permute and $\theta_\alpha \cup \theta_\beta = \iota$. Hence direct union representations of L correspond to decompositions $\omega = \bigcap_\alpha \theta_\alpha$ where $\{\theta_\alpha | \alpha \in A\}$ is a set of permuting, coprime congruence relations. Conversely, any such decomposition of ω leads to a direct union representation of L.

Finally, it should be mentioned that Hashimoto [**10**] has characterized Cartesian product representations in terms of properties of $\Theta(L)$ and a generalized notion of permutability. If $\{\theta_\alpha | \alpha \in A\}$ is a set of congruence relations, let $\theta_\alpha^* = \bigcap \{\theta_\beta | \beta \neq \alpha\}$. The set $\{\theta_\alpha | \alpha \in A\}$ is said to be *completely coprime* if $\theta_\alpha \cup \theta_\alpha^* = \iota$ for all α. Likewise the set $\{\theta_\alpha | \alpha \in A\}$ is said to be completely permutable if $a_\alpha(\theta_\beta^* \cup \theta_\beta^*)a_\beta$ for all α and β, implies that there exists $a \in L$ such that $a\theta_\alpha a_\alpha$ for all $\alpha \in A$. Then if $\{\theta_\alpha | \alpha \in A\}$ is a set of completely coprime and completely permutable congruence relations, L is isomorphic to $\prod_\alpha \theta_\alpha(L)$. It is also easy to see that the congruence relations on a Cartesian product determined by the component lattices form a set of completely coprime and completely permutable congruence relations.

3. **The classical decomposition theorems.** As mentioned in the introduction and in view of the applications to structure problems we shall be interested in the representation of elements of a lattice as meets of indecomposable elements. However, the appropriate definition of indecomposability will depend upon the type of representations under consideration. The two principal definitions are the following.

Definition 3.1. q is (*meet*) *irreducible* if $q = x \cap y$ implies $q = x$ or $q = y$.

Definition 3.2. q is *completely* (*meet*) *irreducible* if $q = \bigcap S$ implies $q = s$ for some $s \in S$.

In general, the first definition is appropriate for finite decompositions while the second is the appropriate concept for infinite decompositions.

The fundamental problems in decomposition theory concern the question of existence and uniqueness of decompositions into irreducibles. If an element of a lattice can be represented as a meet of the set Q of indecomposables and if the removal of one or more elements from the set Q gives a set with the same meet, then the representation $a = \bigcap Q$ can hardly be unique in any reasonable sense. Hence we shall restrict our attention to irredundant decompositions.

Definition 3.3. A decomposition $a = \bigcap Q$ is *irredundant* if $\bigcap (Q - q) > a$ for all $q \in Q$.

If $a = q_1 \cap \cdots \cap q_n$ is a finite decomposition of a, then by deleting superfluous q's this decomposition can always be refined to an irredundant decomposition. For infinite decompositions, this process breaks down, and hence the construction of irredundant infinite decompositions requires a more elaborate procedure.

The classical existence theorem is the following.

Theorem 3.1. *If L satisfies the ascending chain condition, then every element of L has a finite irredundant decomposition into irreducibles.*

For if the theorem is not true, the lattice L contains a maximal element a which cannot be represented as an irredundant meet of irreducibles. a is reducible since otherwise it would have an irredundant decomposition consisting of one irreducible. Hence $a = x \cap y$ where $x > a$ and $y > a$. But then both x and y may be represented as meets of irreducibles and hence a can likewise be so represented. Refining this decomposition into an irredundant one contradicts the definition of a.

The classical uniqueness theorem is due to Birkhoff [**2**] and concerns distributive lattices.

Theorem 3.2. *If an element of a distributive lattice has a finite irredundant decomposition into irreducibles, this decomposition is unique.*

For if $a = q_1 \cap \cdots \cap q_m = q_1' \cap \cdots \cap q_n'$ are two irredundant decompositions into irreducibles, then $q_i = q_i \cup a = q_i \cup (q_1' \cap \cdots \cap q_n') = (q_i \cup q_1') \cap \cdots \cap (q_i \cup q_n')$ and hence $q_i = q_i \cup q_j'$ for some j. Thus $q_i \geq q_j'$ and similarly $q_j' \geq q_k$ for some k. Since the decompositions are irredundant $i = k$ and hence $q_i = q_j'$. Similarly each q_j' is equal to q_k for some k and hence the two decompositions are identical.

Another type of uniqueness theorem is concerned with the replacement of irreducibles in one decomposition by suitably chosen irreducibles in another decomposition. The classical theorem of this type is due to Kurosch [11] and Ore [12].

Theorem 3.3. *Let $a = q_1 \cap \cdots \cap q_m = q_1' \cap \cdots \cap q_n'$ be two irredundant decompositions of an element of a modular lattice. Then for each q_i there exists a q_j' such that $a = q_i \cap \cdots \cap q_{i-1} \cap q_j' \cap q_{i+1} \cap \cdots \cap q_m$ is an irredundant decomposition of a.*

For if $q_i^* = q_1 \cap \cdots \cap q_{i-1} \cap q_{i+1} \cap \cdots \cap q_m$, then it can be easily verified that $[q_i \cup (q_i^* \cap x)] \cap [q_i \cup (q_i^* \cap y)] = q_i \cup (q_i^* \cap x \cap y)$. Repeated application of this formula gives $[q_i \cup (q_i^* \cap q_1')] \cap \cdots \cap [q_i \cup (q_i^* \cap q_n')] = q_i \cup (q_i^* \cap q_1' \cap \cdots \cap q_n') = q_i \cup (q_i^* \cap a) = q_i \cup a = q_i$. Since q_i is irreducible we have $q_i = q_i \cup (q_i^* \cap q_j')$ for some j and hence $q_i \geq q_i^* \cap q_j'$ for some j. But then $a = q_i \cap q_i^* \geq q_i^* \cap q_j' \geq a$ and hence $a = q_i^* \cap q_j'$. It is easy to see that this representation is irredundant.

The property expressed in this theorem we shall call the *replacement property* for irredundant decompositions in modular lattices. Repeated application of the replacement property shows that $m = n$ and hence that the number of irreducibles in the irredundant decompositions of an element of a modular lattice is unique.

We note at this point that Theorem 3.3 can be sharpened to give a simultaneous replacement theorem. Namely, the q_j' may be remembered so that each q_i can be replaced by q_i'.

The methods of proof for the classical uniqueness and replacement theorems do not extend to more general lattices. Thus a new technique is required. We will begin by discussing the finite dimensional case.

4. **Finite dimensional lattices.** Let L be a finite dimensional lattice and let $a = q_1 \cap \cdots \cap q_m$ be an irredundant decomposition into irreducibles. Then $q_i^* = q_i \cap \cdots \cap q_{i-1} \cap q_{i+1} \cap \cdots \cap q_m \neq a$ for each i and hence there exists p_i such that $q_i^* \geq p_i \succ a$ when $p_i \succ a$ signifies that p_i covers a. Thus irredundant decompositions are closely related to properties of the elements covering a. Let u_a denote the join of all elements p covering a. Then $u_a \cap x = a$ if and only if $x = a$. Hence $a = q_1 \cap \cdots \cap q_m$ is an irredundant decomposition of a if and only if $a = (u_a \cap q_1) \cap \cdots \cap (u_a \cap q_m)$ is an irredundant decomposition of a in the quotient lattice u_a/a. Thus the study of the irredundant decompositions of a is reduced to the study of the irredundant decompositions $a = s_1 \cap \cdots \cap s_m$ in u_a/a where $s_i = q_i \cap u_a$ for some irreducible q_i of L. Now the maximal elements of u_a/a always have this form. For if $u_a \succ s$, let q be a maximal element in L such that $q \geq s$, $q \not\geq u_a$. Then q is irreducible and $q \cap u_a = s$ since $u_a \succ s$. In most cases of interest, the maximal elements of u_a/a are the only elements of u_a/a having this form and hence the study of the irredundant decomposition of a is

reduced to the study of the irredundant representations of a as a meet of maximal elements of u_a/a.

Now if the elements of a lattice L have unique irredundant decompositions into irreducibles, the lattice must be semimodular. For if $a \succ a \cap b$ while $a \cup b > c > b$ there exists an irreducible q_1 such that $q_1 \geq c$, $q_1 \not\geq a \cup b$ and an irreducible q_2 such that $q_2 \geq b$, $q_2 \not\geq c$. Hence $a \cap b = a \cap q_1 = a \cap q_2$. If we take any finite decomposition of a (which exists by Theorem 3.1) this leads to two decompositions of $a \cap b$ which may then be refined to irredundant decompositions. Since $q_1 \geq c$ and $q_2 \not\geq c$, it follows that these two irredundant decompositions are distinct contrary to assumption.

Thus in characterizing finite dimensional lattices with unique irredundant representations we may restrict our attention to semimodular lattices. But in a semimodular lattice every maximal independent set of points of u_a/a has the same number of elements, such a set of points generate a Boolean algebra, and the maximal elements of the Boolean algebra are elements covered by u_a in u_a/a. Hence there exist irredundant decompositions having the same number of components as the number of elements in a maximal independent set of points of u_a/a. Now if q is an irreducible of L such that $q \not\geq u_a$, let $p_1, \cdots, p_k$ be a maximal independent set of points of $q \cap u_a/a$. Extend this set to a maximal set $p_i, \cdots, p_k, \cdots, p_n$ of points of u_a/a. If $s_i = p_i \cup \cdots \cup p_{i-1} \cup p_{i+1} \cup \cdots \cup p_n$, then $u_a \succ s_i$ and hence there exists q_i such that $q_i \cap u_a = s_i$. Then it is easily verified that $a = q \cap q_1 \cap \cdots \cap q_n$ is an irredundant decomposition of a. Hence if the number of components is unique it follows that $k + 1 = n$ and hence $u_a \succ q \cap u_a$. Thus the number of components in the irredundant decompositions of a will be unique only if the number of components in the irredundant representations of a as meets of elements covering u_a is unique. But it can be shown that the number of components in such decompositions is unique if and only if u_a/a is modular. Furthermore, if u_a/a is modular it follows from the classical replacement theorem (Theorem 3.3) applied to u_a/a that the replacement property holds for the irredundant decompositions of a. Finally if the irredundant decomposition of a is unique, then u_a/a consists precisely of the elements generated by a maximal independent set of points of u_a/a and hence is a Boolean algebra. Thus we get the following theorems.

Theorem 4.1. *Let L be a finite dimensional lattice. Then the elements of L have unique irredundant decompositions into irreducibles if and only if L is semimodular and u_a/a is distributive for each a.*

Theorem 4.2. *Let L be a finite dimensional semimodular lattice. Then the number of components in the irredundant decompositions of the elements of L is unique if and only if u_a/a is modular for each a in L.*

In this case the replacement property holds for the irredundant decompositions of an element of L.

It should be observed that the semimodularity of the lattice L is equivalent to the semimodularity of u_a/a for each $a \in L$. For if u_a/a is semimodular for each $a \in L$, then it clearly follows that $a,b \succ a \cap b$ implies $a \cup b \succ a,b$. Thus L is weakly semimodular. But it is well known that for finite dimensional lattices weak semimodularity implies semimodularity. Since distributivity implies semimodularity it follows that L is semimodular if u_a/a is distributive for each $a \in L$.

We shall say that a lattice L has a property P *locally* if u_a/a has the property P for each $a \in L$. Hence L is *locally distributive* if u_a/a is distributive for each $a \in L$ and *locally modular* if u_a/a is modular for each $a \in L$. Then the results given in Theorems 4.1 and 4.2 may be stated as follows:

A finite dimensional lattice has unique irredundant decomposition if and only if it is locally distributive.

A finite dimensional semimodular lattice has replaceable irredundant decompositions if and only if it is locally modular.

Finally we note that for semimodular lattices, local distributivity is equivalent to the property that every modular sublattice is distributive.

5. **Lattices satisfying the ascending chain condition.** In relaxing the requirement of finite dimensionality it is most natural to drop the descending chain condition, since by Theorem 3.1 the ascending chain condition alone is sufficient to insure the existence of finite irredundant decompositions into irreducibles. On the other hand, the techniques of §3 for studying irredundant decompositions can no longer be applied since in general covering elements will not exist. However, there is a lattice closely associated with the lattice L in which covering elements always exist; namely, the lattice of dual ideals. A subset A of L is a *dual ideal* if

(1) $a \in A$ and $x \geq a$ imply $x \in A$;

(2) $a,b \in A$ imply $a \cap b \in A$.

The set of dual ideals of L form a complete lattice in which the join of any set of dual ideals is their set intersection while the meet of any set of dual ideals is the dual ideal generated by their set union. A dual ideal A is *principal* if there exist $a \in L$ such that $A = \{x \in L | x \geq a\}$ in which case we write $A = (a)$. The principal dual ideals form a sublattice of the lattice of all dual ideals which is isomorphic to the lattice L. The covering theorem asserts that *if $A > (a)$, there exists a dual ideal P such that $A \geq P \succ (a)$.* Finally we note that when the descending chain condition holds, every dual ideal is principal and hence $P = (p)$ where $p \succ a$ in L.

It is now clear that in case the descending chain condition does not hold, then in place of the quotient lattice u_a/a of L it is appropriate to consider the quotient lattice $U_a/(a)$ of the lattice of dual ideals of L where $U_a = \bigcup \{P | P \succ (a)\}$. In the finite dimensional case u_a/a is always finite dimensional and this property is fundamental in developing the decomposition theory. For lattices satisfying the ascending chain condition, $U_a/(a)$ need not be finite dimensional and this fact is responsible for some of the essential difficulties associated with this more general decomposition theory. However we shall see that $U_a/(a)$ will indeed be finite dimensional in those cases which are relevant to the treatment of uniqueness and replaceability criteria.

The next step in developing the decomposition theory for lattices satisfying only the ascending chain condition is that of formulating an appropriate definition of semimodularity. Perhaps the most natural definition would be to require that the lattice of dual ideals of L be semimodular. However, this requirement is too severe since there exist lattices satisfying the ascending chain condition and having unique irredundant decompositions but for which the lattice of dual ideals is not semimodular. Now it is easy to see that semimodularity for finite dimensional lattices can be put in the form:

$$p \succ a, \; b \geq a, \; b \nleqq p \text{ imply } p \cup b \succ b.$$

We shall call an element a of a lattice semimodular if it satisfies the above implication for all p and b. A finite dimensional lattice is thus semimodular if and only if each of its elements is semimodular.

A lattice satisfying the ascending chain condition is then defined to be (upper) semimodular if for each $a \in L$, (a) is semimodular in the lattice of dual ideals of L.

The principal lemmas related to the finite dimensionality of $U_a/(a)$ are the following.

LEMMA 5.1. *Let L satisfy the ascending chain condition. If each element of L has a unique irredundant decomposition into irreducibles, then L is semimodular and $U_a/(a)$ is finite dimensional.*

LEMMA 5.2. *If $U_a/(a)$ is a Boolean algebra, then it is finite dimensional.*

LEMMA 5.3. *Let L be a semimodular lattice satisfying the ascending chain condition. Then if the number of components in the irredundant decompositions of a is unique, $U_a/(a)$ is finite dimensional.*

LEMMA 5.4. *Let L be a semimodular lattice satisfying the ascending chain condition. Then if $U_a/(a)$ is modular for each a, $U_a/(a)$ is finite dimensional for each $a \in L$.*

By means of these lemmas, the techniques described in §4 can be applied to arbitrary lattices satisfying the ascending chain condition to give characterizations of lattices with unique decompositions and replaceable decompositions. Analogous to the finite dimensional case let us define L to be locally modular (distributive) if for each $a \in L$ $U_a/(a)$ is modular (distributive). The statement of the fundamental theorems then differ only slightly from those described in §4.

A lattice satisfying the ascending chain condition has unique irredundant decompositions if and only if it is semimodular and locally distributive.

A semimodular lattice satisfying the ascending chain condition has replaceable irredundant decompositions if and only if it is locally modular.

6. **Compactly generated atomic lattices.** In the preceding sections it has
been assumed that the lattices under consideration satisfy the ascending
chain condition. This means that the decomposition theorems can only be
applied to give structure theorems in case the lattice of congruence relations
satisfies the ascending chain condition. This is a very restrictive condition
for most algebraic systems and hence it is desirable to have decomposition
theorems which do not depend upon assumption of a chain condition. On
the other hand, it is easy to give examples of lattices having no irreducible
elements. For such lattices a decomposition theory is essentially meaning-
less. Hence some type of restriction is necessary to insure the existence of
irreducibles and decompositions into irreducibles. Lattices of congruence
relations provide the key to such a restriction in that they always have the
property that they are compactly generated. Namely, if $\theta \in \Theta(L)$, then
$\theta = \bigcup\{\theta(a,b) \mid a\ \theta\ b\}$ where $\theta(a,b)$ denotes the congruence relation generated
by identifying a and b. Now suppose that $\bigcup\Phi \geq \theta(a,b)$ for some set Φ of
congruence relations ϕ. Then since $a\ \theta(a,b)\ b$ we have $a\ \bigcup\Phi\ b$ and hence
there exist elements $a = x_0, x_1, \cdots, x_n = b$ such that $x_{i-1}\ \phi_i\ x_i$ where
$\phi_i \in \Phi$. But then $a(\phi_1 \cup \cdots \cup \phi_n)b$ and hence $\phi_1 \cup \cdots \cup \phi_n \geq \theta(a,b)$.
Thus $\theta(a,b)$ has the property that $\bigcup\Phi \geq \theta(a,b)$ implies $\bigcup\Phi' \geq \theta(a,b)$ where
Φ' is a finite subset of Φ. An element of a lattice having this property is said
to be *compact*. If every element of the lattice is a join of compact elements,
the lattice is said to be *compactly generated*. The above argument shows that
$\Theta(L)$ is always compactly generated and a similar line of reasoning shows that
the lattice of congruence relations on an arbitrary algebraic system is compactly generated.

The property of being compactly generated can be viewed as a generalization of the ascending chain condition. For if the ascending chain condition holds, then the join of an arbitrary set of elements is equal to the join
of some finite subset and hence every element of the lattice is compact. Thus
the lattice is trivially compactly generated.

At the present time a satisfactory decomposition theory for arbitrary
compactly generated lattices does not exist, although in the following section
we shall give some indications of the nature of such a general theory. The
principal difficulty lies in the fact that irredundant decompositions into irreducibles need not exist. For lattices satisfying the ascending chain condition
there exist finite decompositions into irreducibles and hence by refinement
there exist irredundant decompositions. Now an element of a compactly
generated lattice can always be represented as a meet of irreducibles. For
if $a > b$, there exists a compact element c such that $a \geq c$ and $b \not\geq c$. Since
c is compact there exists a maximal element q such that $q \geq b$ and $q \not\geq c$.
Then q is completely meet irreducible since if $q = \bigcap X$ where $x > q$ for all $x \in X$
we must have $x \geq c$ for all $x \in X$ and hence $q = \bigcap X \geq c$ contrary to $q \not\geq c$.
Thus $q \geq b$ and $q \not\geq c$. It follows that each element is the meet of the
completely meet irreducibles containing it. This representation is in general

infinite so that refinement cannot be expected to lead to an irredundant representation. In fact it is easy to give examples of compactly generated lattices in which irredundant decompositions do not exist. The situation is quite different if the compactly generated lattice is atomic. A rather subtle argument due to P. Crawley shows *that every element of a compactly generated atomic lattice has an irredundant decomposition into irreducibles.*

Now atomicity is a consequence of the descending chain condition and hence we may look upon compactly generated atomic lattices as generalizations of finite dimensional lattices. A number of familiar properties of finite dimensional lattices carry over to compactly generated atomic lattices. Thus in such lattices every meet irreducible element is completely meet irreducible; a compactly generated atomic lattice which is upper and lower semimodular is modular; a weakly semimodular compactly generated atomic lattice is semimodular.

Crawley's theorem shows that the classical existence theorem for irredundant decompositions in lattices satisfying the ascending chain condition can be carried over to compactly generated atomic lattices. The uniqueness theorem in the distributive case and the replacement theorem in the modular case also carry over, although quite different proofs are required. In spite of this there are some important properties of decompositions in finite dimensional lattices whose analogues do not hold in general for compactly generated atomic lattices. For example, it can be shown that if $a = q_1 \cap \cdots \cap q_n = q_1' \cap \cdots \cap q_m'$ are two irredundant decompositions of an element a in a finite dimensional lattice, then each q_i' can replace a suitable q_j to give an irredundant decomposition. This replacement theorem is, in a sense, the reverse of the replacement theorem given in §3. It is not hard to give examples of compactly generated atomic lattices in which the analogous replacement theorem for infinite decompositions fails to hold.

As in the finite dimensional case u_a will denote the join of all elements covering a. Also decompositions of a into irreducibles correspond to decompositions of a into relative irreducibles in u_a/a. If L is semimodular, u_a/a is complemented, it is a dual point lattice, and an independent set of points generates a complete atomic Boolean algebra. As before L is said to be locally modular or locally distributive if u_a/a is modular or distributive respectively for every $a \in L$. With these notions, the general methods employed in the finite dimensional case can be applied to this more general situation. The lack of a chain condition considerably complicates the derivation of global results from local conditions. However the final theorems are precisely the same as in the finite dimensional case. *Unique decompositions are characterized by local distributivity and replaceable decompositions in semimodular lattices are characterized by local modularity.*

7. **Compactly generated lattices.** Since the decomposition theory for both compactly generated atomic lattices and lattices satisfying the ascending

chain condition is in quite satisfactory form, it might be supposed that the methods used in the two cases could be combined to give a decomposition theory for arbitrary compactly generated lattices. The actual situation is quite the contrary. At the present time, there is no satisfactory existence theorem on irredundant decompositions for general compactly generated lattices. Moreover, it is even not clear what form such an existence theorem should take. In order to indicate the type of difficulties which arise we shall discuss briefly a class of compactly generated lattices which, though severely restricted, includes all lattices satisfying the ascending chain condition and all compactly generated atomic lattices.

If a is an element of a lattice we shall call a dual ideal C *compact over a* if $C \geq (a)$ and $\bigcup S \in C$ where S is a subset of (a) implies $\bigcup S' \in C$ for some finite subset S' of S. If c is a compact element of L it is easy to verify that the principal dual ideal $(a \cup c)$ is compact over a. Finally, L is said to be *semiatomic* if $b > a$ implies that there exists a dual ideal P which contains b, covers a, and is compact over a.

If L is a lattice satisfying the ascending chain condition, then the join of any subset of L is equal to the join of a finite subset and hence every dual ideal B such that $B \geq (a)$ is compact over a. By the covering theorem for dual ideals, it follows that L is semiatomic. Likewise if L is a compactly generated atomic lattice and $b > a$, then there exists p such that $b \geq p \succ a$. Let c be a compact element such that $p \geq c$, $a \not\geq c$. Then $p = a \cup c$ and hence $(p) = (a \cup c)$ is compact over a. Thus L is again semiatomic. A satisfactory decomposition theory for compactly generated semiatomic lattices should therefore include the results described in the preceding sections.

The first problem to be examined is the existence of irredundant decompositions. But this assumes that the type of decompositions to be studied has been specified. The example of compactly generated atomic lattices shows that we must consider infinite decompositions. A modification of the argument of Crawley referred to above shows that each element of a compactly generated semiatomic lattice has an irredundant representation as a meet of irreducibles. Unfortunately, in semiatomic lattices irreducibles need not be completely meet irreducible. Hence it can occur that the components in these representations are not in simplest form but can be represented as a meet of proper divisors. This raises the question as to whether it is appropriate to weaken the requirements for the components or to try to modify the definition of decomposition. The latter procedure can be justified only if it can be proved that each element has an irredundant decomposition into elements which do not have a decomposition into proper divisors. We shall outline a method for accomplishing this objective in the case of semiatomic lattices.

If a is an element of a compactly generated, semiatomic lattice L, we shall say that *a is represented by a subset S of L* (in symbols $a = R(S)$) *if $I(S) \geq (a)$ and $I(S) \geq C \geq (a)$ where C is compact over a implies $C = (a)$.* $I(S)$ denotes the dual ideal generated by S. Note that $a = R(S)$ implies $a = \bigcap S$, since if $\bigcap S > a$, then $(\bigcap S) \geq P \succ (a)$ where P is compact and hence $I(S) \geq P$ contrary to assumption. On the other hand, the example of an infinite descending chain with null element adjoined shows that $a = \bigcap S$ need not imply that $a = R(S)$. An element q is *r-irreducible* if $q = R(S)$ implies $q \in S$. A representation $a = R(S)$ is irredundant if $a \neq R(S - s)$ for all $s \in S$. It can then be proved that *each element of a compactly generated, semiatomic lattice has an irredundant decomposition into r-irreducible elements.* Furthermore with appropriate definitions of semimodularity, local modularity, and local distributivity, the characterization theorems of §6 hold in compactly generated, semiatomic, lattices.

8. **Some applications and problems.** By means of the correspondence between lattice decomposition theorems and structure theorems outlined in §2, the reader can easily translate the decomposition theorems of the preceding sections into structure theorems. For example, from the results of §6 we have

THEOREM 8.1. *An algebraic system with minimal congruence relations can be represented as an irredundant subdirect union of subdirectly irreducible systems.*

From the characterization theorems we derive

THEOREM 8.2. *The lattice of congruence relations of an arbitrary lattice has a unique irredundant representation as a subdirect union of subdirectly irreducible lattices.*

THEOREM 8.3. *The lattice of ideals of a modular lattice has a unique irredundant representation as a subdirect union of subdirectly irreducible lattices.*

Since a lattice is always a sublattice of its lattice of ideals, Theorem 8.3 also gives a subdirect representation for the lattice itself. However the components will not necessarily be subdirectly irreducible. Incidentally, this theorem is closely related to the imbedding theorem of Frink [8] for complemented modular lattices.

A classical theorem of Birkhoff [2] asserts that any algebraic system is a subdirect union of subdirectly irreducible systems. Such representations are, in general, *not* irredundant. From the discussion of §7 it is clear that irredundant subdirect representations of the usual type cannot be expected in arbitrary algebraic systems. On the other hand, the theorems for semi-atomic lattices suggest that there may be such theorems for a generalized notion of irredundancy. At any rate, this emphasizes the importance of the following problem : *Formulate and prove an existence theorem on "irredundant" decompositions for an arbitrary compactly generated lattice.*

For each of the classes of lattices treated in this paper it was possible to characterize completely those lattices having unique irredundant decompositions. This characterization is possible since semimodularity is always a necessary condition for unique decompositions. On the other hand, semimodularity is not a necessary condition for the uniqueness of the number of components or for replaceability. Hence the characterization theorems hold only for the subclass of semimodular lattices. The results of §6 suggest that the characterization of replaceability is the more fundamental question. Since nearly all of the techniques used so far depend heavily upon semimodularity, it is clear that a quite new idea is needed for the general characterization. Since very little has been proved even in the finite dimensional case, the starting point should be the following problem : *Characterize finite dimensional lattices in which irredundant decompositions into irreducibles have the replaceability property.*

Finally we note that the decompositions considered so far in this discussion are those which are appropriate for the study of subdirect representations of algebraic systems. For purposes of application to direct union representations we should study direct decompositions of the elements of a lattice, preferably in conjunction with an abstract permutability relation.

It is clear that the argument which gives the existence of ordinary decompositions under the assumption of the ascending chain condition carries over immediately to the case of direct decomposition. Thus each element of a lattice satisfying the ascending chain condition has a direct decomposition into a finite number of indecomposable elements. Since direct decompositions are always irredundant whenever the components are non-trivial, we do not have to worry about irredundancy in the study of direct decompositions. It is also clear that the classical uniqueness proof for ordinary decompositions in distributive lattices carries over to give the uniqueness of direct decompositions into indecomposables for distributive lattices. The first lattice theoretic theorem on direct decomposition for more general lattices was given by Ore [**12**]. Ore's theorem may be stated as follows :

Let L be a finite dimensional, modular lattice. If $a = q_1 \cap \cdots \cap q_n = q_1' \cap \cdots \cap q_m'$ are two direct decompositions of an element $a \in L$ into indecomposable elements, then $m = n$ and each component in one decomposition may be replaced by a suitable component in the other to give a new direct decomposition.

This theorem does not hold under the assumption of the ascending chain condition alone since counterexamples exist in the theory of modules. However, Ore has also shown that the theorem does hold in a modular lattice satisfying the ascending chain condition provided the lattice is regular. A lattice is regular if $a \cap b = c \cap d = c \cap b = a \cap d$ and $a \cup b = c \cup d$ imply $a \cup b = c \cup d = c \cup b = a \cup d$.

It is clear from Ore's theorem that the problems concerning uniqueness and replaceability for direct decompositions are quite different in character and probably much deeper than the corresponding problems for decompositions into irreducibles. Up to the present time, very little progress has been made in connection with these problems for direct decompositions. Nevertheless in some respects the situation is simpler for direct decompositions. Thus, though a general existence theorem for decompositions into irreducibles has not yet been proved for arbitrary compactly generated lattices, the writer has proved that every element of a compactly generated lattice has a direct decomposition into indecomposable elements. Furthermore this result remains valid under the assumption of an abstract permutability relation. These partial results suggest that considerable attention should be devoted to the following problem.

Develop a uniqueness and replaceability theory for direct decompositions in compactly generated lattices.

References

1. G. Birkhoff, *On the combination of subalgebra*, Proc. Cambridge Philos. Soc. vol. 29 (1933) pp. 441–464.

2. ——, *Lattice theory*, Amer. Math. Soc. Colloquium Publications, vol. 25, 1948.

3. R. P. Dilworth, *Lattices with unique irreducible decompositions*, Ann. of Math. vol. 41 (1940) pp. 771–777.

4. ——, *The arithmetical theory of Birkhoff lattices*, Duke Math. J. vol. 8 (1941) pp. 286–299.

5. ——, *Ideals in Birkhoff lattices*, Trans. Amer. Math. Soc. vol. 49 (1941) pp. 325–353.

6. ——, *Note on the Kurosch-Ore theorem*, Bull. Amer. Math. Soc. vol. 52 (1946) pp. 659–663.

7. R. P. Dilworth and Peter Crawley, *Decomposition theory for lattices without chain condition*, Trans. Amer. Math. Soc. vol. 96 (1960) pp. 1–22.

8. O. Frink, *Complemented modular lattices and projective spaces of infinite dimensions*, Trans. Amer. Math. Soc. vol. 50 (1946) pp. 452–467.

9. N. Funayama and T. Nakayama, *On the distributivity of a lattice of lattice-congruences*, Proc. Imp. Acad. Tokyo vol. 18 (1942) pp. 553–554.

10. J. Hashimoto, *Direct, subdirect decompositions and congruence relations*, Osaka Math. J. vol. 9 (1957) pp. 87–112.

11. A. Kurosch, *Durchschnittsdarstellungen mit irreduziblen Komponenten in Ringen und sogenanten Dualgruppen*, Mat. Sb. vol. 42 (1935) pp. 613–616.

12. O. Ore, *On the foundations of abstract algebra*, Ann. of Math. vol. 36 (1935) pp. 406–437; vol. 37 (1936) pp. 265–292.

13. ——, *Direct decompositions*, Duke Math. J. vol. 2 (1936) pp. 581–596.

California Institute of Technology,
Pasadena, California

Dilworth's Work on Decompositions
in Semimodular Lattices

BJARNI JÓNSSON

In a series of papers written over a period of twenty years, the first one appearing in 1940, Dilworth investigated representations of an element a in a complete lattice as a meet $a = \bigwedge Q$ of a set Q of completely meet irreducible elements. For brevity we refer to such a representation, or to the set Q itself, as a *decomposition* of a. The decomposition is said to be *irredundant* if no proper subset of Q is a decomposition of a. The survey paper [9] by Dilworth provides an excellent summary of the principal results obtained.

These investigations focus on three properties that may or may not be satisfied in a given complete lattice L.

- *The existence property:* Every element of L has an irredundant decomposition.
- *The uniqueness property:* Every element of L has at most one irredundant decomposition.
- *The unique cardinality property:* Any two irredundant decomposition of the same element of L have the same cardinality.

A fourth property, related to the last one, is also considered:

- *The replacement property:* If Q and Q' are irredundant decompositions of the same element a of L, then there exists for each member q of Q a member q' of Q' such that $(Q - \{q\}) \cup \{q'\}$ is also an irredundant decomposition of a.

Research supported by NSF Grant DMS-8800290

Mostly, the lattices are assumed to be semimodular. Birkhoff calls a finite dimensional lattice L semimodular if it satisfies the condition

$$a \prec p, \quad a \leq b, \quad p \not\leq b \quad \text{imply} \quad b \prec b \vee p$$

(using $x \prec y$ for y covers x). For infinite dimensional lattices this way of describing the property is of limited interest, since some of the elements need not have any covers. For such lattices, Mac Lane's pointfree version (Axiom E_5 in [12]) is more useful:

(E_5) $a \wedge c \leq b \leq a$, $c \not\leq a$ imply that c_1 exists such that $a \wedge c < c_1 \leq c$
       and $b = a \wedge (b \vee c_1)$.

However, Dilworth needs a somewhat stronger property. He embeds L in its lattice $\mathcal{L}$ of dual ideals, and defines L to be semimodular if Birkhoff's condition holds in $\mathcal{L}$ whenever a is a principal dual ideal.

The uniqueness property and the unique cardinality property are closely related to distributivity and to modularity, respectively. In Birkhoff [1] it is shown that a modular lattice satisfying the ascending chain condition has the uniqueness property if and only if it is distributive, and the Kurosch-Ore Theorem asserts that in a modular lattice the replacement property, and hence also the unique cardinality property, hold for finite irredundant decompositions. The central idea of the series of papers is to relax the conditions of distributivity and modularity by only requiring these conditions to hold in the intervals u_a/a, where u_a is the join of all the covers of a. Dilworth calls a lattice L *locally distributive* if all the intervals u_a/a with $a \in L$ are distributive, and he says that L is *locally modular* if all these intervals are modular. As with the notion of semimodularity, these concepts have to be modified in order for them to be useful in infinite dimensional lattices. Dilworth therefore considers again the embedding of L into the lattice $\mathcal{L}$ of all its dual ideals. Taking U_a to be the join of all the covers of the principal dual ideal (a), he says that L is locally distributive, or locally modular, if all the intervals $U_a/(a)$ are distributive, or modular, respectively.

The papers consist mostly of detailed investigations of the intervals u_a/a and their relationship to the rest of the lattice, especially to the completely meet irreducible elements that are above a but not above u_a. These investigations are of a quite technical nature, and they provide deep insight into the structure of the lattices, yet the principal results are aesthetically satisfying in their simple formulation and sweeping generality. It is worthwhile to summarize these results in chronological order.

Dilworth [5]. The lattice L is assumed to be finite dimensional and semimodular. The existence property therefore holds, and every irredundant decomposition is finite. Morgan Ward had pointed out to Dilworth that without semimodularity the uniqueness property always fails. The principal result states that the following conditions are equivalent:

(i) L has the uniqueness property.
(ii) L is locally distributive.
(iii) Every modular sublattice of L is distributive.

Dilworth [6]. As before, L is assumed to be finite dimensional and semimodular. The following conditions are shown to be equivalent:

(i) L has the unique cardinality property.
(ii) L is locally modular.
(iii) L has the replacement property.

Dilworth [7]. Here L is assumed to satisfy the ascending chain condition, which insures that the existence property holds, and that every irredundant decomposition is finite. Since atomicity is not guaranteed, the notion of semimodularity must now be understood in the modified sense, and the same is true of local distributivity and local modularity. A stronger version of local distributivity is also considered:

(A) $a \wedge b \leq x \leq a \vee b$, $a \wedge x = b \wedge x = a \wedge b$ imply $x = a \wedge b$.

The principal results are:

The following conditions are equivalent:
(i) L has the uniqueness property.
(ii) L is semimodular and locally distributive.
(iii) L satisfies E_5 and A.

If L is semimodular, then the following conditions are equivalent:
(i) L has the unique cardinality property.
(ii) L is locally modular.
(iii) L has the replacement property.

Dilworth [8]. This note gives a sharper form of the Kurosch-Ore Theorem. Given two finite irredundant decompositions Q and Q' of an element a in a modular lattice, Dilworth shows that there exists for each $q \in Q$ some $q' \in Q'$ such that q and q' can be exchanged: both $(Q - \{q\} \cup \{q'\}$ and $(Q' - \{q'\}) \cup \{q\}$ are irredundant decompositions of a. He also shows that there exists a bijection $q \rightarrow q'$ from Q onto Q' such that each q can be replaced by the corresponding q'. However, there does not in general exist a bijection such that q and q' can be exchanged.

Dilworth and Crawley [10]. Now L is assumed to be compactly generated. The famous Subdirect Product Theorem, in Birkhoff [2], is equivalent to the assertion that if L is the congruence lattice of an algebra, then every member of L is the meet of completely meet irreducible members. In Birkhoff and Frink [3] it was shown that this holds in every complete, upper continuous lattice in which every element is the join of inaccessible elements. These are precisely what are now known as compactly generated lattices, or as algebraic lattices. Birkhoff and Frink showed that every compactly generated lattice L can be represented as the subalgebra lattice of some algebra. It was not until somewhat later that we learned from Grätzer, Schmidt [11] that L can also be represented as the congruence lattice of some algebra. However, enough evidence was in to show that compactly generated lattices are worth investigating and, in particular, that the decomposition theory for these lattices is important.

Since the ascending chain condition is no longer assumed, even the existence property need not hold without further hypothesis. The principal results are as follows. Remember that L is assumed to be compactly generated.

If L is semimodular and strongly atomic, then:
  (i) L has the existence property.
  (ii) L has the uniqueness property if and only if L is locally distributive.
  (iii) L has the replacement property if and only if L is locally modular.

Strong atomicity is a very strong hypothesis, but it is unlikely that it can be much weakened. In fact, for a compactly generated modular lattice, the uniqueness property is shown to imply strong atomicity. In (ii) the assumption of semimodularity could actually be dispensed with, because it is in fact a consequence of the other assumptions. As regards (iii), we note that the replacement property does not imply the unique cardinality property, since the decompositions need not be finite.

Throughout these papers, the lattice L is assumed to be semimodular, and in view of Morgan Ward's observation, this is reasonable as far as the uniqueness property is concerned. However, in Crawley [4] the results (i) and (iii) are extended to nonsemimodular lattices. Crawley's principal results are as follows. Suppose L is a compactly generated strongly atomic lattice. Then

  (i′) L has the existence property.
  (iii′) L has the replacement property iff, for all $a, b \in L$, the condition that $(a \vee b)/a$ has a unique atom implies that $b/(a \wedge b)$ has a unique atom.

Of course, since both the uniqueness property and local distributivity imply semimodularity, one expects no analogue of (ii) in this more general setting.

REFERENCES

1. G. Birkhoff, *Rings of sets*, Duke Math. J. **3** (1937), 442–454.
2. G. Birkhoff, *Subdirect unions in universal algebra*, Bull. Amer. Math. Soc. **50** (1944), 764–768.
3. G. Birkhoff and O. Frink, *Representations of lattices by sets*, Trans. Amer. Math. Soc. **64** (1948), 299–316.
4. P. Crawley, *Decomposition theory for nonsemimodular lattices*, Trans. Amer. Math. Soc. **99** (1961), 246–254.
5. R. P. Dilworth, *Lattices with unique irreducible decompositions*, Ann. of Math. (2) **41** (1940), 771–777. Reprinted in Chapter 3 of this volume.
6. R. P. Dilworth, *The arithmetical theory of Birkhoff lattices*, Duke Math. J. **8** (1941), 286–299. Reprinted in Chapter 3 of this volume.
7. R. P. Dilworth, *Ideals in Birkhoff lattices*, Trans. Amer. Math. Soc. **49** (1941), 325–353. Reprinted in Chapter 3 of this volume.
8. R. P. Dilworth, *Note on the Kurosch-Ore theorem*, Bull. Amer. Math. Soc. **52** (1946), 659–663. Reprinted in Chapter 3 of this volume.

9. R. P. Dilworth, *Structure and decomposition theory of lattices*, in "Lattice Theory," (Proceedings of Symposia in Pure Mathematics, Vol. 2), R. P. Dilworth, ed., Amer. Math. Soc., Providence, Rhode Island, 1961, pp. 3–16. Reprinted in Chapter 3 of this volume.

10. R. P. Dilworth and P. Crawley, *Decomposition theory for lattices without chain conditions*, Trans. Amer. Math. Soc. **96** (1960), 1–22. Reprinted in Chapter 3 of this volume.

11. G. A. Grätzer and E. T. Schmidt, *Characterization of congruence lattices of abstract algebras*, Acta Sci. Math. (Szeged) **24** (1963), 34–59.

12. S. Mac Lane, *A lattice formulation for transcendence degress and p-bases*, Duke Math. J. 4 (1938), 455–468.

Vanderbilt University
Nashville, TN 37235
U. S. A.

The Consequences of Dilworth's Work on Lattices with Unique Irreducible Decompositions

BERNARD MONJARDET

In 1937, G. Birkhoff [6] proved that every element of a finite dimensional distributive lattice L has a "unique irredundant decomposition" as meet of meet-irreducible elements (or as a join of join-irreducible elements). What does this mean? Let us denote by $M(L)$ or simply M (resp. $J(L)$ or J) the set of all meet-irreducible (resp. join-irreducible) elements of a lattice L; then for each element x of a finite dimensional distributive lattice there exists a unique subset of M such that $x = \bigwedge S$ and $x < \bigwedge(S - \{m\})$ for every m in S (and dually for J). It is easy to find non-distributive lattices having this property of unicity of irredundant meet decomposition (the simplest example is given in Figure 1 of the Background for this chapter). Thus a natural question was to characterize such (finite dimensional) lattices; that is exactly what the Dilworth did in [17]. I don't know how [17] was received in 1940 (I was two years old ...), but it seems it was considered as an ending: a natural question was nicely solved, so there was nothing more to say. And indeed nothing more was said on these lattices for twenty years. But in mathematics fire can spring from coals: since 1960, at least fifty papers have been published that directly or indirectly concern or rediscover the lattices characterized in [17] (or their duals); moreover these lattices appear at the core of fast-developing combinatorial theories, like convex geometries (Edelman and Jamison [28]), greedoids (Korte and Lovász [36]), or exchange systems (Brylawski and Dieter [11]).

In this survey, I'll first give an idea of the proof of the main result of [17] (the genesis of this result is described by Dilworth himself in the background in this chapter) and of its applications; then I'll present several other characterizations of the lattices discussed as well as representation theories obtained by different authors in various fields and under various terminologies; finally I'll point out some other domains in which these lattices have appeared. Two remarks before we start:

first, since these lattices have been rediscovered several times, they have received many names; here I'll use the name *upper locally distributive* (ULD) lattice for reasons that will become clear soon. (This name was used by Avann [4], whereas Dilworth and Crawley [21] used the term locally distributive lattice; the adjunction of "upper" allows us to distinguish such lattices from their duals, called here *lower locally distributive* (LLD) lattices.) Second, I'll consider exclusively the finite case. Indeed [17] deals with the finite dimensional case, but the results are the same and require only minor changes in the proofs. Moreover other papers [18,19,20,21] of Dilworth, the last of which contains an illuminating survey of the results, extend the theory to significantly more general cases (along with the study of the unicity of the number of components in the irreducible decompositions). Finally the recent combinatorial developments are in the finite case.

Let us consider the following two conditions for a finite lattice:

(A) Each element of L has a unique irredundant meet-irreducible decomposition.

(B) L is an upper semimodular (USM) lattice in which every modular sublattice is distributive.

The main result (Theorem 1.1) of Dilworth's paper [17] proves their equivalence. Further it describes the decomposition more precisely: for each element a covering the element x there exists a unique meet-irreducible m such that $m \not\geq a$ and $m \geq b$ for all other elements b covering x. These meet-irreducibles form the iredundant meet-irreducible decomposition of x; thus their number is the number of elements covering x. In what follows we use $a \succ x$ for a covers x.

First Dilworth gives a clever direct proof that (A) implies (B); then in order to prove the converse implication, he introduces the following condition:

(C) For each x in L the set of elements covering x generates a convex distributive sublattice of L, or equivalently, for each x in L the interval $[x, x^+]$, where $x^+ = \bigvee\{a : a \succ x\}$ is a Boolean algebra.

This condition is the key to the proof; one can make a trivial observation: for each atom a of a Boolean algebra, there exists a unique meet irreducible not greater than or equal to a (the coatom complement of a) and 0 is the irredundant meet of these irreducibles. Now the proof goes essentially as follows:

(1) a lattice satisfying (B) satisfies (C); so for a given x in L the interval $[x, \bigvee\{a_i : a_i \succ x\}]$ is a Boolean algebra B_x (Lemma 1.3);

(2) for each atom a_i of B_x there exists a unique meet irreducible element m_i such that $m_i \not\geq a_i$ and m_i is greater than or equal to the coatom complement of a_i in B_x (Lemma 1.6);

(3) if $\bigwedge P = x$ is an irredundant meet-irreducible decosition of x, then P is the set of m_i's of (2) (lemma 1.7).

This proof prompts one question: does (C) imply (A)? The answer is positive since the proofs of (2) and (3) above use only a weakened form of (C) and upper semimodularity (which is trivially implied by (C)); so one obtains another characterization of lattices satisfying (A), implicit in the 1940 paper [17] but explicit in the 1961 paper [21] (and in the generalizations in the 1941 papers [18,19]). In particular, (C) justifies the name "upper locally distributive" given to the lattices where the irredundant meet decomposition is unique.

Applications of Theorem 1.1 of [17] are given in Section 2 of that paper. Theorem 1.1 allows us to characterize (finite) Boolean algebras as complemented locally distributive lattices and distributive lattices as ULD lattices satisfying the following condition: $x \geq y$ if and only if each "component" (meet irreducible of the unique meet decomposition) of x is greater of equal to some component of y. Section 2 also contains results on sublattices and decomposability of ULD lattices and USM lattices.

We are going now to present other characterizations of ULD lattices obtained by various authors. First we return to the unique meet decomposition of an element of a such lattice by the meet-irreducible m_i's (see (2) above). We can prove [13] that if $a_i \succ x$, m_i is the greatest element t such that $x \vee t < a_i \vee t$. Let us say the the interval $[t, z]$ of a lattice L is an upper transpose of the interval $[x, y]$, denoted $[x, y] \nearrow [t, z]$, if $z = y \vee t$ and $x = y \wedge t$. Let us restrict the transposition relation to the set of covering intervals $x \prec y$ of L; this set is partially ordered by the upper transposition relation $\nearrow$. We say that a set C of covering intervals of L is a *projective covering class* if $(C, \nearrow)$ is a connected component of this poset. The above results and definitions allow us to show that an ULD lattice L satisfies the following condition [13]:

(D) L is upper semimodular and every projective covering class has a greatest element $[m, \dot{m}]$ where m is meet irreducible (and $\dot{m}$ denotes the unique element covering m).

Conversely, it is easy to see that (D) characterizes ULD lattices [25]. Indeed a similar but more precise result was obtained by Avann [4], who showed that the following condition characterizes these lattices:

(E) Every projective covering class of L satisfies the following conditions:

    a)   it has a greatest element $[m, \dot{m}]$ $(m \in M)$,

    b)   its minimal elements are intervals $[j_\alpha, \dot{j}_\alpha]$, with $j_\alpha \in J$ the set of join irreducibles (and $\dot{j}_\alpha$ the element covered by j_α).

    c) L is the disjoint union of $\{x \in L : x \leq m\}$ and $\{y \in L : y \geq j_\alpha$ for at least one $j_\alpha\}$.

Notice that this result generalizes the canonical bijection between the sets M and J of a distributive lattice.

But there are in fact a lot of other characterizations of ULD lattices, and we list some of them below. We need the following classical notions: let 0 (1) be the least (greatest) element of a lattice L; the height of $x \in L$ is the length of a longest chain from 0 to x (recall that a one-element chain has length 0); L is ranked if $x \succ y$ implies $h(x) = h(y) + 1$ (or equivalently if it satisfies the Jordan-Dedekind chain condition: all the maximal chains between any two elements of L have the same length). Each of the following conditions characterizes ULD lattices:

THE DILWORTH THEOREMS

(F) L is ranked and $h(1) = |M|$.

(G) L is ranked and for each x in L, $h(x) = |\{m \in M : m \not\geq x\}|$.

(H) L is USM (upper semimodular) and $h(1) = |M|$.

(I) L is USM and for each x in L, $h(x) = |\{m \in M : m \not\geq x\}|$.

(J) L is USM and has no M_5 lattice as covering sublattice.

(K) L is USM and every modular sublattice of order 5 of L is distributive.

(L) L is USM and every lower semimodular sublattice of L is distributive.

(M) L is ranked and there exists a join morphism from L into a Boolean algebra preserving the rank.

(N) For all meet-irreducibles $m \neq m'$ and all x in L, $m \not\geq x$, $m' \not\geq x$, and $m > x \wedge m'$ implies $m' \not> x \wedge m$.

Conditions (F) to (L) are due (in dual form) to Avann in his thorough study of LLD lattices [1,2,3]. The dual form of condition (M) has been given by Greene and Markowsky [32]: in a ULD lattice the map $x \to \{m \in M : m \not\geq x\}$ satisfies the conditions of (M). Condition (N) is a dual variant of the "antiexchange property" introduced by Edelman [26] (see below). Notice that the Avann papers contain other characterizations, especially based on properties of projective class of intervals [2]. Finally let us note this other interesting characterization:

(O) For each x, y in L with $x \prec y$ there exists a unique meet irreducible m such that $m \geq x$, $m \not\geq y$.

(One has easily: (A) implies (O) and (O) implies (G)).

Notice also that since the distributive lattices are the upper and lower locally distributive lattices, these characterizations induce characterizations of distributive or Boolean lattices, some of them given in the papers of Dilworth and Avann. We remark finally that Stern [47] has studied other properties of LLD lattices.

We come now to representation theories of ULD lattices. Let us recall the two "canonical" representations of a lattice $L : x \to M^*(x) = \{m \in M : m \not\geq x\}$ and $x \to J(x) = \{j \in J : j \leq x\}$. We begin with the first one; the family $\mathcal{M}(L) = \mathcal{M} = \{M^*(x) : x \in L\}$ is a union-stable closed family of ideals of M, containing $\emptyset$ and M; it induces an anticlosure operator ω on 2^M (the set of all subsets of M). If L is ULD, ω satisfies the following condition:

(P) $A \subseteq M$, $m, l \in \omega(A)$, $m \notin \omega(A - l) \implies l \in \omega(A - m)$.

Conversely, the lattice of the open (anticlosed) sets of an anticlosure operator satisfying (P) on a set X is an ULD lattice (notice that X endowed with the order $x \leq y$ if and only if each open set containing y contains x is isomorphic with the ordered set of the meet irreducible elements of this lattice). Thus we obtain that a lattice is ULD iff it is isomorphic to the lattice of open sets of an anticlosure operator satisfying (P). This result was obtained by Edelman [26] in the dual form where he considered LLD lattices and closure operator φ on X satisfying the "antiexchange" condition:

(P') $A \subseteq X$, $x, y \notin \varphi(A)$, $x \in \varphi(A + y) \implies y \notin \varphi(A + x)$.

(Note that it is equivalent to assume A closed in this condition).

Anticlosure or closure operators on a set X can be equivalently defined by families of open or closed sets. One obtains, for instance, the following "open sets" characterization of anticlosure operator satisfying (P):

(Q) $\mathcal{F}$ is a family of subsets of X containing $\emptyset$ and X, $\mathcal{F}$ is $\cup$-stable, and for each $A \in \mathcal{F}$ there exists $x \in A$ such that $A - x \in \mathcal{F}$.

Dually, the family of closed sets of a closure operator satisfying the antiexchange property is characterized by:

(Q') $\mathcal{F}$ is a intersection-stable family of subsets of X containing $\emptyset$ and X, such that for each $A \in \mathcal{F}$ there exists $x \notin A$ with $A + x \in \mathcal{F}$.

Edelman and Jamison [28] have called such families "convex geometries"; they were previously studied by Jamison (under the name "antimatroids"; see, for instance, [34]). Jamison gave many other characterizations, based on "convexity" properties. We present just the most striking: let φ be a closure operator whose closed sets are called convex and let us say that x is an extreme point of the convex set C if $x \notin \varphi(C - x)$. Then φ is an antiexchange closure operator if and only if it satisfies:

(R') For each convex C, C is the φ-closure of the set of extreme points of C.

In other words convex geometries are the abstract convex spaces satisfying the finite Krein-Milman property; for other convexity properties of these spaces or of the corresponding LLD lattices see Edelman and Jamison [28].

Returning to the classical representation $\mathcal{M}(L)$ of an arbitrary lattice L in 2^M; we know that it is a $\vee$-morphism; moreover to each maximal chain of L corresponds a chain of ideals of M, these ideals being the principal ideals of the complete preorder defined on M by the chain; thus we obtain that $\mathcal{M}(L)$ is obtained as the set of unions of the principal ideals of complete preorders defined on M. Now if L is an ULD lattice, maximal chains have $|M|$ elements and define linear orders on M (one can construct these linear orders by using (O) and part a) of (E); then $\mathcal{M}(L)$ is obtained as the set of unions of (all) principal ideals of a family of linear orders defined on M. Conversely it is easy to see that the set of unions of all principal ideals of a family of linear orders defined on a set X is a ULD lattice. One obtains:

(S) A lattice L is ULD if and only if it is isomorphic to the lattice obtained by taking the unions of all the principal ideals of a family of linear orders.

This representation result has been for instance obtained by Crapo [13] and Edelman and Jamison [28] in the dual form; using the representation in 2^M, it is also clear that the minimum number of linear orders required in the characterization (S) is the width of the the set of join-irreducible elements of L (a fact noticed, in the dual form, by Edelman and Saks [30], and, in an apparently very different context by Matalon [40], see below).

Crapo [13] expresses the representation result (S) in the terms of the theory of "selectors"—a theory equivalent with a part of the theory of greedoids. The theory of greedoids was developed by Korte and Lovász in a series of papers (see, for instance, [36,37,38]) in order to provide a structural framework for the "greedy" algorithm. Both theories deal with (simple) "hereditary languages" defined as an (hereditary) set of "words", *i.e.*, finite sequences of elements of a ground set (the alphabet); to each word (or set of words) one can associate the underlying set of letters appearing in the word(s), called a "partial alphabet" by Crapo or a "feasible subset" by Korte and Lovász. Special classes of greedoids or selectors are defined by imposing axioms on the hereditary language, or equivalently on the set of partial alphabets. Then it turns out that the notions of "normal greedoid with the interval property without upper bounds", "alternative precedence structure greedoid", "shelling structure", "locally free selector", etc. are all equivalent; moreover the corresponding set of partial alphabets is an ULD lattice. (The fact that conversely each ULD lattice is isomorphic to the lattice of partial alphabets of a locally free selector is the representation result (S)).

In this way we obtain two theories, an ordered theory (for the words) and an unordered theory for the feasible subsets equivalent with the theory of ULD lattices. We also obtain additional characterizations of ULD lattices, for instance by means of properties of "circuits" or by exclusion of "minors" (notice that the complementary sets of the feasible subsets are the convex sets of a convex geometry). A nice survey on convex geometries and greedoids is in Duchet [23], whereas recent variations based on the notion of "exchange systems" are in Brylawski and Dieter [11].

We come now to the other representation of a lattice by ideals of its set J of join-irreducibles: $x \rightarrow \{j \in J : j \leq x\}$. Notice that in the case of an ULD lattice this representation does not preserve rank. In his 1968 paper [4], Avann considers a poset endowed with an equivalence relation θ, and defines a closure operator on the (distributive) lattice of ideals of P giving a lattice of closed ideals $\mathcal{P}_\theta(P)$; then he defines an equivalence relation θ on the set J of an ULD lattice by: $j\theta j'$ if and only if $\mu(\mathrm{j} < j) = \mu(\mathrm{j}' < j')$ (μ is the map implicitly defined in condition E, and j the unique element covered by j), or, if and only if j and j' are minimal elements of the filter $\{x \in L : x \not\leq m\}$; finally he claims that $\mathcal{P}_\theta(J)$ is isomorphic with L. (There is, however, an unresolved problem in the definition of his closure operator. Nonetheless the equivalence θ is very natural and could lead to other characterizations of ULD lattices based on the properties of the ordered set $J \cup M$.)

Finally the "core" of an ULD lattice has been characterized; this leads to another representation of these lattices [24,25].

The above characterizations and representations of locally distributive lattices demonstrate that these lattices occur in many fields. Indeed, as noticed in my note in *Order* [41], there are many instances where the set of all specified subobjects of a mathematical object is a (nondistributive) locally distributive lattice: subtrees of trees and connected subgraphs of certain undirected graphs (Boulaye [8-10]); "convex" subsets of vertices of certain undirected graphs (for different definitions of convexity (Jamison [35])); "convex" subgraphs of a directed graph (Pfaltz [43]);

convex subsets of a partial order (Birkhoff and Bennett [7]); partial orders contained in a linear or a partial order (Dean and Keller [16], Avann [5], Stanley [46], and Edelman and Klingsberg [29]); "convex" sets of an acyclic oriented matroid (Edelman [27]); "separating" sets of a graph (Pym and Perfect [44] and Polat [45]); subsemilattices of a semilattice (Jamison [34]) and more generally closure operators on a poset (Duquenne [24], see also Lapscher [39]); Sperner k-families (*i.e.*, subsets containing no chain of cardinality $k+1$) of a ordered set (Greene and Kleitman [31]). Notice that in many of these instances the LD lattices are atomic or coatomic.

Finally I'll mention another appearance of these lattices in data analysis. In 1955 (see Coombs [14]), Coombs and Kao [15] defined multidimensional generalizations of Guttman scale that they called the conjunctive and disjunctive models. These models intend to account for the answers of a subject to dichotomous items which the subject can pass or fail. In both models the set I of items is endowed with several "difficulty" linear orders; in the disjunctive (resp. conjunctive) case a subject passes an item if and only if his "competence" overcomes the item difficulty on at least one (resp. all) difficulty order(s) (in the Guttman case there is only one such order). Now a "complete family" of answer patterns is the set of all possible answer patterns induced by such models; this set corresponds exactly to the subsets of I generated by the union (intersection) of all principal ideals of the difficulty orders; thus it is an ULD (LLD) lattice.

Conversely given a family of answer patterns, the problem is to find a (minimal) model, that is, a (minimal) set of linear orders generating—disjunctively or conjunctively—this family; this is indeed the problem of the Ferrers (or biorder) dimension of a binary relation (see Cogis [12] and Ducamp, Doignon and Falmagne [22]). But in the case where the family is complete, this minimal number of linear orders is clearly—in the conjunctive case—the width of the set of $\cap$-irreducible patterns, a fact noticed by Matalon [40, p. 96], as mentioned earlier.

In conclusion, it seems clear (at least to me) that finite locally distributive lattices are important for many combinatorial problems and theories in much the same way that geometric lattices are for matroids and combinatorial geometries, and also that combinatorists should be more aware of this fact (in both cases many combinatorial properties derive from the corresponding lattice-theoretic structures). So, Dilworth's beautiful 1940 paper, far from having been an end, was the cornerstone of a theory with many facets and applications.

Acknowledgements. The author thanks K. P. Bogart for editorial assistance which has led to a significant improvement in style and V. Duquenne for helpful discussions on locally distributive lattices.

References

1. S. P. Avann, *Application of the join-irreducible excess function to semimodular lattices*, Math. Ann. **142** (1961), 345–354.
2. ————, *Distributive properties in semimodular lattices*, Math. Z. **76** (1961), 283–287.

                                                    THE DILWORTH THEOREMS

3. __________, *Increases in the join-excess function in a lattice*, Math. Ann. **154** (1964), 420–426.

4. __________, *Locally atomic upper locally distributive lattices*, Math. Ann. **175** (1968), 320–336.

5. __________, *The lattice of natural partial orders*, Aequationes Math. **8** (1972), 95–102.

6. G. Birkhoff, *Rings of sets*, Duke Math. J. **3** (1937), 442–454.

7. G. Birkhoff and M. K. Bennett, *The convexity lattice of a poset*, Order **2** (1985), 223–242.

8. G. Boulaye, *Sous-arbres et homomorphisimes à classes connexes dans un arbre,*, in "Theory of Graphs International Symposium," Gordon and Breach, New York, 1967, pp. 47–50.

9. __________, *Sur l'ensemble ordonné des parties connexes d'un graphe connexe*, R.I.R.O. (1968a), 13-25.

10. __________, *Notion d'extension dans les treillis et méthodes booléennes*, Rev. Roum. Math. Pures Appl. **13-9** (1968b), 1225–1231.

11. T. H. Brylawski and E. Dieter, *Exchange systems*, Discrete Math. **69** (1988), 125–151.

12. O. Cogis, *On the Ferrer's dimension of a digraph*, Discrete Math. **38** (1982), 47–52.

13. H. H. Crapo, *Selectors. A theory of formal languages, semimodular lattices, branching and shelling processes*, Adv. Math. **54** (1984), 233–277.

14. C. H. Coombs, "A Theory of Data," Wiley, New York, 1964.

15. C. H. Coombs and R. C. Kao, *Non-matrix factor analysis*, Engineering Research Bulletin, University of Michigan, Vol. 38.

16. R. A. Dean and G. Keller, *Natural partial orders*, Canad. J. Math. **20** (1968), 535-554.

17. R. P. Dilworth, *Lattices with unique irreducible decompositions*, Ann. of Math. (2) **41** (1940), 771–777. Reprinted in Chapter 3 of this volume.

18. __________, *Ideals in Birkhoff lattices*, Trans. Amer. Math. Soc. **49** (1941), 325–353. Reprinted in Chapter 3 of this volume.

19. __________, *The arithmetical theory of Birkhoff lattices*, Duke Math. J. **8** (1941), 286–299. Reprinted in Chapter 3 of this volume.

20. __________, *Structure and decomposition of lattices*, in "Lattice Theory," Proceedings of Symposia in Pure Mathematics, Vol. 2, R. P. Dilworth, ed., Amer. Math. Soc., Providence, Rhode Island, 1961, pp. 3–16. Reprinted in Chapter 3 of this volume.

21. R. P. Dilworth and P. Crawley, *Decomposition theory for lattices without chain conditions*, Trans. Amer. Math. Soc. **96** (1960), 1–22. Reprinted in Chapter 3 of this volume.

22. A. Ducamp, J. P. Doignon and J. C. Falmagne, *On realizable biorders and the biorder dimension of a relation*, J. Math. Psychology **28** (1984), 73–109.

23. P. Duchet, *Convexity in combinatorial structures*, Rendiconti Circ. Mat. Palermo (1986).

24. V. Duquenne, *Contextual implications between attributes and some representation properties for finite lattices,*, in "Beiträge zur Begriffsanalyse," B. Ganter, R. Wille, and Wolff eds., Wissenschaftsverlag, Mannheim, 1987, pp. 213-239.

25. __________, *On the core of finite lattices*, in Centre d'Analyse et de Mathématique Sociales report, p. P.037.

26. P. H. Edelman, *Meet-distributive lattices and the antiexchange closure*, Algebra Universalis **10** (1980), 290–299.

27. __________, *The lattice of convex sets of an oriented matroid*, J. Combin. Theory Ser. B **33** (1982), 239–244.

28. P. H. Edelman and R. E. Jamison, *The theory of convex geometries*, Geometriae Dedicata **19** (1985), 247–270.

29. P. H. Edelman and P. Klingsberg, *The subposet lattice and the order polynomial*, European J. Combin. **2** (1982), 341–346.

30. P. H. Edelman and M. E. Saks, *Combinatorial representations and convex dimension of convex geometries*, Order **5** (1988), 23–32.

31. C. Greene and D. J. Kleitman, *The structure of Sperner k-families*, J. Combin. Theory Ser. A **20** (1976), 41–68.

Exchange Properties for Reduced Decompositions in Modular Lattices

JOSEPH P. S. KUNG

In his paper [3], Dilworth studied "the manner in which the irreducibles of two decompositions can replace each other" in a modular lattice. Thus this paper anticipated papers on the closely related area of basis exchange properties in matroids or geometric lattices. This relationship is most transparent for modular lattices of finite rank. For such lattices, replacement properties are "local" in the sense that they can be decided by looking only at intervals of the form $[a, u_a]$, where u_a is the join all the elements covering a. This follows from the following variant (cf. [5]) of a result in group theory due to Burnside [2] and Frattini [4]:

LEMMA. *Let a be an element in a modular lattice L of finite rank. Then*

$$q_1 \wedge \cdots \wedge q_n$$

is a reduced decomposition of a into meet-irreducibles in L if and only if

$$(q_1 \wedge u_a) \wedge \cdots \wedge (q_n \wedge u_a)$$

is a reduced decomposition of a into meet-irreducibles in $[a, u_a]$.

In particular, because a modular lattice is a point lattice whenever the maximum element is a join of atoms, a replacement property holds in all modular lattices of finite rank if and only if it holds in all modular point lattices of finite rank.

Brualdi [1] (see also [7,9]) has rediscovered Theorem 1 in Dilworth's paper [3] for bases in a matroid. Theorem 2 for matroids is the symmetric basis exchange axiom; its extension to subsets was proved in [6,8,11]. A survey of basis exchange properties in matroids can be found in [10].

32. C. Greene and G. Markowsky, *A combinatorial test for local distributivity*, IBM Research Report RC 5129, T. J. Watson Research Center, Yorktown Heights, New York, 1974.

33. R. E. Jamison, *A development of axiomatic convexity*, Technical Report 48, Clemson University Mathematics Department, 1970, pp. 15–20.

34. __________, *Copoints in antimatroids,*, Congr. Num. **29** (1980), 535-544.

35. __________, *A perspective on abstract convexity: classifying alignments by varieties*, in "Convexity and Related Combinatorial Geometry," D. G. Kay and M. Breen, ed., Marcel Dekker, New York, 1982, pp. 113-150.

36. B. Korte and L. Lovász, *Structural properties of greedoids*, Combinatorica **3** (1983), 359–374.

37. __________, *Shelling structures, convexity and a happy end*, in "Graph Theory and Combinatorics: Proceedings of the Cambridge combinatorial conference in honour of Paul Erdős," B. Bollobás, ed., Academic Press, London, 1984, pp. 219–232.

38. __________, *Homomorphisms and Ramsey properties of antimatroids*, Discrete Appl. Math. **15** (1986), 283–290.

39. F. Lapscher, *Proprietés de longueurs de chaîne dans l'ensemble des fermetures supérieures sur un ensemble ordonné*, Discrete Math. **4** (1974), 129–140.

40. B. Matalon, "L'analyse Hiérarchique," Mouton, Paris, 1965.

41. B. Monjardet, *A use for frequently rediscovering a concept*, Order **1** (1985), 415–417.

42. J. L. Pfaltz, *Semi-homomorphisms of semimodular lattice*, Proc. Amer. Math. Soc. **22** (1969), 418-425.

43. __________, *Convexity in directed graphs*, J. Combin. Theory **10** (1971), 143–162.

44. J. S. Pym and H. Perfect, *Submodular functions and independence structures*, J. Math. Anal. Appl. **30** (1970), 1–31.

45. N. Polat, *Treillis de séparation des graphes*, Canad. J. Math. **28** (1976)), 725–752.

46. R. P. Stanley, *Combinatorial reciprocity theorems*, Adv. Math. **14** (1974), 194–253.

47. M. Stern, *On meet-distributive lattices*, Studia Sci. Math. Hung. (1988).

Université Paris V et Centre d'Analyse et de Mathématique Sociales
54 Boulevard Raspail
75 270 Paris Cedex 06
France

REFERENCES

1. R. A. Brualdi, *Comments on bases in dependence structures*, Bull. Australian Math. Soc. **1** (1969), 161–167.
2. W. Burnside, *Some properties of groups whose orders are powers of primes*, Proc. London Math. Soc. (2) **13** (1913), 6–12.
3. R. P. Dilworth, *Note on the Kurosch-Ore theorem*, Bull. Amer. Math. Soc. **52** (1946), 659–663. Reprinted in Chapter 3 of this volume.
4. G. Frattini, *Intorno alla generazione de gruppi di operazioni*, Rend. Atti Accad. Lincei (4) **1** (1885), 281–285; 455–457.
5. K. M. Gragg and J. P. S. Kung, *Consistent dually semimodular lattices*, preprint.
6. C. Greene, *A multiple exchange property for bases*, Proc. Amer. Math. Soc. **39** (1973), 45–50.
7. C. Greene, *Another exchange property for bases*, Proc. Amer. Math. Soc. **46** (1974), 155–156.
8. C. Greene and T. L. Magnanti, *Some abstract pivot algorithms*, SIAM J. Appl. Math. **29** (1975), 530–539.
9. J. P. S. Kung, *Alternating basis exchanges in matroids*, Proc. Amer. Math. Soc. **71** (1978), 355–358.
10. J. P. S. Kung, *Basis exchange properties*, in "Theory of Matroids," N. L. White, ed., Cambridge Univ. Press, Cambridge, 1986, pp. 62–75.
11. D. R. Woodall, *An exchange theorem for bases of matroids*, J. Combin. Theory Ser. B **16** (1974), 227–228.

University of North Texas
Denton, TX 76203
U. S. A.

The Impact of Dilworth's Work on Semimodular Lattices on the Kurosch-Ore Theorem

Manfred Stern

In his papers on decompositions in lattices, Dilworth clarifies, among other things, the scope of the celebrated Kurosch-Ore Theorem. In particular, he shows that an upper semimodular lattice of finite length has the Kurosch-Ore replacement property (KORP) for meet-decompositions if and only if it is locally modular. This result and some deep generalizations are described in B. Jónsson's article in this chapter.

Dilworth's result led people to think about characterizing those (semimodular) lattices (of finite length) possessing the KORP for join-decompositions. Several authors subsequently wrote on this problem.

As already mentioned in Jónsson's article, it follows from Crawley [1] that a lattice L of finite length has the KORP for join-decompositions if and only if the following condition holds:

$(*)$ for all $a, b \in L$, $b/(a \wedge b)$ has a unique dual atom implies that $(a \vee b)/a$ has a unique dual atom.

In fact, Crawley proves the dual statement for compactly generated atomic lattices. Condition $(*)$ essentially means that the lattice is consistent in the sense of Kung [6]. For a short direct proof of the statement that a lattice of finite length has the KORP for join-decompositions if and only if it is consistent, see Faigle [3] or Reuter [7].

For lattices which are also (upper) semimodular in the sense of Birkhoff, still more equivalent conditions have been given.

Faigle [2] introduces what he calls a strong lattice and shows that a semimodular lattice of finite length is strong if and only if it has the KORP for join-decompositions.

Reuter [7] takes another approach using the so-called arrow relation due to Wille [8] and introducing the concept of a balanced lattice. For the construction of finite balanced lattices, Reuter develops a method of gluing extending the approach of Herrmann [5]. Herrmann's approach, in turn, is a generalization of a construction given by Hall and Dilworth [4]. Reuter shows that for a finite semimodular lattice L the following conditions are equivalent:

(i) L has the KORP for join-decompositions;

(ii) L is glued by geometric lattices;

(iii) L is balanced.

It is natural to ask what happens if a finite semimodular lattice has the KORP both for meet-decompositions and for join-decompositions. By Dilworth's result such a lattice is locally modular and by Reuter's result it is balanced. Since a geometric lattice is locally modular if and only if it is modular, it follows from the preceding theorem that the whole lattice must be modular. This may be viewed as a converse to the Kurosch-Ore theorem for (finite) modular lattices. Thus, a *finite semimodular lattice is modular if and only if it has the KORP both for meet-decompositions and for join-decompositions.* Similarly, one gets that a *finite locally distributive lattice is balanced if and only if it is distributive.*

REFERENCES

1. P. Crawley, *Decomposition theory for nonsemimodular lattices*, Trans. Amer. Math. Soc. **99** (1961), 246–254.

2. U. Faigle, *Geometries on partially ordered sets*, J. Combin. Theory Ser. B **28** (1980), 26–51.

3. U. Faigle, *Exchange properties of combinatorial closure spaces*, Discrete Appl. Math. **15** (1986), 249–260.

4. M. Hall, Jr. and R. P. Dilworth, *The imbedding problem for modular lattices*, Ann. of Math. (2) **45** (1944), 450–456. Reprinted in Chapter 4 of this volume.

5. C. Herrmann, *S-Verklebte Summen von Verbänden*, Math. Z. **130** (1973), 255–274.

6. J. P. S. Kung, *Matchings and Radon transforms in lattices. I. Consistent lattices*, Order **2** (1985), 105–112.

7. K. Reuter, *The Kurosh-Ore exchange property*, Acta Math. Hung. **53** (1989), 119–127.

8. R. Wille, *Subdirect decomposition of concept lattices*, Algebra Universalis **17** (1983), 275–287.

Martin-Luther-Universität
DDR-4010 Halle
German Democratic Republic

Modular and Distributive Lattices

Background

R. P. Dilworth

Imbedding problems and the gluing construction. One of the most powerful tools in the study of modular lattices is the notion of the projectivity of intervals. The intervals $[a, a \vee b]$ and $[a \wedge b, b]$ are *transposes* of one another. Two intervals which can be transformed into each other by a series of transposes are said to be *projective*. For example, in the lattice of Figure 1 every covering interval is projective to the interval $[a, e]$ and hence every two covering intervals are projective. If an interval is collapsed under a homomorphism, then every transpose is also collapsed and hence every interval projective to it is collapsed. Thus a homomorphism of the lattice in Figure 1 is either an isomorphism or it maps the entire lattice onto a one-element lattice. A lattice with this property is said to be *simple*. In particular, if a simple lattice is imbedded in a direct product of lattices, then it can also be imbedded in one of the component lattices since not all of the homomorphisms onto the components can be trivial.

During my graduate years at Caltech I had noticed that two lattices could be combined to form a larger lattice if an upper interval of one lattice was isomorphic to a lower interval of the other lattice. The process was to patch the two lattices together so that corresponding elements under the isomorphism were identified as indicated in Figure 2.

This process of patching (now called 'gluing') had the property of preserving modularity and also distributivity. Also if the isomorphic intervals were non-trivial and the lattices were simple, then the resulting lattice would likewise be simple.

While I was at Yale, it occurred to me that the patching technique might be relevant to one of the then outstanding problems concerning modular lattices,

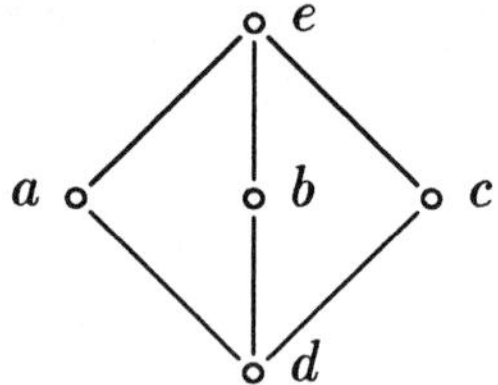

Figure 1

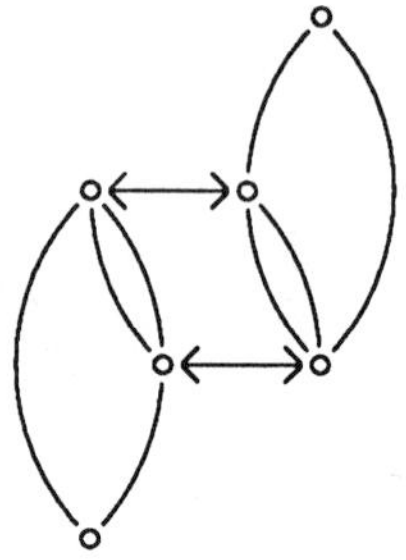

Figure 2

namely, can every modular lattice be imbedded in a complemented modular lattice. It seemed unlikely that there would be an affirmative answer since simple complemented modular lattices were projective geometries which were highly uniform in their structure. I discussed this problem with Marshall Hall who was at Yale at this time and mentioned the patching technique. Almost immediately he came up with the idea of patching the lattice of Figure 1 onto the lattice of subspaces of a non-Desarguesian plane to give the lattice schematically represented in Figure 3.

If this modular lattice could be imbedded isometrically in a complemented modular lattice L, then, since L is a direct product of projective geometries, the lattice could be imbedded in one of the components which would have to be of projective dimension at least 3. But every plane in a projective geometry of projective dimension at least 3 is Desarguesian. Thus such an imbedding is not possible. Hall went on to show that it could not be imbedded in any way in a complemented modular lattice. We then constructed several other examples of modular lattices which could not be isometrically imbedded in a complemented modular lattice.

Meet and join irreducible elements in modular lattices. Meet and join irreducibles have played an important role in many aspects of lattice theory. An element q is meet irreducible if $q = x \wedge y$ implies $q = x$ or $q = y$. Join irreducibles are defined dually. If a finite dimensional lattice is distributive, there is a natural one-to-one correspondence between the meet and join irreducibles, namely if q is a

                                                                 THE DILWORTH THEOREMS

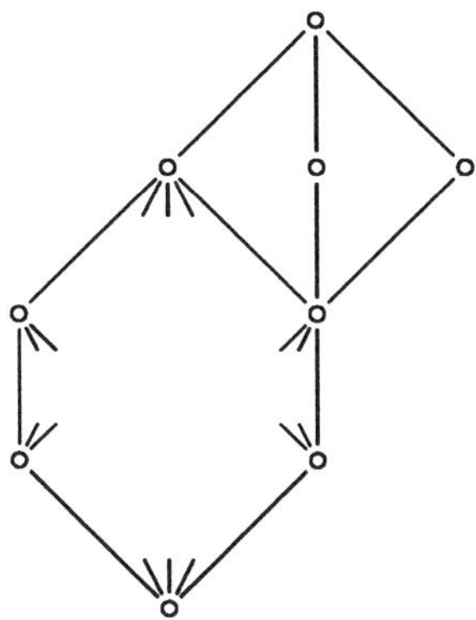

Figure 3

meet irreducible, there is a unique minimal join irreducible p such that $p \nleq q$. In turn, q is the unique maximal meet irreducible not containing p. It follows from this correspondence that a finite distributive lattice has the same number of meet and join irreducibles. Although there is no natural correspondence between meet and join irreducibles in a finite modular lattice it was observed in the early 1940's that the number of meet irreducibles was the same as the number of join irreducibles. This was stated as a conjecture by Schützenberger but for a number of years no proof was forthcoming.

In the early 1950's I decided to make a major attack on the problem. Since meet irreducibles are those elements covered by a single element of the lattice and join irreducibles are those elements covering a single element of the lattice, it was natural to attack the more general problem of showing that $|V_k| = |W_k|$ where V_k is the set of elements of the lattice covered by precisely k lattice elements, W_k is the set of elements covering precisely k elements, and $|S|$ denotes the number of elements in the set S. This equality had been known for some time in the case of the lattice of subspaces of a projective geometry. Since a finite complemented lattice is a direct product of projective geometries, it follows that the equality holds for all finite complemented modular lattices.

I spent some time trying to describe the lattice structure of a finite modular lattice in terms of its complemented sublattices. However, this seemed to be a more complicated problem than the original combinatorial problem. It did become clear from these early investigations that the complemented modular sublattice M generated by the atoms of the lattice L would play a significant role. This led to the idea of reducing the problem from the lattice L to the sublattices L_m, where $m \in M$ and L_m consists of the elements of L containing m. If it turned out to be possible to express $|V_k|$ in terms of the corresponding numbers for lattices L_m and the same expression also works for $|W_k|$, then the result would follow by induction on the dimension of the lattice. Now there was available a standard device for counting elements in an ordered set, namely, the Möbius function. For $m \in M$, $\mu(m)$ is defined by the recurrence relation

$$\sum_{x \leq m} \mu(x) = \begin{cases} 1 & \text{if } m \text{ equals } z, \text{ the null element of the lattice,} \\ 0 & \text{if } m \text{ does not equal } z. \end{cases}$$

From this basic property of the Möbius function it was quite direct to show that $|V_k|$ was expressible as a linear combination of the corresponding numbers for the lattices L_m with Möbius coefficients. With some additional work and making use of the fact that equality holds in the complemented case, I was able to show that the same linear relation held for $|W_k|$ in terms of the corresponding number of the lattices L_m. The theorem then follows by induction.

Distributivity in lattices. A closure operation on a set S is a mapping φ from subsets of S to subsets of S with the following properties:

(1) $T \subseteq \varphi(T)$ for all $T \subseteq S$;

(2) $\varphi(U) \subseteq \varphi(T)$ if $U \subseteq T$;

(3) $\varphi(\varphi(T)) = \varphi(T)$ if $T \subseteq S$;

If S is an ordered set, a closure operation φ is called an imbedding operator if

(4) $\varphi(\{s\}) = (s)$ for all $s \in S$,

where (s) denotes the set of all $x \in S$ such that $x \leq s$. A subset T of S is said to be φ-closed if $\varphi(T) = T$. The φ-closed subsets form a complete lattice under the operation of set intersection and the closure of the set union. If S is an ordered set, the mapping $s \mapsto (s)$ imbeds S in the lattice of φ-closed subsets of S. Imbedding operators have a natural ordering given by $\varphi \leq \psi$ if and only if $\varphi(T) \subseteq \psi(T)$ for all $T \subseteq S$. The minimum operator ω has the defining property that $\omega(T)$ consists of all x such that $x \leq t$ for some $t \in T$. The maximum operator ν is defined by letting $\nu(T)$ be the set of all lower bounds of the upper bounds of T.

The starting point of the paper "Distributivity in Lattices" was the observation that the usual distributivity of meet with respect to join in a lattice L can be expressed in terms of an imbedding operator θ as

$$a \wedge \theta(T) = \theta(a \wedge T)$$

where $a \wedge T = \{a \wedge t : t \in T\}$ and $\theta(T)$ is the ideal of L generated by T, *i.e.*, the set of all elements of L contained in a finite join of elements of T. Replacing θ by an arbitrary imbedding operator φ produces many possibilities for distributivity conditions in lattices. This raised a number of questions concerning the relationships between the different distributivities and the relation to the lattice structure of the imbedding operators. I proposed to investigate these questions and was joined in the study by an innovative young graduate student, Jack McLaughlin.

It is a natural conjecture that if $\varphi \geq \psi$ then φ-distributivity should imply ψ-distributivity. However, this turns out not to be the case. In fact, there exists an imbedding operator φ on the Boolean algebra of order 8 such that the Boolean

THE DILWORTH THEOREMS

algebra is not φ-distributive, while it is also true that every Boolean algebra is ν-distributive. φ-distributivity is related to ν-distributivity in the following way: L is φ-distributive if and only if the lattice of φ-closed subsets of L is ν-distributive. But for complete lattices ν-distributivity is equivalent to ordinary infinite distributivity. Hence, φ-distributivity of L is equivalent to infinite distributivity of the lattice of φ-closed subsets of L. Given an imbedding operator φ, suitable collections of subsets of S provide associated imbedding operators contained in φ. These associated operators have consistency properties for distributivity not found among imbedding operators in general. It turns out that all of the common imbedding operators are associated with the normal imbedding operator ν.

Editors' note. The paper "Aspects of Distributivity," published in 1984, is in part a survey of distributivity and gives much insight into Dilworth's thinking about this area of lattice theory. In addition, Dilworth proved that in a distributive lattice, an irredundant decomposition into irreducibles is unique (if it exists), even when the ascending chain condition fails.

Annals of Mathematics
Vol. 45, No. 3, July, 1944

THE IMBEDDING PROBLEM FOR MODULAR LATTICES

By M. Hall and R. P. Dilworth

(Received November 11, 1943)

1. Introduction

It is trivially true that an arbitrary lattice may be imbedded in a complemented lattice. We need only adjoin a unit and null elements if they do not already exist and a single element which is a complement of each of the elements not the unit of null element. For distributive lattices, the imbedding problem is not so trivial, but is contained in the representation theorem which asserts that any distributive lattice is isomorphic with a ring of sets (Birkhoff (1), Mac Neille (1)). The corresponding problem of imbedding a modular lattice in a complemented modular lattice is an outstanding problem in lattice theory. We exhibit here an example of a modular lattice which cannot be imbedded in any complemented modular lattice. However we will be concerned primarily with the isometric problem for finite dimensional modular lattices; that is, the problem of imbedding a finite dimensional modular lattice in a complemented modular lattice of the same dimensionality.

2. Notation and definitions

We will use the notation and terminology of G. Birkhoff "Lattice Theory" except that proper inclusion will be denoted by $a \supset b$, reserving $a > b$ to indicate that a covers b. It will be recalled that a lattice L is modular if

M: $a \supseteq b$ implies $a \cap (b \cup c) = b \cup (a \cap c)$ and complemented if

C: For each a there is at least one complement a' such that $a \cup a' = u$, $a \cap a' = z$ where u and z are respectively unit and null elements of the lattice.

Any complemented modular lattice is also relatively complemented, that is whenever $a \supseteq x \supseteq b$ there exists a y such that $x \cup y = a$, $x \cap y = b$.

Considerable use will be made of the following fundamental theorem on complemented modular lattices (Birkhoff (1) pp. 60).

Theorem 2.1. *Any finite-dimensional complemented modular lattice is the direct product of projective geometrics.*

It is understood in the statement of the theorem that a Boolean algebra of order two is the projective geometry of the void space and a single point where we may consider the geometric postulates on lines as satisfied by default.

3. Subdirect product decompositions

As a first reduction of the isometric imbedding problem, the lattice L will be represented as a subdirect product of lattices having a simple structure with respect to projectivity.

Definition 3.1. A modular lattice is said to be *projective* if every two prime quotients are projective (Birkhoff (1) pp. 37).

450

The following theorem was proved by Birkhoff for lattices of finite dimension. The proof given here is new and applies to any atomic lattice.

Theorem 3.1. *Let L be an atomic modular lattice. Then L is a subdirect product of projective lattices.*

Proof. Atomicity will be used in the following weak form: $a \supset b$ implies x exists such that $a \supseteq x > b$. Now let $\mathfrak{p}$ be a given prime quotient in L and let $\mathfrak{P}$ denote the set of all prime quotients projective to $\mathfrak{p}$. In L we define $a \sim b \, (\mathfrak{P})$ if and only if there exists a chain of elements $a = a_1, a_2, \cdots, a_n = b$ such that $a_i \cup a_{i+1}/a_i \cap a_{i+1}$ contains *no* quotient projective to $\mathfrak{p}$. The relation $a \sim b$ is clearly an equivalence relation. We show that it is also a congruence relation. Let $a \sim b \, (\mathfrak{P})$ and consider the chain $a \cup c = a_1 \cup c, a_2 \cup c, \cdots, a_n \cup c = b \cup c$. Suppose $(a_i \cup c) \cup (a_{i+1} \cup c)/(a_i \cup c) \cap (a_{i+1} \cup c)$ contains a quotient x/y projective to $\mathfrak{p}$. Then $x > y$ and $y \supseteq (a_i \cup c) \cap (a_{i+1} \cup c) \supseteq (a_i \cap a_{i+1}) \cup c$. Thus $y \supseteq c$ and $y \supseteq a_i \cap a_{i+1}$. Clearly $a_i \cup a_{i+1} \cup c \supseteq x \cup (a_i \cup a_{i+1}) \supseteq y \cup (a_i \cup a_{i+1}) \supseteq a_i \cup a_{i+1} \cup c$. Hence $x \cup (a_i \cup a_{i+1}) = y \cup (a_i \cup a_{i+1})$ and it follows that $x \cap (a_i \cup a_{i+1}) > y \cap (a_i \cup a_{i+1})$. But then $a_i \cup a_{i+1} \supseteq x \cap (a_i \cup a_{i+1}) > y \cap (a_i \cup a_{i+1}) \supseteq a_i \cap a_{i+1}$ and $x \cap (a_i \cup a_{i+1})/y \cap (a_i \cup a_{i+1})$ is projective to x/y and hence is projective to $\mathfrak{p}$ which is contrary to assumption. Thus $(a_i \cup c) \cup (a_{i+1} \cup c)/(a_i \cup c) \cap (a_{i+1} \cup c)$ contains no quotient projective to $\mathfrak{p}$ and we have $a \cup c \sim b \cup c \, (\mathfrak{P})$. In a similar manner $a \cap c \sim b \cap c \, (\mathfrak{P})$.

If $x, y, \cdots$ are elements of L, let $\{x\}_{\mathfrak{P}}, \{y\}_{\mathfrak{P}}, \cdots$ represent the congruence classes determined by $x, y, \cdots$ respectively. Since $a \sim b \, (\mathfrak{P})$ is a congruence relation, if we define $\{x\}_{\mathfrak{P}} \cup \{y\}_{\mathfrak{P}} = \{x \cup y\}_{\mathfrak{P}}$ and $\{x\}_{\mathfrak{P}} \cap \{y\}_{\mathfrak{P}} = \{x \cap y\}_{\mathfrak{P}}$ the congruence classes form a modular lattice $L_{\mathfrak{P}}$. Now let a/b be a prime quotient projective to $\mathfrak{p}$. We shall show that $a \nsim b \, (\mathfrak{P})$. For let $a = a_1, a_2, \cdots, a_n = b$ be a chain of elements such that $a_i \cup a_{i+1}/a_i \cap a_{i+1}$ contains no quotient projective to $\mathfrak{p}$. Now $a \cup (a_1 \cap a_2) = b \cup (a_1 \cap a_2)$ since otherwise $a \cup (a_1 \cap a_2)/b \cup (a_1 \cap a_2)$ is a quotient in $a_1 \cup a_2/a_1 \cap a_2$ which is projective to $\mathfrak{p}$. Hence $a \cap a_1 \cap a_2 > b \cap a_1 \cap a_2$. Now suppose it has been shown that $a \cap a_1 \cap \cdots \cap a_i > b \cap a_1 \cap \cdots \cap a_i$. Then $(a \cap a_1 \cap \cdots \cap a_i) \cup (a_i \cap a_{i+1}) = (b \cap a_1 \cap \cdots \cap a_i) \cup (a_i \cap a_{i+1})$ since otherwise $(a \cap a_1 \cap \cdots \cap a_i) \cup (a_i \cap a_{i+1})/(b \cap a_1 \cap \cdots \cap a_i) \cup (a_i \cap a_{i+1})$ is a prime quotient in $a_i \cup a_{i+1}/a_i \cap a_{i+1}$ which is projective to a/b and hence is projective to $\mathfrak{p}$ which contradicts our assumption. Hence $a \cap a_1 \cap \cdots \cap a_i \cap a_{i+1} > b \cap a_1 \cap \cdots \cap a_i \cap a_{i+1}$. By induction we get $a \cap a_1 \cap \cdots \cap a_n > b \cap a_1 \cap \cdots \cap a_n$. But then $a = b \cup (a_1 \cap \cdots \cap a_n)$. Now $a_n = b$ and hence $a = b$ which contradicts $a > b$. Thus $a \nsim b \, (\mathfrak{P})$. It clearly follows that $\{a\}_{\mathfrak{P}} > \{b\}_{\mathfrak{P}}$ in $L_{\mathfrak{P}}$.

Now let $\{a\}_{\mathfrak{P}} > \{b\}_{\mathfrak{P}}$ and $\{c\}_{\mathfrak{P}} > \{d\}_{\mathfrak{P}}$ in $L_{\mathfrak{P}}$. Since $a \nsim b$, $a \cup b/a \cap b$ contains a quotient x/y which is projective to $\mathfrak{p}$. Then $\{a\}_{\mathfrak{P}} = \{a \cup b\}_{\mathfrak{P}} \supseteq \{x\}_{\mathfrak{P}} > \{y\}_{\mathfrak{P}} \supseteq \{a \cap b\}_{\mathfrak{P}} = \{b\}_{\mathfrak{P}}$ and hence $\{a\}_{\mathfrak{P}} = \{x\}_{\mathfrak{P}}, \{b\}_{\mathfrak{P}} = \{y\}_{\mathfrak{P}}$. In a similar manner $\{c\}_{\mathfrak{P}} = \{v\}_{\mathfrak{P}}, \{d\}_{\mathfrak{P}} = \{w\}_{\mathfrak{P}}$ where $v > w$ and v/w is projective to $\mathfrak{p}$. Thus x/y is projective to v/w and hence $\{a\}_{\mathfrak{P}}/\{b\}_{\mathfrak{P}}$ is projective to $\{c\}_{\mathfrak{P}}/\{d\}_{\mathfrak{P}}$ in $L_{\mathfrak{P}}$. It follows that $L_{\mathfrak{P}}$ is a projective lattice.

Let us set up the correspondence $a \rightarrow (\cdots, \{a\}_{\mathfrak{P}}, \cdots)$ where $\mathfrak{P}$ runs over all sets of projective quotients. Clearly $a \cup b \to (\cdots, \{a\}_{\mathfrak{P}} \cup \{b\}_{\mathfrak{P}}, \cdots)$ and $a \cap b \to (\cdots, \{a\}_{\mathfrak{P}} \cap \{b\}_{\mathfrak{P}}, \cdots)$. Hence the correspondence is a homomorphism. But if $a \neq b$, then $a \cup b \supset a \cap b$ and x exists such that $a \cup b \supseteq x > a \cap b$. Let $\mathfrak{P}$ be the set of quotients projective to $x/a \cap b$. Then $\{x\}_{\mathfrak{P}} \neq \{a \cap b\}_{\mathfrak{P}}$ and hence $\{a \cup b\}_{\mathfrak{P}} \neq \{a \cap b\}_{\mathfrak{P}}$ and clearly $\{a\}_{\mathfrak{P}} \neq \{b\}_{\mathfrak{P}}$. Thus $(\cdots, \{a\}_{\mathfrak{P}}, \cdots) \neq (\cdots, \{b\}_{\mathfrak{P}}, \cdots)$ and the correspondence is an isomorphism. If $a > b$, then $\{a\}_{\mathfrak{P}} > \{b\}_{\mathfrak{P}}$ where $\mathfrak{P}$ is the set of prime quotients projective to a/b, and $\{a\}_{\mathfrak{P}'} = \{b\}_{\mathfrak{P}'}$ where $\mathfrak{P}'$ is any other set of projective prime quotients. Hence $(\cdots, \{a\}_{\mathfrak{P}}, \cdots) > (\cdots, \{b\}_{\mathfrak{P}}, \cdots)$ in the direct product lattice. Thus we have shown that L is isomorphic with an isometric sublattice of the direct product of the projective lattices $L_{\mathfrak{P}}$.

Now if L is a lattice of finite dimensions, by theorem 3.1 it is an isometric sublattice of the direct product $L_1 \times \cdots \times L_n$ where L_i is a projective lattice. Hence if we can imbed L_i isometrically in a complemented modular lattice M_i, then L will be isometrically imbedded in the complemented modular lattice $M = M_1 \times \cdots \times M_n$. Thus the imbedding problem for arbitrary modular lattices of finite dimension is reduced to the consideration of projective lattices of finite dimension.

We remark in this connection that modular lattices of dimension three or less can always be imbedded in a complemented modular lattice (Dilworth (2)).

For projective lattices, the problem is further reduced by the following theorem:

THEOREM 3.2. *If a projective lattice of finite dimensions can be imbedded isometrically in a complemented modular lattice M, then M is a projective geometry.*

PROOF. Since M is a complemented modular lattice of finite dimensions, it is a direct product of projective geometries $M = P_1 \times \cdots \times P_n$. Now let $a > b$ in L. Then $a = (a_1, \cdots, a_n)$ and $b = (b_1, \cdots, b_n)$. Since $a > b$ and the imbedding is isometric we have $a_i > b_i$ for some i and $a_j = b_j$ for $j \neq i$. But since every prime quotient in L is projective to a/b and the correspondence $a \to a_i$ is a homomorphism we have $c > d$ implies $c_i > d_i$. Thus the correspondence $a \to a_i$ is an isomorphism and L is imbedded in P_i. But since the original imbedding is isometric, P_i is the only component which is not null and hence $M = P_i$. That is, M is a projective geometry.

4. Counter-examples

In the construction of the counter examples we shall need the following lemma:

LEMMA 4.1. *Let L_1 and L_2 be lattices with unit elements u_1, u_2 and null elements, z_1, z_2 respectively. Let the quotient lattice u_1/a_1 of L_1 be isomorphic to the quotient lattice a_2/z_2 of L_2. If isomorphic elements are identified, then the set sum L of L_1 and L_2 is a lattice which contains L_1 and L_2 as sublattices and is modular if and only if L_1 and L_2 are modular.*

PROOF. Let a and b be any two elements of L. If both a and b are in L_1 or in L_2, then $a \cup b$ is the union in L_1 or L_2 and $a \cap b$ is cross-cut in L_1 or L_2 respectively. If a is in L_1 and b is in L_2, then $a \cup b = (a \cup z_2) \cup b$ where the first union is in L_1 and the second, in L_2. Similarly $a \cap b = a \cap (u_1 \cap b)$ where the first cross-cut is in L_1 and the second, in L_2. It is readily verified that L is a lattice under these definitions of union and cross-cut. Clearly L_1 and L_2 are sublattices of L. That L is modular if and only if L_1 and L_2 are modular follows from Lemma 4.2 of Dilworth (1).

Consider a lattice L_1 whose diagram is given by Figure 1, where the quotient lattice a/z is the lattice of a non-Desarguesian plane. Since a/z and u/e are modular Lemma 4.1 assures us that L_1 is modular. Since every prime quotient in L_1 is projective to a/e, L_1 is a projective lattice and by Theorem 3.2 if it can be imbedded isometrically in a complemented modular lattice, then the complemented modular lattice is a projective geometry. Here u/z is three dimensional in the customary terminology of projective geometry (four dimensional as a lattice). But every plane in a projective 3-space must be Desarguesian (Veblen and Young (1)) and hence it must be impossible to imbed L_1 isometrically in a

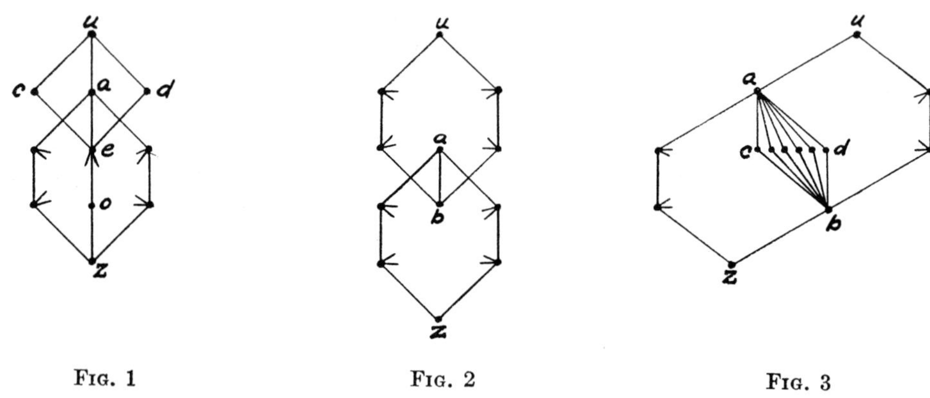

Fig. 1   Fig. 2   Fig. 3

complemented modular lattice. We have been able to show even more, namely that the following statement holds:

L_1 cannot be imbedded in any complemented modular lattice.

For since a/z is non-Desarguesian, there exists "points" O, A_1, B_1, C_1, A_2, B_2, C_2, A_3, B_3, C_3 and "lines" $e = OA_1A_2$, OB_1B_2, $f = OC_1C_2$, $A_1B_1C_3$, $A_2B_2C_3$, $A_1B_3C_1$, $b = A_2B_3C_2$, $A_3B_1C_1$, $A_3B_2C_2$ such that $A_1 \cup B_1 \cup C_1 = A_2 \cup B_2 \cup C_2 = A_3 \cup B_3 \cup C_3 = a$. Thus A_1, B_1, C_1 and A_2, B_2, C_2 are two triangles perspective from O, whose corresponding sides A_1B_1 and A_2B_2, etc. meet in three non-collinear points C_3, A_3, B_3.

Now suppose that u/z is a sublattice of a complemented modular lattice M. Let x be a relative complement of e in d/O and y be a relative complement of e in d/A_2 so that $x \cup e = y \cup d$, $x \cap e = O$, $y \cap e = A_2$. From these relations the following projectivities may be verified: f/z proj. a/A_2 proj. u/y proj. c/A_2 proj. u/b proj. x/z. Furthermore, under the projections $O \to e \to d \to e \to a \to O$. Let O_1 and O_2 be the image of C_1 and C_2 under the series of projections. Then since $C_1 \cup O = C_2 \cup O = C_1 \cup C_2 = f$ and $C_1 \cap O = C_2 \cap O = C_1 \cap C_2 = Z$ we have $O_1 \cup O = O_2 \cup O = O_1 \cup O_2 = x$ and $O_1 \cap O = O_2 \cap O = O_1 \cap O_2 = z$. Let us set

$$A = (A_1 \cup O_1) \cap (A_2 \cup O_2)$$

$$B = (B_1 \cup O_1) \cap (B_2 \cup O_2)$$

$$C = (C_1 \cup O_1) \cap (C_2 \cup O_2).$$

Then $A \cup B = (A_1 \cup B_1 \cup O_1) \cap (A_2 \cup B_2 \cup O_2)$. For $e \cup A \cup B = e \cup A_1 \cup A_2 \cup A \cup B = e \cup [(A_1 \cup A_2 \cup O_1) \cap (A_1 \cup A_2 \cup O_2)] \cup B = [(e \cup O_1) \cap (e \cup O_2)] \cup B = d \cup B = d \cup O_1 \cup O_2 \cup [(B_1 \cup O_1) \cap (B_2 \cup O_2)] = d \cup [(B_1 \cup O_1 \cup O_2) \cap (B_2 \cup O_1 \cup O_2)] = d \cup [(B_1 \cup x) \cap (B_2 \cup x)] = d \cup [(B_1 \cup O \cup x) \cap (B_2 \cup O \cup x)] = d \cup B_1 \cup B_2 = d \cup a = u$ while $e \cap (A_1 \cup B_1 \cup O_1) \cap (A_2 \cup B_2 \cup O_2) = e \cap d \cap (A_1 \cup B_1 \cup O_1) \cap d \cap (A_2 \cup B_2 \cup O_2) = e \cap (A_1 \cup O_1) \cap (A_2 \cup O_2) = [A_1 \cup (e \cap O_1)] \cap [A_2 \cup (e \cap O_2)] = A_1 \cap A_2 = z$. Since $(A_1 \cup B_1 \cup O_1) \cap (A_2 \cup B_2 \cup O_2) \supseteq A \cup B$ the above formula follows from modularity. By symmetry

$$A \cup C = (A_1 \cup C_1 \cup O_1) \cap (A_2 \cup C_2 \cup O_2).$$

Furthermore

$$B \cup C = (B_1 \cup C_1 \cup O_1) \cap (B_2 \cup C_2 \cup O_2).$$

For $f \cup (B \cup C) = f \cup (C_1 \cup C_2) \cup [(C_1 \cup O_1) \cap (C_2 \cup O_2)] \cup B = f \cup [(C_1 \cup C_2 \cup O_1) \cap (C_1 \cup C_2 \cup O_2)] \cup B = [(f \cup O_1) \cap (f \cup O_2)] \cup B = [(f \cup O \cup O_1) \cap (f \cup O \cup O_2)] \cup B = f \cup x \cup B = f \cup O_1 \cup O_2 \cup [(B_1 \cup O_1) \cap (B_2 \cup O_2)] = f \cup [(B_1 \cup O_1 \cup O_2) \cap (B_2 \cup O_1 \cup O_2)] = f \cup [(B_1 \cup O \cup x) \cap (B_2 \cup O \cup x)] = f \cup B_1 \cup B_2 \cup x = a \cup x = a \cup e \cup x = a \cup d = u$ while $f \cap (B_1 \cup C_1 \cup O_1) \cap (B_2 \cup C_2 \cup O_2) = f \cap a \cap (B_1 \cup C_1 \cup O_1) \cap (B_2 \cup C_2 \cup O_2) = f \cap (B_1 \cup C_1) \cap (B_2 \cup C_2) = C_1 \cap C_2 = z$. Since $(B_1 \cup C_1 \cup O_1) \cap (B_2 \cup C_2 \cup O_2) \supseteq B \cup C$, the equality must hold by the modular law.

But then $A \cup B \supseteq (A_1 \cup B_1) \cap (A_2 \cup B_2) \supseteq C_3$ and similarly $A \cup C \supseteq B_3$, $B \cup C \supseteq A_3$. Thus $A \cup B \cup C \supseteq A_3 \cup B_3 \cup C_3 = a$. Hence $A_1 = A_1 \cap a = A_1 \cap (A \cup B \cup C) = A_1 \cap (A_1 \cup B_1 \cup O_1) \cap (A \cup B \cup C) = A_1 \cap [A \cup B \cup ((A_1 \cup B_1 \cup O_1) \cap C)] = A_1 \cap [A \cup B \cup ((A_1 \cup B_1 \cup O_1) \cap (C_1 \cup O_1) \cap (C_2 \cup O_2))] = A_1 \cap [A \cup B \cup ((O_1 \cup [C_1 \cap (A_1 \cup B_1)]) \cap ((C_2 \cup O_2)))] = A_1 \cap [A \cup B \cup (O_1 \cap (C_2 \cup O_2))] = A_1 \cap [A \cup B \cup (O_1 \cap d \cap (C_2 \cup O_2))] = A_1 \cap [A \cup B \cup (O_1 \cap O_2)] = A_1 \cap [A \cup B] = A_1 \cap (A_2 \cup B_2 \cup O_2) = A_1 \cap a \cap (A_2 \cup B_2 \cup O_2) = A_1 \cap (A_2 \cup B_2) = z$. But A_1 is a "point" of a/z and hence is not equal to z. Thus we have a contradiction and u/z cannot be imbedded in any complemented modular lattice.

Let L_2 be a lattice whose diagram is given by Figure 2 where u/b is a Desarguesian plane whose coordinatizing skew-field F is of characteristic p and a/Z is a Desarguesian plane whose coordinatizing skew-field K is of characteristic q and let $q \neq p$. Then, as with L_1, if L_2 can be imbedded isometrically in a complemented modular lattice, this lattice must be a projective 4-space G. But the coordinatizing skew-field of G cannot have subfields F and K of different characteristics and we are led to a contradiction. Hence L_2 cannot be imbedded isometrically in a complemented modular lattice.

The third counter-example is the lattice L_3 whose diagram is Figure 3. Here a/z and u/b are isomorphic Desarguesian planes whose field F contains more than three elements and is not of characteristic 2. We may, for example, let F be the finite field of five elements. In general if F contains n elements, then in the plane which it determines there are $n + 1$ points on every line and $n + 1$ lines through every point. Hence L_3 may be constructed from two isomorphic planes a_1/z and u/b_2 where since a_1/b_1 and a_2/b_2 contain the same number of elements the identifications $a_1 = a_2$, $b_1 = b_2$ and any mapping of the intermediate elements of a_1/b_1 and a_2/b_2 will satisfy the conditions of Lemma 4.1, yielding the modular lattice L_3. Now every prime quotient in L_3 is projective to a/c and hence L_3 is a projective lattice and by Theorem 3.2 if it can be imbedded in a complemented modular lattice, this lattice must be a projective geometry.

Now Lemma 4.1 assures us that an arbitrary mapping of a_1/b_1 on a_2/b_2 will make L_3 modular, but we shall show that if L_3 is imbedded in a projective 3-space then this mapping of a_1/b_1 onto a_2/b_2 cannot be arbitrary, and that therefore the mappings of a_1/b_1 onto a_2/b_2 which are not permissible in this way yield lattices L_3 which cannot be isometrically imbedded in complemented modular lattices. This will depend on the construction of a harmonic line conjugate. If A_1, B_1, C_1 are three lines of the quotient a_1/b_1 then we may construct in a_1/z a line C_1^*, in a_1/b_1 which is the harmonic conjugate of C_1 with respect to A_1 and B_1. Here $C_1^* \neq A_1$, B_1 always and $C_1^* \neq C_1$ if the characteristic of F is not 2. Similarly if A_2, B_2, C_2 are three lines of a_2/b_2, we may construct in u/b_2 a line C_2^* in a_2/b_2 which is the harmonic conjugate of C_2 with respect to A_2 and B_2. Here if u/z is a projective 3-space we must have $C_1^* = C_2^*$ and hence if a_1/b_1 is mapped onto a_2/b_2 so that $A_1 \rightleftarrows A_2$, $B_1 \rightleftarrows B_2$, $C_1 \rightleftarrows C_2$, then we must also map C_1^* onto C_2^* if u/z can be imbedded in a 3-space. But if a/b contains more than four elements and F is not of characteristic 2, this excludes certain mappings of a_1/b_1 onto a_2/b_2.

THEOREM 4.1. *Given a projective 3-space S and a plane π containing a point P, let A, B, C, be any three lines through P lying in π. Let R_1 and S_1 be two lines in π such that R_1, S_1, C are concurrent in a point Q_1 different from P. Construct $M_1 = (A \cap R_1) \cup (B \cap S_1)$, $N_1 = (A \cap S_1) \cup (B \cap R_1)$ and $C_1^* = P \cup (M_1 \cap N_1)$. Let R_2 and S_2 be two lines through P such that R_2, S_2, and C lie in a plane Q_2 different from π. Construct $M_2 = (A \cup R_2) \cap (B \cup S_2)$, $N_2 = (A \cup S_2) \cap (B \cup R_2)$ and $C_2^* = \pi \cap (M_2 \cup N_2)$. Then C_1^*, the harmonic conjugate from below of C with respect to A and B, is independent of the choice of R_1 and S_1; C_2^*, the harmonic conjugate from above of C with respect to A and B is independent of the choice of R_2 and S_2; and $C_1^* = C_2^*$.*

PROOF. We may introduce coordinates in S from the appropriate skew-field F where points are given by homogeneous coordinates (x_1, x_2, x_3, x_4) and planes have equations $u_1x_1 + u_2x_2 + u_3x_3 + u_4x_4 = 0$. By an appropriate choice of the frame of reference we may take P as $(1, 0, 0, 0)$, π as $x_4 = 0$, and A as $x_2 = 0$, $x_4 = 0$, B as $x_3 = 0$, $x_4 = 0$, and C as $x_2 + x_3 = 0$, $x_4 = 0$.

Here we may take R_1 as $x_1 + \alpha x_2 + \beta x_3 = 0$, $x_4 = 0$ and S_1 as $x_1 + (\alpha + \gamma)x_2 + (\beta + \gamma)x_3 = 0$, $x_4 = 0$. Then $A \cap R_1 = (-\beta, 0, 1, 0)$, $B \cap S_1 = (-\alpha -\gamma, 1, 0, 0)$. Hence $M_1 = (A \cap R_1) \cup (B \cap S_1)$ is $x_1 + (\alpha + \gamma)x_2 + \beta x_3 = 0$, $x_4 = 0$. Then $B \cap R_1$ is $(-\alpha, 1, 0, 0)$, $A \cap S_1$ is $(-\beta -\gamma, 0, 1, 0)$ and $N_1 = (A \cap S_1) \cup (B \cap R_1)$ is $x_1 + \alpha x_2 + (\beta + \gamma)x_3 = 0$, $x_4 = 0$. Then $M_1 \cap N_1$ is $(-\alpha -\beta -\gamma, 1, 1, 0)$ and $C_1^* = P \cup (M_1 \cap N_1)$ is $x_2 - x_3 = 0$, $x_4 = 0$. Hence C_1^* is independent of the choice of R_1 and S_1.

We may take

$$R_2 \left\{ \begin{array}{l} x_2 + x_3 + \alpha x_4 = 0 \\ x_3 + \beta x_4 = 0 \end{array} \right. \qquad S_2 \left\{ \begin{array}{l} x_2 + x_3 + \alpha x_4 = 0 \\ x_3 + \gamma x_4 = 0 \end{array} \right.$$

since S is a line coplanar with R_2 and C. Then we have

$$\left. \begin{array}{lll} A \cup R_2 & x_2 + (\alpha - \beta)x_4 = 0 \\ B \cup S_2 & x_3 + \gamma x_4 = 0 \end{array} \right\} M_2 = (A \cup R_2) \cap (B \cup S_2)$$

$$\left. \begin{array}{lll} A \cup S_2 & x_2 + (\alpha - \gamma)x_4 = 0 \\ B \cup R_2 & x_3 + \beta x_4 = 0 \end{array} \right\} N_2 = (A \cup S_2) \cap (B \cup R_2)$$

Hence $M_2 \cup N_2$ is $x_2 - x_3 + (\alpha - \beta - \gamma)x_4 = 0$ and $C_2^* = (M_2 \cup N_2) \cap \pi$ is $x_2 - x_3 = 0$, $x_4 = 0$. Finally C_2^* is independent of the choice of R_2 and S_2 and $C_2^* = C_1^* = C^*$. Note that always $C^* \neq A, B$ and that $C^* \neq C$ if the characteristic of F is different from 2.

Yale University.

REFERENCES

1. G. Birkhoff, *Lattice Theory*, Amer. Math. Soc. Coll. Pub., vol. 25.
2. R. P. Dilworth, *The arithmetical theory of Birkhoff lattices*, Duke Math. Journal vol. 8 (1941) pp. 286–299.
3. ———— *On dependence relations in a semi-modular lattice*, submitted to the Duke Math. Journal.
4. H. M. MacNeille, *Partially ordered sets*, Trans. Amer. Math. Soc. vol. 42 (1937) pp. 416–460.
5. Veblen and Young, *Projective Geometry*, vol. I.

Annals of Mathematics
Vol. 60, No. 2, September, 1954
Printed in U.S.A.

PROOF OF A CONJECTURE ON FINITE MODULAR LATTICES

By R. P. Dilworth

(Received September 18, 1953)

1. Introduction

Since the middle 1930's when the work of Birkhoff and Ore stimulated the modern development of lattice theory, it has been conjectured that in a finite modular lattice the number of meet irreducibles is equal to the number of join irreducibles.[1] I shall prove here the following general combinatorial theorem on finite modular lattices which includes a proof of this conjecture as a special case.

THEOREM. *Let L be a finite modular lattice. If V_k denotes the set of elements of L covered by precisely k elements and W_k denotes the set of elements of L covering precisely k elements, then V_k and W_k contain the same number of elements.*

In particular, V_1 consists of those elements which are covered by exactly one element of L, i.e. the meet irreducibles of L. Similarly W_1 is the set of join irreducibles of L. Hence the theorem implies the equality of the number of meet and join irreducibles.

The critical steps of the proof depend upon properties of Weisner's (Weisner [2]) generalization to partially ordered sets of the Möbius function of number theory. This function is used to reduce the proof to the complemented case where other techniques are available. Curiously the technique of proof by the Weisner-Möbius function fails to work in the complemented case while the methods used in the complemented case do not generalize to the non-complemented case.

The theorem for complemented modular lattices can be deduced from the principle of duality in projective geometries. However, since the proof of the general theorem depends upon the truth of the theorem in this case, it seems appropriate to give a direct lattice-theoretic proof. This is done in Section 2 Section 3 contains the derivation of the necessary properties of the Weisner-Möbius function and Section 4 is devoted to the proof of the theorem.

2. The complemented case

If L is a finite lattice, V_k will denote the set of elements of L having exactly k covering elements. Dually W_k will denote the set of elements which cover exactly k elements. If S is a subset of L, $N(S)$ will denote the number of elements in S. We shall prove now that $N(V_k) = N(W_k)$ for all k if L is a finite complemented modular lattice.

---

[1] This conjecture is stated as a theorem by Schutzenberger, C. R. 218 (1944) 218–219. However, no proof is given and not all finite modular lattices can be obtained by the constructions which he gives. S. Avann also gives a number of results related to this problem in his thesis, California Institute of Tech. (1943).

359

Consider first the case in which L is indecomposable. Let u, z denote the unit and null elements respectively of L and let ρ denote the rank function of L, that is $\rho(a)$ is the length of a complete chain from a to z. From the structure theorems of modular lattices[2] it follows that $\rho(a) = \rho(b)$ if and only if a and b have a common complement. Another characterization of elements of equal rank is also needed.

LEMMA 2.1. *Two elements of L have the same rank if and only if they are covered by the same number of elements of L.*

For let a and b have the same rank. Then by the above remark, a and b have a common complement c. But then u/a is isomorphic[3] to c/z which is in turn isomorphic to u/b. Since u/a and u/b are isomorphic, a and b are covered by the same number of elements. Conversely, let a and b be covered by the same number of elements and suppose that $\rho(a) > \rho(b)$. Then there exists $b_1 > b$ such that $\rho(b_1) = \rho(a)$ and hence by the first part of the proof, b_1 and a are covered by the same number of elements. Let b_1' be a complement of b_1. Then u/b_1 is isomorphic to b_1'/z. Since $b \cap b_1' = z$, we have b_1'/z is isomorphic to $(b \cup b_1')/b$. Thus u/b_1 is isomorphic to $(b \cup b_1')/b$ and hence there is a 1–1 correspondence between the elements covering b_1 and those elements covering b which are contained in $b \cup b_1'$. But there exists x covering b such that $x \le b_1$. Now $x \nleq b \cup b_1'$ since otherwise $x \le b_1 \cap (b \cup b_1') = b \cup (b_1 \cap b_1') = b$. Thus there are more elements covering b than there are covering b_1. But b_1 and a have the same number of covering elements which is a contradiction. Thus a and b must have the same rank.

It follows from Lemma 2.1, that if V_k is non-empty, it consists precisely of all elements of a given rank. The dual of Lemma 2.1 implies that W_k likewise consists of all elements of a given rank.

LEMMA 2.2. *The number of elements of rank r is the same as the number of elements of rank $\rho(u) - r$.*

The lemma is clearly true for lattices of length 1. We proceed by induction and suppose that the lemma holds for all proper quotient lattices of L. We may suppose that $r \le \rho(u) - r$ since otherwise we could consider $r' = \rho(u) - r$. Also the lemma holds if $r = z$ since there is exactly one element of rank zero and one element of rank $\rho(u)$. Hence we may suppose that $r > 0$. Let a be an element of rank r. If a_1 is a second element of rank r, then a and a_1 have a common complement c and hence u/a is projective to c/z which in turn is projective to a/a_1. Thus u/a_1 is isomorphic to u/a, and hence a and a_1 are contained in the same number m of elements of rank $\rho(u) - r$. Now let a' be a complement of a. Then u/a is isomorphic to a'/z. But m, the number of elements of rank $\rho(u) - r$ containing a, is precisely the number of elements of rank $\rho(u) - 2r$ in u/a. Thus by the induction hypothesis m is equal to the number of elements of rank $\rho(u) - r - (\rho(u) - 2r) = r$ in u/a. By the isomorphism m is equal to the number of elements

---

[2] Cf. G. Birkhoff [1].

[3] a/b denotes the quotient lattice of all elements x such that $a \ge x \ge b$.

of rank r contained in a'. Clearly a' is of rank $\rho(u) - r$ and, as in the above argument, every other element of rank $\rho(u) - r$ contains precisely m elements of rank r. Thus if n_r is the number of elements of rank r in L, then $m \cdot n_r$ counts each element of rank $\rho(u) - r = r'$ exactly m times and hence $mn_{r'} = mn_{r'}$. Thus $n_r = n_{r'}$ and the lemma holds for L. Induction completes the proof.

Now let $a \,\epsilon\, V_k$. Then k is the number of elements of rank 1 in u/a. By Lemma 2.2, k is also the number of elements of rank $\rho(u) - \rho(a) - 1$ in u/a. But if a' is a complement of a, then u/a is isomorphic to a'/z and hence a' covers exactly k elements in L. Thus $a' \,\epsilon\, W_k$. By Lemma 2.2, $N(V_k) = N(W_k)$. We have thus shown that the theorem holds when L is a finite, indecomposable, complemented modular lattice.

Consider next the case where L is finite, complemented, and modular but not necessarily indecomposable. Then $L = L_1 \times \cdots \times L_n$ where L_i is a finite indecomposable modular lattice. Now if $a = a_1 \times \cdots \times a_n$ is an element of L, then $b = b_1 \times \cdots \times b_n$ covers a if and only if b_i covers a_i for some i and $b_j = a_j$ for $j \neq i$. It follows that a is covered by k elements of L if and only if $C(a_1) + \cdots + C(a_n) = k$ where $C(a_i)$ is the number of elements covering a_i in L_i. Let $V_{i,k}$ denote the set of elements of L_i which are covered by precisely k elements of L_i. $W_{i,k}$ is similarly defined. Now if $k = k_1 + \cdots + k_n$, it follows that there are $N(V_{1,k_1})N(V_{2,k_2}) \cdots N(V_{n,k_n})$ elements $a = a_1 \times \cdots \times a_n$ of L which are covered by k elements of L and are such that a_i is covered by k_i elements in L_i. Thus

$$N(V_k) = \sum\nolimits_{k=k_1+\cdots+k_n} N(V_{1,k_1})N(V_{2,k_2}) \cdots N(V_{n,k_n}).$$

Similarly

$$N(W_k) = \sum\nolimits_{k=k_1+\cdots+k_n} N(W_{1,k_1})N(W_{2,k_2}) \cdots N(W_{n,k_n}).$$

But by the previous paragraph, $N(V_{i,k}) = N(W_{i,k})$ for all i and k. Hence

$$N(V_k) = N(W_k).$$

We have thus proved that *if L is a finite complemented lattice, then $N(V_k) = N(W_k)$ for every k.*

3. The Weisner-Möbius function

If L is a finite lattice, let $Q_k(a)$ be the number of sets of k distinct points of L whose union is a. Following Weisner (Weisner [2]) we define

DEFINITION 3.2. $\mu(a) = \sum\nolimits_{k=0} (-1)^k Q_k(a)$.

Let P_a denote the number of points of L contained in a.

LEMMA 3.1. $\displaystyle\sum_{b \leq a} Q_k(b) = \binom{P_a}{k}$.

For there are clearly $\dbinom{P_a}{k}$ sets of k distinct points which are contained in a.

Each of these sets has a union b which is contained in a. Furthermore the number of sets having the same union b is exactly $Q_k(b)$. Hence $\sum_{b \leq a} Q_k(b) = \binom{P_a}{k}$.

LEMMA 3.2. $\displaystyle\sum_{b \leq a} \mu(b) \begin{cases} = 1 \ if \ a = z \\ = 0 \ if \ a \neq z. \end{cases}$

For $Q_k(z) = 0$ if $k > 0$ and $Q_0(z) = 1$. Hence $\mu(z) = 1$ and thus $\sum_{b \leq z} \mu(b) = \mu(z) = 1$. If $a > z$, then

$$\sum_{b \leq a} \mu(a) = \sum_{b \leq a} \sum_{k=0}^{\infty} (-1)^k Q_k(b) = \sum_{k=0}^{\infty} (-1)^k \sum_{b \leq a} Q_k(a)$$

$$= \sum_{k=0}^{\infty} (-1)^k \binom{P_a}{k} = (1 - 1)^{P_a} = 0 \qquad \text{since } P_a \geq 1.$$

LEMMA 3.3. $\displaystyle\sum_{x \cup a = b} \mu(x) \begin{cases} = \mu(b) \ if \ a = z \\ = 0 \quad\ if \ a \neq z. \end{cases}$

For if $b = z$, there are terms in the sum only if $a = z$ and in this case the single term if $\mu(z)$. Thus the lemma holds in this case and we now make an induction on the element b. Suppose the lemma holds for all $b < b_1$. If $a = z$, then

$$\sum_{x \cup a = b_1} \mu(x) = \sum_{x = b_1} \mu(x) = \mu(b_1)$$

If $a \neq z$, we may clearly suppose that $a \leq b_1$ since otherwise there are no terms in the sum. But then

$$\sum_{x \cup a = b_1} \mu(x) = \sum_{b < b_1} \sum_{x \cup a = b} \mu(x) + \sum_{x \cup a = b_1} \mu(x) = \sum_{b \leq b_1} \sum_{x \cup a = b} \mu(x)$$

$$= \sum_{x \cup a \leq b_1} \mu(x) = \sum_{x \leq b_1} \mu(x) = 0 \qquad \text{by Lemma 3.2.}$$

Since L is finite, the lemma follows by induction.

4. Proof of the theorem

Throughout this section L will denote a finite modular lattice. If a/b is a quotient lattice of L, the symbols $V_k(a/b)$ and $W_k(a/b)$ will be used to indicate the sets V_k and W_k for the lattice a/b. If $a \, \epsilon \, L$, let a^* denote the join of all elements of L covering a. Dually, let a_* be the meet of all elements covered by a. Clearly $x \cap a^* = a$ implies $x = a$. For if $x \neq a$, then $x \geq y$ where y covers a and hence $a = x \cap a^* \geq y$ which is impossible. Similarly $x \cup a_* = a$ implies $x = a$. Since z^* is a join of points of a modular lattice it follows that $L^* = z^*/z$ is a finite complemented modular lattice. In order to distinguish the sets V_k and W_k for L^* we shall use the symbols V_k^* and W_k^*. Note that the functional μ for L^* is just the functional μ of L restricted to L^*. For, if $a \leq z^*$, the number of sets of k distinct points of L whose join is a is precisely the same as the number of sets of k distinct points of L^* whose join is a.

If $a \, \epsilon \, L$, $C(a)$ will denote the number of elements covering a in L.

LEMMA 4.1. $\displaystyle\sum_{a \, \epsilon \, L^*} \mu(a) N(V_k(u/a)) = \begin{cases} 1 \ if \ C(z) = k \\ 0 \ if \ C(z) \neq k. \end{cases}$

For $V_k(u/a)$ consists of those elements of V_k which contain a. Hence if $b \, \epsilon \, L^*$ the set $T_k(b)$ of those elements $x \, \epsilon \, V_k$ such that $x \cap z^* = b$ is a subset of $V_k(u/a)$ whenever $b \geqq a$. Also if $x \, \epsilon \, V_k(u/a)$, then $x \cap z^* = b$ for some $b \, \epsilon \, L^*$ such that $b \geqq a$. Hence $V_k(u/a) = \vee_{b \geqq a} T_k(b)$ and this representation is clearly disjoint. Hence

$$N(V_k(u/a)) = \sum_{b \geqq a} N(T_k(b)).$$

But then

$$\sum_{a \epsilon L^*} \mu(a) N(V_k(u/a)) = \sum_{a \epsilon L^*} \sum_{b \geqq a} \mu(a) \, N(T_k(b))$$
$$= \sum_{b \epsilon L^*} N(T_k(b)) \sum_{a \leqq b} \mu(a)$$
$$= N(T_k(z)) \qquad \text{by Lemma 3.2}$$

Now $x \, \epsilon \, T_k(z)$ implies $x \cap z^* = z$ which implies $x = z$. Thus $T_k(z)$ is non-empty only if $C(z) = k$ in which case $T_k \, (z)$ consists of the single element z. Thus $N(T_k(z)) = 1$ if $C(z) = k$ and is zero otherwise. Thus the proof is complete.

Now if $x \, \epsilon \, L$, let $S_k(x) = \{a \, \epsilon \, L^* \mid x \, \epsilon \, W_k(u/a)\}$.

LEMMA 4.2. *If* $x \, \epsilon \, L^*$, *then* $\sum_{a \epsilon S_k(x)} \mu(a) = 0$.

We begin by defining a relation over L^* as follows:

$$a \sim b \text{ if and only of } a \, \cup \, (x_* \cap z^*) = b \, \cup \, (x_* \cap z^*).$$

This relation is clearly an equivalence relation over L^*. Let A be an equivalence class. Then if a_0 is a fixed element of A, $a \, \cup \, (x_* \cap z^*) = a_0 \, \cup \, (x_* \cap z^*)$ for all $a \, \epsilon \, A$ and this equality in turn implies that $a \, \epsilon \, A$. Furthermore $x_* \cap z^* \neq z$ since otherwise $x_* = z$ and hence $x \leqq (x_*)^* \leqq z^*$ contrary to $x \, \epsilon \, L^*$. Thus by Lemma 3.3

$$\sum_{a \epsilon A} \mu(a) = 0$$

Next let $S_k(x)$ and A have an element a in common. Let b be any other element of A. Then if x covers y where $y \geqq a$, we have $y \geqq x_*$ and thus $y \geqq (x_* \cap z^*) \, \cup \, a = (x_* \cap z^*) \, \cup \, b \geqq b$. Similarly if x covers y where $y \geqq b$, then $y \geqq a$. It follows that x covers exactly the same set of elements in u/a as in u/b. But $a \, \epsilon \, S_k(x)$ implies x covers k elements in u/a and thus x also covers k elements in u/b. But then $b \, \epsilon \, S_k(x)$. Since b is an arbitrary element of A we have $A \subseteqq S_k(x)$. It follows that $S_k(x)$ is the union of the equivalence classes which it contains. Thus

$$\sum_{a \epsilon S_k(x)} \mu(a) = \sum_{A \subseteqq S_k(x)} \sum_{a \epsilon A} \mu(a) = 0$$

and the proof is complete.

LEMMA 4.3. $\sum_{a \epsilon L^*} \mu(a) \, N(W_k(u/a)) = \begin{cases} 1 \ \textit{if } C(z) = k \\ 0 \ \textit{if } C(z) \neq k. \end{cases}$

Clearly

$$\sum_{a \epsilon L^*} \mu(a) \, N(W_k(u/a)) = \sum_{a \epsilon L^*} \mu(a) \sum_{x \epsilon L} \varphi_k(a, \, x)$$

where $\varphi_k(a, x) = 1$ if x covers k elements in u/a and is zero otherwise. Hence

$$\sum_{a \epsilon L^*} \mu(a) \, N(W_k(u/a)) = \sum_{x \epsilon L} \sum_{a \epsilon L^*} \mu(a)\varphi_k(a, x)$$

$$= \sum_{x \epsilon L} \sum_{a \epsilon S_k(x)} \mu(a)$$

$$= \sum_{x \epsilon L^*} \sum_{a \epsilon S_k(x)} \mu(a) \qquad \text{by Lemma 4.2}$$

$$= \sum_{x \epsilon L^*} \sum_{a \epsilon L^*} \mu(a)\varphi_k(a, x)$$

$$= \sum_{a \epsilon L^*} \mu(a) \sum_{x \epsilon L^*} \varphi_k(a, x)$$

$$= \sum_{a \epsilon L^*} \mu(a) \, N(W_k^*(z^*/a)).$$

Since L^* is complemented, the results of Section 2 imply that

$$N(W_k^*(z^*/a)) = N(V_k^*(z^*/a)).$$

Thus

$$\sum_{a \epsilon L} \mu(a) \, N(W_k(u/a)) = \sum_{a \epsilon L^*} \mu(a) \, N(V_k^*(z^*/a))$$

$$= \begin{cases} 1 \text{ if } C^*(z) = k \\ 0 \text{ if } C^*(z) \neq k \end{cases} \qquad \text{by Lemma 4.1}$$

But $C^*(z) = C(z)$ since z^* contains all of the points of L and hence the proof is complete.

The theorem is deduced from these results as follows: It is trivially true for the lattice of length 1. Hence we proceed by induction and assume that the theorem holds for all proper quotient lattices of L. From Lemmas 4.1 and 4.3 we have

$$\sum_{a \epsilon L^*} \mu(a) \, [N(V_k(u/a)) - N(W_k(u/a))] = 0.$$

By the induction hypothesis, $N(V_k(u/a)) - N(W_k(u/a)) = 0$ for all $a \neq z$. Thus

$$\mu(z) \, [N(V_k(u/z)) - N(W_k(u/z))] = 0.$$

But $\mu(z) = 1$, $V_k(u/z) = V_k$, $W_k(u/z) = W_k$ and hence we conclude that $N(V_k) = N(W_k)$. Thus the proof of the theorem is complete.

CALIFORNIA INSTITUTE OF TECHNOLOGY

REFERENCES

1. G. BIRKHOFF. *Lattice Theory*, Revised edition, Amer. Math. Soc. Coll. Publications, Vol. 25, 1948.
2. L. WEISNER, *Abstract theory of inversion of finite series*, Trans. Amer. Math. Soc., 38 (1935) 474–484.

DISTRIBUTIVITY IN LATTICES

By R. P. Dilworth and J. E. McLaughlin

1. **Introduction.** A lattice L is infinitely (join) distributive if $a \cap \cup_B b = \cup_B (a \cap b)$ whenever the indicated joins exist in L. Clearly infinite distributivity implies ordinary distributivity. On the other hand it is easy to give examples of distributive lattices which are not infinitely distributive. For example, the rational integers under the relation of division form a distributive lattice which is not infinitely distributive. However for complemented lattices, as observed by Tarski [8] and von Neumann [6], distributivity implies infinite distributivity.

Another curious consequence of complementation is the following theorem due to Stone [7] and Glivenko [5]. Let a subset A of a Boolean algebra $\mathfrak{B}$ be called *normally closed* if A contains all lower bounds of its set of upper bounds. Then the collection of closed subsets of $\mathfrak{B}$ is a Boolean algebra under set inclusion and hence is infinitely distributive by the theorem mentioned above. On the other hand, Funayama [4] and Cotlar [2] have given examples of distributive lattices whose closed subsets form non-distributive (indeed, non-modular) lattices. In §6 we give an example of a lattice which is even infinitely distributive and whose lattice of normally closed subsets is non-modular.

The purpose of this paper is to formulate and study the principles upon which these results rest.

We first introduce a general class of distributive laws. Let ϕ be an imbedding operator (Ward [9]) on a lattice L. Thus ϕ is a closure operation (Birkhoff [1]) on L such that $\phi(a) = (a)$ where (a) denotes the set of all $x \leq a$. L is said to be ϕ-*distributive* if

$$a \cap \phi(S) = \phi(a \cap S)$$

for all $a \in L$ and all subsets S of L.

Thus if ϕ is the ideal operator, ϕ-distributivity is ordinary distributivity. Similarly, if ϕ is the complete ideal operator, then ϕ-distributivity is infinite distributivity. (A subset A of L is a *complete ideal* if $S \subseteq A$ and $\cup S$ exists in L, then $\cup S \in A$.)

The imbedding operators on L have a natural partial ordering defined by $\phi \geq \psi$ if and only if $\phi(S) \supseteq \psi(S)$ for all subsets S of L. The null operator under this partial ordering is the operator ω defined by

$$\omega(S) = \{x \mid x \leq s \text{ some } s \in S\}.$$

It is easily shown that every lattice L is ω-distributive. On the other hand, the unit operator ν with respect to the partial ordering is the normal operator

Received February 25, 1952

683

mentioned above. Among the commonly used imbedding operators, the operator ν gives the strongest form of distributivity. Thus the term "normal distributivity" will be used for ν-distributivity. As a strong form of the Tarski-von Neumann theorem we prove that *every Boolean algebra is normally distributive.*

If ϕ is an imbedding operator on L, the collection of ϕ-closed subsets (that is, sets A for which $\phi(A) = A$) form a complete lattice L_ϕ containing L as a sublattice. L_ϕ is called the ϕ-completion of L. That part of the Stone-Glivenko theorem which concerns distributivity is then contained in the following theorem:

Theorem 3.4. *L is ϕ-distributive if and only if L_ϕ is normally distributive.*

Since it will also be shown that in complete lattices normal distributivity is equivalent to infinite distributivity, normal distributivity can be replaced by infinite distributivity in the statement of the theorem. If we take ϕ to be the normal operator ν and apply the general form of the Tarski-von Neumann theorem given above, this theorem gives the distributivity conclusion of the Stone-Glivenko theorem.

If ϕ is a given imbedding operator and $\mathfrak{S}$ is a collection of subsets of L containing all one-element subsets, then ϕ and $\mathfrak{S}$ generate a new imbedding operator $\phi_\mathfrak{S}$ such that A is $\phi_\mathfrak{S}$-closed if and only if $S \subseteq A$ and $S \ \varepsilon \ \mathfrak{S}$ implies $\phi(S) \subseteq A$. The operators $\phi_\mathfrak{S}$ are said to be *associated* with ϕ. A collection $\mathfrak{S}$ is said to be *meet-complete* with respect to ϕ if $a \cap \phi(S) = \phi(a \cap S)$ all $S \ \varepsilon \ \mathfrak{S}$ implies $a \cap S \ \varepsilon \ \mathfrak{S}$ all $S \ \varepsilon \ \mathfrak{S}$. The following theorem is proved:

Theorem 4.1. *Let $\mathfrak{S}$ and $\mathfrak{T}$ be meet-complete collections of subsets of L such that $\phi_\mathfrak{S} \geq \phi_\mathfrak{T}$. Then $\phi_\mathfrak{S}$-distributivity implies $\phi_\mathfrak{T}$-distributivity.*

If $\mathfrak{S}$ is the collection of all subsets of L, then $\phi_\mathfrak{S} = \phi$ and $\phi \geq \phi_\mathfrak{T}$ all $\mathfrak{T}$. Thus ϕ-distributivity implies $\phi_\mathfrak{T}$-distributivity for all meet-complete subsets $\mathfrak{T}$. It is shown that all of the common imbedding operators are associated with the normal operator ν and are generated by meet-complete collections of subsets. Hence normal distributivity implies all of the common types of distributivity.

As a final application, it is proved that *the latice of bounded continuous real functions on a topological space is normally distributive (in the restricted sense).* Hence combining Theorem 3.4 with Theorem 6.1 of Dilworth [3] we have

Theorem 6.5. *The normal completion of the lattice of bounded continuous function on a topological space is infinitely distributive.*

2. Imbedding operators. We begin by recalling some of the basic properties of imbedding operators on lattices. A mapping ϕ which maps the subsets of a given set S into subsets of S is called a *closure operator on S* (Birkhoff [1]) if the following conditions are satisfied:

(1) $\phi(A) \supseteq A$

(2) $A \supseteq B$ implies $\phi(A) \supseteq \phi(B)$

(3) $\phi(\phi(A)) = \phi(A)$.

A set A is ϕ-*closed* if $\phi(A) = A$. The collection of ϕ-closed subsets of S form a complete lattice L_ϕ under set inclusion.

If A consists of a single element a we shall write $\phi(A) = \phi(a)$. Now let L be a lattice with containing relation $x \leq y$. If $a \,\varepsilon\, L$, let (a) denote the set of all x such that $x \leq a$. A closure operator ϕ on L is an *imbedding* operator if

(4) $\phi(a) = (a)$ all $a \,\varepsilon\, L$.

If ϕ is an imbedding operator, the *principal* closed subsets (a) form a partially ordered set within L_ϕ which is isomorphic to L. L_ϕ is called the ϕ-completion of L.

The imbedding operators on L have a natural partial ordering given by $\phi \geq \psi$ if and only if $\phi(A) \supseteq \psi(A)$ all subsets A of L. Under this containing relation the set of imbedding operators on L form a complete lattice Φ. The unit element of Φ is the *normal* operator ν defined by

$$\nu(A) = (A^*)_*$$

where $A^* = \{x \mid a \leq x \mid \text{all } a \,\varepsilon\, A\}$ and $A_* = \{x \mid x \leq a \text{ all } a \,\varepsilon\, A\}$. The null element of Φ is the *minimal* operator ω defined by

$$\omega(A) = \{x \mid x \leq a \text{ some } a \,\varepsilon\, A\}.$$

Finally, we remark that if $\mathfrak{A}$ is a collection of ω-closed subsets of L containing all principal sets (a) and closed under set intersection, then $\mathfrak{A}$ generates an imbedding operator ϕ such that

$$\phi(S) = \cap\{A \,\varepsilon\, \mathfrak{A} \mid A \supseteq S\}.$$

The ϕ-closed subsets are precisely the sets of A. Thus if $\mathfrak{A}$ is the collection of ideals of L, ϕ is the ideal operator. Similarly, the σ-ideals (that is, ideals closed under countable unions whenever they exist) generate the σ-ideal operator and the complete ideals (that is, ideals closed under unrestricted union whenever it exists) generate the complete ideal operator.

3. **Operator distributivity.** If S is a subset of the lattice L, $x \cap S$ will denote the set of elements $x \cap s$ where $s \,\varepsilon\, S$. Now let ϕ be a given imbedding operator defined on L.

DEFINITION 3.1. L is said to be ϕ-*distributive* if

$$x \cap \phi(S) = \phi(x \cap S)$$

for all $x \,\varepsilon\, L$ and all subsets S of L.

Now $x \cap \phi(s)$ is always ϕ-closed. For $x \cap \phi(S) \subseteq \phi(S)$, (x) implies $\phi(x \cap \phi(S)) \subseteq \phi(\phi(S)) = \phi(S)$ and $\phi(x \cap \phi(S)) \subseteq \phi(x) = (x)$. Hence $\phi(x \cap \phi(S)) \subseteq (x) \cap \phi(S) = x \cap \phi(S)$. But then $x \cap S \subseteq x \cap \phi(S)$ implies $\phi(x \cap S) \subseteq \phi(x \cap \phi(S)) = x \cap \phi(S)$ and hence the inclusion $x \cap \phi(S) \supseteq \phi(x \cap S)$ always holds. Thus ϕ-distributivity is equivalent to the formally weaker condition $x \cap \phi(S) \subseteq \phi(x \cap S)$.

The following theorem gives an equivalent form for ϕ-distributivity which is analogous to the formulation $x \leq a \cup b$ implies $x \leq (a \cap x) \cup (b \cap x)$ for ordinary distributivity.

THEOREM 3.1   *L is ϕ-distributive if and only if*

$$x \, \varepsilon \, \phi(S) \; implies \; x \, \varepsilon \, \phi(x \cap S)$$

for all subsets S of L.

Proof. If L is ϕ-distributive and $x \, \varepsilon \, \phi(A)$, then $(x) \subseteq \phi(A)$ and hence $x \, \varepsilon \, (x) = (x) \cap \phi(A) = x \cap \phi(A) = \phi(x \cap A)$. Conversely, if the condition of the theorem is satisfied, let $y \, \varepsilon \, x \cap \phi(S)$. Then $y \, \varepsilon \, \phi(S)$ by (4) and hence $y \, \varepsilon \, \phi(y \cap S)$. But since $y \leq x$, $y \cap s \leq x \cap s$ all $s \, \varepsilon \, A$ and thus $\phi(y \cap S) \subseteq \phi(x \cap S)$. It follows that $y \, \varepsilon \, \phi(x \cap S)$ and hence $x \cap \phi(S) \subseteq \phi(x \cap S)$.

COROLLARY.   *Every lattice L is ω-distributive.*

For $x \, \varepsilon \, \omega(S)$ implies $x \leq s$ some $s \, \varepsilon \, S$. But then $x \leq x \cap s$ some $s \, \varepsilon \, S$ and hence $x \, \varepsilon \, \omega(x \cap S)$.

The concept of *normal* distributivity associated with the normal operator ν can be formulated in the following alternative forms.

THEOREM 3.2.   *Each of the following three conditions is equivalent to normal distributivity.*
$N_1.$ $x \, \varepsilon \, \nu(S)$ *implies* $x = \cup(x \cap S)$
$N_2.$ *If $y \geq s$ all $s \, \varepsilon \, S$ implies $y \geq x$, then $w \geq x \cap s$ all $s \, \varepsilon \, S$ implies $w \geq x$.*
$N_3.$ *$x \cap s \leq y < x$ all $s \, \varepsilon \, S$ implies that c exists such that $s \leq c < c \cup x$ all $s \, \varepsilon \, S$.*

Proof. In order to show that N_1 is equivalent to normal distributivity it will be sufficient to prove that $x \, \varepsilon \, \nu(x \cap S)$ if and only if $x = \cup(x \cap S)$. But if $x \, \varepsilon \, \nu(x \cap S)$, then all elements containing the element of $x \cap S$ also contain x and since x clearly contains all elements of $x \cap S$ we have $x = \cup(x \cap S)$. The argument can clearly be reversed and hence N_1 is equivalent to normal distributivity. N_2 is simply N_1 written out in detail and N_3 is obtained from N_2 by inverting the logic.

For complete lattices normal distributivity reduces to the familiar notion of infinite distributivity.

THEOREM 3.3.   *Let L be complete. Then L is normally distributive if and only if L is infinitely distributive.*

Proof. If L is normally distributive and S is a subset of L having a union $\cup S$, then $y \geq s$ all $s \, \varepsilon \, S$ implies $y \geq \cup S \geq x \cap \cup S$. Thus $x \cap \cup S \, \varepsilon \, \nu(S)$. By N_1, $x \cap \cup S = \cup [(x \cap \cup S) \cap S] = \cup(x \cap S)$ and L is infinitely distributive. Conversely, if L is infinitely distributive and $x \, \varepsilon \, \nu(S)$, then $x \leq \cup(S)$ and hence $x = x \cap \cup(S) = \cup(x \cap S)$. Thus L is normally distributive.

From the first part of the proof of the preceding theorem it is clear that in any lattice, normal distributivity implies infinite distributivity. For an example of an infinitely distributive lattice which is not normally distributive see §6.

ϕ-distributivity for an arbitrary imbedding operator ϕ is closely related to normal distributivity as the following theorem shows.

THEOREM 3.4. *L is ϕ-distributive if and only if L_ϕ is normally distributive.*

Proof. If L_ϕ is normally distributive, let $x \; \varepsilon \; \phi(S)$ where S is a subset of L. Then since $\phi(S)$ is the smallest ϕ-closed set containing all of the elements of S we have $\phi(S) = \cup_s(S)$ where the union symbol refers to L_ϕ. But $x \; \varepsilon \; \phi(S)$ implies $(x) \subseteq \phi(S) = \cup_s(S)$. By Theorem 3.3, L_ϕ is infinitely distributive and hence

$$(x) = (x) \cap \bigcup_s(S) = \bigcup_s[(x) \cap (s)] = \bigcup_s(x \cap s) = \phi(x \cap S).$$

Thus $x \; \varepsilon \; \phi(x \cap S)$ and hence L is ϕ-distributive by Theorem 3.1.

Conversely, let L be ϕ-distributive and let $\mathfrak{A}$ be a collection of ϕ-closed subsets of L. If X is a given ϕ-closed subset of L, let $x \; \varepsilon \; X \cap (\cup\mathfrak{A})$. From the general properties of closure operators $\cup \; \mathfrak{A} = \phi \; (\vee \; \mathfrak{A})$ where $\vee \; \mathfrak{A}$ denote the set union of the sets of $\mathfrak{A}$. Hence $x \; \varepsilon \; X$ and $x \; \varepsilon \; \phi(\vee \; \mathfrak{A})$. Since L is ϕ-distributive, $x \; \varepsilon \; \phi(x \cap (\vee \; \mathfrak{A}))$. But $x \cap (\vee \; \mathfrak{A}) = \vee \; (x \cap \mathfrak{A})$. Hence $x \; \varepsilon \; \phi \; (\vee(x \cap \mathfrak{A})) \subseteq \phi \; (\vee(X \cap \mathfrak{A})) = \cup \; (X \cap \mathfrak{A})$. Thus $X \cap (\cup \; \mathfrak{A}) \subseteq \cup \; (x \cap \mathfrak{A})$. But $X \cap (\cup \; \mathfrak{A}) \supseteq \cup \; (x \cap \mathfrak{A})$ trivially and hence $X \cap (\cup \; \mathfrak{A}) = \cup \; (x \cap \mathfrak{A})$. Thus L_ϕ is infinitely distributive and hence normally distributive by Theorem 3.3.

If ϕ is specialized to be the ideal operator, Theorem 3.4 gives the well-known result that the lattice of ideals of a distributive lattice is infinitely distributive.

Specializing ϕ to be the normal operator ν we get the following corollary:

COROLLARY. *The normal completion of a lattice L is infinitely distributive if and only if L is normally distributive.*

4. Associated imbedding operators.

A survey of the common imbedding operators shows that for these operators, ϕ-distributivity implies ψ-distributivity whenever $\phi \geq \psi$. Nevertheless, this property does not hold in general. For example, let L be the Boolean algebra with three points p_1, p_2, and p_3. Since L is finite the only ν-closed sets are the principal sets. Let ϕ be defined by the requirement that all of the principal sets and the set $\{p_1, p_2, z\}$ be ϕ-closed. Then $\nu \geq \phi$ and L is ν-distributive. On the other hand L is not ϕ-distributive since $(p_1 \cup p_2) \cap \phi(\{p_1 \cup p_3, p_2 \cup p_3\}) = \{p_1 \cup p_2, p_1, p_2, z\}$ while $\phi(p_1 \cup p_2) \cap (\{p_1 \cup p_3, p_2 \cup p_3\}) = \{p_1, p_2, z\}$. It is thus clear that if the above property is to hold, the class of imbedding operators must be restricted in some manner.

Let ϕ be an imbedding operator L and let $\mathfrak{S}$ be a collection of subsets of L containing all of the one-element subsets. Associated with ϕ and $\mathfrak{S}$ is an imbedding operator defined as follows:

DEFINITION 4.1. A subset A of L is $\phi_\mathfrak{S}$-*closed* if

$$(4.1) \qquad S \subseteq A \quad \text{and} \quad S \; \varepsilon \; \mathfrak{S} \quad \text{imply} \quad \phi(S) \subseteq A.$$

It is clear that the intersection of any collection of $\phi_\mathfrak{S}$-closed subsets is again a $\phi_\mathfrak{S}$-closed subset and hence the $\phi_\mathfrak{S}$-closed subsets generate a closure operator

$\phi_\mathfrak{S}$. If A is a ϕ-closed subset containing x, then $\{x\} \subseteq A$ and $\{x\} \, \varepsilon \, \mathfrak{S}$ and hence $(x) = \phi(x) \subseteq A$. On the other hand, $S \subseteq (x)$ implies $\phi(S) \subseteq \phi((x)) = (x)$ all S and hence (x) is $\phi_\mathfrak{S}$-closed. Thus $\phi_\mathfrak{S}(x) = (x)$ and hence $\phi_\mathfrak{S}$ is an imbedding operator.

LEMMA 4.1. *For every collection* $\mathfrak{S}$, $\phi \geq \phi_\mathfrak{S}$. *If* $\mathfrak{S}$ *consists of all subsets of* L, *then* $\phi = \phi_\mathfrak{S}$.

For if $S \subseteq \phi(A)$ and $S \, \varepsilon \, \mathfrak{S}$, then $\phi(S) \subseteq \phi(\phi(A)) = \phi(A)$ and thus $\phi(A)$ is $\phi_\mathfrak{S}$-closed. It follows that $\phi(A) \supseteq \phi_\mathfrak{S}(A)$ all A and hence $\phi \geq \phi_\mathfrak{S}$. If $\mathfrak{S}$ contains all subsets of L, then $A \, \varepsilon \, \mathfrak{S}$ and hence $\phi_\mathfrak{S}(A) \supseteq \phi(A)$.

LEMMA 4.2. *If* $\mathfrak{S} \supseteq \mathfrak{T}$, *then* $\phi_\mathfrak{S} \geq \phi_\mathfrak{T}$.

For if $T \subseteq \phi_\mathfrak{S}(A)$ and $T \, \varepsilon \, \mathfrak{T}$, then $T \, \varepsilon \, \mathfrak{S}$ and hence $\phi(T) \subseteq \phi_\mathfrak{S}(A)$. Thus $\phi_\mathfrak{S}(A)$ is $\phi_\mathfrak{T}$-closed and hence $\phi_\mathfrak{S} \geq \phi_\mathfrak{T}$.

The converse of Lemma 4.2 does not hold in general since different collections $\mathfrak{S}$ and $\mathfrak{T}$ may generate the same imbedding operator. Let us call a collection $\mathfrak{S}$ *maximal* with respect to ϕ if

$$(4.2) \qquad\qquad \phi_\mathfrak{S}(S) = \phi(S) \qquad \text{implies} \qquad S \, \varepsilon \, \mathfrak{S}.$$

LEMMA 4.3. *For every collection* $\mathfrak{S}$, *there exists a collection* $\mathfrak{S}' \supseteq \mathfrak{S}$ *such that* $\mathfrak{S}'$ *is maximal with respect to* ϕ *and* $\phi_{\mathfrak{S}'} = \phi_\mathfrak{S}$.

Let $\mathfrak{S}'$ be the collection of all S such that $\phi_\mathfrak{S}(S) = \phi(S)$. If $S \, \varepsilon \, \mathfrak{S}$, then since $S \subseteq \phi_\mathfrak{S}(S)$ we have $\phi(S) \subseteq \phi_\mathfrak{S}(S) \subseteq \phi(S)$ by lemma 4.1. Thus $\phi(S) = \phi_\mathfrak{S}(S)$ and hence $S \, \varepsilon \, \mathfrak{S}'$. It follows that $\mathfrak{S}' \geq \mathfrak{S}$ and hence $\phi_{\mathfrak{S}'} \geq \phi_\mathfrak{S}$ by lemma 4.2. Now let $S' \subseteq \phi_\mathfrak{S}(A)$ where $S' \, \varepsilon \, \mathfrak{S}'$. Then $\phi(S') = \phi_\mathfrak{S}(S') \subseteq \phi_\mathfrak{S}(\phi_\mathfrak{S}(A)) = \phi_\mathfrak{S}(A)$. Thus $\phi_\mathfrak{S}(A)$ is $\phi_{\mathfrak{S}'}$-closed and hence $\phi_\mathfrak{S} \geq \phi_{\mathfrak{S}'}$. It follows that $\phi_\mathfrak{S} = \phi_{\mathfrak{S}'}$. Clearly $\phi_{\mathfrak{S}'}(S) = \phi(S)$ implies $S \, \varepsilon \, \mathfrak{S}'$ and hence $\mathfrak{S}'$ is maximal with respect to ϕ.

The following lemma is a partial converse to Lemma 4.2.

LEMMA 4.4 *If* $\phi_\mathfrak{S} \geq \phi_\mathfrak{T}$ *and* $\mathfrak{S}$ *is maximal with respect to* ϕ, *then* $\mathfrak{S} \supseteq \mathfrak{T}$.

For if $T \, \varepsilon \, \mathfrak{T}$, then $T \subseteq \phi_\mathfrak{T}(T)$ and hence $\phi(T) \subseteq \phi_\mathfrak{T}(T)$. But then $\phi_\mathfrak{S}(T) \subseteq \phi(T) \subseteq \phi_\mathfrak{T}(T)$. Since $\phi_\mathfrak{S} \geq \phi_\mathfrak{T}$ we have $\phi_\mathfrak{S}(T) = \phi_\mathfrak{T}(T) = \phi(T)$. But then $T \, \varepsilon \, \mathfrak{S}$ since $\mathfrak{S}$ is maximal with respect to ϕ. Hence $\mathfrak{S} \supseteq \mathfrak{T}$.

LEMMA 4.5. *If* $\phi \geq \psi$, *then* $\phi_\mathfrak{S} \geq \psi_\mathfrak{S}$.

For $S \, \varepsilon \, \phi_\mathfrak{S}(A)$ and $S \, \varepsilon \, \mathfrak{S}$ imply $\psi(S) \subseteq \phi(S) \subseteq \phi_\mathfrak{S}(A)$ since $\phi_\mathfrak{S}(A)$ is $\phi_\mathfrak{S}$-closed. Thus $\phi_\mathfrak{S}(A)$ is $\psi_\mathfrak{S}$-closed and the lemma follows.

LEMMA 4.6. *If* $\phi \geq \psi \geq \phi_\mathfrak{S}$, *then* $\phi_\mathfrak{S} = \psi_\mathfrak{S}$.

For let $S \subseteq \psi_\mathfrak{S}(A)$ where $S \, \varepsilon \, \mathfrak{S}$. Then since $\psi_\mathfrak{S}(A)$ is $\psi_\mathfrak{S}$-closed we have $\psi(S) \subseteq \psi_\mathfrak{S}(A)$. On the other hand since $\phi_\mathfrak{S}(S)$ is $\phi_\mathfrak{S}$-closed we have $\phi(S) \subseteq \phi_\mathfrak{S}(S)$. But then $\phi(S) \subseteq \phi_\mathfrak{S}(S) \subseteq \psi(S) \subseteq \psi_\mathfrak{S}(A)$. Hence $\psi_\mathfrak{S}(A)$ is $\phi_\mathfrak{S}$-closed and thus $\psi_{\mathfrak{S}} \geq \phi_{\mathfrak{S}}$. Lemma 4.5 gives the reverse inclusion and the proof of the lemma is complete.

Now if $\mathfrak{S}$ and $\mathfrak{T}$ are such that $\phi_{\mathfrak{S}} \geq \phi_{\mathfrak{T}}$, it is not generally true that $\phi_{\mathfrak{T}}$-distributivity will follow from $\phi_{\mathfrak{S}}$-distributivity. Indeed, the operator described in the first paragraph of this section is $\nu_{\mathfrak{S}}$ where $\mathfrak{S}$ consists of all subsets of L except $\{p_1, p_2\}$ and $\{p_1, p_2, z\}$. Thus the distributivity theorem can be proved only under further restrictions upon the collection $\mathfrak{S}$.

DEFINITION 4.2. A collection $\mathfrak{S}$ is *meet complete* with respect to ϕ if $x \cap \phi(S) = \phi(x \cap S)$ all $S \,\varepsilon\, \mathfrak{S}$ implies $x \cap S \,\varepsilon\, \mathfrak{S}$ all $S \,\varepsilon\, \mathfrak{S}$.

In particular, $\mathfrak{S}$ is meet complete with respect to ω if and only if $S \,\varepsilon\, \mathfrak{S}$ implies $x \cap S \,\varepsilon\, \mathfrak{S}$ for all $x \,\varepsilon\, L$.

Let us call an element $x \,\varepsilon\, L$ ϕ-*distributive* if $x \cap \phi(A) = \phi(x \cap A)$ all subsets A of L. Clearly L is ϕ-distributive if and only if every $x \,\varepsilon\, L$ is ϕ-distributive.

LEMMA 4.7. *If $\mathfrak{S}$ is meet complete with respect to ϕ, then x is $\phi_{\mathfrak{S}}$-distributive if and only if $x \cap \phi(S) = \phi(x \cap S)$ all $S \,\varepsilon\, \mathfrak{S}$.*

For if x is $\phi_{\mathfrak{S}}$-distributive, then $x \cap \phi(S) = x \cap \phi_{\mathfrak{S}}(S) = \phi_{\mathfrak{S}}(x \cap S) \subseteq \phi(x \cap S)$. But $x \cap \phi(S) \supseteq \phi(x \cap S)$ all S and hence $x \cap \phi(S) = \phi(x \cap S)$ all $S \,\varepsilon\, \mathfrak{S}$. Conversely, if the condition of the lemma is satisfied for some $x \,\varepsilon\, L$, let V be an arbitrary subset of L and let A denote the set of all y such that $x \cap y \,\varepsilon\, \phi_{\mathfrak{S}}(x \cap V)$. If $S \subseteq A$ where $S \,\varepsilon\, \mathfrak{S}$, then $x \cap S \subseteq \phi_{\mathfrak{S}}(x \cap V)$ and $x \cap S \,\varepsilon\, \mathfrak{S}$ since $\mathfrak{S}$ is meet complete with respect to ϕ. It follows that

$$x \cap \phi(S) = \phi(x \cap S) \subseteq \phi_{\mathfrak{S}}(x \cap V).$$

Thus $\phi(S) \subseteq A$ and consequently A is $\phi_{\mathfrak{S}}$-closed. Now clearly $V \subseteq A$ and hence

$$x \cap \phi_{\mathfrak{S}}(V) \subseteq x \cap \phi_{\mathfrak{S}}(A) = x \cap A \subseteq \phi_{\mathfrak{S}}(x \cap V).$$

It follows that $x \cap \phi_{\mathfrak{S}}(V) = \phi_{\mathfrak{S}}(x \cap V)$ for all subsets V of L and thus x is $\phi_{\mathfrak{S}}$-distributive. This completes the proof of the lemma.

COROLLARY. *If x is ϕ-distributive and $\mathfrak{S}$ is meet-complete with respect to ϕ, then x is $\phi_{\mathfrak{S}}$-distributive.*

The set $\mathfrak{S}$ associated with the example given at the beginning of this section is clearly not meet complete since $(p_1 \cup p_2) \cap (\{p_1 \cup p_3, p_2 \cup p_3\}) = \{p_1, p_2\}$ $\notin \mathfrak{S}$ while $\{p_1 \cup p_3, p_2 \cup p_3\} \,\varepsilon\, \mathfrak{S}$.

We can prove now the distributivity theorem for operators associated with ϕ.

THEOREM 4.1. *If x is $\phi_{\mathfrak{S}}$-distributive where $\phi_{\mathfrak{S}} \geq \phi_{\mathfrak{T}}$ and $\mathfrak{T}$ is meet complete with respect to ϕ, then x is $\phi_{\mathfrak{T}}$-distributive.*

Proof. Since $\phi \geq \phi_{\mathfrak{S}} \geq \phi_{\mathfrak{T}}$, it follows from Lemma 4.6 that $(\phi_{\mathfrak{S}})_{\mathfrak{T}} = \phi_{\mathfrak{T}}$. Also $\mathfrak{T}$ is meet complete with respect to $\phi_{\mathfrak{S}}$. For if $y \cap \phi_{\mathfrak{S}}(T) = \phi_{\mathfrak{S}}(y \cap T)$ all $T \,\varepsilon\, \mathfrak{T}$, then $y \cap \phi(T) = y \cap \phi_{\mathfrak{T}}(T) \subseteq y \cap \phi_{\mathfrak{S}}(T) = \phi_{\mathfrak{S}}(y \cap T) \subseteq \phi(y \cap T)$ all $T \,\varepsilon\, \mathfrak{T}$ and hence $y \cap T \,\varepsilon\, \mathfrak{T}$ all $T \,\varepsilon\, \mathfrak{T}$ since $\mathfrak{T}$ is meet complete with respect to ϕ. Applying the corollary to Lemma 4.7 we conclude that if x is $\phi_{\mathfrak{S}}$-distributive, then x is $\phi_{\mathfrak{T}}$-distributive.

COROLLARY. *If L is $\phi_{\mathfrak{S}}$-distributive where $\phi_{\mathfrak{S}} \geq \phi_{\mathfrak{T}}$ and $\mathfrak{T}$ is meet complete with respect to ϕ, then L is $\phi_{\mathfrak{T}}$-distributive.*

It should be noted that $\mathfrak{S}$ is always meet complete if $x \cap S \; \varepsilon \; \mathfrak{S}$ all $x \; \varepsilon \; L$ and $S \; \varepsilon \; \mathfrak{S}$.

Finally it will be shown that the property of being meet complete is preserved when $\mathfrak{S}$ is replaced by its maximal extension.

THEOREM 4.2. *If $\mathfrak{S}'$ is the maximal extension of $\mathfrak{S}$ with respect to ϕ and $\mathfrak{S}$ is meet complete with respect to ϕ, then $\mathfrak{S}'$ is meet complete with respect to ϕ.*

Proof. Let $x \cap \phi(S') = \phi(x \cap S')$ all $S' \; \varepsilon \; \mathfrak{S}'$. Then clearly $x \cap \phi(S) = \phi(x \cap S)$ all $S \; \varepsilon \; \mathfrak{S}$ and by Lemma 4.7, x is $\phi_{\mathfrak{S}}$-distributive. It follows that

$$\phi_{\mathfrak{S}}(x \cap S') = x \cap \phi_{\mathfrak{S}}(S') = x \cap \phi(S') = \phi(x \cap S')$$

Thus $x \cap S' \; \varepsilon \; \mathfrak{S}'$ all $S' \; \varepsilon \; \mathfrak{S}'$ and $\mathfrak{S}'$ is meet complete with respect to ϕ.

5. **Examples.** A subset of A of a lattice L is: (1) a J-closed subset, (2) an ideal, (3) a σ-ideal, (4) a complete ideal, or (5) a normal subset, according as it contains all elements bounded above by: (1) elements of A, (2) finite joins of elements of A, (3) countable joins of elements of A, (4) arbitrary joins of elements of A, (5) all upper bounds of the set A. Each of these ideal concepts has played an important role in the development of lattice theory and each generates a corresponding imbedding operator. Furthermore, each of these operators are associated with the normal imbedding operator ν in the sense of §4. One may take for $\mathfrak{S}$: (1) all one element subsets, (2) all finite subsets, (3) all countable subsets for which a join exists, (4) all subsets for which a join exists, (5) all subsets. Note that all of these collections are meet complete with respect to the normal operator ν. This statement is obvious for the first two collections and the last. If the countable set S, for example, has a join $\cup \, S$ and $x \cap \nu \, (S) = \nu(x \cap S)$, then $\cup \, S \; \varepsilon \; \nu(S)$ and hence every upper bound of $x \cap S$ contains $x \cap \cup \, (S)$. Thus $\nu(x \cap S) = (x \cap \cup \, S)$. It follows that $x \cap S$ is countable and has a join $x \cap \cup \, S$. Thus $x \cap S \; \varepsilon \; \mathfrak{S}$. A similar argument holds for the remaining collection. Since each collection is a subcollection of the following, it follows from Theorem 4.1 that the corresponding distributivities are progressively stronger. This can also be easily verified directly.

6. **Applications.** It will be shown in this section that a number of common examples of distributive lattices are indeed normally distributive. Now since the Stone-Glivenko theorem implies that the normal completion of a Boolean algebra is distributive and hence infinitely distributive (Tarski [8], von Neumann [6]), it follows from Theorem 3.4 that every Boolean algebra is normally distributive. This, however, can be verified directly as follows:

THEOREM 6.1. *Every Boolean algebra is normally distributive.*

Proof. Let $x \cap a \leq b < x$ all $a \; \varepsilon \; A$ and let $c = b \cup x'$. Then $a = (a \cap x)$

$\cup (a \cap x') \leq b \cup x' = c$ all $a \, \varepsilon \, A$. Also $c \neq u$ since $b \not\geq x$. Hence $c < u = c \cup x$ and the theorem follows from theorem 3.2.

Since normal distributivity implies infinite distributivity, Theorem 6.1 is a sharp form of the Tarski-von Neumann theorem. Furthermore Theorem 6.1 and Theorem 3.4 give the normal distributivity of the normal completion of a Boolean algebra.

According to Theorem 3.3 if L is complete, then it is always normally distributive if it is infinitely distributive. There are, however, infinitely distributive lattices which are not normally distributive. The following example is an infinitely distributive lattice whose normal completion is non-modular and hence

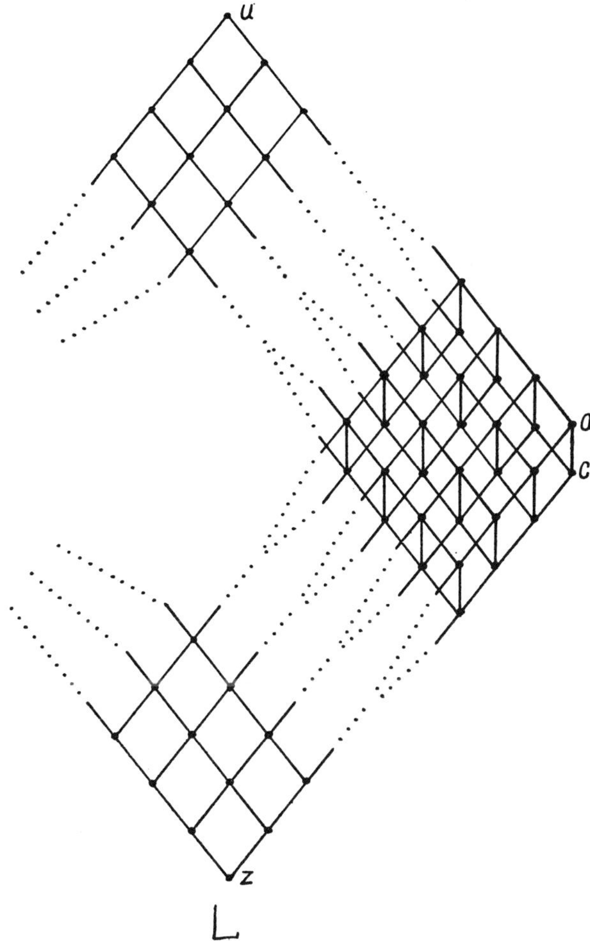

a fortiori is not normally distributive by Theorem 3.4. An inspection of the lattice diagram of L (the dots indicate that the given lattice structure is continued indefinitely and that the represented containing relations between the different parts hold) shows that an ascending chain in L has a join if and only if it has a largest element. Thus since L is clearly distributive, it is also infinitely distributive.

Now let B denote the collection of elements of finite dimension over Z. B is clearly a normal subset of L. But $(a) \cap ((c) \cup B) = (a) \cap ((a) \cup B) = (a)$ while $(c) \cup ((a) \cap B) = (c) \cup ((c) \cap B) = (c)$ and hence L_ν is non-modular.

The direct union of normally distributive lattices need not be normally distributive. Thus, for example, the direct union of two replicas of the integers under their natural ordering is not normally distributive. For $(1,2) \cap (m,1) = (1,1) \subset (1,2)$ all $m \geq 1$, but the set $(m,1)$, $m \geq 1$ has no upper bounds whatever. The reader may readily verify, however, that the following theorem holds.

THEOREM 6.2. *The direct union of normally distributive lattices with unit elements is again normally distributive.*

The anomalous behavior of normal distributivity under direct union arises from the fact that the normal operator has an anomalous behavior on unbounded sets. Hence let us consider the *restricted* normal operator ν' where $\nu'(S)$ consists of all lower bounds of the set of upper bounds of subsets of S which are bounded above. Then it follows that $\nu' = \nu_\mathfrak{S}$ where $\mathfrak{S}$ consists of all subsets of L which are bounded above and the following theorem holds.

THEOREM 6.3. *The direct union of lattices normally distributive in the restricted sense is again normally distributive in the restricted sense.*

Since a chain is clearly normally distributive we have the following corollary.

COROLLARY. *A direct union of chains is normally distributive in the restricted sense. If each chain has a unit element, then the direct union is normally distributive.*

Finally we show that lattices of continuous functions under the natural partial ordering are always normally distributive in the restricted sense.

THEOREM 6.4. *Let $C(X)$ denote the lattice of bounded, real, continuous functions on the topological space X. Then $C(X)$ is normally distributive in the restricted sense.*

Proof. There is no loss in generality if X is assumed to be completely regular. Now let F be a set of functions of $C(X)$ which are bounded above, $f \leq f_0$ all $f \in F$. Let $f \cap g \leq h < g$ all $f \in F$. Then $h(x_0) < g(x_0)$ some $x_0 \in X$ and hence there exists a neighborhood N_{x_0} of x_0 such that $h(y) < g(y)$ all $y \in N_{x_0}$. But then $f(y) \cap g(y) \leq h(y) < g(y)$ all $y \in N_{x_0}$ and $f \in F$ and thus $f(y) \leq h(y)$ all $y \in N_{x_0}$ and $f \in F$. By complete regularity there exists $k \in C(X)$ such that $0 \leq k(x) \leq 1$ all $x \in X$, $k(y) = 0$ all $y \notin N_{x_0}$ and $k(x_0) = 1$. Let

$$r(x) = h(x) \cup (1 - k(x))f_0(x)$$

Then clearly $r(x) \in C(X)$ and if $y \in N_{x_0}$ we have

$$r(y) \geq h(y) \geq f(y) \qquad \text{all} \qquad f \in F.$$

But if $y \notin N_{x_0}$, then $f(y) = h(y) \cup f_0(y) \geq f_0(y) \geq f(y)$. Thus $f \leq r$. Since $r(x_0) = h(x_0) < g(x_0)$ we have $r < r \cup g$. It follows that $C(X)$ is normally distributive in the restricted sense.

From Theorem 4.1 it follows that normal distributivity in the restricted sense implies infinite distributivity and hence we have

COROLLARY 1. *$C(X)$ is infinitely distributive.*

Since it has been shown (Dilworth [3]) that the normal completion (in the restricted sense) of $C(X)$ is isomorphic to $C(Y)$ for a suitable space Y we also have

THEOREM 6.5. *The normal completion (in the restricted sense) of $C(X)$ is infiinitely distributive.*

REFERENCES

1. G. BIRKHOFF, *Lattice Theory*, American Mathematical Society Colloquium Publications, vol. 25, Revised Edition, New York, 1948.
2. MISCHA COTLAR, *Un metodo de construccion de estructuras*, Revista de la Universidad Nacional de Tucuman (A), vol. 4(1944), pp. 105–157.
3. R. P. DILWORTH, *The normal completion of the lattice of continuous functions*, Transactions of the American Mathematical Society, vol. 68(1950), pp. 427–458.
4. NENOSUKE FUNAYAMA, *On the completion by cuts of distributive lattices*, Proceedings of the Imperial Academy, Tokyo, vol. 20(1944), pp. 1–2.
5. V. GLIVENKO, *Sur quelques points de la logique de M. Brouwer*, Bulletin of the Academy of Science, Belgium (5), vol. 15(1929), pp. 183–188.
6. J. VON NEUMANN, *Lectures on continuous geometrics* II, Princeton, 1937.
7. M. H. STONE, *The theory of representations for Boolean algebras*, Transactions of the American Mathematical Society, vol. 40(1936), pp. 37–111.
8. A. TARSKI, *Grundzüge des Systemenkalküls*, I, Fundamenta Mathematicae, vol. 25(1936), pp. 503–526.
9. MORGAN WARD, *The closure operators of a lattice*, Annals of Mathematics, vol. 43(1942), pp. 191–196.

CALIFORNIA INSTITUTE OF TECHNOLOGY.

Modular and Distributive Lattices

Algebra Universalis, **18** (1984) 4-17

0001-5240/84/010004-14$01.50+0.20/0

Aspects of distributivity

R. P. DILWORTH

Introduction

A lattice L is *distributive* if it satisfies the identity

$$a \wedge (b \vee c) = (a \wedge b) \vee (a \wedge c).$$

This identity is equivalent to the dual

$$a \vee (b \wedge c) = (a \vee b) \wedge (a \vee c).$$

It is also equivalent to the self dual identity

$$(a \wedge b) \vee (a \wedge c) \vee (b \wedge c) = (a \vee b) \wedge (a \vee c) \wedge (b \vee c).$$

It is well known that L is distributive if and only if L contains no sublattice isomorphic to either of the following lattices.

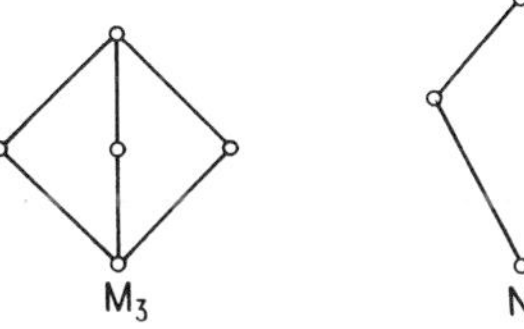

Figure 1.

Distributive lattices were among the earliest lattices studied in some detail. Out of this early work came a variety of representation theorems, embedding theorems, and structure theorems. In fact, so much was known and so many of the basic questions were answered, that it was generally felt that, except possibly for some combinatorial questions, the study of distributive lattices was too easy.

Presented by I. Rival. Received January 26, 1982. Accepted for publication in final form March 29, 1982.

Nevertheless, distributivity began to turn up in some surprising areas of lattice theory and in these areas its role was frequently fundamental. Some examples will be examined in this paper.

Unique decompositions

Let L be a distributive lattice satisfying the ascending chain condition. An element $q \in L$ is *meet irreducible* if

$$q = x \wedge y \quad \text{implies} \quad q = x \text{ or } q = y.$$

An easy inductive argument shows that each element of L is a meet of a finite number of meet irreducibles. If superfluous irreducibles are removed, then each element $x \in L$ has an irredundant representation

$$a = q_1 \wedge q_2 \wedge \cdots \wedge q_n.$$

One of the most obvious consequences of distributivity is the uniqueness of this representation. For if $q \geq a$ where q is meet irreducible, then

$$q = q \vee a = q \vee (q_1 \wedge \cdots \wedge q_n) = (q \vee q_1) \wedge \cdots \wedge (q \vee q_n).$$

Hence $q = q \vee q_i$ and $q \geq q_i$ for some i. If

$$a = q_1' \wedge \cdots \wedge q_m'$$

is another irredundant representation, then each $q_i' \geq q_j$ for some j and $q_j \geq q_k'$ for some k. Thus $q_i' \geq q_k'$ and by the irredundancy $q_i' = q_k'$ so that $q_i' = q_j$. Similarly each $q_j = q_k'$ for some k and hence the two decompositions are identical.

If the ascending chain condition does not hold, then finite irredundant decompositions need not exist. Nevertheless, in the presence of distributivity whenever an irredundant decomposition (possibly infinite) exists, it is unique. The proof for the finite case no longer applies so a different argument is required. Let

$$a = \Lambda Q = \Lambda Q'$$

be two irredundant meet representations of a. For each $q \in Q$, let $x_q = \Lambda (Q - q)$. Then $a = q \wedge x_q$ and by irredundancy $a \neq x_q$. If $q' \in Q'$, then

$$q' = q' \vee a = q' \vee (q \wedge x_q) = (q' \vee q) \wedge (q' \vee x_q).$$

Hence $q' = q' \vee q$ or $q' = q' \vee x_q$. Thus either $q' \geq q$ or $q' \geq x_q$. Now if $q' \not\geq q$ all $q' \in Q'$, then $q' \geq x_q$ all $q' \in Q'$ and

$$a = \Lambda Q' \geq x_q$$

contrary to $a \not\geq x_q$. Thus $q' \geq q$ from some $q' \in Q'$ and similarly $q_0 \geq q'$ for some $q_0 \in Q$. But then $q_0 \geq q$ and by the irredundancy we must have $q = q_0 = q'$. Hence, $Q \subseteq Q'$. If $Q \subset Q'$, then $Q \subseteq Q' - q'$ for some q' and $\Lambda Q \geq \Lambda(Q' - q') > a$ contrary to $\Lambda Q = a$. Thus $Q = Q'$ and the representations are identical.

Although this theorem is quite adequate for uniqueness, it contributes little to the question of existence. Indeed, there is no hope of obtaining an existence of decompositions of this type in general. For example, the lattice of finite subsets of the integers has no irreducibles whatever. A natural condition to insure the existence of irreducibles is compact generation.

DEFINITION 1.1. An element c of L is *compact* if

$$VS \geq c \quad \text{implies} \quad VF \geq c$$

for some finite subset F of S.

L is *compactly generated* or *algebraic* if L is complete and each element of L is a join of compact elements.

If L is compactly generated and $a \not\leq b$, then there exists a compact c such that $c \leq a$ and $c \not\leq b$. A simple application of the Hausdorff Maximal Principle gives the existence of a maximal element q such that $q \geq b$ and $c \not\leq q$. Clearly q is irreducible. In fact q is completely meet irreducible since

$$q = \Lambda S \quad \text{implies} \quad q \in S.$$

Since $a \not\leq q$ and $b \leq q$ it follows easily that each element of L is a meet of irreducibles.

Although compact generation provides the existence of many irreducibles, it does not assure the existence of irredundant decompositions. For example, the lattice of ideals of the boolean algebra of subsets of the integers is compactly generated. Nevertheless, the ideal generated by the finite subsets has no irredundant representation as a meet of irreducible ideals.

Now let L be a distributive lattice in which each element can be represented as an irredundant meet of completely meet irreducibles. If $a < 1$, let $a = \Lambda Q$ be an irredundant representation of a. Then if $q \in Q$, $s = \Lambda(Q - q) > a$. Let $q_1 = \Lambda \{x \in L \mid x > q\}$, then $q_1 > q$ and hence q_1 covers q. Since $q \vee s = q_1 \vee s$ we have

$q_1 \wedge s > q \wedge s = a$. It follows that the quotient lattice $1/a$ contains an atom. An easy extension shows that every proper quotient a/b contains an atom and hence L is strongly atomic.

Conversely, if L is compactly generated and strongly atomic, let P denote the set of atoms of $1/a$. It is easily verified that $VP > V(P-p)$ for each $p \in P$. Let q_p be an irreducible such that $q_p \geq V(P-p)$ and $q_p \not\geq VP$. Then if $Q = \{q_p \mid p \in P\}$ it is easily shown that $a = \Lambda Q$. Since $\Lambda(Q - q_p) \geq p$ the representation is irredundant.

THEOREM [4]. *Every element of a compactly generated strongly atomic distributive lattice has a unique irredundant decomposition into irreducibles.*

It should be noted that a meet irreducible element of a strongly atomic lattice is completely meet irreducible. For if q is irreducible and p covers q, then $x > q$ implies $x \geq p$ and q is completely meet irreducible.

As was observed above, a compactly generated distributive lattice which is not strongly atomic need not have irredundant meet decompositions. Hence a satisfactory decomposition theory for arbitrary compactly generated distributive lattices will require a modification of the notion of meet decomposition.

In formulating an appropriate modification let us note that in a compactly generated lattice $a = \Lambda S$ where each $s \geq a$ if and only if for each compact element c, $s \geq c$ for all $s \in S$ implies $a \geq c$. The appropriate generalization is obtained by replacing compact elements by compact dual ideals.

DEFINITION. A dual ideal C is *compact* if

$$VS \in C \quad \text{implies} \quad VF \in C$$

for some finite subset F of S.

It is easily verified that a dual ideal in a compactly generated lattice is compact if and only if it is generated by its compact elements.

An easy application of the Hausdorff Maximal Principle shows that for $a \neq 1$ there exists a minimal compact dual ideal C such that $a \notin C$.

The following lemma is the key to the decomposition theory of compactly generated distributive lattices.

LEMMA. *Let q be an irreducible of a compactly generated distributive lattice L. Then $C(q) = \{x \in L \mid x \not\leq q\}$ is a compact dual ideal of L.*

In a distributive lattice, $q \geq x \wedge y$ implies $q \geq x$ or $q \geq y$ if q is irreducible.

Hence $C(q)$ is a dual ideal. Since L is compactly generated, if $x \nleq q$, there exists $c \leq x$ such that $c \nleq q$. Hence $c \in C(q)$ and $C(q)$ is generated by its compact elements.

DEFINITION. A set Γ of compact dual ideals of a compactly generated distributive lattice is an *atomic separating* set if $a < b$ implies there exists $C \in \Gamma$ such that

$$a \nleq C, \; b \in C \quad \text{and} \quad (a) \vee C \text{ is minimal over } (a)$$

In this context, minimal means with respect to the lattice ordering of dual ideals, namely, $A \leq B$ if and only if $A \supseteq B$. Hence $(a) \vee C$ is minimal if for each compact dual ideal D such that $a \notin D$, $(a) \vee C \geq (a) \vee D$ implies $(a) \vee C = (a) \vee D$.

The generalized concept of a decomposition will be defined relative to a given atomic separating set γ of compact dual ideals. $J(S)$ will denote the dual ideal generated by S.

DEFINITION. An element a is *represented* by a set S, in symbols, $a = \mathcal{R}(s)$ if $s \geq a$ all $s \in S$ and

$$J(s) \geq C \Rightarrow a \in C \quad \text{all} \quad C \in \Gamma.$$

DEFINITION. An element q is *$\mathcal{R}$-irreducible* if $q = \mathcal{R}(s)$ implies $q \in S$. The next two lemmas show that $\mathcal{R}$-representations are in fact, closely related to ordinary meet representations. As usual, L is a compactly generated distributive lattice.

LEMMA. *If $a = \mathcal{R}(s)$, then $a = \Lambda S$.*

LEMMA. *q is $\mathcal{R}$-irreducible if and only if q is meet irreducible.*

The key to the unicity of $\mathcal{R}$-representations is the following theorem:

THEOREM. *Let L be a compactly generated distributive lattice and let C be a minimal compact dual ideal such that $a \notin C$. Then there is a unique irreducible such that $q \geq a$ and $q \notin C$.*

By the Hausdorff Maximal Principle, there exists a maximal element $q \geq a$ such that $q \notin C$. This element is irreducible and is unique.

Now let $\Gamma_a = \{C \in \Gamma \mid a \notin C\}$. If C and C' are in Γ_a, define $C \sim C'$ if and only if $(a) \vee C = (a) \vee C'$. Let $\tilde{C}$ denote the equivalence class containing C. If $C \in \Gamma_a$, let C_0 be a minimal compact dual ideal such that $C \geq C_0$ and $a \notin C_0$. Let q be the unique irreducible such that $q \geq a$, $q \notin C_0$. It can be shown that q depends only on the equivalence class $\tilde{C}$. Hence we may write $q = q(\tilde{C})$. Let $Q_a = \{q(\tilde{C}) \mid C \in \Gamma_a\}$.

THEOREM. *Let L be a compactly generated distributive lattice. Then each element a has a unique irredundant $\mathscr{R}$-representation in irreducibles given by*

$$a = \mathscr{R}(Q_a).$$

If L is strongly atomic, it is easily shown that the set of principal, compact dual ideals is an atomic separating set and the theorem reduces to the classical decomposition theorem for strongly atomic compactly generated distributive lattices.

The next lemma shows that the set of *all* compact dual ideals is an atomic separating set.

LEMMA. *Let L be a compactly generated distributive lattice and let C be a minimal compact dual ideal such that $a \notin C$. If C' is any compact dual ideal such that $a \notin C'$ and $(a) \vee C' \leq (a) \vee C$, then $(a) \vee C' = (a) \vee C$ and $C \leq C'$.*

If Γ is the set of all compact dual ideals, then the above theorem gives a unique irredundant representation in irreducibles for an arbitrary element of a compactly generated distributive lattice.

If the lattice is strongly atomic, this representation may be quite different from the classical decomposition. For example, if $a = \Lambda Q$ is a classical representation and q_0 is an irreducible such that $q_0 \geq a$, it is possible for $q_0 \not\geq q$ all $q \in Q$. On the other hand, if $a = \mathscr{R}(Q)$ is the general representation, then $q_0 \geq a$ implies $q_0 \geq q$ for some $q \in Q$.

A decomposition theory for an arbitrary distributive lattice will require the addition of new elements, since it may contain no irreducibles. The natural way to get irreducibles is to imbed the lattice in a compactly generated distributive lattice. Furthermore, it is desirable that this imbedding be as economical as possible.

The embedding of a lattice in a complete lattice is accomplished by embedding operators. A mapping φ of subsets of L into subsets of L is an embedding operator if

1) $\varphi(S) \supseteq S$

2) $S \supseteq T$ implies $\varphi(S) \supseteq \varphi(T)$

3) $\varphi(\varphi(S)) = \varphi(S)$

4) $\varphi(\{s\}) = s_*$

where $s_* = \{x \mid x \leq s\}$.

An embedding operator φ is *distributive* [5] if

$$a \wedge \varphi(S) = \varphi(a \wedge S) \quad \text{all} \quad S \subseteq L$$

where $a \wedge S$ denotes the set of elements $a \wedge s$ for $s \in S$.

An embedding operator φ is *compact* if

$$a \in \varphi(S) \quad \text{implies} \quad a \in \varphi(F)$$

for some finite subset F of S.

If L is a distributive lattice, the ideal operator θ given by

$$\theta(S) = \{x \mid x \leq VF, F \subseteq S, F \text{ finite}\}$$

is a compact distributive operator.

THEOREM. *The join of compact distributive embedding operators is compact and distributive.*

Hence there is a unique maximal compact distributive embedding operator γ on a lattice L. Clearly $\gamma \geq \theta$. Since γ is compact and distributive, the lattice L_γ of γ-closed subsets is compactly generated and distributive. Since $\gamma \geq \theta$, each γ-closed subset is an ordinary ideal of L. Hence *each element of L, as a principal ideal, has a unique irredundant representation in terms of ideals of L belonging to L_γ.*

It should be noted that if L is compactly generated, then each γ-closed subset is principal and L_γ is isomorphic to L.

Congruence distributivity

In 1942 Funayama and Nakayama [7] published a short note in which they showed that the lattice $\operatorname{Con} L$ of congruence relations on a lattice L is distributive. The key idea in the proof is the observation that the mapping

$$x \to a \vee (b \wedge x)$$

squeezes the lattice L into the quotient lattice $a \vee b / a \wedge b$. Now the algebraic properties of $\operatorname{Con} L$ are intimately related to structure properties of L. For example, if L is a subdirect product of the lattices L_α and θ_α is the kernel of the projection of L onto L_α, then

$$\omega = \Lambda_\alpha \theta_\alpha$$

where ω is the equality relation on L.

Conversely, if $\omega = \Lambda_\alpha \theta_\alpha$, then L is isomorphic to a subdirect product of the lattices L/θ_α.

Since $\operatorname{Con} L$ is a compactly generated distributive lattice, it follows that every decomposition theorem for such lattices gives a corresponding structure theorem for L. For example, *irredundant subdirect product representations by means of subdirectly irreducible lattices are unique.*

Recall that a lattice has the projectivity property if whenever a quotient is weakly projective into another quotient, it is projective to a subquotient of the latter quotient. Then it is not hard to show that *a weakly atomic lattice having the projectivity property is uniquely an irredundant subdirect product of subdirectly irreducible weakly atomic lattices having the projectivity property* [2].

It was shown above that distributivity implies a strong uniqueness for irredundant meet decompositions into irreducibles. Hence it is not surprising that distributivity also implies strong uniqueness for direct decomposition. For example, let $a = \dot{\Lambda} B$ denote direct meet decomposition, i.e. $a = \Lambda B$ and $(B - b) \vee b = 1$ all $b \in B$. Then if L is a complete distributive lattice the following strong uniqueness theorem holds.

THEOREM. *Let L be a complete distributive lattice. Then*

$$a = \dot{\Lambda} B = \dot{\Lambda} C$$

implies $b = \dot{\Lambda}(b \wedge C)$ all $b \in B$.

Now direct product representations of lattices can be characterized in terms of structure properties and permutability properties of Con L. Thus the above uniqueness theorem yields the following refinement theorem for lattices [8].

THEOREM. *Let* $L \simeq \prod_\alpha L_\alpha \simeq \prod_\beta M_\beta$. *Then there exists lattices* $N_{\alpha\beta}$ *such that* $L_\alpha \simeq \prod_\beta N_{\alpha\beta}$ *and* $M_\beta \simeq \prod_\alpha N_{\alpha\beta}$.

Few completely satisfactory theorems on direct product representations in terms of indecomposable are known. However, congruence distributivity does give the existence and uniqueness for special classes of lattices. Two examples are the following.

THEOREM [6]. *Let L be a relatively complemented lattice satisfying a chain condition. Then L is uniquely a direct product of simple lattices.*

THEOREM [8]. *A complete, weakly atomic, relatively complemented lattice is a direct product of subdirectly irreducible lattices.*

Although congruence distributivity played a significant role in the structure theory of lattices, the most remarkable applications of congruence distributivity occurred in connection with the study of varieties of lattices.

Let $\mathscr{K}$ be a class of algebras of a given type. The variety $V(\mathscr{K})$ generated by $\mathscr{K}$ is the class of algebras satisfying all of the identities holding in all of the algebras of $\mathscr{K}$. A fundamental theorem of Birkhoff [1] asserts that

$$V(\mathscr{K}) = HSP\mathscr{K}.$$

Where $H\mathscr{K}$ denotes the class of algebras isomorphic to homomorphic images of algebras of $\mathscr{K}$. $S\mathscr{K}$ is similarly defined in terms of subalgebras and $P\mathscr{K}$ in terms of direct products.

Now the class of all lattices is a variety in which each member is congruence distributive. The key contribution which opened up the theory of lattice varieties was the observation by Professor Jónsson [9] that for a class $\mathscr{K}$ of algebras belonging to a congruence distributive variety, the Birkhoff representation could be replaced by:

$$V(\mathscr{K}) = P_d HSP_u \mathscr{K}.$$

Where $P_u \mathscr{K}$ denotes the class of algebras isomorphic to ultra products of algebras of $\mathscr{K}$ and P_d denotes the class of algebras isomorphic to subdirect products of algebras of $\mathcal{K}$. The great advantage of this representation arises from the fact that if $\mathcal{K}$ is a finite class of finite algebras, then

$$P_u\mathcal{K} \simeq \mathcal{K}.$$

Thus in this case, a subdirectly irreducible algebra belongs to $V(\mathcal{K})$ if and only if it is a homomorphic image of a subalgebra of an algebra belonging to $\mathcal{K}$.

It is important to note where congruence distributivity comes into play in Jónsson's theorem. Let K be a subdirectly irreducible member of $V(\mathcal{K})$. Then by Birkhoff's theorem $K \simeq L/\theta$ where L is a sublattice of $\prod_\alpha L_\alpha$, each L_α belonging to $\mathcal{K}$. If S is the set of subscripts α and $A \subseteq S$ then $\varphi(A)$ defined by

$$x\varphi(A)y \quad \text{if and only if} \quad A \subseteq \{\alpha \mid x_\alpha = y_\alpha\}$$

is a congruence relation on L. Let $\mathcal{U}$ be a maximal proper filter in S such that

$$A \in \mathcal{U} \Rightarrow \varphi(A) \le \theta.$$

The key step is to show that $\mathcal{U}$ is an ultra filter. If it is not, there exists $D \subseteq S$ such that $D \notin \mathcal{U}$ and $D^c \notin \mathcal{U}$. By the maximality of $\mathcal{U}$ there exists $A \in \mathcal{U}$ such that

$$\varphi(D \cap A) \not\le \theta \quad \text{and} \quad \varphi(D^c \cap A) \not\le \theta.$$

Then we have

$$\theta = \theta \vee \varphi(A) = \theta \vee \varphi((D \cap A) \cup (D^c \cap A)) = \theta \vee [\varphi(D \cap A) \wedge \varphi(D^c \cap A)].$$

By congruence distributivity

$$\theta = (\theta \vee \varphi(D \cap A)) \wedge (\theta \vee \varphi(D^c \cap A))$$

contrary to the subdirect irreducibility of $L/\theta \simeq K$. Thus $\mathcal{U}$ is an ultrafilter and if $\varphi_{\mathcal{U}}$ is the associated congruence relation on L, then $L/\varphi_{\mathcal{U}}$ is a sublattice of an ultraproduct of the lattice L_α. Since $\varphi_{\mathcal{U}} \le \theta$, L/θ is a homomorphic image of $L/\varphi_{\mathcal{U}}$ and $K \simeq L/\theta$ belongs to $HSP_u\mathcal{K}$.

An immediate consequence of the Jónsson representation is the following theorem for lattices.

THEOREM [9]. *If K and L are nonisomorphic subdirectly irreducible lattices, K is finite, and L has at least as many elements as K, then there is an identity that holds in K but not in L.*

Since the intersection of varieties is again a variety, it follows that the varieties of algebras of a given type form a lattice. A further consequence of the congruence distributivity of lattices is the following theorem.

THEOREM. *The lattice of lattice varieties is distributive.*

Since the lattices M_4 and $M_{3,3}$ of Figure 2 are subdirectly irreducible, and the only subdirectly irreducible lattices which are properly homomorphic images of sublattices of M_4 and $M_{3,3}$ are M_3 and the two-element lattice, it follows from the above theorem, that the varieties generated by M_4 and $M_{3,3}$ cover $V(M_3)$ in the lattice of variety of lattices.

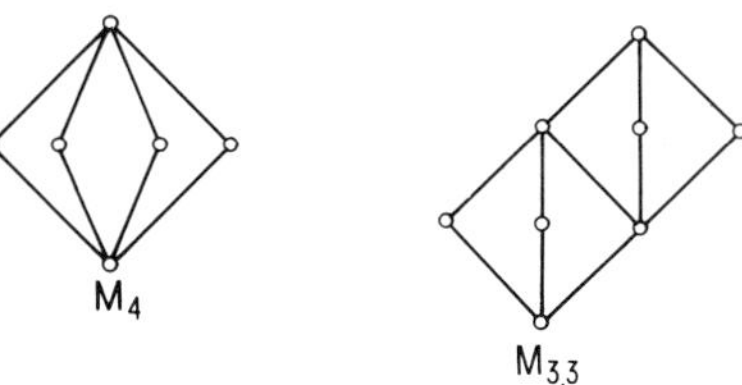

Figure 2.

R. McKenzie [12] constructed fifteen lattices L_1–L_{15} whose varieties are join irreducible covers of $V(N_3)$ in the lattice of varieties.

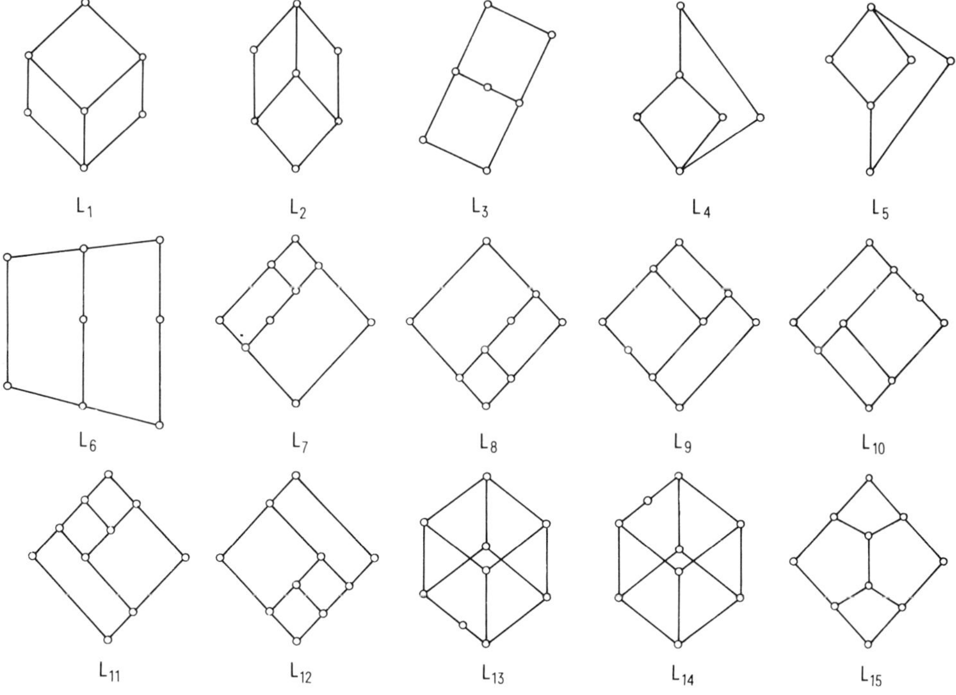

Figure 3.

A deeper result which is related to the subject matter of the next section asserts that these are the only join irreducible covers of $V(N_3)$. Hence the following diagram gives the four lowest levels of the lattice of lattice varieties.

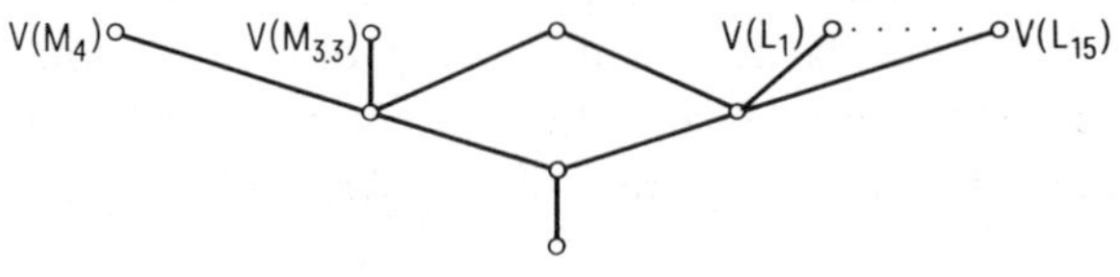

Figure 4.

Semidistributivity

The modular law is, by far, the most important weakening of the distributive law. In fact the study of modular lattices is a major branch of lattice theory. Nevertheless, there have been many other lattice conditions which have been proposed which are weak forms of the distributive law. Most of these were designed for a specific purpose and thus have made only limited contributions to general lattice theory. However, the following weakening of the distributive law, first formulated by Jónsson [10] for the study of free lattices, deserves particular attention.

DEFINITION. A lattice L is *semidistributive* if

1) $u = x \vee y = x \vee z$ implies $u = x \vee (y \wedge z)$

2) $v = x \wedge y = x \wedge z$ implies $v = x \wedge (y \vee z)$

Making use of canonical forms it is relatively easy to verify that 1) and 2) hold in a free lattice. Hence any sublattice of a free lattice is semidistributive. A sublattice of a free lattice also satisfies the condition of Whitman [14].

(W) $x \vee y \geq z \wedge w$ if and only if $x \vee y \geq z$ or $x \vee y \geq w$

$$\text{or} \quad x \geq z \wedge w \quad \text{or} \quad y \geq z \wedge w.$$

Around 1960, Jónsson conjectured that a finite lattice is isomorphic to a sublattice of a free lattice if and only if it is a semidistributive lattice satisfying Whitman's condition. For two decades this conjecture was one of the outstanding problems in lattice theory. In 1980, J. B. Nation, following an approach originally suggested by Jónsson, proved the conjecture.

THEOREM (Nation [13]). *A finite lattice is isomorphic to a sublattice of a free lattice if and only if it is a semidistributive lattice satisfying Whitman's condition.*

Although free lattices provide the outstanding examples of semidistributive lattices, semidistributivity has had a significant role in other areas of lattice theory, in particular, the study of lattice varieties. An examination of the lattices M_3, $L_1, \ldots, L_{15}$ shows that L_6–L_{12} are semidistributive while $M_3, L_1, \ldots, L_5$ are not. In fact, Davey, Poguntke, and Rival [3] showed that a lattice of finite length is semidistributive if and only if it contains no sublattice isomorphic to one of M_3, $L_1, \ldots, L_5$. Making use of ideal and dual ideal lattices. Jónsson and Rival extended this result to varieties of lattices.

THEOREM (B. Jónsson and I. Rival [11]). *A variety of lattices is semi-distributive if and only if it contains none of the lattices M_3, $L_1, \ldots, L_5$.*

Jónsson and Rival then proceed to make a very careful analysis of weak projectivities in semidistributive lattices and show that a finitely generated subdirectly irreducible lattice in a variety containing none of M_3, $L_1, \ldots, L_{15}$ must be N_3. This gives the following theorem which completes the determination of the covers of $V(N_3)$.

THEOREM (B. Jónsson and I. Rival [11]). *Every variety of lattices that properly contains $V(N_3)$ includes one of the lattices M_3, $L_1, \ldots, L_{15}$.*

REFERENCES

[1] G. Birkhoff, *On the structure of abstract algebras*, Proc. Comb. Phil. Soc., *31* (1935), 433–454.
[2] P. Crawley and R. P. Dilworth, *Algebraic theory of lattices*, Englewood Cliffs, N.J., Prentice Hall, *1* (1973).
[3] B. A. Davey, W. Poguntke and I. Rival, *A characterization of semidistributivity*, Alg. Univ. 5 (1975), 72–75.
[4] R. P. Dilworth and P. Crawley, *Decomposition theory for lattices without chain conditions*, Trans. Amer. Math. Soc., (1960), 1–23.
[5] R. P. Dilworth and J. E. McLaughlin, *Distributivity in lattices*, Duke Math. J., *19* (1952), 683–694.
[6] R. P. Dilworth, *The structure of relatively complemented lattices*, Annals of Math., *51* (1950), 348–359.
[7] N. Funayana and T. Nakagama, *On the distributivity of a lattice of lattice congruences*, Proc. Imp. Acad. Tokyo, *18* (1942), 553–554.
[8] J. Hashimoto, *Direct, subdirect decompositions and congruence relations*, Osak Math. J., *9* (1957), 87–112.
[9] B. Jónsson, *Algebras whose congruence lattices are distributive*, Math. Scand., *21* (1967), 110–121.

[10] B. Jónsson, *Sublattices of a free lattice*, Can. J. Math., *13* (1961), 256–264.

[11] B. Jónsson and I. Rival, *Lattice varieties covering the smallest nonmodular variety*, Pacific J. Math., *82* (1979), 463–478.

[12] R. McKenzie, *Equational bases and nonmodular lattice varieties*, Trans. Amer. Math. Soc., *174* (1972), 1–43.

[13] J. B. Nation, *Finite sublattices of a free lattice*, Trans. Amer. Math. Soc., *269* (1982), 311–337.

[14] Ph. Whitman, *Free lattices*, Annals of Math., *42* (1941), 325–329.

California Institute of Technology
Pasadena, California
U.S.A.

The Role of Gluing Constructions
in Modular Lattice Theory

ALAN DAY AND RALPH FREESE

In the 1930's and 1940's lattice theory was often broken into three subdivisions: distributive lattice theory, modular lattice theory, and the theory of all lattices. A question about lattices could usually be formulated for each of these subdivisions. Of the three resulting questions, the one about modular lattices almost always proved to be the most difficult. The problem of embedding a lattice into a complemented lattice was an example of such a problem. It is trivial to see that every lattice can be embedded into a complemented lattice, and Birkhoff's representation theorem [1] shows that every distributive lattice can be embedded in a complemented distributive lattice. However the problem of embedding modular lattices into complemented modular lattices remained open for some time. R. P. Dilworth and Marshall Hall addressed this problem in their 1944 paper [23], showing, in fact, that there are finite modular lattices which cannot be embedded into a complemented modular lattice.

This paper used a construction that has become known as Hall-Dilworth gluing, but is now being called Dilworth gluing since it actually originated in an earlier paper of Dilworth, see below. With this construction Dilworth and Hall produced three examples of modular lattices, none of which can be embedded into a complemented modular lattice. Although other papers of Dilworth (and also Hall) contain deeper results, this paper has proved extremely important in the subsequent development of modular lattice theory. The examples themselves have proved useful in refuting various conjectures. The gluing technique used in constructing these lattices has turned out to be useful in settling some of the deeper questions of modular lattice theory. This gluing technique was the origin of more general gluing, which in turn has proved to be especially fruitful in solving some of the most stubborn problems of modular lattice theory.

This work was supported by the NSF and NSERC.

The Dilworth gluing is simply this: if a nonempty filter $\mathbf{F}$ of a lattice $\mathbf{L_0}$ is isomorphic to an ideal $\mathbf{I}$ of a lattice $\mathbf{L_1}$, let L be the union of L_0 and L_1 with the elements of F and I identified via the isomorphism. L can be ordered with the transitive closure of the union of the orders on $\mathbf{L_0}$ and $\mathbf{L_1}$. It is easy to see that under this order L is a lattice. A schematic representation of this situation is given in Figure 2 of the Background for this chapter.

It is more difficult to see that the lattice $\mathbf{L}$ is modular if both $\mathbf{L_0}$ and $\mathbf{L_1}$ are. This was established by Dilworth in [12]. Since this paper preceded the Hall-Dilworth paper, we now use the term *Dilworth gluing* for this construction. The Dilworth gluing does not preserve equations in general. (However the distributive law is preserved.) Jónsson's Arguesian law (discussed below) is an example of an equation which is not preserved.

The examples and complemented modular lattices. The Hall-Dilworth paper used gluing to construct three types of examples of modular lattices which cannot be embedded into complemented modular lattices. The basic idea behind all of them is that a projective plane can be embedded into a projective geometry of higher dimension if and only if it satisfies Desargues' theorem. Now the subspaces of a projective geometry form a complemented modular lattice and this lattice determines the geoemetry, see Chapter 13 of [2] and [35]. Since a projective geometry is determined by its lattice of subspaces, we identify a projective geometry with its lattice of subspaces. The first example is constructed by gluing the lattice of subspaces of a projective geometry which fails Desargues' theorem and $\mathbf{M_3}$ (the five element modular, nondistributive lattice) over the two element lattice. A schematic representation of this lattice is given in Figure 3 of the Background. By an argument similar to the proof that non-Desarguesian projective planes cannot be embedded into a higher dimensional projective geometry, Hall and Dilworth showed that this lattice could not be embedded into a complemented modular lattice.

The second example was formed by gluing the lattices of subspaces of two finite Desarguesian projective planes over the two element lattice. These planes were coordinatized by fields with different characteristics.

The third example, which was somewhat more subtle, was constructed by gluing two isomorphic copies of a Desarguesian projective plane over a two dimensional interval. The two dimensional intervals in a projective plane are all isomorphic to $\mathbf{M}_n$, where n is the number of points on a line in the plane. There are $n!$ automorphisms of $\mathbf{M}_n$, and hence that many ways of gluing the two planes together over a two dimensional quotient (some of which will be isomorphic as lattices). With the aid of classical coordinatization techniques it can be shown that only some of these lattices can be embedded into complemented modular lattices. The ones that cannot are the third type of Hall-Dilworth example.

Some of these ideas were clarified by the introduction of the *Arguesian* law by Bjarni Jónsson. This is a lattice equation which reflects Desargues' Theorem of projective geometry. In particular, the lattice of subspaces of a projective geometry satisfies this equation if and only if the projective geometry satisfies Desargues' Theorem. It can be shown that a subdirectly irreducible modular lattice of length

THE DILWORTH THEOREMS

at least four which can be embedded into a complemented modular lattice satisfies the Arguesian equation, see Chapter 13 of [2]. The first and the third Hall-Dilworth examples are non-Arguesian and so not embeddable into a complemented modular lattice. In fact, since Arguesian lattices are defined by an equation, it follows that the first and third examples are not even in the variety $\mathcal{K}$ generated by all complemented modular lattices. (It is conceivable that $\mathcal{K}$ equals the class of lattices embeddable into a complemented modular lattice. In fact this is a good unsolved problem: *Is the class of lattices embeddable into a complemented modular lattice closed under the formation of homomorphic images?*)

One might wonder to what extent the nonembeddability of modular lattices into complemented modular lattices is dependent on the failure of the Arguesian law. Indeed the modular law is satisfied by most of the lattices associated with classical algebraic systems. In fact these lattices satisfy stronger equations: Freese and Jónsson [16] have shown that if all the algebras in a variety of algebras have modular congruence lattices, these lattices satisfy the Arguesian equation. Thus the first and the third Hall-Dilworth examples can never lie in such a modular *congruence variety*. In [15] it is shown that the second Hall-Dilworth example also cannot lie in any modular congruence variety.

Is it true that in some restricted class of modular lattices, closer to the class of lattices associated with classical algebraic systems, embedding into complemented lattices might be possible? This was shown not to be the case by Herrmann and Huhn in [28]. They showed that the lattice of subgroups of $(\mathbb{Z}/4\mathbb{Z})^3$ cannot be embedded into any complemented modular lattice.

Some applications of the examples. The Hall-Dilworth examples have been used often in producing counter-examples. In this section we present a few of the important examples. C. Herrmann and W. Poguntke [29] used the second kind of example to show that *the class of all lattices embeddable into the lattice of normal subgroups of a group cannot be defined by finitely many first order axioms.* The idea is to let $\mathbf{L}_p$ be the lattice obtained by gluing (the lattice of subspaces of) a projective plane of characteristic p to a projective plane of characteristic p^+ (the next prime after p) over a 1-dimensional quotient. $\mathbf{L}_p$ is not embeddable into the lattice of normal subgroups of a group In fact it lies in no variety generated by the congruences lattices of a variety of algebras with modular congruence lattices, see [15]. On the other hand it is not hard to see that a nonprincipal ultraproduct of the $\mathbf{L}_p$'s is also one of the Hall-Dilworth examples of the second kind, but the two projective planes have characteristic 0. From this it follows that the whole lattice can be embedded into the lattice of subspaces of a vector space over the rationals. Hence it can be embedded into the lattice of subgroups of an Abelian group. This result also proves that many other classes of modular lattices cannot be defined by finitely many axioms. For example the class of all lattices embeddable into the lattice of subgroups of an Abelian group. Also it shows that the variety generated either of the above two classes cannot be finitely defined.

In [31] Jónsson made a careful investigation of the third type of Hall-Dilworth example. He found necessary and sufficient conditions for this type to be non-Arguesian. These conditions coincided with the Hall-Dilworth conditions for nonembeddability into complemented modular lattices and were equivalent to the lattice not having a representation as a lattice of permuting equivalence relations. This reinforced the idea that the Arguesian equation reflected Desargues' law in geometry.

M. Haiman [20] examined Jónsson's result and, using the correct skewfield, showed that the Arguesian equation required all of its six variables. Let $\mathbf{H}$ be the skewfield of all real quaternions and $\mathbf{L}_0 = \mathbf{L}_1 = \mathbf{L}(\mathbf{H}^3)$. Now $\mathbf{H}$ has a natural antiautomorphism, quaternion conjugation, that is $\mathbf{R}$-linear. Using this antiautomorphism, Haiman constructed a Dilworth gluing of $\mathbf{L}_0$ and $\mathbf{L}_1$ over a 2-dimensional interval that failed to be Arguesian by Jónsson's result which would require that the map be an automorphism. Moreover this lattice has the property that all 5-generated sublattices are Arguesian.

Freese used a modification of the third type of Hall-Dilworth example to settle some of the important previously unsolved problems of modular lattice theory. Let p and q be distinct prime numbers and let $\mathbf{F}$ and $\mathbf{K}$ be countably infinite fields with characteristics p and q, respectively. Let $\mathbf{L}_0$ be the lattice of subspaces of a 4-dimensional vector space over $\mathbf{F}$ and let $\mathbf{L}_1$ be the lattice of subspaces of a 4-dimensional vector space over $\mathbf{K}$. Every 2-dimensional interval in each of these lattices is isomorphic to $\mathbf{M}_\omega$. Let $\mathbf{L}$ be the lattice obtained by gluing these lattices over such an interval. Then $\mathbf{L}$ is not in the variety generated by the finite modular lattices [13]. In particular the variety of modular lattices is not generated by its finite members. The basic idea of the proof is this. If $\mathbf{F}$ and $\mathbf{K}$ were finite fields then $|F| = p^n$ and $|K| = q^m$ for some $n > 0$ and $m > 0$. Since $p^n \neq q^m$, it is impossible to construct $\mathbf{L}$ as above using finite fields. Of course to actually carry out the proof one needs to bring much of $\mathbf{L}$ into the free modular lattice.

Using a similar example, Freese [14] was able to show that the equational theory of modular lattices is undecidable, *i.e.*, there is no algorithm to determine if two lattice terms are equal in all modular lattices. A. Macintyre [34] constructed a skew field interpreting a finitely presented group with unsolvable word problem. If we construct $\mathbf{L}$ as above, but with Macintyre's field for $\mathbf{F}$, we can interpret this group with unsolvable word problem in $\mathbf{FM}(5)$, showing that its word problem is unsolvable. Herrmann, with the aid of his more general gluing construction, has shown that $\mathbf{FM}(4)$ has an unsolvable word problem, see below.

The lattice $\mathbf{L}$ constructed above can also be used to show that there are two lattice terms $v < u$ in five variables such that interval sublattice $[v, u]$ of $\mathbf{FM}(X)$ is distributive for every X which contains the variables of u and v. From this it follows that every free distributive lattice can be embedded into a free modular lattice. A related open problem is this: *Is the class of distributive sublattices of free modular lattices equal to the class of sublattices of free distributive lattices?*

Generalized gluing. In [25], Herrmann significantly generalized the notion of gluing. Herrmann's idea was to consider maximal complemented subintervals of a modular lattice of finite height. These "blocks" can be ordered by means of their least (or equivalently greatest) elements. With respect to this order, this system of blocks, called the *prime skeleton*, forms a lattice. Now if two blocks, $B = a/b$ and $C = c/d$, are comparable in this order, say with $B \leq C$ and $B \cap C \neq \emptyset$ (equivalently $b \leq d \leq a \leq c$) then $B \cup C$ is a sublattice which is isomorphic to the Dilworth gluing of B and C over the interval $[b, c]$ considered as an ideal in B and a filter in C. Thus the original modular lattice is then decomposed into a lattice of complemented modular blocks together with a system of Dilworth gluings between intersecting, comparable blocks. An example of this kind of decomposition is presented below.

For $\mathbf{L}$ a modular lattice of finite length, Herrmann defined two mappings, $x \mapsto x^\sigma$ and $x \mapsto x^\pi$, of L into L by $0^\sigma = 0$ and $1^\pi = 1$ and

$$x^\sigma = \bigwedge \{y \in L : y \prec x\} \qquad x^\pi = \bigvee \{y \in L : y \succ x\}.$$

These mappings are isotone and form a Galois pair in that $x^\sigma \leq y$ if and only if $x \leq y^\pi$. Consequently, L^σ, the range of $x \mapsto x^\sigma$, is a join subsemilattice of $\mathbf{L}$, and L^π is a meet subsemilattice of $\mathbf{L}$, and they are isomorphic. Thus both $\mathbf{L}^\sigma$ and $\mathbf{L}^\pi$ are lattices, though not necessarily sublattices of L. More precisely, $\langle L^\sigma, +, \wedge \rangle$ and $\langle L^\pi, \vee, \cdot \rangle$ are lattices, where $\vee$ and $\wedge$ are the operations of $\mathbf{L}$, and

$$x \cdot y = (x \wedge y)^\sigma \qquad x + y = (x \vee y)^\pi.$$

A fundamental fact discovered by Herrmann was that the maximal complemented subintervals of $\mathbf{L}$ are precisely those of the form x^π / x, for $x \in L^\sigma$, or equivalently y/y^σ for $y \in L^\pi$. We choose the first format and define the *prime skeleton* of $\mathbf{L}$ to be $S(\mathbf{L}) = \mathbf{L}^\sigma$ and for each $x \in S(\mathbf{L})$, we let $L(x) = x^\pi / x$. The intervals $\mathbf{L}(x)$ will be referred to as the blocks of $\mathbf{L}$.

Now if $x \leq y$ in $S(\mathbf{L})$ and $x \leq y \leq x^\pi$, then $\mathbf{L}(x) \cap \mathbf{L}(y) = [y, x^\pi] = M$. The identity map on $\mathbf{M}$ can be viewed as a bijection φ_{xy} from a principal filter of $\mathbf{L}(x)$ to a principal ideal of $\mathbf{L}(y)$. In the case $x \not\leq y$ or $\mathbf{L}(x) \cap \mathbf{L}(y) = \emptyset$, we let $\varphi_{xy} = \emptyset$. These maps satisfy certain natural compatibility constraints, namely

(1) φ_{xx} is the identity map on $\mathbf{L}(x)$.
(2) If $\varphi_{xy} \neq \emptyset$, then it is an isomorphism of a filter of $\mathbf{L}(x)$ onto an ideal of $\mathbf{L}(y)$.
(3) If $x \prec y$ in $\mathbf{S}$, then $\varphi_{xy} \neq \emptyset$.
(4) If $x \leq z \leq y$ in $\mathbf{S}$, them $\varphi_{xy} = \varphi_{xz} \circ \varphi_{zy}$.
(5) $\operatorname{Im} \varphi_{x, x \vee y} \cap \operatorname{Im} \varphi_{y, x \vee y} = \operatorname{Im} \varphi_{x \wedge y, x \vee y}$.
(6) $\operatorname{Dom} \varphi_{x \wedge y, x} \cap \operatorname{Dom} \varphi_{x \wedge y, y} = \operatorname{Dom} \varphi_{x \wedge y, x \vee y}$.

Herrmann's gluing construction [25] (see also [26]) is essentially a converse of the above situation:

Let $\mathbf{S}$ and $\mathbf{L}(x)$, $x \in S$ be lattices of finite length and for $x \leq y$ in $\mathbf{S}$ let $\varphi_{xy} : \mathbf{L}(x) \to \mathbf{L}(y)$ be partial bijections satisfying the above conditions. Let L be the disjoint union of the $\mathbf{L}(x)$'s with elements identified under

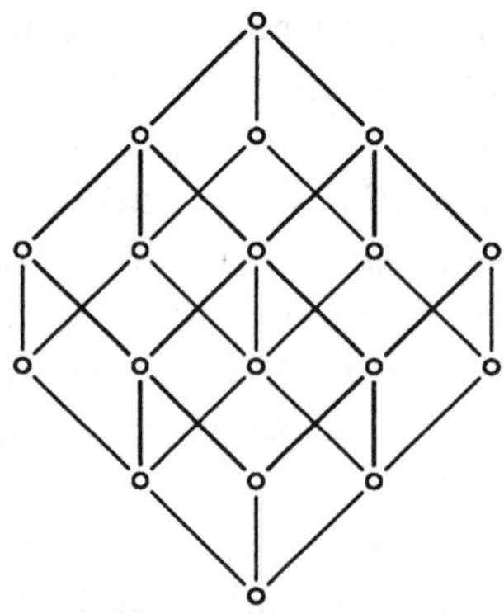

Figure 1

the φ_{xy}'s (i.e., a and b are identified if $\varphi_{xz}(a) = \varphi_{yz}(b)$). Then **L** *is a lattice. If each* **L**(x) *is modular then* **L** *is.*

Herrmann's gluing can be generalized in several ways. The finite length assumptions on **S** and **L**(s) can be relaxed. Moreover the blocks can be *loosely glued* rather than *tightly glued*. (If **F** is a filter of **L**$_0$ which is isomorphic to an ideal **I** of **L**$_1$, we form a lattice on the disjoint union of L_0 and L_1 whose order relation is the transitive closure of the order relations of **L**$_0$ and **L**$_1$ and the relation $x \leq y$ if x is mapped to y under the isomorphism of **F** to **I**. Notice that this construction can be obtained by gluing **L**$_0$, **F** $\times$ **2**, and **L**$_1$ using the ordinary Dilworth gluing.) A related type of gluing was developed by in Graczynska [17] and Gracyznska and Grätzer [18]. Further generalizations of these gluings and of Herrmann's gluing are developed in Day and Herrmann [5]. That paper gives various applications including applications to Maltsev products. Particular applications of this generalized gluing occur in Grätzer and Kelly [19] and Harrison [24].

Applications of generalized gluing. The following example, which is an unpublished result of Jónsson, illustrates how Herrmann's gluing can be used to produce some subtle examples. Let $S = \{0, a, b, 1\} \cong 2^2$ and **L**$_0$, **L**$_a$, **L**$_b$, and **L**$_1$ be four copies of an Arguesian projective plane of order n, **L** $=$ **L**$(\mathbf{F}^3)$, the lattice of subspaces of the 3-dimensional vector space over the field with n elements. We can picture these lattices in the lattice, **L**$(\mathbf{F}^5)$. Let **F**^{5} have as a basis $\{u_1, \ldots, u_5\}$ and consider the following length 3 intervals in **L**$(\mathbf{F}^5)$,

$$\mathbf{L}_0 = [0, \mathbf{F}u_1 + \mathbf{F}u_2 + \mathbf{F}u_3]$$
$$\mathbf{L}_a = [\mathbf{F}u_1, \mathbf{F}u_1 + \mathbf{F}u_2 + \mathbf{F}u_3 + \mathbf{F}u_4]$$
$$\mathbf{L}_b = [\mathbf{F}u_3, \mathbf{F}u_1 + \mathbf{F}u_2 + \mathbf{F}u_3 + \mathbf{F}u_5]$$
$$\mathbf{L}_1 = [\mathbf{F}u_1 + \mathbf{F}u_2 + \mathbf{F}u_3, \mathbf{F}^5]$$

It is easy to see that the union of these intervals is a sublattice of **L**$(\mathbf{F}^5)$. A sublattice of this lattice is given in Figure 1.

     THE DILWORTH THEOREMS

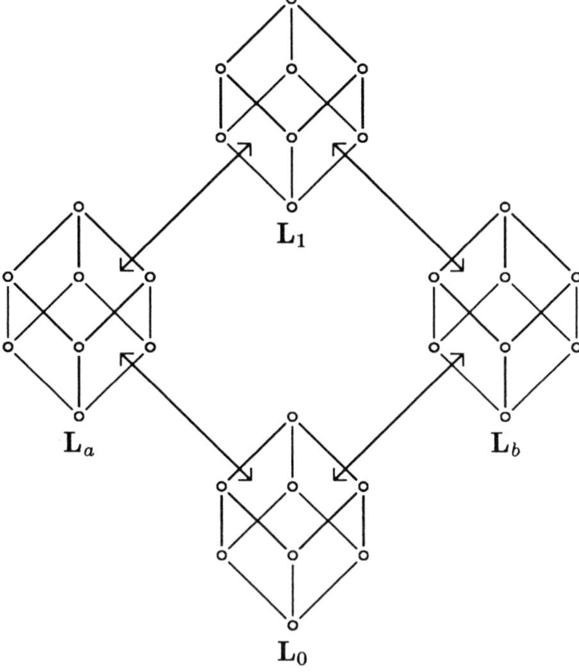

Figure 2

The situation with the blocks pulled apart is represented in Figure 2.

Of course the identification maps for these blocks are the identity maps. Assume that $\mathbf{F}$ has a nontrivial automorphism. What Jónsson did was to modify the map which connects $\mathbf{L}_0$ and $\mathbf{L}_a$ to be the map induced by the automorphism of $\mathbf{F}$. The resulting lattice fails the Arguesian law but the sublattice consisting of the union of $\mathbf{L}_0$ and $\mathbf{L}_a$ is Arguesian. In fact every proper sublattice of this lattice is Arguesian. Herrmann [unpublished] has shown that all non-Arguesian lattices with $S \cong 2^2$ are produced in this way.

Other interesting examples of minimal non-Arguesian lattices were produced by Day, Haiman, Herrmann, Jónsson, and Pickering. For each integer k, Pickering [36] was able to construct a minimal non-Arguesian lattice of length at least k. His construction used Herrmann's gluing with $\mathbf{S} = 2^3$. Using these lattices he was also able to show that *there is a variety of modular, non-Arguesian lattices all of whose finite members were Arguesian*. This guaranteed that obtaining a structural characterization of minimal non-Arguesian lattices would be difficult. Nevertheless some progress has been made. In a series of papers [6], [7], [8], [9], Day and Jónsson conducted a detailed analysis of the Arguesian law in an attempt to understand the structure of minimal non-Arguesian lattices. In analogy to the Desargues configuration, they define a certain lattice configuration which is projective in the class of modular lattices. Accordingly they are called *projectivity configurations*. If the lattice **L** fails the Arguesian equation, one of these configurations is non-Arguesian, and in the ideal lattice of **L** there will be a minimal, non-Arguesian projectivity configuration. Associated with these minimal configurations are several intervals which are projective planes which are glued together in a certain manner. The subsequent analysis split naturally into two cases. For the first case, called *Boolean*, there was a complete classification of models: each was the gluing of projective planes over a skeleton of the form 2^n for $n = 0, 1, 2, 3$. Examples of these have been given above. The results for the second case were not nearly so descriptive and it is clear that this problem is much more intractable.

Herrmann himself of course made frequent use of his gluing. One of his outstanding results was that **FM**(4) has an undecidable word problem, [26]. Although his proof used many ideas from [14], his construction was fundamentally different. He introduced the concept of a *skew frame* and with his gluing, modified the lattice of subgroups of a certain Abelian group producing a 4-generated modular lattice in which the group ring of a finitely presented group with unsolvable word problem could be interpreted. Using the fact that his skew frames are a projective configuration, he could pull this situation back into **FM**(4), proving his result.

Another very important paper of Herrmann which uses gluing is [27]. Let $\mathcal{M}_0$ denote the variety of lattices generated by the subspace lattices of all vector spaces over the rationals. Then Herrmann's result states:

Every variety of modular lattices which contains $\mathcal{M}_0$ either is not generated by its finite dimensional members or does not have a finite equational basis.

This result has several important corollaries:

The variety of Arguesian lattices is not generated by its finite dimensional members. Neither the variety generated by all finite modular lattices nor the variety generated by all finite dimensional modular lattices have a finite equational basis.

As Jónsson noted in [31], lattices of permuting equivalence relations satisfy the Arguesian law. Moreover he showed in [33] that every complemented Arguesian lattice has such a representation. It was an open problem for many years whether complementedness could be dropped from his result. In a beautiful result, M. Haiman [21], [22] used gluings to solve this problem in the negative. A *crown* is an ordered set of the form $C_n = \{a_0, \ldots, a_{n-1}\} \cup \{b_0, \ldots, b_{n-1}\}$ with $a_i < b_i$, b_{i+1} (indices computed modulo n). Let S_n be C_n with a least and greatest element added. A diagram of S_4 is given in Figure 3.

Let **V** be a $2n$-dimensional vector space over a prime field. For each element in S_n Haiman found an interval in **L**(**V**). The union of these intervals is a sublattice of **L**(**V**) whose prime skeleton is S_n. By gluing these intervals back together over S_n, but slightly modifying one of the intersections maps, Haiman produced a lattice which is Arguesian but cannot be represented as a lattice of equivalence relations. Moreover his proof shows in fact that the class of lattices having a representation as a lattice of permuting equivalence relations cannot be defined by finitely many first order axioms, although this class can be defined by an infinite set of Horn

 THE DILWORTH THEOREMS

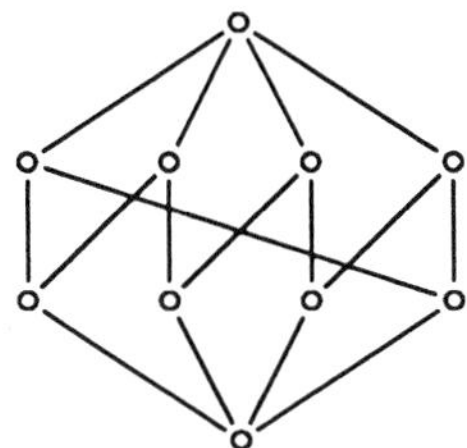

Figure 3

sentences, see Jónsson [32]. A good open problem: *Is the class of lattices having a representation as a lattice of permuting equivalence relations a variety?*

Conclusion. Certainly the results above show that the Hall-Dilworth paper was seminal. Dilworth's gluing and its generalizations together with the examples constructed by Dilworth and Hall are the basis of many of the most important results in modular lattice theory. Along with the papers of Dedekind [10], [11], and von Neumann on coordinatizing complemented modular lattices [35], the Hall-Dilworth paper is among the most influential in the field.

REFERENCES

1. G. Birkhoff, *On the combination of subalgebras*, Proc. Cambridge Phil. Soc. **29** (1933), 441–464.

2. P. Crawley and R. P. Dilworth, "Algebraic Theory of Lattices," Prentice-Hall, Englewood Cliffs, New Jersey, 1973.

3. A. Day, *Geometrical applications in modular lattices*, in "Universal Algebra and Lattice Theory," R. Freese and O. Garcia, eds., Lecture Notes in Mathematics, vol. **1004**, Springer-Verlag, New York, 1983, pp. 111–141.

4. A. Day, *Dimension equations in modular lattices*, Algebra Universalis **22** (1986), 14–26.

5. A. Day and Ch. Herrmann, *Gluings of modular lattices*, Order **5** (1988), 85–101.

6. A. Day and B. Jónsson, *The structure of non-Arguesian lattices*, Bull. Amer. Math. Soc. **13** (1985), 157–159.

7. A. Day and B. Jónsson, *A structural characterization of non-Arguesian lattices*, Order **2** (1986), 335–350.

8. A. Day and B. Jónsson, *Non-Arguesian configurations in a modular lattice*, Acta Sci. Math. (Szeged) **51** (1987), 309–318.

9. A. Day and B. Jónsson, *Non-Arguesian configurations and gluings of modular lattices*, Algebra Universalis (to appear).

10. R. Dedekind, *Über Zerlegungen von Zahlen durch ihre grössten gemeinsamen Teiler*, Festschrift der Herzogl. technische Hochschule zur Naturforscher-Versammlung, Braunschweig (1897). Reprinted in "Gesammelte mathematische Werke", Vol. 2, pp. 103–148, Chelsea, New York, 1968.

11. R. Dedekind, *Über die drei Moduln erzengte Dualgruppe*, Math. Annalen **53** (1900), 371–403.

12. R. P. Dilworth, *The arithmetical theory of Birkhoff lattices*, Duke Math. J. **8** (1941), 286–299. Reprinted in Chapter 3 of this volume.

13. R. Freese, *The variety of modular lattices is not generated by its finite members*, Trans. Amer. Math. Soc. **255** (1979), 277–300.

14. R. Freese, *Free modular lattices*, Trans. Amer. Math. Soc. **261** (1980), 81–91.

15. R. Freese, C. Herrmann and A. P. Huhn, *On some identities valid in modular congruence varieties*, Algebra Universalis **12** (1981), 322–334.

16. R. Freese and B. Jónsson, *Congruence modularity implies the Arguesian identity*, Algebra Universalis **6** (1976), 225–228.

17. E. Graczynska, *On the sums of double systems of lattices and DS-congruences of lattices*, in "Contributions to universal algebra," Colloq., Jozsef Attila Univ., Szeged,1975, 1977, pp. 161–178.

18. E. Graczynska and G. Grätzer, *On double systems of lattices*, Demonstratio Math. **13** (1980), 743–747.

19. G. Grätzer and D. Kelly, *Products of lattice varieties*, Algebra Universalis **21** (1985), 33–45.

20. M. Haiman, *Two notes on the Arguesian identity*, Algebra Universalis **21** (1985), 167–171.

21. M. Haiman, *Arguesian lattices which are not linear*, Bull. Amer. Math. Soc. **16** (1987), 121–124.

22. M. Haiman, *Arguesian lattices which are not type-1*, preprint.

23. M. Hall and R. P. Dilworth, *The imbedding problem for modular lattices*, Ann. of Math. **45** (1944), 450–456. Reprinted in Chapter 4 of this volume.

24. T. Harrison, *A problem concerning the lattice varietal product*, Algebra Universalis **25** (1988), 40–81.

25. Ch. Herrmann, *S-Verklebte Summen von Verbänden*, Math. Z. **130** (1973), 225–274.

26. Ch. Herrmann, *On the word problem for modular lattices with four generators*, Math Ann. **265** (1983), 513–527.

27. Ch. Herrmann, *On the arithmetic of projective coordinate systems*, Trans. Amer. Math. Soc. **284** (1984), 759–785.

28. Ch. Herrmann and A. P. Huhn, *Lattices of normal subgroups generated by frames*, Colloq. Math. Society János Bolyai **14** (1975), 97–136.

29. Ch. Herrmann and W. Poguntke, *The class of sublattices of normal subgroup lattices is not elementary*, Algebra Universalis **4** (1974), 280–286.

30. A. P. Huhn, *Schwach distributive Verbände, I*, Acta Sci. Math. (Szeged) **33** (1972), 297–305.

31. B. Jónsson, *Modular lattices and Desargues theorem*, Math. Scand. **2** (1954), 295–314.

32. B. Jónsson, *Representations of modular lattices and relation algebras*, Trans. Amer. Math. Soc. **92** (1959), 449–464.

33. B. Jónsson, *Representations of complemented modular lattices*, Trans. Amer. Math. Soc. **97** (1960), 64–94.

34. A. Macintyre, *The word problem for division rings*, J. Symbolic Logic **38** (1973), 428–436.

35. J. von Neumann, "Continuous Geometry," Princeton University Press, Princeton, N. J., 1960.

36. D. Pickering, "Minimal non-Arguesian lattices," Ph.D. Thesis, University of Hawaii, 1984.

Lakehead University
Thunder Bay, Ontario P7B 5E1
Canada

University of Hawaii
Honolulu, HI 96822
U. S. A.

Dilworth's Covering Theorem for Modular Lattices

Ivan Rival

Everyone, it seems, has his favourite "Dilworth's Theorem"! At least that is what I was led to believe, almost twenty years ago, when a referee protested that our use of the phrase in an article's title was too vague [6].

"*Which* Dilworth's Theorem (on modular lattices), do you mean?" the anonymous report intoned.

By the time that the proceedings of the *Symposium on Ordered Sets* (1982) appeared it had become clear enough that the current combinatorics generation regarded the "Chain Decomposition Theorem" that,

> *in an ordered set the minimun number of chains whose union is all equals the maximum number of pairwise noncomparable elements* [2]

as "Dilworth's Theorem." Indeed, the Chain Decomposition Theorem played a motivating role in many of the lectures at that symposium—not by design of the scientific programme committee, either. On the other hand, at *Combinatorics and Ordered Sets* (1984) Dilworth's influence was again ubiquitous: "The themes in one day's lectures alone, spanned almost two decades of his fundamental results [Dilworth (1940, 1950, 1954)]," see [19].

It was the last of these [3] that I had had in mind almost twenty years ago as my own favorite "Dilworth's Theorem". To wit,

> *In a finite modular lattice L, let $\mathrm{J}_k(L)$ be the set of elements with k lower covers and let $\mathrm{M}_k(L)$ be the set of elements with k upper covers. Then $|\mathrm{J}_k(L)| = |\mathrm{M}_k(L)|$ for any integer $k \geq 0$.*

It was remarkable for its time. For one thing it settled in the affirmative, as a quite special case, the conjecture of longstanding that, the number of join irreducibles is the same as the number of meet irreducibles, that is to say, $|\mathrm{J}_1(L)| =$

$|\mathrm{M}_1(L)|$. With attention concentrated for many years on this very conjecture, how satisfying it must have been—an understatement!—to formulate and prove this (far more general) combinatorial covering theorem for modular lattices.

The proof itself was a masterful stroke. It used the order-theoretical generalization of the well-known number-theoretical Möbius function developed by Weisner in [21]. For at least a decade this order-theoretical Möbius function was hailed as saviour [20], promoted to give combinatorics a theory, a mathematical continuity and depth, safe from the snipers' fire. Dilworth's Theorem was itself feted as a glorious tribute to this enumeration tool.

This political crossfire notwithstanding, Dilworth's Theorem has survived, and so has combinatorics, although neither because of the Möbius function. Dilworth's Theorem has enriched combinatorial lattice theory and led to important new directions.

Some twenty years passed before any further progress was made on the subject of this Dilworth's Theorem. Ganter and Rival [6] gave a surprisingly simple and elementary proof (independent of the Möbius function). Their proof depends on this basic property of complemented modular lattices.

In a finite complemented modular lattice, the number of k-element subsets of atoms whose supremum is the top equals the number of k-element subsets of coatoms whose infimum is the bottom.

Independently, Kurinnoi [12] also gave an elementary proof of Dilworth's Theorem. Later, Reuter [13], inspired by Wille's "mathematical concept analysis" [22], returned again to the Möbius function theme to give what is perhaps the shortest proof.

Perhaps more important than these simplifications was the "Matching Conjecture" I proposed in 1972 (cf. [16], [17]) that

there is a matching of the join irreducible elements of a finite modular lattice into its meet irreducible elements.

Thus, is there a one-to-one mapping f of the join irreducibles to the meet irreducibles such that $a \leq f(a)$, for every join irreducible a? Actually it was an attack on it that led Ganter and Rival to that first substantial simplification of the proof of Dilworth's Theorem.

The Matching Conjecture was an obvious generalization of that first conjecture that $|\mathrm{J}_1(L)| = |\mathrm{M}_1(L)|$ which Dilworth's Theorem both settled and in turn generalized. But, apart from it, there was yet further evidence lending credibility to it.

In the first place there is always such a matching for a finite distributive lattice L [16]. Indeed, fix any maximal chain $C = \{c_1 < c_2 < \cdots < c_m\}$ in L. Let a_i be a join irreducible satisfying $a_i \leq c_i$ but $a_i \nleq c_{i-1}$ and let b_i be a meet irreducible satisfying $b_i \geq c_i$ but $b_i \ngeq c_{i+1}$. As L is distributive $|\mathrm{J}_1(L)| = m = |\mathrm{M}_1(L)|$

so $J_1(L) = \{a_1, a_2, \ldots, a_m\}$ and $M_1(L) = \{b_1, b_2, \ldots, b_m\}$. Then the function $f(a_i) = b_i$ is a matching of $J_1(L)$ to $M_1(L)$. This fact, in turn, is linked to the correspondence between maximal proper sublattices of a finite distributive lattice and matched pairs of join irreducible and meet irreducible elements.

> *For any maximal proper sublattice S of a finite distributive lattice L there is a join irreducible a and a meet irreducible b with $a \leq b$ and $S = L - \{x | a \leq x \leq b\}$* [15], [16].

Although this description of maximal proper sublattices no longer holds for modular lattices, matching of join irreducibles to meet irreducibles could shed some light on a possible analogue [15].

Greene [7] had shown that every finite geometric lattice has such a matching, thus, for example, so does every finite complemented modular lattice. Rival [17] proved that every modular lattice of breadth at most two does too. Duffus [4] made a serious attack on the conjecture using a "gluing" construction inspired by Herrmann [9] who defined and investigated the *skeleton* of a lattice (which, for a modular lattice, is the set of bottoms of its maximal complemented interval sublattices). Actually this concept has its origin in yet another seminal paper of Dilworth [8]. Duffus did manage to construct a matching as long as the skeleton is the linear sum of lattices L_i such that width$(L_i) \leq 2$ or L_i is the product of two chains or $L_i = 2^3$.

The Matching Conjecture had meanwhile been selected by the editorial board of ORDER as one of the (eight) leading unsolved problems of the subject (cf. Duffus [5]). Finally, Kung [10] succeeded in settling it.

Kung [10,11] called a lattice L *consistent* if for every $x \in L$ and every join irreducible j, $j \vee x$ is join irreducible in the principal filter generated by x. It is easy to see that modular lattices are consistent. He examined a certain linear transformation, whose matrix relative to a certain basis is the incidence matrix with rows indexed by the join irreducible elements of L (that is, $J(L) = J_1(L) \cup J_0(L) = J_1(L) \cup \{0\}$) and columns indexed by the meet irreducible elements. He showed that when L is consistent, the rank of this linear transformation is equal to the number of join irreducible elements. Thus the incidence matrix contains a nonsingular submatrix of size $|J(L)|$. As the determinant is nonzero there is a term in the expansion which supplies the desired matching.

Using Dilworth's result, some of Kung's results, and Hall's marriage theorem, K. Reuter [14] is able to show that if L is linearly indecomposable then there was a matching from $J_1(L)$ to $M_1(L)$.

REFERENCES

1. R. P. Dilworth, *Lattices with unique irreducible decompositions*, Ann. of Math. **41** (1940), 771–777. Reprinted in Chapter 3 of this volume.
2. R. P. Dilworth, *A decomposition theorem for partially ordered sets*, Ann. of Math. **51** (1950), 161–166. Reprinted in Chapter 1 of this volume.

3. R. P. Dilworth, *Proof of a conjecture on finite modular lattices*, Ann. of Math. **60** (1954), 359–364. Reprinted in Chapter 4 of this volume.

4. D. Duffus, *Matching in modular lattices*, J. Combin. Theory, Ser. A **32** (1982), 303–314.

5. D. Duffus, *Matching in modular lattices*, Order **1** (1985), 411–413.

6. B. Ganter and I. Rival, *Dilworth's covering theorem for modular lattices: a simple proof*, Algebra Universalis **3** (1973), 348–350.

7. C. Greene, *A rank inequality for finite geometric lattices*, J. Comb. Theory **9** (1970), 357–364.

8. M. Hall and R. P. Dilworth, *The imbedding problem for modular lattices*, Ann. of Math. **45** (1944), 450–456. Reprinted in Chapter 4 of this volume.

9. Ch. Herrmann, *S-Verklebte Summen von Verbänden*, Math. Z. **130** (1973), 225–274.

10. J. P. S. Kung, *Matchings and Radon transforms in lattices, I. Consistent lattices*, Order **2** (1985), 105–112.

11. J. P. S. Kung, *Matchings and Radon transforms in lattices II. Concordant sets*, Math. Proc. Cambridge Philosophical Society **101** (1987), 221–231.

12. G. C. Kurinnoi, *A new proof of Dilworth's theorem*, Russian, Vestnik Charkov Univ. **93** (1973), 11–15.

13. K. Reuter, *Counting formulas for glued lattices*, Order **1** (1985), 265–276.

14. K. Reuter, *Matchings for linearly indecomposable modular lattices*, Discrete Math. **63** (1987), 245–249.

15. I. Rival, *Maximal sublattices of finite distributive lattices*, Proc. Amer. Math. Soc. **37** (1973), 417–420.

16. I. Rival, "Contributions to combinatorial lattice theory," Ph. D. Thesis, University of Manitoba, Winnipeg, 1974.

17. I. Rival, *Combinatorial inequalities for semimodular lattices of breadth two*, Algebra Universalis **6, no. 3** (1976), 303–311.

18. I. Rival (ed.), "Symposium on Ordered Sets," D. Reidel Publ. Co., Dordrecht, 1982.

19. I. Rival (ed.), "Combinatorics and Ordered Sets," Contempary Math., Amer. Math. Soc., Providence, 1986.

20. G.-C. Rota, *On the foundations of combinatorial theory, I. Theory of Möbius functions*, Z. Wahrscheinlichkeitstheorie und Verw. Gebiete **2** (1964), 340–368.

21. L. Weisner, *Abstract theory of inversion of finite series*, Trans. Amer. Math. Soc. **38** (1935), 474–484.

22. R. Wille, *Tensorial decomposition of concept lattices*, Order **2** (1985), 81–95.

Department of Computer Science
University of Ottawa
Ottawa, Ontario
Canada K1N 6N5

Geometric and Semimodular Lattices

Background

R. P. DILWORTH

Dependence relations. This work began with the general hope that by making use of a suitably defined dependence relation it would be possible to imbed an arbitrary finite dimensional lattice into a semimodular point lattice so that some lattice properties would be preserved. However, some experimentation soon convinced me that at this level of generality too much would be lost to justify the effort.[1] This experimentation did convince me that if the lattice in question was semimodular above its points (atoms), it could be injected into a semimodular point lattice preserving meets and rank and the lattice structure already present above the atoms. Furthermore, this would provide a straightforward lattice construction for imbedding a modular lattice of length 3 in a complemented modular lattice of length 3.

I called these point lattices which satisfied the chain law and were semimodular above atoms *quasi-modular*. Since Mac Lane had shown that a semimodular point lattice can be defined by a suitable dependence relation over its set of points, I decided to use the dependence relation approach. In as much as the lattice was semimodular above its atoms, it was possible to define a rank function $r(a)$ satisfying

(1) $r(0) = 0$
(2) $r(b) = r(a) + 1$ if b covers a
(3) $r(a \vee b) + r(a \wedge b) \leq r(a) + r(b)$ if $a \wedge b \neq 0$.

The advantage in using a dependence relation approach was the fact that independent subsets could be very easily defined, namely, a subset S of the atoms is

---

[1] Later Dilworth was able to show that every finite lattice can be embedded as a sublattice into a finite geometric lattice. See Chapter 8.

independent if and only if for all subsets T of S,

$$r\left(\bigvee T\right) \geq |T|$$

where $|T|$ is the number of elements in T. Then an obvious definition for the dependence relation is $p \bigtriangleup S$ if and only if there exists an independent set T contained in S such that $T \cup \{p\}$ is dependent. If one makes careful use of the properties of the rank function, it can be shown that the dependence relation has the properties:

(1) $p \bigtriangleup S \cup \{p\}$
(2) If $p \bigtriangleup S$ and $S \bigtriangleup T$, then $p \bigtriangleup T$
(3) If $p \bigtriangleup S \cup \{p'\}$, then either $p \bigtriangleup S$ or $p' \bigtriangleup S \cup \{p\}$.

The closed sets under this dependence relation, namely, the sets S such that $p \bigtriangleup S$ if and only if $p \in S$, are closed under set intersection and form a semimodular atomic lattice. As an example of how this process works, let us start with the Boolean algebra of all subset of a four-element set diagrammed in Figure 1.

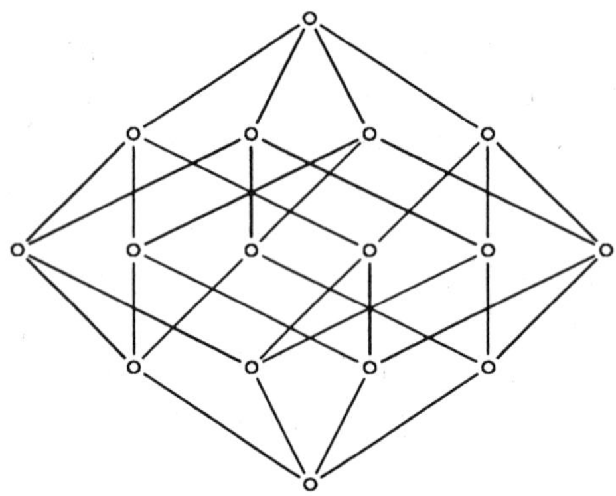

Figure 1

There are 4 one-element subsets, 6 two-element subsets, 4 three-element subsets, and 1 four-element subset as well as the null subset. The null set together with the subsets having at least two elements give the quasi-modular lattice diagrammed in Figure 2.

The construction above adds three new elements a, b, c which restore semimodularity and produce the partition lattice on a four-element set. See Figure 3.[2]

---

[2]The three new elements are the partitions $\{\{1,2\},\{3,4\}\}$, $\{\{1,3\},\{2,4\}\}$, and $\{\{1,4\},\{2,3\}\}$ of the four-element set $\{1,2,3,4\}$. The subsets originally in the quasi-modular lattice are regarded as partitions whose only nontrivial block is the subset itself.

                                                        THE DILWORTH THEOREMS

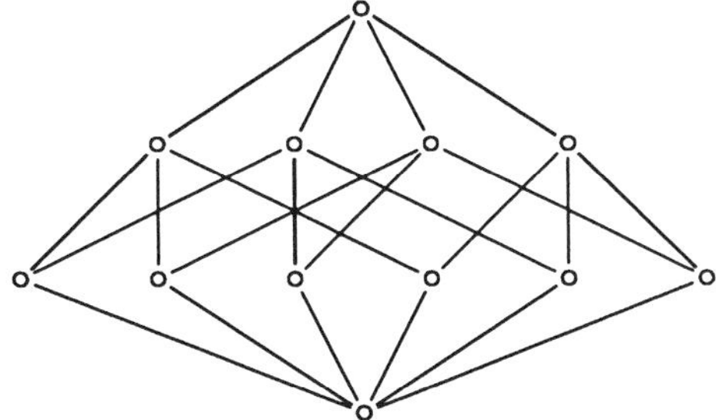

Figure 2

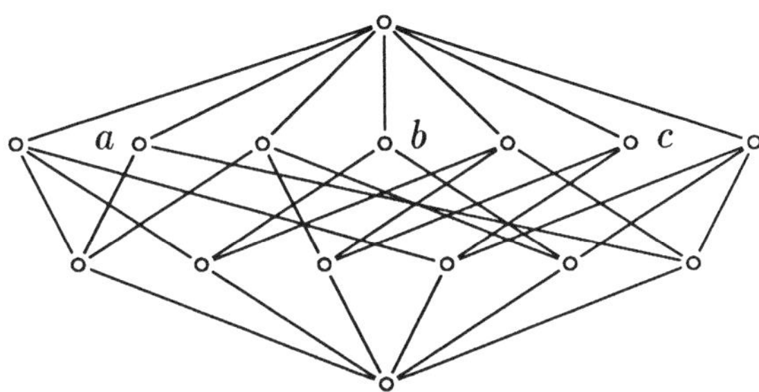

Figure 3

More generally, starting with a Boolean algebra of higher order and applying the process to the quasi-modular lattice of the null set and the sets having at least n elements gives a semimodular lattice with geometric properties generalizing the partition lattice. The process can also be applied to semimodular lattices to get new semimodular lattices.

DEPENDENCE RELATIONS IN A SEMI-MODULAR LATTICE

By R. P. Dilworth

Introduction. Let L be an upper semi-modular lattice of finite dimensions. Then, by definition, L satisfies the Birkhoff condition: If a covers $a \cap b$, then $a \cup b$ covers b. L is also characterized by the existence of a rank function $\rho(a)$ which takes on integer values and has the properties:

$R_1 : \rho(z) = 0$, where z is the null element of L.

$R_2 : \rho(a) = \rho(b) + 1$ if a covers b.

$R_3 : \rho(a \cup b) + \rho(a \cap b) \leq \rho(a) + \rho(b)$.

Let P denote the set of points of L, that is, those elements p such that $\rho(p) = 1$. Then it is well known [3] that, if one defines $p \Delta S$ if and only if $u(S) \supseteq p$, then Δ is a dependence relation over the subsets S of P with the properties:

$D_1 : p \Delta S + p$.

$D_2 :$ If $p \Delta S$ and $S \Delta T$, then $p \Delta T$.

$D_3 :$ If $p \Delta S + p'$, then either $p \Delta S$ or $p' \Delta S + p$.

Conversely, if a dependence relation Δ having properties D_1, D_2, and D_3 is defined over the subsets of a set P, then the closed subsets of P (subsets S such that $p \Delta S$ implies $p \,\varepsilon\, S$) form a semi-modular point lattice. If Δ is the dependence relation defined above, then the lattice of closed subsets is simply the lattice within L of elements which are unions of points.

Now if one defines Δ in a similar manner over the set P_n of elements q such that $\rho(q) = n$, where $n > 1$, then D_3 no longer holds for Δ. Nevertheless, we shall show that by means of a considerable modification of the definition one obtains a dependence relation over P_n such that D_1, D_2, D_3 hold and which reduces to that defined above when $n = 1$. It follows that the elements of P_n are embedded in the semi-modular point lattice of closed subsets.

As an application to embedding problems we prove the following theorem:

Theorem 3.1. *Every modular lattice of length three or less can be embedded isometrically in a complemented modular lattice.*

Since there exist modular lattices of length four which cannot be embedded in complemented modular lattices [2] this is the best possible result.

As an application in a somewhat different direction, let B be the Boolean algebra of subsets of a finite set S. Then we have

Theorem 3.2. *The lattice of closed subsets of the set of elements of B of rank two is isomorphic to the partition lattice of S.*

1. **Quasi-modular lattices.** Let M be a lattice with elements a, b, c, $\cdots$ and containing relation $a \supseteq b$. If $a \supseteq x \supseteq b$ implies $a = x$ or $x = b$ we say that a

Received February 28, 1944.

covers b and write $a > b$. A *chain* of length n joining a to b is a set of elements $a_0 , \cdots , a_n$ such that $a = a_0 \supseteq a_1 \supseteq a_2 \supseteq \cdots \supseteq a_n = b$. If $a_i > a_{i+1}$, the chain is said to be *complete*. Finally, M satisfies the *chain law* if it is of finite dimensions and every two complete chains joining the unit element u to z have the same length.

DEFINITION 1.1. M is *(upper) quasi-modular* if it satisfies the chain law and

$$(\mu) \qquad\qquad a > a \cap b, \; a \cap b \neq z \text{ imply } a \cup b > b.$$

Now let L be a semi-modular lattice of finite dimensions and denote by L_n the set of all elements x of L such that $\rho(x) \geq n$ together with the element z. Then L_n is clearly closed with respect to union and hence is a lattice within L. (If a lattice L' is a subset of a lattice L'', then L' is a lattice *within* L'' if $a \supseteq b$ in L' if and only if $a \supseteq b$ in L''.)

LEMMA 1.1. L_n *is quasi-modular.*

For let $u > a_1 > a_2 > \cdots > a_{k-1} > z$ and $u > b_1 > b_2 > \cdots > b_{l-1} > z$ be two complete chains in L_n joining u to z. Then, in L, $\rho(a_{k-1}) = \rho(b_{l-1}) = n$. Hence there exist in L complete chains $a_{k-1} > a_k > \cdots > a_{k+n-2} > z$ and $b_{l-1} > b_l > \cdots > b_{l+n-2} > z$. But since L satisfies the chain law $k + n - 2 = l + n - 2$ and hence $k = l$. Thus the chain law holds in L_n . Next let $a > a \cap b$ and $a \cap b \neq z$ in L_n . Then $a \cap b$ is also the cross-cut in L and $\rho(a \cap b) \geq n$. But then $a > a \cap b$ in L and $a \cup b > b$ by the semi-modularity of L. Hence $a \cup b > b$ in L_n .

If $n = 1$, then $L_n = L$ and we have

COROLLARY. *Every semi-modular lattice is quasi-modular.*

As with semi-modular lattices, quasi-modular lattices are characterized by the properties of the rank function.

THEOREM 1.1. *If M is a quasi-modular lattice, there is defined over M a rank function $\rho(a)$ with the properties:*
$\quad$ $Q_1 : \rho(z) = 0.$
$\quad$ $Q_2 : \rho(a) = \rho(b) + 1 \; if \; a > b.$
$\quad$ $Q_3 : \rho(a \cup b) + \rho(a \cap b) \leq \rho(a) + \rho(b) \; if \; a \cap b \neq z.$
Conversely, if such a rank function is defined over a lattice M of finite dimensions, then M is quasi-modular.

Proof. If M is quasi-modular, then since the chain law holds, all complete chains from a to z have the same length l. Set $\rho(a) = l$ and $\rho(z) = 0$. Then clearly Q_1 and Q_2 hold. If $a \cap b \neq z$, consider the quotient lattice $u/a \cap b$. Then by (μ) the Birkhoff condition holds in $u/a \cap b$ and hence $u/a \cap b$ is semi-modular. If $r(a)$ denotes the rank function in $u/a \cap b$ we have

$$r(a \cup b) + r(a \cap b) \leq r(a) + r(b).$$

But then, adding $2\rho(a \cap b)$ to both sides of the inequality, we get

$$\rho(a \cup b) + \rho(a \cap b) \leq \rho(a) + \rho(b)$$

and Q_3 holds.

Conversely, if a rank function $\rho(a)$ is defined over M with properties Q_1, Q_2, and Q_3, then by Q_1 and Q_2 every chain from u to z has length $\rho(u)$ and hence the chain law holds in M. If $a > a \cap b$ and $a \cap b \neq z$, then, by Q_2, $\rho(a) = \rho(a \cap b) + 1$ and, by Q_3, $\rho(a \cup b) + \rho(a \cap b) \leq \rho(a \cap b) + 1 + \rho(b)$. Hence $\rho(b) \leq \rho(a \cup b) \leq \rho(b) + 1$. But if $\rho(b) = \rho(a \cup b)$, then $a \cup b = b$ and $a \cap b = a$, contrary to $a > a \cap b$. Thus $\rho(b) < \rho(a \cup b)$ and hence $\rho(a \cup b) = \rho(b) + 1$. Thus $a \cup b > b$ and M is quasi-modular. This completes the proof.

Frequent use will be made of the fact that $a \supseteq b$ and $\rho(a) = \rho(b)$ imply $a = b$ in any quasi-modular lattice. This follows immediately from Q_2.

We shall also need the following property of quasi-modular lattices:

LEMMA 1.2. *Let a_1, a_2, $\cdots$, $a_k \supseteq b \supset z$. Then*

$$\rho(a_1 \cup a_2 \cup \cdots \cup a_k) - \rho(b) \leq \sum_i (\rho(a_i) - \rho(b)).$$

The lemma is trivial if $k = 1$. Suppose that it holds for $k = n - 1$. Then $\rho(a_1 \cup a_2 \cup \cdots \cup a_n) - \rho(b) = \rho(a_1 \cup (a_2 \cup \cdots \cup a_n)) - \rho(b) \leq \rho(a_1) + \rho(a_2 \cup \cdots \cup a_n) - \rho(a_1 \cap (a_2 \cup \cdots \cup a_n)) - \rho(b) = \rho(a_1) - \rho(a_1 \cap (a_2 \cup \cdots \cup a_n)) + (\rho(a_2) - \rho(b)) + \cdots + (\rho(a_n) - \rho(b)) \leq (\rho(a_1) - \rho(b)) + (\rho(a_2) - \rho(b)) + \cdots + (\rho(a_n) - \rho(b))$. Hence the lemma follows by induction.

Now it is clear that in constructing a dependence relation over the elements of rank n of a semi-modular lattice L (which generalizes the usual dependence relation over points) we may restrict our attention to elements of higher rank; that is, elements of L_n. Hence we shall consider the more general problem of constructing a dependence relation for which D_1, D_2, and D_3 hold over the points of an arbitrary quasi-modular lattice.

2. The dependence relation. Let P denote the set of points of a quasi-modular lattice M. If S, T, $\cdots$ are subsets of P, then s, t, $\cdots$ will denote the unions $u(S)$, $u(T)$, $\cdots$ respectively of the elements of S, T, $\cdots$. We will write $S + T$ and ST for the set sum and product respectively of S and T. $n(S)$ will denote the number of elements in S.

DEFINITION 2.1. A subset S of P is *independent* if $\rho(t) \geq n(T)$ for every subset T of S.

A subset S of P which is not independent is said to be *dependent*. Clearly every dependent set S contains a subset T such that $\rho(t) < n(T)$.

It follows from Definition 2.1 that every independent set is finite.

LEMMA 2.1. *Every subset of an independent set is independent. On the other hand, every set which contains a dependent set is dependent.*

The proof is immediate from Definition 2.1.

LEMMA 2.2. *Every subset S of P contains at least one maximal independent subset.*

For every subset consisting of a single point is clearly independent. . On the other hand the number of elements in an independent set cannot be greater than $\rho(u)$.

DEFINITION 2.2. A point p *depends* upon a subset S of P (in symbols $p \, \Delta \, S$) if $p \, \varepsilon \, S$ or if S contains an independent subset T such that $T + p$ is dependent.

LEMMA 2.3. $p \, \Delta \, S + p$.

LEMMA 2.4. *If $p \, \Delta \, S + p'$, then either $p \, \Delta \, S$ or $p' \, \Delta \, S + p$.*

The lemma is trivial if $p = p'$. Hence we may take $p \neq p'$. If $p \, \varepsilon \, S + p'$, then $p \, \varepsilon \, S$ and $p \, \Delta \, S$ holds. Otherwise $S + p'$ contains an independent subset V such that $V + p$ is dependent. Now if $V \subseteq S$, then, according to Definition 2.2, $p \, \Delta \, S$ and the lemma holds. Hence we may assume that $V = W + p'$, where $W \subseteq S$. Since $W \subseteq V$, W is independent by Lemma 2.1. If $W + p$ is dependent, then $p \, \Delta \, S$ and the first conclusion of the lemma holds. But if $W + p$ is independent, then $W + p + p' = V + p$ is dependent and $p' \, \Delta \, S + p$. The proof is thus complete.

Lemmas 2.3 and 2.4 show that properties D_1 and D_3 hold for Δ. Indeed Δ has been so defined that these two properties are almost immediate. Thus the main difficulties are centered in the proof of D_2. Moreover, these difficulties are associated with the structure properties of the independent subsets of P. Hence the lemmas which follow will be concerned with the structure of such sets. Another definition is needed.

DEFINITION 2.3. An independent subset S of P is said to be *normal* if $\rho(s) = n(S)$.

LEMMA 2.5. *If $p \, \Delta \, S$, then $p \, \Delta \, S'$, where S' is a normal subset of S.*

Let $p \, \Delta \, S$ and suppose first that $p \, \varepsilon \, S$. Then $S' = \{p\}$ is normal since $\rho(s') = \rho(p) = 1 = n(S')$ and clearly $p \, \Delta \, S'$. Hence the lemma holds in this case. Now if p is not an element of S, let S' be a minimal independent subset of S such that $S' + p$ is dependent. S' exists since by Definition 2.2 at least one independent subset T of S exists such that $T + p$ is dependent. But then $S' + p$ contains a subset V such that $\rho(v) < n(V)$. Now $V \not\subseteq S'$ since otherwise S' would be dependent. Hence $V = V' + p$, where V' is a subset of S' and hence is independent. But if V' is a proper subset of S', then V' is an independent subset of S such that $V' + p = V$ is dependent, contrary to the minimal property of S'. Hence $V' = S'$ and $V = S' + p$. Thus $\rho(v) = \rho(s' \cup p) < n(S') + 1$. But since S' is independent $\rho(s') \geq n(S')$. Now $n(S') \leq \rho(s') \leq \rho(s' \cup p) < n(S') + 1$. Hence $\rho(s') = n(S')$ and S' is normal by Definition 2.3. p obviously depends upon S'.

LEMMA 2.6. *If S is normal, then $p \, \Delta \, S$ if and only if $s \supseteq p$.*

For if S is normal and $p \Delta S$, then either $p \, \varepsilon \, S$ in which case $s \supseteq p$ or p is not an element of S and $S + p$ is dependent. But if S' is a minimal subset of S such that $S' + p$ is dependent, then S' is normal by the proof of Lemma 2.5. Furthermore, if $T \subset S'$, then $T + p$ is independent by the minimal property of S'. Hence since $S' + p$ is dependent we have $\rho(s' \cup p) < n(S') + 1$ by Definition 2.1. Thus $n(S') = \rho(s') \leq \rho(s' \cup p) < n(S') + 1$. Hence $\rho(s') = \rho(s' \cup p)$ and $s' = s' \cup p$. It follows that $s \supseteq s' \supseteq p$.

Conversely, let $s \supseteq p$. If $p \, \varepsilon \, S$, then trivially $p \Delta S$. If $p \, \varepsilon' \, S$, then $\rho(s \cup p) = \rho(s) = n(s) < n(S + p)$. Hence $S + p$ is dependent and $S \Delta p$ by Definition 2.2.

It follows from Lemmas 2.5 and 2.6 that dependency upon any set can be reduced to a lattice inclusion.

In developing properties of normal sets we shall let $u(N) = z$, where N is the null set. Thus N is trivially normal.

LEMMA 2.7. *Let V and W be normal subsets of an independent set S. Then VW is normal and $v \cap w = u(VW)$. In particular, if $VW = N$, then $v \cap w = z$.*

Let $x = u(VW)$. Then since $V, W \supseteq VW$ we have $v \cap w \supseteq x$. Since S is independent, it follows that $\rho(v \cap w) \geq \rho(x) \geq n(VW)$. Now if $\rho(v \cap w) > n(VW)$, then using Q_2 we get

$$\rho(v) + \rho(w) - \rho(v \cup w) \geq \rho(v \cap w) > n(VW) = n(V) + n(W) - n(V + W).$$

But $\rho(v) = n(V)$ and $\rho(w) = n(W)$ since V and W are normal. Hence we have $\rho(v \cup w) < n(V + W)$. But clearly $v \cup w = u(V + W)$ and hence $V + W$ is dependent, contrary to the independence of S. Thus $\rho(v \cap w) = \rho(x) = n(VW)$. Hence $v \cap w = x$ and VW is normal.

COROLLARY. *Let $V_1, V_2, \cdots, V_k$ be normal subsets of an independent set S. Then $V_1 V_2 \cdots V_k$ is normal and $v_1 \cap v_2 \cap \cdots \cap v_k = u(V_1 V_2 \cdots V_k)$.*

LEMMA 2.8. *Let V and W be normal subsets of an independent set S. Then if $VW \neq N$, $V + W$ is normal and $v \cup w = u(V + W)$.*

For if $VW \neq N$, then by Lemma 2.7 and Q_2 we have

$$\rho(v \cup w) \leq \rho(v) + \rho(w) - \rho(v \cap w) = n(V) + n(W) - n(VW) = n(V + W).$$

But since $V + W$ is a subset of an independent set S, it is also independent and $\rho(v \cup w) \geq n(V + W)$. Hence $\rho(v \cup w) = n(V + W)$ and $V + W$ is normal. It follows from lattice properties that $v \cup w = u(V + W)$.

COROLLARY. *Let $V_1, V_2, \cdots, V_k$ be normal subsets of an independent set S. Then if $(V_1 + V_2 + \cdots + V_i)V_{i+1} \neq N$, $i = 1, \cdots, k - 1$; $V_1 + V_2 + \cdots + V_k$ is normal and $v_1 \cup v_2 \cup \cdots \cup v_k = u(V_1 + V_2 + \cdots + V_k)$.*

LEMMA 2.9. *Two distinct, maximal, normal subsets of an independent set S are always disjoint.*

For if V and W are maximal, normal subsets of an independent set S and $VW \neq N$, then $V + W$ is normal and by the maximal property $V = V + W = W$.

LEMMA 2.10. *An independent set S has a unique representation $S = S_1 + S_2 + \cdots + S_k$ as the sum of mutually disjoint, maximal, normal subsets.*

For if $p \, \varepsilon \, S$, then $\{p\}$ is a normal subset of S and hence $p \, \varepsilon \, S_i$, where S_i is a maximal, normal subset of S. Hence the sum of the maximal, normal subsets of S exhausts S and is clearly disjoint by Lemma 2.9.

LEMMA 2.11. *Let $p \, \Delta \, S$, where S is independent. Then p depends upon exactly one maximal, normal subset of S.*

For suppose that $p \, \Delta \, S_i$ and $p \, \Delta \, S_j$, where $i \neq j$. Then $s_i \supseteq p$ and $s_j \supseteq p$ by Lemma 2.6. Hence $s_i \cap s_j \supseteq p$ and $n(S_i S_j) = \rho(s_i \cap s_j) \geq 1$. Thus $S_i S_j \neq N$, contrary to Lemma 2.9. Hence p depends upon at most one maximal, normal subset of S. That p depends upon at *least* one follows from Lemma 2.5.

It is convenient to extend the notion of dependency to sets. If S and T are subsets of P we shall write $S \, \Delta \, T$ if and only if $p \, \Delta \, T$ for all $p \, \varepsilon \, S$.

LEMMA 2.12. *Let $T \, \Delta \, S$, where T is normal and S is independent. Then $T \, \Delta \, S'$, where S' is a normal subset of S.*

Let $S = S_1 + S_2 + \cdots + S_m$ be the representation of S as a disjoint sum of maximal, normal subsets. Then by Lemma 2.11 each element of T is dependent upon exactly one of the subsets S_i. Let T_i denote those elements of T which depend upon S_i. Furthermore let $T_1, \cdots, T_k$ denote the sets T_i which are non-null. Then $T = T_1 + T_2 + \cdots + T_k$ and $T_1, \cdots, T_k$ are disjoint. Finally, if we let $S' = S_1 + S_2 + \cdots + S_k$, then $T \, \Delta \, S'$. Now, by Lemma 2.6, $s_i \supseteq t_i$ and hence $s_i \cap t \supseteq t_i \supset z$. From Q_2 we get

$$\rho(s_i \cup t) - \rho(t) \leq \rho(s_i) - \rho(s_i \cap t).$$

But $\rho(s_i \cap t) \geq \rho(t_i) \geq n(T_i)$ since T is independent. Thus

$$\rho(s_i \cup t) - \rho(t) \leq \rho(s_i) - n(T_i).$$

Since $T \, \Delta \, S'$ we have $s' = s_1 \cup s_2 \cup \cdots \cup s_k \supseteq t \supset z$ and, by Lemma 1.2,

$$\rho(s') - \rho(t) = \rho(s' \cup t) - \rho(t) = \rho((s_1 \cup t) \cup \cdots \cup (s_k \cup t)) - \rho(t)$$

$$\leq \sum_i (\rho(s_i \cup t) - \rho(t)) \leq \sum_i (\rho(s_i) - n(T_i)) = \sum_i \rho(s_i) - \sum_i n(T_i).$$

But $\rho(s_i) = n(S_i)$ and $\rho(t) = n(T)$ since S_i and T are normal. Furthermore

$$\sum_i n(S_i) = n(S')$$

since $S_1, \cdots, S_k$ are disjoint. Similarly

$$\sum_i n(T_i) = n(T).$$

Thus

$$\rho(s') - n(T) \leq n(S') - n(T)$$

and hence

$$\rho(s') \leq n(S').$$

But since S' is independent $\rho(s') \geq n(S')$ and thus

$$\rho(s') = n(S').$$

Hence S' is normal and the proof is complete.

LEMMA 2.13. *If S is any subset of P, let T be a maximal, independent subset of S. Then $p \, \Delta \, S$ if and only if $p \, \Delta \, T$.*

Let q be any element of S. If $q \, \varepsilon \, T$, then $q \, \Delta \, T$ by Definition 2.2. If $q \, \varepsilon' \, T$, then $T + q$ is a subset of S which properly contains T. Hence $T + q$ is dependent and thus $q \, \Delta \, T$ by Definition 2.2. Hence $S \, \Delta \, T$. Now if $p \, \Delta \, S$, then, by Lemma 2.5, $p \, \Delta \, S'$, where S' is a normal subset of S. But since $S' \, \Delta \, T$ and T is independent it follows from Lemma 2.12 that there is a normal subset T' of T such that $S' \, \Delta \, T'$. Hence $t' \supseteq q$ for all $q \, \varepsilon \, S'$ by Lemma 2.6 and thus $t' \supseteq s' \supseteq p$. Again using Lemma 2.6 we get $p \, \Delta \, T'$ and hence $p \, \Delta \, T$. This proves the necessity. If $p \, \Delta \, T$, then it is trivially true that $p \, \Delta \, S$. Hence the proof is complete.

COROLLARY. *Let T_1 and T_2 be maximal, independent subsets of S. Then $T_1 \, \Delta \, T_2$ and $T_2 \, \Delta \, T_1$.*

LEMMA 2.14. *If $p \, \Delta \, S$ and $S \, \Delta \, T$, then $p \, \Delta \, T$.*

For let T_1 be a maximal, independent subset of T. Then $S \, \Delta \, T_1$ by Lemma 2.13. Now since $p \, \Delta \, S$, by Lemma 2.5, $p \, \Delta \, S'$, where S' is a normal subset of S. Since $S' \, \Delta \, T_1$, by Lemma 2.12 there is a normal subset T_1' of T_1 such that $S' \, \Delta \, T_1'$. But then $p \subseteq s' \subseteq t'$ by Lemma 2.6. Hence $p \, \Delta \, T'$ and thus $p \, \Delta \, T$.

From Lemmas 2.3, 2.4, and 2.14 we have

THEOREM 2.1. *The dependence relation $p \, \Delta \, S$ has the properties* D_1, D_2, *and* D_3.

Now if S is an independent subset of P, by Lemma 2.10, S has a unique representation $S = S_1 + \cdots + S_k$, where the S_i are maximal, normal subsets of S. The elements $s_1, \cdots, s_k$ which are the unions in M of $S_1, \cdots, S_k$ respectively we shall call the *norm* elements associated with S.

LEMMA 2.15. *If s_1 and s_2 are distinct norm elements of an independent set S, then $s_1 \cap s_2 = z$.*

For if $s_1 \cap s_2 \neq z$, then $u(S_1 S_2) = s_1 \cap s_2 \neq z$ by Lemma 2.7.   Hence $S_1 S_2 \neq N$ and $S_1 + S_2$ is normal by Lemma 2.8.   But then $S_1 = S_1 + S_2 = S_2$ by the maximal property, and S_1 and S_2 are not distinct contrary to the hypothesis of the lemma.

Lemma 2.16.   *Let S and T be independent subsets of P.   Then $S \, \Delta \, T$ if and only if for each norm element s_i of S there exists a norm element t_j of T such that $s_i \subseteq t_j$.*

For if $S \, \Delta \, T$, then $S_i \, \Delta \, T$ and hence $S_i \, \Delta \, T_j$ for some maximal, normal subset T_j of T by Lemma 2.12.   But then $s_i \subseteq t_j$ by Lemma 2.6.   Conversely, if each norm element of S is contained in a norm element of T, then if $p \, \varepsilon \, S$ we have $p \subseteq s_i$ for some s_i and hence $p \subseteq s_i \subseteq t_j$ for some t_j .   But then $p \, \Delta \, T_j$ by Lemma 2.6 and hence $p \, \Delta \, T$.   Thus $S \, \Delta \, T$.

Corollary.   *If S and T are independent sets, then $S \, \Delta \, T$ and $T \, \Delta \, S$ if and only if S and T have the same norm elements.*

Now if S is an arbitrary subset of P, according to the corollaries of Lemmas 2.13 and 2.16 the maximal, independent subsets of S all have the same norm elements.   These common norm elements we shall call the norm elements of S.   From Lemmas 2.13, 2.14, and 2.16 we get

Theorem 2.2.   *If S and T are subsets of P, then $S \, \Delta \, T$ if and only if for each norm element s_i of S there exists a norm element t_j of T such that $s_i \subseteq t_j$. S and T are mutually dependent if and only if they have the same norm elements.*

Theorem 2.2 completely characterizes dependency of subsets of P in terms of the lattice properties of M.

If M is semi-modular and S is a subset of P, then (see [1]) $s = p_1 \cup \cdots \cup p_k$, where $p_1 , \cdots , p_k$ generate a Boolean algebra which is complete in M.   Hence $p_1 , \cdots , p_k$ are independent and $\rho(s) = k$.   Thus $p_1 , \cdots , p_k$ is a normal subset of P and s is the only norm element of S.   Hence, by Theorem 2.2, $S \, \Delta \, T$ if and only if $s \subseteq t$ and Δ reduces to the usual dependence relation of semi-modular lattices.

Now, if S is a subset of P, let S^* denote the set of all points p such that $p \, \Delta \, S$. S is said to be *closed* if $S^* = S$.   Let $\mathfrak{M}$ be the set of all closed subsets of P.   Then since $D_1 , D_2 ,$ and D_3 hold for Δ we have [3]

Theorem 2.3 (MacLane).   *$\mathfrak{M}$ is an upper semi-modular point lattice.*

If the quasi-modular lattice M is an upper semi-modular point lattice, then it is well known that $\mathfrak{M}$ is isomorphic to M.   We shall be particularly interested in the case where M is a quasi-modular point lattice.

Lemma 2.17.   *Let M be a quasi-modular point lattice.   Then, if $a \, \varepsilon \, M$, there is a normal set S such that $a = u(S)$.*

If $a = z$, we may take $S = N$.   If a is a point, then S consists of the element a itself.   Now, using induction, let us suppose that the lemma holds for all a such that $\rho(a) < k$. Let $\rho(a) = k$. Since M is finite dimensional there exists an element a_1 such that $a > a_1$. Hence $\rho(a_1) = k - 1$ and by the induction assumption a normal set S_1 exists such that $a_1 = u(S_1)$. Since M is a point lattice, there exists a point p such that $a \supseteq p$, $a_1 \not\supseteq p$. Let $S = S_1 + p$. Now if S is dependent, there is a minimal dependent subset T of S. It follows from Definition 2.1 that $\rho(t) < n(T)$. Since T is dependent, $T \not\subseteq S_1$ and hence $T = T_1 + p$, where $T_1 \subseteq S_1$. But then

$$n(T_1) \leq \rho(t_1) \leq \rho(t_1 \cup p) = \rho(t) < n(T) = n(T_1) + 1.$$

Thus $\rho(t_1) = \rho(t_1 \cup p)$ and $t_1 \cup p = t_1$. Hence $a_1 \supseteq t_1 \supseteq p$, contrary to the definition of p. Thus S is an independent subset of P. Since $a > a_1$ and $a_1 \not\supseteq p$, we have $a = a_1 \cup p$. Hence $a = a_1 \cup p = u(S_1 + p) = u(S)$ and $\rho(a) = \rho(a_1) + 1 = n(S_1) + 1 = n(S)$. Thus S is normal and the proof is complete.

THEOREM 2.4. *If M is a quasi-modular point lattice, then M is isomorphic to a lattice within $\mathfrak{M}$ which is complete in $\mathfrak{M}$.*

A lattice L_1 within L_2 is *complete* in L_2 if $a > b$ in L_1 if and only if $a > b$ in L_2.

Proof. If a is any element of M, let S_a denote the set of all points p such that $a \supseteq p$. Now by Lemma 2.17 there exists a normal set S such that $a = u(S)$. But, by Lemma 2.6, $p \, \Delta \, S$ if and only if $u(S) = a \supseteq p$. Hence $S^* = S_a$ and S_a is an element of $\mathfrak{M}$. Consider the correspondence $a \leftrightarrow S_a$. Since M is a point lattice the correspondence is 1-1. Also $a \cap b \supseteq p$ if and only if $a \supseteq p$ and $b \supseteq p$; that is, if and only if $p \, \varepsilon \, S_a S_b$. Hence $a \cap b \leftrightarrow S_a S_b$. Thus M is isomorphic to a lattice within $\mathfrak{M}$. Now let $a > b$ in M. Since M is a point lattice there exists a point p_0 such that $a \supseteq p_0$, $b \not\supseteq p_0$. Then if T is a normal set such that $b = u(T)$ it follows from the proof of Lemma 2.17 that $T + p_0$ is a normal set such that $a = u(T + p_0)$. Now suppose $S_a \supseteq S^* \supseteq S_b$ for some subset S of P. If $S^* \neq S_b$, there is a point p such that $p \, \varepsilon \, S^*$, p is not an element of S_b. If $p = p_0$, then $p \, \Delta \, T + p_0$ trivially. If $p \neq p_0$, then

$$\rho(b \cup p_0 \cup p) = \rho(a) = \rho(b) + 1 = n(T) + 1 = n(T + p_0) < n(T + p_0 + p)$$

and $T + p_0 + p$ is dependent. Hence in either case $p \, \Delta \, T + p_0$. If $p \, \Delta \, T$, then $p \, \varepsilon \, S_b$, contrary to assumption. Thus, by D_3, we have $p_0 \, \Delta \, T + p$. But then $p_0 \, \Delta \, S^*$ and $S^* \supseteq (T + p_0)^* = S_a$. Thus either $S_a = S^*$ or $S^* = S_b$ and hence $S_a > S_b$ in $\mathfrak{M}$. It follows that M is complete in $\mathfrak{M}$.

THEOREM 2.5. *If M is a quasi-modular point lattice and L is a semi-modular sublattice which is complete in M, then L is isomorphic to a sublattice of $\mathfrak{M}$ which is complete in $\mathfrak{M}$.*

Proof. According to Theorem 2.4, the correspondence $a \leftrightarrow S_a$ takes M into a lattice within $\mathfrak{M}$ which is complete in $\mathfrak{M}$. Hence since L is a complete sublattice of M it takes L into a lattice $\mathfrak{L}$ within $\mathfrak{M}$ which is complete in $\mathfrak{M}$. We shall show that $\mathfrak{L}$ is a sublattice of $\mathfrak{M}$. Let a and b be two elements of L. If $b \supseteq a$, then $S_b \supseteq S_a$ and $a \cup b \leftrightarrow S_a \cup S_b$. If $b \not\supseteq a$, let $a = a_k > a_{k-1} >$

$\cdots > a_1 > a \cap b$ in L. Since L is complete in M, the covering relations hold in M. Also since $b \not\supseteq a_1$, the Birkhoff condition gives $a_1 \cup b > b$. Hence $S_{a_1} \cup b \supseteq S_{a_1} \cup S_b \supset S_b$ and since $S_{a_1 \cup b} > S_b$ we have $S_{a_1 \cup b} = S_{a_1} \cup S_b$. Now suppose that we have shown that $S_{a_i \cup b} = S_{a_i} \cup S_b$. Then since L is semi-modular and $a_{i+1} > a_i$ we have either $a_{i+1} \cup b > a_i \cup b$ or $a_{i+1} \cup b = a_i \cup b$. If $a_{i+1} \cup b > a_i \cup b$ we have $S_{a_{i+1} \cup b} \supseteq S_{a_{i+1}} \cup S_b \supset S_{a_i} \cup S_b = S_{a_i \cup b}$. Since $S_{a_{i+1} \cup b} > S_{a_i \cup b}$ it follows that $S_{a_{i+1} \cup b} = S_{a_{i+1}} \cup S_b$. If $a_{i+1} \cup b = a_i \cup b$, then $S_{a_{i+1} \cup b} \supseteq S_{a_{i+1}} \cup S_b \supseteq S_{a_i} \cup S_b = S_{a_i \cup b} = S_{a_{i+1} \cup b}$ and again we have $S_{a_{i+1} \cup b} = S_{a_{i+1}} \cup S_b$. Hence in either case $S_{a_{i+1} \cup b} = S_{a_{i+1}} \cup S_b$. By induction it follows that $S_{a \cup b} = S_a \cup S_b$. Thus union is preserved and $\mathfrak{L}$ is a sublattice of $\mathfrak{M}$.

We conclude this section with a theorem on the special quasi-modular lattices L_n described in §1.

Theorem 2.6. *If L is a semi-modular point lattice, then L_n is a quasi-modular point lattice.*

Proof. Let a be of positive rank in L_n. Then $\rho(a) = k$, where $k \geq n$. Let $a \supset a_0$, where $\rho(a_0) = n - 1$. Let p be a point of L such that $a \supseteq p$ but $a_0 \not\supseteq p$. Then by the Birkhoff condition $p' = a_0 \cup p > a_0$. Hence $\rho(p') = n$ and p' is a point of L_n. Let S be the set of points p and S', the set of elements p'. Then, since L is a point lattice, $a = u(S) \cup a_0$ and hence $a = u(S')$. Thus L_n is a point lattice. Lemma 1.1 completes the proof.

3. **Applications.** The results of the previous section will be applied first to the imbedding problem for modular lattices. We need two preliminary lemmas.

Lemma 3.1. *Every lattice satisfying the chain law and having length three or less is quasi-modular.*

Let $a > a \cap b$, $a \cap b \neq z$ in L. If $b = a \cap b$, then $a \cup b = a$ and we have $a \cup b > b$. If $b \neq a \cap b$, then $\rho(b) \geq 2$. But $\rho(a \cup b) \leq 3$. Hence $a \cup b > b$ and (μ) holds in L.

We shall make frequent use of the notion of the *dual L'* of a lattice L. L' is the lattice obtained from L by interchanging the operations of union and cross-cut. Clearly the dual of a modular lattice is modular while the dual of an upper semi-modular lattice is lower semi-modular and vice versa. It is also clear that $(L')'$ is isomorphic to L.

Lemma 3.2. *Let L be an upper semi-modular lattice of finite dimensions in which the unit element u is a union of points. Then L' is a complemented point lattice.*

This is essentially Lemma 3.5 of [1].

Theorem 3.1. *Every modular lattice of length three is a sublattice of a complemented modular lattice.*

Proof. Let L be a modular lattice of length 3 and denote by B the Boolean algebra of order 4. Then the direct sum of L and B is a modular lattice of length 5. Hence if we identify elements whose ranks are 3 or greater, we get an upper semi-modular lattice L_1 of length 3 which contains L as a sublattice. Furthermore the union of the points of L_1 is u. By Lemma 3.2, L_1' is a lower semi-modular point lattice and hence is a quasi-modular point lattice by Lemma 3.1. Now since L' is modular, L_1' contains L' as an upper semi-modular sublattice complete in L_1'. Let $\mathfrak{L}$ denote the lattice of closed subsets of the set of points of L_1'. Then, by Theorem 2.5, L_1' is isomorphic to a lattice within $\mathfrak{L}$ while L' is isomorphic to a sublattice of $\mathfrak{L}$. Hence the dual $\mathfrak{L}'$ is a lower semi-modular point lattice of length 3 containing a sublattice isomorphic to L.

Let us set $L_2 \simeq \mathfrak{L}'$ so that L_1 is a lattice within L_2 while L is a sublattice of both L_1 and L_2. If L_3 denotes the lattice of closed subsets of L_2, then L_3 is an upper semi-modular point lattice of length 3 and, by Theorems 2.4 and 2.5, L_2 is isomorphic to a lattice within L_3 while L is isomorphic to a sublattice of L_3. Then L_3' is a quasi-modular point lattice and if we denote by L_4 the dual of the lattice of closed subsets of L_3', L_3 is lattice within L_4 while L is a sublattice of L_4. Continuing in this manner we get a sequence of lattices $L \subseteq L_1 \subseteq L_2 \subseteq L_3 \subseteq \cdots$, where L_{2n} is lower semi-modular, L_{2n+1} is upper semi-modular and L is a sublattice of L_i for each i. Let M denote the set $L_1 + L_2 + L_3 + \cdots$. If $a, b \,\varepsilon\, M$, there is a first L_i such that $a, b \,\varepsilon\, L_i$. Let $a \cup_i b$ denote union in L_i. Then since L_i is a lattice within L_{i+1} we have $a \cup_i b \supseteq a \cup_{i+1} b \supseteq a \cup_{i+2} b \supseteq \cdots$. Since the descending chain condition holds, from some point on the elements of the chain must be equal. If j is the first integer such that $a \cup_j b = a \cup_{j+1} b = \cdots$ we define $a \cup b = a \cup_j b$. Similarly $a \cap_i b \subseteq a \cap_{i+1} b \subseteq \cdots$ and if k is the first integer such that $a \cap_k b = a \cap_{k+1} b = \cdots$ we define $a \cap b = a \cap_k b$. It is clear that M is a lattice under union and cross-cut so defined and that M contains L as a sublattice.

Now let $a > a \cap b$ in M and let $a \cap b = a \cap_k b = a \cap_{k+1} b = \cdots$. Then $a > a \cap_k b$ and $a > a \cap_{k+1} b$. But since either L_k or L_{k+1} is upper semi-modular, we have either $a \cup_k b > b$ or $a \cup_{k+1} b > b$. Hence $a \cup_k b \supseteq a \cup_{k+1} b \supseteq a \cup b \supseteq b$ and $a \cup b > b$ since L_k is complete in M. A dual argument shows that M is lower semi-modular. Hence M is modular. Since L_3 is a point lattice, it follows that M is a point lattice. Hence M is complemented and L is a sublattice of M. This completes the proof.

The next application is in a somewhat different direction. We shall show that the dependence relation of Definition 2.2 gives a direct method for constructing the partition lattice of a set S from the lattice of its subsets.

THEOREM 3.2. *Let L be the Boolean algebra of subsets of a finite set S. Then the lattice of closed subsets of L_2 is isomorphic to the partition lattice of S.*

Proof. The points of L_2 are the pairs $\{x, y\}$ of distinct elements x and y of S. Now if T is a set of pairs, let $m(T)$ denote the number of distinct elements of S occurring in the pairs of T. Then $m(T)$ is the rank of the union of T in the

Boolean algebra L. Hence, in L_2, $\rho(T) = m(T) - 2 + 1 = m(T) - 1$. $n(T)$ is the number of pairs in T. Now let T be a normal subset of pairs and let $T = T_1 + T_2$, where T_1 and T_2 are disjoint and non-null. Then there is an element of S which occurs both in some pair of T_1 and in some pair of T_2. For otherwise the sets of elements in T_1 and T_2 are disjoint and we have

$$\rho(T) = m(T) - 1 = m(T_1) + m(T_2) - 1 = \rho(T_1) + \rho(T_2) + 1$$

$$\geq n(T_1) + n(T_2) + 1 \geq n(t) + 1 > n(T),$$

contrary to the normality of T.

Now if E is an equivalence relation over S and C is an equivalence class, let P_C be the set of all pairs $\{x, y\}$, where $x, y \, \varepsilon \, C$. Set $S_E = \sum P_C$, where the sum is over the equivalence classes C. Hence S_E consists of all pairs $\{x, y\}$ of distinct elements of S such that $x \, E \, y$. Let $p = \{a, b\}$ depend upon S_E. Then, by Lemma 2.5, p depends upon a normal subset T of S_E. Let T_C be the pairs of T in P_C. Then $T = \sum T_C$ and this decomposition is disjoint. But if C and C' are distinct equivalence classes, then the elements in the pairs of T_C are different from the elements in the pairs of $T_{C'}$ since otherwise C and C' would have elements in common. Since T_C is normal we have $T = T_C$ for some C. But $T + p$ is dependent and hence $\rho(t_1 \cup p) < n(T_1) + 1$ for some subset T_1 of T. Thus

$$\rho(t_1) + 1 \geq n(T_1) + 1 > \rho(t_1 \cup p) \geq \rho(t_1)$$

and hence $\rho(t_1) = \rho(t_1 \cup p)$ which implies $m(T_1) = m(T_1 + p)$. It follows that a and b occur in pairs of T and hence that $a, b \, \varepsilon \, C$. But then $p = \{a, b\} \, \varepsilon \, P_C \subseteq S_E$ and S_E is closed under the dependence relation.

Conversely, let T be a set of pairs closed under the dependence relation. Set $x \, E \, y$ if either $x = y$ or the pair $\{x, y\} \, \varepsilon \, T$. Thus $x \, E \, x$ for all x and clearly $x \, E \, y$ implies $y \, E \, x$. Now let $x \, E \, y$ and $y \, E \, w$. If $x = w$, then $x \, E \, w$ and the transitivity holds. We may thus suppose that x, y, and w are distinct. Consider the set T_1 consisting of $\{x, y\}$ and $\{y, w\}$. We have $\rho(T_1) = 3 - 1 = 2$, $n(T_1) = 2$ and hence T_1 is independent. But $\rho(T_1 + \{x, w\}) = 3 - 1 = 2$ while $n(T_1 + \{x, w\}) = 3$. Hence $T_1 + \{x, w\}$ is dependent. Thus $\{x, w\}$ depends upon T_1 and hence upon T. Since T is closed $\{x, w\} \, \varepsilon \, T$ and $x \, E \, w$. Thus E is an equivalence relation over S. Clearly $T = S_E$.

We have shown that there is a 1-1 correspondence between the equivalence relation over S and the closed subsets of pairs. Now let $E \supseteq F$. If C is an equivalence class of F, then $C \subseteq C'$, where C' is an equivalence class of E. Hence $P_{C'} \supseteq P_C$ and thus $\sum P_{C'} \supseteq \sum P_C$. Hence $S_E \supseteq S_F$. Conversely, if $S_E \supseteq S_F$, then $x \, F \, y$ and $x \neq y$ imply $\{x, y\} \, \varepsilon \, S_F$ implies $\{x, y\} \, \varepsilon \, S_E$ implies $x \, E \, y$. Hence $E \supseteq F$. The correspondence preserves order and the lattices are thus isomorphic. This completes the proof of the theorem.

By way of illustration, let $S = \{1, 2, 3, 4\}$ and denote pairs of S by p_{ij}. Then the closed subsets are N, p_{12}, p_{13}, p_{14}, p_{23}, p_{24}, p_{34}, $\{p_{12}, p_{34}\}$, $\{p_{13}, p_{24}\}$,

$\{p_{14}, p_{23}\}$, $\{p_{12}, p_{13}, p_{23}\}$, $\{p_{12}, p_{14}, p_{24}\}$, $\{p_{13}, p_{14}, p_{34}\}$, $\{p_{23}, p_{24}, p_{34}\}$, $\{p_{12}, p_{13}, p_{14}, p_{23}, p_{24}, p_{34}\}$. All of these are the closures of normal sets except $\{p_{12}, p_{34}\}$, $\{p_{13}, p_{24}\}$, $\{p_{14}, p_{23}\}$. Hence the partition lattice is obtained from the Boolean algebra of subsets of S by deleting the points and adjoining the three unions $p_{12'} \cup p_{34}$, $p_{13} \cup p_{24}$, $p_{14} \cup p_{23}$.

REFERENCES

1. R. P. DILWORTH, *The arithmetical theory of Birkhoff lattices*, this Journal, vol. 8(1941), pp. 286–299.
2. M. HALL AND R. P. DILWORTH, *The embedding problem for modular lattices*, Annals of Mathematics, vol. 45(1944).
3. S. MACLANE, *A lattice formulation for transcendence degrees and p-bases*, this Journal, vol. 4 (1938), pp. 455–468.

YALE UNIVERSITY.

Reprinted from JOURNAL OF COMBINATORIAL THEORY Vol. 10, No. 1, January 1971
All Rights Reserved by Academic Press, New York and London *Printed in Belgium*

A Counterexample to the Generalization
of Sperner's Theorem*

R. P. DILWORTH AND CURTIS GREENE

California Institute of Technology, Pasadena, California 91109

Communicated by Gian-Carlo Rota

Received February 21, 1969

It has been conjectured that the analog of Sperner's theorem on non-comparable subsets of a set holds for arbitrary geometric lattices, namely, that the maximal number of non-comparable elements in a finite geometric lattice is max $w(k)$, where $w(k)$ is the number of elements of rank k. It is shown in this note that the conjecture is not true in general. A class of geometric lattices, each of which is a bond lattice of a finite graph, is constructed in which the conjecture fails to hold.

INTRODUCTION

A theorem of Sperner [1] states that $\max_k \binom{n}{k}$ is the maximal number of non-comparable subsets in a set of n elements. The analogous theorem for finite projective geometries asserts that the maximal number of non-comparable subspaces is $\max_k w(k)$, where $w(k)$ is the number of subspaces of dimension k. This theorem has been proved by Harper [2]. Since the lattices of subsets of a set and the lattice of subspaces of a projective space are examples of geometric lattices, it has been conjectured by Rota that the analog of Sperner's theorem holds in any geometric lattice, namely, that the maximal number of noncomparable elements in a finite geometric lattice L is $\max_k w(k)$, where $w(k)$ is the number of elements of rank k in L. It follows from a result of Baker [3] that the conjecture holds for all geometric lattices in which the number of covering elements is constant for elements of a given rank and dually.

It will be shown in this note that the conjecture is not true in general. In fact, we shall construct a class of geometric lattices each of which is a bond lattice of a finite graph, in which the conjecture fails to hold. The

* This research was partially supported by NSF Grant GP 8423

18

lattice of smallest order which is so constructed and which contradicts the conjecture contains 60,073 elements.

This class of geometric lattices also exhibits another unusual property. For each n, there is a bond lattice of dimension n in the class such that, by removing a single point, another geometric lattice is obtained which is also the bond lattice of a graph. In particular, for each n there exist two bond lattices of dimension n which have isomorphic structure above rank 1 but which are not isomorphic. This contrasts with the theorem of McLaughlin [4, Theorem 1.3] that the partially ordered set of points and dual points in a geometric lattice completely characterizes the lattice.

THE COUNTEREXAMPLES

Let G be a finite graph with no trivial cycles and no double edges. A subset E of the set of edges of G is *closed* if it contains every edge whose vertices are joined by a path formed from the edges of E. It is well known that the closed subsets form a geometric lattice (i.e., a semimodular point lattice) $L(G)$ in which the points correspond to the signleton sets formed from the edges of G.

Consider the graph G_n shown in Figure 1. Let $p_0 = \{(a, b)\}$ be the point

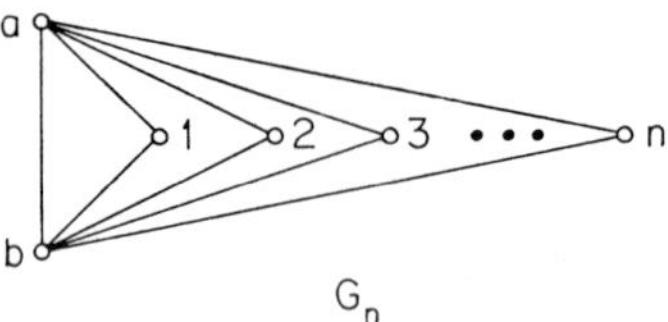

FIGURE 1

of $L(G_n)$ corresponding to the edge (a, b). If A is a closed subset of $E(G_n)$ containing the edge (a, b), then, for each i, A either contains both of the edges (a, i) and (b, i) or neither. Thus A is determined by the subset

$$A^* = \{i \mid (a, i), (b, i) \in A\}.$$

It follows that the quotient lattice $1/p_0$ of $L(G_n)$ is isomorphic to B_n, the Boolean algebra of all subsets of $\{1, 2,..., n\}$. Thus the rank of A in $L(G_n)$ is one more than the number of elements in A^*. Next let B be a closed subset not containing (a, b). Then, for each i, B can contain at most one of the two edges (a, i) and (b, i). Thus B is determined by the subset

$$B^* = \{i \mid \text{either } (a, i) \in B \text{ or } (b, i) \in B\}$$

and a choice function which for each $i \in B^*$ selects one of the two edges (a, i), (b, i). The rank of B in $L(G_n)$ is clearly equal to the number of elements in B^*.

It follows from the observations of the previous paragraph that the number $w(k)$ of elements of rank k in $L(G_n)$ is given by

$$w(k) = 2^k \binom{n}{k} + \binom{n}{k-1}.$$

It is easily verified that the sequence $w(k)$ is unimodal and that the value of k for which $w(k)$ reaches its maximum is approximately $2n/3$. Since $1/p_0$ is isomorphic to B_n, the rank having the most elements containing p_0 is approximately $n/2$.

Now let L be an arbitrary finite geometric lattice. Furthermore let $w(k)$ denote the number of elements of rank k in L and, if p is a point of L, let $w_p(k)$ denote the number of elements of rank k in L which lie above p. Finally let l be such that $w(l) = \max_k w(k)$.

In order to produce a counterexample to the analog of Sperner's theorem it suffices to find a geometric lattice L and a point p such that $w_p(l - 1) > w_p(l)$. For let T_l denote the set of elements of rank l which do not lie above p and let S_{l-1} be the elements of rank $l - 1$ which lie above p. Then $T_l \cup S_{l-1}$ is clearly a non-comparable subset of L and

$$|T_l \cup S_{l-1}| = |T_l| + |S_{l-1}|$$
$$= w(l) - w_p(l) + w_p(l - 1) > w(l) = \max_k w(k).$$

Now, for $L(G_n)$, l is approximately $2/3n$. Since the rank at which the maximum number of elements containing p_0 occurs is approximately $n/2$, it follows that, for large n, $w_{p_0}(l - 1) > w_{p_0}(l)$ and hence the geometric lattices $L(G_n)$ furnish counterexamples for all large n. In particular, for $n = 10$, $l = 7$, we have

$$w_{p_0}(6) = 252 > 210 = w_{p_0}(7)$$

and $L(G_{10})$ is a counterexample. The order of $L(G_{10})$ is 60,073.

Sublattices of $L(G_n)$

Let G_n' denote the graph G_n with the edge (a, b) deleted (Figure 2). If A' is a closed subset of G_n', then A' may be viewed as a subset of G_n. The closure of A' in G_n provides a natural mapping of $L(G_n')$ into $L(G_n)$.

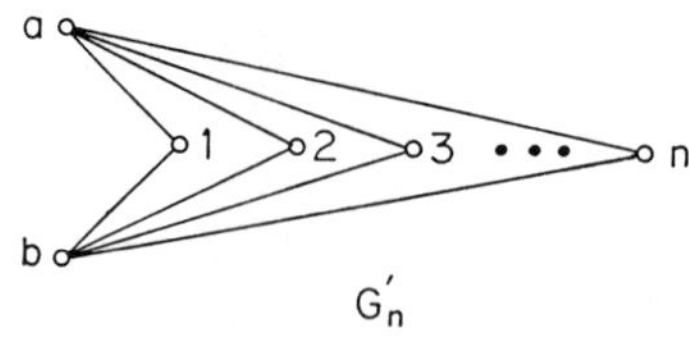

FIGURE 2

Since a closed subset of G_n containing any two of (a, b), (a, i), (b, i) must also contain the third, it is easily seen that the mapping of $L(G_n')$ into $L(G_n)$ is injective and the image set is $L(G_n) - \{(a, b)\}$. Thus $L(G_n')$ is isomorphic to the lattice $L(G_n)$ with the point $\{(a, b)\}$ deleted and hence $L(G_n)$ and $L(G_n')$ are not isomorphic. On the other hand $L_2(G_n) \simeq L_2(G_n')$, where $L_2(G_n)$ and $L_2(G_n')$ denote the partially ordered sets of elements of rank $\geqslant 2$ in $L(G_n)$ and $L(G_n')$ respectively.

It is interesting to note that the graphs G_n are the only graphs whose bond lattices have this property.

REFERENCES

1. E. SPERNER, Ein Satz über Untermengen einer endlichen Menge, *Math. Z.* **27** (1928), 544–548.
2. L. H. HARPER, The Morphology of Geometric Lattices (to appear).
3. K. A. BAKER, A Generalization of Sperner's Lemma (to appear).
4. J. E. McLAUGHLIN, Structure Theorems for Relatively Complemented Lattices, *Pacific J. Math.* **3** (1953), 197–208.

PRINTED IN BELGIUM BY THE ST. CATHERINE PRESS, TEMPELHOF, 37, BRUGES, LTD.

Dilworth's Completion, Submodular Functions, and Combinatorial Optimization

ULRICH FAIGLE

The motivation behind Dilworth's investigation in [5] is the question whether a lattice can be imbedded into a geometric lattice. This paper concentrates on *quasi-modular* point lattices, *i.e.*, point lattices L of finite length such that all maximal chains of L share the same length and the rank function f of L satisfies the weakened submodularity condition for all $a, b \in L$,

$$(1) \qquad a \wedge b \neq 0 \quad \text{implies} \quad f(a \vee b) + f(a \wedge b) \leq f(a) + f(b).$$

The embedding problem is solved in a closure-theoretic sense. First, each lattice element is represented by the set of points it dominates so that L can be viewed as a closure system $\mathcal{L}$ relative to a set of points. This closure system then is completed to the lattice of flats of a matroid $M = M(f)$, where the independent sets of M are specified as follows: The subset X of points is independent if and only if

$$(2) \qquad |S| \leq f(S) \quad \text{for all} \quad S \subseteq X.$$

(Here $f(S)$ is the lattice rank in L of the join of S.) If $f(S)$ were *submodular* relative to the collection of subsets of points, *i.e.*, if for all subsets A and B of points,

$$(3) \qquad f(A \cup B) + f(A \cap B) \leq f(A) + f(B),$$

then one could easily show that

$$(4) \qquad r(X) = \min\{f(S) + |X \backslash S| : S \in \mathcal{L}\}$$

gives rise to the rank function of M and that each $S \in \mathcal{L}$ is closed in M and satisfies $f(S) = r(S)$. In view of (1), however, f is only "almost submodular." Dilworth therefore derives the desired properties for M *via* a detailed analysis of the independence structure induced by (2).

In the following, the term *Dilworth completion* will not only refer to the completion of a closure system by a matroid closure system but, more generally, to the technique of constructing matroids *via* (2) (or (2*) below) from suitable set functions f.

Some time later, Dilworth was able to prove that every finite lattice can be embedded into a finite geometric lattice. The construction is based on the same idea: complete a closure system to the lattice of flats of a matroid. The trick to make this work in general consists in replacing the rank function of the arbitrary finite lattice L with some strictly increasing submodular function on L. An additional advantage of the use of this trick is that now (4) does yield the rank function of the embedding matroid (whose rank, though, may be strictly larger than the length of L). The details are published in Chapter 14 of [3]. Curiously, [5] is not even mentioned in the list of references there!

The Dilworth completion of a submodular function f on a finite lattice L *via* (4) is not the only way to arrive at a matroid such that each $a \in L$ corresponds to a flat of rank $f(a)$ but is, in a sense, the "most free" procedure (see Nguyen [23, 24]). Moreover, using a totally different approach, Pudlák and Tůma [25] have established a much more stringent result: every finite lattice can be embedded into the (geometric) lattice of all partitions of a large enough finite set.

While the Dilworth completion of almost submodular functions has not produced the farthest-reaching embedding result for finite lattices to date, it has led to deep insight into the structure of min-max theorems in combinatorial optimization. The key to those theorems is the fundamental observation in the present paper: *via* (2) a matroid may be obtained if f is "sufficiently submodular." I will outline the development in a bit more detail.

Matroids from Crossing Families. Let $\mathcal{F}$ be a family of subsets of a finite set E and $f : \mathcal{F} \to \mathbf{Z}$ some integer-valued function. With $\mathcal{F}$ associate the family $\mathcal{F}_*$ of all sets that can be partitioned into members of $\mathcal{F}$ and extend f to a function $f_* : \mathcal{F}_* \to \mathbf{Z}$ *via*

$$(5) \qquad f_*(X) = \min \left\{ \sum f(X_i) : X = X_1 \cup \cdots \cup X_k,\ X_i \in \mathcal{F},\ X_i \cap X_j = \emptyset \right\},$$

where $f_*(\emptyset) = 0$ by definition.

Then for each $X \subseteq E$, the following two conditions are equivalent:

$$(2*) \qquad\qquad |X \cap S| \leq f(S) \quad \text{for all} \quad S \in \mathcal{F}$$

and

$$(2**) \qquad\qquad |X \cap S| \leq f_*(S) \quad \text{for all } S \in \mathcal{F}_*.$$

Assume that $\mathcal{F}$ is an *intersecting family*, i.e., for all $A, B \in \mathcal{F}$,

$$(6) \qquad\qquad A \cap B \neq \emptyset \quad \text{implies} \quad A \cap B,\ A \cup B \in \mathcal{F},$$

THE DILWORTH THEOREMS

and that f is submodular in the sense that

$$(7) \qquad A \cap B \neq \emptyset \quad \text{implies} \quad f(A \cup B) + f(A \cap B) \leq f(A) + f(B).$$

Then $\mathcal{F}_*$, is closed under union and intersection. Moreover, f_* is submodular with respect to all pairs of subsets in $\mathcal{F}_*$. Hence, in this case,

$$(4*) \qquad r(X) = \min\{f_*(S) + |X \backslash S| : S \in \mathcal{F}_*\}$$

yields a matroid rank function. In the original situation considered by Dilworth, $\mathcal{F}$ consists of all non-empty subsets of points. The use of f_* instead of f is implicit in Crapo [2] and explicit in Dunstan [7]. Intersecting families are treated in Lovász [20] (see also [21]).

One may try to go one step further and require only that $\mathcal{F}$ be a *crossing family*, i.e., for all $A, B \in \mathcal{F}$,

$$(6*) \qquad A \cap B \neq \emptyset \text{ and } A \cup B \neq E \quad \text{imply} \quad A \cap B, A \cup B \in \mathcal{F}$$

and that f only be submodular on crossing pairs of $\mathcal{F}$. Then the collection of subsets X satisfying condition $(2*)$ need no longer be the system of independent sets of a matroid. However, as Frank and Tardos [14] observe, for each fixed $k \in \mathbb{N}$ the collection of k-element subsets $S \subseteq E$ satisfying $(2*)$ is either empty or the collection of bases of some matroid on E.

To see this, one starts with $\mathcal{F}_0 = \mathcal{F}\backslash\{\emptyset, E\}$ and sets

$$\bar{\mathcal{F}}_0 = \{E\backslash X : X \in \mathcal{F}_0\}.$$

Then $g(E\backslash X) = f(X) - k$ is submodular on the crossing family $\bar{\mathcal{F}}_0$. Define $\mathcal{K} = (\bar{\mathcal{F}}_0)_*\backslash\{E\}$ and $g_* : \mathcal{K} \to \mathbb{Z}$ as above. Thus $\mathcal{F}^* = \bar{\mathcal{K}}$ is an intersecting family and $f^* : \mathcal{F}^* \to \mathbb{Z}$ given by $f^*(X) = g_*(E\backslash X)+k$ is submodular. Hence the k-element sets X satisfying the constraints given by the pair $(\mathcal{F}, f)$ are exactly the k-element independent sets of the matroid obtained as the Dilworth completion of $(\mathcal{F}^*, f^*)$.

Submodular programs. Combinatorial objects may often be viewed as integral $|E|$-dimensional vectors $\mathbf{x}$ satisfying linear inequality constraints

$$(7) \qquad A\mathbf{x} \leq \mathbf{b}.$$

An important special case arises when the row vectors of A are the $(0,1)$-incidence vectors of the members of some family $\mathcal{F}$ of subsets of E and $\mathbf{b}$ then corresponds to some function on $\mathcal{F}$. Generalizing the Dilworth completion expressed *via* (2) or $(2*)$, Edmonds [9] is thus led to introduce *polyhedral matroids* or *polymatroids* as the solution sets of systems of the form

$$(8) \qquad \begin{aligned} \mathbf{x}(e) &\geq 0 && \text{for all } e \in E \\ \mathbf{x}(S) &\leq f(S) && \text{for all } S \in \mathcal{F} \end{aligned}$$

with $\mathbf{x}(S) = \sum\{\mathbf{x}(e) : e \in S\}$, where $\mathcal{F}$ comprises all non-empty subsets of E and f is submodular on $\mathcal{F}$.

Let $\mathbf{c} \in \mathbf{R}^E$ be an arbitrary weight function. A fundamental question now is whether the two optimization problems

$$(9) \qquad \left\{ \begin{array}{c} \max \mathbf{c} \cdot \mathbf{x} \\ A\mathbf{x} \leq \mathbf{b} \end{array} \right\} \quad \text{and} \quad \left\{ \begin{array}{c} \max \mathbf{c} \cdot \mathbf{x} \\ A\mathbf{x} \leq \mathbf{b} \\ \mathbf{x} \text{ integral} \end{array} \right\}$$

have a common optimal solution vector if an optimal solution exists at all.

As is well-known, the question has an affirmative answer if A is *totally unimodular*, i.e., if each subdeterminant of A takes value in $\{-1, 0, +1\}$, and $\mathbf{b}$ is integral. Another sufficient condition is that $A\mathbf{x} \leq \mathbf{b}$ is *totally dual integral*, i.e., that the minimum in the linear programming duality equation

$$(10) \qquad \max\{\mathbf{c} \cdot \mathbf{x} : A\mathbf{x} \leq \mathbf{b}\} = \min\{\mathbf{y} \cdot \mathbf{b} : \mathbf{y} \geq 0, \mathbf{y}A = \mathbf{c}\}$$

has an integer optimum solution for each integral objective function $\mathbf{c}$ for which the minimum exists (Hoffman [17] and Edmonds and Giles [10]).

Edmonds [8] not only shows that the linear optimization problem over a polymatroid can be solved by the analog of the matroid greedy algorithm (see also [9]) and hence has an integral optimal solution. He proves in effect that the defining system (8) for a polymatroid is totally dual integral and that the same is true for the intersection of two polymatroids.

The proof idea consists in "uncrossing dual solutions," which seems to be going exactly the opposite way of deriving matroids from crossing families. To illustrate this idea, let $\mathcal{F}$ be the set of all subsets of E and f_1, f_2 two submodular functions on $\mathcal{F}$ with $f_1(\emptyset) = f_2(\emptyset) = 0$. Consider now the primal linear programming problem over the intersection of the two associated polymatroids:

$$(11) \qquad \begin{array}{lll} \max \mathbf{c} \cdot \mathbf{x} & & \\ \mathbf{x}(e) \geq 0 & \text{for all } e \in E & \\ \mathbf{x}(S) \leq f_1(S) & \text{for all } S \in \mathcal{F} & \\ \mathbf{x}(S) \leq f_2(S) & \text{for all } S \in \mathcal{F}. & \end{array}$$

Denote by $\mathbf{y}_1$ and $\mathbf{y}_2$ the vectors of dual variables in (10) corresponding to the constraints from f_1 and f_2 respectively. Suppose that for some optimal dual solution $(\mathbf{y}_1^*, \mathbf{y}_2^*)$ there exists an intersecting pair $A, B \in \mathcal{F}$ with $A \cap B \neq \emptyset, A \not\subseteq B$ and $B \not\subseteq A$, and $\mathbf{y}_1^*(A) \geq \mathbf{y}_1^*(B) > 0$. Then the solution can be "uncrossed," i.e., $\mathbf{y}_1^*$ can be modified to $\bar{\mathbf{y}}_1^*$, where

$$\bar{\mathbf{y}}_1^*(A) = \mathbf{y}_1^*(A) - \mathbf{y}_1^*(B)$$
$$\bar{\mathbf{y}}_1^*(B) = 0$$
$$\bar{\mathbf{y}}_1^*(A \cap B) = \mathbf{y}_1^*(A \cap B) + \mathbf{y}_1^*(B)$$
$$\bar{\mathbf{y}}_1^*(A \cup B) = \mathbf{y}_1^*(A \cup B) + \mathbf{y}_1^*(B).$$

Because f_1 is submodular, $(\bar{\mathbf{y}}_1^*, \mathbf{y}_2^*)$ also represents an optimal dual solution. It follows that optimal dual solutions exist containing no non-trivial strictly positive intersecting pair relative to $\mathbf{y}_1$ or to $\mathbf{y}_2$. Hence the dual problem may be viewed as coming from a primal problem, whose constraint matrix arises from the union of two families of sets, each of which with no non-trivial intersecting members. Such matrices, however, are totally unimodular, which implies that (11) is totally dual integral.

The argument, in fact, yields more. If $\mathbf{c}$ is integral then (11) has both integral optimal primal and integral optimal dual solutions. Thus (10) gives rise to a min-max relation for primal and dual combinatorial objects. For example, matchings in bipartite graphs can be studied this way, where the matchings are the primal and the cuts are the dual combinatorial objects. The min-max relation (10), interpreted in this context, implies among other min-max theorems Dilworth's decomposition theorem for ordered sets [6].

Edmonds' ideas [8,9] have been generalized. For example, the discussion of matroids arising from crossing families may be repeated to derive from the total dual integrality of (11) also the total dual integrality of the system

$$
\begin{aligned}
\mathbf{x}(E) &= k \\
\mathbf{x}(S) &\le f_1(S) \qquad \text{for all } S \in \mathcal{F}_1 \\
\mathbf{x}(T) &\le f_2(S) \qquad \text{for all } T \in \mathcal{F}_2,
\end{aligned}
\tag{12}
$$

where $k \in \mathbf{Z}$ is fixed and f_1 and f_2 are integer-valued submodular functions with respect to the crossing families $\mathcal{F}_1$ and $\mathcal{F}_2$.

Frank [12] introduces a *generalized polymatroid* as the collection of solution vectors x for the linear system

$$
-f_1(S) \le \mathbf{x}(S) \le f_2(S) \qquad \text{for all } S \in \mathcal{F},
\tag{13}
$$

where f_1 and f_2 are submodular on the collection $\mathcal{F}$ of all subsets of E (see also Frank and Tardos [14]).

The *submodular systems* of Fujishige [15] are the solution vectors $\mathbf{x}$ for the system

$$
\mathbf{x}(S) \le f(S) \qquad \text{for all } S \in \mathcal{F},
\tag{14}
$$

where $\mathcal{F}$ is closed under union and intersection and f is submodular on $\mathcal{F}$. It turns out that generalized polymatroids can be represented as projections of *base polyhedra* of submodular systems, *i.e.*, of solution vectors $\mathbf{x}$ for the system

$$
\mathbf{x}(E) = f(E) \quad \text{and} \quad \mathbf{x}(S) \le f(S) \qquad \text{for all } S \in \mathcal{F}
\tag{14*}
$$

onto a coordinate hyperplane.

Again, these systems (and their intersection versions) are totally dual integral, *i.e.*, these systems are determined by the integral vectors they contain. In other words, these systems essentially are combinatorial structures. Indeed, their combinatorial analysis may be carried out completely within the framework of matroids. (For such an approach, see Faigle [11].) The only tool needed for the analysis is the Dilworth completion.

Let us illustrate the power of the Dilworth completion by deriving Edmonds' matroid intersection theorem [9] combinatorially (see McDiarmid [22]). Let M_1 and M_2 be matroids on E with rank functions r_1 and r_2. The Dilworth completion (2) of $f = r_1 + r_2$ yields a matroid whose independent sets are obtained by taking pairwise unions of independent sets in M_1 and M_2. (This follows directly from the rank formula (4).) We want to know if M_1 and M_2 have a common independent set of size k. So we can assume $r_1(E) = r_2(E) = k$. Denoting by r_2^* the rank function of the matroid dual M_2^* of M_2, the foregoing remarks show the equivalence of our problem with the question whether E is independent in the Dilworth completion of $r_1 + r_2^*$. By (4), the latter is the case if and only if

$$r_1(X) + r_2(E\backslash X) \geq k \quad \text{for all } X \subseteq E,$$

which is Edmonds' theorem (see also Aigner and Dowling [1]).

Submodular flows and generalizations. The submodular linear programs presented so far all have (0,1)-constraint matrices. Edmonds and Giles [10] suggest a model which is seemingly more general.

Let $G = (V, A)$ be a directed graph with set V of vertices and set A of arcs, $\mathcal{F}$ a crossing family of subsets of V and $f : \mathcal{F} \to \mathbf{Z}$ submodular on crossing pairs. A *submodular flow* in G is a vector $\mathbf{x} \in \mathbf{R}^A$ satisfying

$$(16) \qquad \mathbf{x}(\delta^-(S)) - \mathbf{x}(\delta^+(S)) \leq f(S) \qquad \text{for all } S \in \mathcal{F},$$

where $\delta^-(S)[\delta^+(S)]$ is the set of those arcs with just the head [tail] in S. Edmonds and Giles prove directly that (16) is totally dual integral. This fact can, however, also be derived by establishing the equivalence of (16) with a system of the form (12) (see Schrijver [26]).

Further generalizations of the submodular model are obtained by the *lattice polyhedra* of Hoffman and Schwartz [18] and the polyhedra of Grishuhin [16]. These generalizations also lead to totally dual systems since they allow the same proof idea to work: uncross optimal dual solutions and conclude that the optimum is achieved relative to a totally unimodular constraint matrix. (For an in-depth discussion of submodular models, their interrelations and applications, see the survey of Schrijver [26]).

One curiosity, however, seems perhaps noteworthy. Although lattice polyhedra, say, are defined by submodular functions which make the uncrossing technique successful, no matroid-theoretic analysis of their combinatorial structure is apparent. In particular, it is not clear what their "Dilworth completion" should be. This phenomenon is accompanied by the fact that currently no efficient combinatorial algorithm is known to solve optimization problems over lattice polyhedra or more general submodular structures. Such algorithms, namely generalizations of classical network flow algorithms, do exist for the submodular flow model (16) (see, *e.g.*, Cunningham and Frank [4] or Lawler and Martel [19] for an equivalent flow model), which lends itself to a matroid-theoretic combinatorial analysis.

<h1 style="text-align:center">REFERENCES</h1>

1. M. Aigner and T. A. Dowling, *Matching theory for combinatorial geometries*, Trans. Amer. Math. Soc. **158** (1971), 231–245.
2. H. H. Crapo, *Geometric duality and the Dilworth completion*, in "Combinatorial Structures and their Applications," R. Guy *et al.*, eds., Gordon and Breach, New York, 1970, pp. 37–46.
3. P. Crawley and R. P. Dilworth, "Algebraic Theory of Lattices," Prentice-Hall, Englewood Cliffs, New Jersey, 1973.
4. W. H. Cunningham and A. Frank, *A primal-dual algorithm for submodular flows*, Math. Operations Res. **10** (1985), 251–262.
5. R. P. Dilworth, *Dependence relations in a semi-modular lattice*, Duke Math. J. **11** (1944), 575–587. Reprinted in Chapter 5 of this volume.
6. —————, *A decompostion theorem for partially ordered sets*, Ann. of Math **51** (1950), 161–166. Reprinted in Chapter 1 of this volume.
7. F. D. J. Dunstan, *Matroids and submodular functions*, Quart. J. Math., Oxford **27** (1976), 339–348.
8. J. Edmonds, *Matroids and the greedy algorithm*, Math. Programming 1 (1971), 127–136.
9. —————, *Submodular functions, matroids and certain polyhedra*, in "Combinatorial Structures and Their Applications," R. Guy *et al.*, eds., Gordon and Breach, New York, 1970, pp. 69–87.
10. J. Edmonds and R. Giles, *A min-max relation for submodular functions on graphs*, Ann. Discrete Math. 1 (1977), 185–204.
11. U. Faigle, *Matroids in combinatorial optimization*, in "Combinatorial Geometries," N. L. White, ed., Cambridge Univ. Press, Cambridge, 1987, pp. 161–210.
12. A. Frank, *Generalized polymatroids*, in "Finite and Infinite Sets," A. Hajnal *et al.*, eds., North-Holland, Amsterdam, 1984, pp. 285–294.
13. —————, *Matroids from crossing families*, in "Finite and Infinite Sets," A. Hajnal *et al.*, eds., North-Holland, Amsterdam, 1984, pp. 295–304.
14. A. Frank and E. Tardos, *Generalized polymatroids and submodular flows*, Math. Programming **42** (1988), 489–563.
15. S. Fujishige, *Submodular systems and related topics.*, Math. Programming Study **22** (1984), 113–131.
16. V. P. Grishuhin, *Polyhedra related to a lattice*, Math. Programming **21** (1981), 70–89.
17. A. J. Hoffman, *A generalization of max-flow min-cut*, Math. Programming **6** (1974), 352–359.
18. A. J. Hoffman and D. E. Schwartz, *On lattice polyhedra*, in "Combinatorics," A. Hajnal and V. T. Sós, eds., North-Holland, Amsterdam, 1978, pp. 593–598.
19. E. L. Lawler and C. U. Martel, *Computing maximal "polymatroidal" network flows*, Math. Operations Res. **3** (1982), 334–347.
20. L. Lovász, *Flats in matroids and geometric graphs*, in "Combinatorial Surveys," P. Cameron, ed., Academic Press, London, 1977, pp. 45–86.
21. —————, *Submodular functions and convexity*, in "Mathematical Programming—The State of the Art," A. Bachem *et al.*, eds., Springer-Verlag, Berlin and New York, 1983, pp. 235–257.
22. C. J. H. McDiarmid, *Rado's theorem for polymatroids*, Math. Proc. Cambridge Phil. Soc. **78** (1975), 263–281.

23. H. Q. Nguyen, *Semi-modular functions and combinatorial geometries*, Trans. Amer. Math. Soc. **238** (1978), 355–383.

24. _____________, *Semimodular functions*, in "Theory of Matroids," N. L. White, ed., Cambridge Univ. Press, Cambridge, 1986, pp. 272–297.

25. P. Pudlák and J. Tůma, *Every lattice can be embedded in the lattice of all equivalences over a finite set*, Algebra Universalis **10** (1980), 74–95.

26. A. Schrijver, *Total dual integrality from directed graphs, crossing families, and sub- and supermodular functions.*, in "Progress in Combinatorics," W. R. Pulleyblank, ed., Academic Press, New York, 1984, pp. 315–361.

University of Twente
7500 AE Enschede
The Netherlands

THE DILWORTH THEOREMS

Dilworth Truncations of Geometric Lattices

JOSEPH P. S. KUNG

In the paper "Dependence relations in a semi-modular lattice" [5], Dilworth described a construction which represents the elements of a quasimodular point lattice (*i.e.*, a point lattice satisfying the semimodular axiom above points or atoms) as closed sets of a matroid (*i.e.*, a dependence structure satisfying the exchange property). This representation yields an injection of the quasimodular lattice into the geometric lattice of closed sets of the matroid which preserves the rank and meets, but not necessarily joins. Natural examples of quasimodular lattices can be obtained by taking a geometric lattice L of rank n and identifying all the elements of rank less than a fixed positive integer k. Using Dilworth's construction, we obtain a geometric lattice $D_k(L)$ of rank $n - k + 1$ which contains a copy of the upper $n - k$ levels of L. The lattice $D_k(L)$ is now called the kth *Dilworth truncation of L.* For example, the 2nd Dilworth truncation of the Boolean algebra of all subsets of an n-element set S is isomorphic to the lattice of partitions on S.

By looking at the matroid G induced by the geometric lattice L on its points, Crapo and Rota [4, Chap.6] and Mason [9] obtained the following geometric interpretation for the Dilworth truncation. Suppose that the matroid G can be represented as a spanning set G of points in n-dimensional space over a field F. Take a subspace U of dimension $n - k + 1$ in general position. This can be done by extending the field F. Then the matroid induced by $D_k(L)$ is represented by the points $U \cap X$, where X is a k-dimensional flat of G. From this, it follows from that if a matroid G of rank n is representable over a field F, then its Dilworth truncations are representable over extensions of F. A converse to this, stated in Mason [9], is also true: *if G has rank greater than 3 and $D_2(L)$ is representable over F, then G is representable over F.* The proof, due to Brylawski [1], uses the fact that the rank-4 lattice of partitions on a 5-element set is the lattice of flats of the geometric configuration on 10 points arising from Desargues' theorem. See Brylawski [1, 2] and Mason [9,10] for further results.

When the subspace U is not in general position relative to the points of the matroid, then the matroid represented by the points $U \cap X$ can be regarded as a "specialized" Dilworth truncation. This idea has been formulated precisely in two different ways. One way is due to Tůma [13] and extends the 2nd Dilworth truncation. Given a matroid G and a modular cut C, a matroid is constructed which is an analogue of the matroid obtained by taking as points intersections of lines of G with a hyperplane in special position. The positioning of the hyperplane is given by the modular cut C.

A less explicit but more general way uses the notion of a comap due to Crapo [3]. A *normalized comap* from the geometric lattice K to the geometric lattice L is a function preserving the relation of "covers or equals," meets of modular pairs of elements, and the minimum. Given a normalized comap $\gamma : K \to L$, there exists a geometric lattice M having the same rank as K and a modular element U in M satisfying:

(1) K injects into M,
(2) U has the same rank as L and L injects into the lower interval $[O, U]$,
(3) Identifying K and L with their images in M, $\gamma(X) = X \wedge U$ for every element X in K.

The function $\delta : L \to D_k(L)$ that sends x to itself if $\mathrm{rank}(x) \geq k$ and the minimum otherwise is a normalized comap. In the lattice M constructed for δ, the upper interval $[U, I]$ is isomorphic to the upper truncation of L to rank $k - 1$ obtained by identifying all the elements in L of rank greater than or equal to $k - 1$. Since upper truncations are combinatorial analogues of projections by subspaces in general position, this formalizes the geometric interpretation for the Dilworth truncation given earlier. A detailed account of this can be found in Kung [8].

Mason [10] used a variant of the Dilworth truncation to construct Dowling lattices. These lattices are group-labelled analogues of partition lattices. Mazzocca [11] characterized the matroids which are 2nd Dilworth truncations of another matroid. In addition, he has described extensions of matroids by a single line using modular filters in the 2nd Dilworth truncations (see [12]; Halsey [7] has given another description using "parallel classes").

Dilworth also used his construction (in §3 of [5]) to show that *every modular lattice of rank 3 can be embedded (as a sublattice) into a complemented modular lattice of rank 3*. This result complements the examples of modular lattices (of rank at least 4) which cannot be embedded into complemented modular lattices given by Dilworth and Hall in [6].

REFERENCES

1. T. Brylawski, *Coordinatizing the Dilworth truncation*, in "Matroid Theory," Colloq. Math. Soc. János Bolyai, 40, L. Lovász and A. Recski, eds., North-Holland, Amsterdam, 1985, pp. 61–95.
2. ——————, *Constructions*, in "Theory of Matroids," N. L. White, ed., Cambridge Univ. Press, Cambridge, 1986, pp. 127–223.
3. H. H. Crapo, *The joining of exchange geometries*, J. Math. Mech. **17** (1967/68), 837–852.

4. H. H. Crapo and G.-C. Rota, "On the Foundations of Combinatorial Theory: Combinatorial Geometries," M. I. T. Press, Cambridge, Massachusetts, 1970.

5. R. P. Dilworth, *Dependence relations in a semi-modular lattice*, Duke Math. J. **11** (1944), 575–587. Reprinted in Chapter 5 of this volume.

6. M. Hall, Jr. and R. P. Dilworth, *The imbedding problem for modular lattices*, Annals of Math. (2) **45** (1944), 450–456. Reprinted in Chapter 4 of this volume.

7. M. D. Halsey, *Extending a combinatorial geometry by adding a unique line*, J. Combin. Theory Ser. B **46** (1989), 118–120.

8. J. P. S. Kung, *A factorization theorem for comaps of geometric lattices*, J. Combin. Theory Ser. B **34** (1983), 40–47.

9. J. H. Mason, *Matroids as the study of geometrical configurations*, in "Higher Combinatorics," M. Aigner, ed., Reidel, Dordrecht, 1977, pp. 133–176.

10. __________, *Gluing matroids together: a study of Dilworth truncations and matroid analogues of exterior and symmetric powers*, in "Algebraic Methods in Graph Theory (Szeged, 1978)," Colloq. Math. Soc. János Bolyai, 25, L. Lovász and A. Recski, eds., North-Holland, Amsterdam, 1981, pp. 519–561.

11. F. Mazzocca, *On a characterization of Dilworth truncation of combinatorial geometries*, J. Geometry **20** (1983), 63–73.

12. __________, *Extensions of combinatorial geometries by the addition of a unique line*, J. Combin. Theory Ser. A **37** (1984), 32–45.

13. J. Tůma, *Dilworth truncations and modular cuts*, in "Matroid Theory (Szeged, 1982)," Colloq. Math. Soc. János Bolyai, 40, L. Lovász and A. Recski, eds., North-Holland, Amsterdam, 1985, pp. 383–400.

University of North Texas
Denton, TX 76203
U. S. A.

The Sperner Property
in Geometric and Partition Lattices

JERROLD R. GRIGGS

A major field in the study of the combinatorial properties of ordered sets developed from a fundamental result of Sperner concerning the maximum size of an antichain in a Boolean algebra. Investigations by many researchers combined into a theory of surprising elegance, subtlety, and breadth. There are three major thrusts in this field, which is now known as Sperner theory. One approach is to study the maximum antichains (and their generalizations) in *arbitrary* finite ordered sets. The seminal result of this type is Dilworth's decomposition theorem (see [9]), which exposed the profound duality between antichains and partitions into chains. This aspect of Sperner theory is discussed in Chapter 1 of this volume. A second effort has been aimed at generalizing Sperner's theorem to obtain deeper results about families of subsets, *e.g.,* there is Ramsey-Sperner theory [13], in which one assumes that an n-element set has been k-colored, and the problem is to find a maximum collection of subsets that contains no two sets A, B with $A \subset B$ and $B - A$ monochromatic.

We shall concern ourselves here with the third prominent direction in Sperner theory, which deals with maximum antichains in *ranked* ordered sets. A 1967 question of Rota [42] concerning the analogue of Sperner's theorem for the partition lattice inspired the development of this subject. A fundamental contribution to the assault on Rota's conjecture, as it came to be called, was made by Dilworth and his doctoral student, Curtis Greene, in [10]. It is our aim here to place their work in proper perspective by surveying the discoveries inspired by Rota's conjecture.

Rota asked whether a fundamental discovery of Sperner from 1928 [46] about Boolean algebras could be extended to other finite ordered sets. Sperner had determined that a maximum-sized collection of subsets of $[n] = \{1,\ldots,n\}$ with the

This work was supported by the NSF.

property that no set contains any other is given by taking all subsets of the same size k. It is easily checked that k must be $\lfloor n/2 \rfloor$ or $\lceil n/2 \rceil$. In the language of ordered sets, Sperner's theorem states that the *width* (*i.e.*, maximum size of an antichain) of the Boolean algebra B_n of all subsets of $[n]$, ordered by inclusion, is $\binom{n}{\lfloor n/2 \rfloor}$. (In fact, Sperner proved that there are no other maximum-sized antichains.) Rota's conjecture concerns the partition lattice Π_n of all partitions of $[n]$, ordered by refinement. In this lattice, the number of partitions of $[n]$ into exactly k parts is denoted by $S(n,k)$, called the Stirling number. Rota's conjecture is that for all n the width of Π_n equals the maximum Stirling number $S(n,k)$ over all k. An appropriate setting for more general investigations is this large class of finite ordered sets P in which elements can be divided into levels in a natural way: P is *ranked* if there exists a function $r : P \to \{0,1,2,\ldots\}$ such that $r(x) = 0$ for all minimal $x \in P$ and $r(y) = r(x) + 1$ for all $x, y \in P$ with x covered by y. (Equivalently, for every $x \in P$, every unrefinable chain with x as greatest element has the same length, given by its *rank* $r(x)$.) The *rank of* P is $\max_x r(x)$. The general problem suggested by Rota's conjecture is to investigate what ranked ordered sets P have width equal to their maximum *Whitney number* (or rank size) $|P_k|$, where we use P_k to denote the set of elements of rank k, $k \geq 0$. Ordered sets satisfying this condition are said to possess the *Sperner property*. In general P_k is an antichain, so the width is certainly at least as large as the largest Whitney number. Rota and Harper [43] asked whether arbitrary geometric lattices have the Sperner property. This is the question answered by Dilworth and Greene in [10].

A positive result related to Rota's conjecture was obtained much earlier, around 1950, when deBruijn *et al.* [4] proved the analogue of Sperner's theorem for the lattice of divisors D_N of an integer N, ordered by divisibility. The middle rank of D_N is an antichain of maximum size, so D_N has the Sperner property. Sperner's Theorem is itself a special case since D_N is isomorphic to B_n when N is a product of n distinct primes.

Baker [2] extended Sperner's theorem to another large class of ordered sets when he proved that the Sperner property is possessed by any *regular* order, *i.e.*, an ordered set satisfying: for all k the number of covering elements is constant for all elements of rank k and dually the number of elements covered is constant for all elements of rank k. While D_N is not in general regular, B_n certainly is. Another interesting family of ordered sets, the lattices of subspaces of a projective geometry, is regular and, hence, Sperner.

In the course of their search for a new line of attack on the problem of proving Rota's conjecture, Graham and Harper [12,20] introduced the *normalized matching property*, which we abbreviate here by *NMP*. This property holds for an ordered set P if for every rank $k > 0$ and for every collection $S \subseteq P_k$, the set ∂S of elements of P that are covered by some element of S satisfies the inequality (inspired by P. Hall's marriage theorem)

$$\frac{|\partial S|}{|P_{k-1}|} \geq \frac{|S|}{|P_k|}.$$

For example, regular orders are easily seen to possess NMP. Sperner's original proof can be extended naturally to show that any NMP order is Sperner. Thus Baker's result is generalized by this approach.

Harper [22] and, independently but later, Hsieh and Kleitman [25] proved the beautiful result that the direct product $P \times Q$ of NMP orders is NMP provided that both P and Q have log-concave sequences of Whitney numbers (a sequence $s_0, s_1, s_2, \ldots$ is *log-concave*, or *LC*, if for all $k > 0$, $s_k^2 \geq s_{k-1} s_{k+1}$). Log-concavity of a sequence implies unimodality. It should also be noted that $P \times Q$ is LC when P and Q are. Thus by repeatedly applying the product theorem it follows that the following ordered sets are NMP and LC: the Boolean algebra B_n (a product of n chains of size 2), the lattice of divisors D_N (a product of chains), and any modular geometric lattice (a product of subspace lattices).

Kleitman [27] proved that NMP orders have an equivalent characterization which is inspired by Lubell's elegant proof of Sperner's theorem [31]. The characterization is an inequality which must hold for every antichain A in the ranked ordered set P:

$$\sum_{x \in A} \frac{1}{|P_{r(x)}|} \leq 1.$$

Yamamoto [51] and Meshalkin [33] formulated similar inequalities, and it is now standard to refer to ordered sets possessing either (and hence both) properties as *LYM orders*.

One other result involving LYM orders deserves mention here. DeBruijn *et al.* [4] proved that D_N (and consequently its special case, B_n) is Sperner by showing that it is a *symmetric chain order*: This means that it can be partitioned into chains that are saturated and symmetric about middle rank. In such a poset, the middle rank is a maximum antichain, so the Sperner property holds. Griggs [16] proved that any LYM order P that is *rank-symmetric* and *rank-unimodal* (*i.e.*, if n is the rank of P, then $|P_0| = |P_n| \leq |P_1| = |P_{n-1}| \leq \cdots \leq |P_{\lfloor \frac{n}{2} \rfloor}| = |P_{\lceil \frac{n}{2} \rceil}|$) is a symmetric chain order. In particular, this implies that the lattice of subspaces is a symmetric chain order. A long-standing open problem [16] is to extend this chain partition result to LYM orders with arbitrary Whitney numbers.

Let us now return to the partition lattice. Graham and Harper [14] reduced the problem of proving that Π_n is LYM to the smaller problem of showing that the *partition order* P_n, whose elements are the partitions of the *integer* n, is LYM with respect to an appropriate weighting of the elements (not LYM in the usual sense). By this approach they verified by computer that Π_n is LYM, and hence Sperner, for $n \leq 19$. While this seemed to be strong evidence in support of Rota's conjecture in general, it was disappointing that no general proof was evident. Indeed, at about this time, Spencer [45] showed this approach could go no farther when he proved that Π_n is not LYM for all $n \geq 20$.

Mullin [34] showed that it suffices to examine just three ranks in Π_n to settle Rota's conjecture. Let M_n denote the largest k such that $S(n, k)$ is maximum. Then if the width of Π_n exceeds $S(n, M_n)$, Mullin proved that there is some antichain

$A \subseteq \Pi_n$, $|A| > S(n, M_n)$, such that A contains only partitions of $[n]$ into $M_n - 1$, M_n, or $M_n + 1$ blocks.

In [10], Dilworth and Greene settled the stronger problem of Rota and Harper in the negative by producing an elegant construction of geometric lattices *without* the Sperner property. The smallest lattice given by their method has 60,073 elements. Their discovery was yet another indication that Rota's conjecture would not hold up in general. The examples of Dilworth and Greene are bond lattices of graphs. Using designs, Kahn [26] has since constructed non-Sperner *paving* geometric lattices, *i.e.*, geometric lattices in which every proper lower interval is a Boolean algebra.

A counterexample to Rota's conjecture was finally produced in 1978 by Rod Canfield in a series of papers [5,6,7] using sophisticated analytical methods. He showed that Π_n is not Sperner for all sufficiently large n, where "sufficiently large" requires n to be at least around 6×10^{24}. This was soon lowered considerably by Shearer [44] and Peck [35] who reduced the threshold down to around 5×10^6. Their proofs are elementary and considerably simpler.

The location of the maximum Stirling number is another consequence of Canfield's impressive study. An elementary proof of this result is unlikely to be found. It was shown by Lieb [29] that Π_n is strictly log-concave, from which it follows that Π_n has at most two values k which maximize $S(n, k)$. When there are two values, they must be consecutive. Let K_n denote the minimum k such that $S(n, k)$ is maximum. Harper [21] determined the asymptotic behavior of this quantity: $K_n \sim \frac{n}{\ln n}$. Canfield proved that for all sufficiently large n, K_n is either $\lfloor \alpha_n \rfloor$ or $\lfloor \alpha_n \rfloor + 1$, where α_n is the unique solution to the equation

$$\frac{(\alpha_n + 2)\alpha_n \ln(\alpha_n + 2)}{\alpha_n + 1} = n$$

Despite the setback caused by Canfield's counterexample, there continued to be steady theoretical progress in the field as researchers explored the Sperner property itself. Kung [28] introduced linear algebra tools for constructing chains in ordered sets. By this method he reproved the result of Mason [32] that if P is a geometric lattice of rank n, then there exists a family of $|P_1|$ disjoint saturated chains of length $n - 2$ in P. Stanley [47] brought in techniques from algebraic geometry to show that several new classes of ordered sets are Sperner. He solved a long-standing number-theoretic conjecture of Erdős and Moser as a consequence of his study. Griggs [17] noticed that Stanley's ordered sets possess an even nicer property, now called the *strong Sperner property*. This means that for all k, the union of the k largest ranks is a maximum-sized k-family (subset containing no chains of $k + 1$ elements). This generalizes a property of the Boolean algebra which was noted much earlier in 1945 by Erdős [12]. It is easily verified that LYM orders are strongly Sperner.

Much of the nicest theory that subsequently developed concerns Peck orders, which are named in honor of the illustrious imaginary mathematician, G. W. Peck. An ordered set is *Peck* if it is strongly Sperner, rank-symmetric, and rank-unimodal. The Peck orders include Stanley's ordered sets above as well as all symmetric chain orders. There is a linear algebra characterization of Peck orders that was used to determine that the product of Peck orders is Peck [8,40].

Later a Peck quotient theorem was found (in several slightly different forms) that provides nice sufficient conditions for the *quotient order* P/G to be Peck, where P is a ranked ordered set P under the action of a group G of order-preserving automorphisms [21,36,48]. The elements of P/G are the orbits in P under G, and the ordering is inherited in the natural way for P. The motivating application for this result is the ordered set $L(m,n)$, the set of order ideals of the product of an m-chain and an n-chain. The ordered set $L(m,n)$ is a quotient of the Boolean algebra B_{mn}. It can be shown that a Boolean algebra satisfies the conditions of the quotient theorem, so it follows that $L(m,n)$ is Peck. Stanley originally proved $L(m,n)$ is Peck using his tools from algebraic geometry. The quotient approach provided the first natural "combinatorial" proof, although some linear algebra is required. The ordered set $L(m,n)$ continues to be a subject of considerable interest owing to a long-standing conjecture of Stanley that $L(m,n)$ (and, indeed, all of his ordered sets mentioned above) is a symmetric chain order. The conjecture has been verified when m or n is at most 4 [30,41,49].

Another surprising approach to the study of Peck orders is based on Proctor's discovery [37] that Peck orders are characterized by their ability to carry a representation of $sl(2,\mathbb{C})$. He has written a series of papers employing this perspective to provide natural interpretations for the Peck theorems and to derive some new classes of Peck orders [38,39].

Sperner theory has grown to the point where there is now a book devoted to the subject, by Engel and Gronau [11]. It covers in depth the Sperner theory results for arbitrary and for ranked ordered sets but not for families of subsets. For more accessible introductions to Sperner theory and to extremal set theory in general, the following sources are highly recommended: the long survey by Greene and Kleitman [15] (even though it preceded many discoveries), and the more recent texts by Anderson [1] and Bollobás [3]. Other helpful surveys related to this article are [16,19,20,50].

As some indication of how much remains to be understood in this field, we conclude by mentioning three more annoyingly unsettled problems.

First, Erdős and others have been asking for years whether the partition order P_n is Sperner in general. This appears to be unlikely, yet no one has come up with a counterexample since P_n is so difficult to handle.

Second, it remains open to prove that every geometric lattice is rank-unimodal. For example, we have mentioned that many interesting examples are Peck. Another motivating example, the partition lattice, is neither Sperner nor rank-symmetric in general, but it is strictly log-concave and therefore unimodal. We also know some facts about arbitrary geometric lattices, such as Mason's result mentioned above, which support the unimodality conjecture. But this seems to be a very difficult problem.

Finally, we still know very little about the width of the partition lattice. The "asymptotic Rota conjecture" [7,22] states that the ratio, denote it R_n, of the width

THE DILWORTH THEOREMS

of Π_n to the maximum Stirling number, $S(n, K_n)$, tends to 1 as $n \to \infty$. So far we know only that $R_n > 1$ for n sufficiently large. We suspect that in fact $R_n \to \infty$, but there is discouragingly little evidence to support this claim. This problem has not yet received as much attention as it deserves.

REFERENCES

1. I. Anderson, "Combinatorics of Finite Sets," Oxford Univ. Press, Oxford, 1987.
2. K. Baker, *A generalization of Sperner's lemma*, J. Combin. Theory Ser. A **6** (1969), 224–225.
3. B. Bollobás, "Combinatorics," Cambridge Univ. Press, Cambridge, 1986.
4. N. deBruijn, C. A. van Ebbenhorst Tengbergen, and D. R. Kruyswijk, *On the set of divisors of a number*, Nieuw. Arch. Wisk. (2) **23** (1952), 191–193.
5. E. R. Canfield, *On the location of the maximum Stirling number(s) of the second kind*, Stud. Appl. Math. **59** (1978), 89–93.
6. E. R. Canfield, *On a problem of Rota*, Adv. in Math. **29** (1978), 1–10.
7. E. R. Canfield, *Application of the Berry-Esséen inequality to combinatorial estimates*, J. Combin. Theory Ser. A **28** (1978), 17–25.
8. E. R. Canfield, *A Sperner property preserved by product*, Linear Multilinear Alg. **9** (1980), 151–157.
9. R. P. Dilworth, *A decomposition theorem for partially ordered sets*, Ann. of Math. **51** (1950), 161–166. Reprinted in Chapter 1 of this volume.
10. R. P. Dilworth and C. Greene, *A counterexample to the generalization of Sperner's theorem*, J. Combinatorial Theory **10** (1971), 18–21. Reprinted in Chapter 5 of this volume.
11. K. Engel and H.-D. Gronau, "Sperner Theory in Partially Ordered Sets," Teubner, Leipzig, 1985.
12. P. Erdős, *On a lemma of Littlewood and Offord*, Bull. Amer. Math. Soc. **51** (1945), 898–902.
13. Z. Füredi, J. R. Griggs, A. M. Odlyzko, and J. B. Shearer, *Ramsey-Sperner theory*, Discrete Math. **63** (1987), 143–152.
14. R. L. Graham and L. H. Harper, *Some results on matching in bipartite graphs*, SIAM J. Appl. Math. **17** (1969), 1017–1022.
15. C. Greene and D. J. Kleitman, *Proof techniques in the theory of finite sets*, in "Studies in Combinatorics," G.-C. Rota, ed., Mathematical Association of America, Washington, D.C., 1978, pp. 22–79.
16. J. R. Griggs, *Symmetric chain orders, Sperner theorems, and loop matchings*, Ph. D. Dissertation, Massachusetts Institute of Technology (1977).
17. J. R. Griggs, *Sufficient conditions for a symmetric chain order*, SIAM J. Appl. Math. **32** (1977), 807–809.
18. J. R. Griggs, *On chains and Sperner k-families in ranked posets*, J. Combin. Theory Ser. A **28** (1980), 156–168.
19. J. R. Griggs, *The Sperner property*, Ann. Discrete Math. **23** (1984), 397–408.
20. S. Gulati, *A study of Sperner theory and its application to L(m,n)—the Young's lattice*, Master's thesis, University of South Carolina, 1987.
21. L. H. Harper, *Stirling behaviour is asymptotically normal*, Ann. Math. Statist. **38** (1967), 410–414.
22. L. H. Harper, *The morphology of partially ordered sets*, J. Combin. Theory Ser. A **17** (1974), 44–58.
23. L. H. Harper, *Morphisms for the strong Sperner property of Stanley and Griggs*, Linear Multilinear Alg. **16** (1984), 323–337.
24. L. H. Harper, *On a continuous analog of Sperner's problem*, preprint, 1984.
25. W. N. Hsieh and D. J. Kleitman, *Normalized matching in direct products of partial orders*, Stud. Appl. Math. **52** (1973), 258–289.

26. J. Kahn, *Some non-Sperner paving matroids*, Bull. London Math. Soc. **12** (1980), p. 268.

27. D. J. Kleitman, *On an extremal property of antichains in partial orders: the LYM property and some of its implications and applications*, in "Combinatorics," M. Hall, Jr. and J. H. van Lint, eds., Math. Centre Tracts, Amsterdam, 1974, pp. 77–90.

28. J. P. S. Kung, *The Radon transforms of a combinatorial geometry, I*, J. Combin. Theory Ser. A **26** (1979), 97–102.

29. E. H. Lieb, *Concavity properties and a generating function for Stirling numbers*, J. Combin. Theory **5** (1968), 203–206.

30. B. Lindström, *A partition of $L(3,n)$ into saturated symmetric chains*, European J. Combin. **1** (1980), 61–63.

31. D. Lubell, *A short proof of Sperner's theorem*, J. Combin. Theory **1** (1966), p. 299.

32. J. H. Mason, *Maximal families of pairwise disjoint maximal proper chains in a geometric lattice*, J. London Math. Soc. (2) **6** (1973), 539–542.

33. L. D. Meshalkin, *A generalization of Sperner's theorem on the number of subsets of a finite set*, Theory Probability Appl. **8** (1963), 203–204.

34. R. Mullin, *On Rota's problem concerning partitions*, Aequationes Math. **2** (1969), 98–104.

35. G. W. Peck, *On Canfield-type antichains of partitions*, unpublished, 1979.

36. M. Pouzet and I. G. Rosenberg, *Sperner properties for groups and relations*, European J. Combin. **7** (1986), 349–370.

37. R. A. Proctor, *Representations of $\mathfrak{sl}(2,\mathbf{C})$ on posets and the Sperner property*, SIAM J. Alg. Discrete Methods **3** (1982), 275–280.

38. R. A. Proctor, *A Dynkin diagram classification theorem arising from a combinatorial problem*, Adv. in Math. **62** (1986), 103–117.

39. R. A. Proctor, *Solution of a Sperner conjecture of Stanley with a construction of Gelfand*, preprint, 1988.

40. R. A. Proctor, M. Saks, and D. Sturtevant, *Product partial orders with the Sperner property*, Discrete Math. **30** (1980), 173–180.

41. W. Riess, *Zwei Optimierungsprobleme auf Ordungen*, Arbeits. Inst. Math. Masch. Daten., Vol. 11, No. 5, 1978.

42. G.-C. Rota, *Research problem 2-1*, J. Combin. Theory **2** (1967), p. 104.

43. G.-C. Rota and L. H. Harper, *Matching theory, an introduction*, in "Advances in Probability," Vol. 1, P. Ney, ed., Marcel Dekker, New York, 1971, pp. 171–215.

44. J. B. Shearer, *A simple counterexample to a conjecture of Rota*, Discrete Math. **28** (1979), 327–330.

45. J. H. Spencer, *A generalized Rota conjecture for partitions*, Stud. Appl. Math. **53** (1974), 239–241.

46. E. Sperner, *Ein Satz über Untermengen einer endlichen Menge*, Math. Z. **27** (1928), 544–548.

47. R. P. Stanley, *Weyl groups, the hard Lefschetz theorem and the Sperner property*, SIAM J. Alg. Discrete Methods **1** (1980), 168–184.

48. R. P. Stanley, *Quotients of Peck posets*, Order **1** (1984), 29–34.

49. D. B. West, *A symmetric chain decomposition of $L(4,n)$*, European J. Combin. **1** (1980), 379–383.

50. D. B. West, *Extremal problems in partially ordered sets*, in "Ordered Sets (Proceedings of the Banff conference,1982)," I. Rival, ed., Reidel, 1982.

51. K. Yamamoto, *Logarithmic order of free distributive lattices*, J. Math. Soc. Japan **6** (1954), 343–353.

University of South Carolina
Columbia, SC 29208
U. S. A.

Multiplicative Lattices

Background

R. P. DILWORTH

In the middle 1930's Morgan Ward, largely from reading the work of E. Noether, became convinced that much of the basic structure theory of commutative rings could be formulated in lattice theoretic terms provided an appropriate multiplication was defined over the lattice. It was clear that the lattice should be modular and satisfy the ascending chain condition. The multiplication should be commutative, associative, distributive with respect to the lattice join operation, and have the unit of the lattice as a unit element for the multiplication. A lattice endowed with such a multiplication is called a *multiplicative lattice*.

I was a senior undergraduate at Caltech at that time and Morgan Ward suggested that we begin a systematic study of such systems, keeping the ring model in mind. From the ring case it was clear that residuation, a kind of quotient operation, should play a significant role in the theory. Our first efforts were directed toward determining the various formal relationships between multiplication, residuation, and the lattice operations. In particular, we were much interested in how the existence of a multiplication affected the lattice structure. It turned out that, in general terms, there wasn't much effect, but there was one important local effect, namely, if the join of two elements was the unit element of the lattice, then the meet of the two elements was simply their product. Since multiplication distributes with respect to join, this had the effect of making the portion of the lattice near the unit element distributive. As a consequence the only complemented lattices having a multiplication were Boolean algebras and the only multiplication on a Boolean algebra was the meet operation.

The principal aim of these early investigations was to find an abstract version of the Noether decomposition theory for commutative rings. Since the notions of prime and primary carry over immediately to lattices with a multiplication and since in any lattice satisfying the ascending chain condition, each element can be represented as as a meet of irreducibles, the key requirement for a Noether type decomposition is that every irreducible be primary. It soon became clear that this in turn required that the multiplicative lattice have sufficiently many elements which behave like the principal ideals in a commutative ring. The property abstracted from the ring case which seemed an appropriate definition for a principal m was the following:

$$\text{If } m \geq a, \text{ there exists } b \text{ such that } mb = a.$$

Thus a principal element m is a characterized by the property that when it contains an element in the lattice sense, it then divides the element in the multiplicative sense. If every element in the multiplicative lattice is a join of principal elements and the principal elements are closed under multiplication, then it follows that every irreducible is primary and the analogues of the Noether decomposition theorems hold.

At this point in the investigation I turned to the noncommutative case and Morgan Ward turned to more purely arithmetical questions. I was not to return to the commutative theory until many years later.

In the late 1950's, after attending a lecture on rings at UCLA by Irving Kaplansky, I was stimulated to again think about the abstract lattice theoretic approach to commutative ideal theory. It was now apparent to me that the notion of a principal element as defined by Ward and me was simply not strong enough for the deeper results of commutative ideal theory. In fact, it seemed to me that what was needed were some identities on the ideals of a commutative ring involving multiplication, residuation and the lattice operations which do not hold in general but do hold when one of the elements is principal.

After considerable searching, I came upon the following identities:

$$(A \wedge B : M)M = AM \wedge B$$

$$(A \vee BM) : M = A : M \vee B$$

These identities hold for arbitrary ideals A and B in a commutative ring if M is a principal ideal. These identities were particularly satisfying since if multiplication and residuation are considered to be duals of each other, then each of the above identities is the dual of the other. Furthermore, if these identities are taken as the abstract definition of a principal element, then it is a happy consequence that the principal elements are closed under multiplication. Also, they imply that a principal element is principal in the earlier sense of Ward and Dilworth. Accordingly I adopted the two identities as the new definition of a *principal element M* in a multiplicative lattice. The corresponding new definition for a Noether lattice requires that the lattice be a multiplicative, modular lattice satisfying the ascending chain condition such that every element is a join of principal elements. When a Noether lattice is so defined, the Noether decomposition theorems hold in a Noether lattice.

In order to test the effectiveness of the new definition of principal elements, I concentrated on formulating a proof of the abstract version of the intersection theorem. The assumption that every element is a join of principal elements turned out to be exactly what was needed in order to make the proof go through. Encouraged by this success, I turned to the analogue of the Krull principal ideal theorem. The key first step was to show that if L is a local lattice with maximal prime P, then the quotient lattice L/P^k is finite dimensional for each positive integer k. Again the definition of principal elements as given above turned out to be exactly right for proving this result. Then, following an argument due to D. Rees, it was fairly straightforward to formulate a proof of the analogue of the Krull principal ideal theorem.

ABSTRACT RESIDUATION OVER LATTICES*

R. P. DILWORTH

Introduction. The idea of residuation goes back to Dedekind [3],[†] who introduced it in the theory of modules. It has since had extensive applications in the theory of algebraic modular systems [6], in the theory of ideals [8], and in certain topics of arithmetic [9]. On account of its fundamental role in several fields of modern algebra, it is desirable to consider residuation abstractly. A postulational treatment also is a necessary preliminary to the investigation of the structure properties of the residual. We give such an abstract formulation.

In a commutative ring with unit element the residual of an ideal B with respect to an ideal A, written $A:B$, is an ideal with the properties $A \supset (A:B)B$; if $A \supset XB$, then $A:B \supset X$. Although the residual is defined in terms of multiplication, most of its important properties are concerned with the cross-cut and union of ideals. Hence we shall consider a residual defined over a system having only these two operations, that is, over a *lattice* [2]. As an example of a system having a residual but no ordinary multiplication we consider in §5 residuation in a Boolean algebra.

In §1 the postulates for abstract residuation are given. Equality is taken as an undefined relation with cross-cut, union, and residual as undefined connections. In §2 we list a few systems satisfying the postulates. In §3 it is shown that the system defined by the postulates is a lattice and that the residual has all of its important properties which are independent of multiplication. Consistency and independence proofs are given in §4.

I wish to express my thanks to Professors Morgan Ward and E. T. Bell for their many suggestions and helpful criticisms during the preparation of this paper.

1. **Postulates for residuation.** Let Σ be a set of elements A, B, C, $\cdots$; and let $=$, $[\,,\,]$, $(\,,\,)$, and $:$ be relations, satisfying the postulates i–iv; 1–3; I–V. In what follows, $\circ$ denotes an arbitrary one of the relations $[\,,\,]$, $(\,,\,)$, $:$ and the letters A, B, C, $\cdots$, appearing in the statement of the postulates indicate arbitrary elements of Σ.

POSTULATE i. *$A \circ B$ is in Σ whenever A and B are in Σ.*

---

* Presented to the Society, November 27, 1937.

† Numbers in square brackets refer to the bibliography at the end of the paper.

POSTULATE ii. *If $A = B$, then $C \circ A = C \circ B$ and $A \circ C = B \circ C$.* *
POSTULATE iii. *If $A = B$ and $B = C$, then $A = C$.*
POSTULATE iv. *If $A = B$, then $B = A$.* .
POSTULATE 1. $[[A, B], C] = [B, [A, C]]$. †
POSTULATE 2. $[A, A] = A$.
POSTULATE 3. *There is an element I in Σ such that $[A, I] = A$ for all A in Σ.*

As immediate deductions from these postulates we have:

1.1. $[I, A] = [A, I] = A$.

PROOF. $[I, A] = [I, [A, I]] = [[A, I], I] = [A, I] = A$ by ii, 1, 3, iii.

1.2. *The element I in Postulate 3 is unique.*

POSTULATE I. $A:A = I$.
POSTULATE II. $(A:B):C = (A:C):B$.
POSTULATE III. $A:(B, C) = [A:B, A:C]$.
POSTULATE IV. $[A, B]:C = [A:C, B:C]$.
POSTULATE V. *If $A:B = B:A = I$, then $A = B$.*

DEFINITION 1. $A:B = I$ is written $A \supset B$.
DEFINITION 2. $[A, B] = B$ is written $A > B$.

2. **Examples.** We list a few systems satisfying the postulates i–V.

1. Let Σ be the set of ideals in a commutative ring with unit element. Let $[,]$ and $(,)$ be the cross-cut and union respectively. Let $A:B$ be defined by $A \supset (A:B)B$, if $A \supset XB$, $A:B \supset X$.

2. Let Σ be the set of positive integers with $[,]$ and $(,)$ the L. C. M. and G. C. D. respectively. Let $A:B$ be defined by $A/(A, B)$ with $I = 1$.

3. As in 2, let Σ be the set of positive integers with $[,]$ and $(,)$ defined as max $(,)$ and min $(,)$ respectively. If now $A:B$ is defined by max $(0, A - B)$ and 0 is taken to be the element I, the postulates are satisfied.

4. Let Σ be the integers $\leq n$ with $[,]$ and $(,)$ defined by min $(,)$ and max $(,)$, respectively. Define $A:B$ as min $(n, n+A-B)$ with $I = n$.

5. Let Σ be a Boolean algebra with $[,]$, $(,)$ the Boolean operations $\cdot$, $\vee$ respectively. Let $A:B = A \vee B'$.

---

* It is understood that the relations in the postulates hold whenever the respective elements and the indicated combinations are in Σ.

† To facilitate obtaining an independence example, the commutative and associative laws have been combined in one postulate.

3. **Deductions from the postulates.** † We have, first,

2.1. $A = A$ by 3, iv.

2.12. $A \supset B$ and $B \supset A$ is equivalent to $A = B$‡ by V, Definition 1.

2.13. If $M:A = M:B$ for all M in Σ, then $A = B$.

PROOF. $M = A, B$ respectively give

2.14. $A:B = A:A = I$ and $B:A = B:B = I$ by ii, I, iv.

Hence $A = B$ by V.

2.15. $[A, B] = [B, A]$.

PROOF. $[A, B] = [[A, B], I] = [B, [A, I]] = [B, A]$ by 3, 1, 3, ii.

2.16. $(A, B) = (B, A)$ by III, 2.15, 2.13.

2.17. $[[A, B], C] = [A, [B, C]]$ by 2.15, 1.

2.18. $((A, B), C) = (A, (B, C))$ by III, 2.17, III, 2.13.

From 2.17 and 2.18 we may write $[A, B, C]$ for $[[A, B], C]$ and (A, B, C) for $((A, B), C)$. Generally $[A_1, A_2, \cdots, A_n]$, $(A_1, A_2, \cdots, A_n)$ are unambiguous.

*2.19. $M:(A_1, A_2, \cdots, A_n) = [M:A_1, M:A_2, \cdots, M:A_n]$ by induction from III.

*2.2. $[A_1, A_2, \cdots, A_n]:M = [A_1:M, A_2:M, \cdots, A_n:M]$ by induction from IV.

2.22. $(A, A) = A$.

PROOF. $M:(A, A) = [M:A, M:A] = [M, M]:A = M:A$ by III, IV, 2. Hence $(A, A) = A$ by 2.13.

2.23. $(A, B) \supset A$.

PROOF. $(A, B):A = [(A, B):A, I] = [(A, B):A, (A, B):(A, B)]$
$= (A, B):(A, (A, B)) = (A, B):((A, A), B) = (A, B):(A, B) = I$ by
3, I, III, 2.18, 2.22, I.

2.24. If $A > B$, then $A \supset B$.

PROOF. $[A, B] = B$ gives $A:B = [A:B, I] = [A:B, B:B] = [A, B]:B$
$= B:B = I$ by 3, I, IV, I.

2.25. If $[A, B] = B$, then $(A, B) = A$.

PROOF. $A:(A, B) = [A:A, A:B] = [I, A:B] = A:B = I$ by III, I, 3, 2.24. Hence $(A, B) = A$ by 2.23, V.

2.26. $(A, I) = (I, A) = I$ by 2.25, 3, 2.16.

---

† The theorems giving the essential properties of the residual will be starred.

‡ By "equivalent" we mean formal equivalence.

*2.27. $I:A = I.$

PROOF. $I:A = [I, \ I:A] = [I:I, \ I:A] = I:(I, \ A) = I:I = I$ by 1.1, I, III, 2.26, I.

*2.28. $A:I = A.$

PROOF. $(A:(A:I)):I = (A:I):(A:I) = I$ by II, I. $I:(A:(A:I)) = I$ by 2.27. Hence $A:(A:I) = I$ by V. But $(A:I):A = (A:A):I = I:I = I$ by II, I. Hence $A:I = A$ by V.

2.29. $A \supset [A, B]$ by 2.17, 2, 2.24.
2.3. *If $A \supset B$, then $A > B$.*

PROOF. $[A, \ B]:B = [A:B, \ B:B] = [I, \ I] = I$ by IV, I, 2. But $B:[A, B] = I$ by 2.29, 2.15, Definition 1. Hence $[A, B] = B$ by V.

2.31. *$A \supset B$ is equivalent to $A > B$* by 2.24, 2.3.
2.32. $(A, [A, B]) = A$ by 2.29, 2.31, 2.25.
2.33. $[A, (A, B)] = A$ by 2.23, 2.31, Definition 2.
2.34. *Σ is a lattice.*

PROOF. We show that Birkhoff's axioms L1–L4† are satisfied if we take $[,] \equiv \cap$. For L1 is i; L2 is 2.15 and 2.16; L3 is 2.17 and 2.18; L4 is 2.32 and 2.33.

*2.35. $A:(A:B) \supset B$ by II, I.
*2.4. *If $B \supset C$, then $A:C \supset A:B$.*

PROOF. $A:(A:B) \supset B$ and $B \supset C$ by 2.35. Hence $(A:(A:B)):C = I$ by 2.34, Definition 1. Then $(A:C):(A:B) = (A:(A:B)):C = I$ by II.

*2.41. $A:B = A:(A, B)$ by III, I, 1.1.
*2.42. $[A, B]:B = A:B$ by II, I, 3.
*2.43. $I \supset A:B \supset A$ by 2.27, II, I, 2.27.
*2.44. $A:(A:(A:B)) = A:B.$

PROOF. $[A:(A:(A:B))]:(A:B) = [A:(A:B)]:[A:(A:B)] = I$ by II, I. $(A:B):[A:(A:(A:B))] = [A:[A:(A:(A:B))]]:B = I$ by II. Since $A:[A:(A:(A:B))] \supset A:(A:B)$ and $A:(A:B) \supset B$ by 2.35, hence $A:(A:(A:B)) = A:B$ by V.

It will be noted from 2.44 that $A:(A:B)$ and $A:B$ are mutually residual with respect to A.

*2.45. *If $A:C = A:B$, then $A:(A:B) \supset C$* by II, I.

---

† G. Birkhoff, *On the lattice theory of ideals*, this Bulletin, vol. 40 (1934), p. 613. L1, L2, L3, and L4 are his axioms for a lattice.

These theorems are sufficient to show that the usual properties of the residual are deducible from the postulates of §1.

Since the residual as here considered is independent of multiplication, there is a residual completely dual to that defined above. The dual may be defined by the postulates I′–V′:

POSTULATE I′. $A:A = E$ where E is the null element of the lattice.
POSTULATE II′. $(A:B):C = (A:C):B$.
POSTULATE III′. $A:[B, C] = (A:B, A:C)$.
POSTULATE IV′. $(A, B):C = (A:C, B:C)$.
POSTULATE V′. If $A:B = B:A = E$, then $A = B$.

Thus, for the integers $0, 1, 2, \cdots, n$ as the given set, let $[\,,\,]$, $(\,,\,)$ be defined as min $(\,,\,)$, max $(\,,\,)$ respectively with $E = 0$ and $I = n$. Then $A:B \equiv \min (n, n+A-B)$ is a residual satisfying the postulates of §1, while $A:B \equiv \max (0, A-B)$ is a residual satisfying the second set of postulates.

4. Consistency and independence proofs. †

	[,]	(,)	:	
Consist-ency	∣ 1 2 1 ∣ 1 2 2 ∣ 2 2	∣ 1 2 1 ∣ 1 1 2 ∣ 1 2	∣ 1 2 1 ∣ 1 1 2 ∣ 2 1	
Independ-ence 1	∣ 1 2 3 1 ∣ 1 1 2 2 ∣ 2 2 2 3 ∣ 3 3 3	∣ 1 2 3 1 ∣ 1 1 1 2 ∣ 2 2 2 3 ∣ 2 3 3	∣ 1 2 3 1 ∣ 1 1 1 2 ∣ 2 1 1 3 ∣ 3 2 1	$[[1, 3], 2] \neq [3, [1, 2]]$
2	∣ 1 2 1 ∣ 1 2 2 ∣ 2 1	∣ 1 2 1 ∣ 2 1 2 ∣ 1 2	∣ 1 2 1 ∣ 1 1 2 ∣ 2 1	$[2, 2] \neq 2$
3	∣ 1 2 3 1 ∣ 1 1 1 2 ∣ 1 2 1 3 ∣ 1 1 3	∣ 1 2 3 1 ∣ 1 3 3 2 ∣ 3 2 3 3 ∣ 3 3 3	∣ 1 2 3 1 ∣ 1 1 1 2 ∣ 2 1 1 3 ∣ 3 2 1	$[2, 1] \neq 2,\ [3, 2] \neq 3,\ [2, 3] \neq 2$
I	∣ 1 2 1 ∣ 1 2 2 ∣ 2 2	∣ 1 2 1 ∣ 1 1 2 ∣ 1 2	∣ 1 2 1 ∣ 1 1 2 ∣ 2 2	$2:2 \neq 1$

II — $(3:1):2 \neq (3:2):1$

[,]

	1	2	3
1	1	2	3
2	2	2	3
3	3	3	3

(,)

	1	2	3
1	1	1	1
2	1	2	2
3	1	2	3

:

	1	2	3
1	1	1	1
2	3	1	1
3	3	2	1

III — $2:(2,\, 1) \neq [2:2,\, 2:1]$

[,]

	1	2
1	1	2
2	2	2

(,)

	1	2
1	1	1
2	2	2

:

	1	2
1	1	1
2	2	1

IV — $[4,\, 3]:2 \neq [4:2,\, 3:2]$

[,]

	1	2	3	4
1	1	2	3	4
2	2	2	3	4
3	3	3	3	4
4	4	4	4	4

(,)

	1	2	3	4
1	1	1	1	1
2	1	2	2	2
3	1	2	3	3
4	1	2	3	4

:

	1	2	3	4
1	1	1	1	1
2	2	1	1	1
3	3	3	1	1
4	4	2	2	1

V — $2:1 = 1:2 = 1$ but $2 \neq 1$.

[,]

	1	2
1	1	2
2	2	2

(,)

	1	2
1	1	1
2	1	2

:

	1	2
1	1	1
2	1	1

† The independence examples for i–iv are omitted.

5. **Residuation in a Boolean algebra** [1]. If we take Σ to be a Boolean algebra and interpret $[\,,\,]$, $(\,,\,)$ as the Boolean operations $\cdot$, $\vee$ respectively, then it is readily verified that Σ satisfies postulates i–V if we define residuation by $A:B \equiv A \vee B'$. Moreover we have the following theorem:

THEOREM. *Let Σ be a Boolean algebra and let $[\,,\,]$, $(\,,\,)$ be the Boolean operations $\cdot$, $\vee$ respectively. Then the only Boolean operation satisfying postulates* I–V *is* $A:B = A \vee B'$.

PROOF. Write $A:B$ as a general Boolean function of A and B

$$A:B = K_1 AB \vee K_2 AB' \vee K_3 A'B \vee K_4 A'B'.$$

Then

$$1:1 = K_1 = 1, \qquad 0:1 = K_3 = 0,$$

$$1:0 = K_2 = 1, \qquad 0:0 = K_4 = 1.$$

Hence

$$A:B = AB \vee AB' \vee A'B' = A(B \vee B') \vee A'B' = A \vee A'B'$$

$$= (A \vee AB') \vee A'B' = A \vee (A \vee A')B' = A \vee B',$$

and as above this is sufficient that postulates I–V be satisfied.

With this definition of $A : B$, $A \supset B$ becomes the usual inclusion relation of the algebra of classes [5].

REFERENCES

1. E. T. Bell, *Arithmetic of logic*, Transactions of this Society, vol. 29 (1927), pp. 597–611.

2. Garrett Birkoff, *On combination of subalgebras*, Cambridge Philosophical Society Proceedings, vol. 29 (1933), pp. 441–464.

3. R. Dedekind, *Dirichlet, Vorlesungen über Zahlentheorie.*

4. E. Lasker, *Zur Theorie der Moduln und Ideale*, Mathematische Annalen, vol. 60 (1905), pp. 20–115.

5. Lewis and Langford, *Symbolic Logic*, 1932.

6. F. S. Macaulay, *The Algebraic Theory of Modular Systems*, Cambridge Tracts, no. 19, 1916.

7. O. Ore, *Abstract Algebra I, II*, Annals of Mathematics, vol. 36 (1935), pp. 406–437; vol. 37 (1936), pp. 265–292.

8. B. L. van der Waerden, *Moderne Algebra*, vol. 2.

9. M. Ward, *Some arithmetical applications of residuation*, American Journal of Mathematics, vol. 59 (1937), pp. 921–926.

CALIFORNIA INSTITUTE OF TECHNOLOGY

RESIDUATED LATTICES*

BY

MORGAN WARD AND R. P. DILWORTH

I. INTRODUCTION

1. We propose to develop here a systematic theory of lattices† over which an auxiliary operation of multiplication or residuation is defined. We begin by showing that the two operations correspond to one another; under quite general conditions in every lattice over which a multiplication is defined a residuation may be defined and conversely. The residuation and multiplication we introduce have the properties of the like-named operations in the particular instance of polynomial ideal theory.

We next give various necessary conditions and sufficient conditions that such operations may exist in an arbitrary lattice, and apply our results to projective geometries and Boolean algebras.

In the third division of the paper we extend E. Noether's decomposition theorems of the ideal theory of commutative rings to general lattice theory. The introduction of a multiplication is obviously necessary for such a generalization. The surprising result emerges that the decomposition theorems are largely independent of the modular axiom, as we show by specific examples. We take this occasion to correct an error made in the preliminary account of our researches (Ward and Dilworth [1]). Since we wrote this, we have obtained many new results which we give here for the first time.

We plan to describe the main part of our investigations of distributive residuated lattices elsewhere (Ward and Dilworth [1], §§5, 6). Here we settle some questions raised by one of us (Ward [1]) as to the significance of certain auxiliary conditions which a residuation may satisfy by showing in all cases that they imply that the lattice is distributive.

2. It was not until this paper was virtually completed that we learned of the investigation of Krull upon this subject (Krull [1]). There is, however, very little duplication between our results and Krull's. Krull was chiefly concerned with the problem of finding out in what manner the Noether decomposition theorems could be extended to a residuated lattice in which the chain condition was weakened and no connection was assumed between irreducibles and primary elements.

---

* Presented to the Society, March 27, 1937, and April 9, 1938; received by the editors April 21, 1938.

† For a connected account of lattice theory and the literature up to 1937, see Köthe [1].

335

3. We shall use the following terminology and notation. $\mathfrak{S}$ is a fixed lattice with elements $a, \cdots, y$ with or without subscripts. Sublattices of $\mathfrak{S}$ are denoted by German capitals $\mathfrak{A}, \mathfrak{B}$. The letters $\mathfrak{X}, \mathfrak{Y}, \mathfrak{Z}$ are reserved to denote subsets of $\mathfrak{S}$ which are not necessarily sublattices. We write $x \, \varepsilon \, \mathfrak{X}$ for "the set $\mathfrak{X}$ contains the element x." The expressions $x \supset y$ or $y \subset x$, $x \not\supset y$ denote, as usual, x divides y, x does not divide y. We write $x = y$ if $x \supset y$ and $y \supset x$ (Ore [1], p. 42) and $x > y$ or $y < x$ for x covers y (Birkhoff [1]). We use (x, y) and $[x, y]$ for union and cross-cut. If the unit and null elements exist, we denote them by i and z, respectively. Elements covered by i are called divisor-free. If $(a, b) = i$, a and b are said to be co-prime. If every pair of distinct elements of a set $\mathfrak{X}$ are co-prime, the set is said to be co-prime. An element n of $\mathfrak{S}$ is called a node if either $x \supset n$ or $n \supset x$ for every x of $\mathfrak{S}$. A sublattice $\mathfrak{A}$ is said to be dense over $\mathfrak{S}$ if $a_1, a_2 \, \varepsilon \, \mathfrak{A}$ and $a_1 \supset x \supset a_2$ imply $\mathfrak{A}$ contains x. If every set of elements finite or infinite of $\mathfrak{S}$ has a cross-cut (union), $\mathfrak{S}$ is said to be completely closed relative to cross-cut (union). Two properties P and Q which $\mathfrak{S}$ may possess are said to be completely independent if there exist instances of lattices in which both P and Q hold, neither holds, P holds but not Q, Q holds but not P.

We shall find it convenient to use the following conditions for a distributive lattice either of which is equivalent to the usual formulation:

(i) $b \supset [a, c]$ *implies* $b = [(b, a), (b, c)]$.
(ii) $(a, c) \supset b$ *implies* $b = ([b, a], [b, c])$.

II. Residuations and multiplications

4. Assume that $\mathfrak{S}$ contains i. A well-defined one-valued binary operation $x : y$ is called a residuation over $\mathfrak{S}$ if the following conditions are satisfied:

R 1. *If $\mathfrak{S}$ contains a, b, then $\mathfrak{S}$ contains $a : b$.*
R 2. $a : b = i$ *if and only if* $a \supset b$.
R 3. $a \supset b$ *implies that* $a : c \supset b : c$ *and* $c : b \supset c : a$.
R 4. $(a : b) : c = (a : c) : b$.
R 5. $[a, b] : c = [a : c, b : c]$.
R 6. $c : (a, b) = [c : a, c : b]$.

We postpone the consideration of the dual residuation for our second paper.

A well defined binary operation $x \cdot y$ (or xy) is called a multiplication over $\mathfrak{S}$ if the following conditions are satisfied:

M 1. *If $\mathfrak{S}$ contains a, b, then $\mathfrak{S}$ contains $a \cdot b$.*
M 2. *If $a = b$, then $a \cdot c = b \cdot c$.*
M 3. $a \cdot b = b \cdot a$.

M 4. $(a \cdot b) \cdot c = a \cdot (b \cdot c)$.

M 5. *If $\mathfrak{S}$ contains i, then $a \cdot i = a$.*

M 6. $a \cdot (b, c) = (a \cdot b, a \cdot c)$.

It may be shown (Ward [1]) that a residuation exists satisfying R 1–R 6 if a multiplication over $\mathfrak{S}$ exists satisfying M 1–M 6 and the following condition:

M 7. *$\mathfrak{S}$ is completely closed with respect to union, and the product of the unions of any two sets of elements of $\mathfrak{S}$ is the union of the products of all pairs of elements in the sets.*

This residual $a:b$, satisfying R 1–R 6, is defined as follows:

DEFINITION 4.1. (i) $a \supset (a:b)b$; (ii) *if $a \supset xb$, then $a:b \supset x$.*

If we take for $x \cdot y$ the cross-cut $[x, y]$, then conditions M 1–M 6 are all satisfied provided that $\mathfrak{S}$ is distributive. If M 7 holds, the lattice is said to be completely distributive with respect to union. Hence (Ward [2]) *every completely distributive lattice may be residuated in at least one way.*

Another condition* insuring the existence of a residual is the following:

M 8. *For any two elements a, b of $\mathfrak{S}$, the ascending chain condition holds in the set $\mathfrak{X}$ of all x such that $a \supset xb$.*

M 8 insures the existence of a union of the set $\mathfrak{X}$ expressible as the union of a finite number of elements of $\mathfrak{X}$ (Ore [1], §2). This union is the required residual.

We list for reference the more important properties of residuation and multiplication (Ward [1], Dilworth [1]):

(4.1) $a:b \supset a$.

(4.2) $a:(a:b) \supset (a, b)$.

(4.3) $(a:b):c = a:(bc)$.

(4.4) $[a, b]:b = a:b$.

(4.5) $a:(a, b) = a:b$.

(4.51) $c:[a, b] \supset (c:a, c:b)$.

(4.6) *If $a:b = a$, then $a \supset bx$ implies $a \supset x$.*

(4.7) *If $r = a:b$ and $s = a:r$, then $r = a:s$.*

(4.71) *$a \supset b$ implies $ac \supset bc$.*

(4.8) $[a, b] \supset ab \supset [a, b](a, b)$.

(4.81) $ab:a \supset b$.

(4.9) *$a = bc$ implies $b \supset a$.*

(4.10) $(a, b) = i$ *implies* $ab = [a, b]$ *and* $(a, bc) = (a, c)$.

(4.11) $(a, b):c \supset (a:c, b:c)$.

(4.12) *In any chain of powers a, a^2, a^3, $\cdots$ either all elements are distinct or all are equal from a certain point on.*

---

* This axiom is equivalent to the ascending chain condition, as may be seen on taking $a = b = i$. We state it in this manner to emphasize the analogy with R 8 which is *not* equivalent to the descending chain condition.

5. We shall now exhibit a remarkable reciprocity between the operations of residuation and multiplication.

THEOREM 5.1. *If a residuation $x:y$ exists in $\mathfrak{S}$ satisfying conditions R 1–R 6, and if either of the conditions R 7 or R 8 below holds, then a multiplication $x \cdot y$ exists in $\mathfrak{S}$ satisfying* M 1–M 6.

R 7. *$\mathfrak{S}$ is completely closed with respect to cross-cut, and if c is the cross-cut of a set $\mathfrak{X}$, then the cross-cut of the set of all $a:x$, where $a \; \varepsilon \; \mathfrak{S}, \; x \; \varepsilon \; \mathfrak{X}$, equals $a:c$.*

R 8. *For any two elements a, b of $\mathfrak{S}$ the descending chain condition holds in the set $\mathfrak{Y}$ of all elements y such that $y:a \supset b$.*

R 8 is satisfied in many important instances where R 7 does not hold and where the descending chain condition does not hold; for example, in polynomial ideal theory and the classical ideal theory of algebraic rings.

The proof is as follows. Define the "product" $a \cdot b$ of any two elements a and b of $\mathfrak{S}$:

DEFINITION 5.1. (i) $a \cdot b : a \supset b$; (ii) *if $y : a \supset b$, then $y \supset a \cdot b$.*

Postulate M 1 *is satisfied.* For the set $\mathfrak{Y}$ of all y such that $y:a \supset b$ is non-empty, since it includes b by (4.1). If R 7 holds, $\mathfrak{Y}$ has a cross-cut $p = a \cdot b$ satisfying Definition 5.1, (ii), and the cross-cut $[\mathfrak{Y}:a]$ equals $p:a$. Definition 5.1, (i) is therefore satisfied with $p = a \cdot b$ by the definition of cross-cut.

If R 8 holds, then $\mathfrak{Y}$ again has a cross-cut p representable as the cross-cut of a *finite* number of y, $p = [y_1, \cdots, y_k]$. Thus Definition 5.1, (ii) is satisfied, and Definition 5.1, (i) is satisfied by R 5.

Postulate M 2 *is satisfied.* For by R 3, $a = b$ implies $a \supset b$ implies $a \cdot c : b \supset a \cdot c : a$. Hence by Definition 5.1, (i), $a \cdot c : b \supset c$ so that by Definition 5.1, (ii), $a \cdot c \supset b \cdot c$. Similarly $a = b$ implies $b \cdot c \supset a \cdot c$, so that M 2 follows.

Postulate M 3 *is satisfied.* For $b \cdot a$ exists, and by Definition 5.1, if $y:b \supset a$, then $y \supset b \cdot a$. Now by R 4, Definition 5.1, (i) and R 2, $(a \cdot b : b) : a = (a \cdot b : a) : b = i$. Hence by R 2, $a \cdot b : b \supset a$. Hence $a \cdot b \supset b \cdot a$. Similarly, $b \cdot a \supset a \cdot b$, $a \cdot b = b \cdot a$. Condition R 4 is thus seen to insure that multiplication is commutative.

Postulate M 4 *is satisfied.* For by Definition 5.1, (i), $\{a \cdot (c \cdot b)\} : a \supset c \cdot b$. Hence $\{\{a \cdot (c \cdot b)\} : a\} : c \supset c \cdot b : c$ by R 3. But $c \cdot b : c \supset b$ by Definition 5.1, (i). Therefore $\{\{a \cdot (c \cdot b)\} : a\} : c \supset b$ or by R 4, $\{\{a \cdot (c \cdot b)\} : c\} : a \supset b$. Hence $\{a(cb) : c\} \supset ab$ and $a(cb) \supset c(ab)$ by Definition 5.1, (ii). Interchanging a and c, $c(ab) \supset a(cb)$. Hence $a(cb) = c(ab)$, or by M 3 and M 2, $(ab)c = a(bc)$.

Postulate M 5 *is satisfied.* For by R 2, $a:a \supset i$. Hence $a \supset ai$ by Definition 5.1, (ii). Now $ia:i \supset a$ by Definition 5.1, (i). But by (4.10) and M 3, $ia:i = ia = ai$. Hence $ai \supset a$, $a = ai$.

Postulate M 6 *is satisfied.* For since $(b, c) \supset b$, $a(b, c) \supset ab$ by (4.71). Similarly $a(b, c) \supset ac$. Hence $a(b, c) \supset (ab, ac)$. Next $(ab, ac) : a \supset ab : a \supset b$ by R 3 and (4.81). Similarly $(ab, ac):a \supset c$. Hence $(ab, ac):a \supset (b, c)$. Therefore by Definition 5.1, (ii), $(ab, ac) \supset a(b, c)$ giving M 6. This completes the proof.

DEFINITION 5.2. (i) $a \supset (a \circ b)b$; (ii) *if $a \supset xb$, then $a \circ b \supset x$.*

The following theorem further illustrates the reciprocity between multiplication and residuation:

THEOREM 5.2. *If $a \circ b$ is defined as above, where the multiplication xy is defined by Definition 5.1, then $a \circ b = a:b$.*

For since $a:b \supset a:b$, we have $a \supset (a:b)b$ by Definition 5.1, (ii). Therefore by Definition 5.2, (ii), $a \circ b \supset a:b$. Now $a \supset (a \circ b)b$ by Definition 5.2, (i). Therefore by R 3, $a:b \supset \{(a \circ b)b:b\}$. But by M 3 and Definition 5.1, (ii), $(a \circ b)b:b \supset a \circ b$. Hence $a:b \supset a \circ b$, $a:b = a \circ b$.

Hereafter when we speak of a "residuated lattice," we shall mean a lattice in which both a residuation and its associated multiplication are defined satisfying M 1–M 6, R 1–R 6 and the conditions of Definitions 5.1 and 5.2.

6. We may prove by simple examples the following theorem:

THEOREM 6.1. *The Dedekind modular condition and the existence of a residual or a multiplication are completely independent properties of a lattice.*

It is important to observe that a given lattice may usually be residuated in several different ways. To give a simple example, consider the lattice of four elements $i > a > b > z$. The tables for $x:y$ and $x \cdot y$ are as follows:

$x:y$	i	a	b	z		$x \cdot y$	i	a	b	z
i	i	i	i	i		i	i	a	b	z
a	a	i	i	i		a	a	$*$	$*$	z
b	b	$*$	i	i		b	b	$*$	$*$	z
z	z	$*$	$*$	i		z	z	z	z	z

A brief analysis discloses that the combinations denoted by stars may be determined in six ways so as to satisfy R 1–R 8, M 1–M 8:

	I	II	III	IV	V	VI
$b:a$	a	a	b	a	b	b
$z:a$	a	b	b	z	z	z
$z:b$	a	a	a	z	b	z
$a \cdot a$	z	b	a	b	a	a
$a \cdot b$	z	z	z	b	b	b
$b \cdot b$	z	z	z	b	z	b

Cases II and VI are illustrated in the lattice of the ring of integers modulo 8. Here i is the set of residue classes $\{1, 3, 5, 7\}$, a is $\{2, 6\}$, b is $\{4\}$, and z is $\{8\}$. Case II ensues on taking for $x \cdot y$ multiplication modulo 8, and case VI on taking for $x \cdot y$ the L.C.M. operation.

The only other lattice of order four is i, a, b, z with $(a, b) = i$, $[a, b] = z$. This lattice may be residuated in only one way, an illustration of a general theorem on the residuation of Boolean algebras which we prove later.

III. CONDITIONS FOR RESIDUATION

7. In this division of the paper we shall give various sufficient conditions and necessary conditions for the existence of a residuation in a given lattice.

THEOREM 7.1. *A necessary condition that a lattice $\mathfrak{S}$ can be residuated is that any co-prime set of elements of $\mathfrak{S}$, $a_1, a_2, \cdots, a_r$ generates a Boolean algebra $\mathfrak{B}$ of order 2^r.*

This condition is not sufficient for a residuation to exist. It is satisfied, for example, in Dedekind's free modular lattice on three elements of order twenty-eight (Dedekind [1], Birkhoff [1], Ore [1]) which we shall prove later cannot be residuated.

Let $a_1, a_2, \cdots, a_r$ be a co-prime set so that

$$(7.1) \qquad\qquad (a_u, a_v) = i, \qquad\qquad u, v = 1, \cdots, r; u \neq v.$$

The set will remain co-prime if we adjoin i to it. We shall suppose that this has been done, and for definiteness choose our notation so that $a_1 = i$.

Form from the set of a's the "ray" Π of 2^r formally distinct cross-cuts:

$$u = [a_{u_1}, a_{u_2}, \cdots, a_{u_L}], \qquad 1 \leq u_1 < u_2 < \cdots < u_L \leq r; 1 \leq L \leq r.$$

We call the a_u the *constituents* of u. The ray Π is obviously closed under cross-cut. We shall show that Π is the Boolean algebra required.

LEMMA 7.1. *If x is any element of $\mathfrak{S}$, then*

$$(x, [a_u, a_v]) = [(x, a_u), (x, a_v)].$$

This result is trivial if $u = v$. But if $u \neq v$, $(a_u, a_v) = i$. Hence $((x, a_u), (x, a_v)) = i$. Therefore by (4.10) and M 6,

$$[(x, a_u), (x, a_v)] = (x, a_u)(x, a_v) = (x^2, xa_v, a_ux, a_ua_v)$$
$$= (x^2, x(a_u, a_v), a_ua_v) = (x, a_ua_v) = (x, [a_u, a_v])$$

by M 3 and M 6.

The following two corollaries of this lemma may be proved by induction:

LEMMA 7.2. *If $u = [a_{u_1}, \cdots, a_{u_L}]$, then $(x, u) = [(x, a_{u_1}), \cdots, (x, a_{u_L})]$.*

LEMMA 7.3. *If* $u = [a_{u_1}, \cdots, a_{u_L}]$ *and* $v = [a_{v_1}, \cdots, a_{v_M}]$, *then*

$$(u, v) = [(a_{u_1}, a_{v_1}), \cdots, (a_u, a_v), \cdots, (a_{u_L}, a_{v_M})].$$

LEMMA 7.4. *If* x *is any element of* $\mathfrak{S}$ *and if* $(x, b) = (x, c) = i$, *then*

$$(x, [b, c]) = [(x, b), (x, c)].$$

It suffices to show that $(x, [b, c]) = i$. But $(x, [b, c]), \supset (x\, bc) = (x, bx, bc)$ (by (4.10)) $= (x, b(x, c)) = (x, b) = i$.

We return to the proof of our theorem. *The ray* II *is a lattice.* For by Lemma 7.3 and (7.1) it is closed under union. *The lattice is of order* 2^r. It suffices to show that if $u = v$, the constituents of u and v are identical. But if $u = v$, $a_u \supset v$. Hence by Lemma 7.2,

$$a_u = (a_u, v) = [(a_u, a_{v_1}), \cdots, (a_u, a_{v_M})].$$

Since $(a_u, a_v) = a_u$ or i, a_u must be a constituent of v. Thus every constituent of u is a constituent of v. Similarly every constituent of v is a constituent of u, so that u and v are not formally distinct.

The lattice is distributive. For by Lemma 7.3, if $w = [a_{w_1}, \cdots, a_{w_N}]$, then

$$\begin{aligned}
(w, [u, v]) &= [\cdots, (a_w, [u, v]), \cdots] \\
&= [\cdots, [(a_w, u), (a_w, v)], \cdots] \\
&= [[\cdots, (a_w, u), \cdots], [\cdots, (a_w, v), \cdots]] \\
&= [(w, u), (w, v)],
\end{aligned}$$

by Lemma 7.2.

The lattice is complemented. For we assign to the element u the complement

$$u' = [a_{u'_{L+1}}, \cdots, a_{u'_r}]$$

where $u_{L'+1}, \cdots, u_r'$ is the selection complementary to $u_1, \cdots, u_L$ from $1, 2, \cdots, r$. Then $[u, u'] = [a_1, a_2, \cdots, a_r]$, the null element of the lattice $\mathfrak{B}$, and $(u, u') = i$ by Lemma 7.3. Hence $\mathfrak{B}$ is a complemented distributive lattice and thus a Boolean algebra.

THEOREM 7.2. *If* $a_1, \cdots, a_r$ *is a co-prime set of divisor-free elements of a residuated lattice* $\mathfrak{S}$, *then the Boolean algebra* $\mathfrak{B}$ *which they generate is dense over* $\mathfrak{S}$.

For if u lies in $\mathfrak{B}$ and $x \supset u$, then $x = [(x, a_{u_1}), \cdots, (x, a_{u_L})]$ by Lemma 7.2. Since $(x, a_u) = i$ or a_u, the result follows.

This theorem is quite useful in examining finite lattices to see whether or not they can be residuated. We have also found the following exclusion principle useful in this connection. The proof (which we omit) follows from Lemma 7.4.

THEOREM 7.3. EXCLUSION PRINCIPLE. *Let a, b, c, and d be any four elements of a residuated lattice $\mathfrak{S}$ with an ascending chain condition such that $c \supset a$, $c \neq i$; $(b, c) = i$, $d > a$, $d > b$. Then if $m = [b, c]$ we must have $[a, b] = m$ and $a > m$, $b > m$. Furthermore a and b are the only elements covered by d and covering m in the lattice.*

In schematic form (Klein [1], Birkhoff [2]) the lattice must have the following structure, where the dotted lines indicate that the configuration of the remaining lattice parts is irrelevant.

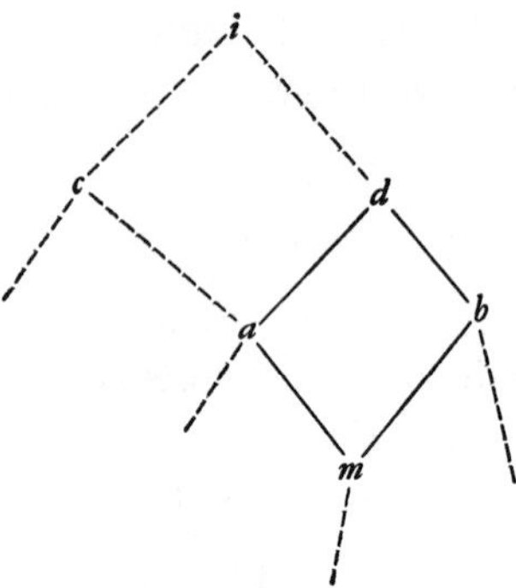

As a simple application, if the reader will diagram the lattice of order nine on three elements b, c, and f where $c \supset f$ (Dedekind [1]) and take $a = [c, (f, b)]$, $d = (f, b)$, he will see that this lattice cannot be residuated.

THEOREM 7.31. *The only complemented lattices which can be residuated are Boolean algebras.*

Since by hypothesis the lattice is complemented, it is sufficient to show that it is distributive. We need the following lemma:

LEMMA 7.5. *If $(b, c) = i$ and $a \supset [b, c]$, then $(a:b, a:c) = i$.*

For we have

$$(a:b, a:c) = (a:b, a:c):(b, c) = [(a:b, a:c):b, (a:b, a:c):c]$$
$$\supset [(a:c):b, (a:b):c] = a:cb = a:[c, b] = i.$$

A complement a' of a is defined by the following conditions:

DEFINITION 7.1. $(a, a') = i$, $[a, a'] = z$, *where z is the null element of* $\mathfrak{S}$.

Let a, b, c be any three elements of $\mathfrak{S}$ and assume that

(i) $\qquad\qquad\qquad\qquad a \supset [b, c].$

Let $u = [(a, b), (a, c)]$ and $v = ([b, c], a')$. It suffices to show that (i) implies $u = a$. We have trivially $u \supset a$ and $v \supset a'$. Hence $(u, v) = i$ by Definition 7.1 so that $uv = [u, v]$. Hence $b:uv = b:[u, v] \supset (b:u, b:v)$ by (4.51). Now $b:u \supset b:a$ by R 3 and $b:v = [b:[b, c], b:a'] = b:a'$. Hence $b:uv \supset (b:a, b:a')$. But by Definition 7.1, $(a, a') = i$ and $b \supset [a, a']$. Hence by Lemma 7.5, $(b:a, b:a') = i$ so that $b \supset uv$. Similarly $c \supset uv$ so that $[b, c] \supset uv$, or by (i), $a \supset uv$, $a:v \supset u$. But $a:v = [a:[b, c], a:a'] = a:a' = a$ by (i) and Definition 7.1. Hence $a \supset u$ so that $a = u$.

COROLLARY. *The only projective geometries (Birkhoff [3]) which can be residuated are Boolean algebras.*

In case the ascending chain condition holds in $\mathfrak{S}$, one can give a much shorter proof by showing that each element may be represented as a cross-cut of a finite number of divisor-free elements and appealing to Theorem 7.1.

THEOREM 7.4. *The only multiplication which can be defined over a Boolean algebra is the cross-cut operation.*

In view of our reciprocity theorems it suffices to show that only one residual is definable. One of us has shown elsewhere (Dilworth [1]) that $a \vee b'$ is a residuation in a Boolean algebra. Suppose that $a:b$ were another. Then

$$(a:b):(a \vee b') = (a:b):b' = a:bb' = i;$$

$$(a \vee b'):(a:b) \supset \{a:(a:b)\} \vee \{b':(a:b)\} \supset b \vee b' = i.$$

Hence $a:b = a \vee b'$.

An interesting consequence of Theorem 7.4 is the following corollary:

COROLLARY. *In the ring of integers modulo a square-free integer, the operations of multiplication and L.C.M. are identical.*

8. We consider in this section some sufficient conditions for residuation. We have the following theorem:

THEOREM 8.1. *Every lattice in which only one divisor-free element exists can residuated in at least one way.*

Let d be the single divisor-free element. We define the residual $a:b$ by the conditions:

(i) $a:i = a$;   (ii) $a:b = i$ if $a \supset b$;   (iii) $a:b = d$ if $a \not\supset b$, $b \neq i$.

Then postulates R 1 and R 2 are obviously satisfied.

R 3 *is satisfied.* For assume $a \supset b$. Then $a:c$ always divides $b:c$ except possibly when $b:c = i$. But then $b \supset c$; so $a \supset c$, $a:c = i$. Similarly $c:b \supset c:a$.

R 4 *is satisfied.* For R 4 obviously holds if a, b, or c equals i. If $a \supset b$, $a \neq i$, then $a:c \supset a \supset b$; so $(a:c):b=(a:b):c=i$. If $a:b \supset c$ but $a \not\supset b$, $b \neq i$, $a \not\supset c$, $c \neq i$, then $a:b=a:c=d$, whence $(a:b):c=d:c=i=d:b=(a:c):b$. If $a \not\supset b$, $a \not\supset c$, $a:b \not\supset c$, $a:c \not\supset b$, then b or $c=i$.

$R\ 5$ *is satisfied.* For if $c=i$, R 5 is trivial. If $a \supset c$, $b \supset c$, then $[a, b] \supset c$ and R 5 obviously holds. If $a \not\supset c$, $c \neq i$, then $[a, b] \not\supset c$ and $[a, b]:c=d=[a:c, b:c]$. Hence R 5 holds in general.

In exactly the same way we show that R 6 is satisfied.

F. Klein has shown (Klein [1]) that the modular or distributive properties of a lattice built up of sublattices connected by nodes ("Schnurstellen") depend upon the modular or distributive properties of the sublattices. We prove a similar result for residuation.

THEOREM 8.2. *A lattice built up out of a set of residuated lattices connected into a chain by nodes can be residuated.*

It will suffice to prove the theorem for the case of two lattices connected by a node.

Let $\mathfrak{S}$ be composed of two lattices $\mathfrak{S}_1$ and $\mathfrak{S}_2$ connected by a node, so that $x_1 \varepsilon \mathfrak{S}_1$ and $x_2 \varepsilon \mathfrak{S}_2$ imply $x_1 \supset x_2$. Let i be the unit element of $\mathfrak{S}_1$. We shall consider the nodal element as belonging to $\mathfrak{S}_1$, and let $x \mathbin{!} y$ denote the residuation in $\mathfrak{S}_1$, $x \circ y$, the residuation in $\mathfrak{S}_2$ when the nodal element is replaced by i.

We now define a residual in $\mathfrak{S}$ by the conditions:

$a:b=a \mathbin{!} b$ if $a, b \varepsilon \mathfrak{S}_1$, $a:b=a \circ b$ if $a, b \varepsilon \mathfrak{S}_2$,
$a:b=i$ if $a \varepsilon \mathfrak{S}_1$, $b \varepsilon \mathfrak{S}_2$, $a:b=a$ if $a \varepsilon \mathfrak{S}_2$, $b \varepsilon \mathfrak{S}_1$.

Then postulates R 1, R 2, and R 3 are obviously satisfied.

Postulate R 4 *is satisfied.* For clearly $a:c \supset a$. Hence if $a \supset b$, then $(a:b):c=(a:c):b$ by R 3. Also if $a \varepsilon \mathfrak{S}_1$, then $(a:b):c=(a:c):b$. Suppose that $a \varepsilon \mathfrak{S}_2$. Then if $b \varepsilon \mathfrak{S}_2$, $c \varepsilon \mathfrak{S}_2$, we have $(a:b):c=(a:c):b$. Similarly if $b \varepsilon \mathfrak{S}_1$, $c \varepsilon \mathfrak{S}_1$, then $(a:b):c=(a:c):b$. Finally if $b \varepsilon \mathfrak{S}_1$, $c \varepsilon \mathfrak{S}_2$, then $(a:b):c=a \circ c=(a:c):b$.

Postulate R 5 *is satisfied.* For R 5 is trivial if a, b, or $c=i$. If $a \supset b$, R 5 follows from R 3. If $a, b \varepsilon \mathfrak{S}_1$ or $a, b \varepsilon \mathfrak{S}_2$, R 5 holds since it holds in $\mathfrak{S}_1$ and $\mathfrak{S}_2$.

In a similar manner one can show that R 6 is satisfied, and the proof is complete.

By the *direct product* (Birkhoff [4]) $\mathfrak{S}$ of the lattices $\mathfrak{S}_1, \cdots, \mathfrak{S}_n$ we mean the set of vectors $a = \{a_1, \cdots, a_n\}$, $(a_i \varepsilon \mathfrak{S}_i)$, where the operations are defined by

$$[a, b] = \{[a_1, b_1], \cdots, [a_n, b_n]\},$$
$$(a, b) = \{(a_1, b_1), \cdots, (a_n, b_n)\},$$

and $a \supset b$ if and only if $a_i \supset b_i$, $(i=1, \cdots, n)$.

If the $\mathfrak{S}_i$ are residuated lattices, then $\mathfrak{S}$ can be residuated, since we may define $a:b$ to be $\{a_1:b_1, \cdots, a_n:b_n\}$.

We shall call two sublattices $\mathfrak{S}_1$, $\mathfrak{S}_2$ of $\mathfrak{S}$ *co-prime* if $a_1 \varepsilon \mathfrak{S}_1$, $a_2 \varepsilon \mathfrak{S}_2$ implies that $(a_1, a_2) = i$. The sublattices $\mathfrak{S}_1, \cdots, \mathfrak{S}_n$ will be called a *co-prime set* if they are co-prime in pairs.

We note that if $\mathfrak{S}$ is the direct product of the sublattices $\mathfrak{S}_1, \cdots, \mathfrak{S}_n$ with unit elements, then $\mathfrak{S}$ contains sublattices $\mathfrak{S}_1', \cdots, \mathfrak{S}_n'$ simply isomorphic to $\mathfrak{S}_1, \cdots, \mathfrak{S}_n$ and such that $\mathfrak{S}_1', \cdots, \mathfrak{S}_n'$ is a co-prime set. Birkhoff (Birkhoff [4]) has defined sublattices $\mathfrak{S}_1, \cdots, \mathfrak{S}_n$ to be "strongly" co-prime if each $\mathfrak{S}_i$ is co-prime to the lattice generated by the remaining lattices. Clearly strong co-primeness implies co-primeness in the ordinary sense. Moreever if $\mathfrak{S}$ is residuated, Lemma 7.4 shows that co-primeness implies strong co-primeness, so that for residuated lattices the notions are identical. We now prove a converse result.

THEOREM 8.3. *Let $\mathfrak{S}_1, \mathfrak{S}_2, \cdots, \mathfrak{S}_n$ be a set of co-prime sublattices of a residuated lattice $\mathfrak{S}$ such that each element of $\mathfrak{S}$ can be expressed as a cross-cut of elements of $\mathfrak{S}_1, \cdots, \mathfrak{S}_n$. Then $\mathfrak{S}$ is the direct product of $\mathfrak{S}_1, \cdots, \mathfrak{S}_n$, and each $\mathfrak{S}_i$ can be residuated.*

Let $a = [a_1, a_2, \cdots, a_n]$, $(a_i \varepsilon \mathfrak{S}_i)$, $b = [b_1, b_2, \cdots, b_n]$, $(b_i \varepsilon \mathfrak{S}_i)$. Then $[a, b] = [[a_1, b_1], \cdots, [a_n, b_n]]$. Furthermore

$$(a, b) = [(a_1, b_1), \cdots, (a_n, b_n)].$$

For by Lemma 7.2,

$$(a, b) = (a, [b_1, \cdots, b_n]) = [(a, b_1), \cdots, (a, b_n)]$$
$$= [\cdots, (a_j, b_k), \cdots] = [(a_1, b_1), \cdots, (a_n, b_n)],$$

since $(a_j, b_k) = i$ if $j \neq k$.

LEMMA 8.1. *If $b_1, b_2, \cdots, b_n$ are a co-prime set, then*

$$a:[b_1, \cdots, b_n] = (\cdots (((a:b_1):b_2):b_3) \cdots):b_n.$$

This result follows by repeated applications of Lemma 7.4, (4.3), and (4.10).

We have now $a:b = a:[b_1, \cdots, b_n] = (\cdots ((a:b_1):b_2) \cdots):b_n$ by Lemma 8.1. But

$$a:b_i = [a_1, \cdots, a_n]:b_i = [a_1:b_i, \cdots, a_n:b_i]$$
$$= [a_1, \cdots, a_{i-1}, a_i:b_i, a_{i+1}, \cdots, a_n].$$

Hence $a:b = [a_1:b_1, a_2:b_2, \cdots, a_n:b_n]$.

If $a = b$, then $(a_i, [a_1, \cdots, a_n]) = (a_i, [b_1, \cdots, b_n])$ or $a_i = (a_i, b_i)$, $a_i \supset b_i$.

Similarly $b_i \supset a_i$. Hence $\mathfrak{S}$ is simply isomorphic with the direct product of the $\mathfrak{S}_i$. We note that if $x = [x_1, \cdots, x_n]$, $(x \supset a_i, a_i \,\varepsilon\, \mathfrak{S}_i)$, then

$$x = (x, a_i) = ([x_1, \cdots, x_n], a_i) = (x_i, a_i) \,\varepsilon\, \mathfrak{S}_i.$$

Since $a_i : b_i \supset a_i$, we see that $\mathfrak{S}_i$ is closed under residuation, which completes the proof.

We conclude with a theorem of a more special character.

THEOREM 8.4. *The free modular lattice of order twenty-eight on three elements cannot be residuated.*

We shall use Dedekind's original notation for the elements of this lattice in the proof of the theorem (Dedekind [1]). Assume that a residuation $x:y$ exists. Then $\mathfrak{d}''''$ is the unit element. Hence $a_0 : \mathfrak{d}' = b_0 : \mathfrak{d}' = c_0 : \mathfrak{d}' = \mathfrak{h} \neq \mathfrak{d}''''$. Now $a_0 = [a', a''']$. Hence $a_0 : a''' = a' : a''' \supset a'$. But $a''' \supset \mathfrak{d}'$. Therefore $a_0 : \mathfrak{d}' \supset a_0 : a'''$ or $\mathfrak{h} \supset a'$. Similarly, $\mathfrak{h} \supset \mathfrak{b}'$, $\mathfrak{h} \supset c'$. Hence $\mathfrak{h} \supset (a', \mathfrak{b}', c')$ or $\mathfrak{h} \supset \mathfrak{d}''''$, $\mathfrak{h} = \mathfrak{d}''''$ giving a contradiction.

It may be observed that the "exclusion principle" of Theorem 7.3 cannot be applied to prove this theorem.

IV. NOETHER LATTICES*

9. Consider any residuated lattice $\mathfrak{S}$. An element c of $\mathfrak{S}$ is *irreducible* if in every decomposition $c = [g, f]$ into a cross-cut of two elements of $\mathfrak{S}$, either $g = c$ or $f = c$. An element p is a *prime* if $p \supset ab$ implies $p \supset a$ or $p \supset b$, and *primary* if $p \supset ab$, $p \not\supset a$ implies $p \supset b^s$ for some integer s. The irreducible elements are thus determined by an intrinsic lattice property, while the primes and primary elements depend upon the particular multiplication introduced into the lattice.

We propose here the name "Noether lattice" for any lattice $\mathfrak{S}$ satisfying the following three conditions:

N 1. *The lattice $\mathfrak{S}$ may be residuated.*
N 2. *The ascending chain condition holds in $\mathfrak{S}$.*
N 3. *Every irreducible element of $\mathfrak{S}$ is primary.*

By N 1 we mean that $\mathfrak{S}$ is closed under operations $x:y$, xy having the properties R 1–R 6, M 1–M 6 and connected by the relationships expressed

---

* Our definition differs from that in Ward and Dilworth [1]. We have found that some of the results stated in §4 of this paper are in error. In postulate D 1, the exponent r must be replaced by 1. The condition $ab = [a, b]$ on the idempotent elements of a finite modular lattice is consequently *necessary* for the truth of D 1 but not sufficient. The postulate M 7 is *not* a sufficient condition for a Noether lattice as stated in the theorem preceding M 7.

in Definitions 4.1, 5.1. By N 2 we mean (Ore [1]) that every chain of lattice elements $a_1 < a_2 < a_3 < \cdots < i$ terminates.

We choose the name in honor of Emmy Noether because the decomposition theorems first proved by her for the ideals of a commutative ring with chain condition all hold. It is to be observed that we do not assume a modular condition.

The proof that the usual decomposition theorems hold may be made by a mere transcription of the proofs given in van der Waerden [1] into lattice language. With each primary q is associated a prime p with the properties $p \supset q$ and $p \supset b$ implies $q \supset b^r$. A cross-cut of primaries is said to be "simple" if no primary in it divides the cross-cut of any of the remaining primaries. *Every element not equal to i of a Noether lattice may be represented as a simple cross-cut of a finite number of primaries each of which is associated with a different prime. The primes themselves and the total number of primaries are uniquely determined by the element.* We obtain from each such representation a representation as the cross-cut of "isolated components" by grouping together the cross-cuts of primaries whose associated primes divide one another. *The isolated components of an element and the corresponding representation as their cross-cut are unique.*

As was pointed out by Krull (Krull [1]), the decompositions into relatively prime ("teilerfremd") elements depend merely upon N 1 and N 2. From our standpoint, they are simple consequences of Theorem 7.1 and the chain condition.

We may specialize our lattice still more by the following assumption:

N 4. *Every prime of $\mathfrak{S}$ is divisor-free.*

Then, since we have trivially from N 1 that every divisor-free element is a prime, we easily see that all primaries associated with a given prime form a lattice which we may say "belongs" to this prime.

The lattices belonging to distinct primes have no elements save i in common. Hence the decomposition theorems in this case are merely an instance of Birkhoff's decomposition of a lattice into direct products relative to cross-cut (Birkhoff [4]).

10. We shall now give some general properties of any Noether lattice.

THEOREM 10.1. *If a and b are any two elements of a Noether lattice, there exists an exponent s such that the following condition holds:*

D 1. $ab \supset [a, b^s]$.

Let $ab = [q_1, \cdots, q_k]$ be a decomposition of ab into a cross-cut of primaries. Then for each q_i, $q_i \supset ab$; hence either $q_i \supset a$ or $q_i \not\supset a$, $q_i \supset b^{s_i}$. With a proper choice of notation, we may assume that $q_i \supset a$, $(i=1, \cdots, l)$, $q_i \supset b^{s_i}$, $(i=l+1, \cdots, k)$. Hence if s is the largest of the s_i, $[q_1, \cdots, q_l] \supset a$, $[q_{l+1}, \cdots, q_k] \supset b^s$ giving D 1.

The following three theorems are immediate corollaries:

THEOREM 10.2. *If b is an idempotent element in a Noether lattice, then $[a, b] = ab$ for any other element a of the lattice, and $b[a, c] = [ba, bc]$ for any elements a and c.*

THEOREM 10.3. *In a Noether lattice, every idempotent element is neutral.**

THEOREM 10.4. *In a Noether lattice, the idempotent elements form a distributive lattice. The product of any two idempotent elements is their cross-cut.*

It is easy to show that this last mentioned property of idempotent elements holds in any lattice in which multiplication is distributive with respect to cross-cut; for if a, b are idempotent,

$$ab \supset [a, b](a, b) = [a(a, b), b(a, b)]$$
$$= [(a^2, ab), (ba, b^2)] = [(a, ab), (b, ab)] = [a, b].$$

The following lattice of order six illustrates how the definition of a Noether lattice depends upon the type of multiplication introduced. The elements are i, j, a, b, k, and z with the coverings $i>j, j>a, j>b; a>k, b>k; k>z$. The lattice is distributive and hence a Noether lattice if multiplication is identified with cross-cut (see §11). Define an operation xy by $ix=xi=x$; $zx=xz=z$; $j^2=j$, $a^2=a$, $b^2=b$, $k^2=z$; $ja=aj=a$; $jb=bj=b$; $jk=kj=z$; $ab=ba=ak=ka=bk=kb=z$. It may be shown that xy is a multiplication satisfying M 1–M 8. But D 1 does not hold; for $ab=z$ and $[a, b]=k$, while a and b are idempotent. Hence N 3 is false by Theorem 10.1.

11. We shall next give some sufficient conditions that a lattice be a Noether lattice.

THEOREM 11.1. *Let $\mathfrak{S}$ be a residuated lattice with ascending chain condition. Then sufficient conditions that $\mathfrak{S}$ be a Noether lattice are as follows:*

D 1. *$ab \supset [a, b^s]$.*
D 2. *$\mathfrak{S}$ is modular.*

It suffices to show that N 3 holds. Let m be irreducible, $m \supset ab$, $m \not\supset a$. Then if $d=(a, m)$, $d \supset m \supset db$. Now by D 1, $db \supset [d, b^s]$ for some s. Hence $d \supset m \supset [d, b^s]$. Therefore by D 2, $m = [(m, d), (m, b^s)]$. Since m is irreducible and $(m, d) = (m, a) \neq m$, $(m, b^s) = m$. Hence $m \supset b^s$ and m is primary.

---

* Following Ore [1], we call an element n of a lattice "neutral" if $[n, (b, c)] = ([n, b], [n, c])$ for every pair of elements b, c of the lattice.

COROLLARY. *Every distributive lattice in which the ascending chain condition holds is a Noether lattice for a suitably defined multiplication.*

We take for the multiplication the cross-cut operation. Then M 1–M 6 and M 8 all hold; so $\mathfrak{S}$ may be residuated. Since $\mathfrak{S}$ is distributive, it is modular. Thus N 1, N 2, and D 2 hold. But D 1 is trivially true. The result now follows from the previous theorem.

We shall next give some conditions enabling us to view the ideal theory of commutative rings from a lattice-theoretic standpoint. It is first necessary to introduce a new concept. Let $\mathfrak{S}$ be a residuated lattice.

DEFINITION 11.1. *An element q of $\mathfrak{S}$ is principal if $q \supset b$ implies that there exists an element c such that $qc = b$.*

Neither c nor b need be principal.

Suppose that a is principal, $a \supset b$. The set $\mathfrak{Z}$ of elements z such that $az = b$ is closed with respect to union. If either postulate M 7 or M 8 holds, the union b/a of $\mathfrak{Z}$ has the properties stated in the following definition:

DEFINITION 11.2. $a \cdot (b/a) = b$; *if $ax = b$ then $b/a \supset x$.*

We call b/a the *quotient* of b by a. It is easily shown (Ward [1]) that *if a is principal and $a \supset b$, then the quotient b/a equals the residual $b : a$ of a with respect to b.*

As a simple consequence, we have the following lemma:

LEMMA 11.1. *If a is principal and if $a \supset b$, then $b = (b : a)a$.*

We may observe that M 8 always holds if the ascending chain condition holds. Hence Lemma 11.1 is true for all principal elements of a residuated lattice with ascending chain condition. We shall now prove the following fundamental theorem:

THEOREM 11.2. *Let $\mathfrak{S}$ be a lattice in which the following conditions hold:*

N 1. *The lattice $\mathfrak{S}$ may be residuated.*
N 2. *The ascending chain condition holds in $\mathfrak{S}$.*
D 2. *$\mathfrak{S}$ is modular.*
D 3. *Every element of $\mathfrak{S}$ is the union of a finite number of principal elements.*
D 4. *The principal elements of $\mathfrak{S}$ are closed under multiplication.*

Then $\mathfrak{S}$ is a Noether lattice.

The instance of ideal theory is obtained by identifying the principal elements of the lattice with the principal ideals or the corresponding ring elements. D 3 is then the basis theorem, and D 4 the closure property of ring multiplication.

It suffices to show that every irreducible element is primary, or inversely that every non-primary element is reducible. Let m be non-primary. Then there exist elements a and b of the lattice such that

$$(11.1) \qquad m \supset ab, \quad m \not\supset a, \quad m \not\supset b^r \text{ any } r.$$

We shall show that m is reducible. By D 3, $b = (b_1, b_2, \cdots, b_k)$ where the b_i are principal. Then $m \supset ab_i$. For at least one b_i, $m \not\supset b_i^r$ for any r. For otherwise, for each b_i there exists an exponent r_i such that $m \supset b_i^{r_i}$. Then if $r > r_1 + r_2 + \cdots + r_l - l$, we have $m \supset b^r$ contrary to hypothesis. Therefore, *we may assume that b in* (11.1) *is principal.*

By N 2, the chain $m:b$, $m:b^2$, $\cdots$, $m:b^k$, $\cdots$ terminates so that $m:b^k = m:b^{k+1}$ for some fixed k. Consider the cross-cut $c = [(m, a), (m, b^k)]$. We have trivially $c \supset m$. Now $(m, b^k) \supset c \supset m$. Hence by D 2 (Ore [1]),

$$(11.2) \qquad c = ([c, m], [c, b^k]).$$

Now $m \supset [c, m]$. We shall show next that $m \supset [c, b^k]$. By D 4, b^k is principal, and $b^k \supset [c, b^k]$. Hence by Lemma 11.1, $[c, b^k] = \{[c, b^k]:b^k\}b^k = (c:b^k)b^k$. Also since $(m, a) \supset c$, $b(m, a) \supset bc$. But $b(m, a) = (bm, ba) \subset m$ by (11.1). Hence $m \supset bc \supset b[c, b^k]$ by (11.2). That is, $m \supset b\{(c:b^k)b^k\}$ or $m:b^{k+1} \supset c:b^k$. But $m:b^{k+1} = m:b^k$. Hence $m:b^k \supset c:b^k$ or $m \supset (c:b^k)b^k$, $m \supset [c, b^k]$. It follows therefore that $m \supset c$. Hence $m = c$ or $m = [(m, a), (m, b^k)]$. But $m \not\supset a$, $m \not\supset b^k$. Hence $(m, a) \neq m$, $(m, b^k) \neq m$, and m is reducible. This completes the proof.

12. To show the significance of the hypotheses of Theorems 11.1 and 11.2, we shall exhibit various lattices in which not all the hypotheses are satisfied.

We first consider the following lattice $\mathfrak{B}_1$; and we define a multiplication xy over $\mathfrak{B}_1$ by the following table:

$x \backslash y$	i	j	a	b	m	z
i	i	j	a	b	m	z
j	j	a	a	z	z	z
a	a	a	a	z	z	z
b	b	z	z	z	z	z
m	m	z	z	z	z	z
z	z	z	z	z	z	z

The reader may verify that M 1–M 8 are satisfied. Thus $\mathfrak{B}_1$ is a residuated lattice in which the ascending chain condition holds. $\mathfrak{B}_1$ is obviously non-modular. Now it is easily verified that D 1 holds in the lattice: $xy \supset [x, y^2]$, for every x, y of the lattice. Nevertheless, *not every irreducible element is primary*. For consider the irreducible m. We have $m \supset ab$ and $m \not\supset b$. But since a is idempotent, $m \not\supset a^r$ for any r.

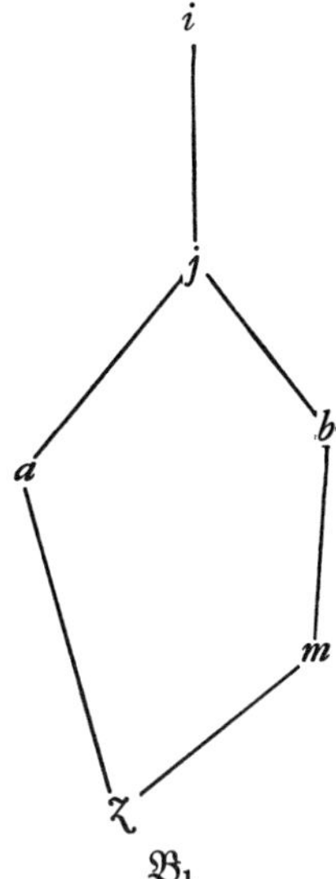

Next, consider the lattice $\mathfrak{B}_2$.

We assign the residuation $x:y$ to $\mathfrak{B}_2$ described in Theorem 8.1. The associated multiplication given by Definition 5.1 is then as follows: $xy=y$, if $x=i$; $xy=x$ if $y=i$; $xy=x$ otherwise.

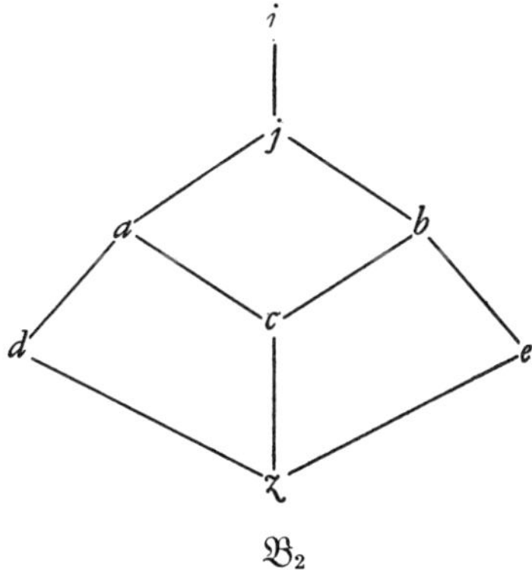

This lattice is non-modular, as it contains the non-modular sublattice j, a, d, e, z. The irreducible elements in it are j, a, b, d, e, and these are all primary since $x \supset y^2$ for any $y \neq i$ and any x. Furthermore, the elements i, c, d, e, and z are principal and $a=(d, c)$, $b=(c, e)$, $j=(a, b)$. Finally the principal elements are closed with respect to multiplication. Thus in this lattice, all hypotheses of Theorem 11.2 hold save modularity; and yet the lattice is a Noether lattice.

Our last example is one in which all the hypotheses of Theorem 11.2 hold save modularity and the lattice is *not* a Noether lattice. We define a multiplication over $\mathfrak{B}_3$ by the following table:

x \ y	i	j	a	b	c	d	m	z
i	i	j	a	b	c	d	m	z
j	j	d	c	b	m	d	m	z
a	a	c	c	z	m	m	m	z
b	b	b	z	b	z	b	z	z
c	c	m	m	z	m	m	m	z
d	d	d	m	b	m	d	m	z
m	m	m	m	z	m	m	m	z
z	z	z	z	z	z	z	z	z

Then it may be verified that the multiplication satisfies M 1–M 8, and that the elements i, a, b, c, m, z are principal and closed under multiplication.

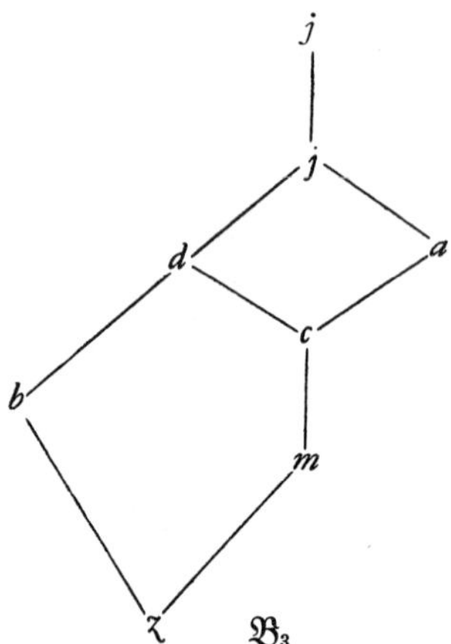

Since $d = (b, m)$, $j = (a, d)$, every element is the union of a finite number of principal elements. The lattice is evidently non-modular. It is not a Noether lattice. For consider the irreducible element m. Then $m \supset ab$, $m \not\supset a$, and $m \not\supset b^s$ for any s, since b is idempotent.

V. Conditions for distributivity

13. We shall conclude by answering some of the questions raised in Ward [4] as to the import of certain auxiliary conditions in a residuated lattice. We consider a residuated lattice in which one or more of the following conditions hold:

R 9. $(a:b, b:a) = i.$ R 10. $a:[b, c] = (a:b, a:c).$ R 11. $(b, c):a = (b:a, c:a).$

THEOREM 13.1. *R 9, R 10, R 11 are equivalent and imply distributivity.*

R 9 *implies* R 11. For

$$(b:a, c:a):\{(b, c):a\} \supset ((b:a):\{(b, c):a\}, (c:a):\{(b, c):a\})$$
$$= ((b:\{(b, c):a\}):a, (c:\{(b, c):a\}):a).$$

But $(b:\{(b, c):a\}):a \supset b:c$ since

$$\{(b:\{(b, c):a\}):a\}:(b:c) = (\{b:(b:c)\}:\{(b, c):a\}):a$$
$$\supset ((b, c):\{(b, c):a\}):a \supset a:a = i.$$

Similarly $(c:\{(b, c):a\}):c \supset c:b.$ Hence $(b:a, c:a):\{(b, c):a\} \supset (b:c, c:b) \supset i$ by R 9. Thus $(b:a, c:a) \supset (b, c):a.$ But $(b, c):a \supset (b:a, c:a)$ trivially.

R 11 *implies* R 10. For by R 11,

$$(a:b, a:c):\{a:[b, c]\} = ((a:b):\{a:[b, c]\}, (a:c):\{a:[b, c]\})$$
$$= ((a:\{a:[b, c]\}):b, (a:\{a:[b, c]\}):c)$$
$$\supset ([b, c]:b, [b, c]:c)$$
$$= (c:b, b:c) = (c:(b, c), b:(b, c)) = (c, b):(b, c) = i$$

by R 11. Hence $(a:b, a:c) \supset a:[b, c].$ But $a:[b, c] \supset (a:b, a:c)$ trivially.

R 10 *implies* R 9. For $(a:b, b:a) = ([a, b]:b, [a, b]:a) = [a, b]:[a, b] = i$ by condition R 10.

R 10 *implies distributivity.* For let $a \supset [b, c].$ Then

$$a:[(a, b), (a, c)] = (a:(a, b), a:(a, c)) = (a:b, a:c) = a:[b, c] = i.$$

Hence $a \supset [(a, b), (a, c)]$ and $[(a, b), (a, c)] \supset a$ trivially. Therefore $a = [(a, b), (a, c)].$

THEOREM 13.2. *If every element of a residuated lattice is principal, then the lattice is distributive.*

Let $(b, c) \supset a.$ We have $a \supset ([a, b], [a, c]).$ Hence

$$a:(b, c) \supset ([a, b], [a, c]):(b, c) = [([a, b], [a, c]):b, ([a, b], [a, c]):c]$$
$$\supset [[a, b]:b, [a, c]:c] = [a:b, a:c] = a:(b, c).$$

Thus $a:(b, c) = ([a, b], [a, c]):(b, c)$. But $(b, c) \supset a \supset ([a, b], [a, c])$. Hence

$$a = (a:(b, c))(b, c) = \{([a, b], [a, c]):(b, c)\}(b, c) = ([a, b], [a, c])$$

by Lemma 13.1.

THEOREM 13.3. *A sufficient condition that a residuated lattice with ascending chain condition be a Noether lattice is that every element in it be principal.*

For by Theorem 13.2, the lattice is distributive and hence modular; so all the hypotheses of Theorem 11.2 are satisfied.

REFERENCES

GARRETT BIRKHOFF

  1. *On the combination of sub-algebras*, Proceedings of the Cambridge Philosophical Society, vol. 39 (1933), pp. 441–464.

  2. *On the structure of abstract algebras*, ibid., vol. 41 (1935), pp. 433–454.

  3. *Combinatorial relations in projective geometries*, Annals of Mathematics, (2), vol. 36 (1935), pp. 743–748.

  4. *On the lattice theory of ideals*, Bulletin of the American Mathematical Society, vol. 40 (1934), pp. 613–619.

R. DEDEKIND

  1. *Ueber die von drei Moduln erzeugte Dualgruppe*, Gesammelte mathematische Werke, vol. 2, 1931, paper 30, pp. 236–271.

R. P. DILWORTH

  1. *Abstract residuation over lattices*, Bulletin of the American Mathematical Society, vol. 44 (1938), pp. 262–268.

FRITZ KLEIN

  1. *Dedekindsche und distributive Verbände*, Mathematische Zeitschrift, vol. 41, pp. 261–280.

G. KÖTHE

  1. *Die Theorie der Verbände* · · · , Jahresbericht der deutschen Mathematiker-Vereinigung, vol. 47 (1937), pp. 125–142.

W. KRULL

  1. *Axiomatische Begründung der allgemeinen Idealtheorie*, Sitzungsberichte der physikalisch-medicinischen Societät zu Erlangen, vol. 56 (1924), pp. 47–63.

O. ORE

  1. *On the foundation of abstract algebra*, I, Annals of Mathematics, (2), vol. 36 (1935), pp. 406–437.

B. L. VAN DER WAERDEN

  1. *Moderne Algebra*, vol. 2, Berlin, 1931.

M. WARD AND R. P. DILWORTH

  1. *Residuated lattices*, Proceedings of the National Academy of Sciences, vol. 24 (1938), pp. 162–164.

M. WARD

  1. *Residuation in structures over which a multiplication is defined*, Duke Mathematical Journal, vol. 3 (1937), pp. 627–636.

  2. *Structure residuation*, Annals of Mathematics, (2), vol. 39 (1938), pp. 558–568.

CALIFORNIA INSTITUTE OF TECHNOLOGY,
  PASADENA, CALIF.

NON-COMMUTATIVE RESIDUATED LATTICES*

BY

R. P. DILWORTH

Introduction and summary. In the theory of non-commutative rings certain distinguished subrings, one-sided and two-sided ideals, play the important roles. Ideals combine under crosscut, union and multiplication and hence are an instance of a lattice over which a non-commutative multiplication is defined.† The investigation of such lattices was begun by W. Krull (Krull [3]) who discussed decomposition into isolated component ideals. Our aim in this paper differs from that of Krull in that we shall be particularly interested in the lattice structure of these domains although certain related arithmetical questions are discussed.

In Part I the properties of non-commutative multiplication and residuation over a lattice are developed. In particular it is shown that under certain general conditions each operation may be defined in terms of the other.

The second division of the paper deals with the structure of non-commutative residuated lattices in the vicinity of the unit element. It is found that this structure may be characterized to a large extent in terms of special types of distributive lattices (arithmetical and semi-arithmetical lattices). The next division contains a discussion of the arithmetical properties of non-commutative residuated lattices. In particular decompositions into primary and semi-primary elements are discussed.

Finally we investigate the case where both the ascending and descending chain conditions hold and prove some structure theorems which are analogous to the structure theorems of hypercomplex systems.

I. Multiplication and residuation

1. **Definitions and notations.** The fixed lattice of elements $a, b, c, \cdots$ will will be denoted by $\mathfrak{S}$. Sublattices will be denoted by German capitals, and Latin capitals will denote subsets of $\mathfrak{S}$ which are not necessarily sublattices. $(,)$, $[,]$, $\supset$ will denote union, crosscut, and lattice division respectively. If $a \neq b$ and $a \supset x \supset b$ implies either $x = a$ or $x = b$, a is said to *cover* b and we write

---

* Presented to the Society in two parts: April 9, 1938, under the title *Non-commutative residuation*, and November 26, 1938, under the title *Archimedian residuated lattices*; received by the editors May 1, 1939.

† Lattices with a commutative multiplication have been investigated by Professor Morgan Ward and the author in a previous paper (Ward-Dilworth [7]).

426

$a > b$. If $\mathfrak{S}$ has a unit element u, the elements covered by u are called *divisor-free* elements of $\mathfrak{S}$. If $\mathfrak{S}$ has a null element it will be denoted by z.

$\mathfrak{S}$ is said to satisfy the ascending chain condition if every chain $a_1 \subset a_2 \subset a_3 \subset \cdots$ has only a finite number of distinct elements. Similarly if every descending chain $a_1 \supset a_2 \supset a_3 \supset \cdots$ has only a finite number of distinct elements, $\mathfrak{S}$ is said to satisfy the descending chain condition. $\mathfrak{S}$ is called *archimedian* if both the ascending and descending chain conditions hold.

The direct product (Birkhoff [1]) of lattices $\mathfrak{L}_1, \mathfrak{L}_2, \cdots, \mathfrak{L}_n$ is defined to be the set of vectors $a = \{a_1, a_2, \cdots, a_n\}$, $a_i \, \varepsilon \, \mathfrak{L}_i$ with division defined by $a \supset b$ if and only if $a_i \supset b_i$. Union and crosscut are given by $(a, b) = \{(a_1, b_1), \cdots, (a_n, b_n)\}$, $[a, b] = \{[a_1, b_1], \cdots, [a_n, b_n]\}$.

2. **Multiplication.** A one-valued, binary operation xy is called a *multiplication* over $\mathfrak{S}$ if the following postulates are satisfied:

M_1. *ab lies in $\mathfrak{S}$ whenever a and b lie in $\mathfrak{S}$.*
M_2. *$a = b$ implies $ac = bc$, $ca = cb$.*
M_3. *$a(b, c) = (ab, ac)$, $(a, b)c = (ac, bc)$.*
M_4. *$a(bc) = (ab)c$.*

From M_2 and M_3 we have
(2.1) $a \supset b$ implies $ac \supset bc$ and $ca \supset cb$;
(2.2) $[ab, ac] \supset a[b, c]$, $[ac, bc] \supset [a, b]c$.

If in addition to M_1–M_4, postulate M_5 below is satisfied, $\mathfrak{S}$ is said to be a *left ideal* lattice.

M_5. *$a \supset ba$.*

In a similar manner if $M_{5'}$ is satisfied, $\mathfrak{S}$ is said to be a *right ideal* lattice.

$M_{5'}$. *$a \supset ab$.*

If a lattice is both a left and right ideal lattice, it is called a *two-sided ideal* lattice, or simply *ideal* lattice.

Consider a lattice with unit element u over which a multiplication satisfying M_1–M_4 is defined and for which M_6 holds.

M_6. *$ua = au = a$.*

Then by M_3, M_5 and $M_{5'}$ hold so that $\mathfrak{S}$ is an ideal lattice. A lattice with unit element in which M_6 holds we call an *ideal lattice with unit.*

$\mathfrak{S}$ is said to be commutative if it satisfies M_7.

M_7. *$ab = ba$.*

3. **Residuation.** Consider now an ideal lattice $\mathfrak{S}$ in which the ascending

chain condition* holds. Let a and b be two elements of $\mathfrak{S}$. Then the set X of all elements $x \, \varepsilon \, \mathfrak{S}$ such that $a \supset xb$ is non-empty and closed with respect to union. Hence by the ascending chain condition X has a unit element $a \cdot b^{-1}$ which we call the *left residual* of b with respect to a. The left residual $a \cdot b^{-1}$ has the fundamental properties:

R_1. $a \supset (a \cdot b^{-1})b$.

R_2. $a \supset xb \rightarrow a \cdot b^{-1} \supset x$.†

In a similar manner the *right residual* $b^{-1} \cdot a$ is defined by the following properties:

$R_{1'}$. $a \supset b(b^{-1} \cdot a)$.

$R_{2'}$. $a \supset bx \rightarrow b^{-1} \cdot a \supset x$.

The two residuals are connected by the relation

$$(3.1) \qquad a^{-1} \cdot (b \cdot c^{-1}) = (a^{-1} \cdot b) \cdot c^{-1}.$$

The residuals are connected with the multiplication by the formulas

$$(3.2) \qquad (ab) \cdot b^{-1} \supset a, \qquad a^{-1} \cdot (ab) \supset b,$$

$$(3.3) \qquad a \cdot (bc)^{-1} = (a \cdot c^{-1}) \cdot b^{-1}, \qquad (ab)^{-1} \cdot c = b^{-1} \cdot (a^{-1} \cdot c).$$

Some of the more important properties of the residuals are the following:

$$(3.4) \qquad a \cdot (b^{-1} \cdot a)^{-1} \supset (a, b), \qquad (a \cdot b^{-1})^{-1} \cdot a \supset (a, b);$$

$$(3.5) \qquad [a, b] \cdot c^{-1} = [a \cdot c^{-1}, b \cdot c^{-1}], \qquad a^{-1} \cdot [b, c] = [a^{-1} \cdot b, a^{-1} \cdot c];$$

$$(3.6) \qquad a \cdot (b, c)^{-1} = [a \cdot b^{-1}, a \cdot c^{-1}], \qquad (a, b)^{-1} \cdot c = [a^{-1} \cdot c, b^{-1} \cdot c];$$

$$(3.7) \qquad (a, b) \cdot c^{-1} \supset (a \cdot c^{-1}, b \cdot c^{-1}), \qquad a^{-1} \cdot (b, c) \supset (a^{-1} \cdot b, a^{-1} \cdot c);$$

$$(3.8) \qquad a \supset b \rightarrow a \cdot c^{-1} \supset b \cdot c^{-1}, \; c^{-1} \cdot a \supset c^{-1} \cdot b;$$

$$(3.9) \qquad a \supset b \rightarrow c \cdot b^{-1} \supset c \cdot a^{-1}, \; b^{-1} \cdot c \supset a^{-1} \cdot c;$$

$$(3.10) \qquad a \cdot b^{-1} \supset a, \; b^{-1} \cdot a \supset a;$$

$$(3.11) \qquad a \cdot b^{-1} \supset c \rightleftarrows c^{-1} \cdot a \supset b.$$

On the other hand, if we start with a lattice $\mathfrak{S}$ in which the descending chain condition‡ holds and over which left and right residuals are defined having the properties given above, then we may define a multiplication over $\mathfrak{S}$ satisfying M_1–$M_{5'}$. For let a and b be two elements of $\mathfrak{S}$ and let X be the

---

* This condition may be replaced by the weaker condition that every set S of elements of $\mathfrak{S}$ have a union $u(S)$ and that $u(S)c = u(Sc)$.

† The symbol $\rightarrow$ indicates formal implication.

‡ As in the previous case this condition may be weakened.

set of elements x such that $x \cdot b^{-1} \supset a$. Then X is non-empty and closed with respect to crosscut, and hence by the descending chain condition has a null element ab. It can be shown that the product so defined satisfies M_1–$M_{5'}$ and moreover is equal to the product similarly defined in terms of the right residual.

II. Residuated lattices with unit

4. Lattice structure. Throughout this and the following section we shall assume that $\mathfrak{S}$ is a lattice in which the ascending chain condition holds and having a multiplication satisfying $M_1, \cdots, M_6$. As a consequence of M_6 the residuals have the following properties:

$$(4.1) \qquad a \supset b \rightleftarrows a \cdot b^{-1} = u \rightleftarrows b^{-1} \cdot a = u;$$

$$(4.2) \qquad a \cdot u^{-1} = u^{-1} \cdot a = a;$$

$$(4.3) \qquad (a, b) = u \rightarrow a \cdot b^{-1} = a, \; b^{-1} \cdot a = a.$$

Conversely, if we start with residuals having property (4.1) and define multiplication in terms of the residuals as in §3, then it is readily verified that the multiplication satisfies M_6.

Of particular importance in the proofs that follow are the properties:

$$(4.4) \qquad (b, c) = u \rightarrow (a, [b, c]) = [(a, b), (a, c)];$$

$$(4.5) \qquad (b, c) = u \rightarrow ([a, b], [a, c]);$$

$$(4.6) \qquad (a, b) = u, \; (a, c) = u \rightarrow (a, [b, c]) = u.$$

As a consequence of (4.4) and (4.6) we have the following property:
(4.7) If $a_1, \cdots, a_n$ are coprime in pairs, then

$$(c, [a_1, \cdots, a_n]) = [(c, a_1), \cdots, (c, a_n)].$$

Two sublattices $\mathfrak{A}$ and $\mathfrak{B}$ are said to be *coprime* if $a \, \varepsilon \, \mathfrak{A}$ and $b \, \varepsilon \, \mathfrak{B}$ imply $(a, b) = u$. We have then

LEMMA 4.1. *Let $\mathfrak{A}$ be the sublattice generated by the sublattices $\mathfrak{A}_1, \mathfrak{A}_2, \cdots, \mathfrak{A}_n$ each of which contains u. Then $\mathfrak{A}$ is the direct product of $\mathfrak{A}_1, \cdots, \mathfrak{A}_n$ if and only if $\mathfrak{A}_1, \cdots, \mathfrak{A}_n$ are coprime in pairs.*

From the definitions of §1 it follows directly that $\mathfrak{A}_1, \cdots, \mathfrak{A}_n$ are coprime in pairs if $\mathfrak{A}$ is the direct product of $\mathfrak{A}_1, \cdots, \mathfrak{A}_n$. Let now $\mathfrak{A}_1, \cdots, \mathfrak{A}_n$ be coprime in pairs and let L denote the set of crosscuts $[a_1, \cdots, a_n]$ where $a_i \, \varepsilon \, \mathfrak{A}_i$. We have clearly

$$[[a_1, \cdots, a_n], [a_1', \cdots, a_n']] = [[a_1, a_1'], [a_2, a_2'], \cdots, [a_n, a_n']].$$

Furthermore

$$([a_1, \cdots, a_n], [a_1', \cdots, a_n']) = [(a_1, [a_1', \cdots, a_n']), \cdots, (a_n, [a_1', \cdots, a_n'])]$$
$$= [(a_1, a_1'), \cdots, (a_n, a_n')]$$

by (4.7). Hence L is a sublattice and is thus equal to $\mathfrak{A}$. If $[a_1, \cdots, a_n] = [a_1', \cdots, a_n']$, then

$$a_i = (a_i, [a_1', \cdots, a_n']) = [(a_i, a_1'), \cdots, (a_i, a_n')] = (a_i, a_i').$$

Whence $a_i \supset a_i'$. Similarly $a_i' \supset a_i$ and hence $a_i = a_i'$. This completes the proof.

If the sublattices $\mathfrak{A}_1, \cdots, \mathfrak{A}_n$ have minimal elements, the conditions of Lemma 4.1 may be simplified.

COROLLARY. *If the sublattices $\mathfrak{A}_1, \cdots, \mathfrak{A}_n$ of Lemma 4.1 have minimal elements $m_1, \cdots, m_n$, then $\mathfrak{A}$ is the direct product of $\mathfrak{A}_1, \cdots, \mathfrak{A}_n$ if and only if $m_1, \cdots, m_n$ are coprime in pairs.*

From Lemma 4.1 we have immediately

LEMMA 4.2. *Any finite set of divisor-free elements generates a finite Boolean algebra.*

If there are only a finite number of divisor-free elements in $\mathfrak{S}$, we may speak of *the* Boolean algebra generated by the divisor-free elements. This is certainly the case when the descending chain condition holds in $\mathfrak{S}$, for we have

LEMMA 4.3. *If the descending chain condition holds in $\mathfrak{S}$, then there are only a finite number of divisor-free elements.*

Let $p_1, p_2, \cdots, p_n, \cdots$ be an infinite sequence of distinct divisor-free elements, and form the descending chain $a_1 \supset a_2 \supset a_3 \supset \cdots$ where $a_i = [p_1, p_2, \cdots, p_i]$. If $a_i = a_i + 1$, then $[p_1, \cdots, p_i] = [p_1, \cdots, p_{i+1}]$ and hence

$$p_{i+1} = (p_{i+1}, [p_1, \cdots, p_i]) = [(p_{i+1}, p_1), \cdots, (p_{i+1}, p_i)] = u,$$

which is impossible. Thus $a_1 \supset a_2 \supset a_3 \supset \cdots$ is an infinite descending chain.

5. We turn now to the study of the structure of a residuated lattice in the vicinity of the unit element and prove first the fundamental

THEOREM 5.1. *Let $\mathfrak{S}$ be a residuated lattice with unit having only a finite number of divisor-free elements $p_1, p_2, \cdots, p_n$. Moreover let $\mathfrak{L}$ be the direct product of chain lattices $\mathfrak{L}_1, \cdots, \mathfrak{L}_n$ where $\mathfrak{L}_i$ is the chain $u \supset p_i \supset a_i \supset \cdots \supset m_i$. Then if $m_k > b$ and b does not belong to $\mathfrak{L}$, the sublattice generated by the elements of $\mathfrak{L}$ and the element b is the direct product of the chain lattices $\mathfrak{L}_1, \cdots, \mathfrak{L}_k', \cdots, \mathfrak{L}_n$ where $\mathfrak{L}_k'$ is the chain lattice $u \supset p_k \supset a_k \supset \cdots \supset m_k \supset b$.*

Proof. In view of the corollary to Lemma 4.1 it is sufficient to show that $(b, m_i) = u, i \neq k$. If $(b, m_i) \neq u$, there exists a divisor-free element p such that

$p \supset (b, m_i)$. Since $p \supset m_i$, we have $p = p_i$. Now $m_k \supset [m_k, p_i] \supset b$ since $p \supset b$. But $m_k \neq [m_k, p_i]$ since otherwise $p_i \supset m_k$ while $(p_i, m_k) = u$. Hence $b = [m_k, b_i]$ and b is contained in $\mathfrak{L}$ which is contrary to assumption. Thus $(b, m_i) = u$, $i \neq k$.

This theorem enables us to construct certain characteristic sublattices with very simple properties. For let $\mathfrak{B}$ be the Boolean algebra generated by the divisor-free elements of $\mathfrak{S}$. If a divisor-free element p of $\mathfrak{B}$ covers an element a_1, not contained in $\mathfrak{B}$, then $\mathfrak{B}$ and a_1 generate a sublattice $\mathfrak{L}_1$ which is a direct product of chain lattices. If a_1 covers a_2 and a_2 does not belong to $\mathfrak{L}_1$, then $\mathfrak{L}_1$ and a_2 generate a sublattice $\mathfrak{L}_2$ which is again a direct product of chain lattices. We may continue in this manner as long as we obtain elements a_i not contained in $\mathfrak{L}_{i-1}$. Having obtained a sublattice $\mathfrak{L}_k$ in this manner, we may further extend it by building chains from other divisor-free elements. Thus if we call lattices which are direct products of chain lattices, *arithmetical* (Ward [5]), we see that the structure of a residuated lattice in the vicinity of the unit element is characterized to a large extent in terms of arithmetical lattices.

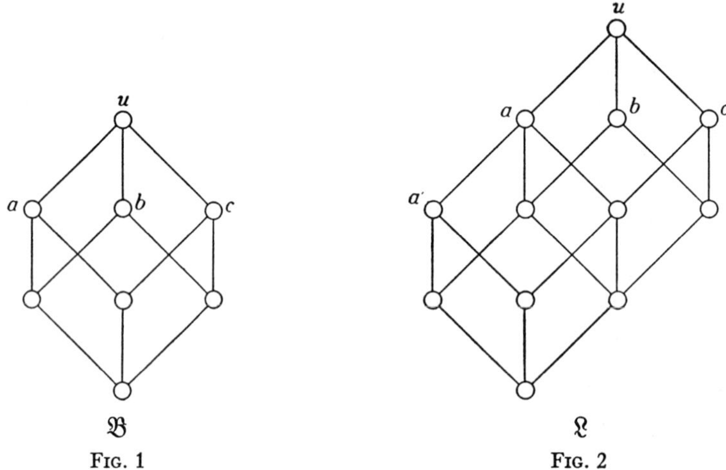

FIG. 1   FIG. 2

This principle is very useful in constructing examples of residuated lattices. For example, suppose we wish to construct a residuated lattice containing three divisor-free elements. We start then with the Boolean algebra $\mathfrak{B}$ of Fig. 1.

Now if we wish to add an element a' covered by a, by Theorem 5.1 we have immediately the sublattice $\mathfrak{L}$ of Fig. 2.

The condition of Theorem 5.1 that each divisor-free element be a member of one of the chain lattices is essential for the truth of the theorem as may be seen by simple examples. However in general a residuated lattice will have an infinite number of divisor-free elements and Theorem 5.1 will no longer apply. It may be generalized as follows:

THEOREM 5.2. *Let $\mathfrak{L}$ be the direct product of chain lattices $\mathfrak{L}_1, \cdots, \mathfrak{L}_n$ of a residuated lattice $\mathfrak{S}$, and let $\mathfrak{B}$ be the lattice generated by $\mathfrak{L}$ and the set of divisor-free elements p which divide at least one element of $\mathfrak{L}$. Furthermore let $m_k > b$. Then either b lies in $\mathfrak{B}$ or the lattice generated by $\mathfrak{L}$ and b is the direct product of the chain lattices $\mathfrak{L}_1, \cdots, \mathfrak{L}_k', \cdots, \mathfrak{L}_n$ where $\mathfrak{L}_k' = \{\mathfrak{L}_k, b\}$.*

Proof. If $(b, m_i) \neq u$, $i \neq k$, there exists a divisor-free element p such that $p \supset (b, m_i)$. Now $m_k \supset [m_k, p] \supset b$ and $m_k \neq [m_k, p]$ since otherwise $p \supset m_k$ while $(p, m_k) = u$. Hence $b = [m_k, b]$ and $b \, \varepsilon \, \mathfrak{B}$. Hence if $b \, \epsilon \, \mathfrak{B}$, $(b, m_i) = u$, $i \neq k$, and the theorem follows by Lemma 4.

The structure of the lattice $\mathfrak{B}$ of Theorem 5.2 is comparatively simple. We shall study its properties in terms of the notion of *semi-arithmetical* lattices introduced by Morgan Ward (Ward [5]). We make the

DEFINITION 5.1. *A distributive lattice $\mathfrak{D}$ is said to be semi-arithmetical if the indecomposable elements divisible by a given divisor-free element form a chain lattice.*

A semi-arithmetical lattice in which the ascending chain condition holds may be characterized as follows:

LEMMA 5.1. *A distributive lattice $\mathfrak{D}$ in which the ascending chain condition holds is semi-arithmetical if and only if the indecomposables occurring in the reduced representation of an element as a crosscut of indecomposables are coprime in pairs.*

From Definition 5.1 it follows trivially that an arithmetical lattice is semi-arithmetical.

We shall show now that the lattice $\mathfrak{B}$ of Theorem 5.2 is semi-arithmetical and to that end prove the

THEOREM 5.3. *Let $\mathfrak{L}$ be a semi-arithmetical sublattice of a residuated lattice $\mathfrak{S}$ and let $\mathfrak{L}$ contain the unit element u. Then if p is a divisor-free element of $\mathfrak{S}$, the sublattice $\mathfrak{L}'$ generated by p and the sublattice $\mathfrak{L}$ is semi-arithmetical.*

Proof. If p is contained in $\mathfrak{L}$, the theorem is trivial and we may thus assume that $p \, \epsilon \, \mathfrak{L}$. Now let U be the set of all elements of the form a or $[p, a]$ where $a \, \varepsilon \, \mathfrak{L}$. The set U is clearly closed with respect to crosscut. We show that U is also closed with respect to union. Let x and y be two members of the set U. If both x and y are contained in $\mathfrak{L}$, (x, y) is obviously in U. Let $x = [p, x_1]$, $p \, \not\supset x_1$ and $y \, \varepsilon \, \mathfrak{L}$. Let $x_1 = [q_1, \cdots, q_s]$ where the q_i are indecomposables and $(q_i, q_j) = u$, $i \neq j$. Then since $p \, \not\supset x_1$, $(p, q_i) = u$ $(i = 1, \cdots, s)$. Hence

$$(x, y) = (y, [p, q_1, \cdots, q_s]) = [(y, p), (y, q_1), \cdots, (y, q_s)]$$

by (4.3). But (y, p) is either p or u hence (x, y) is contained in U. If $x = [p, x_1]$, $p \not\supset x_1$ and $y = [p, y_1]$, $p \not\supset y_1$, then

$$(x, y) = ([p, q_1, \cdots, q_s], [p, q_1', \cdots, q_{s'}'])$$
$$= [p, (q_1, p), \cdots, (q_s, p), \cdots, (q_{s'}', p), (q_1, q_1'), \cdots, (q_s, q_{s'}')] = [p, a]$$

where $a \; \varepsilon \; \mathfrak{L}$. Hence U is identical with $\mathfrak{L}'$.

Now let a, b, c be contained in U. Then in. exactly the same manner as above we find that $(a, [b, c] = [(a, b), (a, c)]$. For example, if $b = [p, b_1]$, $p \not\supset b_1$ and $c \; \varepsilon \; \mathfrak{L}$, then

$$(a, [b, c]) = (a, [p, q_1, \cdots, q_s, q_1', \cdots, q_{s'}'])$$
$$= [(a, p), (a, q_1), \cdots, (a, q_{s'}')] = [(a, b), (a, c)]$$

if $p \not\supset c$; and if $p \supset c$, then

$$(a, [b, c]) = (a, [q_1, \cdots, q_s, q_1', \cdots, q_{s'}'])$$
$$= [(a, q_1), \cdots, (a, q_s), (a, q_1'), \cdots, (a, q_{s'}')]$$
$$= [(a, q_1), \cdots, (a, q_s), (a, c)] = [(a, p), (a, q_1), \cdots, (a, q_s), (a, c)]$$
$$= [(a, b), (a, c)].$$

Hence $\mathfrak{L}'$ is distributive.

Finally let $x \; \varepsilon \; \mathfrak{L}'$; then either $x \; \varepsilon \; \mathfrak{L}$ or $x = [p, x_1]$ where $p \not\supset x_1$. If $x \; \varepsilon \; \mathfrak{L}$, then $x = [q_1, \cdots, q_r]$ where the q_i are indecomposable and $(q_i, q_j) = u$, $i \neq j$. If $x = [p, x_1]$ then $x = [p, q_1, \cdots, q_r]$ where $p, q_1, \cdots, q_r$ are indecomposable and $(q_i, q_j) = u$, $i \neq j$; $(p, q_i) = u$ $(i = 1, \cdots, r)$. Thus $\mathfrak{L}'$ is semi-arithmetical by Lemma 5.1 and the proof is complete.

Now since $\mathfrak{B}$ is obtained from an arithmetical lattice $\mathfrak{L}$ by a successive adjunction of divisor-free elements and since at each stage a semi-arithmetical sublattice is obtained, $\mathfrak{B}$ itself is semi-arithmetical. We have thus proved

THEOREM 5.4. *The lattice $\mathfrak{B}$ of Theorem 5.2 is a semi-arithmetical sublattice of $\mathfrak{S}$.*

In forming the sublattice $\mathfrak{B}$ from the arithmetical lattice $\mathfrak{L}$ only divisor-free elements which are divisors of some element of $\mathfrak{L}$ are considered. If we adjoin a divisor-free element which does not divide any of the elements of $\mathfrak{L}$, the results are even simpler; for we have

THEOREM 5.5. *Let $\mathfrak{L}$ be a direct product of the chain lattices $\mathfrak{L}_1, \cdots, \mathfrak{L}_n$, and let p denote a divisor-free element not contained in $\mathfrak{L}$. Then if p does not divide any of the elements of $\mathfrak{L}$, the sublattice generated by p and $\mathfrak{L}$ is the direct product $\mathfrak{L}'$*

of the chain $\{u, p\}$ and the chain lattices of $\mathfrak{L}$. Furthermore if $\mathfrak{L}$ is dense in $\mathfrak{S}$, then $\mathfrak{L}'$ is dense in $\mathfrak{S}$.

Proof. Since p does not divide a_i if $a_i \, \varepsilon \, \mathfrak{L}_i$, $(a^i, p) = u$. Hence the first part of the theorem follows. Let now $x \supset [p, a_1, \cdots, a_n]$. Then $x = [(x, p), (x, a_1), \cdots, (x, a_n)]$. Now (x, p) is clearly in $\mathfrak{L}'$ and (x, a_i) is in $\mathfrak{L}$ by hypothesis. Hence $x \, \varepsilon \, \mathfrak{L}'$.

We conclude this section with

THEOREM 5.6. *Let $\mathfrak{L}$ be the direct product of the chain lattices $\mathfrak{L}_1, \cdots, \mathfrak{L}_n$ of a residuated lattice $\mathfrak{S}$ and let $m_k > b$ where b is indecomposable. Then $\mathfrak{L}$ and b generate a sublattice $\mathfrak{L}'$ which is the direct product of the chain lattices $\mathfrak{L}_1, \cdots, \{\mathfrak{L}_k, b\}, \cdots, \mathfrak{L}_n$. Furthermore if $\mathfrak{L}$ is dense in $\mathfrak{S}$, then $\mathfrak{L}'$ is dense in $\mathfrak{S}$.*

Proof. The first part follows directly from Theorem 5.2. Let now $x \supset [m_1, m_2, \cdots, b, \cdots, m_n]$. Then $x = [(x, m_1), \cdots, (x, b), \cdots, (x, m_n)]$. Since $\mathfrak{L}$ is dense by hypothesis, $(x, m_1), \cdots, (x, m_n)$ are contained in $\mathfrak{L}$. Now either $(x, b) = b$ in which case $x \, \varepsilon \, \mathfrak{L}'$ or $(x, b) \supset m_i$ since b is indecomposable. But then $(x, b) \, \varepsilon \, \mathfrak{L}$ and x is contained in $\mathfrak{L}'$.

III. ARITHMETICAL PROPERTIES OF IDEAL LATTICES

6. Assume that $\mathfrak{S}$ is an ideal lattice in which the ascending chain condition holds.

DEFINITION 6.1. *An element $p \, \varepsilon \, \mathfrak{S}$ is said to be a prime if $p \supset ab$ and $p \not\supset a$ implies $p \supset b$.*

DEFINITION 6.2. *An element $q \, \varepsilon \, \mathfrak{S}$ is said to be right primary if $q \supset ab$ and $q \not\supset a$ implies $q \supset b^s$ for some whole number s.*

In the theory of commutative residuated lattices a residuated lattice in which the ascending chain condition holds is said to be a *Noether* lattice (Ward-Dilworth [7]) if every irreducible is primary. It is then shown that every element of a Noether lattice may be represented as a simple* crosscut of a finite number of primaries each of which is associated with a different prime. The primes themselves and the total number of primaries are uniquely determined by the element. This result also holds for the non-commutative case although there are certain complications due to the non-commutativity of the multiplication. We shall show how these complications may be avoided.

Let $\mathfrak{S}$ be a non-commutative Noether lattice; that is, assume that every irreducible is right primary. If a and b are elements of $\mathfrak{S}$, the product ab then has the form $ab = [q_1, \cdots, q_r]$ where the q_i are right primary. Let $q_i \supset a$

---

* A crosscut representation is said to be simple if omitting any one of the terms changes the representation.

$(i=1, \cdots, l)$, $q_i \not\supset a$ $(i=l+1, \cdots, r)$. Then since $q_i \supset ab$ we have $q_i \supset b^{s_i}$ $(i=l+1, \cdots, r)$. If we then set $s = \max(s_{l+1}, \cdots, s_r)$, we have

$$(6.1) \qquad\qquad ab \supset [a, b^s] \supset b^s a.$$

Let q be right primary and consider the union p of all elements x such that $q \supset x^s$ for some whole number s. Then $q \supset p^t$ for some whole number t by the ascending chain condition. Furthermore p is a prime. For if $p \supset ab$, then $q \supset p^t \supset (ab)^t \supset a^r b^t$ by (6.1). If $q \supset a^r$, then $p \supset a$. If $q \not\supset a^r$, then $q \supset b^{ts}$ and $p \supset b$. Hence either $p \supset a$ or $p \supset b$. This prime is clearly unique and is called the prime element associated with the right primary q. We have moreover

LEMMA 6.1. *The crosscut of two right primaries associated with the same prime p is also a right primary associated with p.*

Let $[q, q'] \supset ab$, $[q, q'] \not\supset a$. Then either q or q', say q, does not divide a and hence $q \supset b^s$. But then $p \supset b$ and hence $q' \supset b^t$. Hence $[q, q'] \supset b^{s'}$ where $s' = \max(s, t)$. Obviously $[q, q']$ is associated with p.

LEMMA 6.2. *Let q and q' be right primaries associated with p and p' respectively. Then if $p \not\supset p'$, $q \cdot q'^{-1} = q$.*

For $q \supset (q \cdot q'^{-1})q'$. Hence either $q = q \cdot q'^{-1}$ or $q \supset q'^s$. But if $q \supset q'^s$, then $p \supset p'^t$ and hence $p \supset p'$ contrary to hypothesis.

Note that Lemma 6.2 holds only for the right residual. If we were considering left primaries, the left residual would replace the right residual.

The proof from this point on is exactly analogous to the proof in classical ideal theory and will be omitted. We thus obtain

THEOREM 6.1. *Let $\mathfrak{S}$ be a non-commutative Noether lattice. Then every element of $\mathfrak{S}$ may be represented as a simple crosscut of a finite number of right primaries. The primes and the total number of right primaries are uniquely determined by the element.*

The following theorem proved in Ward-Dilworth [7] for the commutative case holds also for non-commutative residuated lattices and is proved in exactly the same manner.

THEOREM 6.2. *The following two conditions are sufficient that $\mathfrak{S}$ be a Noether lattice:*

(i) $\mathfrak{S}$ *is modular,*
(ii) $ab \supset [a, b^s]$.

The distinction between left and right primaries may be removed by weakening the condition of Definition 6.2. We adopt the name *semi-primaries* for these new elements.

DEFINITION 6.3. *An element $a \, \varepsilon \, \mathfrak{S}$ is said to be semi-primary if $a \supset bc$ and $a \not\supset b^s$ for all s implies $a \supset c^t$ for some whole number t.*

Let $\mathfrak{S}$ be an ideal lattice in which every element may be represented as a crosscut of a finite number of semi-primaries. Moreover let x and y be any two elements of $\mathfrak{S}$. Then $xy = [a_1, \cdots, a_r]$ where the a_i are semi-primary. Let $a_i \supset x^{s_i}$ for $i = 1, \cdots, l$ and $a_i \supset y^{t_i}$, $i = l+1, \cdots, r$. Then $xy \supset [x^s, y^t]$ where $s = \max (s_1, \cdots, s_l)$ and $t = \max (t_{l+1}, \cdots, t_r)$. We thus have

THEOREM 6.3. *If every element of a residuated lattice $\mathfrak{S}$ is expressible as a crosscut of a finite number of semi-primaries, then for every x and y in $\mathfrak{S}$, there exist whole numbers s and t such that*

$$(6.2) \qquad\qquad xy \supset [x^s, y^t].$$

If (6.2) holds in a residuated lattice, the semi-primary elements may be simply characterized as follows:

THEOREM 6.4. *Let $\mathfrak{S}$ be a residuated lattice in which (6.2) holds. Then an element a is semi-primary if and only if a prime p exists such that $p \supset a \supset p^s$ for some whole number s.*

Proof. Let a be semi-primary, and let p denote the union of all elements x such that $a \supset x^r$ for some r. Then $a \supset p^t$ for some t. Now let $p \supset xy$. Then $a \supset xy \supset x^m y^n$ for some integers m and n by (6.2). Hence $a \supset x^s$ for some s or $a \supset y^t$ for some t. Hence either $p \supset x$ or $p \supset y$. Clearly $p \supset a \supset p^s$ for some s.

Conversely let $p \supset a \supset p^s$ and suppose that $a \supset bc$. Then $p \supset bc$, and hence either $p \supset a$ or $p \supset b$. Hence either $a \supset b^s$ or $a \supset c^s$.

The converse to Theorem 6.3 does not hold in general. However under the assumption of the distributive law we have

THEOREM 6.5. *The following two conditions are sufficient that every element of a residuated lattice $\mathfrak{S}$ satisfying the ascending chain condition be expressible as a crosscut of a finite number of semi-primaries.*

   (i) *$\mathfrak{S}$ is distributive,*

   (ii) *$xy \supset [x^s, y^t]$ for suitable s and t.*

Every element of $\mathfrak{S}$ is clearly expressible as a crosscut of a finite number of indecomposables. Hence it is sufficient to show that every indecomposable is semi-primary. Let a be indecomposable, and let $a \supset bc$, $a \not\supset b^s$, for any s. Then $a \supset [b^s, c^t]$ by (ii). Hence $a = [(a, b^s), (a, c^t)]$ by (i). But $(a, b^s) \neq a$. Hence since a is indecomposable, $a = (a, c^t)$ and $a \supset c^t$.

The distributive condition is essential in Theorem 6.5 as is shown by the example in Fig. 3.

Let $\mathfrak{L}_a$ denote the sublattice $\{a', b'', a, c'', b''', z', c''', d', e', z\}$, $\mathfrak{L}_b$ the sublattice $\{d, b', b\}$, and $\mathfrak{L}_c$ the sublattice $\{e, c', c\}$. We define a multiplication over $\mathfrak{L}$ as follows: $u^2 = u$, $ux = b$ if $x \, \varepsilon \, \mathfrak{L}_b$, $ux = c$ if $x \, \varepsilon \, \mathfrak{L}_c$, $ux = z$ if $x \, \varepsilon \, \mathfrak{L}_a$. The product of any two elements in $\mathfrak{L}_b$ is b. The product of any two elements in $\mathfrak{L}_c$ is c. The product of any element of $\mathfrak{L}_b$ with an element of $\mathfrak{L}_c$ is z. The product of an element of $\mathfrak{L}$ with an element of $\mathfrak{L}_c$ is z. It is readily verified that the multiplication so defined satisfies $M_1, \cdots, M_{5'}$ and is also commutative. $\mathfrak{L}$ is

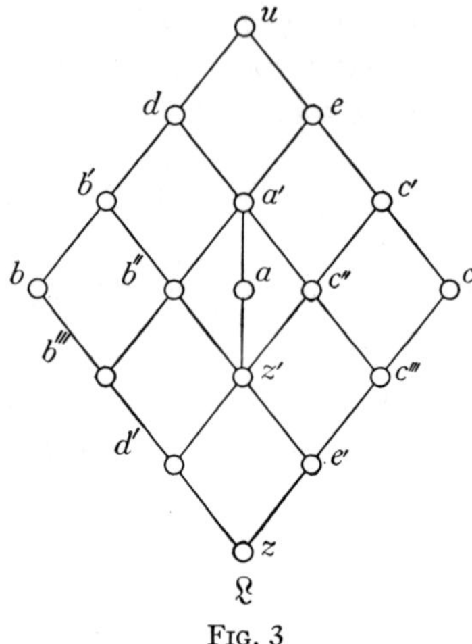

Fig. 3

clearly *not* distributive. It can also be verified that $xy \supset [x^s, y^t]$ for suitable s and t. However it is not true that $xy \supset [x, y^s]$ for some s, since $dc \not\supset [d, c^s]$. Furthermore a is indecomposable but *not* semi-primary since $a \supset bc$, but $a \not\supset b^s$ any s and $a \not\supset c^t$ any t.

7. **Ideal lattices with unit.** We turn now to the study of the properties of divisor-free elements in an ideal lattice with unit. We prove first the

LEMMA 7.1. *Let f be a divisor-free element of $\mathfrak{S}$, and let a be any element not divisible by f. Then one and only one of the following formulas holds:*
(1) $fa \supset af$,
(2) $fa = (fa) \cdot f^{-1}$.

We have $(fa \cdot f^{-1})^{-1} \cdot fa \supset f$ by (4.4). Hence either $(fa \cdot f^{-1})^{-1} \cdot fa = u$ or $(fa \cdot f^{-1})^{-1} \cdot fa = f$. In the first case $fa \supset fa \cdot f^{-1}$. But $fa \cdot f^{-1} \supset fa$ by (3.10). Hence $fa = fa \cdot f^{-1}$. If $(fa \cdot f^{-1})^{-1} \cdot fa = f$, then

$$f = f \cdot a^{-1} = ((fa \cdot f^{-1})^{-1} \cdot fa) \cdot a^{-1} = (fa \cdot f^{-1})^{-1} \cdot (fa \cdot a^{-1}) \supset (fa \cdot f^{-1})^{-1} \cdot f.$$

But $(fa \cdot f^{-1})^{-1} \cdot f \supset f$. Hence $(fa \cdot f^{-1})^{-1} \cdot f = f$. But then $f^{-1} \cdot (fa \cdot f^{-1}) = fa \cdot f^{-1}$ or $(f^{-1} \cdot fa) \cdot f^{-1} = fa \cdot f^{-1}$. Then $fa \cdot f^{-1} \supset a \cdot f^{-1} \supset a$. Hence $fa \supset (fa \cdot f^{-1})f \supset af$.

If both (1) and (2) hold, then $fa = fa \cdot f^{-1} \supset af \cdot f^{-1} \supset a$. But then $f = (f, fa) \supset (f, a) = u$, contrary to the assumption that f is a divisor-free element.

We clearly have a similar result for left residuals.

LEMMA 7.2. *Let f be a divisor-free element of a residuated lattice in which (6.2) holds. Then f commutes with every element which it does not divide.*

Let $a \, \varepsilon \, \mathfrak{S}$ such that $f \not\supset a$. Then by Lemma 7.1 either $fa \supset af$ or $fa = fa \cdot f^{-1}$. If $fa = fa \cdot f^{-1}$, then $fa \supset (a^s f^t) \cdot f^{-1} \supset a^s f^{t-1}$ by (6.2). But then $fa \supset fa \cdot f^{-1} \supset (a^s f^{t-1}) \cdot f^{-1} \supset a^s f^{t-2}$. Continuing in this manner we finally get $fa \supset a^s$. But then $f = (f, fa) \supset (f, a^s) = u$ since $f \not\supset a^s$. Hence $f = u$ which is contrary to our assumption that f is a divisor-free element. We thus have $fa \supset af$. In a similar manner using left residuals we get $af \supset fa$. Hence $af = fa$.

As a corollary to Lemma 7.2 the divisor-free elements in a residuated lattice for which (6.2) holds always commute. In particular we have from Theorem 6.3

LEMMA 7.3. *If in a residuated lattice every element is expressible as a crosscut of semi-primaries, then the divisor-free elements commute.*

Let $\mathfrak{S}$ be an arbitrary residuated lattice in which the ascending chain condition holds and denote by $\mathfrak{S}'$ the set of all elements x which divide a finite product of divisor-free elements. $\mathfrak{S}'$ is clearly closed under union, crosscut, multiplication and residuation and hence a residuated sublattice of $\mathfrak{S}$. Then

LEMMA 7.4. *Every prime in $\mathfrak{S}'$ is divisor-free.*

Let p be a prime in $\mathfrak{S}'$. Then by the definition of $\mathfrak{S}'$, $p \supset f_1 f_2 \cdots f_r$ where $f_1, f_2, \cdots, f_r$ are divisor-free elements of $\mathfrak{S}$. Hence $p = f_i$ for some i.

LEMMA 7.5. *Every element of $\mathfrak{S}'$ divides a finite product of its divisor-free divisors.*

This lemma follows directly from the following lemma due to Krull [3].

LEMMA 7.6. *Let $\mathfrak{S}$ be a non-commutative residuated lattice in which the ascending chain condition holds. Then each element $a \, \varepsilon \, \mathfrak{S}$ has only a finite number of minimal prime divisors $p_1, \cdots, p_n$ and a divides a power of $p_1 \cdots p_n$.*[*]

---

* Krull states this lemma for the more general case where the ascending chain condition is assumed only for prime elements while a residual chain condition holds for all elements. However his proof seems to be in error as he uses the following rule: If $a \supset a_1' a_2'$, then $a \supset a_1 a_2$ where $a_1 = a \cdot a_2'^{-1}$ and $a_2 = a_1'^{-1} \cdot a$. This rule is in general not correct as the following example shows: Let $\mathfrak{S}$ be the lattice defined by the covering relations $u > a > b > c > z$, $b > d > z$. The multiplication is defined by $ux = xu = x$, all $x \, \varepsilon \, \mathfrak{S}$; $a^2 = a$, and all other products are equal to z. Then $z \cdot c^{-1} = a$, $d^{-1} \cdot z = a$ and $z \supset cd$. However $z \not\supset (z \cdot c^{-1})(d^{-1} \cdot z) = a^2 = a$.

The lemma is readily seen to be correct under the assumption of the ascending chain condition since we may take $a_1 = (a, a_1')$ and $a_2 = (a, a_2')$ and the rule stated above holds.

A further consequence of Lemma 7.6 is the result that $\mathfrak{S}'$ is the maximal residuated sublattice all of whose prime elements are divisor-free.

In certain cases $\mathfrak{S}'$ is simply the Boolean algebra $\mathfrak{B}$ generated by the divisor-free elements. For example we have

THEOREM 7.1. *Let $\mathfrak{S}$ be a residuated lattice with only a finite number of divisor-free elements all of which commute among themselves. If the only elements covered by the divisor-free elements are elements of the Boolean algebra $\mathfrak{B}$ generated by them, then $\mathfrak{S}' = \mathfrak{B}$.*

Proof. Under the hypothesis of the theorem, $f^2 \subset [f, f']$ or $f^2 = f$. But if $[f, f'] \supset f^2$, then $f' \supset f$ which is impossible. Hence $f^2 = f$. But then $[f_1, f_2, \cdots, f_n] = f_1 f_2 \cdots f_n$ and $(f_1 \cdots f_n)^2 = f_1 \cdots f_n$.

If the divisor-free elements do not commute, the theorem does not hold in general. Consider the lattice $\mathfrak{L}$ defined by the covering relations $u > b > c > z$, $u > a > c$. The multiplication is given by $ux = xu = x$, $x \ \varepsilon \ \mathfrak{L}$, and $ab = c$, $ba = z$, $ac = ca = bc = cb = c^2 = z$, $a^2 = a$, $b^2 = b$, $zx = z$, all $x \ \varepsilon \ \mathfrak{L}$. Then $\mathfrak{S}' = \mathfrak{L}$ while $\mathfrak{B}$ is the sublattice $\{u, a, b, c\}$.

Applying Theorem 7.1 to hypercomplex systems we obtain

THEOREM 7.2. *A hypercomplex system in which the prime two-sided ideals are commutative is a direct sum of simple two-sided ideals if and only if each irreducible two-sided ideal which is not a prime has at least two prime ideal divisors.*

We conclude this section by giving a variation of a theorem due to Krull.*

THEOREM 7.3. *Each element of $\mathfrak{S}'$ is expressible as a crosscut of a finite number of semi-primaries if and only if the divisor-free elements commute.*

Proof. The second part follows from Lemma 7.3. To prove the first let $a = [a_1, \cdots, a_r]$ be the decomposition of a into coprime indecomposable elements. Then $a_i \supset f_1^{n_1} \cdots f_r^{n_r} = [f_1^{n_1}, \cdots, f_r^{n_r}]$ or $a_i = [(a_i, f_1^{n_1}), \cdots, (a_i, f_r^{n_r})]$ whence $a_i = (a_i, f_j^{n_j})$ for some j. We have then $f_j \supset a_i \supset f_j^{n_j}$. Let $a_i \supset bc$; then $f_j \supset bc$ and hence either $f_j \supset b$ or $f_j \supset c$. Hence either $a_i \supset b^{n_i}$ or $a_i \supset c^{n_i}$. Thus if the divisor-free elements of $\mathfrak{S}$ commute, each element of $\mathfrak{S}'$ may be uniquely represented as a crosscut of coprime semi-primary elements.

IV. ARCHIMEDEAN RESIDUATED LATTICES

8. Throughout this section unless the contrary is explicitly stated it will be assumed that $\mathfrak{S}$ is an ideal lattice in which the ascending and descending chain conditions hold. The unit element of $\mathfrak{S}$ need not be the unit of multiplication.

---

* Krull proves the theorem for "primary" elements where an element is primary if it has only one divisor-free divisor.

DEFINITION 8.1. *An element a of $\mathfrak{S}$ is said to be nilpotent if $a^s = z$ for some whole number s.*

LEMMA 8.1. *The union m of all nilpotent elements of $\mathfrak{S}$ is nilpotent. m is called the radical of $\mathfrak{S}$.*

If $a_1{}^{t_1} = z$ and $a_2{}^{t_2} = z$, then $(a_1, a_2)^t = z$ where $t = t_1 + t_2 - 1$. The result follows from the ascending chain condition.

DEFINITION 8.2. *An element s of $\mathfrak{S}$ is said to be simple if $s > z$ where z is the null element of $\mathfrak{S}$.*

LEMMA 8.2. *A necessary and sufficient condition that the radical be the null element is that each simple element be idempotent.*

Let $m = z$. If s is a simple element, since $s \supset s^2$, either $s = s^2$ or $s^2 = z$. But if $s^2 = z$, then $z \supset m \supset s$ contrary to Definition 8.2. Suppose now that each simple element is idempotent and let $m \neq z$. Then $m \supset s$ where s is simple, whence $z = m^t \supset s^t = s$, which contradicts the definition of s. Hence $m = z$.

DEFINITION 8.3. *If the radical is the null element, $\mathfrak{S}$ is said to be semisimple.*

LEMMA 8.3. *Let $\mathfrak{S}$ be semisimple and s be any simple element of $\mathfrak{S}$. Then $a \supset s \rightleftarrows as = sa = s$, $a \not\supset s \rightleftarrows as = sa = z$.*

Let $a \supset s$. Then $as \supset s^2 = s$ and hence $as = s$. Similarly $sa = s$. If $a \not\supset s$, then $[a, s] = z$ and hence $as = sa = z$.

The position of the radical in the lattice may have important bearing on the arithmetical properties of the lattice. For example, we have the following theorem:

THEOREM 8.1. *Let $\mathfrak{S}$ be an archimedean residuated lattice whose divisor-free elements generate a Boolean algebra with null element m. Then the divisor-free elements are the only primes of $\mathfrak{S}$.*

Proof. Since $\mathfrak{S}$ is archimedean there is only a finite number of divisor-free elements. Let p be a prime of $\mathfrak{S}$. Then $p \supset m^t \supset z$ and hence $p \supset m$. But $m = [f_1, \cdots, f_n]$ where $f_1, \cdots, f_n$ are the divisor-free elements of $\mathfrak{S}$. Hence $p \supset [f_1, \cdots, f_n]$ and hence $p = f_i$ for some i.

The conclusion of Theorem 8.1 may be stated in the form $\mathfrak{S} = \mathfrak{S}'$.

Let $\mathfrak{S}_m$ denote the sublattice of all elements x such that $x \supset m$. The study of the structure of $\mathfrak{S}_m$ may be reduced to the study of the structure of semisimple lattices. For since $\mathfrak{S}_m$ is dense in $\mathfrak{S}$ it is closed with respect to residuation and hence has a multiplication (§3). We call this multiplication the multiplication *in* $\mathfrak{S}_m$ and denote it by $a \cdot b$.

THEOREM 8.2. *Let $a, b \, \varepsilon \, \mathfrak{S}_m$. If $ab \, \varepsilon \, \mathfrak{S}_m$, then $ab = a \cdot b$.*

Proof. $a \cdot b$ is defined by

   (i) $(a \cdot b) \cdot b^{-1} \supset a$,

   (ii) $x \cdot b^{-1} \supset a$, $x \ \varepsilon \ \mathfrak{S}_m \rightarrow x \supset a \cdot b$.

Similarly ab is defined by

   (i′) $(ab) \cdot b^{-1} \supset a$,

   (ii′) $x \cdot b^{-1} \supset a$, $x \ \varepsilon \ \mathfrak{S} \rightarrow x \supset ab$.

Hence if $ab \ \varepsilon \ \mathfrak{S}_m$, then $ab \supset a \cdot b$ by (i′), (ii). On the other hand by (i), (ii′), $a \cdot b \supset ab$. Hence $a \cdot b = ab$.

In general we have

Lemma 8.4. $a \cdot b \supset ab$.

Let now p be a prime element of $\mathfrak{S}$. Then $p \supset m^t = z$ and hence $p \supset m$. Thus $p \ \varepsilon \ \mathfrak{S}_m$. Now let $p \supset a \cdot b$. Then $p \supset ab$ by Lemma 8.4. Hence either $p \supset a$ or $p \supset b$. We thus have

Theorem 8.3. *If p is a prime element of $\mathfrak{S}$, then $p \ \varepsilon \ \mathfrak{S}_m$ and p is a prime in $\mathfrak{S}_m$ with respect to the multiplication in $\mathfrak{S}_m$.*

Theorem 8.4. *$\mathfrak{S}_m$ is semisimple.*

Proof. Let s be a simple element of $\mathfrak{S}_m$. Then $s > m$. Now $s \supset s \cdot s$. Hence $s = s \cdot s$ or $s \cdot s = m$. But if $s \cdot s = m$, $m \supset s^2$ by Lemma 8.4 and hence $s^{2t} = z$. This contradicts the definition of m. Hence each simple element is idempotent and by Lemma 8.1 $\mathfrak{S}_m$ is semisimple.

The most important application of archimedean residuated lattices is in the theory of hypercomplex systems. More generally, let $\mathfrak{S}$ be the set of two-sided ideals of a non-commutative ring R in which the ascending and descending chain conditions hold for left ideals. Then m is the radical of R. Now the quotient ring R/m is isomorphic to $\mathfrak{S}_m$ and hence is semisimple by Theorem 8.4. However from a well known structure theorem, a semisimple ring is a direct sum of simple two-sided ideals. Its lattice of two-sided ideals is thus a Boolean algebra, and Theorem 8.1 gives

Theorem 8.5. *The only prime two-sided ideals in a hypercomplex system are the divisor-free ideals.*

9. Semisimple lattices. In this section we shall be particularly interested in the sublattices generated by the simple elements of a semisimple lattice $\mathfrak{S}$.

Lemma 9.1. *There are only a finite number of simple elements in a semisimple lattice $\mathfrak{S}$.*

Let $s_1, s_2, s_3, \cdots$ be an infinite sequence of simple elements. Consider the chain $a_1 \subset a_2 \subset a_3 \subset \cdots$ where $a_i = (s_1, s_2, \cdots, s_i)$. The members of this chain are distinct. For suppose that $a_i = a_{i+1}$; then $(s_1, \cdots, s_i) = (s_1, \cdots, s_{i+1}.)$

Hence we have

$$s_{i+1} = \overset{2}{s_{i+1}} = (s_1 s_{i+1}, s_2 s_{i+1}, \cdots, \overset{2}{s_{i+1}}) = (s_1, \cdots, s_{i+1}) s_{i+1}$$

$$= (s_1, \cdots, s_i) s_{i+1} = (s_1 s_{i+1}, \cdots, s_i s_{i+1}) = z.$$

This contradicts Definition 8.2. Hence $a_1 \subset a_2 \subset \cdots$ is an infinite ascending chain contradicting the ascending chain condition.

THEOREM 9.1. *Let $\mathfrak{S}$ be a semi-simple lattice. Then if each element of $\mathfrak{S}$ can be expressed as a union of simple elements, $\mathfrak{S}$ is a Boolean algebra.*

Proof. Let $a \,\varepsilon\, \mathfrak{S}$ have the representation

$$(9.1) \qquad\qquad a = (s_1, \cdots, s_n)$$

where $s_1, \cdots, s_k$ are distinct simple elements. The representation (9.1) is unique and $s_1, \cdots, s_k$ are the only simple elements which a divides. For let $a = (s_1, \cdots, s_k) = (s_1', \cdots, s_l')$. Multiplying by s_i' we have $s_i' = (s_1 s_i', \cdots, s_k s_i')$. Hence all of the products are null except one, say $s_i s_i'$. Then $s_i s_i' = s_i'$ and hence $s_i \supset s_i'$ by Lemma 8.3. Thus $s_i = s_i'$ and $k = l$. If $a \supset s$, where s is simple and not equal to any of $s_1, \cdots, s_k$, then $(s_1, s_2, \cdots, s_k) = (s_1, \cdots, s_k, s)$ contrary to the result we have just obtained.

We show now that the product of any two elements is equal to their crosscut.

We clearly have $[a, b] \supset ab$. Let $[a, b] = (s_1, \cdots, s_k)$. Then since $a, b \supset [a, b]$, $a = (s_1, s_2, \cdots, s_k, a')$ and $b = (s_1, \cdots, s_k, b')$. Hence

$$ab = (s_1, \cdots, s_k, a')(s_1, \cdots, s_k, b') = (s_1, \cdots, s_k, a'b') \supset [a, b].$$

Thus $[a, b] = ab$.

Since the product is distributive with respect to union, the crosscut must be distributive and hence $\mathfrak{S}$ is distributive. Furthermore $\mathfrak{S}$ is complemented. For let $a = (s_1, \cdots, s_k)$, $u = (s_1, \cdots, s_n)$ and define $a' = (s_{k+1}, \cdots, s_n)$. Then $(a, a') = u$ and $[a, a'] = aa' = (s_1, \cdots, s_n)(s_{k+1}, \cdots, s_n) = z$. Hence $\mathfrak{S}$ is a Boolean algebra.

In an arbitrary semisimple lattice, the set of elements which can be represented as a union of simple elements need not be closed with respect to crosscut as we shall show by an example. However, if we assume the modular* condition we have the following theorem.

THEOREM 9.2. *Let $\mathfrak{S}$ be a modular semisimple lattice. Then the simple elements of $\mathfrak{S}$ generate a Boolean algebra $\mathfrak{S}_B$. Moreover $\mathfrak{S}_B$ is dense in $\mathfrak{S}$.*

Proof. Let U be the set of all elements of $\mathfrak{S}$ which can be expressed as a

---

* For various statements of the modular axiom see Ore [4].

union of simple elements of $\mathfrak{S}$. The set U is obviously closed with respect to union. We shall show that U is dense in $\mathfrak{S}$ and hence closed with respect to crosscut. Let $(s_1, \cdots, s_n) \supset x$, and let $x \supset s_1, \cdots, s_l, x \not\supset s_{l+1}, \cdots, s_n$. Then $x = [x, (s_1, \cdots, s_n)] = (s_1, \cdots, s_l, [x, (s_{l+1}, \cdots, s_n)])$ by the modular condition. If $[x, (s_{l+1}, \cdots, s_n)] \neq z$, then there is a simple element s such that $[x, s_{l+1}, \cdots, s_n] \supset s$. But then $x \supset s$ and $(s_{l+1}, \cdots, s_n) \supset s$. Hence

$$s = s(s_{l+1}, \cdots, s_n) = (s_{l+1}s, \cdots, ss_n) = s_i s$$

by Lemma 8.3. Thus $s = s_i$ and $x \supset s_i$ contrary to assumption. Hence $[x, (s_{l+1}, \cdots, s_n)] = z$ and $x = (s_1, \cdots, s_l)$.

Since U is dense in $\mathfrak{S}$, it is closed with respect to multiplication and is clearly semisimple. Moreover every element of U can be expressed as a union of simple elements. Hence by Theorem 9.1, $U = \mathfrak{S}_B$ is a Boolean algebra.

To show the significance of the modular condition in the previous theorem we give an example of a non-modular semisimple lattice in Fig. 4.

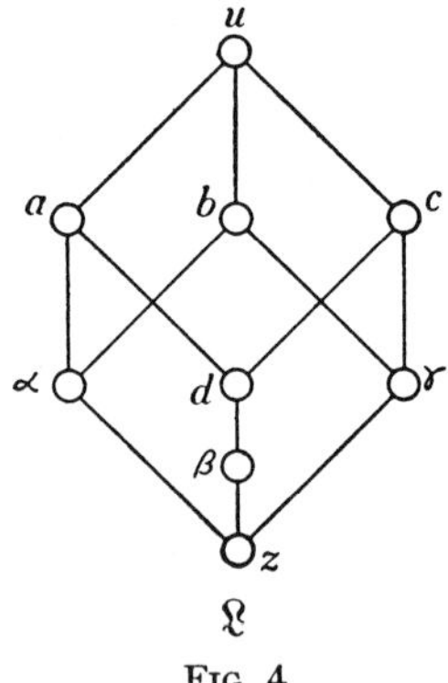

Fig. 4

If U denotes the set of elements of $\mathfrak{L}$ which can be expressed as a union of simple elements, we define a multiplication over $\mathfrak{L}$ as follows: If $x, y \, \varepsilon \, U$, $x \neq a$, $y \neq b$, then $xy = [x, y]$, $ac = \beta$, $dx = \beta$ or z according as $x \supset d$ or $x \not\supset d$. It can be readily verified that all of the multiplication postulates are satisfied. Also $\mathfrak{L}$ is non-modular since it contains the non-modular sublattice $\{a, \alpha, d, \beta, z\}$. The simple elements α, β, γ do not generate a Boolean algebra. In fact, U is not closed with respect to crosscut since $d = [(\alpha, \beta), (\beta, \gamma)]$.

THEOREM 9.3. *Let $\mathfrak{S}$ be a modular semisimple lattice. Then if for each simple element s there exists an element $s' \neq u$ such that $(s, s') = u$, $\mathfrak{S}$ is a Boolean algebra.*

Proof. We may take the s''s to be divisor-free elements since if s_i' is not divisor-free, there exists a divisor-free element f_i such that $f_i \supset s_i'$. But then

$(s_i, f_i) \supset (s_i, s_i') = u$. Let $v = (s_1, \cdots, s_n)$. Then the length of chain from v to z is n. But now $[s_1', s_2', \cdots, s_n'] = z$, since if $[s_1', \cdots, s_n'] \neq z$, there exists an s_i such that $[s_1', \cdots, s_n'] \supset s_i$. But then $s_i' \supset s_i$, which is impossible. Since $[s_1', \cdots, s_n'] = z$, the length of chain from u to z is equal to or less than n. But $u \supset v$. Hence $u = v$.

Theorem 9.3 gives immediately

THEOREM 9.4. *A complemented, modular, semisimple lattice is a Boolean algebra.*

We conclude with the statement of Theorem 9.3 in terms of the two-sided ideals of a non-commutative ring.

THEOREM 9.5. *Let R be a ring without radical in which the ascending and descending chain conditions hold for two-sided ideals. Then if for each two-sided ideal $\mathfrak{a}$ there exists an ideal $\mathfrak{a}' \neq R$ such that $(\mathfrak{a}, \mathfrak{a}') = R$, R is a direct sum of two-sided simple ideals.*

Such an ideal $\mathfrak{a}'$ always exists if $\mathfrak{a}$ has a principle unit. For in that case we may take $\mathfrak{a}'$ to be the set of all elements x such that $\mathfrak{a}x = 0$.

REFERENCES

1. G. Birkhoff, Bulletin of the American Mathematical Society, vol. 40 (1934), pp. 613–619.
2. R. P. Dilworth, Bulletin of the American Mathematical Society, vol. 44 (1938), pp. 262–267.
3. W. Krull, Mathematische Zeitschrift, vol. 28 (1928), pp. 481–503.
4. O. Ore. Annals of Mathematics, (2), vol. 36 (1935), pp. 406–432.
5. M Ward, Annals of Mathematics, (2), vol. 39 (1938), pp. 558–568.
6. M. Ward and R. P. Dilworth, Proceedings of the National Academy of Sciences, vol. 24 (1938), pp. 162–164.
7. ———— , these Transactions, vol. 45 (1939), pp. 335–354.

CALIFORNIA INSTITUTE OF TECHNOLOGY,
    PASADENA, CALIF.

Reprinted from DUKE MATHEMATICAL JOURNAL
Vol. 5, No. 2, June, 1939

NON-COMMUTATIVE ARITHMETIC

BY R. P. DILWORTH

1. Introduction and summary. The problem of determining the conditions that must be imposed upon a system having a single associative and commutative operation in order to obtain unique factorization into irreducibles has been studied by A. H. Clifford [1],[1] König [1], and Ward [2]. The more general problem of determining similar conditions for the non-commutative case has been treated by M. Ward [1]. However, the conditions given by Ward are more stringent than those satisfied by actual instances of non-commutative arithmetic, for example, quotient lattices and non-commutative polynomial theory (Ore [1, 2]). Moreover, in both of these instances the factorization is unique only up to a similarity relation, and instead of a single operation of multiplication the additional operations G. C. D. and L. C. M. are involved.[2] Accordingly, we shall concern ourselves with the arithmetic of a non-commutative multiplication defined over a lattice.

As the decomposition of lattice quotients gives an important instance of non-commutative arithmetic, we shall summarize here a few of the fundamental ideas of Ore's theory (Ore [1]). Let Σ be the set of quotients[3]

$$\alpha = \frac{a_1}{a_2}, \qquad a_2 \supset a_1, \qquad a_1, a_2 \, \epsilon \, L,$$

where L is a lattice in which the ascending chain condition holds. If $\beta = b_1/b_2$, we define $(\alpha, \beta) = (a_1, b_1)/(a_2, b_2)$, $[\alpha, \beta] = [a_1, b_1]/[a_2, b_2]$. With these definitions Σ is a lattice which is modular or distributive if and only if L is modular or distributive. Ore defines the product $\alpha \cdot \beta$ only for elements $\alpha, \beta \, \epsilon \, \Sigma$ such that $a_2 = b_1$, in which case $\alpha \cdot \beta = a_1/b_2$. Let us set $\alpha \asymp \beta$ if and only if $a_2 = b_1$, so that a necessary and sufficient condition for the existence of the product $\alpha \cdot \beta$ is that $\alpha \asymp \beta$. Although the relation $\asymp$ is neither reflexive nor symmetric, it is in a certain sense transitive since

(1) $\qquad$ if $\alpha \asymp \beta$ and $\gamma \asymp \delta$, then $\gamma \asymp \beta$ implies $\alpha \asymp \delta$.

Furthermore, the relation is preserved under union and cross-cut; that is,

(2) $\qquad$ $\alpha \asymp \beta, \gamma \asymp \delta$ implies $(\alpha, \gamma) \asymp (\beta, \delta), [\alpha, \gamma] \asymp [\beta, \delta]$.

Received September 2, 1938.

[1] The numbers in brackets refer to the references at the end of the paper.

[2] In this regard note that the necessary and sufficient conditions for unique factorization in the commutative case are stated in their most elegant form in terms of the G. C. D. operation (König [1]).

[3] Our inclusion is the reverse of Ore's.

270

Also

$$(3) \qquad \alpha \backsimeq \beta, \ \beta \backsimeq \gamma \text{ implies } \alpha \backsimeq \beta \cdot \gamma, \ \alpha \cdot \beta \backsimeq \gamma.$$

In the abstract theory given below we shall choose (1), (2), and (3) as the defining properties of the abstract relation $\backsimeq$.

When we have a commutative multiplication, the connection of the multiplication with the lattice operations automatically makes the lattice modular (in fact, distributive (Ward-Dilworth [1, 2])). If the multiplication is non-commutative, the lattice need not be modular; however, the assumption of the modular condition is essential since we shall prove that it is one of the necessary and sufficient conditions for arithmetic in a non-commutative semigroup. In particular, our results show the importance of the modularity[4] of a non-commutative polynomial domain in determining its arithmetical properties.

We conclude by stating our fundamental decomposition theorem for the elements of a lattice Σ with a multiplication having the properties of §2.

DECOMPOSITION THEOREM. *Each element a of Σ not equal to a unit has a decomposition into irreducible elements. If there are two such decompositions*

$$a = p_r p_{r-1} \cdots p_2 p_1 = q_s q_{s-1} \cdots q_2 q_1 ,$$

then $r = s$ and the p's and q's are similar in pairs.

2. The multiplication.[5] Let Σ be a lattice in which the ascending chain condition holds and let i denote its unit element. Consider in Σ a relation $\backsimeq$ having the following properties:

T1.   For each $a \ \epsilon \ \Sigma$ there are elements a', $a'' \ \epsilon \ \Sigma$ such that $a \backsimeq a'$, $a'' \backsimeq a$.

T2. $a \backsimeq b$, $a = c$, and[6] $b = d \rightarrow c \backsimeq d$.

T3. $a \backsimeq b$, $c \backsimeq d \rightarrow (a, c) \backsimeq (b, d)$, $[a, c] \backsimeq [b, d]$.

T4.   If $a \backsimeq b$, $c \backsimeq d$, then $c \backsimeq b \rightarrow a \backsimeq d$.

DEFINITION 2.1.   Let L_a denote the set of elements x such that $x \backsimeq a$.

By T1 and T3, L_a is non-empty and closed with respect to union and cross-cut. Hence L_a is a sublattice of Σ. Thus with each element a of Σ we associate a sublattice L_a .

In a similar manner we may associate with each element a of Σ a sublattice S_a defined as the set of all x's such that $a \backsimeq x$.

THEOREM 2.1.   *The lattices L_a and L_b are either disjoint or they are identical.*

Proof. If L_a and L_b are not disjoint, they have an element c in common. Let x be an arbitrary element of L_a . Then $x \backsimeq a$, $c \backsimeq a$, $c \backsimeq b$, and hence $x \backsimeq b$ by T3. Similarly, each element of L_b belongs to L_a .

---

[4] That a non-commutative polynomial domain is modular can be easily seen from the fact that the degree of a polynomial is a rank function over the lattice in the sense of Birkhoff (Birkhoff [1], Ore [2]).

[5] See Ward-Dilworth [1, 2] for lattice notation.

[6] We shall use $\rightarrow$ to denote "implies".

Clearly, a similar result holds for S_a and S_b .

Let now L_s and $L_{s'}$ be the L-lattices corresponding to s, $s' \in S_a$. Then $a \in L_s$, $L_{s'}$ and hence L_s and $L_{s'}$ are identical by Theorem 1.1. *Thus to each S-lattice we can associate an L-lattice L, where L is the L-lattice corresponding to an arbitrary element of S; and conversely, to each element of L corresponds the S-lattice S.*

DEFINITION 2.2. We write $a \sim b$ if a and b belong to the same L-lattice.

$\sim$ is clearly an equivalence relation in Σ with the L-lattices as equivalence classes.

We consider now a multiplication over Σ having the following properties:

M00. *To each pair of elements a, b such that $a \asymp b$, there is ordered a unique element ab, the product of a and b.*

M0.[7] $a = b \rightarrow ac = bc, da = db$.

M1. *With each $a \in \Sigma$ there exists an element $u_a \asymp a$ such that $u_a a = a$.*

M2. $a \supset ba$.

M3. $(a, b)c = (ac, bc)$.

M4. $(ab)c = a(bc)$.

M5. $ac = bc \rightarrow a = b$.

M6. $a \asymp b, b \asymp c \rightarrow a \asymp bc, ab \asymp c$.

From M00–M6 follow

$$(2.1) \qquad \begin{aligned} a \asymp b, c \asymp ab &\rightarrow c \asymp a; \\ a \asymp b, ab \asymp c &\rightarrow b \asymp c. \end{aligned}$$

Proof. $u_a \asymp a$ by M1. Hence $u_a \asymp ab$ by M6. But $c \asymp ab$ and thus $c \asymp a$ by T4. A similar proof gives the second statement.

$$(2.2) \qquad u_a \text{ is the unit element of } L_a .$$

Proof. Let $x \in L_a$. Then $a \supset xa \rightarrow u_a a = a = (a, xa) = (u_a a, xa) = (u_a , x)a$ by M1, M2, M3. Hence $u_a = (u_a , x)$ by M5.

$$(2.3) \qquad a \supset b \rightarrow ac \supset bc \text{ if } a \asymp c, b \asymp c \text{ by M3.}$$

Let $_aL$ denote the L-lattice to which a belongs. Let $_au$ denote its unit element. Then we have

$$(2.4) \qquad a \,_au = a.$$

Proof. We have $a \asymp \,_au$ since if $a \asymp x$, then $_au \asymp x$ and $_aux = x$ by (2.2) and M1. But then $a \asymp \,_aux$, and hence $a \asymp \,_au$ by (2.1). Now $ax = a(_aux) = (a \,_au)x \rightarrow a = a \,_au$ by M4 and M5.

DEFINITION 2.3. b is said to *divide a on the right* if there is an element $x \in \Sigma$ such that $a = xb$.

The element x of Definition 2.3 is unique by M5 and is called the quotient of a and b.

---

[7] In this and the remaining postulates the statements are assumed to hold if and only if all the products appearing in the statements exist.

DEFINITION 2.4. If b divides $[a, b]$ on the right, we write $a \ominus b$ and denote the quotient of $[a, b]$ and b by $a \cdot b^{-1}$.

THEOREM 2.2. *The quotient $a \cdot b^{-1}$ has the following properties:*

R1.
$$a \supset (a \cdot b^{-1})b;$$

R2.
$$a \supset xb \to a \cdot b^{-1} \supset x.$$

Proof. R1 is clear from Definition 2.4. Let $a \supset xb$ so that $[a, b] \supset xb$ by M2. Then

$$(a \cdot b^{-1})b \supset xb \to (a \cdot b^{-1}, x)b = ((a \cdot b^{-1})b, xb) = (a \cdot b^{-1})b.$$

Hence $(a \cdot b^{-1}, x) = a \cdot b^{-1}$ by M5.

Since R1 and R2 are the defining properties of the residual (Ward-Dilworth [1, 2]), we shall call $a \cdot b^{-1}$ the residual of b with respect to a. It exists only if $a \ominus b$.

We note that $a \ominus a$ and $ba \ominus a$ since $[a, a] = u_a a$ and $[a, ba] = ba$. Hence the residuals $a \cdot a^{-1}$ and $ba \cdot a^{-1}$ always exist.

$$(2.5) \qquad a \cdot a^{-1} = u_a .$$

$$(2.6) \qquad (ab) \cdot b^{-1} = a.$$

$$(2.7) \qquad a \ominus c, b \ominus c \to [a, b] \ominus c.$$

$$(2.8) \qquad [a, b] \cdot c^{-1} = [a \cdot c^{-1}, b \cdot c^{-1}] \text{ if } a \ominus c, b \ominus c.$$

$$(2.9) \qquad a \supset b \text{ if and only if } a \cdot b^{-1} = u_b .$$

3. **The decomposition theory.** Throughout this section we make the following assumptions:

A1. $\quad a \sim b \to a \ominus b.$[8]

A2. $\quad \Sigma$ *is modular.*

As consequences of A1 we have

$$(3.1) \qquad a \ominus c, b \ominus c \to a \cdot c^{-1} \ominus b \cdot c^{-1}.$$

Proof. Since $a \cdot c^{-1} \approx c, b \cdot c^{-1} \approx c$, we have $a \cdot c^{-1} \ominus b \cdot c^{-1}$ by A1.

$$(3.2) \qquad a \sim b \to a \ominus (a, b).$$

$$(3.3) \qquad (ba) \cdot c^{-1} = (b \cdot (c \cdot a^{-1})^{-1})(a \cdot c^{-1}) \text{ if } b \approx a, c \ominus a, a \ominus c, ba \ominus c.$$

---

[8] A1 is much stronger than necessary. However, since the weaker formulations are more complicated and artificial, and since the methods of proof remain essentially the same, we adopt the present formulation. If the proofs are examined, the various weaker conditions will be readily apparent to the reader (as, for example, the set given in §5). Also A1 is always satisfied in the important instances of the theory.

Proof. $((ba) \cdot c^{-1})c = [ba, c] = [ba, c, a] = ([ba, c] \cdot a^{-1})a = [b, c \cdot a^{-1}]a$ by (2.8). Now $ba \cdot a^{-1} \ominus c \cdot a^{-1}$ by (3.1) and hence $b \ominus c \cdot a^{-1}$ by (2.6). Hence

$$((ba) \cdot c^{-1})c = ((b \cdot (c \cdot a^{-1})^{-1})(c \cdot a^{-1}))a = (b \cdot (c \cdot a^{-1})^{-1})((c \cdot a^{-1})a)$$

$$= (b \cdot (c \cdot a^{-1})^{-1})[a, c] = (b \cdot (c \cdot a^{-1})^{-1})((a \cdot c^{-1})c) = ((b \cdot (c \cdot a^{-1})^{-1})(a \cdot c^{-1}))c$$

by M4 and Definition 2.4.

DEFINITION 3.1. If $a' = a \cdot b^{-1}$, where $a \sim b$ and $(a, b) = {}_a u$, we say that a' is conjugate to a.

DEFINITION 3.2. a is *similar* to b if there exists a chain of elements $a = a_0$, $a_1, \cdots, a_n = b$ such that either a_i is conjugate to a_{i+1} or a_{i+1} is conjugate to a_i (Ore [1]).

The relation of similarity is clearly reflexive, symmetric, and transitive.

DEFINITION 3.3. An element $p \, \epsilon \, \Sigma$ is irreducible if $p \neq {}_p u$, and if $x \supset p$, $x \sim p \rightarrow x = {}_p u$ or $x = p$.

(3.4) *If p is irreducible, then $p \not\supset a$ and $p \sim a \rightarrow (p, a) = {}_p u$.*

(3.5) *If p is irreducible and $p \supset ab$, $p \sim b$, $p \not\supset b$; then $p' \supset a$, where p' is conjugate to p.*

Proof. Take $p' = p \cdot b^{-1}$.

THEOREM 3.1. *An element conjugate to an irreducible element is an irreducible element.*

Proof. Let p be an irreducible element, and let $p' = p \cdot a^{-1}$, where $p \sim a$ and $(p, a) = {}_p u$. Let $x \supset p'$ with $x \sim p'$. Then $x \supset p \cdot a^{-1}$ and $xa \supset (p \cdot a^{-1})a = [a, p]$ by (2.3), Definition 2.4. Hence $xa = (xa, [a, p]) = [a, (xa, p)]$ by A2. Thus $x = (xa) \cdot a^{-1} = [a, (xa, p)] \cdot a^{-1} = (xa, p) \cdot a^{-1}$ by (2.6), (2.8), (2.9). Now $a \eqsim d$ for some d by T1 and hence $p'a \eqsim d$, $xa \eqsim d$ by M6. Thus we have $xa \sim p'a = (p \cdot a^{-1})a = [a, p]$. But since $p \sim a$, $[a, p] \sim p$ and hence $xa \sim p$. $(xa, p) \supset p$ gives $(xa, p) = {}_p u$ or p since p is irreducible; and hence $x = {}_p u \cdot a^{-1} = u_a = {}_p u$ or $x = p \cdot a^{-1} = p'$. This proves the theorem.

THEOREM 3.2. *If an irreducible p is conjugate to an element p', then p' is an irreducible.*

Proof. Let $p = p' \cdot a^{-1}$, where $p' \sim a$ and $(p', a) = {}_{p'} u$, and let $x \supset p'$, $x \sim p'$. Then $x \cdot a^{-1} \supset p' \cdot a^{-1}$ by (2.8), A1 and thus $x \cdot a^{-1} \supset p$. Also $x \cdot a^{-1} \sim p$ since $x \cdot a^{-1} \eqsim a$ and $p \eqsim a$. Hence we have either (i) $x \cdot a^{-1} = {}_p u$ or (ii) $x \cdot a^{-1} = p$. If (i) holds, then $x \cdot a^{-1} = u_a$ since $p \eqsim a$ and hence $x \supset a$ by (2.9). But $x \supset p'$ by hypothesis, hence $x \supset (a, p') = {}_{p'} u$. And since $x \sim p'$, $x = {}_{p'} u$. If (ii) holds, we have $(x \cdot a^{-1})a = (p' \cdot a^{-1})a$ or $[x, a] = [p', a]$ by Definition 2.4. But then $x \supset p' \supset [x, a]$ and by A2, $p' = [x, (p', a)] = [x, {}_{p'} u] = x$. Hence either $x = {}_{p'} u$ or $x = p'$, and thus p' is an irreducible. This completes the proof.

THEOREM 3.3. *Every element similar to an irreducible element is irreducible.*

Proof. The theorem is clear from Theorems 3.1 and 3.2 and Definition 3.2.

Theorem 3.4. *Let a' be conjugate to $a = a_k a_{k-1} \cdots a_2 a_1$, then $a' = a'_k a'_{k-1} \cdots a'_2 a'_1$, where a'_i is conjugate to a_i.*

Proof. Suppose the theorem is true for every product of $k - 1$ elements and let $a' = a \cdot b^{-1}$, $a \sim b$, $(a, b) = {}_b u$. Then $a \ominus b$, $a_1 \ominus b$, $b \ominus a_1$ since $a \sim b$ and $a_1 \sim b$. Thus $a' = ((a_k \cdots a_2) \cdot (b \cdot a_1^{-1})^{-1})(a_1 \cdot b^{-1})$ by (3.3). Now $a_1 \sim b$ and $(a_1, b) \supset (a, b) = {}_b u$ which gives $(a_1, b) = {}_b u$. Hence $a'_1 = a_1 \cdot b^{-1}$ is conjugate to a_1. Let $b' = b \cdot a_1^{-1}$ and $s = a_k \cdots a_2$. Then $b' \sim s$ since $b' \approx a_1$ and $s \approx a_1$. Now $(s, b')a_1 = (sa_1, b'a_1) = (a, (b \cdot a_1^{-1})a) = (a, [b, a_1]) = [a_1, (b, a)] = [a_1, {}_a u] = a_1$ by Definition 2.4 and A2. Hence $(s, b') = a_1 \cdot a_1^{-1} = {}_{b'} u$. Thus $s \cdot b'^{-1}$ is conjugate to s and hence $s \cdot b'^{-1} = a'_k a'_{k-1} \cdots a'_2$ by hypothesis. Substitution gives $a' = a'_k \cdots a'_1$. The theorem is therefore proved.

We now prove the fundamental

Uniqueness Theorem. *If an element $a \in \Sigma$ has two representations as a product of irreducibles*

$$a = p_r p_{r-1} \cdots p_2 p_1 = q_s q_{s-1} \cdots q_2 q_1,$$

then $r = s$ and the p's and q's are similar in pairs.

Proof.[9] Let

$$(1) \qquad a = p_r p_{r-1} \cdots p_2 p_1 = q_s q_{s-1} \cdots q_2 q_1.$$

If $p_1 = q_1$, this factor may be canceled. If $p_1 \neq q_1$, let k be the first number such that $q_1 \supset p_k p_{k-1} \cdots p_2 p_1$; then $q_1 \not\supset p_{k-1} \cdots p_1$ and $p_{k-1} \cdots p_1 \sim q_1$. Hence $(q_1, p_{k-1} \cdots p_1) = {}_{q_1} u$ by (3.4). But then $q_1 \cdot (p_{k-1} \cdots p_1)^{-1} \supset p_k$ and $q'_1 = q_1 \cdot (p_{k-1} \cdots p_1)^{-1}$ is conjugate to q_1 and thus is an irreducible by Theorem 3.2. Hence $q'_1 = p_k$. Now $p_k p_{k-1} \cdots p_1 = (q_1 \cdot (p_{k-1} \cdots p_1)^{-1})p_{k-1} \cdots p_1 = [q_1, p_{k-1} \cdots p_1] = ((p_{k-1} \cdots p_1) \cdot q_1^{-1})q_1$ by Definition 2.4. Hence $p_k p_{k-1} \cdots p_1 = p'_{k-1} \cdots p'_1 q_1$ by Theorem 3.4 and p'_i is conjugate to p_i $(i = 1, \cdots, k - 1)$. Substituting this result in (1) and canceling q_1, we may treat the resulting expression in the same manner. Thus we find $r = s$ and the p's and q's similar in pairs.

Concerning the existence of a decomposition into irreducibles we have the

Existence Theorem. *If the descending chain condition holds for the right factors of $a \neq {}_a u$, then a has a decomposition into irreducible elements.*

Proof. If a is not an irreducible, then there is an element $a_1 \neq {}_a u$ such that $a_1 \supset a$, $a_1 \sim a$ and $a_1 \neq a$. But then $a = (a \cdot a_1^{-1})a_1$ by A1. If a_1 is not an irreducible, we have an $a_2 \neq {}_a u$ such that $a_2 \supset a_1$, $a_2 \sim a_1$, and $a_2 \neq a_1$. Then $a = (a \cdot a_1^{-1})(a_1 \cdot a_2^{-1})a_2$. Thus we get a chain of elements $a \subset a_1 \subset a_2 \subset \cdots$ which must break off giving an irreducible element p_1 such that $a = bp_1$. But if b is not an irreducible, $b = b_1 p_2$. Since $p_1 \supset p_2 p_1 \supset \cdots$ is a descending chain of factors of a, it must break off giving a decomposition $a = p_k p_{k-1} \cdots p_2 p_1$. The proof of the theorem is complete.

[9] This proof is essentially that given by Ore [2] for non-commutative polynomials.

We note that the descending chain condition for the factors of an element of Σ does not follow from the ascending chain condition in Σ, as in the commutative case. However, it does follow from the ascending chain condition in Σ', where Σ' is the lattice of left union and cross-cut if they exist.

As examples of the abstract theory let Σ be the lattice N of a non-commutative polynomial domain. Then for every a and b, $a \asymp b$ and $a \ominus b$ so that T1-T4, M6 and A1 are trivially satisfied. The relations of similarity and conjugacy are identical. Furthermore, in N the irreducible elements are those elements whose only right divisors are the elements themselves and the elements of the fundamental field.

More generally the above results apply to any non-commutative domain of integrity having a Euclidean algorithm.

Again if we interpret Σ to be the quotient lattice Q of a lattice L, we have $A \asymp B$, where $A = a_1/a_2$, $B = b_1/b_2$ if and only if $a_2 = b_1$ in which case $AB = a_1/b_2$. Furthermore $u_A = a_1/a_1$, $_Au = a_2/a_2$. Postulates T1-T4 are clearly satisfied by the relation $\asymp$, and it is readily verified that the multiplication satisfies M00-M6. We have $A \ominus B$ if and only if $a_2 \supset b_2$ in which case $A \cdot B^{-1} = [a_1, b_1]/b_1$. We observe that $A \sim B$ if and only if $a_2 = b_2$ so that A_1 is satisfied. If we start with a modular lattice L, then Σ is modular and A2 is satisfied. The irreducible elements are those quotients p for which p_2 covers p_1.[10]

4. **The arithmetic of a semigroup.** Let S be a semigroup of elements $a, b, c,$ $\cdots$ and unit element i such that each pair of elements a, b has a G. C. D. (a, b). Then if the ascending chain condition holds in S, a and b have an L. C. M. defined as the G. C. D. of those elements which both a and b divide. As in §2 we define $a \cdot b^{-1} = [a, b]/b$.

DEFINITION 4.1. If $a' = a \cdot b^{-1}$, where $(a, b) = i$, we say that a' is conjugate to a.

DEFINITION 4.2. a is *similar* to b if there exists a chain of elements $a = a_0,$ $\cdots,$ $a_n = b$ such that either a_i is conjugate to a_{i+1} or a_{i+1} is conjugate to a_i.

We have then the following fundamental theorem:

THEOREM 4.1. *Let S be a semigroup with G. C. D. and L. C. M. operations. (S is thus a lattice with respect to G. C. D. and L. C. M.) Then the following*

---

[10] As another example, consider the set M of all finite matrices for which the number of rows is greater than or equal to the number of columns over a non-commutative ring R with unit element. A subset A of M is called an ideal if the matrices of A (i) all have the same number of rows and the same number of columns, (ii) are closed under addition, and (iii) are closed under multiplication by all square matrices for which the product exists. We write $A \asymp B$ if the matrices of A have the same number of columns as the matrices of B have rows. The product of A and B is defined only if $A \asymp B$ and is the ideal generated by the products of the matrices of A with those of B. With a suitable definition of union and cross-cut the set Σ of ideals of M satisfies T1-T4, M00-M4, M6, A2. Moreover, if we give a similar definition of left ideals in M and R is a non-commutative domain of integrity for which every left ideal is principal, then the set Σ of left ideals of M satisfies T1-A2 and is an instance of our abstract theory. A detailed account of these systems will be given in another paper.

three conditions are necessary and sufficient that each element not equal to i of S be expressible as a product of irreducibles unique up to similarity:[11]

    (i) *the ascending chain condition in S;*

    (ii) *the descending chain condition for the right factors of each element in S;*

    (iii) *the modular condition in S.*

Proof. The sufficiency of conditions (i)–(iii) follows from the results of §3. For since the product of any two elements always exists, T1-T4 are trivially satisfied. M00-M6 are readily verified and A1 is trivially true since $a \ominus b$ for every a and b. (iii) gives A2. Hence the existence and uniqueness theorems of §3 hold.

Suppose now that each element not equal to i of S is uniquely (up to similarity) expressible as a product of irreducibles. We define $\rho(i) = 0$, $\rho(a) = s$ if $a = p_s p_{s-1} \cdots p_2 p_1$. Then $\rho(a) = 0$ if and only if $a = i$ and $\rho(a) = 1$ if and only if a is an irreducible. Furthermore, $\rho(ab) = \rho(a) + \rho(b)$ since if $a = p_{\rho(a)} \cdots p_1$ and $b = q_{\rho(b)} \cdots q_1$, then $ab = p_{\rho(a)} \cdots p_1 q_{\rho(b)} \cdots q_1$. Hence $a \supset b$ and $a \neq b$ implies that $\rho(a) < \rho(b)$. It follows that the ascending chain condition holds in S and the descending chain condition holds for the factors of each element in S.

We note that if a' is similar to a, $\rho(a') = \rho(a)$.

Let a and b be any two elements of S. We have then $a = a_1(a, b)$, $b = b_1 (a, b)$, where $(a_1, b_1) = i$. Then

$$[a, b] = [a_1(a, b), b_1(a, b)] = [a_1, b_1](a, b) = (a_1 \cdot b_1^{-1}) b_1(a, b) = a_1' b_1(a, b),$$

where a_1' is similar to a_1. Then

$$\rho([a, b]) = \rho(a_1') + \rho(b_1) + \rho((a, b)) = \rho(a_1) + \rho(b_1) + \rho(a, b).$$

But $\rho(a) = \rho(a_1) + \rho((a, b))$, so that $\rho(a_1) = \rho(a) - \rho((a, b))$. Similarly, $\rho(b_1) = \rho(b) - \rho((a, b))$. Hence $\rho([a, b]) = \rho(a) + \rho(b) - \rho((a, b))$ or $\rho([a, b]) + \rho((a, b)) = \rho(a) + \rho(b)$. Thus ρ is a rank function over S in the sense of Birkhoff (Birkhoff [1], p. 447) and S is modular by Birkhoff's result. Hence conditions (i)–(iii) are satisfied.

5. **Properties of the L-lattices.** Using the notations of §3, we make

DEFINITION 5.1. The unit elements of the L-lattices are called the units of Σ.

Let now $a_1, a_2 \, \epsilon \, L$, $a_1', a_2' \, \epsilon \, L'$, where L and L' are any two L-lattices. Then if $a_1, a_2 \backsimeq x_1$, $a_1', a_2' \backsimeq x_2$, we have $(a_1, a_1') \backsimeq (x_1, x_2)$ and $(a_2, a_2') \backsimeq (x_1, x_2)$. Hence (a_1, a_1') and (a_2, a_2') belong to the same L-lattice. We call this L-lattice to which all the unions of elements from L_1 and L_2 respectively belong the union of L_1 and L_2 and write (L_1, L_2). In a similar manner we define the cross-cut $[L_1, L_2]$ of two L-lattices. Hence we make the L-lattices into a lattice Σ_l. Σ_l will be modular if L is modular.

---

[11] By "up to similarity" we mean that the irreducibles appearing in the decompositions of similar elements are similar in pairs.

In general, the union of the unit elements of L_1 and L_2 will not be the unit element of (L_1 , L_2).   However, we prove

THEOREM 5.1.   *If the descending chain condition holds in Σ, then the units of Σ are closed under union and cross-cut and form a lattice isomorphic to Σ_l .*

Proof.   We prove first a necessary lemma.

LEMMA 5.1.   *If the descending chain condition holds in Σ, then the only elements of Σ such that $x \backsimeq x$ are the units of Σ.*

Proof of lemma.   We note that $u \backsimeq u$ for every unit u, since if $u \backsimeq x$, then $x = ux$; and hence $u \backsimeq u$ by M6.   Now let $a \backsimeq a$.   Then the chain $a, a^2, a^3, \cdots$ must break off by the descending chain condition so that $a^{m+n} = a^n$ or $a^m = u_a$ by M5.   But since $a \backsimeq a$, $u_a = {_a}u$ and hence $a^m = {_a}u$.   We have then $a^{m-1} = {_a}u \cdot a^{-1} = {_a}u$, and finally $a = {_a}u$.

We continue with the proof of the theorem.   Let u and u' be two units, so that $u \backsimeq u, u' \backsimeq u'$.   Then $(u, u') \backsimeq (u, u')$ and $[u, u'] \backsimeq [u, u']$.   Whence (u, u') and $[u, u']$ are units by Lemma 5.1.   This completes the proof of the theorem.

The L-lattices have a number of interesting interrelations.   We mention, however, only one:

THEOREM 5.2.   *Let L be an arbitrary L-lattice and let $l \in L$.   Then L has a sublattice isomorphic to L_l with l as the unit element.*

Proof.   Let $x \in L_l$ and set up the correspondence $x \leftrightarrow xl$, where xl is clearly in L.   Then $(x, y) \leftrightarrow (x, y)l = (xl, yl)$ and $[x, y] \leftrightarrow [x, y]l = [xl, yl]$.   Furthermore, the correspondence is 1-1 by M5.   Hence the theorem follows.

We next characterize the irreducibles of Σ in terms of the lattice properties of the L-lattices.

THEOREM 5.3.   *The irreducibles of Σ are the divisor-free elements of the L-lattices.*

Proof.   Let p be a divisor-free element of an L-lattice, and let $x \supset p, x \sim p$; then clearly $x = {_p}u$ or $x = p$.   Conversely, if p is an irreducible, let p' be the divisor-free element of $_pL$ dividing p.   Then $p' \supset p$, $p' \sim p$ and $p' \neq {_p}u$, and hence $p' = p$.

We note that Theorem 5.3 may not hold if we weaken postulate A1.   For example, let us replace A1 by

B1. $$a \ominus b, \quad a \ominus c, \quad b \ominus c \rightarrow a \cdot c^{-1} \ominus b \cdot c^{-1}.$$

B2. $$a \supset b \quad \text{and} \quad b \ominus c \rightarrow a \ominus c.$$

B3. $$a \ominus b \quad \text{and} \quad a \sim b \rightarrow a \ominus (a, b).$$

We define conjugate elements and irreducible elements by

DEFINITION 5.1.   If $a' = a \cdot b^{-1}$, where $a \sim b, a \ominus b, b \ominus a$, and $(a, b) = {_a}u$, we say that a' is conjugate to a.

DEFINITION 5.2. p is an irreducible if $p \neq {}_p u$ and if $x \supset p, x \sim p, p \ominus x \rightarrow$ $x = p$ or $x = {}_p u$.

With these definitions the proofs of the existence and uniqueness theorems follow, with some modifications as in §3.   However, there may be irreducibles which are not divisor-free elements since we may have $x \supset p, x \neq p, {}_p u; x \sim p$, but $x \ominus p$ is not true.

We conclude this section with the investigation of the special case where each L-lattice and its corresponding S-lattice are identical.   Then $\asymp$ is an equivalence relation and the equivalence classes are multiplicatively closed sublattices. Let $a = a_s \cdots a_1 = b_r \cdots b_1$ be two decompositions of a.   Then $a \asymp a_1$, and $a \asymp b_1$.   But $a_s \cdots a_2 \asymp a_1$ and $a_s \cdots a_2 \asymp a_2$.   Hence $a \asymp a_s \asymp \cdots \asymp a_1$ $\asymp b_r \asymp \cdots \asymp b_1$.

Thus this case reduces to that of Σ closed under multiplication.

6. The commutative case.

6. **The commutative case.**   In this section we investigate the consequences of assuming that the multiplication is commutative.   Explicitly we assume

A3. $$a \asymp b \rightarrow b \asymp a, \; ab = ba.$$

We have then

(6.1) $$a \asymp b, \; b \asymp c \rightarrow a \asymp c.$$

Proof.   Since $a \asymp b$, ab exists and $ab \asymp c$.   But $ab = ba \asymp c$, whence $a \asymp c$.

(6.2) $$a \asymp a.$$

Hence $\asymp$ is an equivalence relation giving equivalence classes $\{a\}, \{b\}, \cdots$. The L-lattices and the equivalence classes coincide.   Furthermore, we note that each equivalence class is closed with respect to union, cross-cut, multiplication and residuation.   We note also that $u_a = {}_a u$.

THEOREM 6.1. a' *is conjugate to* a *if and only if* $a' = a$.

Proof.   We prove first a series of lemmas.

LEMMA 6.1.   *If* $a \sim b \sim c$ *and* $b \supset c$, *then* $a \cdot c^{-1} \supset a \cdot b^{-1}$.

Proof.   The residuals exist by A1.   Furthermore, $a \supset (a \cdot b^{-1})b \supset (a \cdot b^{-1})c$ $\rightarrow a \cdot c^{-1} \supset a \cdot b^{-1}$ by R1 and R2.

LEMMA 6.2. $a \sim b \sim c \rightarrow a \cdot (b, c)^{-1} = [a \cdot b^{-1}, a \cdot c^{-1}]$.

Proof. $[a \cdot b^{-1}, a \cdot c^{-1}] \supset a \cdot (b, c)^{-1}$ by Lemma 6.1.   But $a \supset ((a \cdot b^{-1})b,$ $(a \cdot c^{-1}) c) \supset [a \cdot b^{-1}, a \cdot c^{-1}](b, c)$.   Hence $a \cdot (b, c)^{-1} \supset [a \cdot b^{-1}, a \cdot c^{-1}]$ and thus $a \cdot (b, c)^{-1} = [a \cdot b^{-1}, a \cdot c^{-1}]$.

LEMMA 6.3. $a \cdot {}_a u^{-1} = a \cdot u_a^{-1} = a$.

Proof. $a \supset a u_a \rightarrow a \cdot u_a^{-1} \supset a$.   But $a = a u_a \supset (a \cdot u_a^{-1})a \rightarrow a \supset a \cdot u_a^{-1}$. Hence $a = a \cdot u_a^{-1}$.   Since $u_a = {}_a u$, the lemma follows.

We continue with the proof of the theorem. Let $a' = a \cdot b^{-1}$, $a \sim b$, $(a, b) = {}_au$. Then

$$a = a \cdot {}_au^{-1} = a \cdot (a, b)^{-1} = [a \cdot a^{-1}, a \cdot b^{-1}] = [{}_au, a \cdot b^{-1}] = a \cdot b^{-1} = a'$$

by Lemmas 6.2 and 6.3. The proof of the theorem is complete.

Now obviously any irreducible factor of an element belongs to the same equivalence class as the element itself. Furthermore, since multiplication is commutative, the ascending chain condition implies the descending chain condition for the factors of an element $a \epsilon \Sigma$. Hence by the uniqueness and existence theorems and Theorem 6.1 each element not a unit in Σ is uniquely expressible as a product of irreducibles, the irreducibles belonging to the same equivalence class. Thus considered as a lattice, each equivalence class is a direct product of chain lattices; i.e., an arithmetical lattice (Ward [3]).

REFERENCES

GARRETT BIRKHOFF.
1. *On combination of subalgebras*, Proc. Cambridge Phil. Soc., vol. 29(1933), pp. 441–464.

A. H. CLIFFORD.
1. *Arithmetic and ideal theory of abstract multiplication*, Bull. Am. Math. Soc., vol. 40(1934), pp. 326–330.

J. KÖNIG.
1. *Algebraischen Grossen*, Leipzig, 1903, Chapter I.

O. ORE.
1. *Abstract algebra, I, II*, Annals of Math., vol. 36(1935), pp. 406–437; vol. 37(1936), pp. 265–292.
2. *Theory of non-commutative polynomials*, Annals of Math., vol. 34(1933), pp. 480–508.

M. WARD.
1. *Postulates for an abstract arithmetic*, Proc. Natl. Acad. Sci., vol. 14(1928), pp. 907–911.
2. *Conditions for factorization in a set closed under a single operation*, Annals of Math., vol. 36(1933), pp. 36–39.
3. *Structure residuation*, Annals of Math., vol. 39 (1938), pp. 558-568.

M. WARD AND R. P. DILWORTH.
1. *Residuated lattices*, Proc. Natl. Acad. Sci., vol. 24(1938), pp. 162–165.
2. *Residuated lattices*, to appear in the Trans. of the Amer. Math. Soc.

J. H. M. WEDDERBURN.
1. *Non-commutative domains of integrity*, Journal für Math., vol. 167(1931), pp. 129–141.

CALIFORNIA INSTITUTE OF TECHNOLOGY.

ABSTRACT COMMUTATIVE IDEAL THEORY

R. P. Dilworth

1. Introduction. Several years ago, M. Ward and the author [4] began a study in abstract form of the ideal theory of commutative rings. Since it was intended that the treatment should be purely ideal-theoretic, the system which was chosen for the study was a lattice with a commutative multiplication. For such multiplicative lattices, analogues of the Noether decomposition theorems for commutative rings were formulated and proved. However the theorems corresponding to the deeper results on the ideal structure of commutative rings were not obtained; the essential difficulty being the problem of formulating abstractly the notion of a principal ideal. This difficulty occurred in a mild form in treating the Noether theorem on decompositions into primary ideals. In fact, in the above mentioned paper, a weak concept of "principal element" was introduced which sufficed for the proof of the decomposition theorem into primaries. Nevertheless, the definition had serious defects and it was immediately obvious that it was not adequate for the further development of the abstract theory.

In this paper, I give a new and stronger formulation for the notion of a "principal element", and, in terms of this concept, prove an abstract version of the Krull Principal Ideal Theorem. Since there are generally many non-principal ideals of a commutative ring which are "principal elements" in the lattice of ideals, the abstract theorem represents a considerable strengthening of the classical Krull result.

It seems appropriate at this point to include a brief description of the new "principal elements" and to sketch their relationship to principal ideals.

Let L be a lattice with a multiplication and an associated residuation. The product of two lattice elements A and B will be denoted by AB and the residual, by $A : B$. An element M of L is said to be *meet principal* if

$$(1.1) \qquad (A \cap B : M)M \supseteq AM \cap B \qquad \text{all} \quad A, B \in L.$$

Similarly, M is said to be *join principal* if

$$(1.2) \qquad (A \cup BM) : M \subseteq A : M \cup B \qquad \text{all} \quad A, B \in L.$$

Finally, M is said to be *principal* if it is both meet and join principal.

Now let L be the lattice of ideals of a commutative ring R and let $M = (m)$ be a principal ideal of R. If $x \in AM \cap B$, then $x \in B$ and

---

Received April 20, 1961.

481

$x = am$ where $a \in A$. It follows that $a \in B : M$ and hence that $a \in A \cap B : M$. Thus $x = am \in (A \cap B : M)M$. According to (1.1), M is meet principal. Next let $y \in (A \cup BM) : M$. Then $ym \in A \cup BM$ and hence $ym = a + bm$ where $a \in A$ and $b \in B$. Thus $a = (y - b)m$, $y - b \in A : M$, and hence $y = (y - b) + b \in A : M \cup B$. It follows that M is also join principal. Hence every principal ideal is indeed a principal element in the lattice of ideals.

It will be shown in §7 that in polynomial rings, properties (1.1) and (1.2) in fact characterize the principal ideals. In more general rings there may be many nonprincipal ideals which are principal elements of the lattice of ideals. Thus if R is the ring of integers in an algebraic number field, then $A : B$ is the quotient $(A \cap B)/B$ and every ideal satisfies (1.1) and (1.2). We note that since distinct prime ideals of R are noncomparable, the conclusion of the Krull Principal Ideal Theorem also holds for all ideals in this ring.

2. Preliminary definitions and results. Throughout the paper, elements of a multiplicative lattice will be denoted by latin capitals, $\cup$ and $\cap$ will denote the lattice operations and $\supseteq$ will denote the latice inclusion relation with $\supset$ being reserved for proper inclusion. The lattices to be studied will be complete with unit element I and with null element 0.

A lattice is said to be *multiplicative* if there is defined on the lattice a commutative, associative, join distributive, multiplication. We shall also require that I be the identity element for multiplication. Such a multiplicative lattice also has a residuation satisfying the basic relations

$$(2.1) \qquad A \supseteq (A : B)B$$

$$(2.2) \qquad A \supseteq XB \quad \text{implies} \quad A : B \supseteq X .$$

Further important properties of the residuation and multiplication are the following

$$(2.3) \qquad A \supseteq B \text{ if and only if } A : B = I$$

$$(2.4) \qquad (A \cap B) : C = (A : C) \cap (B : C)$$

$$(2.5) \qquad A : (BC) = (A : B) : C$$

$$(2.6) \qquad A : B \supseteq A$$

$$(2.7) \qquad A : I = A$$

$$(2.8) \qquad (AB) : B \supseteq A$$

$$(2.9) \qquad (A \cup B) : C \supseteq (A : C) \cup (B : C)$$

$$(2.10) \qquad (A \cap B)C \subseteq (AC) \cap (BC)$$

$$(2.11) \qquad (A \cap B) : B = A : B$$

$$(2.12) \qquad A : (A \cup B) = A : B$$

$$(2.13) \qquad A : (B \cup C) = A : B \cap A : C$$

(2.14)   If $A \cup C = B \cup C = I$,   then $(AB \cup C) = I$

(2.15)   If $A \cup C = I$,   then $(A \cap B) \cup C = B \cup C$

(2.16) $(A_1 \cup \cdots \cup A_n)^{k_1 + \cdots + k_n} \subseteq A_1^{k_1} \cup \cdots \cup A_n^{k_n}$.

Now let L denote a multiplicative lattice satisfying the ascending chain condition. An element $P \in L$ is *prime* if the following condition is satisfied for all $A, B \in L$

(2.17) $P \supseteq AB$ implies $P \supseteq A$ or $P \supseteq B$.

The element I is trivially a prime according to this definition. However, ''prime'' will normally refer to prime elements distinct from I.

An element $Q \in L$ is *primary* if for all $A, B \in L$

(2.18) $Q \supseteq AB$ implies $Q \supseteq A$ or $Q \supseteq B^k$ for some integer k .

Many of the results relating prime and primary ideals in a commutative ring carry over directly to multiplicative lattices. Thus if Q is a primary element of L, the join P_Q of all elements X such that $X^s \subseteq Q$ for some integer s is a prime element containing Q. It is easily verified that P_Q is the minimal prime containing Q. Furthermore, $Q \supseteq P_Q^k$ for some integer k. The prime P_Q is called the prime *associated* with Q and it is characterized by the properties

(2.19) $P_Q \supseteq Q \supseteq P_Q^k$ for some integer k

(2.20) $Q \supseteq AB$ implies $Q \supseteq A$ or $P_Q \supseteq B$.

If $P_Q \not\supseteq A$, it is easily verified that $Q : A = Q$. Also if Q_1 and Q_2 are primary elements associated with the same prime P, then $Q_1 \cap Q_2$ is likewise a primary element associated with P.

An element A is said to have a primary decomposition if there exists primaries $Q_1, \cdots, Q_m$ such that

(2.21) $$A = Q_1 \cap \cdots \cap Q_m .$$

If superfluous Q_i are removed and the primaries associated with the same prime are combined we obtain a reduced primary decompsition in which distinct primaries are associated with distinct primes. Such a primary decomposition is called a *normal* primary decomposition. The fundamental theorem on primary decompositions is proved exactly as in the commutative ring case and may be stated as follows:

Any two normal decompositions of an element A have the same number of components and the same set of associated primes.

Let $A = Q_1 \cap \cdots \cap Q_n$ be a normal decomposition of A and let $P_1, \cdots, P_n$ denote the associated primes. A subset $\mathfrak{S}$ of $\{P_1, \cdots, P_n\}$ is *isolated* if $P_i \in \mathfrak{G}$ implies $P_j \in \mathfrak{G}$ whenever $P_i \supseteq P_j$. The element $A_\mathfrak{G} = \cap \{Q_i \,|\, P_i \in \mathfrak{G}\}$ is called an *isolated component* of A. The fundamental result on isolated components is the following:

The element $A_\mathfrak{G}$ depends only upon A and $\mathfrak{G}$ and not upon the particular normal decomposition.

For let $A = Q_1' \cap \cdots \cap Q_n'$ be a second normal decomposition and let $A_\mathfrak{G}' = \cap \{Q_1' \,|\, P_i \in \mathfrak{G}\}$. If we set $B' = \cap \{Q_j' \,|\, P_j \notin \mathfrak{G}\}$, then $Q_i \supseteq A = A_\mathfrak{G}' \cap B' \supseteq A_\mathfrak{G}' B'$. Now if $P_i \in \mathfrak{G}$, then $P_i \not\supseteq B'$ since otherwise $P_i \supseteq Q_j' \supseteq P_j^k$ and hence $P_i \supseteq P_j$ so that $P_j \in \mathfrak{G}$ contrary to $P_j \notin \mathfrak{G}$. It follows that $Q_i \supseteq A_\mathfrak{G}'$ for all i such that $P_i \in \mathfrak{G}$. Thus $A_\mathfrak{G} = \cap \{Q_i \,|\, P_i \in \mathfrak{G}\} \supseteq A_\mathfrak{G}'$. Similarly $A_\mathfrak{G}' \supseteq A_\mathfrak{G}$ and hence $A_\mathfrak{G} = A_\mathfrak{G}'$.

There are multiplicative lattices satisfying the ascending chain condition in which elements fail to have primary decompositions.[1] On the other hand it is well known that every element of a lattice satisfying the ascending chain condition has a decomposition into meet irreducibles. Since a meet irreducible can have a primary decomposition only if it is itself primary, it follows that the elements of L will have primary decomposition if and only if every meet irreducible element of L is primary. Suitable condition which insure that meet irreducibles will be primary involve properties of principal elements. These will be discussed in the following section.

It should be pointed out that the relationship between prime and primary elements can be developed in multiplicative lattices which are compactly generated. These include multiplicative lattices satisfying the ascending chain conditions as a very special case.

3. Noether lattices. In §1, the notions of meet principal elements, join principal elements, and principal elements were introduced. We begin by developing some elementary properties of these elements.

LEMMA 3.1 *An element M is meet principal if and only if*

$$(3.1) \qquad (A \cap B : M)M = AM \cap B \qquad all\ A, B \in L .$$

For $(A \cap B : M)M \subseteq AM \cap (B : M)M \subseteq AM \cap B$ by 2.10 and 2.1

LEMMA 3.2 *An element M is join principal if and only if*

$$(3.2) \qquad (A \cup BM) : M = A : M \cup B \qquad all\ A, B \in L .$$

For $(A \cup BM) : M \supseteq A : M \cup (BM) : M \supseteq A : M \cup B$ by 2.9 and 2.8. Setting $A = I$ in 3.1 and $A = 0$ in 3.2 gives the following corollaries

<hr>

[1] Ward-Dilworth [**4**] contains several examples.

COROLLARY 3.1. *If M is meet principal, then*

$$(3.3) \qquad (B:M)M = B \cap M \qquad all \ B \in L .$$

COROLLARY 3.2. *If M is join principal, then*

$$(3.4) \qquad (BM):M = B \cup (0:M) \qquad all \ B \in L .$$

In particular, if $M \supseteq B$, then $(B:M)M = B$ and if $B \supseteq 0:M$, then $(BM):M = B$ respectively.

LEMMA 3.3. *If M_1 and M_2 are meet principal, then M_1M_2 is meet principal.*

For

$$[A \cap B:(M_1M_2)]M_1M_2 = [A \cap (B:M_2):M_1]M_1M_2$$
$$= (AM_1 \cap B:M_2)M_2 = AM_1M_2 \cap B$$

by (2.5) and (3.1).

LEMMA 3.4. *If M_1 and M_2 are join principal, then M_1M_2 is join principal.*

For

$$(A \cup BM_1M_2):(M_1M_2) = [(A \cup BM_1M_2):M_2]:M_1$$
$$= [A:M_2 \cup BM_1]:M_1 = (A:M_2):M_1 \cup B = A:(M_1M_2) \cup B$$

by (2.5) and (3.2).

COROLLARY 3.3. *If M_1 and M_2 are principal, then M_1M_2 is principal.*
The conditions which insure that every meet irreducible is primary are formulated in the following theorem.

THEOREM 3.1. *Let L be a modular, multiplicative lattice satisfying the ascending chain condition. Furthermore let every element of L be a join of meet principal elements. Then every meet irreducible element is primary.*

Proof. Let Q be meet irreducible and let $Q \supseteq AM$ where $Q \not\supseteq A$ and M is meet principal. Then by (2.13) $(A \cup Q):M \subseteq (A \cup Q):M^2 \subseteq \cdots \subseteq (A \cup Q):M^k \subseteq \cdots$. By the ascending chain condition there exists k such that

$$(A \cup Q):M^k = (A \cup Q):M^{k+1} .$$

Now let $C = (A \cup Q) \cap (M^{k+1} \cup Q)$.

Then $C : M^{k+1} = (A \cup Q) : M^{k+1} \cap (M^{k+1} \cup Q) : M^{k+1} = (A \cup Q) : M^{k+1} \cap I = (A \cup Q) : M^{k+1} = (A \cup Q) : M^{k}$ by (2.4).

By Lemma 3.3 M^k and M^{k+1} are meet principal and hence by (3.3) $(C : M^{k+1})M^{k+1} = C \cap M^{k+1}$ and $[(A \cup Q) : M^k]M^k = (A \cup Q) \cap M^k$. Thus by modularity we have $C = (M^{k+1} \cup Q) \cap C = Q \cup (C \cap M^{k+1}) = Q \cup (C : M^{k+1})M^{k+1} = Q \cup [(A \cup Q) : M^k]M^{k+1} = Q \cup [(A \cup Q) \cap M^k]M \subseteq Q \cup (A \cup Q)M = Q \cup AM \cup QM \subseteq Q \subseteq C$.

Hence

$$Q = C = (A \cup Q) \cap (M^{k+1} \cup Q).$$

Since Q is meet irreducible and $Q \neq A \cup Q$ we must have $Q = M^{k+1} \cup Q$ and hence $Q \supseteq M^{k+1}$. Now let $Q \supseteq AB$ where $Q \not\supseteq A$. Then B is a join of meet principal elements, say $B = M_1 \cup \cdots \cup M_r$. Thus $Q \supseteq AM_i$ for $i = 1, \cdots, r$. Hence there exists k_i such that $Q \supseteq M_i^{k_i}$. Let $k = k_1 + \cdots + k_r$. Then by (2.16) $Q \supseteq B^k$ and it follows that Q is primary.

In Ward-Dilworth [4], a multiplicative lattice satisfying the ascending chain condition in which every irreducible is primary was called a "Noether lattice." Since these conditions are not sufficient for the results of commutative ideal theory we shall strengthen the definition as follows:

DEFINITION 3.1. A multiplicative lattice L is a *Noether lattice* if the following conditions are satisfied:
  (1) L is modular
  (2) L satisfies the accending chain condition
  (3) Every element of L is a join of principal elements.

From the results of §1 we conclude *the lattice of ideals of a Noetherian ring is a Noether lattice.*

Theorem 3.4 implies that *every irreducible element of a Noether lattice is primary.* From the results of §2, it follows that *every element of a Noether lattice has a normal primary decomposition.*

Now let $A = Q_1 \cap \cdots \cap Q_n$ be a normal decomposition of A and let $\{P_1, \cdots, P_n\}$ be the associated primes. If B is an arbitrary element of L, then $\{P_i \,|\, P_i \cup B \neq I\}$ is clearly an isolated set of primes. Let A_B denote the corresponding isolated component of A. We prove now an abstract version of the intersection theorem.

THEOREM 3.2. *(Intersection Theorem) If A and B are elements of a Noether lattice, then $\cap_k (A \cup B^k) = A_B$.*

Proof. According to the definition of isolated components $A_B = \cap \{Q_i \,|\, P_i \cup B \neq I\}$. Let $A_B^* = \cap \{Q_j \,|\, P_j \cup B = I\}$. Now $P_j \cup B = I$

implies $P_j^k \cup B = I$ for all k by (2.14) and hence $Q_j \cup B = I$. But then

$$A_B^* = \cap \{Q_j \mid P_j \cup B = I\} \supseteq \varPi \{Q_j \mid P_j \cup B = I\}$$

and hence $A_B^* \cup B = I$ by (2.14). From (2.14) we get $A_B^* \cup B^k = I$ all k. Hence $A \cup B^k = (A_B \cap A_B^*) \cup B^k = A_B \cup B^k$ by 2.15. Thus $A \cup B^k \supseteq A_B$ all k and hence $\cap_k (A \cup B^k) \supseteq A_B$.

Next let $\cap_k (A \cup B^k) \supseteq M$ where M is principal. Let $A \cup BM = R_1 \cap \cdots \cap R_m$ be a normal decomposition of $A \cup BM$ with associated primes $P_1', \cdots P_m'$. Then $R_i \supseteq BM$ and hence either $R_i \supseteq M$ or $P_i' \supseteq B$. But if $P_i' \supseteq B$, then $R_i \supseteq P_i'^{k_i} \supseteq B^{k_i}$ and hence $R_i \supseteq A \cup B^{k_i} \supseteq \cap_k (A \cup B^k) \supseteq M$. In either case we get $R_i \supseteq M$ and hence $A \cup BM = R_1 \cap \cdots \cap R_m \supseteq M$. It follows that $A \cup BM = A \cup M$. Since M is principal we have

$$A : M \cup B = (A \cup BM) : M = (A \cup M) : M = I.$$

Thus $P_i \cup B \neq I$ implies $P_i \not\supseteq A : M$. Since $Q_i \supseteq A \supseteq (A : M)M$ it follows that $Q_i \supseteq M$ for all i such that $P_i \cup B \neq I$. Hence $A_B = \cap \{Q_i \mid P_i \cup B \neq I\} \supseteq M$.

Since L is a Noether lattice, $\cap_k (A \cup B^k)$ is a join of principal elements. Thus

$$A_B \supseteq \underset{k}{\cap} (A \cup B^k)$$

and the proof of the theorem is complete

Setting $A = 0$ in Theorem 3.2 we get

COROLLARY 3.1. *If B is an element of a Noether lattice L, then*

$$\cap B^k = 0_B .$$

If $A \neq I$ is an element of a Noether lattice, then there exists a maximal element $P \neq I$ such that $P \supseteq A$. By (2.14) and (2.17) such an element is prime and clearly is a maximal prime of L.

DEFINITION 3.2. A Noether lattice L is *local* if it contains precisely one maximal prime.

Now suppose that L is local with the unique maximal prime P_0. If $P_i \neq I$ and $B \neq I$, then $P_0 \supseteq P_i \cup B$ and hence $P_i \cup B \neq I$. Thus $A_B = \cap \{Q_i \mid P_i \cup B \neq I\} = A$. Accordingly we get

COROLLARY 3.2. *If L is a local Noether lattice and $B \neq I$ is an element of L, then for each $A \in L$*

$$\underset{k}{\cap} (A \cup B^k) = A .$$

Setting $A = 0$, we have

COROLLARY 3.3.   *If $B \neq I$ is an element of a local Noether lattice, then*

$$\bigcap_k B^k = 0 \, .$$

4.   Quotient lattices. If L is a Noether lattice and $D \in L$, then $L/D = \{A \in L \mid A \supseteq D\}$ is clearly a sublattice of L. However, L/D need not be closed with respect to multiplication. We make L/D into a multiplicative lattice by defining

$$A \circ B = AB \cup D \, .$$

This multiplication is easily seen to be commutative, associative, distributive with respect to join, and has I as an identity element. Thus with this multiplication L/D is a multiplicative lattice.

Since $A : B \supseteq A$, it follows that L/D is closed with respect to residuation. This residuation is also the residuation associated with $A \circ B$ since

$$A \supseteq X \circ B \text{ if and only if } A \supseteq XB \, .$$

LEMMA 4.1.   *If L is a Noether lattice, then L/D is a Noether lattice. Furthermore if M is a principal element of L, then $M \cup D$ is a principal element of L/D.*

For if L is modular and satisfies the ascending chain condition, then clearly L/D is modular and satisfies the ascending chain condition.

Now let M be a principal element of L and let $A, B \in L/D$. Then

$$[A \cup B \circ (M \cup D)] : (M \cup D) = (A \cup BM \cup D) : (M \cup D)$$
$$= (A \cup BM \cup D) : M = (A \cup BM) : M$$
$$= A : M \cup B = A : (M \cup D) \cup B$$

and $M \cup D$ is thus join principal.   On the other hand

$$[A : (M \cup D) \cap B] \circ (M \cup D) = (A : M \cap B) \circ (M \cup D)$$
$$= (A : M \cap B)(M \cup D) \cup D$$
$$= (A : M \cap B)M \cup (A : M \cap B)D \cup D = (A : M \cap B)M \cup D$$
$$= (A \cap BM) \cup D = A \cap (BM \cup D)$$
$$= (A \cap (BM \cup BD) \cup D) = A \cap B \circ (M \cup D) \, .$$

and $M \cup D$ is thus meet principal.   Note that modularity is required in the sixth step of this argument.   Now if $A \in L/D$, then $A = A \cup D = M_1 \cup \cdots \cup M_k \cup D = (M_1 \cup D) \cup \cdots \cup (M_k \cup D)$ and hence every element of L/D is a join of principal elements.   Thus L/D is a Noether lattice.

The next lemma shows that the arithmetical properties of L are preserved in L/D.

LEMMA 4.2. *An element is prime or primary in L/D if and only if it is prime or primary in L.*

Let $A, B, C \in L/D$. Then $A \supseteq B \circ C$ if and only if $A \supseteq BC$. Hence A is prime or primary if and only if it is prime or primary in L.

COROLLARY 4.1. *If L is a local Noether lattice, and $D \neq I$, then L/D is a local Noether lattice.*

5. Congruence lattices. Congruence relations provide a second method for constructing new Noether lattices from a given Noether lattice. This construction is the abstract version of the construction of the generalized ring of quotients introduced by Chevalley [1]. Let L be a Noether lattice and let D be an arbitrary element of L. If $P_1, \cdots, P_n$ are the primes associated with a normal decompsition of A, then $\{P_i \mid D \supseteq P_i\}$ is an isolated set of primes and hence determines an isolated component A_D of A. For a given normal decomposition $A = Q_1 \cap \cdots \cap Q_n$ we have $A_D = \cap \{Q_i \mid D \supseteq P_i\}$. We will let A_D' denote $\{Q_j \mid D \not\supseteq P_j\}$ so that $A = A_D \cap A_D'$. We define $A \equiv B(D)$ if and only if $A_D = B_D$.

LEMMA 5.1. *If $A \supseteq B$, $A_D \supseteq B_D$.*

For let $A = Q_1 \cap \cdots \cap Q_m$ be a normal decomposition of A and let $P_1, \cdots, P_m$ be the associated primes. If $D \supseteq P_i$, then $Q_i \supseteq A \supseteq B \supseteq B_D \cap B_D'$. If $P_i \supseteq B_D'$, then $P_i \supseteq P_j'$ where $D \not\supseteq P_j'$ contrary to $D \supseteq P_i$. Hence $P_i \not\supseteq B_D'$ and thus $Q_i \supseteq B_D$. It follows that $A_D = \cap \{Q_i \mid D \supseteq P_i\} \supseteq B_D$.

LEMMA 5.2. *If $A \equiv B(D)$, then $A \cap C \equiv B \cap C(D)$.*

For let $A \cap C = Q_1 \cap \cdots \cap Q_n$ be a normal decomposition of $A \cap C$ with associated primes $P_1, \cdots P_n$. Then if $D \supseteq P_i$ we have $Q_i \supseteq A \cap C = A_D \cap C \cap A_D'$ and $P_i \not\supseteq A_D'$. Thus $Q_i \supseteq A_D \cap C = B_D \cap C \supseteq B \cap C$. Thus $(A \cap C)_D = \cap \{Q_i \mid D \supseteq P_i\} \supseteq B \cap C$. But then $(A \cap C)_D = ((A \cap C)_D)_D \supseteq (B \cap C)_D$ by Lemma 5.1. Similarly $(B \cap C)_D \supseteq (A \cap C)_D$ and thus $(A \cap C)_D = (B \cap C)_D$. It follows that $A \cap C \equiv B \cap C(D)$.

LEMMA 5.2. *If $A \equiv B(D)$, then $A \cup C \equiv B \cup C(D)$.*

For let $A \cup C = Q_1 \cap \cdots \cap Q_n$ be a normal decomposition of $A \cup C$ with associated primes $P_1, \cdots, P_n$. Then if $D \supseteq P_i$ we have $Q_i \supseteq A_D \cap A_D'$ and $Q_i \supseteq C$. Since $P_i \not\supseteq A_D'$ we have $Q_i \supseteq A_D$. Thus $Q_i \supseteq B_D \supseteq B$ and this $Q_i \supseteq B \cup C$. But then $(A \cup C)_D = \cap \{Q_i \mid D \supseteq P_i\} \supseteq B \cup C$ and hence $(A \cup C)_D \supseteq (B \cup C)_D$ by Lemma 5.1. Similarly $(B \cup C)_D \supseteq (A \cup C)_D$ and hence $A \cup C \equiv B \cup C(D)$.

LEMMA 5.3. *If* $A \equiv B(D)$, *then* $AC \equiv BC(D)$.

For let $AC = Q_1 \cap \cdots \cap Q_n$ be a normal decomposition of AC with associated primes $P_1, \cdots, P_n$. Then if $D \supseteq P_i$ we have $Q_i \supseteq AC = (A_D \cap A'_D)C \supseteq A_D C A'_D$ where $P_i \not\supseteq A'_D$. Thus $Q_i \supseteq A_D C = B_D C \supseteq BC$. It follows that $(AC)_D \supseteq BC$ and hence by Lemma 5.1 that $(AC)_D \supseteq (BC)_D$. Similarly $(BC)_D \supseteq (AC)_D$ and hence the conclusion of the lemma follows.

LEMMA 5.4. *If* $A \equiv B(D)$, *then* $A : C \equiv B : C(D)$ *and* $C : A \equiv C : B(D)$.

For let $X : Y = Q_1 \cap \cdots \cap Q_n$ be a normal decomposition of $X : Y$ with associated primes $P_1, \cdots, P_n$. Then if $D \supseteq P_i$ we have $Q_i \supseteq X : Y = (X_D \cap X'_D) : Y = (X_D : Y) \cap (X'_D : Y) \supseteq (X_D : Y)(X'_D : Y)$. Now $P_i \not\supseteq X'_D : Y$ since otherwise $P_i \supseteq X'_D : Y \supseteq X'_D$. Thus $Q_i \supseteq X_D : Y \supseteq X_D : Y_D$ since $Y_D \supseteq Y$. Hence $(X : Y)_D = \cap \{Q_i \mid D \supseteq P_i\} \supseteq X_D : Y_D$.

On the other hand, $X \supseteq (X : Y)Y$ and hence

$$X_D \supseteq [(X : Y)Y]_D = (X : Y)_D Y_D \text{ by Lemma 5.1 and 5.3 .}$$

Hence $X_D : Y_D \supseteq (X : Y)_D$ and thus $(X : Y)_D = X_D : Y_D$ by the result of the previous paragraph.

Now if $A \equiv B(D)$, then $(A : C)_D = A_D : C_D = B_D : C_D = (B : C)_D$ and thus $A : C \equiv B : C(D)$. Similarly $C : A \equiv C : B(D)$.

Lemma 5.1–5.4 show that congruence mod D is a congruence relation on L preserving meet, join, multiplication, and residuation. In view of the fact that principal elements are defined in terms of equations involving meet, join, multiplication, and residuation it follows that the congruence class of a principal element is principal in the lattice of congruence classes.

Now let L_D denote the multiplicative lattice of congruence classes. Then we have

THEOREM 5.1. *If L is a Noether lattice, then L_D is a Noether lattice.*

The arithmetical properties of L_D may be described as follows:

THEOREM 5.2. *The primes and primaries of L_D are precisely the congruence classes determined by the primes and primaries of L. The proper prime elements of L_D are the congruence classes determined by primes P such that $D \supseteq P$.*

Proof. Let the congruence class $\{C\}$ be a prime in L_D. Now $C \supseteq P_1^{k_1} \cdots P_n^{k_n}$ where $P_1, \cdots, P_n$ are primes of L such that $P_i \supseteq C$. But then $\{C\} \supseteq \{P_1\}^{k_1} \cdots \{P_n\}^{k_n}$ and hence $\{C\} \supseteq \{P_i\} \supseteq \{C\}$ for since i. Thus $\{C\} = \{P_i\}$ where P_i is a prime of L. Conversely, if P is a prime of L, then, by definition, $P_D = P$ or $P_D = I$ according as $D \supseteq P$ or $D \not\supseteq P$. Hence $\{P\} \supseteq \{A\}\{B\}$ implies $P_D \supseteq A_D B_D$ and hence $P_D \supseteq A_D$ or

$P_D \supseteq B_D$ which implies $\{P\} \supseteq \{A\}$ or $\{P\} \supseteq \{B\}$. Thus the primes of L_D have the form $\{P\}$ where P is a prime of L and $\{P\}$ is proper if and only if $D \supseteq P$.

Next let $\{C\}$ be a primary element of L_D associated with the prime $\{P\}$. Now $C_D = Q_1 \cap \cdots \cap Q_r$ when $D \supseteq P_i$ and P_i is the prime associated with Q_i. We may suppose that $\{C\} \neq \{I\}$ and thus that $D \supseteq P$. Now $\{C\} \supseteq \{P\}^k$ for some k. Hence $C_D \supseteq P_D^k = P^k$. But then $P_i \supseteq P$ for all i. Since $P \supseteq C_D$ we have $P \supseteq P_j$ for some j and hence $P = P_j$. On the other hand $P_1, \cdots, P_r$ are distinct and thus $P \not\supseteq P_i$ for $i \neq j$. Since $\{C\} \supseteq \{Q_1\} \cdots \{Q_r\}$ and $\{P\} \not\supseteq \{Q_i\}$ for $i \neq j$. We have $\{C\} \supseteq \{Q_j\}$ since $\{C\}$ is primary in L_C. But then $\{C\} = \{Q_j\}$ where Q_j is a primary of L. Conversely if Q is a primary element of L, then, by definition, $Q_D = Q$ or $Q_D = I$ according as $D \supseteq P_Q$ or $D \not\supseteq P_Q$. Hence $\{Q\} \supseteq \{A\}\{B\}$ implies $\{Q\} \supseteq \{A\}$ or $\{P\} \supseteq \{B\}$ and $\{Q\}$ is a primary element of L_D. This completes the proof of the theorem.

It should be noted that distinct primes contained in D determine distinct congruence classes.

THEOREM 5.3. *If P is a proper prime of L, then L_P is a local Noether lattice.*

Proof. For if $\{P'\}$ is a proper prime of L_P, then $P \supseteq P'$ and hence $\{P\} \supseteq \{P'\}$. On the other hand $\{P\}$ is a proper prime of L_P. Thus $\{P\}$ is the unique maximal proper prime of L_P.

6. The principal element theorem. The critical steps in the proof of the principal element theorem will depend upon the finite dimensionality of certain quotient lattices. The finite dimensionality rests in turn upon the following theorem on modular lattices.

THEOREM 6.1. *Let S be a subset of a modular lattice L which generates L under finite joins. Then if S satisfies the descending chain condition, L also satisfies the descending chain condition.*

We begin by proving two lemmas on quotient lattices in a modular lattice.

LEMMA 6.1. *Let L be a modular lattice. If a/b and b/c are quotient lattices satisfying the descending chain condition, then a/c satisfies the descending chain condition.*

For let $x_1 \geq x_2 \geq \cdots \geq x_n \geq \cdots$ be a descending chain in a/c. Then $x_1 \cup b \geq x_2 \cup b \geq \cdots \geq x_n \cup b \geq \cdots$ and $x_1 \cap b \geq x_2 \cap b \geq \cdots \geq x_n \cap b \geq \cdots$ are descending chains in a/b and b/c respectively. Hence there exists an integer k such that $x_n \cup b = x_k \cup b$ and $x_n \cap b = x_k \cap b$ all $n \geq k$.

Since $x_k \geqq x_n$ all $n \geqq k$ it follows from the modular law that $x_k = x_n$ all $n \geqq k$ and hence a/c satisfies the descending chain condition.

LEMMA 6.2. *Let L be a modular lattice. If a/b and c/d satisfy the descending chain condition, then $a \cup c/b \cup d$ and $a \cap c/b \cap d$ satisfy the descending chain condition.*

For $a/a \cap (b \cup d)$ is a subquotient of a/b and hence satisfies the descending chain condition. But $a/a \cap (b \cup d)$ is isomorphic to $a \cup d/b \cup d$ and hence $a \cup d/b \cup d$ satisfies the descending chain condition. A similar argument shows that $a \cup c/a \cup d$ satisfies the descending chain condition. But then $a \cup c/b \cup d$ satisfies the descending chain condition by Lemma 6.1. A dual proof shows that $a \cap c/b \cap d$ likewise satisfies the descending chain condition.

Repeated application of Lemma 6.2 gives the following corollary.

COROLLARY 6.1. *Let L be a modular lattice. If $a_1/b, \cdots, a_n/b$ satisfy the descending chain condition, then $(a_1 \cup \cdots \cup a_n)/b$ satisfies the descending chain condition.*

We now give the proof of the theorem.

Proof of Theorem 6.1. If s_1 and s_2 are elements of S, then $s_1 \cap s_2$ is a join of elements of S. Hence there exists $s_3 \in S$ such that $s_1 \geqq s_3$ and $s_2 \geqq s_3$. Thus S is a directed set and hence has a null element z by the descending chain condition. z is clearly the null element of L. We shall show first that the quotient lattices s/z satisfy the descending chain condition by making an induction upon the elements of S. Clearly z/z satisfies the descending chain condition. Let us suppose that s/z satisfies the descending chain condition for all $s < s_0$. If s_0/z does not satisfy the descending chain condition, there exists an infinite descending chain

$$a_0 > a_1 > \cdots > a_n > \cdots$$

in s_0/z. Then $s_0 \geqq a_0 > a_1$ and $a_1 = s_1 \cup \cdots \cup s_n$. But then s_i/z satisfies the descending chain condition by the induction assumption and hence $a_1/z = (s_1 \cup \cdots \cup s_n)/z$ satisfies the descending chain condition by Corollary 6.1. But $a_1 > \cdots > a_n > \cdots$ is an infinite descending chain in a_1/z. Hence s_0/z satisfies the descending chain condition. Since S itself satisfies the descending chain condition it follows by induction that s/z satisfies the descending chain condition for all $s \in S$. Now let $a \in L$ and let $a = s_1 \cup \cdots \cup s_m$. Since s_i/z satisfies the descending chain condition for each i, it follows from Corollary 6.1 that a/z satisfies the descending chain condition and hence that L itself satisfies the descending chain condition. This completes the proof of the theorem.

The following theorem describes the finite dimensionality properties of local Noether lattices.

THEOREM 6.2. *Let L be a local Noether lattice with maximal prime P. Then L/P^k is finite dimensional.*

Proof. The elements $P^k \cup M$ where M is principal generate L/P^k under finite joins. Hence by Theorem 6.1 it will suffice to show that these elements satisfy the descending chain condition. Let

$$P^k \cup M_1 \supset P^k \cup M_2 \supset \cdots \supset P^k \cup M_n \cdots$$

We may clearly assume that $P^k \cup M_1 \neq I$. Since L has exactly one maximal prime element P it follows that $P \supseteq P^k \cup M_1 \supseteq M_1$. Now $(P^k \cup M_{r+1}) : M_r \neq I$ since otherwise $M_r = [(P^k \cup M_{r+1}) : M_r]M_r = (P^k \cup M_{r+1}) \cap M_r$ by Corollary 3.1 and hence $P^k \cup M_{r+1} \supseteq M_r$ which implies $P^k \cup M_{r+1} \supseteq P^k \cup M_r$ contrary to $P^k \cup M_r \supset P^k \cup M_{r+1}$. Thus

$$P \supseteq (P^k \cup M_{r+1}) : M_r \text{ for all } r \ .$$

Now $P \supseteq M_1$. Suppose that $P^r \supseteq M_r$ where $r < k$. Then

$$P^{r+1} \supseteq [(P^k \cup M_{r+1}) : M_r]P^r \supseteq [(P^k \cup M_{r+1}) : M_r]M_r$$
$$= (P^k \cup M_{r+1}) \cap M_r \ .$$

Thus $P^{r+1} \supseteq P^k \cup [(P^k \cup M_{r+1}) \cap M_r] = (P^k \cup M_{r+1}) \cap (P^k \cup M_r) = P^k \cup M_{r+1} \supseteq M_{r+1}$ by the modular law. Hence by induction we get

$$P^k \supseteq M_k \ .$$

Thus $P^k = P^k \cup M_k \supset P^k \cup M_{k+1} \supseteq P^k$ which is impossible. It follows that the elements $P^k \cup M$ satisfy the descending chain condition and hence L/P^k is finite dimensional.

A lemma on lattice isomorphism will be needed in the proof of the main theorem.

LEMMA 6.3. *Let L be a Noether lattice in which 0 is a prime. Then if M is a nonzero principal element the quotient lattice AM/BM is lattice isomorphic to A/B.*

For if 0 is a prime, then $0 \supseteq (0 : X)X$ where $X \neq 0$ implies $0 \supseteq 0 : X$. Hence $0 : X = 0$ whenever $X \neq 0$. But then $(XM) : M = (0 \cup XM) : M = 0 : M \cup X = X$. Now the mapping $X \to XM$ maps A/B into AM/BM. It is order preserving since $X \supseteq Y$ implies $XM \supseteq YM$. It is 1-1 since $XM = YM$ implies $X = (XM) : M = (YM) : M = Y$. Finally if $V \in AM/BM$, then $AM \supseteq V$ and hence $V = V \cap AM = (V : M \cap A)M$ since M is principal. But $A \supseteq A \cap V : M \supseteq A \cap (BM) : M \supseteq A \cap B = B$ and hence

$X = A \cap V : M$ maps onto V. Thus $X \to XM$ is a 1-1 order preserving map of A/B onto AM/BM and hence A/B and AM/BM are lattice isomorphic:

In formulating the proof of the Principal Element Theorem, we will make use of a modification of an argument due to D. Rees [3].

THEOREM 6.3. *Let L be a local Noether lattice in which 0 is a prime. Let P be the maximal prime of L. Then if the principal element M is primary with associated prime P, P is the only non zero prime of L.*

Proof. Let P' be a nonzero prime of L. Then $P' \supseteq N$ where N is a nonzero principal element of L. Now by (2.13)

$$N : M \subseteq N : M^2 \subseteq \cdots \subseteq N : M^n \subseteq \cdots .$$

Since the ascending chain condition holds in L, it follows that $N : M^k = N : M^{k+1}$ for some k. But then

$$N \cap M^{k+1} = (N : M^{k+1})M^{k+1} = (N : M^k)M^k M = (N \cap M^k)M .$$

Then if $d(A/B)$ denotes the dimension of the quotient A/B we have

$$\begin{aligned}
d((N \cup M^{k+1})/M^{k+1}) &= d(N/(N \cap M^{k+1})) \\
&= d(N/(N \cap M^k)M) \\
&= d(N/NM) + d(NM/(N \cap M^k)M) \\
&= d(I/M) + d(N/(N \cap M^k)) \\
&= d(M^k/M^{k+1}) + d((N \cup M^k)/M^k) \\
&= d((N \cup M^k)/M^{k+1})
\end{aligned}$$

making use of modularity and Lemma 6.3. Note that all of the quotients are finite dimensional since $M \supseteq P^l$ for some l and hence $M^{k+1} \supseteq P^{(k+1)l}$. But $L/P^{(k+1)l}$ is finite dimensional by Theorem 6.2. Since $N \cup M^k \supseteq N \cup M^{k+1}$ it follows from the equality of the dimension that $N \cup M^k = N \cup M^{k+1}$.
Then

$$M^k = M^k \cap (N \cup M^k) = M^k \cap (N \cup M^{k+1}) = (M^k \cap N) \cup M^{k+1}$$
$$= (N : M^k)M^k \cup M^{k+1} = (N : M^k \cup M)M^k .$$

Hence $I = M^k : M^k = [(N : M^k \cup M)M^k] : M^k = N : M^k \cup M$. Now $P \not\supseteq N : M^k$ since otherwise $P \supseteq N : M^k \cup M = I$. Since L is local with maximal prime P it follows that $N : M^k = I$ and hence $N \supseteq M^k$. But then

$$P' \supseteq N \supseteq M^k \supseteq P^{kl} .$$

Thus $P' \supseteq P$ and since P is maximal we have $P' = P$. This completes the proof of the theorem.

DEFINITION 6.1. Let P be a prime element of a Noether lattice L. The least upper bound of the length of chains $P \supset P_1 \supset P_2 \supset \cdots \supset P_k$ where the P_i are primes of L is called the *rank* of P.

The abstract version of the Krull Principal Ideal Theorem (Krull [2]) is formulated as follows:

THEOREM 6.4. (*Principal Element Theorem*) *Let P be a minimal prime containing the principal element M of a Noether lattice. Then the rank of P is at most one.*

Proof. Let L denote the Noether lattice. Without any loss of generality we may suppose that L is local and that P is the maximal prime of L. For L_P is a local Noether lattice in which the maximal prime is the congruence class of P. By Theorem 5.2, the rank of the congruence class of P in L_P is the same as the rank of P in L. But if L is a local Noether lattice and P is the maximal prime of L, then M is primary with P as the associated prime. For let P_i be a prime associated with a normal decomposition of M. Then $P \supseteq P_i$ since P is maximal. But since P is minimal over M we must have $P = P_i$ and hence M is primary associated with P.

Now let $P \supset P_1 \supseteq P_2$. Then L/P_2 is a local Noether lattice with maximal prime P. The zero element of L/P_2 is P_2 which is prime by Lemma 4.2. Also by Lemma 4.2, the principal element $M \cup P_2$ of L/P_2 is primary with P as associated prime. Hence by Theorem 6.3, P is the only nonzero prime of L/P_2. But by Lemma 4.2, P_1 is a prime of L/P_2 which is distinct from P. Thus $P_1 = P_2$ and the rank of P is at most one.

We will conclude this section by proving the theorem which extends the result of Theorem 6.4 to nonprincipal elements. A preliminary lemma is required.

LEMMA 6.4. *Let L be a Noether lattice and let $P = P_0 \supset P_1 \supset \cdots \supset P_{r-1} \supset P_r$ be a chain of prime elements of L. If M is a principal element of L such that $P \supseteq M$, then there exists a chain of prime elements $P = P_0^* \supset P_1^* \supset \cdots P_{r-1}^* \supset P_r$ such that $P_{r-1}^* \supseteq M$.*

For let us set $P_0^* = P_0 = P$. If $P_1 \supseteq M$ we set $P_1^* = P_1$, otherwise we let P_1^* be a minimal prime such that $P \supseteq P_1^* \supseteq M \cup P_2$. Then P_1^* is a minimal prime containing the principal element $M \cup P_2$ of L/P_2. Hence by Theorem 6.4 we conclude that $P \neq P_1^*$ since P is of rank two in L/P_2. Also since $P_1 \not\supseteq M$ we have $P_2 \not\supseteq M$ and hence $P_1^* \neq P_2$. Thus

$P = P_0^* \supset P_1^* \supset P_2$ and $P_1^* \supseteq M$. Repeating this argument $r - 1$ times gives the chain $P = P_0 \supset P_1^* \supset \cdots \supset P_{r-1}^* \supset P_r$ with $P_{r-1}^* \supseteq M$.

THEOREM 6.5. *Let L be a Noether lattice and let P be a minimal prime containing $M_1 \cup \cdots \cup M_k$ where each M_i is principal. Then the rank of P is at most k.*

Proof. Let $P = P_0 \supset P_1 \supset \cdots \supset P_{r-1} \supset P_r$ be an arbitrary chain of primes beginning with P. Then $P \supseteq M_k$ and hence by Lemma 6.4 there exists a chain $P = P_0^* \supset P_1^* \supset \cdots \supset P_{r-1}^* \supset P_r$ such that $P_{r-1}^* \supseteq M_k$. Now P is a minimal prime containing $(M_1 \cup M_k) \cup \cdots \cup (M_{k-1} \cup M_k)$ where each $M_i \cup M_k$ is principal in L/M_k. Since $P = P_0^* \supset \cdots \supset P_{r-1}^*$ is a prime chain in L/M_k we may suppose by induction that $r - 1 \leq k - 1$ and hence conclude that $r \leq k$. This completes the proof of the theorem.

7. Polynomial ideals. In this section it will be shown that the principal elements of the lattice of ideals of a unique factorization integral domain are precisely the principal ideals. In particular, this result holds for ideals in a polynomial ring.

THEOREM 7.1. *Let R be a unique factorization integral domain. Then every principal element in the lattice of ideals of R is a principal ideal.*

Proof. Let $M = (m_1, \cdots, m_k)$ be a principal element in the lattice of ideals of R. Then $M \supseteq (m_i)$ and hence $(m_i) = (m_i) \cap M = ((m_i) : M)M = X_i M$ where $X_i = (m_i) : M$. Then there exist elements $x_{i1}, \cdots, x_{ik}$ in X_i such that

$$m_i = x_{i1}m_1 + \cdots + X_{ik}m_k \qquad\qquad i = 1, \cdots, k \, .$$

Let

$$\Delta = \begin{vmatrix} 1 - x_{11} & - x_{12} & \cdots & - x_{1k} \\ - x_{21} & 1 - x_{22} & \cdots & - x_{2k} \\ \cdot\ \cdot\ \cdot\ & \cdot\ \cdot\ \cdot\ \cdot & \cdot\ \cdot & \cdot\ \cdot \\ - x_{k1} & - x_{k2} & \cdots & 1 - x_{kk} \end{vmatrix}$$

Then $\Delta m_i = 0$ for $i = 1, \cdots, k$ and hence $\Delta = 0$ since R is an integral domain. But $\Delta = 1 - x$ where $x \in X_1 \cup \cdots \cup X_x$. Hence $X_1 \cup \cdots \cup X_k = (1)$ and thus there exist elements $x_i \in X_i$ such that

$$x_1 + \cdots + x_n = 1 \, .$$

Since $(m_i) = X_i M$ there exist elements r_{ij} such that

$$x_i m_j = r_{ij} m_i \, .$$

Let b be the g. c. d. of $m_1, \cdots, m_k$. Then since R is a unique factorization domain it follows that x_i is a multiple of m_i/d. Thus $dx_i = y_i m_i$. Hence

$$d = dx_1 + \cdots + dx_k = y_1 m_1 + \cdots + y_k m_k \in M .$$

On the other hand it is clear that $M \subseteq (d)$. Hence $M = (d)$ and it follows that M is a principal ideal.

It should be noted that if L is a Noether lattice which satisfies the cancellation law for nonzero elements and in which every element has a unique factorization into primes (for example, the lattice of ideals of the ring of integers in an algebraic number field), then $A : B = (A \cap B)/B$ and hence 1.1 and 1.2 hold identically in L. Thus every element of L is principal. R. S. Pierce notes that the converse is also true. *If L is a Noether lattice in which 0 is prime and in which every element is principal, then the cancellation law holds in L and every element of L has a unique decomposition into primes.* The cancellation law follows from Lemma 6.3 and the unique decomposition into primes can be deduced from Corollary 3.1.

8. Distributive lattices. Let L be a distributive lattice satisfying the ascending chain condition and with a null element 0. Then if the meet operation is taken as the multiplication it is well known (Ward [5]) that L is a multiplicative lattice with residuation. Since L is modular and satisfies the ascending chain condition, L will be a Noether lattice if and only if the principal elements of L generate L under finite joins.

LEMMA 8.1. *M is a principal element of L if and only if M has a complement in L.*

For let M be principal. Then $0 \supseteq (0 : M) \cap M$ and hence $(0 : M) \cap M = 0$. On the other hand, $0 : M \cup M = [0 \cup (M \cap M] : M = M : M = I$ and hence $0 : M$ is a complement of M. Conversely, suppose that M is an element with a complement M'. It is easily verified that $X : M = X \cup M'$. Hence

$$[A \cup (B \cap M)] : M = A \cup (B \cap M) \cup M' = A \cup M' \cup B = A : M \cup B$$

$$[A \cap (B : M)] \cap M = [A \cap (B \cup M')] \cap M = (A \cap M) \cap B$$

and M is principal. The proof of the lemma is thus complete.

Now the complemented elements of a distribution lattice are closed under meet and join and form a Boolean algebra. Thus we get

THEOREM 8.1. *A distributive lattice is a Noether lattice if and only if it is a finite Boolean algebra.*

It should be noted that the prime elements of a Boolean algebra are the elements covered by the unit element. Any two such elements are non comparable and hence the rank of every proper prime is zero.

REFERENCES

1. C. Chevalley, *On the notion of the ring of quotients of a prime ideal*, Bull. Amer. Math. Soc., **50** (1944), 93–97.

2. W. Krull, *Dimensionstheorie in Stellenringe J. reine angew*, Math. **179** (1938), 204–226.

3. D. Rees, *Two classical theorems of ideal theory*, Proc. Camb. Phil. Soc., **52** (1956), 155–157.

4. M. Ward and R. P. Dilworth, *Residuated lattices*, Trans. Amer. Math. Soc., **45** (1939)' 335–354.

5. M. Ward, *Residuated distributive lattices*, Duke Math. Journal **6** (1940),641–651.

CALIFORNIA INSTITUTE OF TECHNOLOGY

Dilworth's Early Papers on Residuated and Multiplicative Lattices

D. D. Anderson

The idea of residuation goes back to Dedekind, who introduced it into ideal theory. Residuation now plays an important role in several fields of mathematics, especially commutative ring theory. Let R be a commutative ring with identity. The *residual* of an ideal B with respect to an ideal A is the ideal $A\!:\!B = \{x \in R \mid xB \subseteq A\}$. Here $A\!:\!B$ is the largest ideal X of R with the property that $BX \subseteq A$. Residuation may be defined in other algebraic structures and may be defined independently of multiplication.

In his paper "Abstract residuation over lattices" [8], Dilworth began an abstract study of residuation over lattices independently of multiplication. Let L be a lattice with greatest element I. Residuation is defined as a binary operation $:$ on L satisfying

(1) $A\!:\!A = I$,
(2) $(A\!:\!B)\!:\!C = (A\!:\!C)\!:\!B$,
(3) $A\!:\!(B \vee C) = (A\!:\!B) \wedge (A\!:\!C)$,
(4) $(A \wedge B)\!:\!C = (A\!:\!C) \wedge (B\!:\!C)$,
(5) $A\!:\!B = B\!:\!A = I$ implies that $A = B$

It is straightforward to show that $A\!:\!B = I$ if and only if $B \leq A$. In [8] Dilworth gives essential properties of residuation along with consistency and independence proofs of the axioms. He shows that for a Boolean algebra, the only residuation is the one given by $A\!:\!B = A \vee B$.

The idea of an abstract study of residuation in multiplicative structures was initiated by Ward [17,18]. Ward and Dilworth collaborated to write the paper "Residuated lattices" [21] ([20] is an announcement of the results which appeared in [21]). The first part of this paper exhibits a reciprocity between the operations of residuation and multiplication. Let L be a lattice with greatest element I. If L

has a commutative associative multiplication that has I as a multiplicative identity and distributes over finite joins, then a residuation may be defined on L if either

    (1) L is join-complete and the multiplication distributes over arbitrary joins, or

    (2) L satisfies the ascending chain condition on sets of the form

$$\{X \in L \mid XB \leq A\}.$$

In either case, the residuation may be defined by $A{:}B = \bigvee\{X \in L \mid BX \leq A\}$. Dually, suppose that there is defined a residuation on the lattice L satisfying the five previously given properties of residuation. If either

    (1) L is meet-complete and $A{:}(\bigvee A_i) = \bigwedge(A{:}A_i)$, or

    (2) L satisfies the descending chain condition on sets of the form

$$\{Y \in L \mid B \leq Y{:}A\},$$

then one can define a product on L with the previously mentioned properties by defining AB to be $\bigwedge\{C \in L \mid B \leq C{:}A\}$.

The paper [21] gives various conditions for a lattice to be residuated. For example, the complemented lattices which can be residuated are the Boolean algebras. The paper also introduces the notion of a weak meet principal element. The importance of this concept in abstract ideal theory is discussed in the article by E. W. Johnson in this chapter. The paper ends by showing that the conditions:

$$(A{:}B) \vee (B{:}A) = I,$$
$$A{:}(B \wedge C) = (A{:}B) \vee (A{:}C),$$
$$(B \vee C){:}A = (B{:}A) \vee (C{:}A)$$

are all equivalent and imply distributivity.

The importance of [21] is two-fold. Besides laying the foundation for Dilworth's future paper "Abstract commutative ideal theory" [11] which is discussed in Johnson's article, this paper points out the "duality" between multiplication and residuation. This duality is, however, better exhibited using lattice modules. This point of view is discussed in [2] where this lattice module duality is used to give a unified treatment of primary and coprimary decomposition and of grade and cograde. This same point of view is also used by W. H. Rowan [15] to study tertiary and cotertiary decompositions in lattice modules. Rowan's paper contains some nice applications to congruence modular varieties.

Two other papers of Ward and Dilworth involving commutative multiplicative lattices are "Evaluations on residuated structures" [23] and "The lattice theory of ova" [22]. In today's language, an "ovum" is a commutative monoid with zero. In [22], it is observed that the lattice of ideals of a commutative monoid with zero forms a complete distributive multiplicative lattice and that the principal ideals are weak meet principal elements. Also considered is the lattice of divisorial ideals. The theory of distributive Noether lattices and its relationship to the lattice of ideals of a semigroup is investigated in [1] and a fairly thorough survey of the ideal theory of a commutative monoid with zero and its relationship to abstract ideal theory is carried out in [3].

The last two papers of Dilworth that we will discuss both involve noncommutative multiplicative lattices. Just as the lattice of ideals of a commutative ring forms a commutative multiplicative lattice, the lattice of left ideals (or right ideals or two-sided ideals) of a noncommutative ring forms a complete lattice with a noncommutative multiplication. In "Non-commutative residuated lattices" [10], the reciprocity between residuation and commutative multiplication given in [21] is extended to the noncommutative setting. Many other results from [21] including primary decompositions are also extended to noncommutative multiplicative lattices. Results from noncommutative ring theory are generalized to these multiplicative lattices. In "Non-commutative arithmetic" [9], Dilworth studies the problem of determining conditions on a noncommutative multiplicative lattice in order to obtain unique factorization into irreducibles.

The study of noncommutative multiplicative lattices or "noncommutative residuation" has received much less attention than the commutative case. Recently, I. Fleischer [13] has used the setting of a noncommutative multiplicative lattice to give a lattice-theoretic study of the various radicals defined on a ring. As previously mentioned, [15] involves noncommutative multiplication and residuation. In a series of papers [5,6,7], M. R. da Costa Reis, has considered partial residuation in groupoid-lattices. Related work, due to O. Steinfeld, is in [16].

We have only barely scratched the surface of papers related to residuation. For the general theory of residuation (or more properly, residuated mappings), the reader is referred to the monograph *Residuation Theory* by T. S. Blyth and M. F. Janowitz [4]. Curiously, there seems to be little interaction between the work done in this general setting and lattice-theoretic work. Partially ordered algebraic systems are thoroughly covered in a book by that title by L. Fuchs [14]. Both books contain extensive bibliographies. There is also a French connection in the theory of residuated semigroups and lattices. The interested reader is referred to the treatise [12] by M. L. Dubriel-Jacotin, L. Lesieur, and R. Croisot.

REFERENCES

1. D. D. Anderson, *Distributive Noether lattices*, Michigan Math. J. **22** (1975), 109–115.
2. ________, *Fake rings, fake modules, and duality*, J. Algebra **47** (1977), 425–432.
3. ________ and E. W. Johnson, *Ideal theory in commutative semigroups*, Semigroup Forum **30** (1984), 127–158.
4. T. S. Blyth and M. F. Janowitz, "Residuation Theory," Pergamon Press, Oxford, 1972.
5. M. Raquel P. da Costa Reis, *Partial residuals in groupoid-lattices-I*, Math. Japonica **25** (1980), 19–26.
6. ________, *Partial residuals in groupoid-lattices-II*, Math. Japonica **25** (1980), 607–618.
7. ________, *Partial residuals in groupoid-lattices-III*, Math. Japonica **26** (1981), 467–474.

8. R. P. Dilworth, *Abstract residuation over lattices*, Bull. Amer. Math. Soc. **44** (1938), 262–267. Reprinted in Chapter 6 of this volume.

9. ________, *Non-commutative arithmetic*, Duke Math. J. **5** (1939), 270–280. Reprinted in Chapter 6 of this volume.

10. ________, *Non-commutative residuated lattices*, Trans. Amer. Math. Soc. **46** (1939), 426–444. Reprinted in Chapter 6 of this volume.

11. ________, *Abstract commutative ideal theory*, Pacific J. Math. **12** (1962), 481–498. Reprinted in Chapter 6 of this volume.

12. M. L. Dubriel-Jacotin, L. Lesieur, and R. Croisot, " Leçons sur la Théorie des Treillis des Structures Algébrique Ordonnés et des Treillis Géométriques," Gauthier-Villars, Paris, 1953.

13. I. Fleischer, *A lattice theoretic look at some ring theoretic radicals*, preprint.

14. L. Fuchs, "Partially Ordered Algebraic Systems," Pergamon Press, Oxford, 1963.

15. W. H. Rowan, *Tertiary decomposition in lattice modules*, Algebra Universalis **22** (1986), 27–49.

16. O. Steinfeld, *On groupoid-lattices*, in "Contributions to General Algebra, Proceedings of the Klagenfurt conference," 1978, pp. 357–372.

17. M. Ward, *Residuation in structures over which a multiplication is defined*, Duke Math. J. **3** (1937), 627–636.

18. ________, *Structure residuation*, Ann. of Math. (2) **39** (1938), 558–568.

19. ________, *Residuated distributive lattices*, Duke Math. J. **6** (1940), 641–651.

20. ________ and R.P. Dilworth, *Residuated lattices*, Proc. Nat. Acad. Sci. **24** (1938), 162–164.

21. ________ and ________, *Residuated lattices*, Trans. Amer. Math. Soc. **35** (1939), 335–354. Reprinted in Chapter 6 of this volume.

22. ________ and ________, *The lattice theory of ova*, Ann. of Math. (2) **40** (1939), 600–608.

23. ________ and ________, *Evaluations over residuated structures*, Ann. of Math. (2) **40** (1939), 328–338.

University of Iowa
Iowa City, IA 52242
U. S. A.

Abstract Ideal Theory:
Principals and Particulars

E. W. Johnson

The subject of abstract ideal theory has developed along two more or less distinct though related lines since its beginnings in the twenties and thirties. One line involves an ideal operator x which is applied to subsets of a semigroup S to produce a lattice ordered semigroup L of ideals. The other, and the one of primary interest to us in this discussion, begins immediately with the lattice ordered semigroup L, dispensing entirely with the assumption of underlying elements.

To a large extent, the differences between the two approaches are cosmetic. Identification of an element E in the lattice setting with the set $S(E)$ of elements less than or equal to E integrates the second approach with the first [12,26]. However, the points of view are sufficiently different to recognize them as distinct. The first approach has its origins in the work of Prüfer [30]. The second has its origins in the work of Ward [31,32,33] and (independently) Krull [24,25]. Dilworth uses the latter approach in his paper "Abstract commutative ideal theory" [11], following the same style used in the earlier Ward [31,32,33], Ward-Dilworth [34,35], and Dilworth [9,10] papers in the area.

Working directly with ideals, without the availability of underlying elements, leads to a very elegant presentation style. However, there is no longer an obvious idea of what a principal ideal (element) is or should be. In the ideal-generated-by approach of Prüfer [30], Lorenzen [27], Aubert [5], etc., the elements (a) of the ideal lattice are natural candidates (the degree of naturalness depending on the axioms, of course). If, for example, it is assumed that $E = (a) = aS$ in a commutative semigroup S with 1, then the weak meet principal (WMP) identity:

$$E \cap B = (B{:}E)E$$

holds.

The WMP identity has long been recognized as an important property of principal ideals and was chosen by Ward and Dilworth for the definition of a principal element in [34], for example. However, the perspective of time has shown that this property, while sufficient to yield the Noether decomposition theorems [34], is insufficient to stand on its own as a definition of a principal element. In fact, a multiplicative lattice L in which every element has this property (a principal element lattice if you will) need not even be Noetherian. Nor need it satisfy the cancellation law if 0 is prime, even if L is assumed Noetherian.

If A, B and E are ideals in an integral domain R, $AE = BE$ implies $A = B$ if E is a nonzero principal ideal. Dilworth observed that in general the weak join principal (WJP) identity:

$$(AE){:}E = A + (0{:}E)$$

holds for E a principal ideal, and that from this it follows that $AE = BE$ implies $A + (0{:}E) = B + (0{:}E)$. It can be argued that it is the recognition of the importance of this identity that makes Dilworth's paper [11] a cornerstone in the development of abstract ideal theory.

In a modular multiplicative lattice, elements satisfying both the WMP identity and the WJP identity satisfy the (apparently) stronger identities:

(MP) $$A \cap BE = ((A{:}E) \cap B)E$$
(JP) $$A \cup (B{:}E) = (AE \cup B){:}E.$$

These two identities together form the definition of a principal element and the foundation for the concept of a Noether lattice introduced by Dilworth in "Abstract commutative ideal theory" [11]. (See also the Background for this chapter.)

Dilworth was content in [11] to lay out the framework he had discovered and point out a few of the more major consequences of the introduction of his new principal elements (*e.g.*, the Principal Element Theorem and the Intersection Theorem), leaving the more detailed implications of his discovery to others.

When I first ran upon Dilworth's paper, some of the questions that came immediately to my mind were: To what extent do the principal identities characterize principal ideals? Are all Noether lattices isomorphic to lattices of ideals of Noetherian rings? Which properties of ideal lattices do they have to have and which may they fail to have? At the time, I was working on a problem for my dissertation having to do with multiplicities but I found that these new questions continually pushed thoughts of multiplicities out of my mind. It seemed to me that the flexibility of this setting might just make it of consequence equal to, say, category theory. Certainly it provided a natural setting in which to look at some of the questions that Rees had investigated. In fact, it seemed tailor made as a way to handle homogeneous ideals more cleanly. I found my thinking of the possible applications going in all directions at once and my enthusiasm grew unboundedly, as is more or less typical of a student at that stage of development who feels he has hit upon something of great potential. And as far as I could tell, I was among the first to find what might be a potential gold mine.

After some preliminary spade work which produced some promising results, I consulted with L. J. Ratliff and we agreed that I should continue digging and see what could be unearthed.

Some twenty-five years after the appearance of Dilworth's paper, it is interesting to reflect on what has been found and what might yet be found.

It follows from the Intersection Theorem that principal elements in a local Noether lattice (L, M) are join irreducible. And it is immediate from this that the only principal elements in the lattice of ideals of a local ring are the principal ideals [6,13]. On the other hand, every ideal in a Dedekind domain is a principal element, though not necessarily a principal ideal. In the lattice of ideals of any commutative ring with 1, the principal elements are characterized by being compact and locally principal [28]. All this suggests that Dilworth's definition of a principal element is at the same time very close to characterizing principal ideals but just far enough away to naturally extend a variety of theorems or show them in a new light (*e.g.*, a Dedekind domain is a principal element domain, but not necessarily a principal ideal domain).

If (R, M) is a local ring with nonzero maximal ideal M, if H is the lattice of homogeneous ideals of $R[x]$ and B is the ideal $MR[x] \cap (x)$, then the Noether lattice $L = H/B$ is one dimensional with unique maximal element $M + (x)$, but every principal element contained in $M + (x)$ is contained in either M or (x). In other words, L does not have enough principal elements to be the lattices of ideals of a ring. This observation settled the narrowest version of the representation question (which is addressed in some detail elsewhere in this chapter by D. D. Anderson) and led to the formulation of two conditions which a Noether lattice would have to satisfy to be representable as the lattice of ideals of a ring. A Noether lattice L is said to satisfy the *union condition on primes* if, given any finite set F of primes of L and any element A of L not contained in any of the elements of F, it follows that there exists a principal element E contained in A which is not contained in any element of F. A Noether lattice L is said to satisfy the *weak union condition* if, given any elements A, B, C with A contained in neither B nor C, it follows that there exists a principal element E contained in A which is contained in neither B nor C. It is a trivial exercise to see that the lattice of ideals of any ring satisfies the weak union condition. The union condition on primes is only slightly only harder to verify and is well known in commutative ring theory [37, p. 215].

In the lattice of ideals of a local ring (R, M) there are three equivalent measures of dimension: the Krull dimension k, the smallest number s of principal elements which generate an M-primary ideal, and the degree h of the Hilbert polynomial. The example L above shows that, in general, the first two need not agree in a local Noether lattice, whereas it is not even obvious at the outset that the third is meaningful. However, the dimension of the quotient L/M^n is indeed given by a polynomial for large n [13]. This raises the question of the relation of the three dimensions. Under the hypothesis of the weak union condition, variations of the usual ring theoretic arguments yield $k = h = s$. In general, as it turns out, $k = h \leq s$ [11,13].

Since a Noether lattice need not have a system of parameters of the proper size, it is natural to consider the extreme case of a local Noether lattice L of Krull dimension d in which the maximal element is actually generated by d principal elements, *i.e.*, a regular local Noether lattice. If $E_1, \ldots, E_k$ are a regular system of parameters, then the collection $RL_k = RL(E_1, \ldots, E_k)$ of joins of the E_i is a distributive complete multiplicative sublattice of L. Moreover, RL_k is again a regular local Noether lattice of Krull dimension k and is independent of L [6]. The significance of RL_k is twofold: every distributive regular local Noether lattice is an RL_k (depending only on Krull dimension) and every distributive local Noether lattice is a quotient of an RL_k by a special type of congruence relation [7]. We note that the construction of RL_k can be made with any prime sequence $E_1, \ldots, E_k$ in a local Noether lattice [22].

In $RL_k = RL(E_1, \ldots, E_k)$, every principal element is contained in one of the primes E_i, so the union condition on primes fails wildly. The distributive local Noether lattices which are representable as the lattice of ideals of a ring are those that satisfy the union condition on primes [19]. Any distributive Noether lattice (local or not) is embeddable in the lattice of ideals of a ring [3].

The number of primes in RL_k is finite. If we call a Noether lattice with a finite number of primes *small*, then any small regular local Noether lattice is an RL_k [8], and hence satisfies the unique factorization property on principal elements [4].

It is interesting to note that in a Noether lattice (but not in an arbitrary multiplicative lattice), the principal elements are characterized by the simpler notion of principalness used in the earlier Ward and Ward-Dilworth papers. In particular, then, any meet-principal element (*i.e.*, element satisfying the (MP) identity above) is principal. The join-principal (JP) elements are both more general and more exotic.

A join-principal element in a local Noether lattice (L, M) has a minimal base $F_1, \ldots, F_s$ with $F_i F_j = 0$ whenever i and j are distinct [18]. If M itself is join-principal, then every element of L is join-principal [14] and M has a base of elements $F_1, \ldots, F_s$ with $F_i \cap F_j = 0$ whenever i and j are distinct [18]. If M is join-principal and not a prime of 0, then L is distributive and pieced together out of chains in such a way that it can be embedded in the lattice of ideals of a homomorphic image of a regular local ring [20]. Join-principal elements satisfy the Principal Element Theorem [18].

Noether lattices admit an A-adic type topology which has been considered in some detail (*e.g.*, [23]). A local Noether lattice (L, M) complete in the M-adic topology is completely determined by the quotients L/M^n [16]. A semilocal Noether lattice $(L, M_1, \ldots, M_n)$ complete in the $J = (M_1 \cap \ldots \cap M_n)$-adic topology is the direct product of its localizations, and an arbitrary Noether lattice is determined by its semi-localizations [17]. Hence, representation problems for Noether lattices are partially reduced to the local Krull dimension 0 case.

Abstract module theory has been considered in some detail and by a number of people (*e.g.*, [23,36]), especially in regard to completions. One of the most interesting applications is in the area of duality. If M is a lattice module over L, then the lattice dual M^* of M is again a lattice module over L under the multiplication

$A * K = K : A$. This dual can be used, for example, to obtain the theory of coprimary decompositions from the theory of primary decompositions [2]. If M is principally generated and satisfies the ascending chain condition, then M^* is also principally generated. In essence, if L is a complete semilocal Noether lattice, then the category of algebraic Artinian principally generated lattice modules over L is dual to the category of complete Noetherian principally generated lattice modules over L. Moreover, the completeness of L is a necessary condition for this duality [14].

The definition of a Noether lattice has been generalized to provide a natural framework in which to do abstract ideal theory without chain conditions [1,29].

At the time of writing, the general question of unique factorization in regular local Noether lattices remains open. It is perhaps the proverbial gold at the end of the rainbow that has eluded everyone who has worked in the field to date. I remain optimistic that the problem has an affirmative solution, and that the solution, when found, will open new ways of thinking about ideal theory.

A second question (or group of questions) which is of a great deal of interest and into which a fair amount of work has gone is the question of embeddability and representability. This is addressed by D. D. Anderson elsewhere in this chapter.

In closing, I feel the need to comment that the papers I have mentioned represent only a portion of the literature on Noether lattices and related topics (r-lattices, m-lattices, lattice modules, etc.). This article focuses on the outcome of R. P. Dilworth's contribution to abstract ideal theory and some directly related areas of the literature into which interested readers might look. The lack of a reference to a particular paper or author should not be construed to indicate a lack of appreciation for the work on my part.

REFERENCES

1. D. D. Anderson, *Abstract commutative ideal theory without chain condition*, Algebra Universalis **6** (1976), 131–145.
2. __________________, *Fake rings, fake modules and duality*, J. Algebra **47** (1977), 425–432.
3. __________________, *Distributive Noether lattices*, Michigan Math. J. **22** (1975), 109–115.
4. __________________ and E. W. Johnson, *Unique factorization and small regular local Noether lattices*, J. Algebra **59** (1979), 260–263.
5. K. E. Aubert, *Theory of x-ideals*, Acta Math. **107** (1962), 1–52.
6. K. P. Bogart, *Structure theorems for regular local Noether lattices*, Michigan Math. J. **15** (1968), 167–176.
7. __________, *Distributive local Noether lattices*, Michigan Math. J. **16** (1969), 215–223.
8. __________, *Unique factorization and small regular local Noether lattices*, J. Algebra **59** (1979), 260–263.
9. R. P. Dilworth, *Abstract residuation over lattices*, Trans. Amer. Math. Soc. **44** (1938), 262–268. Reprinted in Chapter 6 of this volume.
10. __________, *Non-commutative residuated lattices*, Trans. Amer. Math. Soc. **46** (1939), 426–444. Reprinted in Chapter 6 of this volume.
11. __________, *Abstract commutative ideal theory*, Pacific J. Math. **12** (1962), 481–498. Reprinted in Chapter 6 of this volume.
12. I. Fleischer, *Equivalence of x-systems and m-lattices*, in "Contributions to Lattice Theory (Szeged, 1980)," Colloq. Math. Soc. János Bolyai 33, North-Holland, Amsterdam, 19830, pp. 381–400.

13. E. W. Johnson, *A-transforms and Hilbert functions in local lattices*, Trans. Amer. Math. Soc. **137** (1969), 125–139.

14. ______________, *Modules: duals and principally generated fake duals*, Algebra Universalis **24** (1987), 111–119.

15. ______________, *Join-principal element lattices*, Proc. Amer. Math. Soc. **57** (1976), 202–204.

16. ______________ and J. A. Johnson, *M-primary elements of a local Noether lattice*, Canad. J. Math. **22** (1970), 327–331.

17. ______________, ______________ and J. P. Lediaev, *A structural approach to Noether lattices*, Canad. J. Math. **22** (1970), 657–665.

18. ______________, *Join-principal elements in Noether lattices*, Proc. Amer. Math. Soc. **36** (1972), 73–78.

19. ______________ and J. P. Lediaev, *Representable distributive Noether lattices*, Pacific J. Math. **28** (1969), 561–564.

20. ______________ and __________, *Structure of Noether lattices with join-principal maximal elements*, Pacific J. Math. **37** (1971), 101–107.

21. ______________ and __________, *Join-principal elements in Noether lattices*, Proc. Amer. Math. Soc. **36** (1972), 73–78.

22. ______________ and M. Detlefsen, *Prime sequences and distributivity in local Noether lattices*, Fund. Math. **86** (1974), 149–156.

23. J. A. Johnson, *A-adic completions of Noether lattice modules*, Fund. Math. **66** (1970), 347–373.

24. W. Krull, *Axiomatische Begrundung der allgemeinen Idealtheorie*, Sitz. der Physikalisch-Medicinische Societät zu Erlangen **56** (1924), 47–63.

25. ________, *Zur Theorie der zweiseitigen Ideale in nictkommutativen Bereichen*, Math. Z. **28** (1927), 481–503.

26. J. P. Lediaev, *Relationship between Noether lattices and x-systems*, Acta Math. Acad. Sci. Hungar. **21** (1970), 323–325.

27. P. Lorenzen, *Abstrakte Begrundung der multiplikativen Idealtheorie*, Math. Z. **45** (1939), 533–553.

28. P. J. McCarthy, *Note on abstract commutative ideal theory*, Amer. Math. Monthly **74** (1967), 707–707.

29. H. M. Nakkar, *Localization in multiplicative lattice modules*, (Russian), Mat. Issled., Kishinev **9** (1974), 88–108.

30. H. Prüfer, *Untersuchungen über Teilbarkeitseigen-schaften in Korpern*, J. Reine Angew. Math. **168** (1932), 1–36.

31. M. Ward, *Residuation in structures over which a multiplication is defined*, Duke Math. J. **3** (1937), 627–636.

32. ________, *Structure residuation*, Ann. of Math. (2) **39** (1938), 555–568.

33. ________, *Residuated distributive lattices*, Duke Math. J. **6** (1940), 641–651.

34. ________ and R. P. Dilworth, *Residuated lattices*, Proc. Nat. Acad. Sci. **24** (1938), 162–164.

35. ________ and ___________, *Residuated lattices*, Trans. Amer. Math., Soc. **45** (1939), 335–354. Reprinted in Chapter 6 of this volume.

36. G. Whitman, *On ring theoretic lattice modules*, Fund. Math. **70** (1971), 221–229.

37. O. Zariski and P. Samuel, "Commutative Algebra," Vol. I, Van Nostrand, Princeton, New Jersey, 1960.

The University of Iowa
Iowa City, IA 52242
U. S. A.

 THE DILWORTH THEOREMS

Representation and Embedding Theorems for Noether Lattices and r-Lattices

D. D. Anderson

The purpose of this article is to discuss Dilworth's embedding problem for Noether lattices. While this problem does not appear explicitly in Dilworth's papers, he made it clear in personal interactions [13]. We phrase the problem as follows:

Are there significant classes of Noether lattices, defined in lattice theoretic terms, which can be embedded in rings in such a way that their essential structure is inherited from the lattice of ideals of the ring?

The following definitions formalize what is meant by "their essential structure is inherited from the lattice of ideals of the ring." We say that a Noether lattice L is *embeddable* if there exist a Noetherian ring R with lattice of ideals $L(R)$ and a product-preserving lattice monomorphism $\theta : L \to L(R)$ that sends principal elements to principal elements and 0 to 0. If the map θ also sends prime elements to prime ideals and primary elements to primary ideals, we say that θ is a *Noether lattice embedding* and that L is *Noether lattice embeddable*. If the map θ is a lattice isomorphism, L is said to be *representable*.

The first example of a Noether lattice L that is not Noether lattice embeddable was given by K. P. Bogart [10]. Here is a sketch of his example. The elements of L are the symbol I and the elements of the lattice of subspaces of a non-Desarguesian projective plane. Denote the maximum element of the lattice of subspaces by M. Make L into a Noether lattice by taking I as the greatest element of L and defining the product of two subspaces to be 0. Then (L, M) is a local Noether lattice with $M^2 = 0$. If (L, M) is Noether lattice embeddable, then there is actually such an embedding $\theta : L \to L(R)$, where (R, P) is a local ring with $P^2 = 0$. Then $\theta(M) = P$ is a vector space over R/P and hence $\theta(L)$ is Arguesian, contradicting the fact that L is non-Arguesian.

Not surprisingly, the sharpest results concerning embeddability and representability are in the distributive case. In fact, the representable distributive Noether lattices have been completely characterized. A commutative ring R satisfies the property that if A, B, and C are ideals of R with $A \subseteq B \cup C$, then $A \subseteq B$ or $A \subseteq C$. The lattice analog of this property is the *weak union condition*: if $A, B, C \in L$ with $A \not\leq B$ and $A \not\leq C$, then there exists a principal element $E \leq A$ but with $E \not\leq B$ and $E \not\leq C$.

Clearly a representable Noether lattice must satisfy the weak union condition. In the distributive case, the converse is true. The following representation theorem is due to E. W. Johnson and J. P. Lediaev [15].

THEOREM. *For a Noether lattice L the following conditions are equivalent.*

(1) *L is distributive and representable.*

(2) *L is distributive and satisfies the weak union condition.*

(3) *For each maximal element $M \in L$, L_M is a chain.*

Outside of the distributive case, the weak union condition need not imply representability. For example, the lattice of homogeneous ideals of $C[X, Y]$, the polynomial ring in two variables over the complex numbers, is a two-dimensional regular local Noether lattice satisfying the weak union condition, but is not representable [3, Example 2].

Let k be a field. The subset of $L(k[X_1, \dots, X_n])$ consisting of all ideals generated by monomials in $X_1, \dots, X_n$ forms a multiplicative sublattice. This sublattice, which is denoted by RL_n, is a distributive n-dimensional regular local Noether lattice. Moreover, any distributive n-dimensional regular local Noether lattice is isomorphic to RL_n. Certainly RL_n is Noether lattice embedded into $L(k[X_1, \dots, X_n])$, but for $n > 1$, RL_n is not representable since RL_n does not satisfy the weak union condition. Moreover, every distributive local Noether lattice is a certain type of congruence lattice of RL_n. For these results due to K. P. Bogart, see [11,12].

The Noether lattice RL_n has come to play a central role in the theory of Noether lattices. RL_n has n rank one (principal) primes $(X_1), \dots, (X_n)$ and each nonzero principal element of RL_n is a product of powers of these principal primes. Either of these two properties actually characterizes RL_n as is shown by the following theorem [5, Theorem 3].

THEOREM. *For a local Noether lattice L the following conditions are equivalent.*

(1) *L is isomorphic to RL_n.*

(2) *L is an n-dimensional regular local Noether lattice with a finite number of rank one prime elements.*

(3) *L is an n-dimensional local Noether lattice domain with a finite number of principal prime elements and every principal element is a product of principal prime elements.*

(4) *L is a regular local Noether lattice with n rank one prime elements.*

(5) *L is a local Noether lattice domain with n principal elements and every principal element is a product of principal prime elements.*

Noether lattices arise not only as sublattices of the lattice of ideals of a ring, but also as the lattice of ideals of semigroups. (For a survey of the ideal theory of commutative semigroups and its connection with multiplicative lattice theory, see [6].) This point of view, that a distributive local Noether can be considered as the lattice of ideals of a certain type of commutative semigroup, is put forth in [1].

Let S be a commutative semigroup with 0 and 1. An ideal J of S is a nonempty subset $J \leq S$ with $JS \leq J$. S is called a *weak-cancellation semigroup* if $xb = xc \neq 0$ implies $b = \lambda c$ for some unit $\lambda \in S$. If S is a Noetherian (*i.e.*, every ideal of S is finitely generated) weak-cancellation semigroup, then the lattice of ideals $L(S)$ of S is a distributive local Noether lattice. The converse is also true. If L is a distributive local Noether lattice, then L is isomorphic to the lattice of ideals of a Noetherian weak-cancellation semigroup. Moreover, we can suppose that this semigroup has 1 as its only unit and hence satisfies the stronger cancellation property: $xa = xb \neq 0$ implies $a = b$.

Let L be a distributive local Noether lattice. Then $L \approx L(S)$ where S is a Noetherian weak-cancellation semigroup whose only unit is 1. Let $\{x_1, \ldots, x_n\}$ be a minimal basis for the maximal ideal of S. We can define a congruence $\sim$ on $Z_0 \times \ldots \times Z_0$, the direct product of n copies of the additive semigroup of nonnegative integers, by

$$(a_1, \ldots, a_n) \sim (b_1, \ldots, b_n) \quad \Longleftrightarrow \quad x_1^{a_1} \ldots x_n^{a_n} = x_1^{b_1} \ldots x_n^{b_n}.$$

Let $\mathfrak{F}$ be the semigroup $T/\sim$. For a field k, let $k[X, \mathfrak{F}]$ be the semigroup ring of $\mathfrak{F}$ with coefficients from k. Note that $k[X, \mathfrak{F}]$ is Noetherian since it is a homomorphic image of $k[X_1, \ldots, X_n]$. Denote the basis elements of $k[X, \mathfrak{F}]$ by $X\overline{(a_1, \ldots, a_n)}$ where $\overline{(a_1, \ldots, a_n)} \in \mathfrak{F}$. The map $\theta : S \rightarrow k[X, \mathfrak{F}]$ determined by $x_1^{a_1} \ldots x_n^{a_n} \rightarrow X\overline{(a_1, \ldots, a_n)}$ induces a map $\theta : L(S) \rightarrow L(k[X, \mathfrak{F}])$. The map θ is a product preserving monomorphism. Thus every distributive local Noether lattice is embeddable [1, Theorem 3]. If $\mathfrak{F}$ is a torsionless cancellation monoid, then this embedding is actually a Noether lattice embedding. (In particular, if we apply this embedding to RL_n, we just get Bogart's embedding of RL_n into $L(k[X_1, \ldots, X_n])$.) In certain other special cases (see, for example, [8] or [16]), a distributive local Noether lattice is Noether lattice embeddable. This raises the question of whether a distributive local Noether lattice is always Noether lattice embeddable. While writing this article, I observed that the local distributive Noether lattice domain given in [9, Example 4.2] is not Noether lattice embeddable. It is interesting to note that this Noether lattice has dimension one and its maximal element is a join of two principal elements.

On the positive side, distributive regular Noether lattices have been completely characterized and they are all Noether lattice embeddable. In fact, a distributive regular Noether lattice is Noether lattice embeddable into a regular Noetherian ring of the same Krull dimension. For these results, the reader is referred to [1]. Moreover, every distributive Noether lattice is a certain type of congruence lattice of a distributive regular Noether lattice [7, Theorem 4].

A distributive Noether lattice L of the form $L = \hat{L}/K$ where $\hat{L}$ is a distributive regular Noether lattice is Noether lattice embeddable because $\hat{L}$ is. The converse is not always true because $\hat{L}/M$ is embeddable for any maximal element M of $\hat{L}$. It is interesting to speculate to what extent the converse is true. A related question is what Noether lattices can be embedded in the lattice of ideals of a Noetherian ring whose lattice of ideals is distributive. A characterization of such Noether lattices may be found in [4].

In summary, a distributive local Noether lattice is embeddable into the lattice of ideals of a Noetherian ring. Hence in the global case, a distributive Noether lattice may be embedded into the lattice of ideals of a product of Noetherian rings. A distributive regular Noether lattice, or more generally, the quotient of a distributive regular Noether lattice, is Noether lattice embeddable into the lattice of ideals of a Noetherian ring. But there is a distributive local Noether lattice domain that is not Noether lattice embeddable into the lattice of ideals of a Noetherian ring.

A natural question is which of the previously mentioned results concerning embeddings and representations have analogs for multiplicative lattices without the ascending chain condition. In [2], an *r-lattice* is defined to be a complete modular multiplicative lattice that is principally and compactly generated, and has greatest element I compact. If R is any commutative ring, then $L(R)$ is an r-lattice. An r-lattice is a Noether lattice if and only if it satisfies the ascending chain condition. As in the Noetherian case, a distributive quasi-local r-lattice is the lattice of ideals of a special type of commutative semigroup with 0 and 1, called an r-semigroup, which satisfies the cancellation condition:

$$xb = xc \neq 0 \Rightarrow b = \lambda c \quad \textit{for some unit } \lambda.$$

Using this characterization, it is shown that any distributive r-lattice is embeddable into the lattice of ideals of a commutative ring. In fact, the distributive r-lattice domains which are representable have been characterized by the following theorem [12, Theorem 3.4].

THEOREM. *For an r-lattice domain, the following conditions are equivalent.*

(1) *L is distributive and satisfies the weak union condition.*
(2) *For each maximal element M of L, L_M is a chain.*
(3) *$(A : B) \vee (B : A) = I$ for all compact elements $A, B \in L$.*
(4) *$(A \vee B) : C = (A : C) \vee (B : C)$ for $A, B, C \in L$ with C compact.*
(5) *$C : (A \wedge B) = (C : A) \wedge (C : B)$ for $A, B, C \in L$ with A, B compact.*
(6) *$A(B \wedge C) = AB \wedge AC$ for $A, B \in L$.*
(7) *$(A \vee B)(A \wedge B) = AB$ for $A, B \in L$.*
(8) *Every compact element of L is principal.*
(9) *L is representable as the lattice of ideals of a Prüfer domain.*

Another related question is what sublattices of the lattice $L(R)$ that can be defined in a natural way are actually Noether lattices or r-lattices. Let R be a commutative ring. An element of R is called *regular* if it is a non-divisor of zero and an ideal J of R is called *regular* if it contains a regular element. Consider the following four ordered subsets of $L(R)$:

$$L_{rg}(R) = \{0\} \cup \{J \in L(R) \mid J \text{ is generated by regular elements}\}$$
$$L_r(R) = \{0\} \cup \{J \in L(R) \mid J \text{ is regular}\}$$
$$L_{sr}(R) = \{0\} \cup \{J \in L(R) \mid J^* \text{ contains a f.g. ideal } J^* \text{ with } \operatorname{ann}(J^*) = 0\}$$
$$L_f(R) = \{0\} \cup \{J \in L(R) \mid \operatorname{ann}(J) = 0\}.$$

These four ordered subsets are thoroughly investigated in [9]. Here, $L_r(R)$, $L_{sr}(R)$, and $L_f(R)$ are all multiplicative sublattices of $L(R)$ and are also complete, but not necessarily complete sublattices. However, $L_{rg}(R)$ need not even be a sublattice of $L(R)$. A natural question is when any of these ordered subsets are representable as the lattice of ideals of a ring. Again, only in the distributive case is there a definite answer [9, Theorem 3.7 and 3.11].

THEOREM. *For a ring R the following conditions are equivalent.*

(1) *R is a (strong) Prüfer ring, i.e., every finitely generated (semi-) regular ideal of R is invertible.*

(2) *$L_r(R)$ $\left(L_{sr}(R)\right)$ is distributive.*

(3) *There is a Prüfer domain D with $L(D)$ and $L_r(R)$ $\left(L_{sr}(R)\right)$ isomorphic as multiplicative lattices.*

In general, the sublattice $L_f(R)$ and ordered subset $L_{rg}(R)$ may be badly behaved. For example, $L_f(R)$ may be distributive without being representable [9, Example 3.15] and need not even be compactly generated [9, Example 3.16]. Also, the previously mentioned example of a distributive local Noether lattice domain L that is not Noether lattice embeddable has the form $L = L_{rg}(R)$ for some ring R where $L_{rg}(R)$ is not a sublattice of $L(R)$. On the positive side, B. G. Kang [17] has shown that if the lattice $L_t(D)$ of t-ideals of an integral domain D is distributive, then $L_t(D) \approx L(D)'$ for some Prüfer domain D'.

We end with the following open questions.

Question 1. What Noether lattices are embeddable into the lattice of ideals of a Noetherian ring, or more generally, into any commutative ring? (I know of no r-lattice that is not embeddable into the lattice of ideals of a commutative ring.)

Question 2. What Noether lattices are Noether lattice embeddable into the lattice of ideals of a Noetherian ring? (Here even the distributive local case remains open.)

Question 3. What classes of Noether lattices, outside the distributive case, are representable?

Question 4. Is a distributive r-lattice representable if it satisfies the weak union condition? (This is the case if it is a domain.)

Question 5. Let R be a commutative ring. When does there exist an integral domain D with $L_r(R)$ isomorphic to $L(D)$? (This is the case when $L_r(R)$ is distributive.)

REFERENCES

1. D. D. Anderson, *Distributive Noether lattices*, Michigan Math. J. **22** (1975), 109–115.
2. ______, *Abstract commutative ideal theory without chain condition*, Algebra Universalis **6** (1976), 131–145.
3. ______, *Multiplicative lattices in which every principal element is a product of prime elements*, Algebra Universalis **8** (1978), 330–335.
4. ______ and E. W. Johnson, *Arithmetically embeddable local Noether lattices*, Michigan Math. J. **23** (1976), 177–184.
5. ______ and ______, *Unique factorization and small regular local Noether lattices*, J. Algebra **59** (1979), 260–263.
6. ______ and ______, *Ideal theory in commutative semigroups*, Semigroup Forum **30** (1984), 127–158.
7. ______, ______, and J. A. Johnson, *Noether lattices representable as quotients of the lattice of monomially generated ideals of polynomial rings*, Canad. J. Math. **31** (1979), 789–799.
8. ______, ______, and ______, *Structure and embedding theorems for small strong-π-lattices*, Algebra Universalis **16** (1983), 147–152.
9. ______ and J. Pascual, *Regular ideals in commutative rings, sublattices of regular ideals, and Prüfer rings*, J. Algebra **111** (1987), 404–426.
10. K. P. Bogart, *Nonimbeddable Noether lattices*, Proc. Amer. Math. Soc. **22** (1962), 129–133.
11. ______, *Structure theorems for regular local Noether lattices*, Michigan Math. J. **15** (1968), 167–176.
12. ______, *Distributive local Noether lattices*, Michigan Math. J. **16** (1969), 215–223.
13. ______, *Personal communication*.
14. R. P. Dilworth, *Abstract commutative ideal theory*, Pacific J. Math. **12** (1962), 481–498. Reprinted in Chapter 6 of this volume.
15. E. W. Johnson and J. P. Lediaev, *Representable distributive Noether lattices*, Pacific J. Math. **28** (1969), 451–564.
16. ______ and ______, *Structure of Noether lattices with join-principal maximal elements*, Pacific J. Math. **37** (1971), 101–108.
17. B. G. Kang, "*-Operations on Integral Domains," Thesis, University of Iowa, 1987.

The University of Iowa
Iowa City, IA 52242
U. S. A.

Miscellaneous Papers

Background

R. P. DILWORTH

Relatively complemented lattices and lattice structure theory. One of the first fundamental structure theorems of lattice theory was the Birkhoff-Menger theorem which states that a finite dimensional complemented modular lattice is a direct product of projective geometries. An examination of the proof makes it clear that projectivity of intervals, congruence relations, as well as modularity play a significant role. Now a complemented modular lattice is, in fact, relatively complemented, namely, if $a \leq x \leq b$, there exists y such that $a = x \wedge y$ and $x \vee y = b$. Furthermore, relatively complementation introduces many projectivities into a lattice. This suggests that relative complementation in a lattice with unit and null elements may well provide the basic structure which underlies the Birkhoff-Menger theorem.

Let us recall that a congruence relation θ on a lattice is a relation between elements of the lattice which has the following properties:

(1) $a\,\theta\,a$ for all a in L
(2) $a\,\theta\,b$ implies $b\,\theta\,a$ for all a, b in L
(3) $a\,\theta\,b$ and $b\,\theta\,c$ imply $a\,\theta\,c$ for all a, b, c in L
(4) $a\,\theta\,b$ implies $a \wedge c\,\theta\,b \wedge c$ and $a \vee c\,\theta\,b \vee c$ for all a, b, c.

It is easily verified that if $[a,b]$ and $[c,d]$ are projective intervals then $a\,\theta\,b$ if and only if $c\,\theta\,d$. Hence projectivities and congruence relations are closely related. Every lattice has two trivial congruence relations α and ω where $a\,\alpha\,b$ for all a, b in L and $a\,\omega\,b$ if and only if $a = b$. The congruence relations on a lattice have a natural order defined by $\theta \leq \psi$ if and only if $a\,\theta\,b$ implies $a\,\psi\,b$ for all a, b in L. Funayama and Nakayama showed in the early 1940's that the congruence relations on a lattice form a distributive lattice under this ordering.

Let us also recall that a lattice is *simple* if it has only the trivial congruence relations α and ω. The simple complemented modular lattices of finite dimension are simply the projective geometries. Two congruence relations θ and ψ are said to *permute* if $a\,\theta\,c$ and $c\,\psi\,b$ imply that there exists d such that $a\,\psi\,d$ and $d\,\theta\,b$. It was known that a finite dimensional lattice is a direct product of simple lattices if and only if the congruence relations permute and the lattice of congruence relations is a finite Boolean algebra.

I had observed that the congruence relations on a relatively complemented lattice always permute. Hence, in order to prove the analogue of the Birkhoff-Menger theorem for a relatively complemented lattice with unit and null element it was only necessary to show that under the ascending chain condition the lattice of congruence relations was indeed a finite Boolean algebra. This was the principal objective of this paper.

The normal completion and lattices of continuous functions. In the late 1940's I was attending a seminar on functional analysis in which considerable attention was devoted to the lattice of continuous real functions on a topological space. Since the lattice of real continuous functions on a compact Hausdorff space determines the space, it occurred to me that lattice constructions on the continuous function lattice should have some kind of analogue as constructions on the corresponding topological space. The normal completion of a lattice is one of the most important constructions in lattice theory, being the generalization to an arbitrary ordered set of the Dedekind construction of the real numbers from the rational numbers. This seemed to be a particularly appropriate construction to study for the lattice of continuous real functions. In particular, the objective would be to determine if the normal completion of the lattice of continuous real functions could itself be represented as the lattice of continuous functions on a suitable topological space. The first step would be to represent the normal completion as a lattice of real functions on the given topological space. There is no loss in generality in assuming that the topological space is completely regular.

If A is a bounded set of continuous real functions on a completely regular topological space, the normal closure $\overline{A}$ of A is obtained as the set of all lower bounds of the set of all upper bounds of the functions of A. Then A is normal if $\overline{A} = A$. The obvious choice for a function representation of a normal set A would be the function φ_A which is the pointwise infimum of the functions in A. It turns out that these functions can be characterized as the bounded functions on the topological space which satisfy

$$(\varphi_*)^* = \varphi$$

where φ_* is the lim inf of φ and φ^* is the lim sup of φ. The bounded real functions satisfying this relation are said to be normal and they form a lattice isomorphic to the lattice of all normal sets and hence isomorphic to the normal completion of the lattice of continuous functions on the space.

The normal functions on a topological space are in general not continuous. How do we go about finding a suitable topological space whose lattice of bounded continuous functions is isomorphic to the lattice of normal functions? A first step in getting an answer to this question is to find those subsets of the given topological space whose characteristic function (the function which takes the value 1 on the set and is 0 elsewhere) is normal. It turns out that this happens if and only if the complement of the set is a regular open set. Now it was well known that the regular open sets form a Boolean algebra and associated with the Boolean algebra is the topological space of its maximal filters. This appears to be a likely candidate for the space we are looking for and further investigation showed this to be indeed the case.

Annals of Mathematics
Vol. 51, No. 2, March, 1950

THE STRUCTURE OF RELATIVELY COMPLEMENTED LATTICES

By R. P. Dilworth

(Received October 28, 1948)

1. Introduction

In the initial development of lattice theory considerable attention was devoted to the structure of modular lattices. Two of the principal structure theorems which came out of this early work are the following:

Every complemented modular lattice of finite dimensions is a direct union of a finite number of simple[1] complemented modular lattices (Birkhoff [1], Menger [4]).

Every finite dimensional modular lattice is a subdirect union of a finite number of simple modular lattices (Birkhoff [2]).

The standard proofs of these theorems depend heavily upon the modular law. Nevertheless, I shall show in this paper that the essential part of the hypothesis of these theorems is not the modularity but rather the complementation. Thus corresponding to the first of the above results the following theorem is proved.

THEOREM 4.4. *Every relatively complemented lattice satisfying the ascending chain condition is a direct union of a finite number of simple relatively complemented lattices.*

Since modularity and complementation imply relative complementation, Theorem 4.2 contains the Birkhoff-Menger theorem as a special case.

As in the case of modular lattices, simplicity can be characterized in terms of projectivity.

THEOREM 4.5. *A relatively complemented lattice satisfying a chain condition is simple if and only if every two prime quotients are projective.*

Let L be a finite dimensional lattice. If a is an element of L let u_a denote the join of all elements covering a and let z_a denote the meet of all elements covered by a. L is said to be *locally* relatively complemented if the quotient lattices u_a/a and a/z_a are relatively complemented for each $a \epsilon L$. The following structure theorem is proved.

THEOREM 5.1. *Every locally relatively complemented lattice of finite dimensions is a subdirect union of a finite number of simple locally relatively complemented lattices.*

If L is a finite dimensional modular lattice, then u_a/a and a/z_a are point lattices and hence are relatively complemented. Thus the structure theorem for finite dimensional modular lattices is contained in Theorem 5.1.

As in the previous case, a locally relatively complemented lattice of finite dimensions is simple if and only if every two prime quotients are projective.

The technique of lattice congruence relations is used systematically in the proofs of the theorems. Accordingly, a brief account, and some new results, of the general theory of lattice congruences is given in Sections 2 and 3.

---

[1] A lattice is *simple* if it has only trivial congruence relations.

348

2. Congruence relations

If a and b are elements of L such that $a \geq b$ we shall let a/b denote the formal quotient of a and b.   Since the chance of confusion is slight we shall also use the same symbol to denote the quotient lattice of all elements x such that $a \geq x \geq b$. The quotient a/b is said to be *contained* in the quotient c/d if $c \geq a$ and $b \geq d$. Each of the quotients $a/a \cap b$ and $a \cup b/b$ is called a *transpose* of the other. Finally, two quotients a/b and c/d are *projective* if there exists a sequence of quotients $a/b = x_1/y_1, \cdots, x_n/y_n = c/d$ such that x_i/y_i is a transpose of x_{i+1}/y_{i+1} .

We shall begin by determining the congruence relation generated on L by identifying certain pairs of elements of L. Now if θ is a congruence relation on L and if $a \equiv b(\theta)$, then $a \cup b \equiv b \cup b \equiv b \equiv b \cap b \equiv a \cap b(\theta)$. Also if c belongs to the quotient lattice $a \cup b/a \cap b$, then $c \equiv (a \cup b) \cap c \equiv (a \cap b) \cap c \equiv a \cap b(\theta)$. Hence a congruence relation identifies a and b if and only if it identifies $a \cup b$ and $a \cap b$. We shall say that θ collapses a quotient c/d if $c \equiv d(\theta)$. From the previous remark it follows that we need only consider the problem of determining the congruence relation generated by collapsing the quotients belonging to a certain set Q.

The theory of congruence relations in a modular lattice is based upon the projectivity of quotients. For arbitrary lattices a somewhat more general notion of projectivity is required.

DEFINITION 2.1. A quotient a/b is *weakly projective* into c/d if there exists a sequence of quotients $a/b = x_1/y_1, \cdots, x_n/y_n = c/d$ such that x_i/y_i is contained in a transpose of x_{i+1}/y_{i+1} .

For modular lattices it can easily be verified from the isomorphism theorem, that a/b is weakly projective into c/d if and only if a/b is projective to a quotient c_1/d_1 contained in c/d.

It is clear from Definition 2.1 that weak projectivity is a transitive relation. Also if x/y is contained in a transpose of u/v where $u \equiv v(\theta)$ for some congruence relation θ on L, an easy calculation shows that $x \equiv y(\theta)$. Hence if a/b is weakly projective into c/d where $c \equiv d(\theta)$, then $a \equiv b(\theta)$.

Let Q be a set of quotients of L and let us set $a \equiv b(\theta_Q)$ if and only if there exists a sequence of elements $a \cup b = x_1 \geq x_2 \geq \cdots \geq x_n = a \cap b$ such that x_i/x_{i+1} is weakly projective into some quotient of Q.

LEMMA 2.1. θ_Q *is the congruence relation generated on L by collapsing the quotients of Q.*

We observe first that if θ is any congruence relation which collapses the quotients of Q, then $a \cup b \equiv x_1 \equiv x_2 \equiv \cdots \equiv x_n \equiv a \cap b(\theta)$ and hence $a \equiv b(\theta)$. Thus to prove the lemma it will be sufficient to show that θ_Q is a congruence relation. Now $a \equiv a(\theta_Q)$ since the quotient a/a is weakly projective into every quotient. Also if $a \equiv b(\theta_Q)$ from the symmetry of the definition we have $b \equiv a(\theta_Q)$.

We shall show next that if c/d is contained in a transpose a_1/b_1 of a/b where $a \equiv b(\theta_Q)$, then $c \equiv d(\theta_Q)$. Let then $a \cup b_1 = a_1, a \cap b_1 = b$ and let $a = x_1 \geq \cdots \geq$

$x_n = b$ where x_i/x_{i+1} is weakly projective into a quotient of Q. If we set $y_i = c \cap (d \cup x_i)$, then $c = y_1 \geq \cdots \geq y_n = d$. Also y_i/y_{i+1} is a transpose of $y_i \cup d \cup x_{i+1}/d \cup x_{i+1}$ which is contained in $d \cup x_i/d \cup x_{i+1}$ which is a transpose of $x_i/x_i \cap (d \cup x_{i+1})$. But $x_i/x_i \cap (d \cup x_{i+1})$ is contained in x_i/x_{i+1} which is weakly projective into a quotient of Q. Thus y_i/y_{i+1} is weakly projective into a quotient of Q and hence $c \equiv d(\theta_Q)$. A similar argument gives the result if a_1/b_1 is a lower transpose of a/b.

Now let $a \equiv b(\theta_Q)$ and $b \equiv c(\theta_Q)$. If $a \geq b \geq c$, then it follows directly from the definition that $a \equiv c(\theta_Q)$. In the general case, we have $a \cup b \cup c \equiv a \cup b(\theta_Q)$ since $a \cup b \cup c/a \cup b$ is a transpose of $b \cup c/(a \cup b) \cap (b \cup c)$ which is contained in $b \cup c/b \cap c$. Also $a \cap b \cap c \equiv a \cap b(\theta_Q)$ since $a \cap b/a \cap b \cap c$ is a transpose of $(a \cap b) \cup (b \cap c)/b \cap c$ which is contained in $b \cup c/b \cap c$. But since $a \cup b \equiv a \cap b(\theta_Q)$ we have $a \cup b \cup c \equiv a \cap b \cap c(\theta_Q)$. From the previous paragraph it follows that $a \cup c \equiv a \cap c(\theta_Q)$ and hence that $a \equiv c(\theta_Q)$.

Finally if $a \equiv b(\theta_Q)$, then since $a \cup b \cup c/b \cup c$ is a transpose of $a \cup b/(a \cup b) \cap (b \cup c)$ which is contained in $a \cup b/a \cap b$ we have $a \cup b \cup c \equiv b \cup c(\theta_Q)$. Similarly $a \cup b \cup c \equiv a \cup c(\theta_Q)$ and hence $a \cup c \equiv b \cup c(\theta_Q)$. A dual argument shows that $a \cap c \equiv b \cap c(\theta_Q)$ and the proof of the lemma is thus complete.

As a corollary of Lemma 2.1, it follows that L is simple if and only if for every pair of proper quotients a/b and c/d, there exists a chain from a to b whose consecutive quotients are all weakly projective into c/d.

Let $\Theta(L)$ denote the set of congruence relations over L. $\Theta(L)$ is partially ordered in the usual way by defining $\theta \leq \varphi$ if and only if $x \equiv y(\theta)$ implies $x \equiv y(\varphi)$. The unit congruence relation ι is defined by $x \equiv y(\iota)$ all x and y. The null congruence relation ω is defined by $x \equiv y(\omega)$ if and only if $x = y$. Clearly $\omega \leq \theta \leq \iota$ all $\theta \in \Theta(L)$. $\Theta(L)$ is a complete lattice under the partial ordering. For if Φ is a class of congruence relations it can be easily verified that $\bigcap \Phi$ and $\bigcup \Phi$ are given by

$$x \equiv y(\textstyle\bigcap \Phi) \quad \text{if and only if} \quad x \equiv y(\varphi) \text{ all } \varphi \in \Phi$$

$$x \equiv y(\textstyle\bigcup \Phi) \quad \text{if and only if} \quad x = x_1, \cdots, x_n = y \text{ exist such that}$$

$$x_i \equiv x_{i+1}(\varphi_i) \text{ for some } \varphi_i \in \Phi.$$

We need the following result due to Funayama and Nakayama

LEMMA 2.2. $\Theta(L)$ *is a distributive lattice.*

For if $\theta \in \Theta(L)$ and Φ is a class of congruence relations $\varphi \in \Theta(L)$, let $\theta \cap \Phi$ denote the set of congruence relations $\theta \cap \varphi$. Now suppose that $x \equiv y(\theta \cap (\bigcup \Phi))$. Then $x \equiv y(\theta)$ and $x \equiv y(\bigcup \Phi)$. Thus elements $x = x_1, x_2, \cdots, x_n = y$ exist such that $x_i \equiv x_{i+1}(\varphi_i)$ where $\varphi_i \in \Phi$. Let us set

$$y_i = (x \cup y) \cap (x_i \cup x_{i+1} \cup \cdots \cup x_n).$$

Then $x \cup y = y_1 \geq y_2 \geq \cdots \geq y_n = y$. Since $x \equiv y(\theta)$ we have $x \cup y \equiv y(\theta)$ and hence $y_i \equiv y_{i+1}(\theta)$. But

$$y_i \equiv (x \cup y) \cap (x_{i+1} \cup \cdots \cup x_n) \equiv y_{i+1}(\varphi_i).$$

Hence $y_i \equiv y_{i+1}(\theta \cap \varphi_i)$ and thus $y_i \equiv y_{i+1}(\bigcup(\theta \cap \Phi))$. By transitivity, $x \cup y \equiv y(\bigcup(\theta \cap \Phi))$. From the symmetry of the argument we have $x \cup y \equiv x(\bigcup(\theta \cap \Phi))$ and hence $x \equiv y(\bigcup(\theta \cap \Phi))$, It follows that.

$$\theta \cap (\bigcup \Phi) \leqq \bigcup(\theta \cap \Phi).$$

But trivially
$$\theta \cap (\bigcup \Phi) \geqq \bigcup(\theta \cap \Phi).$$

Hence $\theta \cap (\bigcup \Phi) = \bigcup(\theta \cap \Phi)$ and the lemma is proved.

If θ is a congruence relation of $\Theta(L)$, the lattice of congruence classes will be denoted by L_θ .

3. General structure theorems

We will collect in this section some results on general lattice structure which will be needed in the following sections. Let us recall that a lattice L is the *direct union* of lattices L_α (in symbols $L = \vee_\alpha L_\alpha$) if L consists of all vectors $\{x_\alpha\}$ where meet and join are defined componentwise. In order to connect direct union representations of a lattice with the properties of congruence relations on L we make the following definition.

DEFINITION 3.1. Two congruence relations θ and φ on L *permute* if $a \equiv b(\theta)$ and $b \equiv c(\varphi)$ imply that $d \in L$ exists such that $a \equiv d(\varphi)$ and $d \equiv c(\theta)$.

It is clear that when θ and φ permute, then $a \equiv b(\theta \cup \varphi)$ if and only if c exists such that $a \equiv c(\theta)$ and $c \equiv b(\varphi)$. We also have

LEMMA 3.1. *If θ permutes with all congruence relations φ of a set Φ, then θ permutes with $\bigcup \Phi$.*

For if $a \equiv b(\theta)$ and $b \equiv c(\bigcup \Phi)$ then $b = b_1, b_2, \cdots, b_n = c$ exist such that $b_i \equiv b_{i+1}(\varphi_i)$ where $\varphi_i \in \Phi$. But then by $n-1$ applications of Definition 3.1 we have $a \equiv a_1(\varphi_1)$, $a_1 \equiv a_2(\varphi_2)$, $\cdots$, $a_{n-2} \equiv a_{n-1}(\varphi_{n-1})$ and $a_{n-1} \equiv c(\theta)$. Hence $a \equiv a_{n-1}(\bigcup \Phi)$, and $a_{n-1} \equiv c(\theta)$ and the lemma follows.

The set of congruence relations γ of $\Theta(L)$ which permute with θ for all $\theta \in \Theta(L)$ is called the *center* of $\Theta(L)$ and will be denoted by $\Gamma(L)$.

Now if $L = N \vee M$, the mapping $\{x_N, x_M\} \to x_N$ is a homomorphism of L onto N and hence generates a congruence relation θ_N . Clearly $L_{\theta_N} \cong N$. We will call θ_N a *decomposition* congruence relation on L. Such congruence relations can be determined intrinsically as follows:

LEMMA 3.2. *A congruence relation θ on L is a decomposition congruence relation if and only if θ belongs to $\Gamma(L)$ and has a complement in $\Theta(L)$.*[2]

To prove the necessity, let $\theta = \theta_N$ where $L = N \vee M$. Now $x \equiv y(\theta_N)$ and $x \equiv y(\theta_M)$ imply that $x_N = y_N$ and $x_M = y_M$ which imply $x = y$. Hence $\theta_N \cap \theta_M = \omega$. If x and y are arbitrary elements of L, let $v = \{x_N, y_M\}$. Then $x \equiv v(\theta_N)$ and $v \equiv y(\theta_M)$ and hence $\theta_N \cup \theta_M = \iota$. It follows that θ has a complement in $\Theta(L)$. Now let φ be an arbitrary congruence relation of $\Theta(L)$ and suppose that $x \equiv w(\theta_N)$, $w \equiv y(\theta_M \cap \varphi)$. Then $x_N = w_N$, $w_M = y_M$ and $w \equiv y(\varphi)$. Let $v = \{y_N, x_M\}$,

---

[2] This lemma and its corollary sharpen for lattices a similar result of Birkhoff for general algebras (Birkhoff [2] page 87).

$r = \{x_N \cup y_N , x_M\}$ and $s = \{x_N \cap y_N , x_M \cup y_M\}$. Then clearly $x \equiv v(\theta_M)$ and $v \equiv y(\theta_N)$. But an elementary calculation shows that $x = r \cap (s \cup w)$ and $v = r \cap (s \cup y)$ and hence that $x \equiv v(\varphi)$. Thus $x \equiv v(\theta_M \cap \varphi)$, $v \equiv y(\theta_N)$. It follows that θ_N permutes with $\theta_M \cap \varphi$. But θ_N trivially permutes with $\theta_N \cap \varphi$. Hence by Lemmas 2.2 and 3.3, $\theta = \theta_N$ permutes with $(\theta_M \cap \varphi) \cup (\theta_N \cap \varphi) = \varphi$. Thus $\theta \in \Gamma(L)$ and the proof of necessity is complete.

For the sufficiency proof, let θ have a complement $\theta' \in \Theta(L)$ and suppose that θ permutes with θ'. If $x \in L$, let x_θ and $x_{\theta'}$ denote the congruence classes of θ and θ' respectively to which x belongs. We set up the correspondence

$$x \to \{x_\theta , x_{\theta'}\}$$

which maps L into $L_\theta \vee L_{\theta'}$. Since θ and θ' are congruence relations we clearly have $x \cup y \to \{x_\theta \cup y_\theta , x_{\theta'} \cup y_{\theta'}\}$ and $x \cap y \to \{x_\theta \cap y_\theta , x_{\theta'} \cap y_{\theta'}\}$. Also $\{x_\theta , x_{\theta'}\} = \{y_\theta , y_{\theta'}\} \to x \equiv y(\theta), x \equiv y(\theta') \to x \equiv y(\omega) \to x = y$. Hence L is isomorphic to a sublattice of $L_\theta \vee fL_{\theta'}$. Since θ and θ' permute and $\theta \cup \theta' = \iota$ we have $x \equiv w(\theta)$ and $w \equiv y(\theta')$. But then $\omega_\theta = x_\theta$ and $w_{\theta'} = y_{\theta'}$. Hence $w \to \{x_\theta, y_{\theta'}\}$ and L is thus isomorphic to $L_\theta \vee L_{\theta'}$. The mapping $L \to L_\theta$ clearly generates θ and the proof of the lemma is thus complete.

From the proof of Lemma 3.2 we have the following corollary:

COROLLARY: *Let θ have a complement θ' in $\Theta(L)$. Then $\theta \in \Gamma(L)$ if and only if θ and θ' permute.*

As an application of Lemma 3.2 we have

THEOREM 3.1. *A lattiec L is a direct union of a finite number of simple lattices if and only if $\Gamma(L) = \Theta(L)$ and $\Theta(L)$ is a finite Boolean algebra.*

PROOF. Let $L = L_1 \vee \cdots \vee L_n$ where each L_i is simple and let θ_1 be the congruence relation generated by the mapping $L \to L_i$. Since L_i is simple, θ_i is a maximal element of $\Theta(L)$. If $x \equiv y(\theta_1 \cap \cdots \cap \theta_n)$, then $x_i = y_i, i = 1, \cdots, n$ and hence $x = y$. Thus $\theta_1 \cap \cdots \cap \theta_n = \omega$ and it follows that $\Theta(L)$ is a finite Boolean algebra. Let π_i be the complement of θ_i in $\Theta(L)$. By Lemma 3.2, π_i and θ_i permute and hence by the corollary to Lemma 3.2, $\pi_i \in \Gamma(L)$. But then $\Theta(L) - \Gamma(L)$ by Lemma 3.1.

The sufficiency of the conditions of the theorem follow immediately from Lemma 3.2.

We turn now to subdirect representations. Recall that L is a subdirect union of lattices L_α if L is a sublattice of $\vee_\alpha L_\alpha$ and each element of L_α occurs as a component in some element of L.

DEFINITION 3.2. A lattice L is *irreducible* if there exists a pair of distinct elements a and b in L such that $a \equiv b(\theta)$ for every congruence relation θ not equal to ω.

G. Birkhoff [3] has proved the following general structure theorem.

THEOREM 3.2. *Every lattice L is a sub-direct union of irreducible lattices.*

For if a and b are distinct elements of L, by the Maximal Principal there exists a maximal congruence relation $\theta(a, b)$ such that $a \not\equiv b(\theta(a, b))$. But then $L_{\theta(a , b)}$ is irreducible and the mapping $x \to \{x_{\theta(a,b)}\}$ is an isomorphic mapping of L into the direct union $\vee_{a \neq b} L_{\theta(a, b)}$.

It is clear from Theorem 3.2, that L is irreducible if and only if L is subdirectly irreducible in the sense of Birkhoff [3].

Now if L is a subdirect union of lattices L_α where the L_α are irreducible, let θ_α be the congruence relation determined by the mapping $L \to L_\alpha$. If $\theta_\alpha = \cap\Phi$ where $\varphi > \theta_\alpha$ all $\varphi \in \Phi$ and if a and b are such that a_α and b_α are the elements characterizing the irreducibility of L_α, then each φ generates a non-trivial congruence relation on L_{θ_α} and hence $a \equiv b(\varphi)$ all $\varphi \in \Phi$. But then $a \equiv b(\theta_\alpha)$ and hence $a_\alpha = b_\alpha$ contrary to assumption. Thus $\theta_\alpha = \varphi$ for some $\varphi \in \Phi$ and θ_α is a completely meet-irreducible element of $\Theta(L)$. It is also clear that $\cap_\alpha \theta_\alpha = \omega$. Conversely, if $\omega = \cap_\alpha \theta_\alpha$ where the θ_α are completely meet-irreducible, then L is a subdirect union of the L_{θ_α} and each L_{θ_α} is irreducible. It follows (Birkhoff [2]) that representations of L as a subdirect union of irreducible lattices are associated with representations of ω as a meet of completely meet-irreducible elements of $\Theta(L)$. In particular, L is a subdirect union of simple lattices if and only if ω can be represented as a meet of maximal elements of $\Theta(L)$. Thus we have (Birkhoff [2]).

THEOREM 3.3. *A lattice L is a subdirect union of a finite number of simple lattices if and only if $\Theta(L)$ is a finite Boolean algebra.*

4. Relatively complemented lattices

We shall prove first some general structure theorems for arbitrary relatively complemented lattices from which the structure theorem stated in the introduction can then be easily derived. The unit and null elements of a relatively complemented lattice will be denoted by u and z respectively.

THEOREM 4.1. *Let θ be a congruence relation on a relatively complemented lattice L. Then the following three statements concerning θ are equivalent.*

1) *There exists a maximal element $m \in L$ such that $m \equiv z(\theta)$.*

2) *There exists an element $a \in L$ such that $x \equiv y(\theta)$ if and only if $x \cap a = y \cap a$.*

3) *θ is a decomposition congruence relation.*

PROOF. We will show that 2) $\to$ 1), 1) $\to$ 3), and finally that 3) $\to$ 2). Suppose then that 2) holds and let m be an arbitrary complement of a. Then since $m \cap a = z = z \cap a$ we have $m \equiv z(\theta)$. If $m_1 \geqq m$ and $m_1 \equiv z(\theta)$, let w be a relative complement of m_1 in u/m. Since $m_1 \equiv z \equiv m(\theta)$ we have $u \equiv w(\theta)$ and hence $a \equiv a \cap w(\theta)$. But then $a = a \cap w$ by 2). Hence $w \geqq a$ and since $w \geqq m$ it follows $w \geqq a \cup m = u$. But then $m = w \cap m_1 = m_1$ and m is thus a maximal element congruent to z. Hence 1) follows from 2).

Next let us suppose that 1) holds and let m be a maximal element such that $m \equiv z(\theta)$. Let m' be an arbitrary complement of m. If x is an arbitrary element of L, let $y = (m \cap x) \cup (m' \cap x)$. Clearly $x \geqq y$. Let v be a relative complement of y in x/z. Now

$$y = (m \cap x) \cup (m' \cap x) \equiv (z \cap x) \cup (m' \cap x) \equiv m' \cap x = (m' \cup z) \cap x$$

$$\equiv (m' \cup m) \cap x \equiv x(\theta)$$

and hence $v \equiv z(\theta)$. But then $m \cup v \equiv z(\theta)$ and by the maximal property of m

we must have $m \cup v = m$. But then $v = m \cap v = m \cap x \cap v = (m \cap x) \cap y \cap v = z$.
But then $x = y \cup v = y$ and hence for all $x \in L$ we have

(i) $$x = (m \cap x) \cup (m' \cap x)$$

Now let $x = x_1 \cup x_2$ where $m \geqq x_1$ and $m' \geqq x_2$. Then $m \cap x \geqq x_1$ and $m' \cap x \geqq x_2$. We shall show that $m \cap x = x_1$ and $m' \cap x = x_2$. For let y_1 be a relative complement of $m \cap x$ in x/x_1. Then $m \cap x/x_1$ is projective to x/y_1. Now

$$y_1 \cup (m' \cap x) = y_1 \cup x_1 \cup (m' \cap x) = y_1 \cup x_1 \cup x_2 \cup (m' \cap x)$$

$$= y_1 \cup x \cup (m' \cap x) = x.$$

Hence x/y_1 is projective to $m' \cap x/m' \cap y_1$. Let ω_1 be a relatively complement of $m' \cap y_1$ in $m' \cap x/z$. Then $m' \cap x/m' \cap y_1$ is projective to w_1/z and hence w_1/z is projective to $m \cap x/x_1$. But since $m \equiv z(\theta)$ we have $m \cap x \equiv x_1(\theta)$ and hence $w_1 \equiv z(\theta)$. Thus $m \cup w_1 \equiv z(\theta)$ and since m is a maximal element congruent to z we have $m \geqq w_1$. But then $z = m \cap m' \geqq w_1$ and since $m \cap x/x_1$ is projective to w_1/z we have $m \cap x = x_1$. Now let w_2 be a relative complement of x_2 in $m' \cap x/z$. Then $m' \cap x/x_2$ is projective to w_2/z. But $m' \cap x = m' \cap (x_1 \cup x_2) \equiv m' \cap (z \cup x_2) \equiv m' \cap x_2 = x_2(\theta)$. Hence $w_2 \equiv z(\theta)$. By the maximal property of m we have $m \geqq w_2$. But then $z = m \cap m' \geqq w_2$ and hence $m' \cap x = x_2$.

Consider the correspondence

(ii) $$x \to \{m \cap x, m' \cap x\}$$

which maps L into $m/z \vee m'/z$. If $x_1 \in m/z$ and $x_2 \in m'/z$ then by the preceding paragraph $x_1 \cup x_2 \to \{x_1, x_2\}$ and hence the correspondence (ii) maps L onto $m/z \vee m'/z$. From formula (i) it follows that distinct elements in L have distinct images in $m/z \vee m'/z$. Since the order relation is clearly preserved we have

$$L \simeq m/z \vee m'/z.$$

Let φ be the congruence relation determined by the mapping

$$x \to m' \cap x.$$

If $x \equiv y(\varphi)$, then $m' \cap x = m' \cap y$ and taking joins with m we have $m \cup x = m \cup y$. But then $x \equiv m \cup x = m \cup y \equiv y(\theta)$. Conversely, if $x \equiv y(\theta)$, let v_1 be a relative complement of $m' \cap x$ in $m' \cap (x \cup y)/z$. Since $m' \cap (x \cup y) \equiv m' \cap x(\theta)$ we have $v_1 \equiv z(\theta)$. By the maximal property of m we conclude that $m \geqq v_1$. But then $z = m \cap m' \geqq v \geqq z$ and hence $m' \cap x = m' \cap (x \cup y)$. Similarly $m' \cap (x \cup y) = m' \cap y$ and thus $x \equiv y(\varphi)$. Hence $\theta = \varphi$ and the implication 1) $\to$ 3) follows.

Finally the implication 3) $\to$ 2) is trivial. For if φ is determined by the mapping $L \to a/z$ where $L \simeq a/z \vee b/z$. Then $x \equiv y(\varphi)$ if and only if $a \cap x = a \cap y$. The proof of the theorem is thus complete.

THEOREM 4.2. *Every two congruence relations on a relatively complemented lattice permute.*

PROOF. Let θ, φ be two congruence relations on a relatively complemented lattice L. Let $a \equiv b(\theta)$ and $b \equiv c(\varphi)$. Furthermore let a_1 be a relative complement of $a \cup b$ in $a \cup b \cup c/a$ and let c_1 be a relative complement of $b \cup c$ in $a \cup b \cup c/c$. Finally set $d = a_1 \cap c_1$. Now $a \cup b = a \cup b \cup b \equiv a \cup b \cup c(\varphi)$ and since a_1/a is a transpose of $a \cup b \cup c/a \cup b$ we have $a_1 \equiv a(\varphi)$. Similarly $c_1 \equiv c(\theta)$. Also $a \cup b \equiv a \cup a = a(\theta)$ and since $a \cup b/a$ is a transpose of $a \cup b \cup c/a_1$ we have $a \cup b \cup c \equiv a_1(\theta)$. But $a \cup b \cup c \geqq a_1 \cup c_1 \geqq a_1$ and hence $a_1 \cup c_1 \equiv a_1(\theta)$. Taking meets with c_1 we have $c_1 \equiv a_1 \cap c_1 = d(\theta)$. Similarly $a_1 \equiv d(\varphi)$. But then by transitivity $a \equiv d(\varphi)$ and $d \equiv c(\theta)$. Thus θ and φ permute.

The effect of a chain condition in a relatively complemented lattice upon the structure of the lattice can be described as follows.

THEOREM 4.3. *If a relatively complemented lattice L satisfies a chain condition, then $\Theta(L)$ satisfies the same chain condition.*

PROOF. Let us suppose that L satisfies the ascending chain condition. Then if θ is a congruence relation on L, there exists a maximal element m_θ such that $m_\theta \equiv z(\theta)$. Now let $\theta_1 > \theta_2$. Then $m_{\theta_2} \equiv z(\theta_2)$ implies $m_{\theta_2} \equiv z(\theta_1)$ and hence $m_{\theta_1} \cup m_{\theta_2} \equiv z(\theta_1)$. By the maximal property of m_{θ_1} we have $m_{\theta_1} \cup m_{\theta_2} = m_{\theta_1}$ and hence $m_{\theta_1} \geqq m_{\theta_2}$. Let a, b be such that $a \equiv b(\theta_1)$ but $a \not\equiv b(\theta_2)$. Let w be a relative complement of $a \cap b$ in $a \cup b/z$. Then $w \equiv z(\theta_1)$ but $w \not\equiv z(\theta_2)$. Hence $m_{\theta_1} \geqq w$ and $m_{\theta_2} \not\geqq w$. We conclude that $m_{\theta_1} > m_{\theta_2}$. It follows then from the ascending chain condition in L that the ascending chain condition also holds in $\Theta(L)$. A dual argument holds for the descending chain condition.

We give now the proof of the structure theorem for relatively complemented lattices satisfying a chain condition.

THEOREM 4.4. *A relatively complemented lattice satisfying the ascending chain condition is a direct union of a finite number of simple relatively complemented lattices.*

PROOF. Since the ascending chain condition holds in L. property 1) of Theorem 4.1 is automatically satisfied. Hence every congruence relation on L is a decomposition congruence and thus by Lemma 3.2, $\Theta(L)$ is complemented. By Lemma 2.2, $\Theta(L)$ is distributive and by Theorem 4.3 $\Theta(L)$ satisfies the ascending chain condition. Hence $\Theta(L)$ is a finite Boolean algebra. But by Theorem 4.2, $\Gamma(L) = \Theta(L)$ for any relatively complemented lattice L. Hence by Theorem 3.1, L is a direct union of a finite number of simple lattices which clearly must be relatively complemented.

It should be remarked that it is possible to derive Theorem 4.4 directly from Theorem 4.1 using the decomposition of L into indecomposable lattices. However the proof given seems to show more clearly the role of relative complementation in structure theorems.

Let L be a relatively complemented lattice and let x/y be contained in the quotient $a/a \cap b$. Let x_1 be a relative complement of y in $x/a \cap b$. Then x/y is projective to $x_1/a \cap b$. Since $x_1 \cap b = (x_1 \cap a) \cap b = a \cap b$, $x_1/a \cap b$ is projective to $x_1 \cup b/b$ which is contained in $a \cup b/b$. Thus x/y is projective to a quotient contained in $a \cup b/b$. Repeated application of this result shows that if x/y is weakly projective into a/b, then x/y is projective to a quotient contained in a/b. Let us recall that x "covers" y (in symbols $x \succ y$) if the quotient lattice x/y

consists of only x and y. a/b is a *prime* quotient if $a \succ b$. It follows then that x/y is weakly projective into a prime quotient a/b if and only if x/y is projective to a/b.

Now let L be a simple relatively complemented lattice and let x/y and a/b be prime quotients of L. Let θ be the congruence relation generated on L by collapsing a/b. Then $\theta \neq \omega$ and since L is simple we have $\theta = \iota$. But then $x \equiv y(\theta)$ and hence by Lemma 2.1, x/y is weakly projective into a/b. Since a/b is prime, x/y and a/b are projective. Conversely, if L satisfies the ascending chain condition and every two prime quotients are projective, let θ be a congruence relation on L. If $\theta \neq \omega$, there exist distinct elements a and b such that $a \equiv b(\theta)$. Let $a \cup b \succ a_1 \geqq a \cap b$. Then $a \cup b \equiv a_1(\theta)$. Now by the ascending chain condition there exists a maximal element m such that $m \equiv z(\theta)$. Let m' be a complement of m. If $m \neq u$, then $m' \neq z$ and again by the ascending chain condition x exists such that $m' \succ x \geqq z$. By hypothesis m'/x and $a \cup b/a_1$ are projective and hence $m' \equiv x(\theta)$. Let x_1 be a relative complement of x in m'/z. Then x_1/z is projective to m'/x and hence $x_1 \equiv z(\theta)$. But by the maximal property of m we must have $m \geqq x_1$. Hence $z = m \cap m' \geqq x_1 \geqq z$ and thus $m' = x$ contrary to $m' \succ x$. We conclude that $m = u$ and hence $u \equiv z(\theta)$. It follows that $\theta = \iota$ and L is therefore simple. This proves

THEOREM 4.5. *Let L be a relatively complemented lattice satisfying a chain condition. Then L is simple if and only if every two prime quotients of L are projective.*

It should be noted that for general relatively complemented lattices one chain condition does not imply the other as in the case with complemented modular lattices. For example, consider the lattice constructed as follows. Let I denote the set of positive integers and let S_k denote the subset of I consisting of the two integers $2k - 1$ and $2k$ where $k = 1, 2, 3, \cdots$. Let L consist of I and all finite subsets A of I which contain at most one integer of S_k for each k. Clearly if $A \neq I$ and $A \in L$, then every subset of A also belongs to L and hence the intersection of any collection of sets belonging to L also belongs to L. Thus L is a lattice. Now let $A \supset B \supset C$ in L. If $A \neq I$, let $B_1 = A \cap (C \cup B')$ where B' is the complement of B. Then $B_1 \in L$ and $B_1 \cup B = A$, $B_1 \cap B = C$. If $A = I$, let $n \in B$ but $n \notin C$. Then $n \in S_k$ for some k. Let n' be the other member of S_k. Then $n' \notin B$ since otherwise $S_k \subseteq B$ and hence $n' \notin C$. Let $B_1 = C + n'$. Then since B_1 does not contain n and $C \in L$, it follows that $B_1 \in L$. Clearly $B_1 \cap B = C$. On the other hand, $S_k \subseteq B_1 + B$ and hence $B_1 \cup B = I = A$. Thus L is relatively complemented. Now the descending chain condition obviously holds in L while

$$\{1\} \subset \{1, 3\} \subset \{1, 3, 5\} \subset \cdots$$

is an infinite ascending chain.

5. Locally relatively complemented lattices

Throughout this section the discussion will be restricted to lattices of finite dimension. If a is an element of such a lattice, let u_a denote the join of all elements covering a and let z_a denote the meet of all elements covered by a.

DEFINITION: 5.1. A lattice of finite dimensions is *locally* relatively complemented if u_a/a and a/z_a are relatively complemented for each a.

We shall show first that weak projectivity of prime quotients in a locally relatively complemented lattice can be reduced to ordinary projectivity. A preliminary lemma is needed.

LEMMA 5.1. *Let a and b be elements of a locally relatively complemented lattice of finite dimensions. Then every prime quotient contained in $a/a \cap b$ is projective to a prime quotient contained in $a \cup b/b$.*

For suppose that the conclusion of the lemma does not hold, then by the descending chain condition there exists a minimal element m containing two elements for which the conclusion of the lemma does not hold. We may suppose that the two elements are a and b. By the minimal property of m we have $m = a \cup b$. Let x/y be a prime quotient contained in $a/a \cap b$ which is not projective to any prime quotient contained in $a \cup b/b$. Now $(a \cup b) \cap x = x \neq y = (a \cup b) \cap y$. Hence by the descending chain condition there exists a minimal element c contained in $a \cup b/b$ such that $c \cap x \neq c \cap y$. Since $b \cap x = b \cap a = b \cap y$ we have $c > b$. Hence d exists such that $c > d \geq b$. By the minimal property of c we have $d \cap x = d \cap y$. If $c = c \cap x$, then $a \geq c \cap x = c \geq b$ and x/y itself is a quotient contained in $a \cup b/b$ which is projective to x/y contrary to assumption. Hence $c > c \cap x$ and there exists an element e such that $c > e \geq c \cap x$.

Now suppose that x/y is projective to a quotient x_1/y_1 contained in $(e \cap d) \cup c \cap x)/e \cap d$. Then since $e \cap d \in c/z_c$ and c/z_c is relatively complemented by hypothesis, there exists a relative complement w_1 of y_1 in $x_1/e \cap d$. If $d \geq w_1$, then $e \cap d \geq w_1$ and $x_1 = y_1$ which is impossible. Hence $d \cup w_1 = c$ and clearly $d \cap w_1 = d \cap e \cap w_1 = d \cap e$. Thus $w_1/d \cap e$ is projective to c/d. But then x/y is projective to c/d which is contained in $a \cup b/b$ which is contrary to hypothesis. Hence we conclude that x/y is projective to no quotient contained in $(e \cap d) \cup (c \cap x)$ $/e \cap d$.

Next let $c \cap x \geq v > c \cap y$. Then $y \not\geq v$ since otherwise $v \cap y \geq v$. Hence $y \cup v = x$ and clearly $y \cap v = y \cap c \cap v = c \cap y$. Thus $v/c \cap y$ is projective to x/y and hence is projective to no quotient contained in $(e \cap d) \cup (c \cap x)/e \cap d$. But $(e \cap d)(\cap c \cap x) = e \cap d \cap x = e \cap d \cap y = e \cap d \cap c \cap y = c \cap y$. Thus $c \cap x/c \cap y$ contains the prime quotient $v/c \cap y$ which is projective to no prime quotient contained in $(e \cap d) \cup (c \cap x)/e \cap d$. Now $m = a \cup b \geq c > e \geq (e \cap d) \cup (c \cap x)$ and this contradicts the minimal property of m. Hence the conclusion of the lemma holds for all a and b.

Recalling the definition of weak projectivity and making repeated use of Lemma 5.1 and its dual we get immediately.

THEOREM 5.1. *Let L be a locally relatively complemented lattice of finite dimensions. Then a prime quotient is weakly projective into a quotient a/b if and only if it is projective to a prime quotient contained in a/b.*

The next lemma shows that local relative complementation is preserved under homomorphisms.

LEMMA 5.2. *Let θ be a congruence relation on a locally relatively complemented lattice L of finite dimensions. Then L_θ is locally relatively complemented.*

For if A is any congruence class of L_θ, then A is a sublattice of a finite dimensional lattice and hence has a unit element a. Now we shall denote by $C(x)$ the congruence class to which x belongs. Hence $A = C(a)$. Now let $B > A$ in L_θ and let $b \in B$. Then since $B > A$, $a \cup b \in B$ and $a \cup b \neq a$. Hence b_1 exists such that $a \cup b \geq b_1 > a$. But then $B = C(a \cup b) \geq C(b_1) \geq C(a) = A$. But if $C(b_1) = A$, then $a \geq b_1$ contrary to $b_1 > a$. Hence we have $B = C(b_1)$ where $b_1 > a$. It follows that $u_A \leq C(u_a)$. On the other hand if $b_1 > a$, let $C(b_1) \geq X \geq A$. If x is an element of X, let $x_1 = a \cup (b_1 \cap x)$. Then $C(x_1) = A \cup (C(b_1) \cap X) = X$. But $b_1 \geq x_1 \geq a$ and hence $b_1 = x_1$ or $x_1 = a$. Thus $C(b_1) = X$ or $X = A$. It follows that $C(b_1) > A$ and hence that $u_A = C(u_a)$. Now let $u_A \geq Y \geq A$ and let $y \in Y$. If $y_1 = a \cup (u_a \cap y)$, then $C(y_1) = A \cup (u_A \cap Y) = Y$ and y_1 clearly belongs to u_a/a. Hence the mapping $x \to C(x)$ is a homomorphic (indeed, isomorphic!) mapping of u_a/a onto u_A/A. Since u_a/a is relatively complemented for each $a \in L$ it follows that u_A/A is relatively complemented for each $A \in L_\theta$. A dual argument gives the relative complementation of A/z_A for each $A \in L_\theta$ and L_θ is thus locally relatively complemented.

We can now prove the basic structure theorem for locally relatively complemented lattices.

THEOREM 5.2. *Every locally relatively complemented lattice of finite dimensions is a subdirect union of a finite number of simple locally relatively complemented lattices.*

PROOF. Let $u = a_1 > a_2 > \cdots > a_{n+1} = z$ be a complete chain joining u to z. Let θ_i be the congruence relation generated by collapsing the quotient a_i/a_{i+1}. Suppose $\theta_i \geq \theta > \omega$ for some θ. Then there exists a prime quotient a/b such that $a \equiv b(\theta)$. But then $a \equiv b(\theta_i)$ and hence a/b is weakly projective into a_i/a_{i+1}. Thus from Theorem 5.1, a/b and a_i/a_{i+1} are projective. It follows that $a_i \equiv a_{i+1}(\theta)$ and hence that $x \equiv y(\theta_i)$ implies $x \equiv y(\theta)$. Thus $\theta_i \leq \theta$ and hence $\theta_i = \theta$. We conclude that each θ_i is a point of $\Theta(L)$. But clearly $u \equiv z(\theta_1 \cup \cdots \cup \theta_n)$ and hence $\theta_1 \cup \cdots \cup \theta_n = \iota$. Since $\Theta(L)$ is distributive by Lemma 2.2 it follows that $\Theta(L)$ is a finite Boolean algebra. By Theorem 3.3, L is a subdirect union of a finite number of simple lattices. By Lemma 5.2, these simple lattices are locally relatively complemented and the proof of the theorem is complete.

COROLLARY 1. *Every irreducible locally relatively complemented lattice of finite dimensions is simple.*

COROLLARY 2. *Let L be a locally relatively complemented lattice of finite dimensions and let $u = a_1 > a_2 > \cdots > a_{n+1} = z$ be a complete chain joining u to z. Then every prime quotient is projective to one of the quotients a_i/a_{i+1}.*

For if a/b is a prime quotient of L, let θ be the congruence relation generated by collapsing a/b. Then from the proof of Theorem 5.2, $\theta \geq \theta_i$ for some i and hence $a_i \equiv a_{i+1}(\theta)$. Thus a_i/a_{i+1} is weakly projective into a/b and by Theorem 5.1, a_i/a_{i+1} and a/b are projective.

COROLLARY 3. *A locally relatively complemented lattice of finite dimensions is simple if and only if every two prime quotients are projective.*

CALIFORNIA INSTITUTE OF TECHNOLOGY

References

1. G. Birkhoff. *Combinational relations in projective geometries.* Ann. of Math. 36 (1935), 743–748.
2. G. Birkhoff. *Lattice Theory*, Revised edition, Amer. Math. Soc. Colloquium Publications, Vol. 25, 1948.
3. G. Birkhoff. *Subdirect unions in universal algebra.* Bull. Amer. Math. Soc. 50 (1944), 764–768.
4. K. Menger. *New foundations of projective and affine geometry.* Ann. of Math. 37 (1936), 456–482.
5. N. Funayama and T. Nakayama. *On the distributivity of a lattice of lattice-congruences* Proc. Imp. Acad. Tokyo 18 (1942), 553–554.

THE NORMAL COMPLETION OF THE LATTICE OF CONTINUOUS FUNCTIONS

BY

R. P. DILWORTH

1. Introduction. Let S be a topological space[1] and let $C(S)$ denote the set of all real valued, bounded, continuous functions on S. It is well known that $C(S)$ is a distributive lattice under the operations sup (f, g) and inf (f, g). In general, however, $C(S)$ is not a complete lattice; that is, an arbitrary bounded set of continuous functions in $C(S)$ need not have a least upper bound in the lattice $C(S)$. Furthermore, the structure of the minimal completion of $C(S)$ by means of normal subsets has not been determined even in the simple case where S is the real interval $[0, 1]$.

The first part of the paper will be devoted to the construction of a set of functions which form a complete lattice isomorphic to the normal completion of $C(S)$. We use for this purpose a class of bounded, upper semicontinuous functions (called *normal*) which are characterized by the following property[2].

$$(f_*)^* = f.$$

It is proved that *the normal completion of $C(S)$ is isomorphic with the lattice of all normal, upper semicontinuous functions on a suitably determined completely regular space S_0. If S is completely regular, then S_0 is simply S itself.*

As an application we deduce the Stone-Nakano theorem on spaces S for which $C(S)$ is lattice complete.

In the second part of the paper it is shown that the normal completion of $C(S)$ is itself isomorphic to the lattice of all continuous functions on some compact Hausdorff space. The precise theorem is the following.

Let S be completely regular. Then the normal completion of $C(S)$ is isomorphic with the lattice of all continuous functions on the Boolean space associated with the Boolean algebra of regular open sets of S.

Birkhoff has shown that if S is a completely regular space without isolated points and satisfying the second countability axiom, then the Boolean algebra of regular open sets is isomorphic with the normal completion of the free Boolean algebra with a countably infinite set of generators. Hence specializing

---

Presented to the Society, September 10, 1948 and September 1, 1949; received by the editors August 16, 1948 and, in revised form, August 10, 1949.

[1] The term "topological space" is used in the sense of Alexandroff and Hopf, *Topologie*, Berlin, 1936. I am indebted to Professors Bohnenblust and Karlin for their advice in connection with the topological questions arising in the work.

[2] f^* and f_* represent respectively the upper and lower limit functions of f. See formulas (3.1) and (3.2) for the precise definitions.

427

we get the following theorem:

If S is a completely regular space without isolated points and satisfying the second countability axiom, then the normal completion of $C(S)$ is isomorphic with the lattice of all continuous functions on the Boolean space associated with the normal completion of the free Boolean algebra with a countably infinite set of generators.

Thus the lattices of continuous functions on spaces satisfying the conditions of the theorem all have the same normal completion. In particular, this theorem gives a simple representation of the normal completion of the lattice of continuous functions on the real interval $[0, 1]$.

PART I. NORMAL UPPER SEMICONTINUOUS FUNCTIONS

2. **Preliminary reduction.** Since we shall be interested in lattice properties of $C(S)$, we may, if we wish, assume that S is completely regular[3] (Čech [2][4]). The reduction to the completely regular case can be accomplished as follows: Define

$$x \sim y \quad \text{if} \quad f(x) = f(y) \qquad \text{for all } f \in C(S).$$

The relation $x \sim y$ is clearly an equivalence relation and hence separates S into equivalence classes $X, Y, Z, \cdots$. Let S_0 denote the set of equivalence classes. To each $f \in C(S)$ there corresponds a function F on S_0 defined by $F(X) = f(x)$ where x is an element of X. If A_0 is a subset of S_0, let the closure of A_0 consist of all X such that for every F, $F(X) = 0$ whenever $F(Y) = 0$ for all Y contained in A_0. Then S_0 becomes a completely regular topological space under this definition of closure and the mapping

$$f \to F$$

is a lattice isomorphism of $C(S)$ onto $C(S_0)$.

By appealing to the Stone-Čech compactification theorem we could also assume that S is compact. However, little is gained from the additional assumption and it seems desirable that the results of part I should not depend upon transfinite methods.

We shall frequently use the fact that every completely regular space is regular; that is, if N is any open set containing x, there is an open set A containing x whose closure is contained in N.

3. **Properties of normal upper semicontinuous functions.** Let $B(S)$ denote the set of all bounded, real functions on S. If x is a point of S, let N_x denote an arbitrary open set containing x. Then the two basic unary operations on $B(S)$ which we shall use are defined as follows:

---

[3] A topological space S is completely regular if for each x and open set A containing x, there is a continuous function f having the value 1 at x and vanishing outside A. Replacing f by sup $(0, \inf (1, f))$ if necessary one may assume that the values of f lie between 0 and 1.

[4] Numbers in brackets refer to the references cited at the end of the paper.

$$(3.1) \qquad \phi^*(x) = \inf_{N_x} \sup_{y \in N_x} \phi(y),$$

$$(3.2) \qquad \phi_*(x) = \sup_{N_x} \inf_{y \in N_x} \phi(y).$$

LEMMA 3.1. *The operations ϕ^* and ϕ_* have the following properties:*

$$(3.3) \qquad \phi^* \geq \phi \geq \phi_*,$$

$$(3.4) \qquad \phi \geq \psi \;\rightarrow\; \phi^* \geq \psi^* \quad and \quad \phi_* \geq \psi_*,$$

$$(3.5) \qquad (\phi^*)^* = \phi^*, \qquad (\phi_*)_* = \phi_*,$$

$$(3.6) \qquad (((\phi^*)_*)^*)_* = (\phi^*)_*, \qquad (((\phi_*)^*)_*)^* = (\phi_*)^*.$$

Properties (3.3), (3.4), and (3.5) follow immediately from (3.1) and (3.2). Also by (3.3), $((\phi^*)_*)^* \geq (\phi^*)_*$ and hence $(((\phi^*)_*)^*)_* \geq ((\phi^*)_*)_* = (\phi^*)_*$ by (3.4) and (3.5). On the other hand $(\phi^*)_* \leq \phi^* \rightarrow ((\phi^*)_*)^* \leq (\phi^*)^* = \phi^* \rightarrow (((\phi^*)_*)^*)_* \leq (\phi^*)_*$ by (3.3) and (3.4). Thus the first part of (3.6) is proved and the second part follows in a similar manner.

DEFINITION 3.1. ϕ is *upper semicontinuous* on S if $\phi^* = \phi$.

Lower semicontinuous functions are defined dually. Clearly ϕ is continuous if and only if $\phi^* = \phi_*$.

The functions of $B(S)$ which will be used to characterize the normal completion of $C(S)$ are defined as follows:

DEFINITION 3.2. An upper semicontinuous function ϕ on S is *normal* if $(\phi_*)^* = \phi$. Clearly every continuous function is normal.

Normality can be characterized as follows:

THEOREM 3.1. *An upper semicontinuous function ϕ on S is normal if and only if for each $\epsilon > 0$, $x \in S$, and open set N containing x, there exists a nonempty open set $A \subseteq N$ such that $\phi(y) > \phi(x) - \epsilon$ all $y \in A$.*

For the proof let ϕ be an upper semicontinuous function on S and let us suppose first that ϕ is normal. Let $\epsilon > 0$ and let N be an open set containing x. By (3.1),

$$\sup_{z \in N} \phi_*(z) \geq (\phi_*)^*(x) = \phi(x).$$

For some $z \in N$

$$\phi_*(z) > \phi(x) - \epsilon.$$

By (3.2) there is a neighborhood A of z contained in N such that

$$\inf_{y \in A} \phi(y) > \phi(x) - \epsilon.$$

This gives the necessity of the condition of the theorem.

Conversely, if the condition is satisfied for all $\epsilon > 0$ and N containing x, let $z \in A$. Then

$$\phi_*(z) \geqq \inf_{v \in A} \phi(y) \geqq \phi(x) - \epsilon.$$

Hence

$$\sup_{z \in N} \phi_*(z) \geqq \phi(x) - \epsilon.$$

Thus

$$(\phi_*)^*(x) = \inf_{N_x} \sup_{z \in N_x} \phi_*(z) \geqq \phi(x) - \epsilon.$$

Since ϵ is arbitrary we have

$$(\phi_*)^* \geqq \phi.$$

Since ϕ is upper semicontinuous we have by (3.3) and (3.4)

$$(\phi_*)^* \leqq \phi^* = \phi.$$

Hence $(\phi_*)^* = \phi$ and the proof is complete.

Now a lower semicontinuous function can be characterized by the condition that $\{x \mid \phi(x) > \lambda\}$ is open for each real λ. A dual result holds for upper semicontinuous functions. Normal upper semicontinuous functions can also be characterized in a similar manner.

THEOREM 3.2. *An upper semicontinuous function ϕ on S is normal if and only if for each real λ, $\{x \mid \phi(x) > \lambda\}$ is a union of closures of open sets.*

Let us suppose first that $\phi = (\phi_*)^*$ and let $A = \{x \mid \phi(x) > \lambda\}$. Let x_0 be an arbitrary element of A. Then $\phi(x_0) > \lambda$ and hence $\phi(x_0) > \lambda + \delta$ for some $\delta > 0$. Let $B = \{x \mid \phi_*(x) > \lambda + \delta\}$. Clearly B is open since ϕ_* is lower semicontinuous. If N is an arbitrary open set containing x_0, then

$$\sup_{v \in N} \phi_*(y) \geqq (\phi_*)^*(x_0) = \phi(x_0) > \lambda + \delta.$$

Hence $\phi_*(y) > \lambda + \delta$ for some $y \in N$. Thus $B \cap N \neq 0$ for all N and hence $x_0 \in \overline{B}$. Moreover, if $y_0 \in \overline{B}$, then $B \cap N \neq 0$ for every open set N containing y_0 and thus

$$\sup_{v \in N} \phi_*(y) > \lambda + \delta \quad \text{all } N \text{ containing } \quad y_0.$$

Hence $\phi(y_0) = (\phi_*)^*(y_0) \geqq \lambda + \delta > \lambda$ and thus $y_0 \in A$. But then $x_0 \in \overline{B} \subseteq A$ and it follows that A is a union of closures of open sets.

On the other hand, suppose that ϕ is upper semicontinuous and that

$\{x \mid \phi(x) > \lambda\}$ is a union of closures of open sets for each real λ. Let $\epsilon > 0$, $x_0 \in S$, and N be an arbitrary open set containing x_0. Then $\{x \mid \phi(x) > \phi(x_0) - \epsilon\}$ is a union of closures of open sets and hence there exists an open set $A_1 \subseteq \{x \mid \phi(x) > \phi(x_0) - \epsilon\}$ such that $x_0 \in \overline{A}_1$. But then $A = A \cap N$ is a nonempty open set contained in N such that $\phi(y) > \phi(x_0) - \epsilon$ all $y \in A$. Thus ϕ is normal by Theorem 3.1. This completes the proof of the theorem.

COROLLARY. *Every normal upper semicontinuous function on S is continuous if and only if the closure of every open subset of S is open.*

For by Theorem 3.2 the characteristic function of the closure of an open set is upper semicontinuous and normal. Hence if every normal upper semicontinuous function is continuous, the closure of every open set is open. Conversely, if the closure of every open set is open and ϕ is any normal upper semicontinuous function on S, then by Theorem 3.2, ϕ is lower semicontinuous and hence continuous.

4. **Normal subsets of $C(S)$.** Before applying these results to the completion problem we shall recall some relevant facts from the theory of partially ordered sets[5]. A subset S of a partially ordered set P is normal if S contains all a for which $a \geq x$ for every x such that $y \geq x$ for all $y \in S$. If X is an arbitrary subset of S, the set of all x containing all elements of X is normal. In particular, for each a the set of all $x \geq a$ is a normal subset called the *principal* normal subset generated by a. The collection of normal subsets of P form a complete lattice containing P as the partially ordered set of principal normal subsets and preserving sup and inf whenever they exist in P. This normal completion is minimal in the sense that if P is imbedded in any other complete lattice L, the lattice of normal subsets is isomorphic with a lattice within L.

In the present case P is the lattice $C(S)$ of continuous functions on S. If $\phi \in B(S)$, let L_ϕ denote the set of all $f \in C(S)$ such that $f \geq \phi$.

LEMMA 4.1. *If $\phi \in B(S)$, then* $\inf(L_\phi) = \phi^*$.

Since $\phi^*(x) = \inf_{N_x} \sup_{y \in N_x} \phi(y)$, for $\epsilon > 0$ there exists an open set N containing x such that $\phi^*(x) > \sup_{y \in N} \phi(y) - \epsilon$. By complete regularity, $g \in C(S)$ exists such that $g(x) = 1$, $g(y) = 0$ all $y \in {}'N$ and $g \leq 1$. Let $m = \sup_{y \in S} \phi(y)$ and let

$$f = m - (m - \sup_{v \in N} \phi(y))g.$$

Clearly $f \in C(S)$. If $y \in N$, then $f(y) \geq m - (m - \sup_{y \in N} \phi(y)) \geq \phi(y)$. If $y \in {}'N$, then $f(y) = m \geq \phi(y)$. Hence $f \geq \phi$ and thus $f \in L_\phi$. We have then

$$\phi^*(x) > \sup_{v \in N} \phi(y) - \epsilon = f(x) - \epsilon \geq \psi(x) - \epsilon$$

---

[5] The reader is referred to Birkhoff [1] for an account of this theory.

where $\psi = \inf (L_\phi)$. Since ϵ is arbitrary, $\phi^*(x) \geq \psi(x)$ for all x. On the other hand, $f \geq \phi$ implies $f = f^* \geq \phi^*$, which implies $\psi \geq \phi^*$. Hence $\phi^* = \psi = \inf (L_\phi)$.

LEMMA 4.2. *Let ϕ be a normal, upper semicontinuous function on S. Then L_ϕ is a normal subset of $C(S)$.*

For let $f \geq g$ for all g contained in the functions of L_ϕ. We must show that $f \in L_\phi$. Let $x \in S$ and let $\epsilon > 0$. Since f is continuous, there exists an open set N containing x such that $f(y) - f(x) < \epsilon/2$ all $y \in N$. Since ϕ is normal there exists a non-empty open set $A \subseteq N$ such that $\phi(y) > \phi(x) - \epsilon/2$ all $y \in A$. Let y_0 be a point of A. By complete regularity, there exists a continuous function $h(y)$ such that $h \leq 1$, $h(y_0) = 1$, and $h(y) = 0$ all $y \in {}'A$. Let $m_\epsilon = \inf_{y \in S} \phi(y) - \epsilon/2$ and set $g = m_\epsilon + (\phi(x) - \epsilon/2 - m_\epsilon)h$. Now if $y \in A$, then

$$g(y) \leq m_\epsilon + (\phi(x) - \epsilon/2 - m_\epsilon) = \phi(x) - \epsilon/2 < \phi(y).$$

But if $y \in {}'A$, then $g(y) = m_\epsilon < \phi(y)$. Hence $g \leq \phi$ and thus g is a continuous function contained in all of the functions of L_ϕ. It follows that $f \geq g$. But then

$$f(y_0) \geq g(y_0) = m_\epsilon + (\phi(x) - \epsilon/2 - m_\epsilon) = \phi(x) - \epsilon/2.$$

Since $y_0 \in A \subseteq N$ we have

$$f(x) = f(y_0) + (f(x) - f(y_0)) > \phi(x) - \epsilon.$$

Since ϵ is arbitrary, $f(x) \geq \phi(x)$ for all x and hence $f \in L_\phi$. This completes the proof of the lemma.

We need also a converse result.

LEMMA 4.3. *Let $\mathfrak{A}$ be a normal subset of $C(S)$. Then $\inf (\mathfrak{A})$ is a normal, upper semicontinuous function on S.*

For let $\phi = \inf (\mathfrak{A})$ and let $\phi_* \leq f$ where $f \in C(S)$. Then if g is contained in all of the functions of $\mathfrak{A}$, we have $g \leq \phi$ and hence $g \leq \phi_* \leq f$. Hence $f \in \mathfrak{A}$ since $\mathfrak{A}$ is normal. But then by Lemma 4.1

$$(\phi_*)^* = \inf (L_{\phi_*}) = \inf (\mathfrak{A}) = \phi.$$

Thus ϕ is a normal, upper semicontinuous function and the lemma follows.

With these lemmas we are ready to prove the fundamental isomorphism theorem.

THEOREM 4.1. *Let S be a completely regular topological space. Then the completion of $C(S)$ by normal subsets is isomorphic with the lattice of all normal, upper semicontinuous real functions on S.*

For the proof let us recall that $B(S)$ is a complete lattice containing $C(S)$ as a sublattice and hence it follows from the general theory of the normal completion of a partially ordered set that if $\mathfrak{A}$ is a normal subset $C(S)$ the mapping $\mathfrak{A} \to \inf (\mathfrak{A})$ is an isomorphism. By Lemma 4.3, $\mathfrak{A}$ is mapped into the set of normal, upper semicontinuous real functions on S. But by Lemmas 4.1 and 4.2 every normal upper semicontinuous function is an image of a normal subset of $C(S)$. The proof is thus complete.

If $\psi \in B(S)$ and $\phi = (\psi_*)^*$, then $(\phi_*)^* = \phi$ by (3.6). Conversely, if $(\phi_*)^* = \phi$, then ϕ trivially has the form $(\psi_*)^*$ with $\psi \in B(S)$. Hence Theorem 4.1 can also be stated in the following way.

COROLLARY. *If S is completely regular, then the normal completion of $C(S)$ is isomorphic to the lattice of all functions of the form $(\psi_*)^*$ where ψ is a bounded real function on S.*

Now it is clear from Theorem 3.2 that sup (ϕ_1, ϕ_2) where ϕ_1 and ϕ_2 are normal upper semicontinuous functions is also upper semicontinuous and normal. Hence sup (ϕ_1, ϕ_2) is the lattice union of ϕ_1 and ϕ_2. However, if $\mathfrak{A}$ is a bounded class of normal upper semicontinuous functions, sup $(\mathfrak{A})$ need not be normal. For example, let ϕ be defined over the real interval $[0, 1]$ by $\phi(x) = 1$ when $x \neq 1/2$ and $\phi(1/2) = 0$. Let $\mathfrak{A}$ be the set of all continuous functions f such that $f \leq \phi$. Then sup $(\mathfrak{A}) = \phi$ and ϕ is not normal. Also it should be noted that inf (ϕ_1, ϕ_2) need not be normal if ϕ_1 and ϕ_2 are normal. For example, let ϕ_1, ϕ_2 be the characteristic functions of the closed intervals $[0, 1/2]$ and $[1/2, 1]$ respectively. Then $\{x \mid \text{inf } (\phi_1, \phi_2) > 0\}$ consists of the single point $x = 1/2$ and hence is not a union of closures of open sets.

The general determination of the lattice operations in the set of normal upper semicontinuous functions is contained in the following theorem.

THEOREM 4.2. *Let S be an arbitrary topological space and let $\mathfrak{A}$ be a bounded collection of normal upper semicontinuous functions on S. Then the unique minimal normal upper semicontinuous function containing the functions of $\mathfrak{A}$ is* $(\text{sup } \mathfrak{A})^*$, *while the unique maximal normal upper semicontinuous function contained in the functions of $\mathfrak{A}$ is* $((\text{inf } \mathfrak{A})_*)^*$.

For by (3.3), (3.4), and (3.5) we have $(((\text{sup } \mathfrak{A})^*)_*)^* \leq (\text{sup } \mathfrak{A})^*$. On the other hand, since sup $\mathfrak{A} \geq \phi$ all $\phi \in \mathfrak{A}$, we have $(((\text{sup } \mathfrak{A})^*)_*)^* \geq ((\phi^*)_*)^*$ $= (\phi_*)^* = \phi$ for all $\phi \in \mathfrak{A}$. Hence $(((\text{sup } \mathfrak{A})^*)_*)^* \geq \text{sup } \mathfrak{A}$ and thus $(((\text{sup } \mathfrak{A})^*)_*)^*$ $\geq (\text{sup } \mathfrak{A})^*$. We conclude that $(\text{sup } \mathfrak{A})^*$ is normal. If ψ is a normal upper semicontinuous function such that $\psi \geq \phi$ all $\phi \in \mathfrak{A}$, then $\psi \geq \text{sup } \mathfrak{A}$ and hence $\psi = \psi^* \geq (\text{sup } \mathfrak{A})^*$. Thus the first conclusion of the theorem holds. Now if $\psi \leq \phi$ all $\phi \in \mathfrak{A}$, then $\psi \leq \text{inf } \mathfrak{A}$ and hence $\psi = (\psi_*)^* \leq ((\text{inf } \mathfrak{A})_*)^*$ and $((\text{inf } \mathfrak{A})_*)^*$ is a normal, upper semicontinuous function by (3.6). The proof is thus complete.

5. **An application.** We show now that the results of §§3 and 4 contain as a special case the theorem of Stone [5, 6] and Nakano [3] on complete lattices of continuous functions.

THEOREM 5.1 (Stone-Nakano). *If S is a topological space in which the* *closure of every open set is open, then $C(S)$ is complete. Conversely, if $C(S)$ is complete and S is completely regular, then the closure of every open set is open.*

For if the closure of every open set is open, by the corollary to Theorem 3.2, every normal upper semicontinuous function is continuous and by Theorem 4.2, $C(S)$ is complete. Conversely, if $C(S)$ is complete and S is completely regular, by Theorem 4.1 every normal upper semicontinuous function is continuous and hence by the corollary to theorem 3.2, the closure of every open set is open.

PART II. THE BOOLEAN SPACE ASSOCIATED WITH THE NORMAL COMPLETION

6. **The second representation theorem.** In this section it will be shown that the normal completion of the lattice of continuous functions on a topological space is isomorphic to the lattice of all continuous functions on another suitably determined topological space. Now it is well known (Birkhoff [1]) that the regular open sets[6] of a topological space form a complete Boolean algebra under set inclusion. Furthermore, with any Boolean algebra there is associated the Boolean space of minimal dual ideals. The precise theorem to be proved is the following:

THEOREM 6.1. *Let S be completely regular. Then the normal completion of $C(S)$ is isomorphic with the lattice of all continuous functions on the Boolean space[7] associated with the Boolean algebra of regular open sets of S.*

Let $\mathfrak{S}$ denote the Boolean space associated with the Boolean algebra Σ of regular open sets of S. Thus $\mathfrak{S}$ is the set of all minimal dual ideals[8] of Σ. The topology in $\mathfrak{S}$ is such that the closure of a subset $\mathfrak{A}$ of $\mathfrak{S}$ consists of all minimal dual ideals $\mathfrak{p}$ of $\mathfrak{S}$ for which $\cup\mathfrak{A}\supseteq\mathfrak{p}$ in the lattice of dual ideals.

We next define a pair of correspondences, σ and τ, one of which maps $B(S)$ into $B(\mathfrak{S})$ while the other maps $B(\mathfrak{S})$ into $B(S)$. The mapping σ is defined by

$$(6.1) \qquad \sigma f(\mathfrak{p}) = \inf_{P\in\mathfrak{p}} \sup_{y\in P} f(y).$$

Thus for each regular open set $P\in\mathfrak{p}$, the upper bound of f on P is calculated and the lower bound of these values for all $P\in\mathfrak{p}$ is $\sigma f(\mathfrak{p})$. The mapping τ is defined by

$$(6.2) \qquad \tau F(x) = \inf_{x\in A} \sup_{A\in\mathfrak{q}} F(\mathfrak{q}).$$

Thus for each regular open set A containing x, the upper bound of $F(\mathfrak{q})$ for all

---

[6] See Birkhoff [1, p. 177].

[7] See Stone [4].

[8] The minimal dual ideals of the lattice Σ are in one-to-one correspondence with the maximal ring ideals of Σ as a Boolean ring.

$\mathfrak{q}$ containing A is calculated and the lower bound of these values for all A containing x is $\tau F(x)$.

The proof of Theorem 6.1 will rest on a series of lemmas concerning the mappings σ, τ.

LEMMA 6.1. *If $f^* \geq g$, then $\sigma f \geq \sigma g$. Dually, if $F^* \geq G$, then $\tau F \geq \tau G$.*

For if N is any open set, we have

$$\sup_{x \in N} f(x) = \sup_{x \in N} f^*(x).$$

Hence

$$\sigma f(\mathfrak{p}) = \inf_{P \in \mathfrak{p}} \sup_{x \in P} f(x) = \inf_{P \in \mathfrak{p}} \sup_{x \in P} f^*(x) \geq \inf_{P \in \mathfrak{p}} \sup_{x \in P} g(x) = \sigma g(\mathfrak{p}).$$

If A is any regular open set of S, then the set of all $\mathfrak{q}$ containing A is both open and closed and hence

$$\sup_{A \in \mathfrak{q}} F(\mathfrak{q}) = \sup_{A \in \mathfrak{q}} F^*(\mathfrak{q}).$$

Thus if $F^* \geq G$, we have

$$\tau F(x) = \inf_{x \in A} \sup_{A \in \mathfrak{q}} F(\mathfrak{q}) = \inf_{x \in A} \sup_{A \in \mathfrak{q}} F^*(\mathfrak{q}) \geq \inf_{x \in A} \sup_{A \in \mathfrak{q}} G(\mathfrak{q}) = \tau G(x).$$

LEMMA 6.2. *σf and τF are upper semicontinuous for each $f \in B(S)$ and $F \in B(\mathfrak{S})$.*

For let $\sigma f(\mathfrak{p}) < \lambda$. Then $P \in \mathfrak{p}$ exists such that $\sup_{y \in P} f(y) < \lambda$. If $P \in \mathfrak{q}$, then $\sigma f(\mathfrak{q}) \leq \sup_{y \in P} f(y) < \lambda$. Since the set of all $\mathfrak{q}$ containing P is both open and closed, it follows that $\{\mathfrak{p} \mid \sigma f(\mathfrak{p}) < \lambda\}$ is open and hence σf is upper semicontinuous.

Similarly if $\tau F(x) < \lambda$, then there exists a regular open set A containing x such that $\sup_{A \in \mathfrak{p}} F(\mathfrak{p}) < \lambda$. Hence if $y \in A$, we have $\tau F(y) \leq \sup_{A \in \mathfrak{p}} F(\mathfrak{p}) < \lambda$. Since A is open $\{x / \tau F(x) < \lambda\}$ is open and τF is thus upper semicontinuous.

LEMMA 6.3. *If f is a normal, upper semicontinuous function on S, then σf is a continuous function on $\mathfrak{S}$.*

For let $\sigma f(\mathfrak{p}) > \lambda$ and suppose that for each $P \in \mathfrak{p}$ there exists a $\mathfrak{q}$ containing P such that $\sigma f(\mathfrak{q}) \leq \lambda$. Let $\sigma f(\mathfrak{p}) > \lambda_1 > \lambda$. Then $\sigma f(\mathfrak{q}) < \lambda_1$, and hence there exists $Q \in \mathfrak{q}$ such that $f(y) < \lambda_1$ all $y \in Q$. Now $P \cap Q$ belongs to $\mathfrak{q}$ and hence is non-empty. Let

$$W = \{y \mid f(y) < \lambda_1\}.$$

Then $P \cap Q \subseteq W \subseteq \overline{W}$ and if B denotes the interior of $\overline{W}$ we have $P \cap Q \subseteq B$. Hence $B \cap P \neq 0$ for every $P \in \mathfrak{p}$ and thus $(B) \cap \mathfrak{p} \neq 0$. But then $B \in \mathfrak{p}$. On the other hand, since $B \subseteq \overline{W}$, $f(y) < \lambda_1$, on a set dense in B. Hence $f_*(y) \leq \lambda_1$

for all $y \in B$. But then $(f_*)^*(y) \leq \lambda_1$ all $y \in B$. From the normality of f we get

$$f(y) = (f_*)^*(y) \leq \lambda_1 \qquad\qquad \text{all} \quad y \in B.$$

Hence $\sigma f(\mathfrak{p}) \leq \sup_{y \in B} f(y) \leq \lambda_1 < \sigma f(\mathfrak{p})$ which is impossible. Thus for some $P \in \mathfrak{p}$, $\sigma f(\mathfrak{q}) > \lambda$ all A containing P. It follows that $\{\mathfrak{p} \mid \sigma f(\mathfrak{p}) > \lambda\}$ is open for each λ and hence σf is lower semicontinuous. But then Lemma 6.2 implies that σf is continuous.

LEMMA 6.4. *If S is regular and F is a lower semicontinuous function on $\mathfrak{S}$, then τF is a normal, upper semicontinuous function on S.*

Now τF is upper semicontinuous by Lemma 6.2. Hence if τF is not normal, by Theorem 3.1 there exists $\epsilon > 0$, $x \in S$ and open set N containing x such that $U = \{y \mid \tau F(y) \leq \tau F(x) - \epsilon\}$ is dense in N. By regularity there exists a regular open set A such that $x \in A \subseteq N$. It follows that $A \cap U$ is dense in A. Let $\tau F(x) > \lambda > \tau F(x) - \epsilon$. If $y \in A \cap U$ then $\tau F(y) < \lambda$, and hence a regular open set A_y containing y exists such that $F(\mathfrak{p}) < \lambda$ all $\mathfrak{p}$ containing A_y. Let $\mathfrak{A}$ be the collection of all $\mathfrak{p}$ for which $A_y \in \mathfrak{p}$ for some $y \in A \cap U$. Let $B \in \cup \mathfrak{A}$. Then $B \supseteq A_y$ all $y \in A \cap U$ and hence $B \supseteq A \cap U$. Since $A \cap U$ is dense in A we have $\overline{B} \supseteq A$. But B is a regular open set and hence $B \supseteq A$. Thus $\cup \mathfrak{A} \supseteq \mathfrak{p}$ all $\mathfrak{p}$ containing A. But $F(\mathfrak{q}) < \lambda$ all $\mathfrak{q} \in \mathfrak{A}$ and hence by the lower semicontinuity of F, $F(\mathfrak{p}) \leq \lambda$ all $\mathfrak{p}$ containing A. But then

$$\tau F(x) \leq \sup_{A \in \mathfrak{p}} F(\mathfrak{p}) \leq \lambda < \tau F(x),$$

which is impossible. It follows that τF is a normal, upper semicontinuous function on S.

LEMMA 6.5. *If S is regular and $f \in B(S)$, then $\tau \sigma f \leq f^*$.*

For all regular open sets A containing x we have

$$\sup_{A \in \mathfrak{p}} \sigma f(\mathfrak{p}) \geq \tau \sigma f(x).$$

Thus if $\epsilon > 0$, for each A containing x, there exists a p containing A such that $\sigma f(\mathfrak{p}) > \tau \sigma f(x) - \epsilon$. But then

$$\sup_{y \in A} f(y) \geq \sigma f(\mathfrak{p}) > \tau \sigma f(x) - \epsilon.$$

Thus $f(y) > \tau \sigma f(x) - \epsilon$ for some y in each A containing x. Hence if S is regular, $f^*(x) \geq \tau \sigma f(x) - \epsilon$. Since ϵ is arbitrary we have $f^* \geq \tau \sigma f$.

LEMMA 6.6. *If f is a normal, upper semicontinuous function on S, then $\tau \sigma f \geq f$.*

For if $\epsilon > 0$ and x is any element of S, there exists a regular open set A such that $\sigma f(\mathfrak{p}) < \tau \sigma f(x) + \epsilon$ for all $\mathfrak{p}$ containing A. But then for some $P \in \mathfrak{p}$

we have $f(y) < \tau \sigma f(x) + \epsilon$ all $y \in P$. Let

$$W = \{y \mid f(y) < \tau \sigma f(x) + \epsilon\}.$$

Then for each $\mathfrak{p}$ containing A, there is a $P \in \mathfrak{p}$ such that $W \supseteq P$. Thus if B denotes the interior of $\overline{W}$, $B \supseteq P$ and hence $B \in \mathfrak{p}$ all $\mathfrak{p}$ containing A. But then $B \supseteq A$ and hence $x \in B$. Now $f(y) < \tau \sigma f(x) + \epsilon$ on a set dense in $\overline{W}$ and hence dense in B. Thus $f_*(y) \leq \tau \sigma f(x) + \epsilon$ for all $y \in B$. By the normality of f we have

$$f(y) = (f_*)^*(y) \leq \tau \sigma f(x) + \epsilon \qquad \text{all} \quad y \in B.$$

In particular, $f(x) \leq \tau \sigma f(x) + \epsilon$ for each $\epsilon > 0$. Thus $f \leq \tau \sigma f$.

LEMMA 6.7. *If* $F \in B(\mathfrak{S})$, *then* $F_* \leq \sigma \tau F \leq F^*$.

For let $\epsilon > 0$ and let $\mathfrak{p}$ be an arbitrary element of $\mathfrak{S}$. Then $P \in \mathfrak{p}$ exists such that $\tau F(y) < \sigma \tau F(\mathfrak{p}) + \epsilon$ all $y \in P$. But for each $y \in P$ there exists a regular open set A_y containing y, such that $F(\mathfrak{q}) < \sigma \tau F(\mathfrak{p}) + \epsilon$ all $\mathfrak{q} \in A_y$. Let $\mathfrak{A}_1$ be the set of all $\mathfrak{q}$ containing A_y for some y. If $B \in \mathsf{U}\mathfrak{A}_1$, then $y \in A_y \subseteq B$ for all y and hence $P \subseteq B$. But then $B \in \mathfrak{p}$ and hence $\mathsf{U}\mathfrak{A}_1 \supseteq \mathfrak{p}$. Since $\mathfrak{p} \in \overline{\mathfrak{A}}_1$, we have

$$F_*(\mathfrak{p}) \leq \sigma \tau F(\mathfrak{p}) + \epsilon.$$

But ϵ is arbitrary, and hence $F_* \leq \sigma \tau F$.

On the other hand, for every $P \in \mathfrak{p}$ we have

$$\sup_{y \in P} \tau F(y) \geq \sigma \tau F(\mathfrak{p}).$$

Hence if $\epsilon > 0$, there is a $y \in P$ such that $\tau F(y) > \sigma \tau F(\mathfrak{p}) - \epsilon$ and thus $\sup_{P \in \mathfrak{q}} F(\mathfrak{q}) > \sigma \tau F(\mathfrak{p}) - \epsilon$. Let $\mathfrak{A}_2 = \{\mathfrak{q} \mid F(\mathfrak{q}) > \sigma \tau F(\mathfrak{p}) - \epsilon\}$. Then $\mathsf{U}\mathfrak{A}_2 \cap (P) \neq 0$ for every $P \in \mathfrak{p}$. Thus $\mathsf{U}\mathfrak{A}_2 \cap \mathfrak{p} \neq 0$ and hence $\mathsf{U}\mathfrak{A}_2 \supseteq \mathfrak{p}$. Since $\mathfrak{p}$ is a limit point of $\mathfrak{A}_2$, we have

$$F^*(\mathfrak{p}) \geq \sigma \tau F(p) - \epsilon.$$

But ϵ is arbitrary, and hence $F^* \geq \sigma \tau F$.

Proof of Theorem 6.1. By Lemma 6.3, σ maps normal, upper semicontinuous functions on S into continuous functions on $\mathfrak{S}$. By Lemmas 6.5 and 6.6, distinct normal semi-continuous functions on S map into different continuous functions. By Lemmas 6.4 and 6.7, every continuous function on $\mathfrak{S}$ is an image of a normal, upper semicontinuous function on S. Finally, Lemma 6.1 shows that the mapping is an isomorphism. Hence the theorem follows from Theorem 4.1 of Part I.

It should be noted that if $C(S)$ is lattice complete, then the regular open sets are simply the open and closed set of S and Boolean space of Theorem 6.1 is the Stone-Čech compactification of S.

7. **Special cases.** Birkhoff [1, p. 177] has shown that if S is a completely regular space without isolated points and satisfying the second

countability axiom, then the Boolean algebra of regular open sets is isomorphic with the normal completion of the free Boolean algebra with a countably infinite set of generators. Applying Theorem 6.1 to this case we obtain the following theorem.

THEOREM 7.1. *Let S be a completely regular space without isolated points and satisfying the second countability axiom. Then the normal completion of $C(S)$ is isomorphic with the lattice of all continuous functions on the Boolean space associated with the normal completion of the free Boolean algebra with a countably infinite set of generators.*

As an immediate consequence we have the following corollary.

COROLLARY. *All completely regular spaces without isolated points and satisfying the second countability axiom have the same normal completion for their lattices of continuous functions.*

In particular, Theorem 7.1 gives a simple representation of the normal completion of the lattice of continuous functions on the interval $[0, 1]$. According to the corollary, the Cantor set and the real line also have lattices of continuous functions with this same normal completion.

REFERENCES

1. G. Birkhoff, *Lattice theory*, rev. ed., Amer. Math. Soc. Colloquium Publications, vol. 25, 1949.

2. E. Čech, *On bicompact spaces*, Ann. of Math. vol. 38 (1937) pp. 823–844.

3. H. Nakano, *Über das System aller stetigen Funktionen auf Linem topologischen Raum*, Proc. Imp. Acad. Tokyo vol. 17 (1941) pp. 308–310.

4. M. H. Stone, *Applications of the theory of Boolean rings to general topology*, Trans. Amer. Math. Soc. vol. 41 (1937) pp. 375–481.

5. ———, *A general theory of spectra*. I, Proc. Nat. Acad. Sci. U.S.A. vol. 26 (1940) pp. 280–283.

6. ———, *Boundedness properties in function lattices*, Canadian Journal of Mathematics vol. 1 (1949) pp. 176–186.

CALIFORNIA INSTITUTE OF TECHNOLOGY,
PASADENA, CALIF.

A GENERALIZED CANTOR THEOREM

A. M. GLEASON AND R. P. DILWORTH

A well known theorem of Cantor asserts that the cardinal of the power-set of a given set always exceeds the cardinal of the original set. An analogous result for sets having additional structure is the well known theorem that the set of initial segments of a well ordered set always has order type greater than the original set. These two theorems suggest that there should be a similar result for general partially ordered sets. In formulating such a theorem an extension to partially ordered sets of the notion of an initial segment of a well ordered set is required. Of the several possibilities for this choice, the most natural one is the concept of an order ideal. If P is a partially ordered set with order relation $\leqq$, then a subset I of P is an order ideal if $a \leqq b \in P$ implies $a \in P$. The set $\mathcal{I}(P)$ of all order ideals of P is easily seen to be a complete partially ordered set when ordered by set inclusion, since the union and intersection of any set of order ideals is again an order ideal. Note that the empty set is specifically included among the order ideals.

The general theorem can then be formulated as follows:

THEOREM A. *If P is any partially ordered set and $\mathcal{I}(P)$ is the set of all order ideals of P, then $\mathcal{I}(P)$ is not order isomorphic to any subset of P.*

The authors found two independent proofs of this theorem which upon analysis indicated that an even stronger theorem holds.

THEOREM B. *Let P be a partially ordered set and let $\mathcal{I}(P)$ denote the set of order ideals of P. Then if ϕ is a one-to-one map of $\mathcal{I}(P)$ into P, neither ϕ nor ϕ^{-1} is order preserving.*

In the terminology of Theorem B, Theorem A asserts that ϕ and ϕ^{-1} are not both order preserving.

The Cantor theorem follows from this result by assigning to an arbitrary set P the trivial order relation (no unequal elements are comparable). Then $\mathcal{I}(P)$ is the power set of P and the inverse ϕ^{-1} of a one-to-one map ϕ of $\mathcal{I}(P)$ into P would necessarily be order preserving. On the other hand, any infinite well ordered set is a partially ordered set for which $\mathcal{I}(P)$ and P have the same cardinal.

More generally, if P is complete, linearly ordered, and infinite, then $\mathcal{I}(P)$ and P have the same cardinal. If P is linearly ordered but not complete (for example, the set of rationals where $\mathcal{I}(P)$ is isomorphic

---

Received by the editors September 1, 1961.

704

to the Cantor set) the cardinal of $\mathcal{I}(P)$ may be greater than the cardinal of P.

PROOF OF THEOREM B. (1) ϕ *is not order preserving.* For if ϕ preserves order, let $\mathcal{I}_0$ the set of all order ideals I for which $\phi(I) \in I$ and let A be the intersection of these ideals. If $I \in \mathcal{I}_0$, then $A \subseteq I$ and hence $\phi(A) \leq \phi(I) \in I$ which implies $\phi(A) \in I$. It follows that $\phi(A) \in \bigcap \mathcal{I}_0 = A$ and hence $A \in \mathcal{I}_0$. Consider the ideal $B = \{p \in P \mid p < \phi(A)\}$. Clearly $B \subset A$ so that $\phi(B) < \phi(A)$ and therefore $\phi(B) \in B$. But then $B \in \mathcal{I}_0$ and hence $A \subseteq B$ contrary to $B \subset A$.

(2) ϕ^{-1} *is not order preserving.* For if ϕ^{-1} preserves order, let $\mathcal{I}_1$ denote the set of all order ideals I for which $\phi(I) \notin I$ and let

$$C = \{p \mid p \leq \phi(I) \text{ for some } I \in \mathcal{I}_1\}.$$

Then C is an order ideal. If $\phi(C) \notin C$, then $C \in \mathcal{I}_1$; and hence $\phi(C) \in C$ by the definition of C. On the other hand, if $\phi(C) \in C$, then choose $I \in \mathcal{I}_1$ so that $\phi(C) \leq \phi(I)$. By the order preserving property of ϕ^{-1} we have $C \subseteq I$, and since $\phi(I) \in C$ it follows that $\phi(I) \in I$ contrary to the definition of $\mathcal{I}_1$.

It may be noted that the argument of part (2) also holds for quasi-orderings, in which case the map need not be one-to-one.

HARVARD UNIVERSITY,
    CALIFORNIA INSTITUTE OF TECHNOLOGY, AND
        INSTITUTE FOR DEFENSE ANALYSES, PRINCETON, NEW JERSEY

Generators of lattice varieties

R. P. Dilworth* and Ralph Freese*

Introduction

Although it is well known that the variety of all lattices is generated by the subclass of finite lattices, there are lattice varieties which are not generated by their finite members. In fact, there are modular varieties which are not even generated by their finite dimensional members [3]. At the present time it is not known if the variety of all modular lattices is generated by its finite or even its finite dimensional members. This raises the question: Do there exist generators for lattice varieties which satisfy some kind of finiteness conditions? In this note we will be concerned with generators satisfying atomicity conditions. Since any lattice variety is generated by its subdirectly irreducible members, we shall be particularly interested in generators which are also subdirectly irreducible. The main results are the following. The notation and terminology for this paper is taken from [1].

THEOREM 1. *Every lattice variety is generated by its strongly atomic members.*

THEOREM 2. *Every modular lattice variety is generated by its weakly atomic, subdirectly irreducible members.*

THEOREM 3. *Every modular lattice can be imbedded in a strongly atomic, subdirectly irreducible modular lattice.*

It follows from Theorem 3, that the variety of all modular lattices is generated by its strongly atomic, subdirectly irreducible members. Whether a similar result holds for modular varieties in general is an open question.

Finally, we give some applications of Theorem 2.

Presented by B. Jónsson. Received April 14, 1975. Accepted for publication in final form September 9, 1975.

* This research supported in part by NSF Grants GP-35678 and GP-37772.

263

Proofs of the Theorems. If L is a lattice, the lattice of filters (dual ideals) of L will be denoted by $\mathcal{F}(L)$. Principal filters will be identified with the corresponding elements of L, so that $L \subseteq \mathcal{F}(L)$. Inductively, we define $\mathcal{F}^n(L) = \mathcal{F}(\mathcal{F}^{n-1}(L))$.

Proof of Theorem 1. For any lattice L we have

$$L \subseteq \mathcal{F}(L) \subseteq \cdots \subseteq \mathcal{F}^n(L) \subseteq \cdots$$

We note that if $a > b$ in $\mathcal{F}^n(L)$, then $a > b$ in $\mathcal{F}^{n+1}(L)$. For if $(a) > C \geq (b)$ in $\mathcal{F}(\mathcal{F}^n(L))$ then there exists $c \in C$ such that $c \not\geq a$. But then $a > a \cap c \geq b$ and hence $b = a \cap c$. Thus b is contained in C and hence $C = (b)$. Now let $L_0 = \bigcup_n \mathcal{F}^n(L)$. Let $a, b \in L_0$ such that $a > b$. Then $a, b \in \mathcal{F}^n(L)$ for some n and hence there exists $c \in \mathcal{F}^{n+1}(L)$ such that $a \geq c > b$ in $\mathcal{F}^{n+1}(L)$. But then $c > b$ in L_0 and hence L_0 is strongly atomic. Finally, any identity satisfied by L is satisfied by $\mathcal{F}(L)$ and hence by $\mathcal{F}^n(L)$ for all n. But then L_0 satisfies every identity satisfied by L and thus belongs to the variety generated by L. It follows that every variety is generated by its strongly atomic members.

Proof of Theorem 2. Let L be a subdirectly irreducible modular lattice and let

$$\mathcal{F}(L) \subseteq \prod_\alpha L_\alpha$$

be a representation of $\mathcal{F}(L)$ as a subdirect product of subdirectly irreducible lattices L_α. Let π_α denote the projection $\prod_\alpha L_\alpha \to L_\alpha$. Since L is subdirectly irreducible, there exists elements $a, b \in L$ such that $a > b$ and $\theta a = \theta b$ for every nontrivial congruence relation θ on L. Let $a \geq c > b$ where $c \in \mathcal{F}(L)$. Then there exists a subscript α such that $\pi_\alpha c \neq \pi_\alpha b$. Since π_α is a surjection, it follows that $\pi_\alpha c > \pi_\alpha b$ in L_α. By hypothesis L_α is subdirectly irreducible and modular. It follows that $\pi_\alpha c / \pi_\alpha b$ is weakly projective into every proper quotient of L_α. Since $\pi_\alpha c / \pi_\alpha b$ is prime, every proper quotient of L_α contain a prime quotient and L_α is weakly atomic. Finally since $\pi_\alpha c \neq \pi_\alpha b$ we have $\pi_\alpha a \neq \pi_\alpha b$ and the mapping $x \to \pi_\alpha x$ is one-to-one on L. Thus L is isomorphic to a sublattice of L_α. Now every identity holding in L, also holds in $\mathcal{F}(L)$ and hence holds in L_α. Clearly every identity holding in L_α also holds in every sublattice and hence holds in L. It follows that L belongs to a variety if and only if L_α belongs to the variety and hence every modular variety is generated by its weakly atomic, subdirectly irreducible members.

Proof of Theorem 3. Let L be a modular lattice and let $L_0 = \mathcal{F}(L)$. Let a_α / b_α, $\alpha < \lambda$, be a well ordering of the prime quotients of L_0. For each $\alpha < \lambda$ we shall inductively construct a lattice L_α such that if $\beta < \alpha < \lambda$, L_β is a sublattice of L_α, coverings in L_β are preserved in L_α, and a_α/b_α is projective to a_0/b_0 in L_α. Let us suppose that L_β has been constructed for $\beta < \alpha < \lambda$. Let $A = \mathscr{F}(\bigcup_{\beta < \alpha} L_\beta)$. By the dual of (6.2) of [1], we can find a completely join-irreducible element a_0' in A such that $a_0' \le a_0$ and $a_0' \nleq b_0$. If we set $b_0' = b_0 \wedge a_0'$, then a_0/b_0 and a_0'/b_0' are projective.

We will find it convenient to make repeated use of the following construction from [5]. Let X and Y be modular lattices such that a principal ideal Z of X is isomorphic to a principal dual ideal Z' of Y. Then the set $X \cup Y$ with Z and Z' identified can be made into a modular lattice by defining $a \le b$ if and only if $a \le b$ in X or $a \le b$ in Y or if $a \le z'$, $z \le b$ where z and z' are corresponding elements under the isomorphism of Z to Z'.

Let $B = \{a \in A : a \le b_0'\}$. Let x, y, z be the atoms of a copy of M_5 with unit and null elements u and v respectively. Applying the above construction to $B \times 2$ and M_5 where $(b_0', 1)/(b_0', 0)$ is identified with x/v we get a modular lattice C. Since a_0' is completely join irreducible in A, the ideal $a_0'/0$ in A is isomorphic to the dual ideal $u/(0, 1)$ of C. Identifying isomorphic elements and using the above construction gives a lattice D. If we let $a_0'' = (0, 1)$ and $b_0'' = (0, 0)$, then in D we have

$$a_0/b_0 \searrow a_0'/b_0' \searrow y/(b_0', 0) \nearrow u/z \searrow x/v \searrow a_0''/b_0''$$

Note that A is isomorphic to the principal dual ideal of D generated by $(0, 1)$.

We now repeat the entire process starting with a_α/b_α in D rather than a_0/b_0 in A. In the resulting lattice E, a_α/b_α is projective to $(0, 1)/(0, 0)$. Thus in E there are two distinct atoms c_0 and c_α such that a_0/b_0 is projective to $c_0/0$ and a_α/b_α is projective to $c_\alpha/0$.

Let x_i, y_i, z_i for $i = 1, 2, 3$ be the atoms of three copies of M_5 with null element v_i and unit element u_i, $i = 1, 2, 3$. Let F be the lattice obtained from the first two copies of M_5 by identifying x_2 with v_1 and u_2 with z_1. Let G be the lattice obtained from the union of E and the third copy of M_5 by identifying c_0 with u_3 and 0 with x_3. Finally let L_α be the lattice obtained from $F \cup G$ by identifying u_1 with c_α, $u_2 = z_2$ with $0 = x_3$, and z_1 with v_3. Clearly L_β is isomorphic to a sublattice of L_α for all $\beta < \alpha$. Since coverings are preserved in the natural imbedding in the dual ideal lattice, coverings in L_β are preserved in L_α. Finally since $c_0/0$ and $c_\alpha/0$ in E are projective in L_α to prime quotients of F, it follows that a_α/b_α and a_0/b_0 are projective in L_α.

Let $M_1 = \bigcup_{\alpha < \lambda} L_\alpha$. We repeat the entire procedure with M_1 in place of L. In this manner we obtain M_2 and inductively M_2 and inductively $M_3, M_4, \ldots$. Let $M = \bigcup M_i$. Then, as in the proof of Theorem 1, M is strongly atomic. Since all prime quotients of M are projective, M is subdirectly irreducible. This completes the proof of Theorem 3.

Although the techniques of the proof of Theorem 3 do not apply to all varieties of modular lattices, there are certain lattice identities which are preserved under this construction. The most important of these is the Arguesian law.

Application

The study of varieties of modular lattices is often facilitated if the variety is generated by subdirectly irreducible lattices containing prime quotients. Theorem 2 shows that we may always assume this to be the case. For example, the proofs of the results in [2] concerning modular varieties can be considerably shortened with the aid of Theorem 2 (see Appendix B of [2]). On the other hand, it should be pointed out that the full structural results of [2] cannot be obtained in this way. Theorem 2 may also be used to shorten the proofs of some of the results of [6].

As another application of Theorem 2, consider the result of B. Jónsson [7] which states that a modular lattice L not in the variety generated by M_5 has a homomorphic image of a sublattice isomorphic to one of the two lattices of Figure 1.

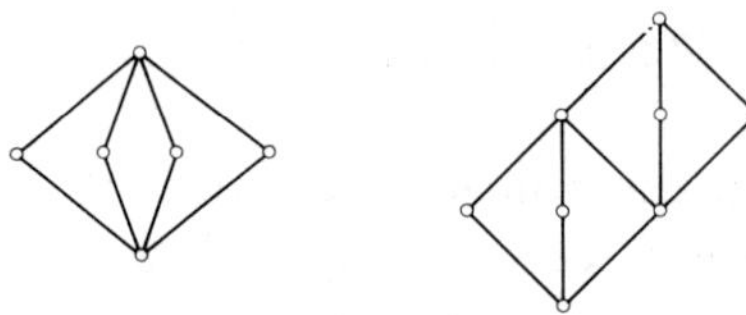

Figure 1

G. Grätzer [4] obtained this result under the assumption that L is finite. It follows from the more general result of Jónnson that the variety generated by M_5 is strongly covered by two modular varieties. Since the Grätzer proof depends only on L having a prime quotient, the covering property for the variety generated by M_5 follows from Grätzer's result and Theorem 2. Again, it should be noted that the full structural result of Jónsson cannot be recovered in this manner.

REFERENCES

[1] P. Crawley and R. P. Dilworth, *Algebraic Theory of Lattices*, Prentice-Hall, Englewood Cliffs, N.J. 1973.

[2] R. Freese, *The structure of modular lattices of width four with applications to varieties of lattices*, Memoir, Amer. Math. Soc. (to appear).

[3] ——, *Some varieties of modular lattices not generated by their finite dimensional members*, Proc. of the Szeged Universal Algebra Conference.

[4] G. GRÄTZER, *Equational classes of lattices*, Duke Math. J. *33* (1966), 613–622.
[5] M. HALL and R. P. DILWORTH, *The embedding problem for modular lattices*, Annals of Math. *45* (1944), 450–456.
[6] D. X. HONG, *Covering relations among lattice varieties*, Pacific J. Math. *40* (1972), 575–603.
[7] B. JÓNSSON, *Equational classes of lattices*, Math. Scand. *22* (1968), 187–196.

California Institute of Technology
Pasadena, California
U.S.A.
University of Hawaii
Honoululu, Hawaii
U.S.A.

Lattice Congruences and Dilworth's Decomposition
of Relatively Complemented Lattices

GEORGE F. MCNULTY

At the focus of "The structure of relatively complemented lattices," which is Dilworth [6], is the following theorem:

THE BASIC STRUCTURE THEOREM FOR RELATIVELY COMPLEMENTED LATTICES. *Every relatively complemented lattice with the ascending chain condition is isomorphic to a direct product of simple relatively complemented lattices.*

This theorem alone would secure for this paper by Dilworth an important place in the theory of lattices. The paper is significant for lattice theory in another and perhaps more far-reaching way: the techniques employed in proving this theorem are fundamental for dealing with congruences on lattices. They have been used and elaborated in the development of lattice theory in contexts which have little other connection to direct decomposition theorems.

As with other direct decomposition theorems, for example the Fundamental Theorem for Finitely Generated Abelian Groups, there are essentially three features to consider.

- The existence of a direct decomposition into directly indecomposable factors.
- The uniqueness, up to isomorphism and the rearrangement of factors, of this direct decomposition.
- The information which may be obtained concerning the directly indecomposable factors.

Loosely speaking, the first feature is usually obtained by the imposition of a finiteness condition—often a chain condition on the *congruence lattice* of the algebra in question. In the paper considered here, Dilworth proves that if **L** is a relatively complemented lattice satisfying a chain condition *on its elements*, then the congruence lattice of **L** satisfies the same chain condition. Since the ascending chain condition on the congruence lattice of **L** is enough to ensure the existence of the decomposition, Dilworth obtains the decomposition from the ascending chain condition on **L**. The power, however, of such direct decomposition theorems can be traced mostly to the other two features. Uniqueness implies that the direct decomposition yields a test for isomorphism—the systems of directly indecomposable factors are complete systems of invariants. A complete classification of the direct indecomposables offers a strategy for proving theorems about a whole class of algebras.

At the time this paper of Dilworth's was being written, a number of general results were known that entail the uniqueness of the direct decompositions Dilworth considered. For example, the Birkhoff-Ore Theorem (see Birkhoff [2] pages 95–96) asserts unique direct factorization for any algebra which has permutable congruence relations, a one element subalgebra, and a congruence lattice of finite height. In the paper we are considering, Dilworth proves that

THEOREM. *Any relatively complemented lattice has permutable congruences.*

Indeed, this is one of the key insights underlying Dilworth's direct decomposition theorem—although its importance concerns mostly the third feature of direct decompositions. The Birkhoff-Ore Theorem is a result of some difficulty and the uniqueness we desire can be arrived at more simply. The congruence lattice of any lattice is distributive. This important result was first published by Funayama and Nakayama in [7]. In [6], Dilworth offers a different and elegant proof which originated at about the same time as that of Funayama and Nakayama. It is not difficult to prove that any algebra with a distributive congruence lattice has the refinement property (or even the strict refinement property of Chang, Jónsson and Tarski [3]). The refinement property in turn implies that the algebra can have no more than one direct decomposition into directly indecomposable factors. That lattices have the refinement property was, perhaps, first noted in the literature by Hashimoto in [11].

The principal conclusion to be drawn from Dilworth's structure theorem for relatively complemented lattices concerns the third feature of direct decompositions. Indeed, the theorem can be recast as:

THEOREM. *Every directly indecomposable relatively complemented lattice with the ascending chain condition is simple.*

Generally, it is no easy task to discover whether an algebra—even one so nice as a finite lattice—is directly indecomposable. Apparently, one must search through the *congruences* of the algebra for a complementary pair of nontrivial congruences which permute with each other; such congruences would be the kernels of the projection functions associated with a direct factorization. On the other hand, to discover whether an algebra is simple, one need only search through the *elements* of the algebra for a pair of distinct elements which generate a nontrivial congruence; such a congruence denies the simplicity of the algebra. Of course, to accomplish such a search one must be able to generate (principal) congruences or at least determine whether the congruence generated by a pair of elements is different from the largest congruence. Dilworth offers the following result along these lines:

THE DILWORTH THEOREMS

THEOREM. *A relatively complemented lattice satisfying a chain condition is simple if and only if every two prime quotients are projective.*

This test for simplicity provides a handle for using Dilworth's basic structure theorem.

Dilworth's paper [6] addresses the question:

Which lattices can be decomposed as direct products of simple lattices?

At the time this paper by Dilworth was being written, it was known from general algebraic considerations (and the fact that congruence lattices of lattices are distributive) that these were exactly the lattices **L** with the following two properties:

(1) **L** is congruence permutable
(2) The congruence lattice of **L** is Boolean (i.e. distributive and complemented).

But see also Dilworth's Lemma 3.2 of the paper before us. Thus Dilworth's insight that relatively complemented lattices have permuting congruence relations played a key role in establishing the decomposition into simple lattices. In proving that (2) holds when **L** is a relatively complemented lattice with the ascending chain condition, Dilworth makes essential use of the chain condition.

The techniques that Dilworth uses to obtain these decomposition results are built around his characterization of the congruence relation generated by a given set of pairs of lattice elements. This characterization uses the notion of weak projectivity, which Dilworth introduces in the paper before us in Definition 2.1. This characterization, which is Lemma 2.1, has since been put to a multitude of uses and should rightly be counted among the most useful tools in lattice theory. Dilworth promptly displays its usefulness in giving a proof that the congruence lattice of a lattice is always distributive. Dilworth's characterization of congruence generation provides the means to effectively calculate the congruence relations using the lattice operations. Here is a way to render Dilworth's characterization so that this aspect becomes more apparent. Call a unary polynomial $s(x)$ on the lattice **L** a *projectivity polynomial* provided

$$s(x) = (((x \wedge a_0) \vee a_1) \wedge \cdots) \vee a_{n-1}$$

where $a_0, a_1, \ldots, a_{n-1} \in L$. It is easily seen that the quotient a/b is weakly projective into the quotient c/d iff there is a projectivity polynomial $s(x)$ such that

$$s(d) \leq b \leq a \leq s(c)$$

Thus, Dilworth's characterization may be recast as follows:

THEOREM. *Let $X \subseteq L^2$ where **L** is a lattice. The congruence of **L** generated by X is the binary relation θ on L defined for all $a, b \in L$ by*

$a\theta b$ iff there is a finite sequence $e_0, e_1, \ldots, e_n \in L$, a sequence $s_0(x), \ldots,$ $s_{n-1}(x)$ of projectivity polynomials, and a sequence of pairs $(c_0, d_0), \ldots,$ $(c_{n-1}, d_{n-1}) \in X$ such that $a \wedge b = e_0 \leq e_1 \leq \cdots \leq e_n = a \vee b$ and $s_i(c_i \wedge d_i) \leq e_i \leq e_{i+1} \leq s_i(c_i \vee d_i)$ for all $i < n$.

One easy consequence of Dilworth's characterization that has proved very useful is that if a covers b in a lattice L, then there is a *unique* maximal congruence of L which separates a and b. It is worth mentioning that there is a related, but less powerful, characterization of congruence generation applying to algebras generally. This result, detailed by Grätzer [9] page 53 or by McKenzie, McNulty, and Taylor in [14] page 155 is generally credited to Maltsev [13]. Dilworth's characterization is substantially more useful than the restriction of the general result to lattices, because the unary polynomials used in the characterization are much more sharply controlled.

The paper before us contains a number of related results concerning subdirect decompositions of locally relatively complemented lattices. These are gathered in the last section of the paper.

In the mid-1930's Birkhoff in [1] and Menger in [15] had established a fundamental structure theorem for finite dimensional complemented modular lattices:

THE BIRKHOFF-MENGER THEOREM. *Each finite dimensional complemented modular lattice is isomorphic to a direct product of simple finite dimensional complemented modular lattices. The simple finite dimensional complemented modular lattices are the two-element chain and the lattices of subspaces of finite dimensional projective geometries.*

This theorem was already a classic at the time paper before us was being written. In view of the fact that every complemented modular lattice is relatively complemented, the first part of the Birkhoff-Menger Theorem is a special case of Dilworth's structure theorem. The second part of the Birkhoff-Menger Theorem draws a sharper description of the simple lattices under the strong hypothesis of modularity. Until this paper by Dilworth, all the proofs of the first part of the Birkhoff-Menger Theorem made substantial use of the modular law. Dilworth's result is surprising, since it reveals that modularity can be avoided and that the conclusion follows from the existence of relative complements.

Dilworth's structure theorem has been extended to relatively complemented lattices with conditions weaker than the ascending chain condition. Perhaps the most notable result obtained in this direction is the following theorem of Hashimoto from [12]:

THEOREM OF HASHIMOTO. *Each complete weakly atomic relatively complemented lattice is a direct product of subdirectly irreducible lattices.*

Recall that a lattice is called weakly atomic if every interval contains a covering. The conclusion that the directly indecomposable factors are subdirectly irreducible

                                             THE DILWORTH THEOREMS

is weaker than the conclusion of simplicity in Dilworth's theorem. Dilworth's theorem can be obtained from Hashimoto's theorem and the fact (due to Dilworth [6]) that the congruence lattice of a relatively complemented lattice with the ascending chain condition is Boolean. Crawley and Dilworth [5] provides a nice account of Hashimoto's results.

In problem 72 on page 153 of [2], Birkhoff asks for a characterization of those lattices whose congruence lattices are Boolean. Of course, this paper by Dilworth offers an advance on this problem. The first characterization was given in 1952 by Tanaka in [18] with later characterizations in 1958 by Grätzer and Schmidt in [10] and in 1960 by Crawley in [4]. Crawley's characterization is the most intrinsic— being phrased entirely in terms of the arithmetic of the lattice operations applied to elements. Both the works of Crawley and of Grätzer and Schmidt just cited make essential use of Dilworth's characterization of congruences generated by sets of pairs of lattice elements. Schmidt [16], [17], and Grätzer [8] provide ample evidence of just how useful Dilworth's techniques for dealing with congruences on lattices have become.

Finally, a remark about the lucid way in which this paper is written is in order. Reading it some forty years after it was written, it strikes this reader as if it were a fresh result, conveyed with simplicity in the most current notation and speaking to issues now at the breaking edge of research in lattice theory. This is certainly a testament to the lasting impact that Dilworth's contributions to lattice theory have had. Today we are speaking and writing and thinking about lattices in the manner of Robert Dilworth.

REFERENCES

1. G. Birkhoff, *Combinatorial relations in projective geometries*, Ann. of Math. **36** (1935), 743–748.
2. G. Birkhoff, "Lattice Theory," rev. ed., Colloquium Publications, Amer. Math. Soc., Providence, R.I., 1948.
3. C. C. Chang, B. Jónsson, and A. Tarski, *Refinement properties for relational structures*, Fund. Math. **55** (1964), 249–281.
4. P. Crawley, *Lattices whose congruences form a Boolean algebra*, Pacific J. Math. **10** (1960), 787–795.
5. P. Crawley and R. P. Dilworth, "Algebraic Theory of Lattices," Prentice-Hall, Englewood Cliffs, New Jersey, 1973.
6. R. P. Dilworth, *The structure of relatively complemented lattices*, Ann. of Math. **51** (1950), 348–359. Reprinted in Chapter 7 of this volume.
7. N. Funayama and T. Nakayama, *On the distributivity of a lattice of lattice-congruences*, Proc. Imp. Acad. Tokyo **18** (1942), 553–554.
8. G. Grätzer, "General Lattice Theory," Series on Pure and Applied Mathematics, Academic Press, New York, N.Y.; Mathematische Reihe, Band 52, Birkhauser Verlag, Basel; Akademie Verlag, Berlin., 1978.
9. G. Grätzer, "Universal Algebra," second edition, Springer-Verlag, New York, 1979.
10. G. Grätzer and E. T. Schmidt, *Ideals and congruence relations in lattices*, Acta. Math. Acad. Sci. Hungar. **9** (1958), 137–175.
11. J. Hashimoto, *On direct product decomposition of partially ordered sets*, Ann. of Math. **54** (1951), 315–318.

12. J. Hashimoto, *Direct, subdirect decompositions and congruence relations*, Osaka J. Math. **9** (1957), 87–112.

13. A.I. Maltsev, *On the general theory of algebraic systems*, (Russian), Mat. Sbornik **77** (1954), 3–20.

14. Ralph McKenzie, George McNulty and Walter Taylor, "Algebras, Lattices, Varieties, Volume I," Wadsworth and Brooks Cole, Monterey, California, 1987.

15. K. Menger, *New foundations of projective and affine geometry*, Ann. of Math. **37** (1936), 456–482.

16. E. T. Schmidt, "Kongruenzrelationen Algebraischer Strukturen," VEB Deutscher Verlag der Wissenschaften, 1969.

17. E. T. Schmidt, "A Survey on Congruence Lattice Representations," 115 pages, Teubner-Texte zur Mathematik, vol. 42, 1982.

18. T. Tanaka, *Canonical subdirect factorizations of lattices*, J. Sci. Hiroshima Univ. **16** (1952), 239–246.

University of South Carolina
Columbia, SC 29208
U. S. A.

THE DILWORTH THEOREMS

The Normal Completion of the Lattice
of Continuous Functions

Gerhard Gierz

Dilworth's paper [10] on the order completion of the Banach lattice $C(S)$ of all continuous bounded functions on a topological space S was one of the first papers written on this topic and more is known nowadays. It might be worthwhile to follow Dilworth's ideas in modern terminology and see how a proof of his results could look were this paper written in 1989 instead of 1949.

Firstly, unlike Dilworth, we will restrict our attention to compact Hausdorff spaces only. There is no loss of generality, since every space of the form $C(S)$ is isometrically isomorphic with $C(\beta S)$, where βS denotes the Stone-Čech compactification of S (see [7, 28]). Easy examples show that bounded subsets $A \subseteq C(S)$ need not have suprema (*i.e.*, $C(S)$ is not "order complete"). As a matter of fact, $C(S)$ is order complete if and only if S is extremally disconnected (*i.e.*, if the closure of every open subset $U \subseteq S$ is again open, see [22,29,30]). Therefore, one might want to construct an "optimal" completion of $C(S)$. There are various different ways to do this. One of those completions is the ideal completion of $C(S)$, *i.e.*, the collection of all lattice ideals of $C(S)$. However, certainly this ideal completion would not be satisfactory, because most of the structure of $C(S)$ is lost. Since $C(S)$ is a Banach lattice, one would like to complete $C(S)$ in such a way that the resulting space is again a Banach lattice. The bidual of $C(S)$ would be such a lattice, which certainly is much too large in general. A smaller completion is the completion by Dedekind cuts (see [6,20,31]), a method which was used to construct the real numbers from the rational numbers.

Let P be an arbitrary ordered set. A pair (N, M) of subsets $M, N \subseteq P$ is a *Dedekind cut* if M is the set of all lower bounds of N, and N is the set of all upper bounds of M. We order the Dedekind cuts by $(M_1, N_1) \leq (M_2, N_2)$ if and only if $M_1 \subseteq M_2$. With this order, the set of all Dedekind cuts becomes a complete lattice $D(P)$, and we can embed P into $D(P)$ by sending an element $x \in P$ to the cut $(\downarrow x, \uparrow x)$, where $\downarrow x$ [respectively, $\uparrow x$] is the set of elements y such that $y \leq x$ [respectively, $y \geq x$]. This embedding preserves all existing suprema and infima. In our situation, $D(P)$ has the disadvantage that it has a largest and a smallest element, and clearly no Banach lattice can have elements of this nature. Hence we let $D_0(P) = \{(M, N) \in D(P) : M \text{ and } N \text{ nonempty}\}$. Here we found the right kind of object:

THEOREM (VULIKH [31]). *If V is an Archimedean vector lattice, then $D_0(V)$ is an order complete vector lattice.*

(An *Archimedean* vector lattice is a vector lattice satisfying the condition:

if $x, y \in E$ and $n \cdot x \leq y$ for all natural numbers n, then $x \leq 0$.

The Archimedean axiom is essential in showing that $D_0(V)$ is actually an additive group—as a matter of fact this axiom is equivalent to the statement that $D_0(V)$ is a vector lattice.)

We now would like to further investigate the structure of $D_0(V)$ in the case where $V = C(S)$. Let $e \in C(S)$ be the constant function with value 1. Then e is a *strong order unit*, i.e., for every $f \in C(S)$ there is a number $r > 0$ so that $-r \cdot e \leq f \leq r \cdot e$. Moreover,

$$\sup\{|f(s)| : s \in S\} = \|f\|_\infty = \inf\{r > 0 : -r \cdot e \leq f \leq r \cdot e\}$$

It is easy to see that the same element $e \in C(S) \subseteq D_0(C(S))$ serves as a strong order unit for the normal completion $D_0(C(S))$. Therefore, we can define a norm on $D_0(C(S))$ by

$$\|\xi\|_\infty = \inf\{r > 0 : -r \cdot e \leq \xi \leq r \cdot e\}.$$

Is $D_0(C(S))$ complete in this norm? In order to verify this, we have to show that every Cauchy sequence $(\xi_n)_{n \geq 0}$ in $D_0(C(S))$ has a cluster point. By selecting a subsequence, if necessary, we may assume that $\|\xi_m - \xi_n\| < (1/2)^n$ whenever $m > n$. From $\xi_0 - e \leq \xi_n \leq \xi_0 + e$ we conclude that the set $\{\xi_n : 0 \leq n\}$ is order bounded, and therefore the limit inferior

$$\xi_\infty = \sup_{n > 0} \left(\inf_{m > n} \xi_m \right)$$

exists in $D_0(C(S))$. Hence, the point ξ_∞ would be a good candidate for a limit point of $(\xi_n)_{n \geq 0}$. Indeed, since

$$\xi_n - \frac{1}{2^n} \cdot e \leq \xi_m \leq \xi_n + \frac{1}{2^n} \cdot e$$

for $m > n$, it follows that

$$\xi_n - \frac{1}{2^n} \cdot e \leq \xi_\infty \leq \xi_n + \frac{1}{2^n} \cdot e$$

and therefore $\|\xi_n - \xi_\infty\| \leq (1/2)^n$. It should be noted at this point that in general the notions of being complete with respect to a norm and being complete with respect to an order are independent of each other.

From the previous discussion we conclude that $D_0(C(S))$ is itself a Banach lattice with a strong order unit, hence of the form $C(T)$ for a certain extremally disconnected compact space T (see [22,29,30]). We would like to identify T. There are two notions we have to introduce at this point. The first notion concerns essential extensions (see [3,23]).

Let $\mathcal{C}$ be a category. A $\mathcal{C}$-morphism $\phi : A \to B$ is called an *essential extension* if for every $\mathcal{C}$-morphism $\psi : B \to C$ the composition $\psi \circ \phi$ is a monomorphism if and only if ψ is a monomorphism. Obviously, every essential extension is a monomorphism.

THEOREM. *Let V be an Archimedean vector lattice. Then the embedding $\iota_V : V \to D_0(V)$ is an essential extension in the category of all vector lattices with linear lattice homomorphisms as morphisms.*

Indeed, let $\psi : D_0(V) \to W$ be a linear lattice homomorphism that is not injective. Then there is a positive $0 < \xi \in D_0(V)$ such that $\psi(\xi) = 0$. Pick any $x \in V$ such that $0 < \iota_V(x) \leq \xi$. Since ψ is order preserving, it follows that $\psi(\iota_V(x)) = 0$, hence $\psi \circ \iota$ is not injective.

We conclude that $D_0(C(S)) = C(T)$ is an essential extension of $C(S)$ in the category $\mathcal{BL}_o$ of all Banach lattices with strong order unit, and linear lattice homomorphisms preserving the strong order unit as morphisms.

The second notion concerns injectivity. An object I in a category $\mathcal{C}$ is called *injective* if for every $\mathcal{C}$-monomorphism $\epsilon : A \to B$ and every $\mathcal{C}$-morphism $\phi : A \to I$ there is a morphism $\psi : B \to I$ such that $\phi = \psi \circ \epsilon$. If T is extremally disconnected, then $C(T)$ is injective in the category of all Banach spaces with contractions as morphisms (see [12,18,21]), in the category of all Banach lattices with positive linear maps as morphisms (see [9, 25]), as well as in $\mathcal{BL}_o$. The following paragraphs contain a proof of this statement.

Let B be a Banach lattice with strong order unit (*i.e.*, a $\mathcal{BL}_o$-object). Let $Spec(B) = \{I \subseteq B : I$ is a maximal vector lattice ideal of $B\}$. We introduce a topology on $Spec(B)$ by declaring sets of the form $h(J) = \{I : J \subseteq I\}$ to be closed, where J is an arbitrary ideal of B. With this topology, $Spec(B)$ becomes a compact Hausdorff space. If $B = C(X)$, then $X \cong Spec(B)$. We extend $Spec$ to functions in the following way: if $\phi : B_0 \to B_1$ is a linear mapping preserving the lattice structure and the strong order unit, then define

$$Spec(\phi) : Spec(B_1) \to Spec(B_0)$$
$$I \mapsto \phi^{-1}(I)$$

Then $Spec$ becomes a functor from $\mathcal{BL}_o$ with values in the category $Comp$ of all compact Hausdorff spaces with continuous maps as morphisms. In the other direction, we have a functor C that assigns the Banach lattice $C(S)$ to every compact Hausdorff space S. If $f : S_0 \to S_1$ is continuous map between compact spaces, define

$$C(f) : C(S_1) \to C(S_0)$$
$$g \mapsto g \circ f$$

The functors $Spec$ and C define a duality between the categories $\mathcal{BL}_o$ and $Comp$.

Applying these procedures and functors to our situation, we arrive at the following conclusions:

(i) Extremally disconnected spaces are exactly the retracts of Stone-Čech compactifications of discrete sets, and these are exactly the projective objects in $Comp$. Hence Banach lattices of the form $C(T)$, T extremally disconnected, are exactly the injective objects in $\mathcal{BL}_o$. (See [13,14,26].)

(ii) It follows that $C(T) = D_0(C(S))$ is isometrically isomorphic to the injective hull of $C(S)$. Hence T is homeomorphic to the projective cover (or projective resolution) of S. (For more details on injective hulls, see [3,23]. The idea of projective covers is discussed in [2,12,24].)

(iii) For any compact Hausdorff space S, the projective cover of S is homeomorphic to the maximal ideal space of the complete Boolean algebra of all regular open sets of S. (See [12,24].)

Here we finally arrive at Dilworth's theorem (Theorem 6.1 in [10]): *The normal completion of $C(S)$ is isomorphic with the lattice of all continuous functions on the maximal ideal space of the Boolean algebra of regular open sets of S.* Of course, Dilworth's beautiful concrete representation of $D_0(C(S))$ *via* functions on S (see Theorem 4.1 in [10]) is lost in this general treatment.

We end by briefly describing three papers which arose directly from [10]. In [16], Horn described directly the completion of a subset C of elements in a complete lattice B as a subset of B. He used this description to obtain another proof of Theorem 4.1 in [10]. Cohen refined the main results to "k-completions" in [8]. Finally, August and Byrne [1] proved analogues of Dilworth's results for lattices of semi-continuous functions on ordered topological spaces.

REFERENCES

1. J. August and C. Byrne, *Completion of lattices of semi-continuous functions*, J. Australian Math. Soc. Ser. A **26** (1978), 453–464.

2. B. Banaschewski, *Projective covers in categories of topological spaces and topological algebras*, in "Proceedings of the Kanpur Topology Conference, 1968," Academic Press, New York, 1971, pp. 63–91.

3. —————————, *Injectivity and essential extensions in equational classes of algebras*, in "Proceedings of the Conference on Universal Algebras, Queen's University, Kingston, Ontario, 1970," pp. 131–147.

4. B. Banaschewski and G. Bruns, *Categorical characterization of the MacNeille completion*, Arch. Math. (Basel) **18** (1967), 369–377.

5. —————————, *Injective hulls in the category of distributive lattices*, J. Reine Angew. Math. **232** (1968), 102–109.

6. G. Birkhoff, "Lattice Theory," 3rd edn., Amer. Math. Soc., Providence, Rhode Island, 1967.

7. E. Čech, *On bicompact spaces*, Ann. Math. **38** (1937), 823–844.

8. H. B. Cohen, *The k-normal completion of function lattices*, Canad. J. Math. **17** (1965), 669–675.

9. M. M. Day, "Normed Linear Spaces," 2nd edn., Springer-Verlag, Berlin and New York, 1973.

10. R. P. Dilworth, *The normal completion of the lattice of continuous functions*, Trans. Amer. Math. Soc. **68** (1950), 427–438. Reprinted in Chapter 7 of this volume.

11. J. Flachsmeyer, *Dedekind-MacNeille extensions of Boolean algebras and of vector lattices of continuous functions and their structure spaces*, General Topology and Appl. **8** (1978), 73–84.

12. A. M. Gleason, *Projective topological spaces*, Illinois J. Math. **2** (1958), 482–489.

13. D. B. Goodner, *Projections in normed linear spaces*, Trans. Amer. Math. Soc. **69** (1950), 89–107.

14. P. Halmos, *Injective and projective Boolean algebras*, in "Lattice Theory," Proc. Symp. Pure Math., Vol. 2, R. P. Dilworth, ed., Amer. Math. Soc., Providence, Rhode Island, 1961, pp. 114–122.

THE DILWORTH THEOREMS

15. _____, "Lectures on Boolean Algebras," Princeton Univ. Press, Princeton, New Jersey, 1963.

16. A. Horn, *The normal completion of a subset of a complete lattice and lattices of continuous functions*, Pacific J. Math. **3** (1953), 137–152.

17. W. Luxemburg and A. C. Zaanen, "Riesz Spaces I/II," North-Holland, Amsterdam and New York, 1971/1983.

18. J. L. Kelley, *Banach spaces with the extension property*, Trans. Amer. Math. Soc. **72** (1952), 323–326.

19. H. P. Lotz, *Extensions and liftings of positive mappings on Banach lattices*, Trans. Amer. Math. Soc. **211** (1975), 85–100.

20. H. MacNeille, *Partially ordered sets*, Trans. Amer. Math. Soc. **42** (1937), 416–460.

21. L. A. Nachbin, *A theorem of the Hahn-Banach type for linear transformations*, Trans. Amer. Math. Soc. **68** (1950), 28–46.

22. H. Nakano, *Über das System aller stetigen Funktionen of einem topologischen Raum*, Proc. Imperial Acad. Tokyo **17** (1941), 308–310.

23. B. Pareigis, "Kategorien und Funktoren," Teubner-Verlag, Stuttgart, 1969.

24. J. R. Porter and R. G. Woods, "Extensions and Absolutes of Hausdorff Spaces," Springer-Verlag, Berlin and New York, 1988.

25. H. H. Schaefer, "Banach Lattices and Positive Operators," Springer-Verlag, Berlin and New York, 1974.

26. Z. Semadeni, "Projectivity, Injectivity and Duality," Rozprawy Matematycyne 35, Warsaw, 1963.

27. _____, "Banach Spaces of Continuous Functions," Polish Scientific Publishers, Warsaw, 1971.

28. M. H. Stone, *Applications of the theory of Boolean rings to general topology*, Trans. Amer. Math. Soc. **41** (1937), 375–481.

29. _____, *A general theory of spectra I*, Proc. Nat. Acad. Sci. U.S.A. **26** (1940), 280–283.

30. _____, *Boundedness properties in function lattices*, Canad. J. Math. 1 (1949), 176–186.

31. B. C. Vulikh, "Introduction to the Theory of Partially Ordered Vector Spaces," Wolters-Noordhoff, Groningen, 1967.

University of California at Riverside
Riverside, CA 92521
U. S. A.

Cantor Theorems for Relations

Joseph P. S. Kung

In [1], Dilworth and Gleason proved the following Cantor-type theorem for ordered sets: *There exists no onto order-preserving map from an ordered set P into the set of ideals of P ordered by inclusion.* Fajtlowicz [2] has generalized this theorem to directed graphs or binary relations. A discussion of the Dilworth-Gleason theorem in the context of embedding theorems for ordered sets can be found in Chapter 5 of [3].

References

1. R. P. Dilworth and A. M. Gleason, *A generalized Cantor theorem*, Proc. Amer. Math. Soc. **13** (1962), 704–705. Reprinted in Chapter 7 of this volume.
2. S. Fajtlowicz, *A graph-theoretical generalization of a Cantor theorem*, Proc. Amer. Math. Soc. **63** (1977), 177–179.
3. R. Fraïssé, "Theory of Relations," North-Holland, Amsterdam, 1986.

University of North Texas
Denton, Texas 76203
U. S. A.

Ideal and Filter Constructions in Lattice Varieties

J. B. NATION

In [5], Dilworth and Freese introduced an important method for dealing with lattice varieties. The basic argument is contained in the proof of Theorem 2; we give here a slight variation.

It is well-known that the ideal lattice and filter lattice, $\mathcal{I}(L)$ and $\mathcal{F}(L)$ respectively, of a lattice L are in the variety generated by L ([20], cf. [1], [2]). ($\mathcal{F}(L)$ is ordered by reverse set inclusion.) Of course, L is embedded in each of these lattices, and they are weakly atomic. Now assume that L is subdirectly irreducible and a/b is a critical quotient, *i.e.*, a and b are collapsed by every nonzero congruence. In $\mathcal{F}(L)$ there is a prime quotient c/d with $a \geq c \succ d \geq b$. By Dilworth's characterization of lattice congruences [4], there is a unique largest congruence ψ on $\mathcal{F}(L)$ such that $(c, d) \notin \psi$. Let $K = \mathcal{F}(L)/\psi$. Then K is subdirectly irreducible, and because ψ cannot collapse a/b, L is embedded in K. Clearly K is in the variety generated by L, so $\mathbf{V}(K) = \mathbf{V}(L)$. However, K has the additional property that its critical quotient is a covering pair. We conclude that every variety of lattices is generated by its subdirectly irreducible members containing a prime critical quotient.

The basic idea (which Freese credits to Dilworth) is to use filter or ideal lattices to obtain covering relations without leaving the variety; there are as many variations of the details of the argument as there are applications. The idea is so simple as to seem obvious in retrospect, but it was new at the time, and immediately became a standard technique for working with varieties of lattices. The advantages of having covering relations to work with in lattices are well-known. For example, a subdirectly irreducible modular lattice with a prime critical quotient is weakly atomic. The last section of the paper gives two instances of important results whose proofs are greatly simplified by Theorem 2.

This paper is in a sense a sequel to Ralph Freese's paper on ideal lattices of lattices [7]. There it is shown using similar methods that every compactly generated

This work was supported by the NSF.

lattice is a subdirect product of subdirectly irreducible lattices which are complete and upper continuous. This result was proved in answer to a question of Dilworth's; the application to lattice varieties followed naturally. Indeed, as soon as Freese's result was announced, it was used by Christian Herrmann and Rudolf Wille to find all subdirectly irreducible modular lattices generated by two complemented pairs [11], [9].

In [14], Bjarni Jónsson and Ivan Rival considered the *location* of critical quotients in a subdirectly irreducible lattice. They showed that every nonmodular subdirectly irreducible lattice L contains a critical pair a/b which also sits as the critical quotient of some five-element sublattice of L isomorphic to the pentagon. The corresponding theorem for modular nondistributive lattices and the diamond was also proved. As they observed, these properties are preserved by the above construction to obtain prime critical quotients given above.

These techniques played a critical role in Jónsson and Rival's proof that there are exactly sixteen lattice varieties covering the variety generated by the pentagon [15]. To appreciate this, consider their lemma (which has been of enduring usefulness) that a variety V is semidistributive if and only if it does not contain any of six specific minimally non-semidistributive lattices. An earlier version of this lemma involving chain conditions had been proved by Davey, Poguntke and Rival [3]. The hypotheses involving chain conditions are removed by using ideals and filters to obtain covering relations instead.

If V and W are lattice varieties with $V \succ W$, then V contains a finitely generated subdirectly irreducible lattice K not in W. For clearly there is a finitely generated lattice $L \in V$ witnessing the failure of some equation ε holding in W, and we may take K to be a subdirect factor of L which also fails ε. Thus it is important to observe that the property of being finitely generated can be preserved by a modification of the construction to obtain a prime critical quotient. To do this, we simply replace $\mathcal{F}(L)$ by the sublattice S of $\mathcal{F}(L)$ generated by $L \cup \{c, d\}$ at the appropriate place in the construction. If L is finitely generated, then so will S be, and hence also $K = S/\psi$. The author has found this technique very useful in studying covers of lattice varieties [17], [19].

Of course, the main impetus for all modern work on lattice varieties is Jónsson's Lemma on subdirectly irreducible algebras in congruence distributive varieties [12]. This result provided an approach to questions concerning lattice varieties which were heretofore inaccessible, which in turn necessitated the development of other new techniques. All these results should be viewed in this context.

Another basic tool developed for studying lattice varieties is the use of limit tables and the associated (generalized) splitting equations. This method lends itself to a more equational (as opposed to structural) approach. The method was developed independently in slightly different forms by Ralph McKenzie and Bjarni Jónsson; both versions appear in McKenzie's classic paper on nonmodular lattice varieties [16]. Examples of this approach are contained in [6] (cf. [13]) and [18].

A primary motivating question for [5] was whether the variety $\mathcal{M}$ of all modular lattices was generated by its finite members. Freese later showed that this is not

                                                          THE DILWORTH THEOREMS

the case [8], and in fact Christian Herrmann proved that $\mathcal{M}$ is not even generated by its finite dimensional members [10].

The author would like to thank Ralph Freese and Bjarni Jónsson for several interesting discussions on these topics, which contributed substantially to the remarks above.

References

1. K. Baker and A. W. Hales, *From a lattice to its ideal lattice*, Algebra Universalis **4** (1974), 250–258.
2. P. Crawley and R. P. Dilworth, "Algebraic Theory of Lattices," Prentice-Hall, Englewood Cliffs, New Jersey, 1973.
3. B. A. Davey, W. Poguntke, and I. Rival, *A characterization of semidistributivity*, Algebra Universalis **5** (1975), 72–75.
4. R. P. Dilworth, *The structure of relatively complemented lattices*, Ann. of Math. **51** (1950), 348–359. Reprinted in Chapter 7 of this volume.
5. R. P. Dilworth and R. Freese, *Generators for lattice varieties*, Algebra Universalis **6** (1976), 263–267. Reprinted in Chapter 7 of this volume.
6. R. Freese, *Breadth two modular lattices*, in "Proc. Univ. of Houston Lattice Theory Conference," Univ. of Houston, 1973, pp. 409–451.
7. R. Freese, *Ideal lattices of lattices*, Pacific J. Math. **57** (1975), 125–133.
8. R. Freese, *The variety of modular lattices is not generated by its finite members*, Trans. Amer. Math. Soc. **255** (1979), 277–300.
9. Ch. Herrmann, *On modular lattices generated by two complemented pairs*, Houston Jour. Math. **2** (1976), 513–523.
10. Ch. Herrmann, *On the arithmetic of projective coordinate systems*, Trans. Amer. Math. Soc. **284** (1984), 759–785.
11. Ch. Herrmann and R. Wille, *On modular lattices with four generators II*, Notices Amer. Math. Soc. **22** (1975), A-54.
12. B. Jónsson, *Algebras whose congruence lattices are distributive*, Math. Scand. **21** (1967), 110–121.
13. B. Jónsson, *Equational classes of lattices*, Math. Scand. **22** (1968), 187–196.
14. B. Jónsson and I. Rival, *Critical edges in subdirectly irreducible lattices*, Proc. Amer. Math. Soc. **66** (1977), 194–196.
15. B. Jónsson and I. Rival, *Lattice varieties covering the smallest non-modular variety*, Pacific J. Math. **82** (1979), 463–478.
16. R. McKenzie, *Equational bases and non-modular lattice varieties*, Trans. Amer. Math. Soc. **174** (1972), 1–43.
17. J. B. Nation, *Some varieties of semidistributive lattices*, in "Universal Algebra and Lattice Theory," S. Comer, ed., Lecture Notes in Mathematics, vol. **1149**, Springer Verlag, New York, 1985, pp. 198–223.
18. J. B. Nation, *Lattice varieties covering $V(L_1)$*, Algebra Universalis **23** (1986), 132–166.
19. J. B. Nation, *An approach to lattice varieties of finite height*, Algebra Universalis (to appear).
20. D. Sachs, *Identities in finite partition lattices*, Proc. Amer. Math. Soc. **12** (1961), 944–945.

University of Hawaii
Honolulu, HI 96822
U. S. A.

Two Results from "Algebraic Theory of Lattices"

Editors' note: In addition to the theorems in the earlier chapters, there are two theorems of Dilworth which are unpublished but widely known and used since the 1940's. These theorems finally appeared in 1973 in his book with Peter Crawley, "Algebraic Theory of Lattices." The two theorems state:

> *Every finite distributive lattice is isomorphic to the congruence lattice of some finite lattice.*
>
> *Every finite lattice can be embedded into a finite geometric lattice.*

A motivated account of how the congruence theorem was first proved is given in Dilworth's background. Other proofs and extensions of this theorem are surveyed in the article by George Grätzer. A sketch of Dilworth's proof of the embedding theorem is also included.

Background

R. P. Dilworth

Congruence lattices. One of the basic tools in the study of algebraic systems is the concept of a homomorphism, *i.e.*, a mapping h from the elements of an algebra to the elements of another algebra of the same type which preserves the basic operations of the algebra. Associated with a homomorphism h is a congruence relation θ_h on the algebra defined by $a\ \theta_h\ b$ if and only if $h(a) = h(b)$. θ_h is an equivalence relation on the elements of algebra which preserves the basic operations of the algebra. Conversely, any congruence relation θ on the algebra, *i.e.*, an equivalence relation which preserves the basic operations of the algebra, determines a natural homomorphism h_θ such that $h_\theta(a)$ is the congruence class of a under the congruence relation θ. There is a natural ordering defined on the congruence relations of an algebra given by $\theta \leq \psi$ if and only if $a\ \theta\ b$ implies $a\ \psi\ b$ for all elements a and b in the algebra. Under this ordering the congruence relations on an algebra form an algebraic (compactly generated) lattice. For a lattice, the lattice of congruence relations is distributive. This immediately raises the question: *Is every algebraic distributive lattice isomorphic to a lattice of congruence relations of a suitable lattice?* I began a study of this question by looking into the case of a finite distributive lattice.

I had felt for some time that the conjecture was true. Thus when I began to work on the problem, I started with the simplest non-trivial example namely, the three element chain. A little experimenting showed that the second lattice in Figure 1 did the job.

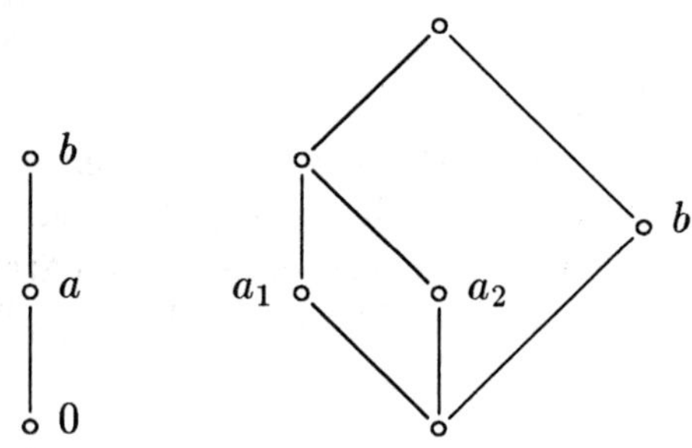

Figure 1

I next tried the four element chain. The representing lattice should be some combination of the lattice which represents the three element chain. Several trials then led to the lattice of Figure 2.

I tried one further example—the distributive lattice diagrammed on the left of Figure 3. This led quickly to the representing lattice of Figure 3.

From these three examples it was clear how to construct the representing lattice in general. Start with the set of join irreducibles. Replicate every irreducible which is properly contained in another irreducible. From this new set select all subsets which have the property that if two comparable irreducibles belong to the subset then the replicate of the lower one also belongs. These subsets are closed under meets and hence form a lattice. This lattice has the property that every interval has a lower transpose whose first element is the null element of the lattice. This makes it easy to determine its congruence lattice which turns out to be the original distributive lattice.

It is clear that this method also handles the case of a complete distributive lattice in which each element is a join of compact join irreducible elements.

This work was never published since I had hoped to have time to do some definitive work on the general question.

                                                         THE DILWORTH THEOREMS

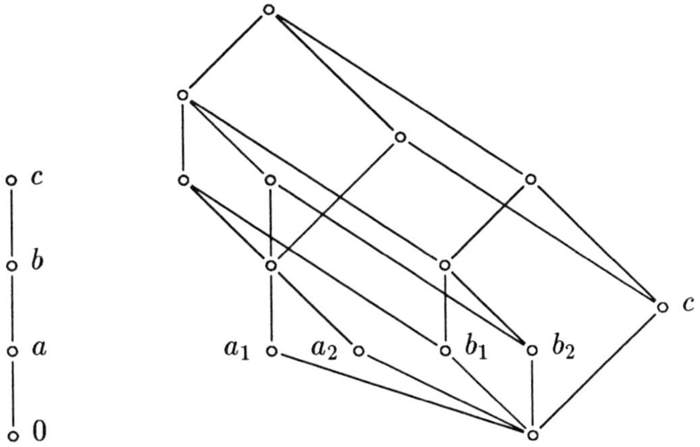

Figure 2

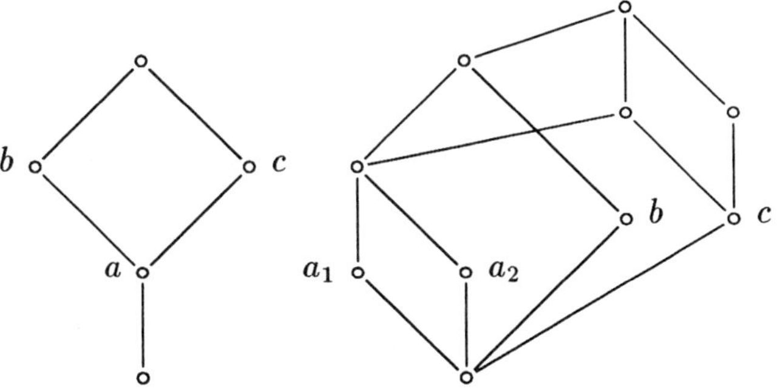

Figure 3

For background on the embedding theorem, see Chapter 5.

Dilworth's Proof of the Embedding Theorem

JOSEPH P. S. KUNG

The following embedding theorem was found by Dilworth in the 1940's.

THEOREM 1. *Every finite lattice L can be embedded (as a sublattice) in a finite geometric lattice.*

(In general, the geometric lattice has greater rank than the original lattice.) Dilworth wanted to use this result to prove Whitman's conjecture: *Every finite lattice can be embedded into a finite partition lattice.* His strategy was to break up Whitman's conjecture into two parts: embedding a finite lattice into a finite geometric lattice and embedding a finite geometric lattice into a finite partition lattice. He felt that the proof of the second part would be easier than the proof of the first. Thus, when he obtained Theorem 1, he held off publishing it until he has also proved the second part. This turned out to be harder than he anticipated. Whitman's conjecture was eventually proved by Pudlák and Tůma in [5].

Theorem 1 is a close relative of the main theorem in [1] and some of the essential ideas for its proof can be found in [1]. A detailed proof of this theorem was published in 1973 in [2]. We shall sketch that proof here.

The first step in the proof is to define by induction an integer-valued function σ on the lattice L such that

(1) $\sigma(0) = 0$

(2) $\sigma(a) > \sigma(b)$ whenever $a > b$

(3) $\sigma(a \vee b) + \sigma(a \wedge b) \geq \sigma(a) + \sigma(b)$ for all $a, b \in L$.

The second step is to represent L as a lattice of subsets. A *lattice $\mathcal{L}$ of subsets* is a collection of subsets containing the null set and closed under set-intersection. A *rank function ρ* on $\mathcal{L}$ is an integer-valued function defined on $\mathcal{L}$ satisfying:

(1) $\rho(\emptyset) = 0$

(2) $\rho(A) \geq \rho(B)$ whenever $A \supseteq B$

(3) $\rho(A \vee B) + \rho(A \cap B) \leq \rho(A) + \rho(B)$ for all $A, B \in \mathcal{L}$

(4) $\rho(A) - \rho(B) \leq |A - B|$ whenever $A \supseteq B$.

If, in addition, $\rho(A) > \rho(B)$ whenever $A \supset B$, ρ is said be *strictly increasing*.

Now let I be the set of all nonzero join-irreducibles in L. Assign to each nonzero join-irreducible r a set X_r such that X_r and X_s are d isjoint sets if $r \neq s$, and,

$$|X_r| = \sigma(r) - \sigma(r_*),$$

where r_* is the unique element covered by r. Set $X = \bigcup X_r$. For each $a \in L$, define

$$S_a = \bigcup_{r \in I,\, r \leq a} X_r.$$

The function $a \to S_a$ is an isomorphism of L with the lattice of subsets $\mathcal{L}$ consisting of the subsets S_a, $a \in L$. The rank function ρ, defined by $\rho(S_a) = \sigma(a)$, is a strictly increasing rank function on $\mathcal{L}$.

The third step is to extend the rank function on $\mathcal{L}$ to all subsets. The is done using the following theorem which is of independent interest in matroid theory.

THEOREM 2. *Let $\mathcal{L}$ be a lattice of subsets on the finite set X and let ρ be a strictly increasing rank function on $\mathcal{L}$. Then the function π defined by*

$$\pi(S) = \min\{\rho(A) + |S - A| : A \in \mathcal{L}\}$$

is a matroid rank function on the set X (that is, a rank function on the lattice of all subsets of X). Moreover, $\pi(A) = \rho(A)$ for every subset A in $\mathcal{L}$.

Using this theorem, we can now finish the proof by showing that $\mathcal{L}$, and hence L, is a sublattice of the geometric lattice of closed sets of the matroid defined on X by the rank function π.

Although it was unpublished, the third step in the proof was a seminal influence in the theory of submodular functions. A survey of this area is given in Faigle's article in Chapter 5. We also note two related papers [3,4].

REFERENCES

1. R. P. Dilworth, *Dependence relations in a semi-modular lattice*, Duke Math J. **11** (1944), 575–587. Reprinted in Chapter 5 of this volume.
2. P. Crawley and R. P. Dilworth, "Algebraic Theory of Lattices," Prentice-Hall, Englewood Cliffs, New Jersey, 1973, pp. 125–131.
3. D. T. Finkbeiner, *A general dependence relation for lattices*, Proc. Amer. Math. Soc. **2** (1951), 756–759.
4. —————————, *A semimodular imbedding of lattices*, Canad. J. Math. **12** (1960), 582–591.
5. P. Pudlák and J. Tůma, *Every finite lattice can be embedded in the lattice of all equivalences over a finite set*, Algebra Universalis **10** (1980), 74–95.

University of North Texas
Denton, TX 76203
U. S. A.

On the Congruence Lattice of a Lattice

Georg Grätzer

1. Dilworth's Theorem. Let L be a lattice. It is proved in N. Funayama and T. Nakayama [10] that the congruence lattice of L is distributive. For a finite lattice L the converse of this result is:

Dilworth's Theorem. *Every finite distributive lattice D can be represented as the lattice of congruence relations of a finite lattice L.*

R. P. Dilworth never published this result. He wanted to solve the more general problem:

Dilworth's Conjecture. *Every distributive algebraic lattice D be represented as the congruence lattice of a lattice L.*

Dilworth undoubtedly did not fully realize the difficulty of this problem, which remains unresolved to this day.

Dilworth's Theorem appears as an exercise (*albeit* one marked as difficult) in G. Birkhoff [3]; the first published proof is in G. Grätzer and E. T. Schmidt [20]. Schmidt and I were honored that the proof in P. Crawley and R. P. Dilworth [4] follows our proof closely.

Although the general problem is still unresolved, several significant theorems have since been obtained. These are discussed below. Dilworth's Theorem is a testimony how far ahead of his time Dilworth was.

See [11, pp.81–84] for a proof of Dilworth's Theorem I found with H. Lakser; see also the proof with H. Lakser referenced in §2.

2. Stronger results: L and D finite. The proof in G. Grätzer and E. T. Schmidt [20] constructs a finite lattice L of between $O(2^n)$ and $O(4^n)$ elements and of length $2n - 1$, where n is the number of join-irreducible elements of D.

S.-K. Teo [39] constructs a finite lattice L *of length* $5m$, where m is the number of dual atoms of D; for $m = 1$ and the conjecture that lead to Teo's result, see E. T. Schmidt [31]. S.-K. Teo also proves that his result is best possible.

I have been working with H. Lakser on representing any complete lattice as the lattice of complete congruence relations of a complete lattice; see the references [14]–[18] (see also [19]). I proved that every complete lattice K can be represented as the lattice of complete congruence relations of a complete lattice L. With H. Lakser we proved a stronger result: Every $\mathfrak{m}$-algebraic lattice K in which the unit element is $\mathfrak{m}$-compact can be represented as the lattice of $\mathfrak{m}$-complete congruence relations of a bounded $\mathfrak{m}$-complete lattice L. It turns out that the techniques developed in these papers gave us a new insight into Dilworth's Theorem. We found a new proof with a "very small" L that can be drawn. L has only $O(n^3)$ elements and it is *planar*. In addition, L is *rigid*, that is, it has no proper automorphisms.

3. Stronger results: D finite. If we wish to construct a modular lattice L with a given congruence lattice, then we must drop the condition that L be finite. Indeed, the congruence lattice of a finite modular lattice is Boolean.

E. T. Schmidt [29] constructs a *modular* L; R. Freese [6] shows that L can be chosen to be *finitely generated*. Schmidt improves his result in [35]: Every finite distributive lattice D can be represented as the congruence lattice of a suitable *complemented modular* lattice L.

4. D infinite. Dilworth's Conjecture is attacked in a series of papers by E. T. Schmidt (see references). His best result is in [33]: If D is isomorphic to the ideal lattice of a (distributive) lattice with zero (or, equivalently, if the meet of any two compact elements is compact), then the conjecture is true.

P. Pudlák [24] attacks the problem "from below." The conjecture is true for D finite. Now suppose there is a stronger proof of the finite case: If S_1 and S_2 are distributive join-semilattices and S_1 is a $\{0\}$-subsemilattice of S_2, then the corresponding L_1 is a sublattice of L_2, and the congruences from L_1 extend naturally. It would then follow that the conjecture holds. This is actually done by P. Pudlák under the assumption of Schmidt's result quoted above. A. Huhn introduces the "reduced free product" of distributive lattices in [21]. In [22], he succeeds in applying his results to prove the conjecture for $|D| \leq \aleph_0$, a result he attributes to Bauer. In a posthumous paper [23], he extends this to $|D| \leq \aleph_1$.

In [5], H. Dobbertin proves that the conjecture holds if every compact element of D contains at most countably many compact elements.

The result of R. Freese, W. A. Lampe, and W. Taylor in [7] also seems relevant. For every fixed type τ of algebras a modular algebraic lattice K is constructed that cannot be represented as the congruence lattice of an algebra of type τ. Their method does not apply to the distributive case. An affirmative answer to Dilworth's Conjecture would show that in the distributive case the type $\langle 2, 2 \rangle$ will always do.

Presently, it is not known whether any one type would work for all distributive algebraic lattices.

5. Enhanced results. These results can be enhanced in many ways. We can represent D as the congruence lattice of a lattice satisfying some condition not related to D. For instance, we may require that the automorphism group be known.

There is such a result for finite D in G. Grätzer and H. Lakser [17]: For a given finite D, the lattice L can be constructed with a prescribed automorphism group G (and if G is finite, L can be chosen to be finite), combining the result of Dilworth with the result (on automorphism groups of lattices) of R. Frucht [8] and [9] and G. Sabidussi [25]. Another way of putting this: the congruence lattice and the automorphism group of a finite lattice are independent.

If L_1 is a finite lattice and L_2 is an ideal of L_1, then the map

$$\varphi : \Theta \to \Theta|_{L_2}$$

is a $\{0,1\}$-homomorphism from the congruence lattice of L_1 into the congruence lattice of L_2. In G. Grätzer and H. Lakser [14], we prove the converse: Let D_1 and D_2 be finite distributive lattices, and let a $\{0,1\}$-homomorphism $\varphi : D_1 \to D_2$ be given. Then there exists a finite lattice L_1 and an ideal L_2 of L_1 such that D_1 and D_2 are the congruence lattices of L_1 and L_2, and φ is represented by the restriction map $\Theta \to \Theta|_{L_2}$. See also E. T. Schmidt [36]. The method we developed also applies to this problem: L_1 can be chosen to be a planar lattice.

Technically, the best result I proved with H. Lakser is the following (see [18]): Let D, D' be finite distributive lattices, and let $\psi : D \to D'$ be a $\{0,1\}$-preserving lattice homomorphism. Let G, G' be groups, and let $\eta : G \to G'$ be a group homomorphism. Then there are a lattice L, an ideal L' in L, lattice isomorphisms

$$\varrho : D \to \operatorname{Con} L, \qquad \varrho' : D' \to \operatorname{Con} L',$$

and group isomorphisms

$$\tau : G \to \operatorname{Aut} L, \qquad \tau' : G' \to \operatorname{Aut} L'$$

such that, for each $x \in D$, the congruence relation $x\psi\varrho'$ on L' is the restriction to L' of the congruence relation $x\varrho$ on L, and, for each $g \in G$, the automorphism $g\eta\tau'$ of L' is the restriction of the automorphism $g\tau$ of L.

REFERENCES

1. J. Berman, *On the length of the congruence lattice of a lattice*, Algebra Universalis **2** (1972), 18–19.
2. G. Birkhoff, *Universal Algebra*, in "Proc. First Canadian Math. Congress, Montreal, 1945," University of Toronto Press, Toronto, 1946, pp. 310–326.
3. __________, "Lattice Theory," Amer. Math. Soc. Colloq. Publ. vol. 25, Revised Edition, Amer. Math. Soc., New York, N. Y., 1948.

4. P. Crawley and R. P. Dilworth, "Algebraic Theory of Lattices," Prentice-Hall, Englewood Cliffs, N. J., 1973.

5. H. Dobbertin, *Vaught measures and their applications in lattice theory*, J. of Pure and Applied Algebra **43** (1986), 27–51.

6. R. Freese, *Congruence lattices of finitely generated modular lattices*, in "Proceedings of the Lattice Theory Conference, Ulm," 1975, pp. 62–70.

7. R. Freese, W. A. Lampe, and W. Taylor, *Congruence lattices of algebras of fixed similarity types. I.*, Pacific J. Math. **82 (1)** (1979), 59–68.

8. R. Frucht, *Herstellung von Graphen mit vorgegebener abstrakter Gruppe*, Compos. Math. **6** (1938), 239–250.

9. ________, *Lattices with a given group of automorphisms*, Canad. J. Math. **2** (1950), 417–419.

10. N. Funayama and T. Nakayama, *On the distributivity of a lattice of lattice-congruences*, Proc. Imp. Acad. Tokyo **18** (1942), 553–554.

11. G. Grätzer, "General Lattice Theory," Academic Press, New York, N. Y.; Birkhäuser-Verlag, Basel; Akademie Verlag, Berlin, 1978.

12. ________, "Universal Algebra," Second Edition, Springer-Verlag, New York, Heidelberg, Berlin, 1979.

13. ________, *On the the automorphism group and the complete congruence lattice of a complete lattice*, Abstracts of papers presented to the Amer. Math. Soc. 88T-06-215.

14. G. Grätzer and H. Lakser, *Homomorphisms of distributive lattices as restrictions of congruences*, Canad. J. Math. **38** (1986), 1122–1134.

15. ________, *On the $\mathfrak{m}$-complete congruence lattice and the automorphism group of an $\mathfrak{m}$-complete lattice*, Abstracts of papers presented to the Amer. Math. Soc. 88T-08-253.

16. ________, *On the complete congruence lattice and the automorphism group of a complete lattice. I. and II*, Manuscript.

17. ________, *On the automorphism group and the congruence lattice of a finite lattice*, Abstracts of papers presented to the Amer. Math. Soc. 89T-06.

18. ________, *Homomorphisms of distributive lattices as restrictions of congruences. II. Planar lattices and automorphism groups*, Abstracts of papers presented to the Amer. Math. Soc. 89T-06.

19. G. Grätzer, H. Lakser, and B. Wolk, *On the lattice of complete congruences of a complete lattice: On a result of K. Reuter and R. Wille*, Acta Sci. Math. (Szeged) (1988) (to appear).

20. G. Grätzer and E. T. Schmidt, *On congruence lattices of lattices*, Acta Math. Acad. Sci. Hungar. **13** (1962), 179–185.

21. A. Huhn, *A reduced free product of distributive lattices. I*, Acta Math. Hungar. **42** (1983), 349–354.

22. ________, *On the representation of distributive algebraic lattices. I*, Acta Sci. Math. (Szeged) **45** (1983), 239–246.

23. ________, *On the representation of distributive algebraic lattices. II and III*, Acta Sci. Math. (Szeged) (to appear).

24. P. Pudlák, *On the congruence lattices of lattices*, Algebra Universalis **20** (1985), 96–114.

25. G. Sabidussi, *Graphs with given infinite groups*, Monatsch. Math. **68** (1960), 64–67.

26. E. T. Schmidt, *Über die Kongruenzverbände der Verbände*, Publ. Math. Debrecen **9** (1962), 243–256.

27. ________, "Kongruenzrelationen Algebraischer Strukturen," Math. Forschungberichte, XXV. VEB Deutcher Verlag der Wissenschaften, Berlin, 1967.

28. ________, *Zur Characterisierung der Kongruenzverbände der Verbände*, Math. Časopis **18** (1968), 3–20.

29. ________, *Every finite distributive lattice is the congruence lattice of a modular lattice*, Algebra Universalis **4** (1974), 49–57.

30. ________, *Über die Kongruenzverbände der modularen Verbände*, Beiträge zur Algebra und Geometrie, Halle **3** (1974), 59–68.

Two Results from "Algebraic Theory of Lattices"

31. __________, *On the length of the congruence lattice of a lattice*, Algebra Universalis **5** (1975), 98-100.

32. __________, *On the characterization of congruence lattices of lattices*, in "Proceedings of the Lattice Theory Conference in Ulm," 1975, pp. 162–179.

33. __________, *The ideal lattice of a distributive lattice with 0 is the congruence lattice of a lattice*, Acta Sci. Math. (Szeged) **43** (1981), 153–168.

34. __________, "A Survey on Congruence Lattice Representations," Teubner-Texte zur Mathematik, 42, BSB B. G. Teubner Verlagsgesellschaft, Leipzig, 1982.

35. __________, *Congruence lattices of complemented modular lattices*, Algebra Universalis **18** (1984), 386-395.

36. __________, *Homomorphisms of distributive lattices as restrictions of congruences*, Acta Sci. Math. (Szeged) **51** (1987), 209–215.

37. __________, *On a representation of distributive lattices*, Periodica Math. Hungar. **19** (1988), 25–31.

38. S.-K. Teo, *Representing finite lattices as complete congruence lattices of complete lattices*, Abstracts of papers presented to the Amer. Math. Soc. 88T-06-207.

39. __________, *On the length of the congruence lattice of a lattice*, Manuscript (1988), 1–9.

University of Manitoba
Winnipeg, Manitoba R3T 2N2
Canada

     THE DILWORTH THEOREMS

Permissions

We would like to thank the original publishers of the papers of Robert P. Dilworth for granting permission to reprint the articles in this volume. Articles are referred to by their numbers in the chronological list of publications.

Academic Press
[28] Reprinted from *J. Combin. Theory* **10**, ©1971 by Academic Press.

The American Mathematical Society
[1] Reprinted from *Bull. Amer. Math. Soc.* **38**, ©1938 by the American Mathematical Society.
[3] Reprinted from *Trans. Amer. Math. Soc.* **45**, ©1939 by American Mathematical Society.
[4] Reprinted from *Trans. Amer. Math. Soc.* **46**, ©1939 by the American Mathematical Society.
[11] Reprinted from *Trans. Amer. Math. Soc.* **49**, ©1941 by the American Mathematical Society.
[14] Reprinted from *Trans. Amer. Math. Soc.* **57**, ©1945 by the American Mathematical Society.
[15] Reprinted from *Bull. Amer. Math. Soc.* **52**, ©1946 by the American Mathematical Society.
[20] Reprinted from *Trans. Amer. Math. Soc.* **68**, ©1950 by the American Mathematical Society.
[23] Reprinted from *Trans. Amer. Math. Soc.* **96**, ©1960 by the American Mathematical Society.
[24] Reprinted from *Combinatorial Analysis*, ©1960 by the American Mathematical Society.
[25] Reprinted from *Lattice Theory*, ©1961 by the American Mathematical Society.
[27] Reprinted from *Proc. Amer. Math. Soc.* **13**, ©1962 by the American Mathematical Society.

Annals of Mathematics
[8] Reprinted from *Ann. of Math. (2)* **41**, ©1940 by Annals of Mathematics.
[13] Reprinted from *Ann. of Math. (2)* **45**, ©1944 by Annals of Mathematics.
[18,19] Reprinted from *Ann. of Math. (2)* **51**, ©1950 by Annals of Mathematics.
[22] Reprinted from *Ann. of Math. (2)* **60**, ©1954 by Annals of Mathematics.

Birkhäuser-Verlag
[30] Reprinted from *Algebra Universalis* **6**, ©1976 by Birkhäuser-Verlag.
[32] Reprinted from *Algebra Universalis* **18**, ©1984 by Birkhäuser-Verlag.

Duke University Press
[5] Reprinted from *Duke Math. J.* **5**, ©1939 by Duke University Press.
[10] Reprinted from *Duke Math. J.* **8**, ©1941 by Duke University Press.
[12] Reprinted from *Duke Math. J.* **11**, ©1944 by Duke University Press.
[21] Reprinted from *Duke Math. J.* **19**, ©1952 by Duke University Press.

Pacific Journal of Mathematics
[26] Reprinted from *Pacific J. Math.* **12**, ©1962 by Pacific Journal of Mathematics.

Tôhoku Mathematical Journal
[7] Reprinted from *Tôhoku Math. J.* **47**, ©1940 by Tôhoku Mathematical Journal.